CONVERSIONS BETWEEN U.S. CUSTOMARY UNITS AND SI UNITS

U.S. Customary unit		Times co		Equals SI unit	
		Accurate	Practical		
Acceleration (linear)					
foot per second squared	ft/s^2	0.3048*	0.305	meter per second squared	m/s^2
inch per second squared	in./s^2	0.0254*	0.0254	meter per second squared	m/s^2
Area					
circular mil	cmil	0.0005067	0.0005	square millimeter	mm^2
square foot	ft^2	0.09290304*	0.0929	square meter	m^2
square inch	in.^{2}	645.16*	645	square millimeter	mm^2
Density (mass)					
slug per cubic foot	slug/ft^3	515.379	515	kilogram per cubic meter	kg/m^3
Density (weight)					
pound per cubic foot	lb/ft^3	157.087	157	newton per cubic meter	N/m^3
pound per cubic inch	lb/in.^{3}	271.447	271	kilonewton per cubic meter	kN/m^3
Energy; work					
foot-pound	ft-lb	1.35582	1.36	joule (N·m)	J
inch-pound	in.-lb	0.112985	0.113	joule	J
kilowatt-hour	kWh	3.6*	3.6	megajoule	MJ
British thermal unit	Btu	1055.06	1055	joule	J
Force					
pound	lb	4.44822	4.45	newton (kg·m/s^2)	N
kip (1000 pounds)	k	4.44822	4.45	kilonewton	kN
Force per unit length					
pound per foot	lb/ft	14.5939	14.6	newton per meter	N/m
pound per inch	lb/in.	175.127	175	newton per meter	N/m
kip per foot	k/ft	14.5939	14.6	kilonewton per meter	kN/m
kip per inch	k/in.	175.127	175	kilonewton per meter	kN/m
Length					
foot	ft	0.3048*	0.305	meter	m
inch	in.	25.4*	25.4	millimeter	mm
mile	mi	1.609344*	1.61	kilometer	km
Mass					
slug	lb-s^2/ft	14.5939	14.6	kilogram	kg
Moment of a force; torque					
pound-foot	lb-ft	1.35582	1.36	newton meter	N·m
pound-inch	lb-in.	0.112985	0.113	newton meter	N·m
kip-foot	k-ft	1.35582	1.36	kilonewton meter	kN·m
kip-inch	k-in.	0.112985	0.113	kilonewton meter	kN·m

CONVERSIONS BETWEEN U.S. CUSTOMARY UNITS AND SI UNITS (Continued)

U.S. Customary unit		Times conversion factor		Equals SI unit	
		Accurate	Practical		
Moment of inertia (area)					
inch to fourth power	in.^{4}	416,231	416,000	millimeter to fourth power	mm^4
inch to fourth power	in.^{4}	0.416231×10^{-6}	0.416×10^{-6}	meter to fourth power	m^4
Moment of inertia (mass)					
slug foot squared	slug-ft^2	1.35582	1.36	kilogram meter squared	kg·m^2
Power					
foot-pound per second	ft-lb/s	1.35582	1.36	watt (J/s or N·m/s)	W
foot-pound per minute	ft-lb/min	0.0225970	0.0226	watt	W
horsepower (550 ft-lb/s)	hp	745.701	746	watt	W
Pressure; stress					
pound per square foot	psf	47.8803	47.9	pascal (N/m^2)	Pa
pound per square inch	psi	6894.76	6890	pascal	Pa
kip per square foot	ksf	47.8803	47.9	kilopascal	kPa
kip per square inch	ksi	6.89476	6.89	megapascal	MPa
Section modulus					
inch to third power	in.^{3}	16,387.1	16,400	millimeter to third power	mm^3
inch to third power	in.^{3}	16.3871×10^{-6}	16.4×10^{-6}	meter to third power	m^3
Velocity (linear)					
foot per second	ft/s	0.3048*	0.305	meter per second	m/s
inch per second	in./s	0.0254*	0.0254	meter per second	m/s
mile per hour	mph	0.44704*	0.447	meter per second	m/s
mile per hour	mph	1.609344*	1.61	kilometer per hour	km/h
Volume					
cubic foot	ft^3	0.0283168	0.0283	cubic meter	m^3
cubic inch	in.^{3}	16.3871×10^{-6}	16.4×10^{-6}	cubic meter	m^3
cubic inch	in.^{3}	16.3871	16.4	cubic centimeter (cc)	cm^3
gallon (231 in.^{3})	gal.	3.78541	3.79	liter	L
gallon (231 in.^{3})	gal.	0.00378541	0.00379	cubic meter	m^3

*An asterisk denotes an *exact* conversion factor

***Note:* To convert from SI units to USCS units, *divide* by the conversion factor**

Temperature Conversion Formulas

$$T(°C) = \frac{5}{9}[T(°F) - 32] = T(K) - 273.15$$

$$T(K) = \frac{5}{9}[T(°F) - 32] + 273.15 = T(°C) + 273.15$$

$$T(°F) = \frac{9}{5}T(°C) + 32 = \frac{9}{5}T(K) - 459.67$$

Thermodynamics *for* Engineers

Thermodynamics *for* Engineers

SI Edition

Kenneth A. Kroos

Villanova University

Merle C. Potter

Michigan State University

SI Edition Adapted by

Shaligram Tiwari

Indian Institute of Technology Madras

CENGAGE Learning®

Andover • Melbourne • Mexico City • Stamford, CT • Toronto • Hong Kong • New Delhi • Seoul • Singapore • Tokyo

Thermodynamics for Engineers, SI Edition
Kenneth A. Kroos, Merle C. Potter
SI edition by Shaligram Tiwari

ISBN-13: 978-81-315-2046-8
ISBN-10: 81-315-2046-3

Cengage Learning India Private Limited
418, F.I.E., Patparganj
Delhi 110092
India

Tel: 91-11-43641111
Fax: 91-11-43641100
Email: asia.infoindia@cengage.com

Cengage Learning is a leading provider of customized learning solutions with office locations around the globe, including Andover, Melbourne, Mexico City, Stamford (CT), Toronto, Hong Kong, New Delhi, Seoul, Singapore, and Tokyo. Locate your local office at: www.cengage.com/global

Cengage Learning Products are represented in Canada by Nelson Education, Ltd.

For product information, visit our website at **www.cengage.co.in**

Printed in India
Fourth Indian Reprint 2018

To my wife Kathleen.
Ken

To my wife Gloria.
Merle

Contents

Nomenclature List

a	Constant,
a	Acceleration,
a	Helmholtz function
A_0,	Constant
AF	Air/fuel mass ratio
B	Bulk modulus
BDC	Bottom dead center
BWR	Back work ratio
C	Specific heat,
C	A constant
c	Speed of light
COP_{HP}	Coefficient of performance of a heat pump
COP_R	Coefficient of performance of a refrigerator
C_p	Constant pressure specific heat
C_v	Constant pressure specific heat
$dE_{c.v.}/dt$	Change in energy per unit time
ds	Differential length segment
f	A representative function
F	Force
$\mathbf{F}$	Force vector
FA	Fuel/air mass ratio
F_n	Normal component of a force
G	Gibbs function
g	Gravity,
g	Gibbs function
g_c	Gravitational constant
h	Specific enthalpy,
h	Height,
h	Planck's constant
$\overline{h}^\circ$	Enthalpy at the reference state
$\overline{h}_f^\circ$	Enthalpy of formation
$\overline{h}_{fg}$	Enthalpy of vaporization
h_c	Convection heat transfer coefficient
H	Enthalpy
$\overline{H}_p, H_P$	Enthalpy of the products
$\overline{H}_R, H_R$	Enthalpy of the reactants
HHV	Higher heating value
i	Current,
i	Irreversibility per unit mass
I	Irreversibility
k	Ratio of specific heats,
k	Thermal conductivity
kg_a	A kilogram of air
kg_w	A kilogram of water
kg_v	A kilogram of water vapor
K	Spring constant
K_P	Equilibrium constant
KE	Kinetic energy
L	Thickness
LHV	Lower heating value
m	Mass
m_a	Mass of dry air
m_f	Mass of liquid,
m_f	The final mass
m_g	Mass of vapor
m_i	The initial mass
m_v	Mass of water vapor contained in the air
$\dot{m}$	Mass flow rate (mass flux)
M	A general function,
M	Molar mass
M_i	Molar mass of substance
m_i	Mass of component i of a substance
MEP	Mean effective pressure
n	A constant
N	A general function,
N	Number of moles
N_i	Number of moles of component i
P	Pressure
P_a	Partial pressure of the dry air
P_{cr}	Critical-point pressure
P_f	The final pressure
P_i	Partial pressure of component i,
P_i	The initial pressure
P_r	Relative pressure
P_R	Reduced pressure
P_v	Partial pressure of water vapor
PE	Potential energy
q	Specific heat transfer

Q	Heat transfer
Q_H	Heat transfer from a high-temperature reservoir
Q_L	Heat transfer from a low-temperature reservoir
$\dot{Q}$	Rate of heat transfer
$\dot{Q}_B$	Boiler rate of heat transfer
$\dot{Q}_C$, $\dot{Q}_{\text{Cond}}$	Condenser rate of heat transfer
$\dot{Q}_{\text{Evap}}$	Evaporator rate of heat transfer
r	Compression ratio,
r	Radius
r_c	cutoff ratio
r_p	pressure ratio
R	Gas constant,
R	Electrical resistivity,
R	*R*-factor
R_u	Universal gas constant
s	Specific entropy
$\bar{s}$	Specific entropy per mol
s^o	An entropy function
S	Entropy
SG	Specific gravity
S_{gen}	Entropy generated
S_{surr}	Entropy of surroundings
$\dot{S}_{\text{prod}}$	Entropy production
t	Time
T	Temperature, Torque
T_0	Dead state temperature
T_{cr}	Critical-point temperature
T_{db}	Dry-bulb temperature
T_{dp}	Dew-point
T_f	The final temperature
T_H	Temperature of a high-temperature reservoir
T_i	The initial temperature
T_L	Temperature of a low-temperature reservoir
T_R	Reduced temperature
T_{wb}	Wet-bulb temperature
TDC	Top dead center
u	Specific internal energy
U	Internal energy
$\overline{U}_P$, U_P	Internal energy of products
$\overline{U}_R$, U_R	Internal energy of reactants
v	Specific volume
$\bar{v}$	Molar specific volume
v_f	Volume of liquid
v_g	Volume of vapor
v_r	Relative specific volume
v_R	Pseudo-reduced specific volume
V	Velocity
V	Volume,
V	Voltage
$\dot{V}$	Volumetric flow rate (flow rate)
w	Work per unit mass
w_s	Shaft work per unit mass
W	Work
$\dot{W}$	Work rate
$\dot{W}_C$	Compressor work rate
$\dot{W}_{\text{Comp}}$	Compressor work rate
$\dot{W}_T$	Turbine work rate
W_S	Shaft work
W_a	Actual work
x	Quality
x_i	The mass fraction of a component
y_i	The mole fraction
z	Elevation
Z	Compressibility factor
ΔS_{gen}	Generated entropy
ΔS_{net}	Net entropy change
ΔS_{univ}	Entropy change in the universe
β	A characteristic constant,
β	Volume expansivity
δ	Signifies an inexact differential
ε	Emissivity,
ε	Utilization factor
η	Efficiency
η_{II}	Second law efficiency
μ_J	Joule-Thomson coefficient
ν	Stoichiometric coefficient,
ν	Light frequency
ρ	Density
ρ_x	Density of an unknown substance
σ	Stefan-Boltzmann constant
τ	Shear stress
ϕ	Equivalence ratio,
ϕ	Relative humidity
Ψ	A special property
ψ	Exergy per unit mass
ψ	A special property
ω	Specific humidity (humidity ratio)

Preface

The motivation to write this text on thermodynamics is due to the huge tomes that other texts on the subject have become. The concern and detail with tangential subjects has created difficulty in the introduction of this subject to beginning engineering students. It's a challenge for a student to know what is important and what is of tangential interest. We have attempted to provide an introduction to thermodynamics by focusing on the material that is essential and have included only sufficient related material to provide insight as to how thermodynamics can be used to explain examples of everyday phenomena.

Thermodynamics, which involves the storage, transfer, and transformation of energy, is the first course in the thermal sciences for engineering students. It provides the foundation for the basic concepts and problem solving skills that are later used in fluid mechanics, heat transfer, and the design of thermo-fluid systems. This textbook will serve to develop the essential skills in thermodynamics, primarily in a one-semester course, but it will also have sufficient content for a second semester. The text is designed to provide a solid understanding of the principles, terminology, and methodology needed to thoroughly understand this subject.

The language of thermodynamics will be explained in careful detail so that students can quickly understand the concepts presented and the analysis techniques used. Extensive use of practical examples will demonstrate the proper set-up and solution of problems. These skills will then be further developed by providing a wide variety of homework problems. The homework problems are presented with an increasing degree of complexity to allow the solution of basic problems and also more challenging problems.

The structure of the book is such that it can be effectively used to support a single course in basic thermodynamics or a two-semester sequence of basic concepts and applications. The text is divided into three parts. In Part I, *Concepts and Basic Laws*, the terminology, concepts, and basic laws used in the subject of thermodynamics are presented, explained, and illustrated. In Part II, *Applications*, power and refrigeration cycles are presented in detail along with an introduction to mixtures, psychrometrics, and combustion. In Part III, *Contemporary Topics*, alternative energy sources and thermodynamics of living organisms are presented.

Part I includes the thermodynamic properties of materials and how they are used in the solution of engineering problems. Emphasis is placed on common working fluids used in industry in addition to air and water. Special attention is placed on using a structured problem solving procedure designed to understand the problem presented, organize the information given, and develop a solution to obtain the required results. This procedure is emphasized in numerous examples in the text and is intended to develop good problem solving skills in students. Topics covered include properties of substances, the first law of thermodynamics, work integrals, engineering devices, the second law of thermodynamics, and nonideal gas effects.

Part II applies thermodynamic principles to a number of engineering devices and cycles. If desired, selected topics in this part can be included in a first course. In this part we also analyze power plants, combustion engines, refrigeration systems, psychrometrics, and combustion which is foundational for subsequent courses in energy conversion, engines, and HVAC.

In Part III alternative energy is reviewed. The use of fossil fuel is not sustainable over the centuries to come, so sustainable sources of energy will be required. Several such energy sources are presented. Finally, the thermodynamics of living organisms is reviewed.

Properties of a number of substances are included in tables in the Appendix. Interpolation, a time-consuming procedure, is often required to determine the required properties. To avoid numerous interpolations, we have included the steps necessary to use the IRC Fluid Property Calculator, introduced in Chapter 2, which provides the most efficient method known by the authors to quickly determine material properties of several often encountered substances. You will greatly appreciate this Internet tool, which is officially referred to as the IRC (2012) *Fluid property calculator*, developed and maintained by the Industrial Refrigeration Consortium of the University of Wisconsin-Madison.

Thermodynamics is one of the first problem-solving courses in the mechanical or chemical engineering curricula. It also forms the foundation to the field of thermal sciences. Students learn how transferring energy to or from a substance can change the basic properties of the substance. It is equally important for them to develop skills in interpreting physical descriptions while solving practical engineering problems of interest.

It is assumed that students have completed courses in integral and differential calculus although algebra is the primary mathematical tool used to solve the majority of the problems in Thermodynamics. Some of the derivations do require some calculus concepts.

Many students take the Fundamentals of Engineering (FE) exam, the first step in becoming a professional engineer, at the end of their senior year. The problems in the FE exam are all four-part, multiple choice. Consequently, we have included this type of problem at the end of the appropriate chapters. These FE-type questions cover the material in the entire chapter so it may be best to respond to those questions when the chapter is concluded, or during the review for the chapter exam. Those FE-type questions could also be used as examples for multiple-choice exams. In fact, Thermodynamics is an excellent course in which to use multiple-choice exams; engineering students do not experience such exams in most, if not all of their other engineering courses and, since national exams are all multiple-choice exams, the experience is very beneficial to the students. Multiple-choice problems will be presented using SI units exclusively since the FE and GRE/Engineering exams use only SI units. Additional information on the FE exam can be obtained from a website at www.ppi2pass.com, or at www.ncees.org/Exams/FE_exam.php, or by Googling "NCEES."

The introductory material included in Part I, Chapters 1 through 7, has been selected carefully to introduce students to the fundamental areas of Thermodynamics. Not all of the material in each chapter need be covered in an introductory course. The instructor can fit the material to a selected course outline. A section or two at the end of several chapters may be omitted without loss of continuity in later chapters. After the introductory material has been presented, there is sufficient material to present an additional course, which could include material that was omitted in Chapters 1 through 7 and selected sections from Chapters 8 through 14.

We have included examples worked out in detail to illustrate each important concept presented. Numerous home problems, many having multiple parts for better homework assignments, provide the student with ample opportunity to gain experience solving problems of various levels of difficulty. All parts in problems with parts labeled i), ii), iii), etc., are expected to be worked, as in the examples. But, it is expected that only the selected part will be assigned in problems with parts labeled with lower-case italic letters [for example, *a*), *b*), and *c*)]. Answers to selected home problems are presented just prior to the Index.

After studying the material, reviewing the examples, and working several of the home problems, students should gain the needed capability to work many of the problems encountered in actual engineering situations in each topic presented. Of course, there are numerous classes of problems that are extremely difficult to solve, even for an experienced engineer. To solve these more difficult problems, the engineer must gain considerably more information than is included in this introductory text. There are, however, many problems of interest to the professional that can be solved successfully using the material and concepts presented herein.

The U.S. text was written with a combined SI and US Customary (English) system of units with emphasis on the SI system (approximately 20% of examples and home problems use English units), while this text uses only SI units.

The authors are very much indebted to both their former professors and to their present colleagues. Rickey Caldwell reviewed all of the text and made numerous helpful edits. We would also like to thank our reviewers who helped immensely in arriving at the final manuscript. They are:

Mahesh Chand Aggarwal, *Gannon University*
William Bathie, *Iowa State University*
Carlos F. M. Coimbra, *University of California, San Diego*
S. Mostafa, Ghiaasian, *Georgia Institute of Technology*
Pei-feng Hsu, *Florida Institute of Technology*
Melina Keller, *California Polytechnic University, San Luis Obispo*
John Kramlich, *University of Washington*
Edward Lumsdaine, *Michigan Technological University*
Sameer Naik, *Purdue University, West Lafayette*
Than Ke Nguyen, *California State University, Pomona*
Steven G. Penoncello, *University of Idaho*
Laura Schaefer, *University of Pittsburgh*
Elisa Toulson, *Michigan State University*

KENNETH KROOS
MERLE C. POTTER

The metrication of this text was done by *Shaligram Tiwari*, Indian Institute of Technology Madras.

About the Au

Kenneth A. Kroos

Education: BS in Physics: University of Toledo
MS in Mechanical Engineering: University of Toledo
PhD in Chemical and Biological Transport Phenomena: University of Toledo

Experience:

- Taught at Christian Brothers College and Villanova University
- Served as Student Section Advisor and Chair of the Memphis – Mid-south Section of ASME
- Taught thermodynamics, fluid mechanics, and several other courses
- Authored numerous publications in the fields of fluid mechanics, heat transfer, engineering education, and computer graphics for flow visualization
- Fellow of the ASME and a member of the American Society for Engineering Education. Received the ASME Dedicated Service Award
- Served as Vice President of ASME in 2001 and served a three-year term on the Council for Member Affairs.

Merle C. Potter

Education: BS in Mechanical Engineering: Michigan Technological University
MS in Engineering Mechanics: Michigan Technological University
MS in Aerospace Engineering: University of Michigan
PhD in Engineering Mechanics: University of Michigan

Experience:

- Taught at Michigan Tech, The U of Michigan, and Michigan State U
- Served as Student Section Advisor of ASME
- Authored and co-authored 35 textbooks, help books, and exam review books
- Performed research in fluid mechanics and energy conservation
- Received numerous awards that include:

  Teacher-Scholar Award
  ASME Centennial Award
  Member of Michigan Tech's Mechanical Engineering Academy
  James Harry Potter Gold Medal (Thermodynamics-ASME)
- Courses taught were on the subjects of mechanics, thermal sciences, and applied math

Thermodynamics *for* Engineers

Part I

Concepts and Basic Laws

In Part I, the terminology, concepts, and basic laws used in the subject of thermodynamics will be derived, explained, and illustrated. At least one example problem will demonstrate the application of each concept introduced, and numerous practice problems at the end of each chapter will allow students to reinforce the information presented. These concepts and basic laws will be applied to devices of interest to engineers; the devices will then be organized into several simple power and refrigeration cycles. In Part II, Applications, the power and refrigeration cycles will be studied in much more detail along with an introduction to psychrometrics and combustion. Part III, Contemporary Topics, will present alternative energy sources and thermodynamics of living organisms.

CHAPTER

1

Basic Concepts and Systems of Units

Gtranquillity/Shutterstock.com

Outline

Nomenclature

The following nomenclature is introduced in this chapter:

a	Acceleration
f	A representative function
F	Force
F_n	Normal component of a force
g	Gravity
g_c	Gravitational constant
h	Height
KE	Kinetic energy
m	Mass
P	Pressure
PE	Potential energy
R	Electrical resistance
SG	Specific gravity
T	Temperature
U	Internal energy
V	Volume
$\mathcal{V}$	Velocity
v	Specific volume
z	Elevation
α_T	Coefficient of thermal expansion
β	A characteristic constant
δ	Signifies an inexact differential
ρ	Density
ρ_x	Density of an unknown substance

Learning Outcomes

- ❑ **Understand the basic concepts of thermodynamics**
- ❑ **Understand the basic quantities in thermodynamics**
- ❑ **Work with SI and English systems of units**
- ❑ **Become familiar with basic properties**

Motivational Example—A Lost Orbiter

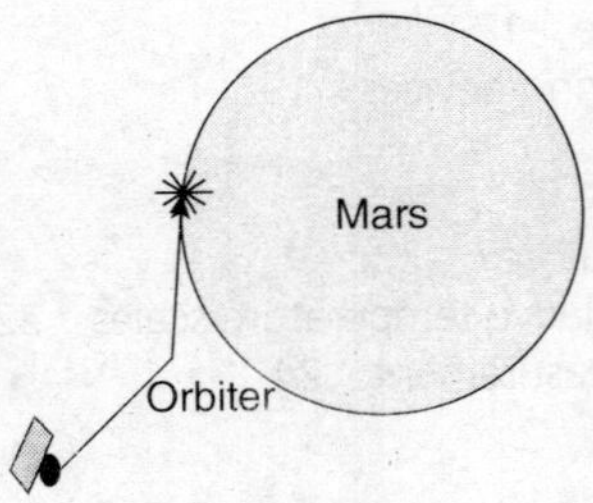

On September 23, 1999, NASA instructed the Mars Climate Orbiter (MCO) to perform a thruster burn that would put it into orbit around Mars. The purpose of this satellite was to monitor conditions in the atmosphere of Mars over an extended period of time. Measurements were to be taken to study the daily weather conditions, the atmospheric temperature profile, and water vapor, and dust content in the Martian atmosphere. Shortly after the command for orbital insertion was given, NASA lost contact with the orbiter. It was later determined that the orbiter crashed due to an error in the thrust measurement. The design called for a thruster impulse to send the orbiter into a Mars orbit. Because of a mix-up in the units between pounds of thrust and newtons of thrust, the proper orbit was not attained and the orbiter plunged into the planet. The proper use of units must be understood.

1.1 Introduction

1.1.1 What is thermodynamics?

Thermodynamics is a word that most people find difficult to comprehend. It is founded on basic physical laws that are very straightforward to present and apply, although some unusual words for physical properties are used, such as enthalpy and exergy. One of the purposes of this text will be to define these words in a way that gives their meaning real significance and to show how they are used by engineers.

Thermodynamics: The storage, transfer, and transformation of energy.

Thermodynamics involves the storage, transfer, and transformation of energy. Energy can be added to a mass or taken away from it. This is accomplished by several physical processes that we will study in depth in this text. For example, when we burn gasoline in a vehicle engine, the heat created by burning a mixture of air and fuel in a cylinder dramatically increases the pressure and temperature in the cylinder. The high pressure is used to push a piston in the cylinder, which leads to the production of power. This example demonstrates two types of thermodynamic processes. First, the chemical energy contained in the gasoline is released as heat when the gasoline in the air undergoes combustion. Second, the subsequent high pressure moves the piston, thereby doing work. Heat transfer and work—to be carefully defined in the thermodynamics context in Chapter 3—are the two most important ways that energy is transferred by processes of interest in our study.

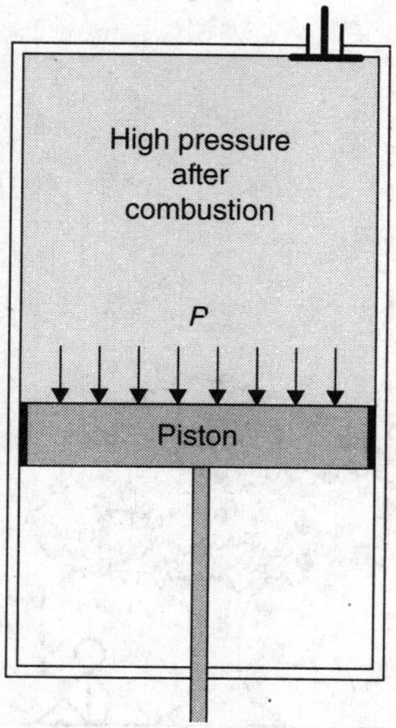

Another example of the transformation of energy is the process of photosynthesis in plants, described in more detail in Chapter 14. Light energy from the sun is absorbed by chlorophyll in the leaves of plants to manufacture sugars that are used to feed the plant. A similar example is the use of a photovoltaic solar cell, introduced in Chapter 13, which transforms energy in sunlight to electricity.

In all of these examples, energy is being used to produce a desirable outcome. But energy doesn't always increase temperatures; it can be used to decrease temperatures as in refrigeration systems. When you think of thermodynamics, think of energy.

Thermodynamics is the first in a series of subjects that constitutes thermal sciences. Thermodynamics is usually followed by courses in fluid mechanics and heat transfer. In some curricula, these courses are followed by a comprehensive design course in thermal-fluid systems, which integrate all three subjects into a design-oriented experience. Elective courses in energy conversion, engine design, power-plant design, and propulsion may follow.

1.1.2 How do we use thermodynamics?

Comment

Thermodynamics is foundational in the design of conventional power plants.

Anytime energy is stored, transferred, or transformed from one form of energy to another, we are applying thermodynamics. Thermodynamics is foundational in the design of conventional power plants, both large and small. In these plants that burn fossil fuels or use nuclear energy, energy in the form of heat is converted to energy in the form of work by transforming water into steam and using this steam to power turbines, which in turn transform mechanical power into electrical power using generators. Electrical engineers assume responsibility for the power after it passes from the turbines to the generators.

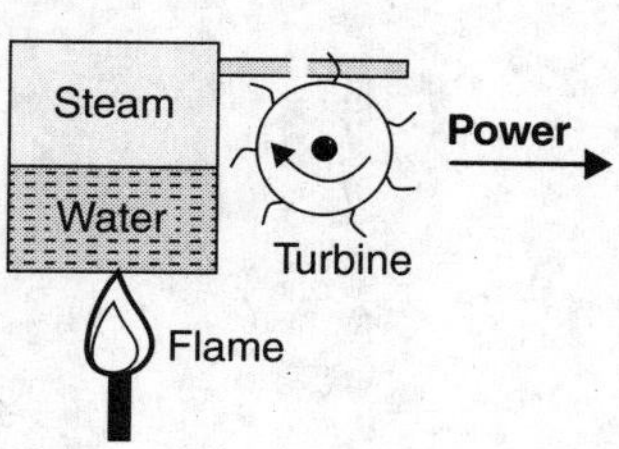

Thermodynamics is used in the design of engines ranging from small engines that power a model airplane to automobile engines, to jet engines, and to the largest engines that power ships. It is also used to analyze alternative energy in hydroelectric plants, wind machines, and solar power generators. Thermodynamics finds application in the analysis of a myriad of industrial processes that produce chemicals, medicines, and food. It also provides an overview of how nature converts the energy in food into energy in animals. Thermodynamics can also be used to understand how solar energy is converted into nutrition for plant life.

1.1.3 How do engineers use thermodynamics?

Thermodynamics is used by engineers who work in a wide variety of fields, mainly by mechanical and chemical engineers. Mechanical engineers are responsible for the design, construction, and operation of major power plants. They also design engines used in cars, boats, airplanes, and ships. Both mechanical and chemical engineers apply thermodynamics in the design of manufacturing processes to produce consumer products, pharmaceuticals, and food products. Engineers, using thermodynamics, are developing new sources of energy to replace fossil fuels, which are a major source of pollutants. They are working in the fields of wind energy, hydropower, biofuels, hydrogen generation, and solar energy, to name a few, in an attempt to eventually develop a sustainable energy system.

1.1.4 What is the history of thermodynamics?

Law: Developed from direct observation under strict constraints.

Humans have had an interest in thermodynamics ever since our species first used fire to keep warm and cook food. From about 3000 BC to 500 BC, Egyptian and then Greek scientists speculated that heat energy was a fluid that they called phlogiston. Fire was one of the basic elements of nature along with earth, wind, and water. Around 500 BC, Greek scientists were debating whether a vacuum could possibly exist or be created. After hundreds of years, the seventeenth and eighteenth centuries saw the transformation from the primitive magic of alchemy to the subject of thermochemistry. In this period, many of the first scientific laws predicting how gases behaved were developed. A *law* is developed from direct observation and has strict constraints. In 1662, Boyle's law, a special case of the ideal-gas law, was published. In 1802 both Charles's law and Guy-Lussac's law, both of which are related to the ideal-gas law, were published.

The first engine, a steam engine, was patented in 1769 by James Watt. In 1824, Sadi Carnot published his work on the equivalence of work and heat. For this work he is known as the "father of thermodynamics." The industrial revolution brought about great changes in engineering and in society. Products that were once made in "cottage" industries, like blacksmith shops, were now manufactured in large industrial plants. One of the side effects of this industrial growth was the need for mechanical engineers who could design and operate the power plants necessary to run factories.

1.1.5 What is the future of thermodynamics?

One of the most important problems that we face today is the diminishing availability of usable energy. In the 1800s and early 1900s, energy was readily available, plentiful, and cheap. Industry was able to thrive and expand because energy supplies were abundant. Today this picture is changing. We know that our sources of fossil fuels, notably coal, petroleum, and natural gas, are limited and will run out someday. Alternative sources of energy, such as solar power, wind power, and hydrogen, are being developed but are far from being able to replace fossil fuels. Nuclear power is another energy source, but public concern about radiation leaks and how to dispose of "spent" fuel rods has slowed this industry to a standstill in the past several decades. In addition, accidents at Chernobyl, Three Mile Island, and Japan have raised serious concerns about the safety of nuclear reactors. Because nuclear power produces little if any CO_2, and public concern for its safety is becoming less problematic, it is regaining an interest in the United States. Nuclear power has been continually increasing in other parts of the world. France and Japan account for over 50% of the world's nuclear power-generating capacity, although that has been decreasing because of recent accidents due to earthquakes and other natural disasters. More effective safeguards are required.

Questions: Is nuclear power once again in our future? Is it safe? Did the Japan containment failure in 2011 doom it as a safe power source?

The study of how processes and our life style affect the long-term availability of the natural resources required to perform the processes and sustain our life style is called *sustainability*. When we review our energy, water, food, and mineral resources we have to consider the long-term effects of our consumption of these resources. Future generations must be able to meet their needs. Even today, many communities in the world suffer from severe lack of drinkable water and available firewood; they live an unsustainable way of life. We will attempt to incorporate sustainability into the various chapters of this book.

Sustainability: The capacity to endure.

1.1.6 What are the fundamental concepts and assumptions?

The term *thermodynamics* is something of a misnomer since "dynamics" suggests motion. Even though the process of adding or removing energy from a substance is by definition a dynamic process, we will not be interested in the instantaneous steps required to change from an initial state to a final state. Instead, we will consider the "snapshot" pictures of the state of a system before and after a process has occurred. Properties of a stable state will be defined prior to the process, and those same properties will be identified at the end of the process. This exercise will yield "before" and "after" pictures of the system but will not detail what is happening during the energy transfer process. The objective of the problem will be to determine the final state, given the initial state. We will specify the initial state and describe the process that occurs. From this information we will then solve for the final state. Occasionally, the final state will be identified and the initial state will be sought.

A *system* is defined to be a fixed mass that occupies a space, a space that may or may not be changing in volume or shape. The mass of a system, such as the helium in a helium balloon, does not change. We then analyze how a particular system changes as energy is added or removed. The *surroundings* include everything external to the system. If the system does not exchange energy with the surroundings, it is an *isolated system*.

In many situations, it is not reasonable to focus attention on a fixed mass; rather, the focus should be on a fixed volume into which and/or from which a fluid may flow, such as a pump, a turbine, or an emptying propane tank. Such a volume is called a *control volume*, and the surface that surrounds the control volume is the *control surface*. The surroundings then include everything external to the control volume. The form that the basic equations in thermodynamics take for

System: A fixed identified quantity of mass.

Surroundings: Everything external to the system.

Isolated system: A system that does not exchange energy with the surroundings.

Control volume: A fixed volume into which and/or from which a fluid flows.

Control surface: The surface that surrounds the control volume.

a system differ from those for a control volume, so it's important to know which is being analyzed.

Continuum: A substance occupies all points in a region of interest.

Note: There are about 3×10^{16} molecules in a cubic millimeter of air. The assumption of a continuum is quite reasonable for all substances of interest considered in this text.

There are two approaches to the study of thermodynamics. Classical thermodynamics is a macroscopic, or global, approach in which we assume that a substance is a *continuum* in that it occupies all points in a region of interest. There are about 3×10^{16} molecules in a cubic millimeter of air at sea level, so the assumption that air occupies all points in a volume is quite reasonable when considering problems of interest. The properties of a finite quantity of matter, the system, are treated as spatially averaged properties. For example, when you are in a room, you think in terms of the room temperature or pressure. This is a temperature or pressure averaged over the entire room, as opposed to properties measured at many points in the room.

Another approach to thermodynamics is encountered in statistical thermodynamics, where the motions of molecules are analyzed using statistical methods to predict how a substance will react to the addition or subtraction of energy. Statistical mechanics relates molecular activity to macroscopic thermodynamic quantities. Properties of bulk materials are related to the spectroscopic data of individual molecules. Statistical thermodynamics is a specialized subject offered either as an undergraduate elective or as a graduate course. In this text, the motion of individual molecules will not be of interest.

PD-USGOV

A power plant.

In this introduction to thermodynamics, we will utilize the concepts and methodology of classical thermodynamics to solve problems encountered in common engineering systems. Even though simplifying assumptions will be made about the systems being analyzing, classical thermodynamics is a very powerful tool in understanding and designing the numerous devices utilized in the production of energy, such as in engines, power plants, and refrigerators.

1.1.7 What are the phases of matter?

Solid: Does not take on the shape of its container.

Liquid: Takes on the shape of its container but does not expand to fill the entire volume available.

Gas: Fills the entire volume available. Its molecules are relatively far apart.

In thermodynamics we study the three basic phases, or states, of matter: the solid phase, the liquid phase, and the gas phase. A *solid* does not flow to take on the shape of its container, nor does it expand to fill the entire volume available. Solid molecules may shift relative to each other when subjected to a stress, but they do not move continuously or independently with respect to neighboring molecules. A *liquid* flows to take on the shape of its container but does not expand to fill the entire volume available, whereas a *gas* expands to fill the entire volume available. Liquids and gases move independently when subjected to a stress. In fact, they will move continuously as long as the stress is applied. Water is a good example in thermodynamics of the three phases: ice (solid), liquid, and vapor (gas). Courses in fluid mechanics deal extensively with analysis of the motion of liquids and gases.

☑ You have completed Learning Outcome (1)

1.2 Dimensions and Units

It's quite easy to confuse dimensions of a quantity with the units used to measure those dimensions. Dimensions are used to describe a quantity, whereas units provide the magnitude of those dimensions. There are two types of dimensions, primary (or fundamental) and derived. *Primary dimensions* are mass, length, time, and temperature (force, length, time, and temperature could have been selected). Other dimensions that describe electric and magnetic properties could also be included in a list of primary dimensions, but these are not of interest in our study. *Derived dimensions* are a combination of primary dimensions. For example, Newton's second law of motion defines a force as the product of a mass times its acceleration; this law is stated as

Key Concept: A unit is used to measure a dimension. Mass is a dimension, kg is a unit.

Primary dimensions: Mass, length, time, and temperature.

Derived dimensions: Combination of primary dimensions.

$$F = ma \tag{1.1}$$

We have selected mass m to be a primary dimension. Acceleration a has the derived dimensions of length divided by time squared. The dimensions on force F are a combination of length, mass, and time as required by Eq. 1.1. Velocity is measured by the dimensions of length divided by time.

Two major systems of units are in use today. The U.S. Customary System of Units is used by the United States, and the SI system (Système international d'unités), a particular metric system, is used by most other nations. Both systems will be used in the U.S. version of this text, but only the SI system will be used in the international version.

An important concept involving dimensions is that of *dimensional homogeneity*. It demands that all terms in an equation must have the same dimensions. A quick check on the validity of an equation is to make sure that the dimensions on all terms are the same. If the dimension is force on one term in an equation, then all terms must have the dimension of force. Then, when units are assigned to the quantities in an equation, make sure the units on each term are the same; that is, if the unit is kN on one term, it cannot be N on another term in the same equation.

Dimensional homogeneity: All terms in an equation must have the same dimensions.

We finish this section with comments on significant digits. In almost every calculation, a material property is involved or a number is the result of a measurement. Material properties are seldom known to four significant digits and often only to three, and measurements are made to three and possibly four significant digits. So, it is not appropriate to express answers to five or six significant digits. Calculations are only as accurate as the least accurate number in our equations. For example, we use gravity as 9.81 m/s^2, only three significant digits; a diameter may be stated as 2 cm, which is assumed to be 2.00 or 2.000, three or four significant digits. It is usually acceptable to express answers using four significant digits, but not five or six. The use of calculators may even provide eight. The engineer does not, in general, provide results to five or six significant digits.

Comment

Calculations are only as accurate as the least accurate number in a calculation.

Observation: "Digits" and "figures" are synonyms.

1.2.1 The SI system

The SI system of units was created in 1793 by the French government as a decimalized alternative to the English system. Use of the SI system spread throughout Europe as a result of Napoleon Bonaparte's military conquests. The primary dimensions and their units are shown in Table 1.1. The unit of force in the SI system is the newton. To obtain the force F in newtons, multiply the mass m in

Table 1.1 Primary Dimensions and Units for the SI System

Dimension	Unit	Abbreviation
Length	meter	m
Time	second	s
Mass	kilogram	kg
Temperature	degree	K or °C

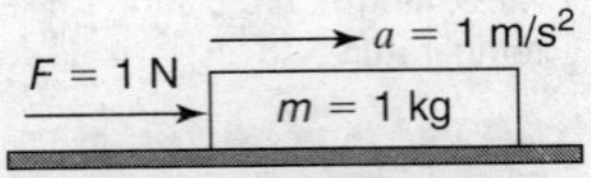

Figure 1.1.

A force accelerating a mass on a horizontal frictionless surface.

kilograms by the acceleration a in meters per second squared, as stated by Eq. 1.1. One newton accelerates a one-kilogram mass one meter per second squared when acting on a frictionless horizontal surface, as shown in Fig. 1.1. So, Eq. 1.1 takes the form

$$1\ \text{N} = 1\ \text{kg} \times 1\ \text{m/s}^2 \tag{1.2}$$

showing that $\text{N} = \text{kg}\cdot\text{m/s}^2$. One newton is equivalent to $1\ \text{kg}\cdot\text{m/s}^2$.

When expressing a quantity in SI units, certain letter prefixes, shown in Table 1.2, may be used to represent multiplication by a power of 10. So, rather than writing 30 000 N (commas are not used in the SI system) or 30×10^3 N, we may simply write 30 kN.

Comment

We write 10 newtons, not 10 Newtons, following the National Institute of Standards and Technology (NIST) rules. The SI system follows a very detailed set of rules.

The SI system of measurement is used by the entire world except Burma, Liberia, and the United States. Products manufactured in the United States using the English system of units are often incompatible with designs developed using the SI system. Engineers practicing in the United States are encouraged to be familiar with both systems of units, although a particular industry may have its own set of units, which may not be either SI or English units. In a world where trade is taking place between most countries, it is imperative that we all have a uniform set of units. The SI system provides such a system, and the United States should move more quickly to adopting it in all its industries.

Comment

Recall our Motivational Example: A NASA spacecraft was lost because the conversions between units were confused.

The derived units for other properties, such as work, thermal energy, and power, will be presented in future chapters where these terms are defined.

This sub-section is included to inform the reader of the U.S. Customary system of units that remain in use in several countries, including the United States. Only the SI system of units will be used in the examples and problems in this text so this sub-section and Example 1.1 may be omitted if desired.

Table 1.2 Prefixes for SI Units

Multiplication factor	Prefix	Symbol
10^{12}	tera	T
10^{9}	giga	G
10^{6}	mega	M
10^{3}	kilo	k
10^{-2}	centi[1]	c
10^{-3}	milli	m
10^{-6}	micro	μ
10^{-9}	nano	n
10^{-12}	pico	p

[1] Discouraged except when measuring length, area, and volume: cm, cm^2, or cm^3.

1.2.2 The English system

The U.S. Customary System of Units is based on a traditional English system of units and will hereafter be referred to as the English system, even though the English no longer use it. The primary dimensions in the English system are listed in Table 1.3.

Table 1.3 Primary Dimensions and Units for the English System

Dimension	Unit	Abbreviation
Length	foot	ft
Time	second	s
Mass	slug or pound	slug or lbm
Temperature	degree	°R or °F

The traditional unit of mass in the English system is the slug. The unit of force is the pound. One pound of force is needed to accelerate a one-slug mass at a rate of one foot per second squared on a horizontal frictionless surface. It is represented by Newton's second law of motion:

$$1 \text{ lbf} = 1 \text{ slug} \times 1 \text{ ft/s}^2 \tag{1.3}$$

which provides the relation slug = lbf-s^2/ft.

This system worked very well until force was equated to mass. If an object weighs one pound, and the acceleration of gravity is $g = 32.2$ ft/s^2, then it has a mass of one pound. At some point, people started describing a mass that weighed one pound as having a mass of one pound, even though mass and force are two very different quantities. All this resulted in a system that defines two units of pounds. There is the pound force, lbf, and the pound mass, lbm. One pound mass (lbm) weighs one pound force (lbf) assuming gravitational acceleration of 32.2 ft/s^2. If it is not 32.2 ft/s^2, the lbf and the lbm are not equal in magnitude. However, for most problems considered in this book, $g =$ 32.2 ft/s^2, so 1 lbm does weigh 1 lbf.

In order to make the English system using lbm and lbf work, we define a *proportionality constant* $g_c = 32.2$ lbm-ft/lbf-s^2. Apply Newton's second law, assuming a mass of 1 slug (1 slug = 32.2 lbm), and Eq. 1.1 provides

$$F = \frac{32.2 \text{ lbm}}{32.2 \text{ lbm-ft/lbf-s}^2} \times 1 \frac{\text{ft}}{\text{s}^2} = 1 \text{ lbf} \tag{1.4}$$

So, we see that a force of 1 lbf accelerates a mass of 32.2 lbm at an acceleration of 1 ft/s^2. Consequently, when using English units in thermodynamics with the mass measured in lbm, Newton's second law, Eq. 1.1, can be written as

$$F = \frac{m}{g_c} a \tag{1.5}$$

In the SI system, the proportionality constant is not needed when using the force in newtons, the mass in kilograms, and the acceleration in m/s^2. An example will illustrate the English system.

Conversion:

$$\text{slug} = \frac{\text{lbf-s}^2}{\text{ft}}$$

Comment

To convert a unit, simply Google it. For example, if you Google "10 lbf in N" you will get "10 pounds force = 44.4822162 newtons." Or, check for the proper conversion in Appendix A. Mr. Google is not an engineer, so nine significant digits were used.

Proportionality constant:

$$g_c = 32.2 \frac{\text{lbm-ft}}{\text{lbf-s}^2}$$

Comment

Thermodynamics is one of the few courses, if not the only course, to use the lbm and the lbf, often a rather confusing combination. It's slowly being phased out as SI units are phased in.

Example 1.1 What force is needed to accelerate a car?

Consider the force required to accelerate a car on a horizontal road, as sketched in Fig. 1.2. What force do the wheels of a car that weighs 9000 N need to apply to accelerate the car at a rate of 3 m/s^2?

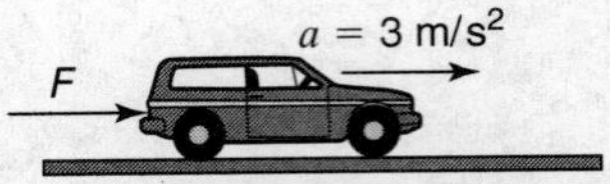

Figure 1.2

Solution

If the car weighs 9000 N, it has a mass of 917.4 kg, assuming $g = 9.81$ m/s^2, which will be the assumption if the local gravitational acceleration is not given. Newton's second law provides

$$F = ma = \frac{W}{g}a = \frac{9000\text{ N}}{9.81\text{ m/s}^2} \times 3\text{ m/s}^2 = \underline{2752\text{ N}}$$

Comment

Gravity g is essentially a constant anywhere on the Earth's surface. It varies by less than 1% from the top of the highest mountain to the bottom of the deepest ocean trench. So, using $g = 9.81$ m/s^2 everywhere on the Earth's surface is quite acceptable.

☑ You have completed Learning Outcome (2)

1.3 Properties, Processes, and Equilibrium

1.3.1 Properties and state of a system

Properties: Characteristics of a substance that can be measured or calculated from measurements.

State: Defined by the specific values of the properties of a substance.

Properties are characteristics of a substance that can be measured, or calculated from direct measurements. Common properties such as size, volume, pressure, and temperature are used in everyday life; they are relatively easy to determine. Other properties, such as the amount of energy contained in a specified mass of a substance or the moisture content in the air or even acceleration, are not easy to measure. Properties are coupled to what is called the state of a substance. The *state* is defined by specified values assigned to the properties of a substance. For example, an ice cube is defined by its size and shape as well as its temperature. For the ice cube to be ice, its temperature must be at or below the freezing temperature of water. If we change the properties of a substance, its state will change. If we increase the temperature of the ice cube sufficiently, it will change to liquid water. As a result, not only will the temperature change but also its shape will change as it becomes a liquid.

In this chapter we will consider thermodynamic properties that are measurable or directly determined from measurable properties. Properties such as size, pressure, and temperature can be measured using a variety of devices, but properties such as volume and density are directly determined from the measured properties.

In the next chapter we will consider several unfamiliar thermodynamic properties that can be determined from the measurable properties.

There is circular reasoning here, which is sometimes hard to follow. The properties of a substance are defined by its state. The state of a substance is defined by specific values of its properties. The two concepts are coupled together. In thermodynamics, we want to know the effect of adding energy to or subtracting energy from a substance. This effect will be shown by a change in the properties of the substance, that is, by a change of state. If the properties of a state are determined, that state is completely described.

A property has a unique value when a system is in a particular state; its value does not depend on the states through which the system passed—that is, it is not a path function. If f represents a property, then

$$\int_{f_1}^{f_2} df = f_2 - f_1 \tag{1.6}$$

and df is referred to as an exact differential of f. The quantity $f_2 - f_1$ represents the change in the property as the system changes from state 1 to state 2.

Quantities, such as work W, will be encountered that are path functions and whose differentials are not exact. Its integral from state 1 to state 2 would be represented by

$$\int_{\text{state 1}}^{\text{state 2}} \delta W = W_{1\text{-}2} \tag{1.7}$$

The differential δW signifies an inexact differential. Such path-dependent quantities are not referred to as properties and do not have a value at a given state; they are never represented by W_1 or W_2.

Key Concept: The differential of a property is exact and is represented by df, whereas the differential of a path function is inexact, and is represented by δf.

Properties are also classified as extensive and intensive. An *extensive property* is a property whose value depends on the mass of the system, such as volume and weight. An *intensive property* is a property whose value is independent of the mass of the system. The temperature of a substance is the same no matter how much of it you have; it is intensive since it does not depend on the mass. A good way to decide if a property is extensive or intensive is to divide the system in half, delete one half, and determine if the property decreases by a half. If it does, the property is extensive; if it remains the same, it is intensive.

Extensive property: A property whose value depends on the mass of the system.

Intensive property: A property whose value is independent of the mass of the system.

Consider the air in the room you are occupying. If the room is divided in two and you now have a smaller space, both the volume and the mass of the smaller space are reduced. This makes volume and mass extensive properties. Consider the density, pressure, and temperature of the air; they remain the same for both spaces; none of them change when the room is divided, so they are intensive properties.

1.3.2 Density and specific volume

When identifying the volume of an object or empty space, we might select Cartesian, cylindrical, or spherical coordinates. Volume V is a physical property that is treated as a variable in thermodynamics. The volume of a gas will decrease if it is compressed and will increase if the gas is allowed to expand. A good example is the air inside an engine's cylinder. In one process, the air–fuel mixture is compressed by the piston decreasing the volume of the mixture. As a result, the pressure and temperature of the air increase. During the power stroke, the piston moves out due to the high pressure after combustion, thereby increasing the volume of the combusted mixture.

Density: The mass divided by the volume occupied by the mass.

The *density* ρ of a substance is the mass m divided by the volume V occupied by the mass:

$$\rho = \frac{m}{V} \tag{1.8}$$

The unit of mass is the kilogram (kg). Density has units of kg/m^3. The higher the density, the "heavier" a substance. Solids and liquids, especially metals, will have relatively high densities. Gases, on the other hand, have relatively low densities.

In thermodynamics, we treat density as a variable, one that can change during a process. Density can vary with the temperature of a substance and, especially for gases, with the pressure of the substance. When solving problems involving liquids or solids, density is usually treated as a constant.

Specific volume: The volume occupied by a substance divided by its mass.

Specific volume v is defined as the volume occupied by a substance divided by its mass:

$$v = \frac{V}{m} \tag{1.9}$$

It is the inverse of the density of a substance. The units on specific volume are m^3/kg.

Specific property: It is formed by dividing an extensive property by the mass of the system.

Note: In thermodynamics, the word *specific* has a special meaning. It means "per unit mass." Thus, specific volume is volume per unit mass. In general, an upper-case letter denotes the property, and a lower-case letter denotes the specific property. An exception is the mass m.

Comment

In thermodynamics we most often use specific volume rather than density.

In most engineering courses, the density of a substance is used, whereas in thermodynamics we most often use specific volume, the reciprocal of density. It takes a little "getting used to" to become comfortable with using specific volume. A substance with a high density like liquid water has a low specific volume. A substance like air at sea level has a density and a specific volume that are of the same order of magnitude when measured using the SI system of units. Table 1.4 shows a comparison of these values.

Specific gravity: The ratio of the density of a liquid to the density of standard water. It is seldom used in our study of thermodynamics and is included because it does occur occasionally in certain courses.

Another term you should be familiar with is specific gravity. *Specific gravity* (SG) is defined as the ratio of the density of a substance to the density of standard water:

$$SG = \frac{\rho_x}{\rho_{\text{water}}} \tag{1.10}$$

Table 1.4 Density and Specific Volume for Water and Air at Standard Conditions (15°C, 101 kPa, 59°F, 14.7 psia)

Substance	Density	Specific Volume
Liquid water	1000 kg/m^3 62.4 lbm/ft^3	0.001 m^3/kg 0.0160 ft^3/lbm
Air	1.22 kg/m^3 0.0765 lbm/ft^3	0.820 m^3/kg 13.1 ft^3/lbm

A substance with SG greater than one is heavier than water; if SG is less than one, it is lighter than water. Many handbooks will publish the specific gravity of liquids rather than densities or specific volumes.

Properties associated with volume — Example 1.2

A substance occupying a volume of 50 cm^3 has a mass of 0.1 kg. Determine its density, specific volume, and specific gravity.

Solution

The density is the mass divided by the volume:

$$\rho = \frac{m}{V} = \frac{0.1 \text{ kg}}{50 \text{ cm}^3 \times 10^{-6} \text{ m}^3/\text{cm}^3} = \underline{2000 \text{ kg/m}^3}$$

Its specific volume is the volume divided by the mass or the reciprocal of the density:

$$v = \frac{V}{m} = \frac{50 \times 10^{-6} \text{ m}^3}{0.1 \text{ kg}} = \underline{0.0005 \text{ m}^3/\text{kg}} \text{ or } v = \frac{1}{\rho} = \frac{1}{2000} = \underline{0.0005 \text{ m}^3/\text{kg}}$$

The specific gravity is found using Eq. 1.10 to be

$$SG = \frac{\rho_x}{\rho_{\text{water}}} = \frac{2000 \text{ kg/m}^3}{1000 \text{ kg/m}^3} = \underline{2.00}$$

Note: Specific gravity is dimensionless.

1.3.3 Processes and equilibrium

When we solve problems, the initial state of a substance will be given by specifying its properties. The addition or subtraction of energy will cause a change of state. From the information given for the initial state and the *process* that a substance passes through as it changes states, we can determine the final state of the substance. A system that exists in an initial state, undergoes two or more processes, and returns to its initial state completes a *cycle*.

In classical thermodynamics, we assume that a system is in *thermodynamic equilibrium* if it is in thermal equilibrium, mechanical equilibrium, phase equilibrium, and chemical equilibrium. That means the temperature is assumed to be the same, or essentially the same, at all points in a system or control volume; the pressure does not change from point to point in a relatively small volume even as the pressure may be changing with time, although it may change with elevation in a large volume, as in the atmosphere or a lake; the mass of each phase remains constant; and the chemical composition does not change with time. If a small change is made to the state of a system in thermodynamic equilibrium, the system will return to its prior state. If a large change occurs due to a small change to the

Process: The states that a substance passes through as it changes from an initial state to a final state.

Cycle: A system that exists in an initial state, undergoes two or more processes, and returns to its initial state.

Thermodynamic equilibrium: Requires thermal (uniform temperature), mechanical (constant pressure at each point), phase (constant mass of each phase), and chemical (no reactions) equilibriums.

state of a system that is in equilibrium, such as a small displacement to a bowling ball on the top of a larger sphere, the system is in metastable equilibrium; such systems are not of interest in our study.

A quasi-equilibrium process: If a system, in passing from one equilibrium state to another equilibrium state, deviates from equilibrium by a very small amount, each state in between the two end states is idealized as an equilibrium state. The quasi-equilibrium process refers to the system only. It matters not what is happening to the surroundings.

One of the most important equilibrium concepts in our study is that of the *quasi-equilibrium process*: If a system, in passing from one equilibrium state to another equilibrium state, deviates from equilibrium by a small amount, each state in between the two end states is idealized as an equilibrium state. If a property changes at a point, such as the pressure in the air in a cylinder being compressed by a piston, all pressures in the volume are assumed to change simultaneously. A piston can be moving at 3000 rpm, yet it is slow thermodynamically (molecules are extremely fast) and the process is modeled as a quasi-equilibrium process. If the process represents combustion in a cylinder during which the pressure changes abruptly, it is a *nonequilibrium process* and a dashed line may be used, as shown in Fig. 1.3 from state 2 to state 3, since we do not know what path it takes. A quasi-equilibrium process from state 1 to state 2 is represented by a solid line.

Finally, we should introduce the *state postulate*, which states that in order to establish the equilibrium state of a simple compressible system, we must know two independent, intensive properties. Two properties are independent if one can vary while the other is held constant. A simple compressible system is free of magnetic, electrical, gravitational, and surface tension as well as effects due to motion. The two properties that establish the state cannot depend on each other; they must be independent.

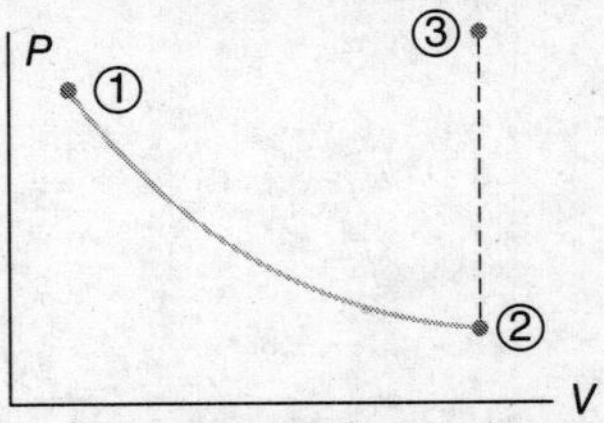

Figure 1.3
A quasi-equilibrium process from 1 to 2 and a nonequilibrium process from 2 to 3.

Example 1.3 Is it a quasi-equilibrium process?

A particular process could be considered a quasi-equilibrium process, or it could be a nonequilibrium process. Consider the weight W to be placed on a frictionless piston in a cylinder containing air. Explain how it could be one or the other.

Solution

To be a quasi-equilibrium process, the weight W would be divided into numerous weights whose total weight is W. The weights would be placed, one at a time, on top of each other with time in between placements until the total weight W is reached, as displayed in Fig. 1.4*a*.

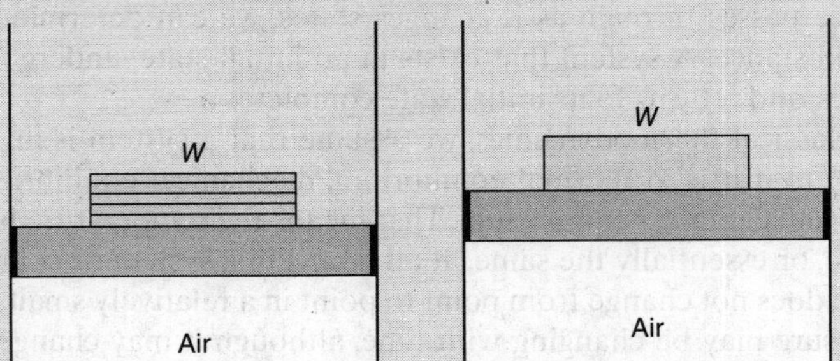

Figure 1.4

A nonequilibrium process would occur if the total weight W were suddenly dropped on the piston, as shown in Fig. 1.4*b*. The position of the piston would not be the same after each process occurred due to the energy that would be lost in the nonequilibrium process.

☑ You have completed Learning Outcome (3)

1.4 Pressure

1.4.1 What is pressure?

Pressure is a thermodynamic property of liquids and gases. As molecules of air move randomly in a room, they strike surfaces in the room, creating an impact-momentum reaction that produces a normal impulse force on the surface. The pressure force is the sum of the multitude of these reactions that occur over the surface. *Pressure* P is the normal component F_n of a force F acting on an area, displayed in Fig. 1.5, divided by the surface's area A, written as

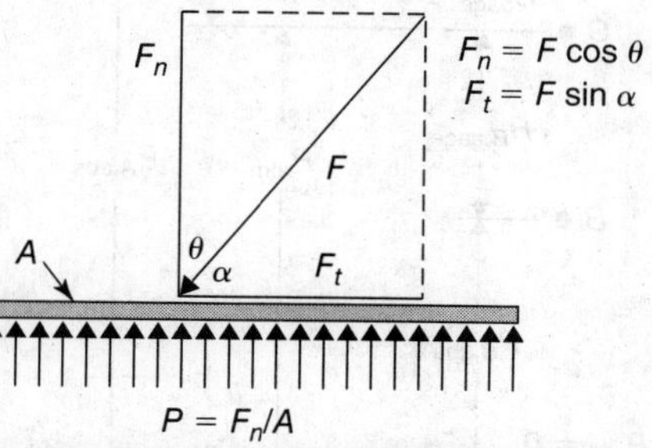

Figure 1.5

The relationship between a pressure and a force.

$$P = \frac{F_n}{A} \tag{1.11}$$

The tangential component F_t does not influence the pressure.

A good example of pressure is the force on our bodies due to atmospheric pressure. The pressure in the atmosphere is a function of altitude above the Earth. In outer space, there is no pressure since there are no air molecules to create pressure. As we travel from the outer edge of the atmosphere to the Earth's surface, the atmospheric pressure will increase due to the weight of the air. The pressure at a point on the Earth's surface is due to the weight of the air above that surface. Mountain climbers who scale the highest peaks in the world must carry oxygen tanks because the air pressure gets too low to support proper respiration (at the top of Mount Everest, atmospheric pressure is about 25% of the pressure at sea level).

Pressure: The normal component of a force acting on a surface area divided by the area of the surface.

A device that measures pressure in the atmosphere is called a *barometer*. Atmospheric pressure is seldom constant: It changes significantly with ground elevation and slightly with the weather. A drop in air pressure is an indicator of stormy weather ahead. A rise in air pressure usually indicates fair weather ahead. This is why a good barometer was an essential tool for ship captains. It allowed them to determine if a change in the weather was expected.

Barometer: A device that measures pressure in the atmosphere.

1.4.2 Absolute and gage pressure

There are two types of pressure used in engineering. The first is called *absolute pressure* since it is measured relative to absolute zero pressure. Absolute zero

Absolute pressure: The pressure measured relative to absolute zero pressure.

Gage pressure: The pressure measured relative to atmospheric pressure.

Vacuum: A negative gage pressure.

pressure occurs when there is no molecular activity resulting in a complete absence of any pressure force. Outer space is a good example of absolute zero pressure since it is essentially devoid of molecules. For many engineering calculations we must use absolute pressure. It is impossible to have an absolute pressure less than zero.

The second type of pressure is *gage pressure*, which is measured relative to the current atmospheric pressure. Pressure above atmospheric pressure is positive, and pressure below atmospheric pressure is negative gage pressure. Negative gage pressure is also called a *vacuum*. When we analyze car engine performance, we often use vacuum gages to measure the suction pressure, a negative gage pressure, in engine cylinders during intake.

Absolute pressure can be calculated from gage pressure by adding the current atmospheric pressure:

$$P_{\text{absolute}} = P_{\text{gage}} + P_{\text{atmospheric}} \tag{1.12}$$

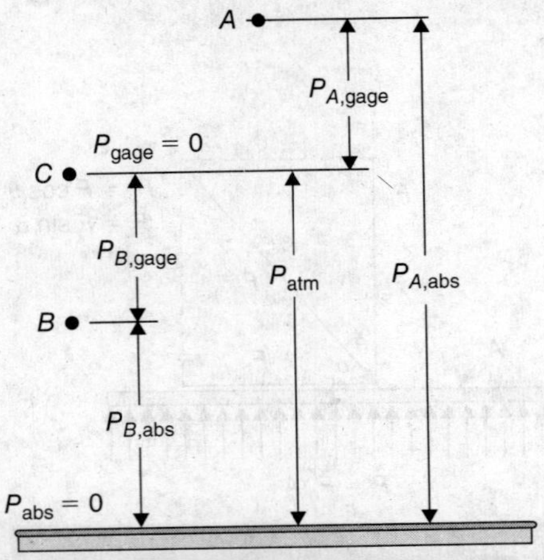

Figure 1.6
Absolute and gage pressures.

Thus, if you require absolute pressure for a calculation, and only a device that measures gage pressure is available, an accurate barometer would be needed to obtain an accurate value of the atmospheric pressure, although near sea level, the atmospheric pressure can be assumed to be 100 kPa. Figure 1.6 illustrates absolute and gage pressures. The gage pressure at A would be positive, whereas at B it would be negative, a vacuum. The gage pressure at C would be zero since it represents atmospheric pressure.

It is always important to know which type of pressure is given. If a problem states that a gage was used to measure a pressure, it is a gage pressure. In thermodynamics if the word "gage" is not stated in the problem, it is always assumed that the pressures given are absolute pressures.

Comment

If the word "gage" is not stated, we will always assume an absolute pressure.

1.4.3 Units of pressure

Pressure has units of force divided by area. The basic unit of pressure in the SI system is the force in newtons divided by the area in square meters. We refer to 1 N/m^2 as 1 Pa, where Pa represents a pascal. One pascal of pressure is very small, so the common unit of pressure is the kilopascal, or kPa. In this textbook most SI pressures will be given in kPa. Some extreme pressures may be given in megapascals (e.g., 10^7 Pa would be stated as 10 MPa).

Numerous units of pressure are used in various fields of science. One common unit of pressure is the *atmosphere*, which is the dry (no humidity) atmospheric pressure at sea level at a temperature of 15°C. The *standard atmosphere* is 101.3 kPa, although 100 kPa is most often used: A 1% error in engineering calculations is normally tolerated since material properties are most often not known to within 1%. Thus, if a diver is experiencing 5 atmospheres of pressure, the pressure on his body is 506 kPa. Again, the average atmospheric pressure varies with altitude, as shown in Table B-1 in the Appendix.

Standard atmosphere:
15°C
101.3 kPa
760 mm Hg
760 Torr
1.013 bar

1.4.4 Pressure-measuring devices

The mercury barometer

The first device used to measure atmospheric pressure is the inverted tube barometer developed by Evangelista Torricelli. A glass tube that has one end closed and one end open is filled with mercury. The open end is closed, turned upside down, and placed in an open container of mercury, as shown in Fig. 1.7. The open end is then released into the dish of mercury. The weight of the mercury pulls it down, leaving a vacuum in the closed end of the tube that holds up the column of height H. Atmospheric pressure acts on the mercury in the container as shown. The height H of the column of mercury in the tube is used to calculate the atmospheric pressure according to

$$P_{atm} = \rho g H \tag{1.13}$$

where ρ is the density of the mercury and g is the gravitational acceleration.

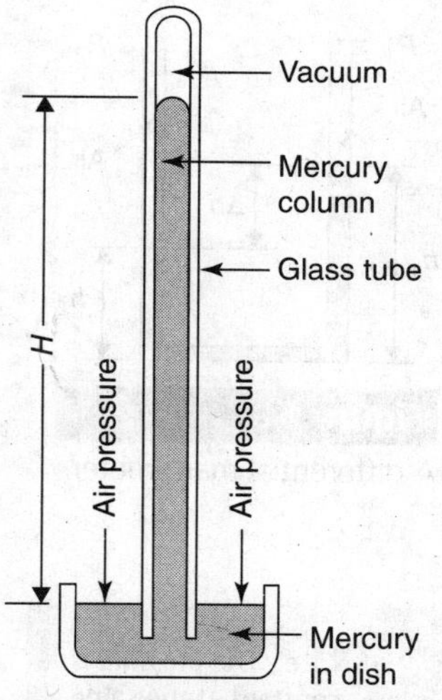

Figure 1.7

A Torricelli barometer.

Convert height of mercury to kPa — Example 1.4

The meteorologist reports that the air pressure during a storm is 70 cm of mercury. Calculate the air pressure in kPa. Assume $SG_{Hg} = 13.6$.

Solution

The density of mercury is calculated (see Eq. 1.10) to be

$$\rho_{Hg} = SG \times \rho_{water} = 13.6 \times 1000\ \text{kg/m}^3 = 13\,600\ \text{kg/m}^3$$

The pressure is now calculated using the acceleration of gravity as 9.81 m/s^2 with $H = 0.70$ m:

$$P_{atm} = \rho g H$$

$$= 13\,600 \frac{\text{kg}}{\text{m}^3} \times 9.81 \frac{\text{m}}{\text{s}^2} \times 0.70\ \text{m} = 93\,400\ \text{Pa}$$

$$\text{Pressure in kPa} = \frac{93\,400\ \text{Pa}}{1000\ \text{Pa/kPa}} = 93.4\ \text{kPa}$$

The pressure is absolute since the 70 cm of mercury given in the example statement was the pressure on the Earth's surface surface.

Comment

Spaces rather than commas are used in the SI system since commas in many countries represent decimal points.

Comment

An answer is given to either three or four significant digits, never five or more.

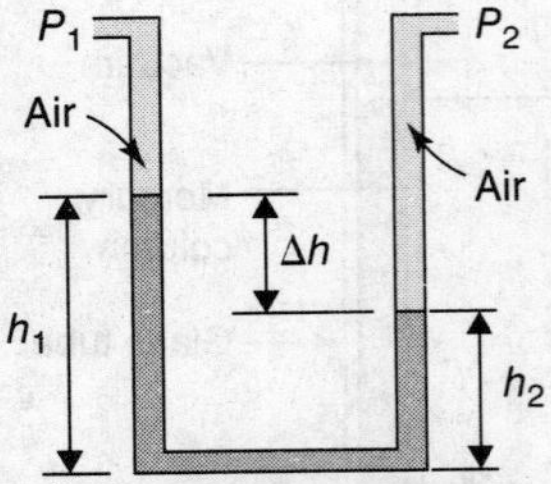

Figure 1.8
A differential manometer.

Example 1.4 points out the pressure measured in inches of mercury. When the meteorologist says that the barometric pressure is 76 cm of mercury, it is a barometer reading, a pressure reading. All meteorologists do this. It is also reported whether the barometer reading is rising, falling, or holding steady. A falling barometer reading indicates a trend toward stormy weather; a rising barometer reading indicates fair weather ahead; and a steady barometer reading indicates no change in the weather.

Mercury barometers are in very restricted use today due to the high toxicity of mercury and mercury vapors. The sale of both mercury barometers and mercury thermometers is becoming illegal in many countries.

The differential manometer

The differential manometer is an adaptation of the single-tube Torricelli barometer. Figure 1.8 shows the basic configuration of a differential manometer. It consists of two long vertical tubes connected by a common tube at the bottom. Thus, the pressure at the bottom of the two vertical tubes, which are at the same elevation, is the same. The tops of the tubes are exposed to two different pressures P_1 and P_2. In many cases, one of these tubes is left open to the atmosphere so that the pressure at the top of that tube is atmospheric pressure, that is, zero gage pressure. The tube with the higher pressure (P_2 in the figure) at the top will push the liquid in that tube down, and the level of the liquid on the other side will rise. This creates a Δh, which is equal to $h_1 - h_2$, as shown in Fig. 1.8. We can use

Note: The pressure is constant at the same elevation in a static fluid, such as 2 m below the surface of a lake. It only changes with elevation.

$$P_2 - P_1 = \rho g \Delta h \tag{1.14}$$

to calculate the pressure difference.

The sensitivity of the differential manometer is mainly a function of the density of the fluid used in the tubes. A high-density liquid, such as mercury, will have a low sensitivity. This means that a relatively large difference in pressure between the two tubes will cause a small Δh. Lighter fluids like water or alcohol will produce a larger Δh for the same pressure difference.

Differential manometers are good for laboratory use. They are cheap and easy to make. The main disadvantage of using these manometers is that they have to be read and then a calculation must be made to obtain a pressure reading. Other pressure-measuring devices can give a direct reading of pressure, and some digital devices can be used for remote data acquisition.

Example 1.5 Use a manometer to measure pressure

A water differential manometer has a reading of $\Delta h = 58$ cm. The left leg of the tube is open to the atmosphere. What is the absolute pressure P in the air shown in Fig. 1.9? Air can be considered weightless over relatively small heights.

Solution

The density of water is always taken as 1000 kg/m^3, unless otherwise stated. The Δh reading (see Fig. 1.9) is converted from 58 cm to 0.58 m. The acceleration of gravity in the SI system is 9.81 m/s². The pressure difference over the 58-cm height is calculated using Eq. 1.14:

$$\begin{aligned}\Delta P &= \rho g \Delta h \\ &= 1000\ \text{kg/m}^3 \times 9.81\ \text{m/s}^2 \times 0.58\ \text{m} = 5690\ \text{kg/m·s}^2 \\ &= 5690\ (\text{N·s}^2/\text{m})/\text{m·s}^2 = 5690\ \text{N/m}^2 \quad \text{or} \quad 5.69\ \text{kPa}\end{aligned}$$

where Eq. 1.3 shows that kg = N · s²/m. Also, $\Delta P = P$ since $P_1 = 0$.

To find the absolute pressure P, the atmospheric pressure (assumed to be 100 kPa) is added to the gage pressure:

$$P = P_{\text{gage}} + P_{\text{atm}} = 5.69\ \text{kPa} + 100\ \text{kPa} = \underline{105.7\ \text{kPa}}$$

retaining four significant digits.

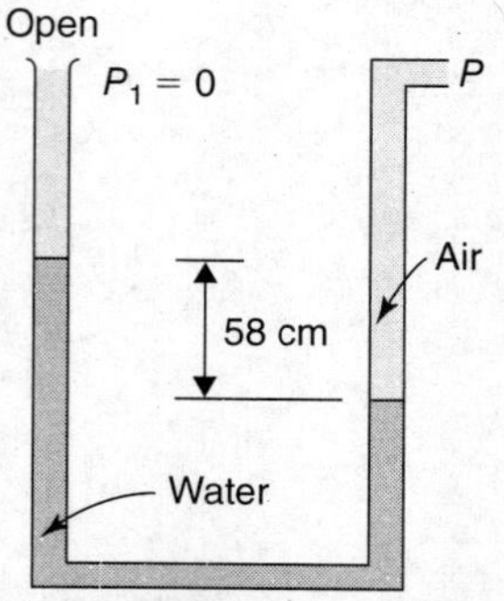

Figure 1.9

Note: We use P_{atm} = 100 kPa, unless otherwise stated.

The Bourdon gage

A Bourdon gage is a mechanical device used to measure gage pressure. Figure 1.10 shows the typical construction of the gage. A pressure tap at the bottom leads into an elliptical tube that is bent in a circular arc. Atmospheric pressure acts on the outside of the tube. If the pressure inside the tube is greater than atmospheric pressure, the pressure forces try to straighten out the tube. This bending is magnified through a linkage and gear pair that turns an indicator needle. If the inside pressure is less than atmospheric, the tube tends to curl up and the needle is turned in the opposite direction. Bourdon gages are very durable and easy to read. The one drawback to these devices is that they have to be read by a person. They are not useful for remote data acquisition.

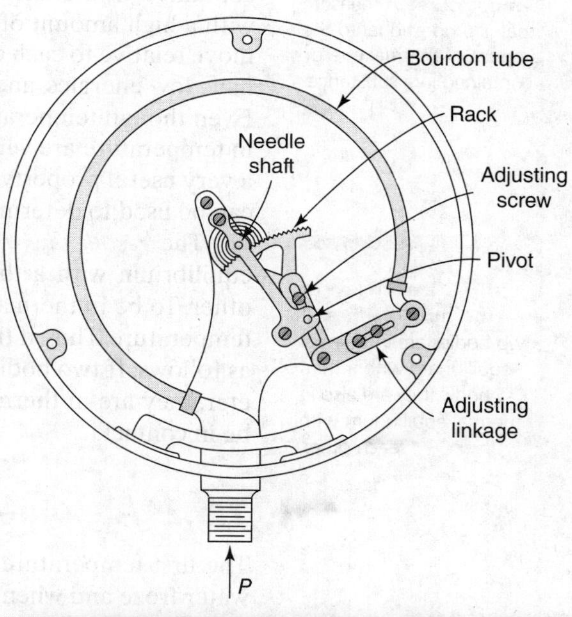

Based on J.P. Holman, Experimental Methods for Engineers, McGraw-Hill, 1971, p.171

Figure 1.10
A Bourdon gage.

Pressure transducers

A *transducer* is a measuring device that converts one type of physical signal to another type. A pressure transducer is one of a variety of devices that converts a pressure to an analog electrical output.

There are many different types of pressure transducers. Some utilize a thin metal diaphragm with an attached strain gage to measure pressure. The flexing of the diaphragm under pressure will stretch the strain gage. Since small changes in electrical resistance are difficult to measure, a Wheatstone bridge circuit is used to convert the resistance change into a measurable voltage change.

Transducer: A measuring device that converts one type of physical signal to another type.

Since pressure transducers output an analog electrical voltage, an analog-to-digital converter can be used to convert this voltage to a digital number that can be transmitted to a remote readout device. This makes the pressure transducer ideal for data acquisition and remote sensing applications. They are widely used in many applications.

1.5 Temperature

1.5.1 What is temperature?

Temperature is a thermodynamic property we are accustomed to in our daily lives. We are familiar with temperature through our sense of touch. This sense, though, can be misleading. For example, if you pick up a piece of wood in one hand and a piece of steel in the other, the steel feels colder. In truth, the wood and the steel are at the same room temperature. What we sense through our fingers is not temperature but the flow of heat. Something feels hot if thermal energy flows from an object into our fingers. It feels cold if thermal energy flows from our fingers into the object.

Temperature: A measure that is proportional to the amount of thermal energy contained in a substance.

Temperature is a measure that is proportional to the amount of thermal energy contained in a substance, manifested as molecular kinetic energy. A molecule with a high amount of thermal energy moves with very high speed. Gas particles move relative to each other with random high velocities. Solids, on the other hand, have low energies, and the molecules are held in place by intermolecular forces. Even though temperature is defined at the molecular level, the effects of change in temperature are very noticeable and easily measured. This makes temperature a very useful property. Temperature is a basic property, it can be measured, and it can be used to determine other thermodynamic properties.

Zeroth law of thermodynamics: If two bodies are in thermal equilibrium with a third body, they are also in thermal equilibrium with each other.

The *zeroth law of thermodynamics* states that if two bodies are in thermal equilibrium with a third body, they are also in thermal equilibrium with each other. To be in thermal equilibrium requires that the three bodies have the same temperature. Thus, if the third body is a thermometer, the zeroth law can be stated as follows: If two bodies have the same temperature, as measured by a thermometer, they are in thermal equilibrium with each other even though they may not be in contact.

1.5.2 Absolute and relative temperature scales

The first temperature scales were based on the difference in the states at which water froze and when it boiled. At a pressure of 101 kPa, these phenomena occur at very precise temperatures. The SI temperature scale measures temperature in degrees Celsius (°C), named for Swedish astronomer Anders Celsius. The temperature at which water freezes at 101 kPa is selected as 0°C. The temperature at which water boils at 101 kPa is selected as 100°C. A common lab exercise in physics is to take an unmarked thermometer, measure the point at which ice melts and the point at which water boils, to establish where on the thermometer 0°C and 100°C exist. Then a temperature scale is marked between these two points. Since this temperature scale measures temperature relative to these phenomena, it is called a *relative temperature scale*.

Relative temperature scale: The scale selected between the boiling point and freezing point of water at standard atmospheric pressure.

The relative temperature scale used in the English system is the Fahrenheit scale, named for the German instrument maker Daniel Fahrenheit. The symbol used for these units is degrees Fahrenheit (°F). The same two physical phenomena are used to define the scale. Water freezes at 32°F and boils at 212°F at a pressure of 14.7 psia. To convert degrees Celsius to degrees Fahrenheit, or degrees Fahrenheit to degrees Celsius, we use

$$T(°\text{F}) = 32 + \frac{9}{5}T(°\text{C}) \tag{1.15}$$

$$T(°\text{C}) = \frac{5}{9}[T(°\text{F}) - 32] \tag{1.16}$$

Convert Fahrenheit to Celsius — Example 1.6

Normal body temperature is 98.6°F. Calculate this temperature in degrees Celsius.

Solution

Use Eq. 1.16 to discover

$$T(°\text{C}) = \frac{5}{9}(98.6 - 32) = \underline{37.0°\text{C}}$$

These relative temperature scales serve well for most engineering applications. The property tables in the appendices are based on degrees Celsius or Fahrenheit. Some applications, however, require that temperature be based on a reference temperature where no thermal energy exists in the substance. This reference temperature is called *absolute zero temperature*. The substance is in the solid form and there is an absence of motion. At this temperature even hydrogen exists as a solid. The SI scale of absolute temperature is named for Lord Kelvin, and the units are labeled simply as K. Absolute 0 K corresponds to a relative temperature of −273.15°C. For engineering calculations, we round this number off to −273°C. The English scale of absolute temperature is named for William Rankine. Absolute zero corresponds to a relative temperature of −459.67°F. For engineering calculations, we round this number off to −460°F. The relationships between absolute temperature and relative temperature are

Absolute zero temperature: The temperature where no thermal energy exists.

Comment

In the SI system we do not use the degree symbol when writing absolute temperature, such as 0 K.

$$T(\text{K}) = T(°\text{C}) + 273 \tag{1.17}$$

$$T(°\text{R}) = T(°\text{F}) + 460 \tag{1.18}$$

The most common use of the absolute temperature scale is that of the ideal-gas law. This law is described in Chapter 2, but you undoubtedly have used it in earlier chemistry and physics courses. To use the ideal-gas law properly, all physical properties must be entered in absolute values. Thus, either degrees Rankine or degrees kelvin must be used.

Example 1.7 Express a temperature in various degrees

Normal room temperature is 22°C. Calculate this temperature in °F, K, and °R.

Solution

Use Eq. 1.15 through 1.18:

$$T(°\text{F}) = 32 + \frac{9}{5} \times 22°\text{C} = \underline{71.6°\text{F}}$$

$$T(\text{K}) = 22°\text{C} + 273 = \underline{295\ \text{K}}$$

$$T(°\text{R}) = 71.6°\text{F} + 460 = \underline{531.6°\text{R}}$$

Note: In the SI system, we do not use the degree symbol when writing 295 K. We would write it out as 295 kelvins, analogous to 10 ohms (not 10 Ohms) when using electrical units. We abide by the NIST rules.

Comment

The National Institute of Standards and Technology (NIST) was known between 1901 and 1988 as the National Bureau of Standards (NBS).

1.5.3 Temperature measurement

A wide variety of temperature-measuring devices are available for use in applications. Characteristics to be noted about these devices are the range of temperatures that can be measured, their durability in various environments, and the ability to communicate temperature data to remote locations using data acquisition. A device that measures temperature is often called a *thermometer*. Several such devices will be described in the following for informational purposes.

Liquid expansion thermometers

The most familiar thermometer is the liquid expansion thermometer. It utilizes a mechanical property of materials called the *coefficient of thermal expansion* α_T, which is defined as the change in volume of a substance per degree increase in the temperature of the substance, which is expressed as

$$\Delta V = \alpha_T V \Delta T \tag{1.19}$$

Coefficient of thermal expansion α_T: The change in volume of a substance per degree increase in the temperature of the substance.

A positive coefficient α_T indicates a substance that expands with increasing temperature. A negative coefficient indicates a substance that contracts or gets smaller with increasing temperature. A substance that has a zero coefficient of thermal expansion is one that does not change size or shape with changing temperature. Such a characteristic would be useful in a device that has to operate over a wide range of temperatures, such as the engine block in a car. The effect of this property is readily noticed when observing the shape of telephone lines suspended between telephone poles. In the summer, the wires sag low because the copper in the wire expands with higher temperatures. In the winter, the telephone lines are tighter since the wire contracts with cooler temperatures.

Note: Most substances (water is an exception) contract when freezing. Water expands when freezing, so ice particles float; this allows for ice fishing on the surface, rather than the bottom, of a lake!

A liquid expansion thermometer consists of a reservoir of liquid attached to a vertical capillary tube, as shown in Fig 1.11. As the temperature of the liquid

Table 1.5 Coefficients of Thermal Expansion for Common Liquids

Substance	Coefficient of Thermal Expansion α_T
Water	0.0002 °C^{-1}
Alcohol	0.0011 °C^{-1}
Mercury	0.00018 °C^{-1}

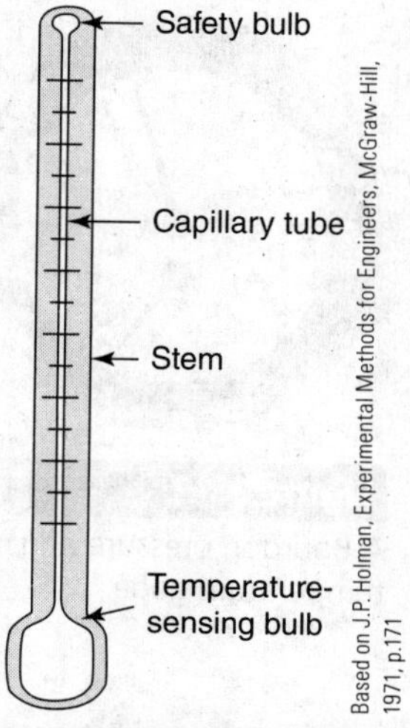

Figure 1.11
A liquid expansion thermometer.

increases, the liquid expands in volume. Since the reservoir has a fixed volume, the liquid can only expand up the capillary tube. Thus the level of the liquid in the tube is an indicator of the temperature of the liquid, which is at the same temperature as the environment whose temperature is being measured.

Higher coefficients of thermal expansion will make the thermometer more sensitive to temperature change. The coefficients of thermal expansion α_T for several common liquids are shown in Table 1.5. Metals that are in the liquid state at normal temperatures make the most effective thermometers. The traditional liquid used in thermometers is mercury, but it is becoming illegal to use in thermometers sold to the public due to its high toxicity.

Liquid expansion thermometers can be very accurate. This type of thermometer cannot be used for data acquisition since it must be read by the user. They are also not considered ideal for industrial use since they are easily broken.

Bi-metallic strips

Another temperature-measuring device makes use of the coefficient of thermal expansion of a bi-metallic strip. The strip, sketched in Fig. 1.12, consists of two different metals that are bonded together at room temperature. At room temperature, both strips are horizontal. As the temperature increases, both strips expand according to their expansion coefficient. The metal with the larger coefficient of thermal expansion will expand further, causing the strip to bend. If the temperature is decreased, the strip will bend in the other direction.

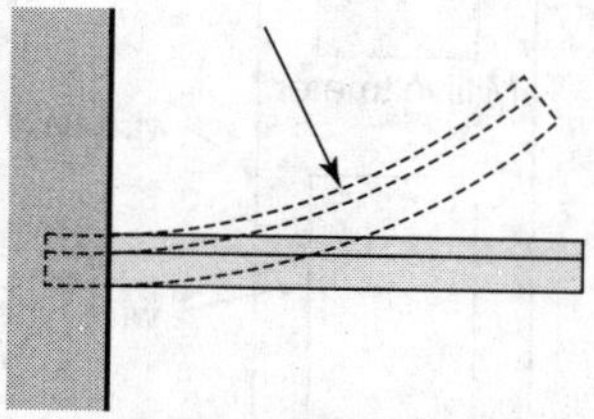

Figure 1.12
A bi-metallic strip thermometer.

The bi-metallic strip is not the most accurate way to measure temperature. It does not lend itself to data acquisition. It is, however, very durable due to its solid nature. Bi-metallic strips have been commonly used as temperature controls for heating and cooling systems. In this application, they are called *thermostats*. As the strip bends in one direction as temperature decreases, it can close an electrical contact and turn on a furnace. As it heats up and bends in the other direction, it can close a contact and turn on an air conditioner. This type of thermostat has largely been replaced by programmable electronic thermostats that use thermistors to measure temperature. Occasionally one will find a kitchen thermometer that is a bi-metallic strip bent in a helix to measure temperatures.

Bourdon gage thermometers

For a fixed volume of gas, the pressure will vary linearly with the temperature. A Bourdon gage thermometer uses a fixed volume of air in a tube attached to a Bourdon gage pressure-measuring device. As the temperature of the contained gas increases, the pressure will also rise. If the temperature drops, the pressure will drop accordingly. We can replace the face of the Bourdon gage of Fig. 1.10 with a temperature scale. A Bourdon gage thermometer has all the advantages and disadvantages

Nomad_Soul/Shutterstock.com

Figure 1.13

A Bourdon pressure and temperature gage.

of a Bourdon pressure gage. Figure 1.13 shows a device that uses one Bourdon gage to measure both temperature and pressure. The tap to the pressure gage is open, and the stem leading to the temperature gage is sealed off.

Thermocouples

A thermocouple is a type of transducer that converts a temperature to an analog voltage. One characteristic of a metal is the existence of free electrons, which allow electricity to be conducted. Different metals have different numbers of free electrons. If the copper and iron wires are connected together, electrons will flow from the copper side that has many electrons to the iron side that has fewer. This establishes a voltage between the two wires, and current, even though very small, will flow through an electrical load connected to these wires. A characteristic of this device is that the greater the temperature of the junction between the two wires, the greater the voltage produced. The voltage increases linearly with temperature. This is called the Seebeck effect. The voltage produced by a thermocouple is very small, requiring a very precise millivoltmeter to read it.

There are many types of thermocouples that use different combinations of wires. In some cases, it is desirable to use metals that won't rust or corrode. In other cases, metals that won't melt must be used at the temperatures that are being measured. Table 1.6 lists several of the commercially available thermocouple combinations, as established by the American National Standards Institute (ANSI).

Thermocouples cannot be used singly since the connection to a voltage readout device effectively creates another thermocouple. The first use of thermocouples was in pairs of identical thermocouples. The two thermocouples have like wires connected, which means the voltages produced by each thermocouple oppose each other. One thermocouple was used as a measuring probe. The other thermocouple was immersed in a bath of melting ice. The thermocouple in the ice bath produces a voltage associated with a temperature of 0°C. This voltage is subtracted from the voltage of the test thermocouple. The millivoltmeter connected to these thermocouples reads the voltage differential between the two thermocouples. Tables are published for various types of thermocouples that relate the measured differential voltage to the temperature of the test probe. Figure 1.14 shows the traditional configuration for a thermocouple pair using an ice bath and millivoltmeter. The use of an ice bath has been replaced with a precise voltage source that produces the exact voltage that the ice bath thermocouple would produce. These "electronic ice baths" are built into most thermocouple readers today.

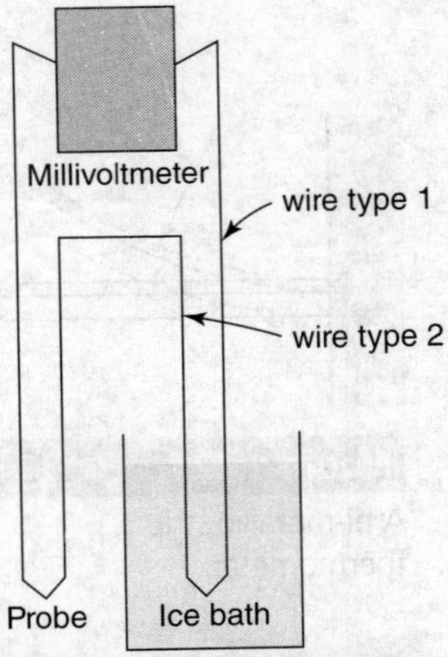

Figure 1.14

A thermocouple configuration.

Thermocouples are widely used in industry because they are accurate and not easily damaged. As long as the two thermocouple wires are in electrical contact, they will work. The response time of a thermocouple is dependent on the size of

Table 1.6 Thermocouple Wire Combinations

ANSI Code	Wire #1	Wire #2
J	Iron	Copper–Nickel Alloy
K	Nickel–Chromium Alloy	Nickel–Aluminum Alloy
T	Copper	Copper–Nickel Alloy
N	Nickel–Chromium-Silicon Alloy	Nickel–Silicon–Magnesium Alloy
R	Platinum–10%Rhodium Alloy	Platinum

the electrical junction between the two wires. If they are simply twisted together, they form a large junction that will respond relatively slowly to changes in temperature. Precision thermocouples are made from very thin wires that are welded together, end to end, to make the smallest connection.

Thermistors

A thermistor is a solid-state device whose electrical resistance changes noticeably with temperature. They are usually constructed as a piece of semiconductor substance with electrodes and lead wires connected to each end. The size and shape of these devices vary widely. The most effective are small spherical beads covered with a protective layer of glass. This is done because the semiconductor substance is easily contaminated by water and other liquids. The relationship between electrical resistance and temperature for these substances is

$$R = R_0 e^{\beta(T_0 - T)/T_0 T} \tag{1.20}$$

where R is the electrical resistance at temperature T, R_0 is the base electrical resistance at reference temperature T_0, and β is a characteristic constant of the semiconductor substance.

What makes these devices very useful is the fact that the electrical resistance of these thermistors is large and measurable and that the change in resistance with temperature is also large enough to be measured easily with an ohmmeter. Thermistors are used widely as temperature-measuring devices and as temperature sensors in electronic thermostats. The main disadvantage of thermistors is that they are more easily damaged than thermocouples. Larger, more durable thermistors tend to have slower response times.

1.6 Energy

There are various forms of energy: kinetic, potential, internal, magnetic, electrical, nuclear, and chemical. In thermodynamics, however, we are primarily concerned with the first three and, most often, even the first two are negligible. So, our attention will be focused on the internal energy of a system. We will also consider situations in which kinetic energy and potential energy are of interest, as in a nozzle or a hydroelectric plant. Kinetic energy, KE, and potential energy, PE, have been of interest in physics courses and are represented, respectively, by

$$KE = \frac{1}{2}mV^2 \quad \text{and} \quad PE = mgz \tag{1.21}$$

where m is the mass, V is the velocity, g is gravity, and z is the elevation above a selected datum. *Internal energy* has probably not been of serious interest in previous courses; it is the sum of all the energy associated with the molecular structure and the molecular activity of the molecules in a system. In combustion, energy is released when chemical bonds between atoms are rearranged; in nuclear reactions, energy is released when changes occur between subatomic particles. Our interest in thermodynamics will focus on the influence of temperature (the motion of the molecules) on internal energy, although we will consider the combustion process in Chapter 12. Other forms of internal energy will not be of interest. Internal energy will be represented by U and specific internal energy

Internal energy: The energy associated with the molecular structure and the molecular activity.

Comment

In thermodynamics, interest will be focused on the influence of temperature on internal energy.

by u. It will be analyzed in detail in later sections of this book. The total energy E of a system will be represented by

$$E = KE + PE + U \qquad (1.22)$$
$$= \frac{1}{2}mV^2 + mgz + U$$

The conservation of energy: The total amount of energy in an isolated system remains constant over time.

A very important law is *the conservation of energy*, which states that the total amount of energy in an isolated system remains constant over time; that is, it is conserved over time. Energy is not created or destroyed; it can only be transformed from one form to another in an isolated system, that is, in a system that does not exchange energy with the surroundings. So, for an isolated system, the conservation of energy takes the form

$$\frac{1}{2}mV_1^2 + mgz_1 + U_1 = \frac{1}{2}mV_2^2 + mgz_2 + U_2 \qquad (1.23)$$

Very few systems are indeed isolated. But assumptions are usually made so that an approximate solution can be obtained.

Example 1.8 The velocity of a falling object

An object is released from rest at an elevation h above the ground. Find a relationship for the velocity of the object when it is a distance z above the ground. See Fig. 1.15.

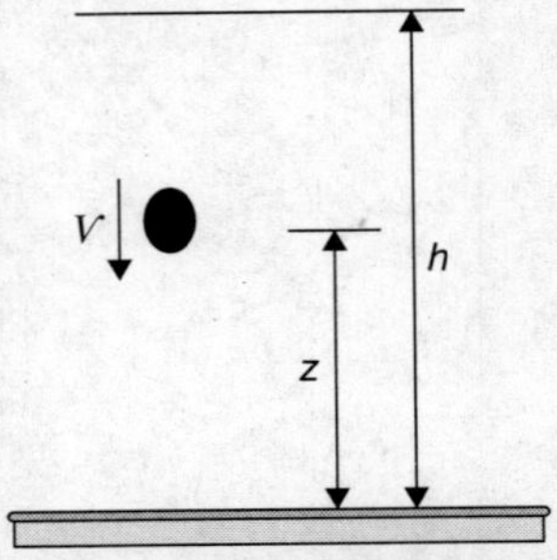

Figure 1.15

Solution

The conservation of energy requires

$$\frac{1}{2}mV_1^2 + mgz_1 + U_1 = \frac{1}{2}mV_2^2 + mgz_2 + U_2$$

Using $V_1 = 0$, $z_1 = h$, $z_2 = z$ and assuming that the temperature of the object does not change so that $U_2 = U_1$, we can simplify the equation by striking out the appropriate terms:

$$\frac{1}{2}m\cancel{V_1^2} + mgh + \cancel{U_1} = \frac{1}{2}mV_2^2 + mgz + \cancel{U_2}$$

This yields the relationship

$$V_2 = \sqrt{2g(h - z)}$$

Note that we did not find the usual $\sqrt{2gh}$ because of the way the problem was stated.

☑ You have completed Learning Outcome (4)

1.7 Summary

In this chapter we introduced the subject of thermodynamics: the study of how energy affects matter. Energy can be stored, transformed, and transferred. Properties such as density and specific volume and their dimensions and units were introduced. Both SI units, with all their prefixes, and English units were presented. Pressure and temperature, the two properties of primary importance in our study, were defined along with their measurement techniques. Finally, a general introduction to energy conservation was reviewed.

Numerous terms were defined:

Coefficient of thermal expansion: *The change in volume of a substance per degree increase in its temperature.*

Control volume: *A fixed volume into which and/or from which a fluid flows.*

Density: *The mass divided by the volume occupied by the mass.*

Extensive property: *A property whose value depends on the mass of the system.*

Independent properties: *Two properties are independent if one can vary while the other is held constant.*

Intensive property: *A property that does not depend on the mass of the system.*

Internal energy: *The energy associated with the molecular structure and the molecular activity of the molecules in a system.*

Law: *Developed from direct observation under strict constraints.*

Pressure: *The normal component of a force acting on a surface area divided by the area of the surface.*

Process: *The states that a substance passes through as it changes from an initial state to a final state.*

Properties: *Characteristics of a substance that can be measured or calculated from measurements.*

Quasi-equilibrium process: *If a system, in passing from one equilibrium state to another equilibrium state, deviates from equilibrium by a very small amount, each state in between the two end states is idealized as an equilibrium state.*

State: *Defined by the specific values of the properties of a substance.*

State postulate: *The equilibrium state of a simple, compressible system is established by two independent, intensive properties.*

Surroundings: *Everything external to the system.*

Sustainability: *The capacity to endure.*

System: *A fixed quantity of mass.*

Temperature: *A measure that is proportional to the amount of thermal energy contained in a substance.*

Several important equations were presented:

Density and specific volume:	$\rho = \frac{m}{V}$ and $v = \frac{V}{m}$
Newton's second law:	$F = ma$
Pressure:	$P = \frac{F_n}{A}$
Absolute pressure:	$P_{\text{absolute}} = P_{\text{gage}} + P_{\text{atmospheric}}$
Temperature:	$T(°\text{F}) = 32 + \frac{9}{5}T(°\text{C})$
Expansion of a substance:	$\Delta V = \alpha_T V \Delta T$
Thermistor resistance:	$R = R_0 e^{\beta(T_0 - T)/T_0 T}$
Conservation of energy:	$\frac{1}{2}mV_1^2 + mgz_1 + U_1 = \frac{1}{2}mV_2^2 + mgz_2 + U_2$

Problems

FE Exam Practice Questions[2]

1.1 Which of the following is not of interest in our study of thermodynamics?
- **(A)** The transmission of energy
- **(B)** The utilization of energy
- **(C)** The storage of energy
- **(D)** The transformation of energy

1.2 During a quasi-equilibrium process, the temperature:
- **(A)** Would remain constant
- **(B)** Could vary slowly from point to point
- **(C)** Is assumed to be the same at all points in the system
- **(D)** Must be substantially above absolute zero

1.3 The father of thermodynamics is:
- **(A)** Jacques Charles
- **(B)** Robert Boyle
- **(C)** Joseph Guy-Lussac
- **(D)** Sadi Carnot

1.4 Which of the following would utilize a control volume rather than a system?
- **(A)** The gas in a cylinder during compression
- **(B)** The gas in a cylinder during exhaust
- **(C)** The melting of ice cubes in a container
- **(D)** A global analysis of the atmosphere

1.5 Which of the following is not an extensive property?
- **(A)** Mass
- **(B)** Volume
- **(C)** Temperature
- **(D)** Weight

1.6 Which of the following cannot be modeled as a quasi-equilibrium process?
- **(A)** The air escaping from a pressurized balloon being let free
- **(B)** The expansion of the air in a piston operating at 4000 rpm
- **(C)** A bike tire being pumped up
- **(D)** The ignition process when the piston is at top-dead center

[2] Passing the Fundamentals of Engineering exam (FE exam) is the first step in becoming a registered engineer.

1.7 Which of the following is not a unit used to measure a primary dimension in our study of thermodynamics?

(A) Kilogram
(B) Newton
(C) Second
(D) Meter

1.8 A unit of power is the watt, W. An equivalent set of units is:

(A) $N \cdot m/s^2$
(B) $N \cdot m^2/s$
(C) $kg \cdot m/s^2$
(D) $kg \cdot m^2/s^3$

1.9 We could write 34 000 000 000 newtons as:

(A) 34 GN
(B) 340 MN
(C) 3400×10^6 N
(D) 3400 MN

1.10 Ten kg of a liquid occupies 8000 cm^3. Its density, specific volume, and specific gravity are (ρ, v, SG):

(A) (1250 kg/m^3, 0.0008 m^3/kg, 1.25)
(B) (0.0008 kg/m^3, 1250 m^3/kg, 0.8)
(C) (1250 kg/m^3, 0.0008 m^3/kg, 0.8)
(D) (0.0008 kg/m^3, 1250 m^3/kg, 1.25)

1.11 A force of 100 N acts on the 10-kg mass of Fig. 1.16. The mass accelerates at:

(A) 10 m/s^2
(B) 1 m/s^2
(C) 0.1 m/s^2
(D) Can't tell

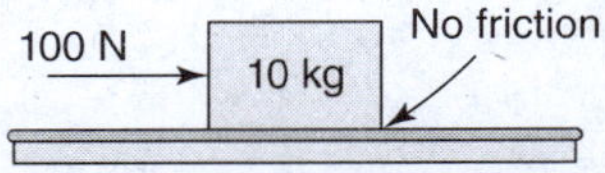

Figure 1.16

1.12 A resultant force of 36 000 N acts on a 200 cm^2 area at an angle of 30° with respect to a normal to the area due to a pressure and friction. The pressure on the area is nearest:

(A) 900 kPa
(B) 1200 kPa
(C) 1560 kPa
(D) 1800 kPa

1.13 An atmospheric pressure on a mountain is measured as 420 mm of mercury. Using $SG_{HG} = 13.6$, the pressure in SI units is nearest:

(A) 56 kPa
(B) 62 kPa
(C) 68 kPa
(D) 72 kPa

1.14 A spring with spring constant 400 N/m is attached to a 40-kg-frictionless piston and is stretched a distance of 20 cm, as shown in Fig. 1.17. The pressure in the cylinder is nearest:

(A) 30 kPa gage
(B) 35 kPa gage
(C) 40 kPa gage
(D) 45 kPa gage

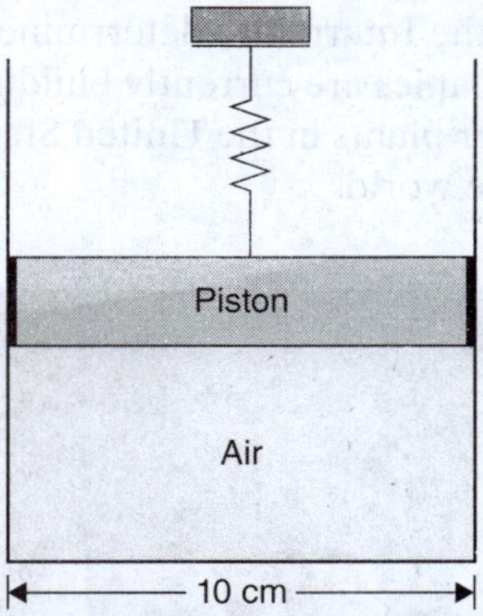

Figure 1.17

1.15 As we take a shower, the water feels cooler than when we first entered the shower, so we turn up the temperature. Why is that?

(A) It takes time for the water to heat up due to the pipes.
(B) Our body heats up, so the water feels cooler.
(C) We have to get used to the feel of the hot water.
(D) The water in the pipes cools down with time, so a hotter water is needed.

1.16 A 2000-kg vehicle is to be crashed into a rigid barrier supported by a very stiff spring (see Fig. 1.18) that simulates a crash into a smaller 1400-kg vehicle. What should be the spring constant K if the spring is to compress a maximum distance of 10 cm when the vehicle is traveling at 80 kph? The potential energy stored in a spring is $Kx^2/2$.

(A) 140 MN/m

(B) 100 MN/m

(C) 60 MN/m

(D) 20 MN/m

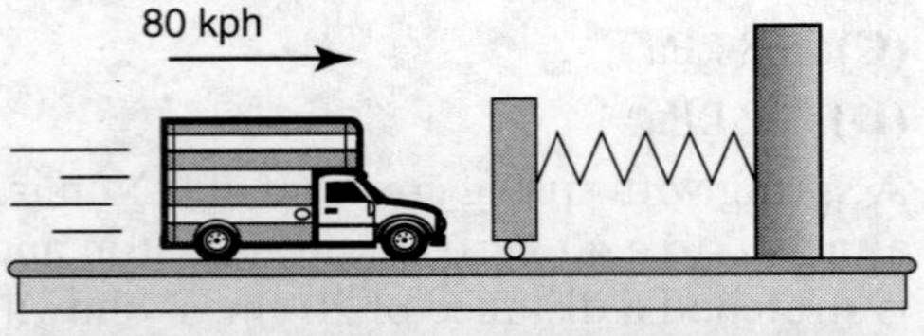

Figure 1.18

Introduction

1.17 Use the Internet to find the top five countries that supply the most oil.

1.18 Use the Internet to deterrmine if any companies are currently building nuclear power plants in the United States. In the rest of the world.

1.19 Use the Internet to find out the different ways in which water can be purified.

1.20 In thermodynamics, we really are not interested in how energy is used. True or false? Comment.

1.21 Select the energy source that is not sustainable.

i) Hydropower from dams
ii) Energy from wind turbines
iii) Electricity from solar photovoltaic panels
iv) Power plants powered by wood
v) Power plants powered by coal

Instructor: It is intended that students respond to all parts of problems listed as i), ii), iii), etc.

1.22 Hydrogen is considered a sustainable energy source for engines and home furnaces in the future. Use the Internet and describe how hydrogen can be manufactured and delivered to customers.

1.23 Select the general time period when the need was felt for engineers in large numbers:

i) When fire was discovered to heat caves
ii) As aqueducts were built to distribute water
iii) When the industrial revolution occurred
iv) When automobiles, trucks, tractors, harvesting machines, and industrial machines were introduced in the late 1800s

1.24 Select a major responsibility of mechanical engineers in the mid-1800s:

i) To design power plants
ii) To oversee the operation of trains
iii) To design mining equipment used to mine coal
iv) To design automobiles

1.25 What is dry ice? Why was it invented? What is its primary use today? What is its lowest temperature if used commercially?

1.26 Identify if a system or a control volume should be selected for each of the following:

i) A shock absorber (see Fig. 1.19)
ii) The turbine in a dam
iii) Heating the water in a pressure cooker from cool water to just before boiling
iv) A piston/cylinder arrangement during the power stroke

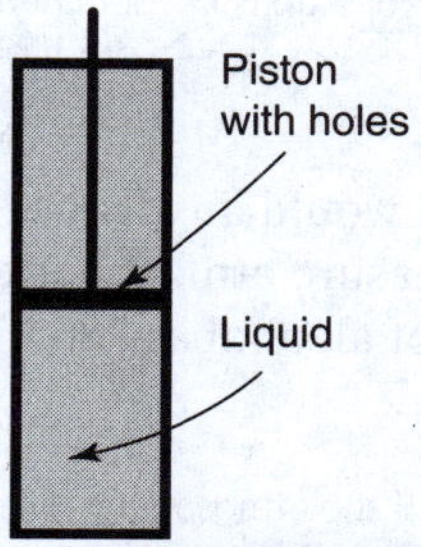

Figure 1.19

1.27 A balloon, the control volume, is pressurized with air, the system, and held fixed immediately after the air is let free to go. Sketch the system and the control volume at $t = 0$ and a moment (more than a nanosecond) after the air is released.

1.28 How large would a sphere have to be at sea level to hold one trillion molecules of air at standard conditions?

1.29 Is ketchup, or catsup, a fluid or a solid? What is it? Check with Mr. Google.

Shake and shake the ketchup bottle, it won't come and then a lot'll!

Dimensions and Units

1.30 In England, weight used to be measured in "stones." How many kilograms are equivalent to 6.3 stones?

Hint: Google "6.3 stone = kg?" if you can't find "stone" in Appendix A.

1.31 What is the advantage of using slugs as the unit of mass in the English system?

A slug!

Properties, Processes, and Equilibrium

1.32 Select the extensive property from the following:

i) Temperature
ii) Pressure
iii) Density
iv) Volume

1.33 If the density of water changes from 992 kg/m^3 to 1002 kg/m^3, what is the percent change in its specific volume?

1.34 The specific volume of ice at 0°C is 1.09×10^{-3} m^3/kg. What is its density? Will it float in water? If it didn't float, what would be the consequences?

A hint!

1.35 Calculate the specific gravity of mercury if $\rho_{HG} = 13\,600$ kg/m^3. How much would 2 m^3 of mercury weigh?

1.36 Determine the entries in the following table if $g = 9.81$ m/s^2 and $V = 2$ m^3.

	v (m^3/kg)	ρ (kg/m^3)	m (kg)	W (N)
a)	5			
b)		2		
c)			1000	
d)				1000

Instructor: It is intended that each part with a lower-case italic letter be assigned as a separate problem.

1.37 Determine the entries in the following table if $g = 9.81$ m/s^2 and $V = 0.54$ m^3.

	v (m^3/kg)	ρ (kg/m^3)	m (kg)	W (N)
(*a*)	3.0			
(*b*)		3.35		
(*c*)			450	
(*d*)				2000

1.38 A substance occupies 8 m^3 and has a specific volume of 4 m^3/kg. Determine its density, specific gravity, mass, and weight.

1.39 A substance, with a density of 3.35 kg/m^3, occupies 0.54 m^3. Calculate its specific volume, specific gravity, mass, and weight.

1.40 Select the process from the following that can be considered a quasi-equilibrium process:
 i) A baseboard heater in a room of a house
 ii) The compression of air by the piston in the cylinder of a diesel engine
 iii) The removal of a membrane that separates a high-pressure region from a low-pressure region

1.41 Calculate the specific gravity of air at i) standard conditions, ii) 5000 m altitude, and iii) 10 000 m altitude. Refer to Table B-1.

Pressure

1.42 If a tire gage reads 2.1 kg/cm^2 (a former European measure), determine the absolute pressure in the tire in kPa at i) sea level, ii) 1000 m elevation, and iii) 5000 m elevation. Refer to Table B-1.

Comment

Some companies continue to use kg/cm^2 as a measure of pressure since the kg is considered a force as well as a mass, which is unacceptable in the SI system. (It is similar to using the pound as both force and mass.)

1.43 A tire gage measures a pressure of 3.4 kg/cm^2 (a former European measure). If a differential manometer using mercury were used to measure this pressure, what would its reading be in millimeters? Refer to Fig. 1.8.

Observation: Strange combinations of units remain in various corporations. We will conform to SI units. You may have used metric units that were not accepted SI units in your chemistry and/or physics classes.

1.44 How tall would an alcohol barometer have to be to measure atmospheric pressure? Use the density of alcohol as 786 kg/m^3.

Note: If the atmospheric pressure is not given, we assume standard pressure to be P_{atm} = 100 kPa.

1.45 The air in the container of Fig. 1.20 has a pressure intended to be 10 atmospheres. Calculate this pressure in kPa gage and kPa.

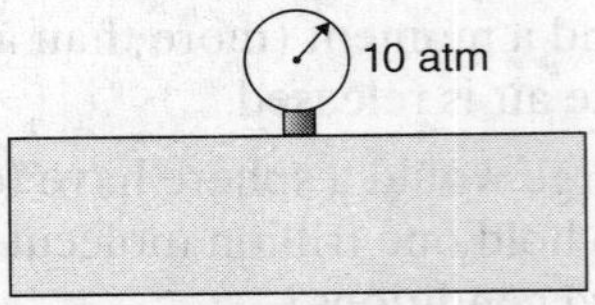

Figure 1.20

1.46 A differential water manometer (see Fig. 1.8) shows a Δh of 25 cm. Air is above the water on both sides. What is the difference in pressure between the two sides in kPa gage? In inches of Hg?

1.47 A mercury differential manometer (see Fig. 1.8) has a Δh of *a*) 10 cm and *b*) 28 cm. If air is above the mercury on both sides, what is the difference in pressure in kPa?

Note: Each part with a lower-case italic letter is independent of the other parts and can be assigned separately.

1.48 The pressure in a container on a mountain at an elevation of 7500 m is measured to be *a*) 35 kPa and *b*) 140 kPa. What is the absolute pressure in kPa? In bar?

1.49 Use the Internet to find the postal addresses and URL addresses of five companies that sell pressure transducers.

1.50 Use the Internet to find five companies that sell Bourdon type pressure gages. Compare the prices for pressure gages that measure pressures from −70 kPa gage to 700 kPa gage.

Temperature

1.51 The air temperature in Phoenix, Arizona is 49°C. Convert this temperature to K.

1.52 The satellite in Fig. 1.21 has an internal temperature of 3 K. What is this temperature in °C?

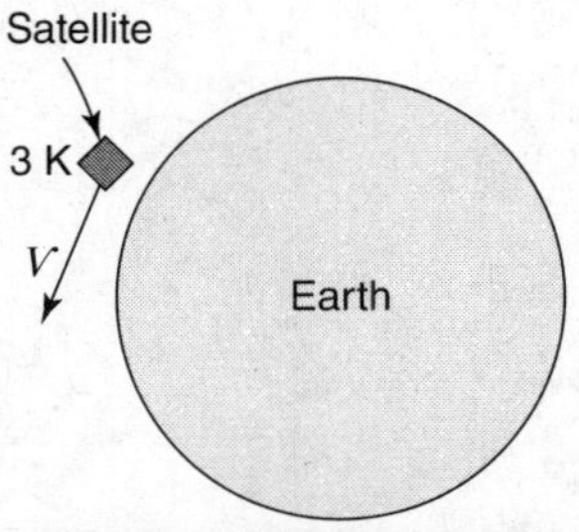

Figure 1.21

1.53 A pipe supplies steam at 204°C. What is the steam temperature in K?

1.54 Normal human body temperature is 37°C. What is this temperature in K?

1.55 A thermistor has a resistance of 3000 Ω at 25°C. If $\beta = 4220$ K, determine its resistance at *a*) 60°C, *b*) 120°C, and *c*) 180°C.

1.56 The mercury thermometer of Fig. 1.22 has a 6-mm-diameter rigid sphere that holds the mercury. If the temperature increases by *a*) 20°C, *b*) 40°C, and *c*) 60°C, estimate the height the mercury will rise in a 0.18-mm-diameter open tube attached to the sphere. Ignore the expansion of the small amount of mercury in the tube compared to the amount in the sphere.

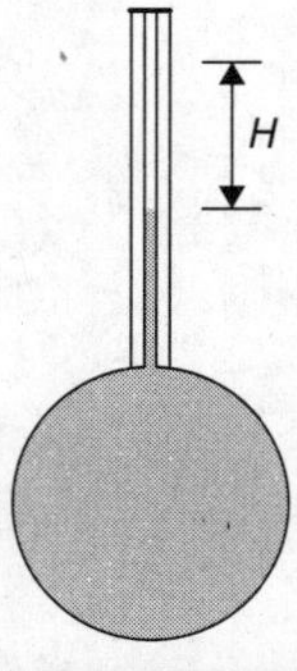

Figure 1.22

Energy

1.57 An automobile with a mass of 1100 kg is traveling at 90 km/h. What is the kinetic energy of the car?

1.58 An airplane with a mass of 5000 kg is flying at an altitude of 1000 m with a speed of 80 m/s. What energy does this airplane possess with respect to the ground?

1.59 A good approximation to gravity above the surface of the Earth is $g = 9.81 - 3.32 \times 10^{-6}h$ m/s^2, where h is the distance above the Earth's surface in meters. The commercial airplane in Fig. 1.23, with a mass of 140 000 kg, flies at a height of 10 km above the Earth's surface. Calculate its weight on the Earth's surface and at 10 km. Also calculate its potential energy at 10 km.

Figure 1.23

CHAPTER

2

Zhabska Tetyana/Shutterstock.com

Properties of Pure Substances

Outline

Nomenclature

The following nomenclature is introduced in this chapter:

A_0, B_0, a, b, c, d	Constants
C_p	Constant-pressure specific heat
C_v	Constant volume specific heat
H	Enthalpy
h	Specific enthalpy
k	Ratio of specific heats
M	Molar mass
m_f	Mass of liquid
m_g	Mass of vapor
N	Number of moles
P_{cr}	Critical-point pressure
P_R	Reduced pressure
R	Gas constant
R_u	Universal gas constant
T_{cr}	Critical-point temperature
T_R	Reduced temperature
u	Specific internal energy
v_f	Volume of liquid
v_g	Volume of vapor
$\overline{v}$	Molar specific volume
x	Quality
v_R	Pseudo-reduced specific volume
Z	Compressibility factor

Learning Outcomes

- ❑ **Understand the phases of a substance**
- ❑ **Calculate properties for various phases of water**
- ❑ **Learn how to use the IRC property calculator**
- ❑ **Understand phase diagrams**
- ❑ **Understand internal energy and enthalpy**
- ❑ **Understand refrigerants**
- ❑ **Apply the equations of state for a gas**
- ❑ **Calculate internal energy and enthalpy changes using specific heats**

Homogeneous: This word is used in two ways: chemically, as a pure substance and mechanically, as a substance whose properties are not dependent on position in the volume of interest.

In this chapter, we will consider different phases in which a pure substance exists and learn how to find important properties of engineering substances of interest in thermodynamics. Substances used in power generation and refrigeration systems will be of primary importance. The thermodynamic properties of internal energy and enthalpy will be defined and used throughout the remainder of the text. They will be calculated for several pure substances at various states so that you can become familiar with the terms and learn how to determine values of these properties. All substances of interest are *pure substances*. That is, they have the same chemical composition throughout; they are *chemically* homogeneous. Water is a good example of a homogeneous substance, even if it is an ice–water mixture, but water mixed with oil is not homogeneous since oil is not soluble in water. They must also be *simple substances;* that is, they must be free from gravitational, electrical, magnetic, and surface tension effects.

Motivational Example—The Eddystone Generating Station

The Eddystone Generating Station, shown in Fig. 2.1, is located near the Philadelphia airport in Eddystone, Pennsylvania. It has been producing 325 MW of power since it began operations in February of 1960. This power plant uses supercritical steam in its operation. Supercritical steam exists at such a high pressure and temperature that there is no difference between a gas and a liquid. The steam entering the high-pressure turbine in this plant is delivered at 34 500 kPa and 620°C. This pressure is 345 times the pressure we feel from the Earth's atmosphere. The pressure of the steam entering the turbine is so great that the steam pipes are made of stainless steel with pipe walls 150 mm or 15 cm thick. Considerable research had to be performed to determine the properties of steam under these conditions. The density of the steam being produced in this plant is two-thirds the density of liquid water. Also, at this temperature and pressure, the stainless steel pipes would display the phenomenon of creep. Creep is a slow deformation of the metal that can occur when the temperature of the pipe nears the melting point of the metal.

Why would we want to create such a dangerous environment in a power plant? As Chapter 8 will show, the efficiency of a power plant increases as the pressure and temperature of the steam passing through the turbine increase. The amount of power that a plant can produce is limited by the maximum temperature of the steam produced in the boiler. The many techniques engineers use to increase the efficiencies of power plants will also be presented in Chapter 8. The Eddystone power plant is an example of engineers pushing the limit to develop new and more effective ways to serve the public more efficiently. By being more efficient, less coal, oil, or wood is used so that our society is more sustainable. We must continue to increase efficiencies in all modes of power production.

Figure 2.1
Eddystone Generating Station, Philadelphia, PA.

Comment
By increasing the efficiencies of all our power-producing devices, our society will become more sustainable by saving energy for future generations.

2.1 Phases of a Substance

In Chapter 1 it was noted that a substance may exist in any of the three phases: a solid, a liquid, or a gas. The term *vapor* is also used to indicate the gas phase, although in some situations it is used to indicate a gas that is very near condensing into a liquid. One of the most common substances of interest in the study of thermodynamics is water. It is nontoxic and readily available. Water, both hot and cold, in the liquid phase is what comes out of our faucets at home, and it is often referred to simply as water. In a sauna, water in the gaseous phase is usually referred to as steam. Water vapor exists in the atmosphere as humidity. In thermodynamics, we primarily study substances that are in the liquid phase or the gas phase. A substance that is either a liquid or a gas is called a *fluid*, a substance that flows when subjected to a shear stress (ketchup is not a fluid). A *working fluid* is a fluid to which energy is added or subtracted while it is undergoing a process. Examples of working fluids would be the steam in a power plant or the air–fuel mixture in the cylinder of an automobile engine.

Vapor: The gas phase of a substance, usually when it is near condensation.

Fluid: A substance that flows when subjected to a shear stress, no matter how small that stress may be. A very small force can move a large ship.

Working fluid: A fluid to which we add or subtract energy while it undergoes a process.

A substance can exist in more than one phase simultaneously. A good example is the glass of ice water served at most restaurants. Water in the liquid phase

Phase equilibrium: When a substance exists in two phases simultaneously.

and water in the solid phase exist in the same container. When a substance exists in two phases simultaneously, it is said to exist in *phase equilibrium*. Another example would be a covered kettle of boiling water. In the kettle you have water in the liquid phase and water in the gaseous phase (steam). To analyze a substance that exists in phase equilibrium, both phase properties are averaged according to the mass of each phase, as will now be discussed.

2.1.1 Phase-change process

Fusion: The process of a liquid becoming a solid.

Dependent property: A property that cannot vary if a second property is held constant.

When a solid is heated, a temperature will be reached where some of the molecules have enough kinetic energy to break away from the solid lattice. This is the beginning of the phase change from solid to liquid. During the phase change, some substances will expand during the melting process and some will contract. Water (ice) is one of the few substances that contracts in volume while melting. Conversely, water will expand slightly when it goes through the *fusion* process of solidifying into ice. That is why ice, fortunately, floats and doesn't form from the bottom up.

Another unique phenomenon of the phase-change process is that pressure and temperature become dependent properties. *Dependent properties* are properties that are coupled. If one dependent property is held constant, the other also stays constant. Two properties are independent if one can vary while the other is held constant. Pressure and temperature are independent properties for pure solids and pure liquids. For example, you can hold the atmospheric pressure constant on a kettle of water, or a block of copper, and change the temperature by heating it up.

Boiling: The phase change from liquid to gas that is accompanied by the existence of bubbles.

Vaporization: The phase change from a liquid to a vapor that is not accompanied by the existence of bubbles.

Condensation: The phase change of a vapor into a liquid.

Sublimation: The phase change from a solid directly into a gas.

Saturation pressure and temperature: The pressure and temperature during the vaporization or condensation process.

During the ice phase-change process of melting, all the energy added to the ice goes to increasing the kinetic energy of ice molecules; some molecules may break away and become liquid. All the energy added to or taken from a phase-change process goes to changing the phase, not the temperature.

Another phase-change process occurs when liquid molecules have enough kinetic energy to break away from the loose liquid molecular structure and become gas molecules. The phase change from liquid to gas is called *boiling* and is accompanied by the existence of bubbles. *Vaporization*, or *evaporation*, also refers to a change of phase from a liquid to a vapor (a gas), but it is not accompanied by the existence of bubbles. For example, sweat evaporates from our skin; it does not boil.

If the phase change occurs from a gas to a liquid, it is referred to as *condensation*. The vapor condenses into a liquid. Steam condenses into water.

When a solid changes directly into a gas, it is undergoing *sublimation*. Sublimation occurs when clothes are hung on a line in the winter. The water in the clothes freezes, and surprisingly the next day the clothes are dry; the water has sublimated.

For water at atmospheric pressure, boiling occurs at 100°C. Pressure and temperature are dependent during this phase change. If a substance undergoes a phase change at a given temperature, the pressure at which this phase change occurs is the *saturation pressure*. If the substance undergoes a phase change at a given pressure, the temperature at which this occurs is the *saturation temperature*.

Observation: Water boils at about 80°C on the top of Pike's peak in Colorado.

Table 2.1 shows the saturation temperature corresponding to a given pressure for water. A more detailed table is included in Appendix C-2. For example, if water is boiled in a kettle at an atmospheric pressure of 100 kPa, the water will remain at the corresponding saturation temperature, which is 100°C. On the top of Pike's Peak in Colorado, where the pressure is about 47 kPa, water would boil at about 80°C; potatoes would take a long time to cook in an open kettle.

Table 2.1 Saturation Temperatures of Water at Various Pressures

Pressure (kPa)	Saturation Temperature (°C)
1	7
10	46
47	80
100	100
400	144
1000	180
5000	264
10 000	311

If we heat the water in a pressure cooker that maintains a constant pressure of 400 kPa, the water will boil at a temperature of 144°C; the water will exist entirely in the liquid phase since the small amount of steam that is created is allowed to escape, thereby maintaining the pressure. If the temperature increases above 144°C while the pressure is maintained at 400 kPa, the water will exist as pure vapor. Similarly, water at a pressure of 10 MPa will boil (will change phase) at 311°C.

Table 2.2 shows how the saturation pressure varies with temperature. A table with more entries is included as Appendix C-1. If water is to be boiled at 40°C, the pressure in the container needs to be 7.4 kPa, which is less than one-thirteenth of an atmosphere. If we want to boil water at 200°C, the pressure needs to be 1555 kPa. If the temperature is 200°C and the pressure is greater than 1555 kPa, the higher pressure will "squeeze" the water into the liquid phase. Likewise, if the pressure drops below 1555 kPa, the water will exist as a vapor (water vapor and steam are used as synonyms).

Comment

Water vapor and steam are synonyms. Water vapor can sometimes be treated as an ideal gas.

The phase-change process is difficult to understand at first. Between the concepts and the terminology used, it is easy to get confused. These concepts, though, form the basis of being able to determine the thermodynamic properties of substances. This information is needed to understand the analysis of processes and cycles presented in later chapters. One way to describe this process is to consider in some detail liquid water being heated at constant pressure to form steam.

Table 2.2 Saturation Pressures of Water at Various Temperatures

Temperature (°C)	Saturation Pressure (kPa)
10	1.2
40	7.4
80	47
100	101
140	362
200	1555
300	8588

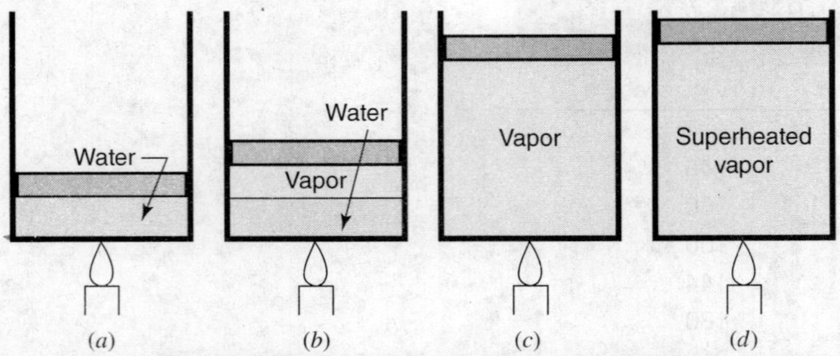

Figure 2.2

The heating of water at constant pressure in a cylinder with a frictionless piston.

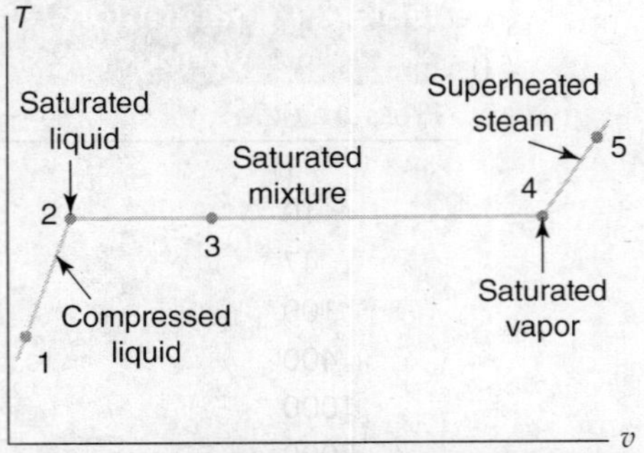

Figure 2.3

A sketch of T vs v for the heating process of water at constant pressure.

Compressed (subcooled) liquid: A liquid that does not have enough internal energy to create gas molecules. It is not boiling.

When liquid water is at standard atmospheric pressure with a temperature between 1°C and 99°C, it exists as a *compressed liquid*, also called a *subcooled liquid*: a liquid that does not have enough internal energy to create gas molecules. That is, its temperature is not sufficiently high to cause the liquid to boil. When you turn on a faucet, subcooled water flows out, whether it's cold or hot.

Saturated liquid: The state in which a substance is just starting to create vapor molecules.

Saturated mixture: A mixture consisting of both liquid and vapor. It is also referred to as the wet region, the saturated region, or the quality region.

Saturated vapor: The state in which a mixture is just completely vaporized.

Superheated vapor: A state in which the vapor is not ready to condense.

Freezing: The phase change of a liquid into a solid.

Quality: The ratio of the mass of the vapor in the mixture to the total mass of the mixture. It is used only in the region between a saturated liquid and a saturated vapor.

Let's mentally run an experiment and heat water in a cylinder fitted with a frictionless piston so the pressure is held constant, as shown in Fig. 2.2; the process is sketched in Fig. 2.3 as T vs v. Compressed water begins its journey in Fig. 2.2*a* and at state 1 in Fig. 2.3. When the water is heated to the saturation temperature at state 2, it is still 100% liquid but has enough internal energy to just start creating steam. This state is called a *saturated liquid*, as shown as Fig. 2.2*a* and at state 2 in Fig. 2.3. The water is still liquid but is just ready to start vaporizing (also called boiling or evaporating). Once the water is heated to the saturation temperature and is vaporizing, it is called a *saturated mixture*, as shown as Fig. 2.2*b* and at state 3 in Fig. 2.3. A saturated mixture consists of both liquid and vapor. As more heat is added to the mixture, more vapor is produced. Since the pressure of the mixture is being held constant, the temperature will also remain constant since pressure and temperature are dependent properties during a phase change. When the mixture is completely vaporized, it is a *saturated vapor*, as shown as Fig. 2.2*c* and at state 4 in Fig. 2.3. If heat continues to be added at constant pressure, the vapor becomes *superheated*, as shown as Fig. 2.2*d* and in Fig. 2.3 at state 5. Because the pressure is held constant, the specific volume increases as the mixture changes from the saturated liquid state to the saturated vapor state and into the superheated state.

If the experiment were run in the opposite direction, condensation of the vapor into liquid would occur, followed by *freezing* (solidification) of liquid into a solid. The solid phase is not shown in Fig. 2.3.

To analyze the properties of a saturated mixture we define a property called quality, denoted with the letter x. *Quality* is defined as the ratio of the mass m_g of the vapor in the mixture to the total mass m of the mixture:

$$x = \frac{m_g}{m} \tag{2.1}$$

Quality, by definition, varies between 0 and 1, as identified by state 3 in Fig. 2.4. If $x = 0$, none of the mass is vapor; it is all liquid, as at state 2 in the figure. As vapor is generated, the quality will increase above 0. A quality of 0.5 means that half of the mass is liquid and half of the mass is vapor. If $x = 1$, the mass is all vapor, as shown in Fig. 2.4 at state 4. But don't confuse mass with volume; quality is the ratio of masses, not the ratio of volumes. The total mass is m and the total volume is V, related to the liquid and vapor components by

$$m = m_f + m_g \qquad V = V_f + V_g \tag{2.2}$$

where m_f and V_f are the liquid components, and m_g and V_g are the vapor components. In terms of specific volumes, the total volume $V = mv$ can be written as

$$mv = m_f v_f + m_g v_g \tag{2.3}$$

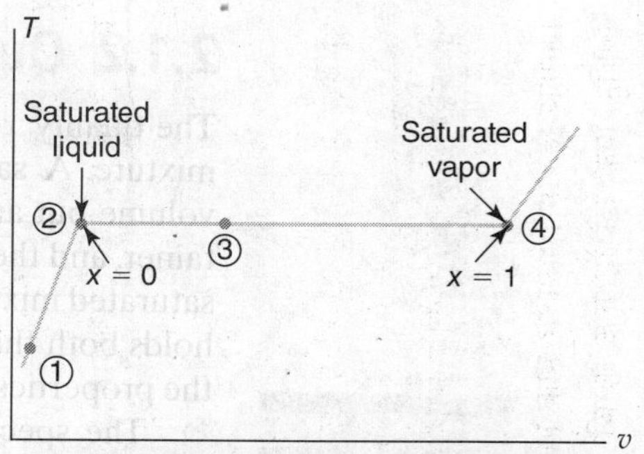

Figure 2.4
Quality x is defined only in the region between states 2 and 4.

where v is the average value of the specific volume for the combined mass, that is, for the entire volume. Substitute $m_g = xm$ from Eq. 2.1 and use $m_f = m - m_g$ from Eq. 2.2 to obtain

Note: The subscript f refers to liquid (a "fluid"), and the subscript g refers to a vapor (a "gas").

$$mv = (m - m_g)v_f + xmv_g \tag{2.4}$$

Now, divide by m and find that, using Eq. 2.1,

$$v = (1 - x)v_f + xv_g$$

Comment

A variable, such as v, that is used for a mixture, is the average value of the variable for the components that make up the mixture. Some books use v_{av} or $\bar{v}$ to denote this average.

or, rewritten in the preferred form,

$$v = v_f + x(v_g - v_f) \tag{2.5}$$

The difference between a gas-phase property and a liquid-phase property is a term labeled with a subscript fg. For specific volumes, it would be

$$v_{fg} = v_g - v_f \tag{2.6}$$

Many thermodynamics tables include values of the difference between the gas and liquid properties. These speed up the calculation of the properties. Equation 2.5 can then take the simplified form

$$v = v_f + xv_{fg} \tag{2.7}$$

This equation requires one multiplication and one addition. With fewer numbers of operations, calculations can be made quicker and with fewer mistakes.

Quality is sometimes referred to as a percentage. A quality of 0% is a quality of 0, a quality of 100% is a quality of 1, and so forth. It is important to remember that quality is defined as a mass ratio. In some problems you may be given the volume of the vapor and the volume of the liquid. The ratio of these is not the quality, as will be illustrated in an example.

Note: A quality of 0.25 could also be stated as a quality of 25%.

☑ You have completed Learning Outcome (1)

2.1.2 Quality and compressed liquid calculations

The quality x of a saturated mixture is used to calculate other properties of the mixture. A saturated mixture contains liquid and gas, which exist in the same volume but are not actually mixed. The liquid will rest on the bottom of the container, and the gas will exist separately above it. When we calculate a property of a saturated mixture, we calculate the average property, such as v, in the container that holds both the liquid and the vapor. This averaged property is very different from the properties of the liquid or the gas unless the state is very close to $x = 0$ or $x = 1$.

Remember: The specific volume of a compressed liquid is approximately v_f at the specified temperature. Ignore the pressure.

The specific volume of a compressed liquid is not significantly affected by pressure, as observed in Table C-4 in the Appendix. It depends primarily on the temperature. Observe in that table that there is approximately a 3% change in the specific volume between $P = 5$ MPa and $P = 30$ MPa at 240°C. So, it is acceptable to use the specific volume v_f from Table C-1 in the Appendix at a specified temperature.

Example 2.1 How do we calculate volumes?

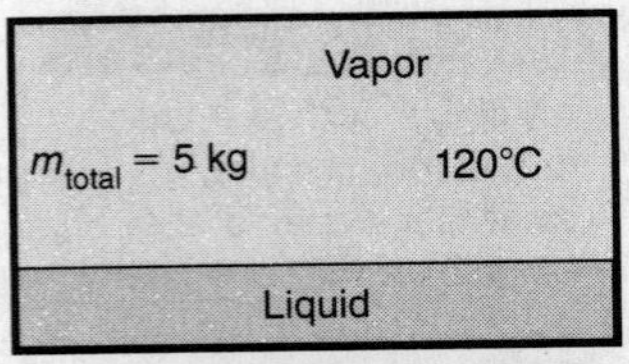

Figure 2.5

A container holds 5 kg of saturated water at 120°C with a quality of 0.6, as sketched in Fig. 2.5. Determine the volume of the liquid water, the volume of the vapor in the container, the total volume, and the pressure.

Solution

A quality of 0.6 defines the state as a saturated mixture. Since quality is defined as the ratio of the mass of vapor to the total mass, we can determine the mass of each phase as follows:

$$m_g = x \cdot m = 0.6 \times 5 \text{ kg} = 3 \text{ kg}$$

$$m_f = m - m_g = 5 - 3 = 2 \text{ kg}$$

To determine the volume of each phase, we need to find the specific volume of both the liquid and the vapor. The liquid phase in the container will have the properties of a saturated liquid at 120°C. The steam in the container will have the properties of a saturated vapor at 120°C. These properties can be found in Table C-1 in the Appendix. The specific volume for a saturated liquid at 120°C is 0.001060 m^3/kg, and for a saturated vapor at 120°C it is 0.8919 m^3/kg. The volume of each phase is now calculated to be

$$V_f = m_f v_f = 2 \text{ kg} \times 0.001060 \frac{\text{m}^3}{\text{kg}} = \underline{0.00212 \text{ m}^3}$$

$$V_g = m_g v_g = 3 \text{ kg} \times 0.8919 \frac{\text{m}^3}{\text{kg}} = \underline{2.676 \text{ m}^3}$$

The total volume of the container is

$$V = 0.00213 \text{ m}^3 + 2.676 \text{ m}^3 = \underline{2.678 \text{ m}^3}$$

The pressure in the container is the saturation pressure at 120°C, which is 198.5 kPa. The liquid, which is 40% of the mass, is only about 0.08% of the volume. That's because the density of the liquid is 840 times that of the vapor. Figures 2.3 and 2.4 are not to scale.

Comment

If Figs. 2.3 and 2.4 were sketched to scale using volumes, the liquid phase would not be noticed. Problem 2.46 requests that the sketches be drawn to scale.

The volume of a mixture in a container — Example 2.2

A container holds 5 kg of water at 120°C with a quality of 0.6. Determine the total volume of the water in the container. This is the same volume as in Example 2.1, but it will be calculated using the average specific volume v.

Solution

The average specific volume of the saturated mixture, using Eq. 2.5 and the values listed in Example 2.1, is

$$v = v_f + x(v_g - v_f)$$
$$= 0.001060 + 0.6 \times (0.8919 - 0.001060) = 0.5356 \text{ m}^3/\text{kg}$$

We can now calculate the volume of the container using this average specific volume and the total mass:

$$V = m \cdot v = 5 \times 0.5356 = \underline{2.678 \text{ m}^3}$$

Note: The specific volume v of the water combines the liquid and the vapor phases.

The pressure and quality of a mixture — Example 2.3

A 2000-cm^3 container holds 0.4 kg of water at 300°C. Determine the pressure and the quality of the mixture in the quality region at state 1 indicated in Fig. 2.6.

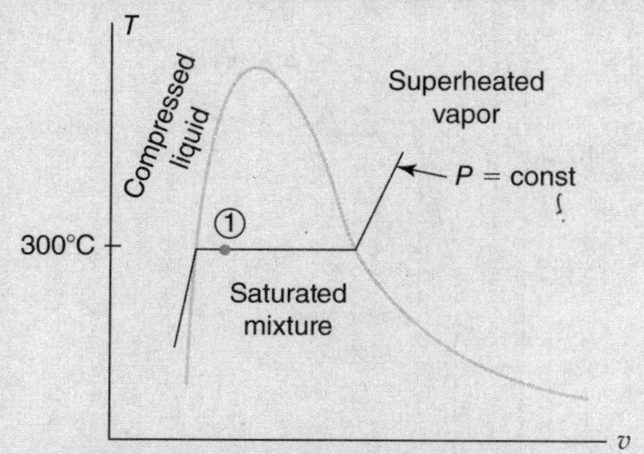

Figure 2.6

(Continued)

Example 2.3 (Continued)

Solution

The pressure and temperature are coupled in the saturated mixture region. So, the pressure can be found next to the temperature in Table C-1 in the Appendix. It is

$$P_1 = \underline{8.58 \text{ MPa}}$$

Comment

If $P > 8.58$ MPa, the water would be compressed. If $P < 8.58$ MPa, the water would be superheated.

Recognizing that 2000 cm³ is 2000×10^{-6} m³, we find that the specific volume of the saturated mixture is

$$v_1 = \frac{V}{m} = \frac{2000 \times 10^{-6}}{0.4} = 0.005 \text{ m}^3/\text{kg}$$

Note: Quantities given in a problem statement are assumed known to four significant digits. So the mass of 0.4 kg is assumed to be 0.4000 kg. Probably three significant digits would be closer to the truth, but assuming four significant digits is usually acceptable.

This is less than v_g but greater than v_f, so it is a mixture of liquid and vapor. We can now calculate the quality:

$$v_1 = v_f + x(v_g - v_f)$$

$$0.005 = 0.001404 + x_1(0.02168 - 0.001404) \quad \therefore x_1 = \underline{0.1774}$$

The state is relatively close to the liquid saturation point. Its quality is 17.74%.

Example 2.4 The change in volume of a compressed liquid

A cylinder holds 200 kg of water at 120°C at a pressure of 20 MPa. The temperature is increased to 340°C while holding the pressure constant, as shown in Fig. 2.7. Determine the volume of the water at both temperatures and the percent change in the volume. Use i) Table C-1 and ii) Table C-4.

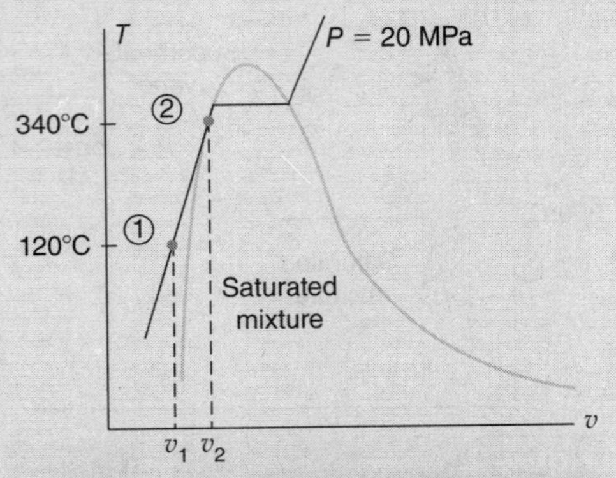

Figure 2.7

Solution

i) Assuming the specific volume depends on temperature only, the volumes are found, using Table C-1, to be

$$V_1 = mv_1 = 200 \times 0.00106 = \underline{0.212 \text{ m}^3}$$

$$V_2 = mv_2 = 200 \times 0.001638 = \underline{0.3276 \text{ m}^3}$$

The percent change in the volumes is

$$\%\text{change} = \frac{0.3276 - 0.212}{0.212} \times 100 = \underline{54.5\%}$$

ii) Using the compressed liquid Table C-4 provides

$$V_1 = mv_1 = 200 \times 0.0010496 = \underline{0.2099 \text{ m}^3}$$

$$V_2 = mv_2 = 200 \times 0.001633 = \underline{0.3266 \text{ m}^3}$$

The percent change in the volumes is

$$\%\text{change} = \frac{0.3266 - 0.2099}{0.2099} \times 100 = \underline{55.6\%}$$

Observe that the values using Table C-1 are quite acceptable for engineering calculations.

2.1.3 Superheated vapor

When the last drop of liquid in a saturated mixture vaporizes, the state becomes a saturated vapor with a quality of $x = 1$. If heat continues to be added to the saturated vapor, it becomes a superheated vapor for which quality has no meaning. In this state, the substance exists as a gas, and temperature and pressure now become independent properties. We once visited a major power plant that, as is common, used steam as the working fluid that drove the turbines. The steam generator (boiler) had a viewport on the side with a sign that read, "superheated steam." When we looked through the viewport, we saw absolutely nothing. A superheated vapor is invisible, as is the water vapor in the atmosphere on a nice sunny day. The properties of superheated water vapor are presented in Table C-3 in the Appendix.

Remember: Steam, vapor, and gas are often used synonymously.

Interpolate to find specific volume — **Example 2.5**

Find the specific volume of superheated steam at 323°C and 2 MPa. Assume a linear relationship between temperature and specific volume in the vicinity of the desired state.

(Continued)

Example 2.5 (*Continued*)

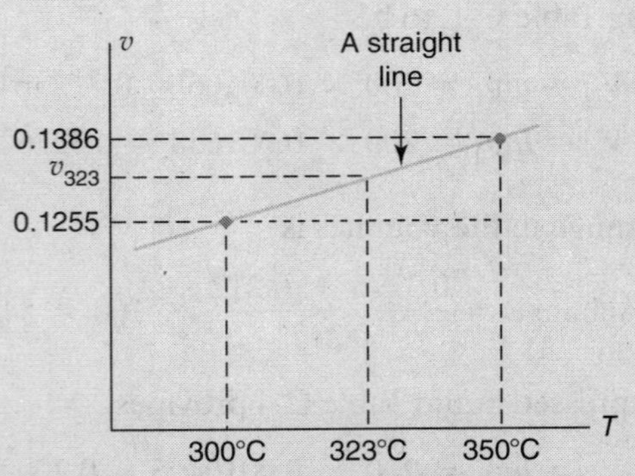

Figure 2.8

Solution

The specific volume at 300°C and 350°C is listed in Table C-3, with two entries displayed as Table 2.3. A straight-line interpolation, sketched in Fig. 2.8, provides the specific volume at 323°C. The ratios from the triangles provide

$$\frac{v_{323} - 0.1255}{0.1386 - 0.1255} = \frac{323 - 300}{350 - 300} \qquad \therefore v_{323} = \underline{0.1315 \text{ m}^3/\text{kg}}$$

There are computer programs that provide properties at any temperature and pressure, but it is important that one can interpolate for values in tables of numbers. Tabulated sets of numbers are encountered in numerous areas of engineering, and interpolation or extrapolation is often necessary; computer programs are not always available. Practice interpolating until you are fairly good at it. It is necessary when you are using tables in various subject areas and may be required during an exam!

Table 2.3 $P = 2$ MPa

T (°C)	v (m³/kg)
300	0.1255
350	0.1386

Example 2.6 Double interpolation of water vapor properties

Water exists at 1120 kPa and 470°C. Determine the specific volume of this superheated water vapor.

Solution

Appendix C-3 presents the properties for superheated water vapor. Tables exist for pressures of 1.0 MPa and 1.2 MPa, so an interpolation for $P = 1.12$ MPa is necessary. The tables also list properties for steam at temperatures of 400°C and 500°C, so an interpolation for $T = 470$°C is also necessary. This is referred to as a *double interpolation*. The following table shows the specific volume at the four listed states from Table C-3:

	$P = 1.0$ MPa	$P = 1.2$ MPa
$T = 400°C$:	$v = 0.3066\ m^3/kg$	$v = 0.2548\ m^3/kg$
$T = 500°C$:	$v = 0.3541\ m^3/kg$	$v = 0.2946\ m^3/kg$

The interpolation for v_1 at $P = 1.12$ MPa and 400°C is performed first, although the temperature interpolation could be selected first. At $T = 400°C$:

$$\frac{1.2 - 1.12}{1.2 - 1.0} = \frac{0.2548 - v_1}{0.2548 - 0.3066} \qquad \therefore v_1 = 0.2755\ m^3/kg$$

Next, interpolate for v_2 at $P = 1.12$ MPa and 500°C:

$$\frac{1.2 - 1.12}{1.2 - 1.0} = \frac{0.2946 - v_2}{0.2946 - 0.3541} \qquad \therefore v_2 = 0.3184\ m^3/kg$$

Next we interpolate between 400°C and 500°C to get the specific volume v at 470°C and 1120 kPa:

$$\frac{470 - 400}{500 - 400} = \frac{v - 0.2755}{0.3184 - 0.2755} \qquad \therefore v = \underline{0.306\ m^3/kg}$$

Note: Three rather than four significant digits are written since the straight-line interpolation method loses at least one significant digit.

☑ You have completed Learning Outcome (2)

2.1.4 Properties using the IRC Property Calculator

Note: You can also Google "IRC Fluid Property Calculator."

In the several examples of this section, we have used the tables in Appendix C to find the properties of water. Several other substances will also be of interest in our study of thermodynamics. Rather than using the tables in the Appendix to find properties, the IRC (Industrial Refrigeration Consortium), in conjunction with the University of Wisconsin,[1] developed the IRC Fluid Property Calculator, displayed in Fig. 2.9, and located at http://www.irc.wisc.edu/properties where properties of interest can be found rather easily. This site will be used in some examples and solutions to problems if it is not requested that the tables in the Appendix be used, and especially if interpolation is required. It should be pointed out that, on this Web site, density is used as an input rather than specific volume, so remember to use

$$\rho = \frac{1}{v} \tag{2.8}$$

when converting from specific volume to density. The specific volume is one of the properties calculated, although it is called "Volume."

The tables in the appendix are generated not by using measured values of properties but by using curve fits of measured values, allowing properties to be carried

[1]There may be periods of time when the IRC Fluid Property Calculator, referred to as the IRC Calculator, is not operating due to computer system updates and/or changes being made in the operation of the calculator.

Comment

The Web site automatically selects English units. If SI units are desired, check the appropriate box on the site.

Choose a fluid.

Enter two conditions (properties).

Click on "Calculate Properties."

Read the values of the properties of interest. Not all properties are shown here, and many shown are not of interest in our study.

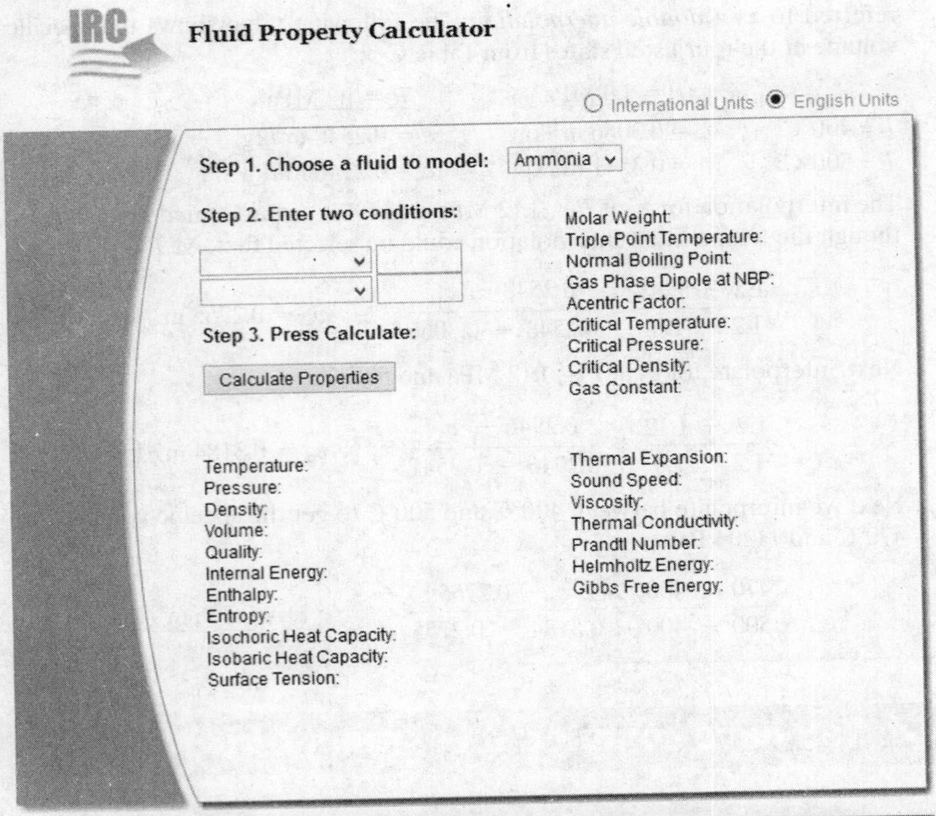

Courtesy of IRC. www.irc.wisc.edu/properties/

Figure 2.9

A display of the IRC Property Calculator.

Comment

It takes practice to become familiar with this IRC Calculator. Check the solutions in Example 2.7 and make sure they are correct.

out to many significant numbers. This is an undesirable feature for engineering students since exaggerated accuracy is implied. Several sets of curve fits are available, so numerical values from one table may not match all the significant digits of the values from another table. The values generated with the IRC Calculator may be slightly different from the values in the tables in this book. The answers should, however, be the same to within acceptable engineering accuracy, which is usually taken as about 3%. The three-digit accuracy on the Web page is more realistic than the multidigit accuracy in the tables and will hopefully prevent both students and faculty from presenting answers to five and six significant digits.[2] The following example will illustrate the use of the Web page, which resembles the reproduction above.

[2] It is suggested that one point be deducted per answer on a partial-credit exam when five or more significant digits are given in an answer. Measured properties (temperature, pressure, etc.) are not known to four significant digits unless extreme care is taken, and probably are never known to five significant digits unless you're a scientist working at the National Bureau of Standards. An answer to five or more significant digits is not acceptable.

Use the IRC Calculator to find properties Example 2.7

i) Find the volume of 5 kg of water if it is at 120°C with $x = 0.6$.
ii) A 2000-cm^3 container holds 0.4 kg of water at 300°C. Find the quality.
iii) Determine the volume of 200 kg of water at 340°C and 20 MPa.
iv) Find the specific volume of superheated steam at 470°C and 1120 kPa.

Solution

i) From the website www.irc.wisc.edu/properties select "International units" and "Water." Select "Temperature" as 120°C and "Quality" as 0.6. Press "Calculate Properties" and under "Volume" find $v = 0.535$ m^3/kg. The volume is

$$V = mv = 5 \text{ kg} \times 0.535 \text{ m}^3/\text{kg} = \underline{2.68 \text{ m}^3}$$

This compares with 2.678 m^3 in Example 2.2.

ii) Since specific volume is not an input, the density is found to be

$$\rho = \frac{m}{V} = \frac{0.4 \text{ kg}}{2000 \text{ cm}^3 \times 10^{-6} \text{ m}^3/\text{cm}^3} = 200 \text{ kg/m}^3$$

Follow the directions in (i) with $T = 300$°C, and find $x = \underline{0.178}$. This compares with $x = 0.1774$ in Example 2.3.

iii) At 340°C and 20 000 kPa, following the directions given in (i), we find $v = 0.00157$ m^3/kg so that

$$V = mv = 200 \text{ kg} \times 0.00157 \text{ m}^3/\text{kg} = \underline{0.314 \text{ m}^3}$$

This compares with a value of 0.328 m^3, based on a linearly interpolated value for v_2 in Example 2.4, a difference of 4%. A more accurate interpolation method for v_2 would reduce the difference. The value using the IRC Calculator is more accurate.

iv) At 470°C and 1120 kPa, following the directions given in (i), the Web site provides $v = \underline{0.303 \text{ m}^3/\text{kg}}$, which compares with 0.306 m^3/kg in Example 2.6.

Comment

The number 0.535 has three significant digits so the answer is expressed using three significant digits, not four.

☑ You have completed Learning Outcome (3)

2.1.5 Phase diagrams

A phase diagram is a two- or three-dimensional graph that shows how a pure substance (a substance that has the same chemical composition throughout) changes state as temperature, pressure, and volume of the substance vary. If numerous experiments were run, similar to the mental experiment described earlier (see Fig. 2.3), and the results plotted on one graph, all the saturated liquid states and the saturated vapor states could be connected to form the *T-v* phase diagram sketched in Fig. 2.10. The diagram is divided into three regions: the compressed liquid region, the saturated mixture region, and the superheated vapor region. A *P-v* diagram

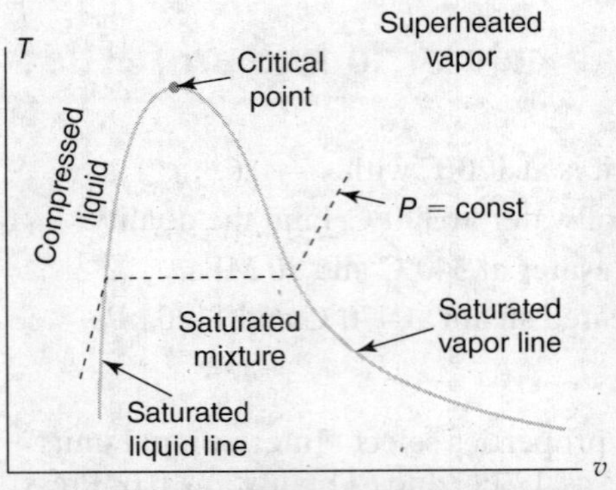

Figure 2.10

A *T-v* phase diagram for a pure substance.

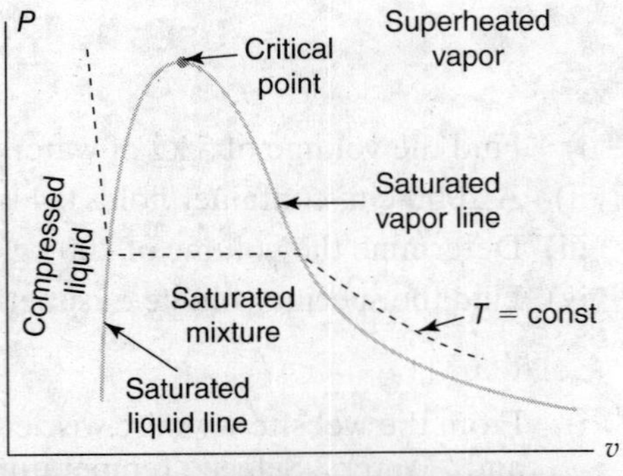

Figure 2.11

A *P-v* phase diagram for a pure substance.

created with constant-temperature experiments, similar to those that created Fig. 2.4, would appear quite similar and, of course, exhibit the same three regions; it is shown in Fig. 2.11.

Saturated liquid line: The solid line separating the compressed liquid from the saturated mixture.

Saturated vapor line: The line separating the superheated vapor from the saturated mixture.

Note: The saturated mixture region is also referred to as the "wet region" or the "quality region."

Critical point: Where the saturated liquid line and the saturated vapor line intersect.

Supercritical region: At a pressure and temperature above the critical point values in a state where a liquid is no longer distinguished from a vapor.

The region on the left represents the range of pressures and temperatures where the substance behaves as a compressed liquid. The region under the solid curve shows the range of pressures and temperatures where the substance behaves as a saturated mixture of liquid and vapor. The solid line separating the compressed liquid from the saturated mixture is called the *saturated liquid line*. It is on this line that the phase-change process from a liquid to a vapor begins. To the left of the line, the fluid is a compressed liquid. To the right of the line it is a saturated mixture. On the line, the fluid exists as a saturated liquid with a quality of zero. The region to the far right is the range of pressures and temperatures where the fluid exists as a superheated vapor. The line separating the superheated vapor from the saturated mixture is the *saturated vapor line*. On this line, the fluid exists as a saturated vapor with a quality of one. The saturated mixture region that appears as a dome between the saturated liquid line and the saturated vapor line is also referred to as the *wet region* or the *quality region*. All pure substances have phase diagrams like the ones shown in Figs. 2.10 and 2.11.

The saturated liquid line and the saturated vapor line intersect at a very high pressure at a point called the *critical point*. For water, the pressure and temperature at the critical point are $P_{cr} = 22.09$ MPa and $T_{cr} = 374.1°C$. The critical temperatures and pressures for other common substances are included in Appendix B-3. The superheat region in which the pressure and temperature are above the critical point values is often referred to as the *supercritical region*; the substance can no longer be distinguished as a liquid or a vapor.

Very few engineering processes outside of power plants operate in the supercritical region. The cost is often too high to safely contain the working fluid. The Eddystone power plant, described at the beginning of this chapter, operates in the supercritical region, as do about 150 other power plants in the United States.

If we combined the two diagrams into one and included the solid phase as well, it would appear as the three-dimensional sketch in Fig. 2.12 for a substance that contracts on freezing. The dashed lines are constant-temperature lines.

Another phase diagram is the pressure-temperature diagram, where the pressure of the substance is plotted as a function of the temperature, as in Fig. 2.13. Three lines separate the three phases. The phase change occurs as a line is crossed.

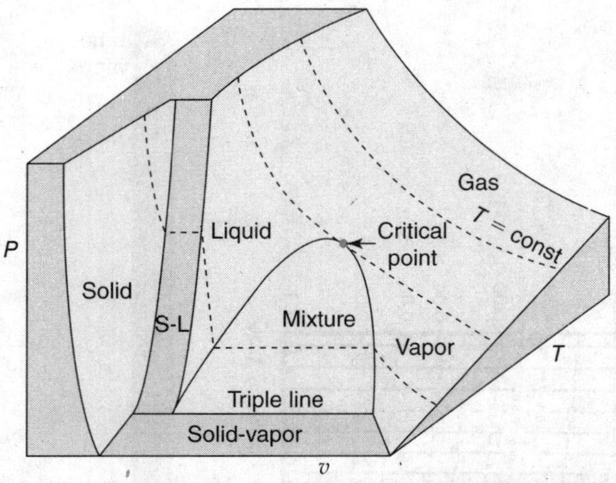

Figure 2.12

The *P-v-T* diagram for a substance that contracts on freezing.

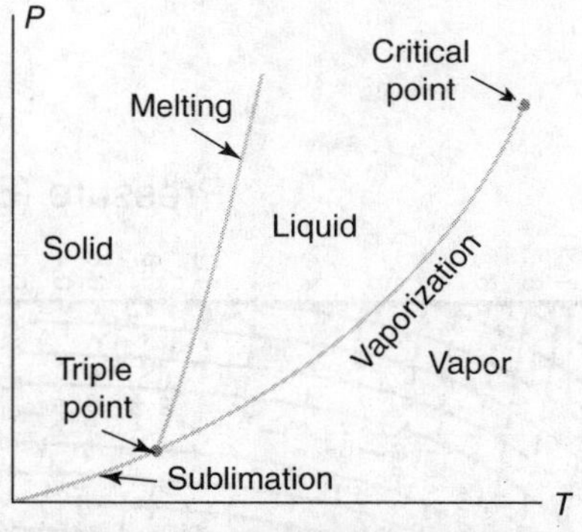

Figure 2.13

The *P-T* phase diagram.

This can be observed if the *P-v-T* diagram of Fig. 2.12 is viewed from the right; the saturated-mixture dome would appear as the curved vaporization curve of Fig. 2.13. It may be difficult to visualize Fig. 2.13, but with some effort it is possible. The *T-v* or *P-v* diagrams are used exclusively in the examples and problems.

The melting line (or fusion line) separates the solid from the liquid. The vaporization line separates the liquid from the vapor, and the sublimation line separates the solid from the vapor. The point where all three lines come together is the *triple point*, where all three phases coexist. (The solid-vapor region wasn't displayed on the bottom, as is common, of the phase diagrams of Figs. 2.10 and 2.11.)

Triple point: Where all three phases coexist.

Sublimation: When a solid changes phase directly to a gas without passing through the liquid phase.

We refer to a liquid changing phase to a vapor as *vaporization* (or boiling or evaporation) and a vapor changing phase to a liquid as *condensation*. Melting and freezing is commonly known in the phase change between solids and liquids. *Sublimation*, which occurs when a solid changes phase directly to a gas without passing through the liquid phase, is less well known and not of significant interest in our study of thermodynamics. Water vapor can become a solid directly if the temperature drops suddenly, resulting in frost. If we could generate power during that process, we'd definitely be interested in it!

None of the diagrams are drawn to scale. In order to display the various regions, the diagrams are skewed much to the right so that the various regions are observable. The saturated liquid lines in Figs. 2.10 and 2.11 are actually very close to the *P*-axis, and the triple point in Fig. 2.13 is actually very close to the origin. A good exercise would be to draw the *T-v* and *P-T* diagrams for water to scale; see Problem 2.46.

Comment

None of the phase diagrams are drawn to scale in order to make the various regions more apparent.

Before extensive tables or computer software were used to determine thermodynamic properties, phase diagrams displayed the values of several properties, and numerical values were read from the diagrams. They are often used in problem solving to help visualize a process, much like the free-body diagram used in statics. Some engineers find them very useful in understanding a problem, whereas others have no interest in them.

One phase diagram that may be provided when taking the Fundamentals of Engineering Exam (FE Exam) is the pressure-enthalpy diagram (enthalpy will be defined in the next section). Fig. 2.14 shows a pressure-enthalpy diagram for the

Comment

A phase diagram can be constructed using any two independent properties. The *T-v* and *P-v* diagrams are usually selected. The *P-h* diagram shown in Fig. 2.14 is unnecessary in this text since the information is provided in tables in the Appendix.

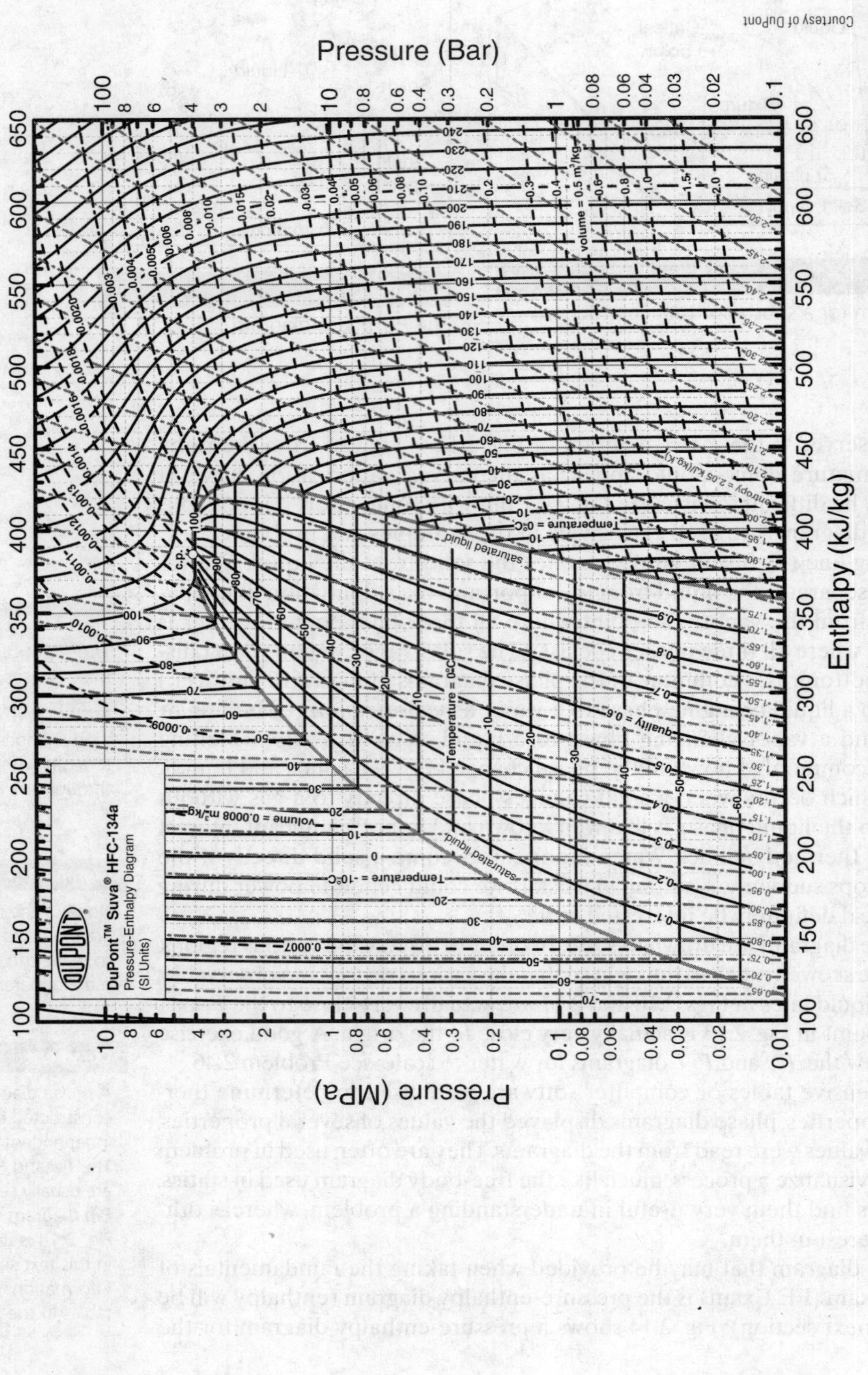

Figure 2.14 Pressure-enthalpy diagram for refrigerant R134a.

refrigerant R134a. When presented as part of a page in a textbook, it is too small to be read accurately; if a much larger diagram is available, it can be used to read values to three significant digits.

A phase diagram — Example 2.8

The properties of the following three states are provided as (T, P, v). Indicate where the states are on a T-v phase diagram provided in Fig. 2.15 and determine the unknown property.

① (340°C, 2000 kPa, v)
② (140°C, P, 0.5 m^3/kg)
③ (T, 5000 kPa, 0.00108 m^3/kg)

Solution

The T-v phase diagram is sketched as shown in Fig. 2.15. By scanning Tables C in the Appendix, state 1 is found in the superheat Table C-3 at 2 MPa with 340°C between 300°C and 350°C. That's 80% of the difference between those two temperatures. Interpolation provides

$$v_1 = (0.13857 - 0.12547) \times 0.8 + 0.12547 = \underline{0.136 \text{ m}^3/\text{kg}}$$

State 2 is observed to be very close to the saturated vapor state, so it is in the mixture region. The pressure is then found next to the temperature in Table C-1 to be $P_2 = 0.3613$ MPa or $\underline{361.3 \text{ kPa}}$.

Now, for state 3. In Table C-2 at 5 MPa (between 4 MPa and 6 MPa), observe that v_3 is less than v_f. Hence, it is a compressed liquid. Its temperature can be found in Table C-1 or Table C-4. From Table C-1 the temperature is $\underline{140°\text{C}}$. Table C-4 would provide essentially the same temperature.

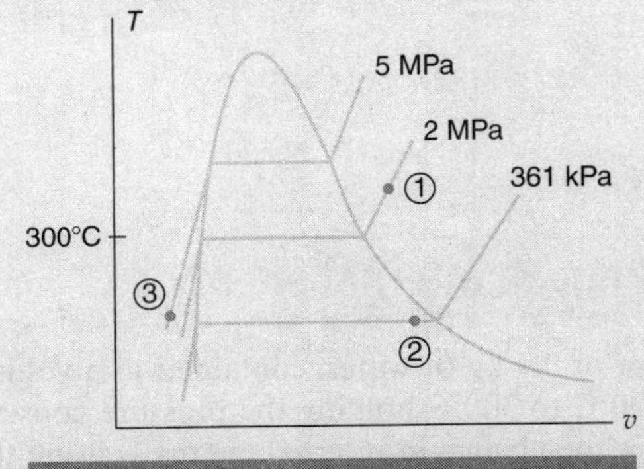

Figure 2.15

Observe: The water you drink in a glass, be it ice water or hot tea, is compressed liquid. It's compressed liquid if it is below the boiling point.

☑ **You have completed Learning Outcome (4)**

2.2 Internal Energy and Enthalpy

2.2.1 Internal energy

In Chapter 1 we considered some of the basic properties of substances, such as density, specific volume, pressure, and temperature. In this section we will add new properties to this list,which will give us information about the energy level of a substance. A key step in being able to solve problems in thermodynamics is the ability to determine numerous properties of a substance. In Chapter 1 we observed that temperature is proportional to the amount of thermal energy contained in a substance. In our study of thermodynamics, with the exception of combustion, the thermal energy contained in a substance is the *internal energy*. Internal energy is a derived property in that it is a function of basic properties, primarily temperature. The concept of internal energy is easier to understand if you consider the effects of changing it. Molecules move according to the amount of thermal energy contained in a substance. As the motion of the molecules increases, the internal energy increases, as does the temperature. In fact, *absolute zero temperature* is defined as the state in which there is no motion of the molecules. For this reason, there are no negative energy levels in a substance.

Internal energy: The amount of thermal energy contained in a substance.

Absolute zero temperature: The state in which there is no motion of the molecules.

In the wet region, specific internal energy u (often referred to simply as internal energy) is calculated in the same way volume was calculated by using

$$u = u_f + x(u_g - u_f) \tag{2.9}$$

Comment

We often omit "specific" when referring to u. In the tables in the Appendix, it may be titled simply "Energy."

The units of internal energy are the same as the units for any energy term. Consider potential energy mgz; it has units in the SI system of (kg)(m/s^2)(m), or using N = kg·m/s^2 (see Eq. 1.2), the units are N·m, which is called a joule (J). A joule is a small amount of energy, so we mostly use units of kilojoules (kJ). All forms of energy in the SI system use this unit. The standard symbol for internal energy is U and for specific internal energy the letter u. Specific internal energy has units of kJ/kg in the SI system and Btu/lbm in the English system.

Note: A joule is a relatively small energy unit whereas the kJ is relatively large so that 1 kJ = 1000 J.

Example 2.9 Find the change in internal energy

The temperature of 0.9 kg of water, contained in a volume of 0.03 m^3, is increased from 100°C to 300°C holding the pressure constant, as illustrated by Fig. 2.16. What is the change in internal energy i) using the equations and ii) using the IRC Calculator?

Solution

i) First, let's determine the initial state of the water. The specific volume is

$$v_1 = \frac{V_1}{m} = \frac{0.03\ \text{m}^3}{0.9\ \text{kg}} = 0.0333\ \text{m}^3/\text{kg}$$

At 100°C, state 1 is in the wet region since $0.0010 < 0.0333 < 1.6730$ (see Table C-1). To find the internal energy at state 1, we must use the quality. It is found to be

$$v_1 = v_f + x_1(v_g - v_f)$$
$$0.0333 = 0.0010 + x_1(1.670 - 0.0010) \qquad \therefore x_1 = 0.0193$$

The specific internal energy is then

$$u_1 = u_f + x_1 u_{fg}$$
$$= 418.9 + 0.0193(2507 - 418.9) = 459.2 \text{ kJ/kg}$$

The pressure at state 1 is found next to the temperature in Table C-1 to be 101.3 kPa. State 2 with $P_2 = 101.3$ kPa and $T_2 = 300$°C is found in the superheat Table C-3, where we read $u_2 = 2810$ kJ/kg. The change in internal energy is calculated to be

$$\Delta U = m(u_2 - u_1)$$
$$= 0.9 \text{ kg}(2810 - 459.2) \text{ kJ/kg} = \underline{2116 \text{ kJ}}$$

ii) Follow the steps outlined in Example 2.7 and find at $T_1 = 100$°C and $\rho_1 = 1/v_1 = 1/0.0333 = 30$ kg/m³ [using the specific volume from part (i)], $u_1 = 459$ kJ/kg and $P_1 = 101$ kPa. At $T_2 = 300$°C and 101.3 kPa, we obtain $u_2 = 2810$ kJ/kg. We have

$$\Delta U = m(u_2 - u_1)$$
$$= 0.9(2810 - 459) = \underline{2120 \text{ kJ}}$$

The difference between the two answers is negligible.

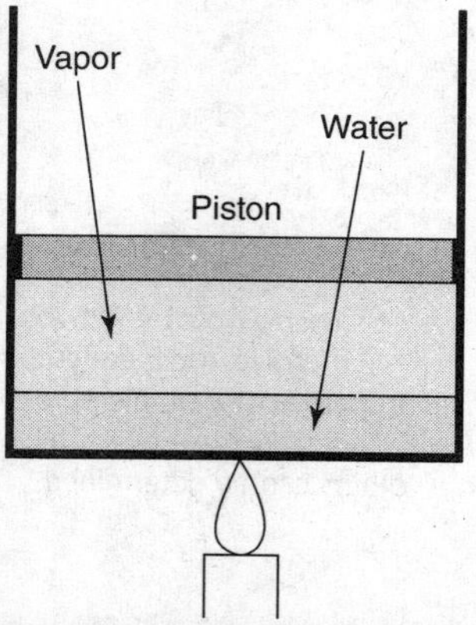

Figure 2.16

Comment

We often refer to the saturated-mixture region as the wet region.

Remember: Density is input using the IRC Calculator, not the specific volume.

2.2.2 Enthalpy

When working with equations in thermodynamics, we often encounter the combination $u + Pv$ of properties. Because this happens so often, especially when analyzing devices involving a flowing working fluid, such as steam flowing through a turbine, we define *enthalpy* H to be

$$H = U + PV \tag{2.10}$$

Specific enthalpy, often referred to simply as *enthalpy*, is then found by dividing by the mass:

$$h = u + Pv \tag{2.11}$$

It simplifies calculations since three properties are combined into one. In fact, in engineering practice, enthalpy is used more often than internal energy, so a table may list enthalpy but not specific internal energy. If that is the case, Eq. 2.11 is used to calculate specific internal energy.

Enthalpy is also found to be a measure of the amount of energy in a substance in some instances, especially in a flowing fluid. If high-pressure vapor or air is allowed to flow over a set of turbine blades, the flow will cause the blades to rotate and

Enthalpy: The combination of properties $(U + PV)$ occurs quite often. It is particularly useful when analyzing devices through which a fluid flows, such as a turbine. It simplifies calculations.

Units: The units on Pv are (kN/m²)(m³/kg), which are equivalent to kJ/kg, the same as the units on u.

create work, which will be shown to be related to the enthalpy change. Enthalpy is the amount of thermal energy contained in a substance plus the potential of that substance to do work. Enthalpy has the same units as internal energy, kJ in the SI system. Similarly, specific enthalpy has units of kJ/kg.

State postulate: If two intensive, independent properties of a simple pure substance are given, all others can be determined.

Solving problems in thermodynamics is like solving a mystery. You are given some clues, and from those clues you have to determine what happened and in what order it happened. The *state postulate* maintains that if two intensive, independent properties of a simple pure substance are given, all others can be determined. So we need two "clues" about a state in order to determine all other properties of that state. In textbooks, you are typically given the exact information needed to solve a problem. In real life, you may have too little or too much information. But what information is not of interest, or what information is necessary that is not available? Experience is often required to make the assumptions required for a possible solution.

Example 2.10 The change in enthalpy

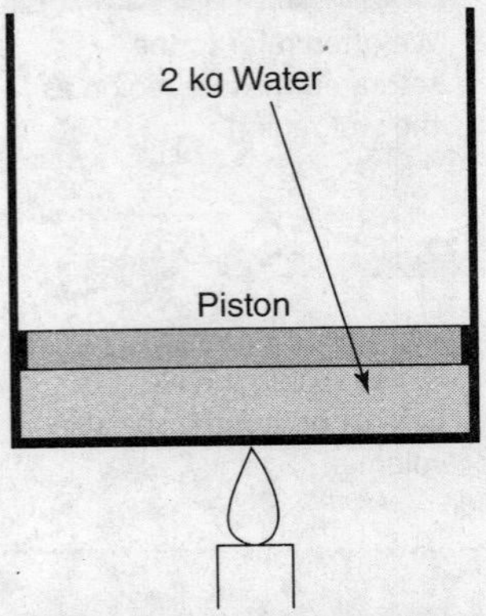

Figure 2.17

The temperature of 2 kg of saturated liquid water contained in a volume of 9000 cm^3 is increased from 60°C to 600°C holding the pressure constant, as illustrated by Fig. 2.17. What is the change in enthalpy?

Solution

The initial enthalpy of the saturated liquid water is found in Table C-1 in the Appendix at $T_1 = 60°\text{C}$ as $h_1 = h_f = 251.1$ kJ/kg. The pressure is noted to be $P_1 = 19.94$ kPa, which is close enough to 20 kPa. At 600°C a pressure of 0.02 MPa is not a table entry, so an interpolation is required to find h_2. It is

$$h_2 = \frac{0.02 - 0.01}{0.05 - 0.01}(3705.1 - 3705.4) + 3705.1 = 3705 \text{ kJ/kg}$$

Notice that the enthalpy is insensitive to the pressure in the superheat region and we could have observed that its value is 3705 kJ/kg without interpolating. Finally, the change in enthalpy is

$$\Delta H = m(h_2 - h_1) = 2 \text{ kg} \times (3705 - 251.1) \text{ kJ/kg} = \underline{6908 \text{ kJ}}$$

2.2.3 Internal energy and enthalpy for liquids and solids

Note: We are using water as the working fluid to establish the relationships that will be used and the techniques used to determine properties of phase-change substances. In the problems, refrigerants R134a and ammonia will also be of interest.

When liquid water is compressed with a pump, as it is in most power plants, it is important to determine its state. As mentioned earlier, Table C-4 provide the properties of compressed water. By studying the entries in those tables, we observe that all the entries, with the exception of enthalpy, change very little with pressure when the temperature is held constant. For example, at 80°C, the percentage changes in v, u, and s (s will be introduced in Chapter 6) between 5 MPa and 50 MPa are at most 2.8%, a relatively small percentage change and within engineering accuracy. Only h experiences a significant change between those two pressures, which is expected since h depends directly on the pressure ($h = u + Pv$). Consequently, we can assume that the properties, except h, are dependent only on the temperature and assume that, for liquids, $v = v_f$, $u = u_f$, and $s = s_f$ where the saturated liquid values with the f subscript come from Tables C-1 and C-1E when water is of interest. The enthalpy is then calculated using $h = u + Pv$. In fact, because the compressed liquid table is not needed, it is often omitted in tables of properties for phase-change substances.

Comment

The properties v, u, and s change very little with pressure for a liquid when the temperature is held constant, except use Eq. 2.11 for enthalpy.

The properties needed to analyze the change of phase of water between a saturated solid and a saturated vapor are found in Table C-5. For ice, this represents ice changing phase directly into a vapor; an example is snow changing phase directly into vapor without melting. It's a very slow process but real. How do clothes hung on a clothesline in the winter dry when the temperature is below freezing? First, the water freezes and then the ice changes phase into a vapor and dry clothes appear the next day. The ice does not melt; it sublimates, that is, changes phase, into a vapor.

Examples will illustrate use of tables for liquids and solids. Water will be the working fluid.

Properties of a liquid — Example 2.11

Determine u and h for the water in Fig. 2.18 at 60°C and 10 MPa using i) Table C-4, ii) the saturated liquid values from Table C-1, and iii) the IRC Calculator.

Solution

i) At a temperature of 60°C and a pressure of 10 MPa, the water is a compressed liquid, so Table C-4 is used to find the desired properties. To four significant digits, they are

$$u = \underline{249.4 \text{ kJ/kg}} \quad \text{and} \quad h = \underline{259.5 \text{ kJ/kg}}$$

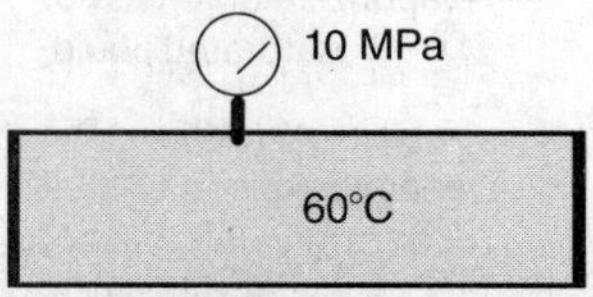

Figure 2.18

ii) Using the saturated liquid values from Table C-1, we find that the properties are

$$u \cong u_f = \underline{251.1 \text{ kJ/kg}} \quad \text{and} \quad h \cong h_f = 251.1 \text{ kJ/kg}$$

Note: If $P = 5$ MPa, the Pv term must be included in the calculation of h.

A more accurate value for h must include the Pv term. It is

$$h \cong u_f + Pv_f$$
$$= 251.1\frac{\text{kJ}}{\text{kg}} + 10\,000\frac{\text{kN}}{\text{m}^2} \times 0.001017\frac{\text{m}^3}{\text{kg}} = \underline{261.3 \text{ kJ/kg}}$$

Units: kPa = kN/m^2

iii) The IRC Calculator provides $u = \underline{249 \text{ kJ/kg}}$ and $h = \underline{260 \text{ kJ/kg}}$.

Obviously, when determining u values and h values for a compressed liquid, using any of the three sources provides acceptable values, but when calculating h using Table C-1, the pressure term Pv must be included, especially if the pressure is relatively high.

Example 2.12 Find a property change of a solid

Determine the change in internal energy of 0.9 kg of water at −17°C between saturated vapor and ice.

Solution

The saturated ice-vapor Table C-5 is not sensitive to the pressure, so only the temperature is needed. At a temperature of −17°C, interpolation is required since −17°C is not a direct entry. Also, u_{ig}, which is the change between ice and saturated vapor, is an entry in the table so interpolation provides, at −17°C,

$$u_g - u_i = u_{ig}$$
$$= \frac{17 - 16}{20 - 16}(2722 + 2719) + 2719 = 2719.75 \text{ kJ/kg}$$

Actually, we observe that u_{ig} is insensitive to temperature, so a value of 2720 kJ/kg could have been selected by observation. That value would be accurate enough for engineering calculations. Then

$$\Delta U = m\Delta u = 0.9 \text{ kg} \times 2720 \text{ kJ/kg} = \underline{2448 \text{ kJ}}$$

Latent heat: The change in enthalpy during a change of phase.

Heat of vaporization: The energy required to vaporize a unit mass of saturated liquid.

Heat of fusion: The energy required to melt (or freeze) a unit mass of a substance.

Heat of sublimation: The energy required to completely vaporize a unit mass of a solid.

2.2.4 Latent heat

The change in enthalpy at the saturated conditions of the two phases experienced during the change of phase of a substance at constant pressure is called *latent heat*. The *heat of vaporization* is the energy required to completely vaporize a unit mass of saturated liquid, or condense a unit mass of saturated vapor; it is equal to $h_{fg} = h_g - h_f$. The energy that is required to melt or freeze a unit mass of a substance at constant pressure is the *heat of fusion* and is equal to $h_f - h_i$, where h_f is the enthalpy of saturated liquid and h_i is the enthalpy of saturated solid. Sublimation occurs when a solid changes phase directly to a vapor; the *heat of sublimation* is equal to $h_{ig} = h_g - h_i$.

The heat of fusion and the heat of sublimation are relatively insensitive to pressure or temperature changes. For ice, the heat of fusion is approximately 330 kJ/kg and the heat of sublimation is about 2840 kJ/kg. The heat of vaporization of water is very dependent on temperature and is included as h_{fg} in Tables C-l and C-2.

Example 2.13 What's the energy needed to melt an ice cube?

The ice cube in Fig. 2.19 is 25 mm on a side. How much energy is needed to melt the ice cube if its initial temperature is 0°C?

Solution

First, let's determine the mass of the cube of ice. In Table C-5, using v_i at 0°C, we find

$$m = \frac{V}{v_i} = \frac{(25 \times 25 \times 25 \times 10^{-9}) \text{ m}^3}{1.091 \times 10^{-3}/\text{kg}} = 0.0156 \text{ kg}$$

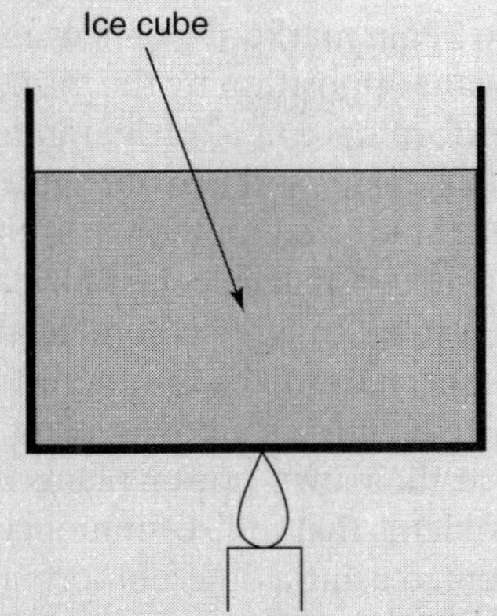

Figure 2.19

Now, what is the heat of fusion for ice, that is, what is the energy to melt the ice? It is the difference between the enthalpy of water h_f at 0°C and the enthalpy of ice h_i at 0°C. The enthalpy of water h_f is found in Table C-1 to be 0.0 kJ/kg. The enthalpy of ice h_i is found in Table C-5 to be −333.4 kJ/kg. The difference is

$$h_f - h_i = 0.0 - (-333.4) = 333.4 \text{ kJ/kg}$$

The energy needed to melt the ice is found to be

$$m(h_f - h_i) = 0.0156 \text{ kg} \times 333.4 \text{ kJ/kg} = \underline{5.20 \text{ kJ}}$$

Suggestion: Alternately, the value of 330 kJ/kg given in the text material could have been used with less than 2% error.

☑ **You have completed Learning Outcome (5)**

2.3 Refrigerants

The working fluid considered in the previous two sections has been water. Refrigerants are another class of working fluids that change phase. We have no interest in the solid phase, only the liquid and the vapor phases. Refrigerant 134a is a common refrigerant for automobile air conditioners, household refrigerators, and air conditioners. It is marketed under the names R134a, HFC-134a, Genetron134a, and Suva134a; it is the chemical compound Tetrafluoroethane and has the chemical formula CH_2FCF_3. R134a is a more environmentally friendly replacement for Freon, the refrigerant R12 (CCl_2F_2), which was found to have harmful effects on the ozone layer of the atmosphere. It is now illegal in the United States to use or design equipment that uses R12. The property tables for R134a are in Appendix D and are organized in the same manner as the steam tables, with the exception that the compressed liquid table and the solid-vapor table are not of interest.

Note: Freon was used for a number of years in air conditioners and refrigerators but was replaced in the early 1990s by R134a.

R134a has also been found to have negative environmental effects, so in a few years HFO-1234yf (Tetrafluoropropene) will be the refrigerant replacing R134a in automotive air conditioners. An advantage in using HFO-1234yf is that it is much

Observation: Engineers can make cool out of heat! Google "absorption refrigeration" to investigate.

less likely to catch fire in the event of an accident. Europe has required its use since 2011. General Motors plans to begin marketing cars using HFO-1234yf, possibly in 2013, although there is significant opposition by the air-conditioning sector, so that date may be postponed. Its performance as a refrigerant is similar to that of R134a. It tends to have a higher vapor density and a lower saturated vapor enthalpy.

The first refrigeration system utilized ammonia as the refrigerant. The system was invented by Dr. John Gorrie in Apalachicola, Florida, to help patients recover from yellow fever in a local hospital. In large commercial applications, ammonia is often used as the refrigerant, especially in absorption refrigeration systems in which cooling is generated using waste heat from a power plant. This is especially popular on college campuses that have their own power plants. Appendix E provides the properties needed to solve problems that utilize ammonia as the refrigerant. The superheated properties are presented using a different format that may be encountered in other property tables. It's wise to become familiar with various forms of tables.

Comment

Ammonia poses a high health hazard, so it is not used in residential and vehicle applications.

Example 2.14 The enthalpy change of R134a

A piston maintains a constant pressure of 500 kPa, as shown in Fig. 2.20. Determine the temperature and the enthalpy change required to completely vaporize the 2 kg of saturated liquid R134a. i) Use the tables and ii) the IRC Calculator.

Solution

i) From Table D-2, the temperature at 500 kPa is 15.74°C, which stays constant until vaporization is complete.

The enthalpy change required to completely vaporize (boil) a kilogram of R134a at 500 kPa is h_{fg} = 184.74 kJ/kg. For 2 kg, the enthalpy change is $2 \times 184.74 = 369$ kJ.

ii) Use the IRC Calculator: Select International Units, R134a, Pressure, and Quality. At 500 kPa and $x = 0$, the temperature is 15.7°C and h_f = 73.4 kJ/kg. At 500 kPa and $x = 1$, h_g = 259 kJ/kg. For 2 kg, the enthalpy change for complete vaporizations is

$$\Delta H = m\Delta h$$
$$= 2 \text{ kg} \times (259 - 73.4) \text{ kJ/kg} = 371 \text{ kJ}$$

Remember: The numbers in the tables are curve fits, and expressing an answer to three significant digits is quite acceptable.

Note: To an engineer, this is essentially the same answer as in (i).

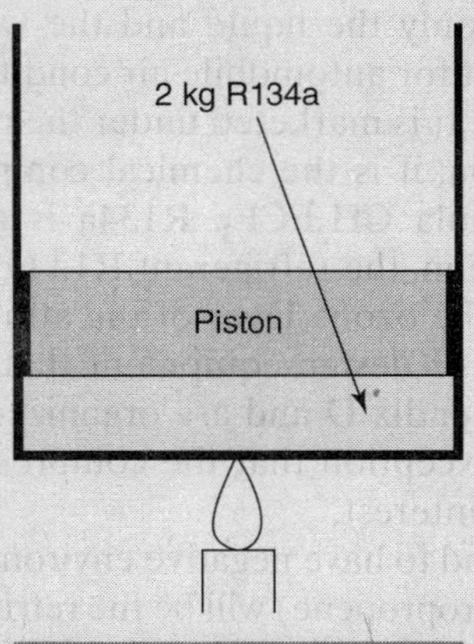

Figure 2.20

The enthalpy change of ammonia — Example 2.15

Determine the temperature and the enthalpy change required to completely vaporize the 2 kg of saturated liquid ammonia shown in Fig. 2.21 at a constant pressure of 500 kPa. i) Use the tables, and ii) the IRC Calculator. These are the same conditions as those of Example 2.14.

Solution

i) From Table E-1, the temperature at 500 kPa is <u>4°C</u>, which stays constant until vaporization is complete.

The enthalpy change required to completely vaporize a kilogram of ammonia at 500 kPa (4°C) is h_{fg} = 1248 kJ/kg. For 2 kg, the enthalpy change is 2 × 1248 = <u>2500 kJ</u>.

ii) Use the IRC Calculator: Select International Units, Ammonia, Pressure, and Quality. At 500 kPa and $x = 0$, the temperature is <u>4.14°C</u> and h_f = 200 kJ/kg. At 500 kPa and $x = 1$, h_g = 1450 kJ/kg. For 2 kg, the enthalpy change for complete vaporization is

$$\Delta H = m\Delta h$$
$$= 2\text{ kg} \times (1450 - 200)\text{ kJ/kg} = \underline{2500\text{ kJ}}$$

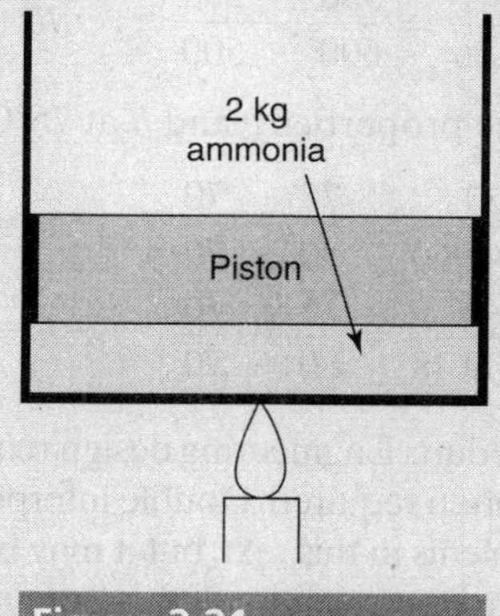

Figure 2.21

Comment

At 4°C the pressure is very close to 500 kPa.

Comment

We used h_g at 4°C since the pressure is so close to 500 kPa.

Double interpolation of R134a properties — Example 2.16

Refrigerant R134a exists at 530 kPa and 75°C. Determine the specific volume and the specific enthalpy. Use the tables.

Solution

Appendix D-3 lists the properties for superheated R134a. Tables exist for pressures of 500 kPa and 600 kPa, so interpolation for 530 kPa is necessary. The tables also list properties for R134a at temperatures of 70°C and 80°C, so another interpolation for 75°C is also necessary, resulting in a double interpolation.

(Continued)

Example 2.16 (*Continued*)

The following table shows the properties at the four listed states from Appendix D-3:

	$P = 500$ kPa	$P = 600$ kPa
$T = 70°C$	$v = 0.05240\ m^3/kg$	$v = 0.04304\ m^3/kg$
	$h = 309.92$ kJ/kg	$h = 308.48$ kJ/kg
$T = 80°C$:	$v = 0.05432\ m^3/kg$	$v = 0.04469\ m^3/kg$
	$h = 319.96$ kJ/kg	$h = 318.67$ kJ/kg

First, at $T = 70°C$, interpolate for a pressure of 530 kPa to find v_1 and h_1:

$$\frac{v_1 - 0.05240}{0.04304 - 0.05240} = \frac{530 - 500}{600 - 500} \qquad \therefore v_1 = 0.04960 m^3/kg$$

$$\frac{h_1 - 309.92}{308.48 - 309.92} = \frac{530 - 500}{600 - 500} \qquad \therefore h_1 = 309.48 \text{ kJ/kg}$$

Units: All units on the left are m^3/kg, and all units on the right are kPa.

Then, at 80°C, interpolate for a pressure of 530 kPa to find v_2 and h_2:

$$\frac{v_2 - 0.05432}{0.04469 - 0.05432} = \frac{530 - 500}{600 - 500} \qquad \therefore v_2 = 0.05143\ m^3/kg$$

$$\frac{h_2 - 319.96}{318.67 - 319.96} = \frac{530 - 500}{600 - 500} \qquad \therefore h_2 = 319.57 \text{ kJ/kg}$$

Next, interpolate to find the properties v and h at 75°C:

$$\frac{v - 0.04960}{0.05143 - 0.04960} = \frac{75 - 70}{80 - 70} \qquad \therefore v = \underline{0.0505\ m^3/kg}$$

$$\frac{h - 309.48}{319.57 - 309.48} = \frac{75 - 70}{80 - 70} \qquad \therefore h = \underline{315 \text{ kJ/kg}}$$

Note: Three rather than four significant digits are written since the straight-line interpolation method loses at least one significant digit.

This is a rather tedious procedure. Engineering design requires the use of numerous tables, and interpolation is often required. Double interpolation is not encountered very often in the home problems in this text, but it may be needed when working a problem on the FE Exam, an exam required for becoming a professional engineer. Programmable calculators and computers are not allowed on that exam.

The IRC Calculator described in Section 2.1.4 is a much quicker way to obtain properties. The software does all the interpolation for you. It provides $v = 0.052\ m^3/kg$, $h = 316$ kJ/kg, which is more accurate than the double interpolation results. However, both are acceptable for engineering calculations.

☑ You have completed Learning Outcome (6)

2.4 Ideal-Gas Law

Comment

This law is based on experimental observations; it is not derived.

The *ideal-gas law* is undoubtedly a familiar law from courses in chemistry and physics. It is an empirical equation of state based on experimental observations of the behavior of gasses at relatively low pressure. An equation of state is an algebraic equation that relates two or more thermodynamic properties. The ideal-gas law relates pressure, temperature, and specific volume in a very simple algebraic equation. The most common form of the ideal-gas law is

$$Pv = RT \tag{2.12}$$

Air: $R = 0.287$ kJ/kg·K
$= 53.34$ ft-lbf/lbm-°R

where P must be the absolute pressure of the gas, v is the specific volume, R is the gas constant, and T is the absolute temperature. The ideal-gas law is also called the ideal-gas equation of state, or the ideal-gas equation.

Nitrogen: $R = 0.297$ kJ/kg·K
$R = 55.15$ ft-lbf/lbm-°R

The gas constant is different for each gas and is related to the universal gas constant R_u, which is the same for all gases, by

$$R = \frac{R_u}{M} \tag{2.13}$$

$R_u = 8.314$ kJ/kmol·K
$= 1545$ ft-lbf/lbmol-°R
$= 1.986$ Btu/lbmol-°R
$= 10.73$ psia-ft^3/lbmol-°R

where M is the molar mass of the gas, tabulated in Table B-2 in the Appendix for a number of gases. The molar mass of oxygen is 32, meaning that there are 32 kg in one kmol.

Other forms of the ideal-gas law that are used are

$$PV = mRT \qquad P = \rho RT \qquad PV = NR_uT \tag{2.14}$$

where N is the number of moles (or kmols).

Pressure of an ideal gas — Example 2.17

Ten kilograms of oxygen exist at 20°C in the 1.2-m^3 container of Fig. 2.22. Determine the pressure in the container.

Solution

The pressure and temperature are given so that the ideal-gas law of Eq. 2.14 (the first version) provides

Note: In thermodynamics, pressure is absolute, unless otherwise stated.

$$P = \frac{mRT}{V}$$

$$= \frac{10 \text{ kg} \times 0.260 \text{ kN·m/kg·K} \times (20 + 273) \text{ K}}{1.2 \text{ m}^3} = \underline{635 \text{ kPa}}$$

Units: Use kJ = kN·m when writing the units on R.

The temperature must be absolute, and the pressure determined is also absolute. If R were used as 260 J/kg·K, the pressure would be in Pa rather than kPa.

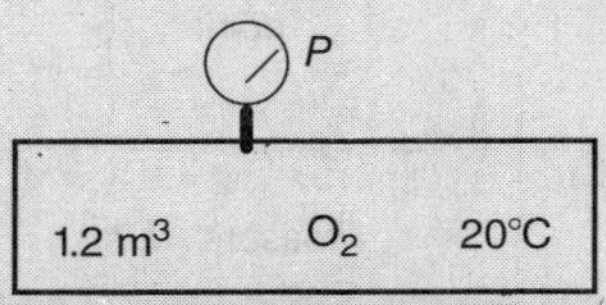

Figure 2.22

Example 2.18 The volume of an ideal gas

Four kg mass of helium exist in a container at 800 kPa and 16°C. What is the volume of the container?

Solution

The gas constant for helium is found in Table B-2 to be 2.077 kJ/kg·K. The ideal gas law of Eq. 2.14 is used to find the volume:

$$V = \frac{mRT}{P}$$

$$= \frac{4\ \text{kg} \times 2.077\ \dfrac{\text{kJ}}{\text{kg·K}} \times (273 + 16)\,\text{K}}{800\ \text{kPa}} = 3\ \text{m}^3$$

2.5 Real Gas Equations of State

Compressibility factor: A factor Z that accounts for real gas behavior.

Generalized compressibility chart: It provides the factor Z for all gases.

The ideal-gas law can be modified to better model real gases by the use of a compressibility factor. The standard symbol for the *compressibility factor* is the letter Z and is defined by

$$Z = \frac{Pv}{RT} \qquad (2.15)$$

For an ideal gas, $Z = 1$. A graph of the compressibility factor for nitrogen as a function of the pressure and temperature is shown in Fig. 2.23. A *generalized*

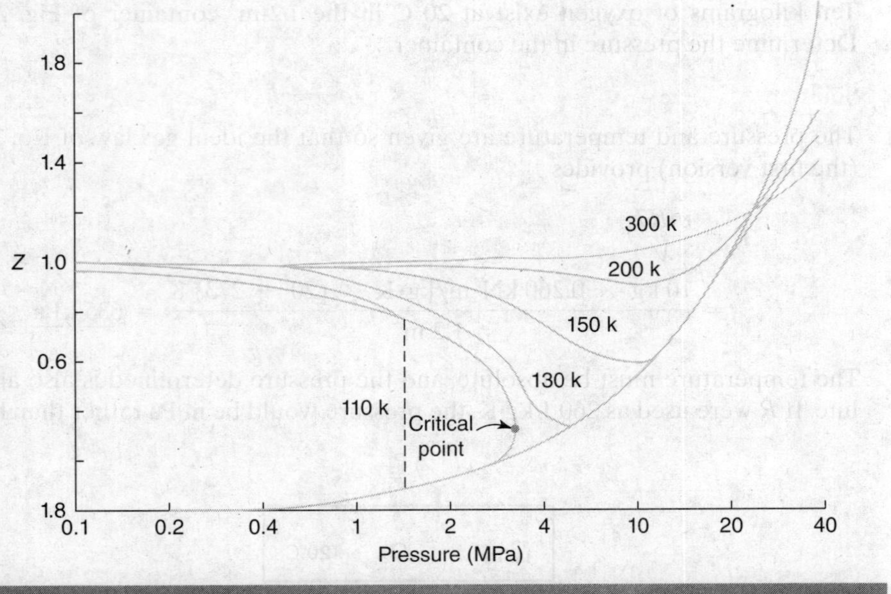

Figure 2.23

Compressibility factor for nitrogen.

compressibility chart that provides the factor Z for all gases is included as Figs. H-1 and H-2 in Appendix H. The generalized charts use *reduced temperature* T_R, *reduced pressure* P_R, and *pseudo-reduced specific volume* v_R, defined by

$$P_R = \frac{P}{P_{cr}} \qquad T_R = \frac{T}{T_{cr}} \qquad v_R = \frac{v}{RT_{cr}/P_{cr}} \tag{2.16}$$

where the critical-point constants P_{cr} and T_{cr} are found in Table B-3. The quantity v_R is called a pseudo-reduced specific volume because it is not defined as v/v_{cr}.

There are more sophisticated real gas equations of state that also relate pressure, temperature, and specific volume but use algebraic equations that require more than one constant. A relatively simple equation is the *van der Waals equation*,

$$P = \frac{RT}{v-b} - \frac{a}{v^2} \tag{2.17}$$

where the constants a and b are related to the critical-point values found in Table B-3 by

$$a = \frac{27R^2T_{cr}^2}{64P_{cr}} \qquad b = \frac{RT_{cr}}{8P_{cr}} \tag{2.18}$$

The numerical values of the constants a and b are presented in Table B-10.

Another real gas equation of state is the Beattie–Bridgeman equation, which offers more precise modeling of gases:

$$P = \frac{R_uT}{\overline{v}^2}\left(1 - \frac{c}{\overline{v}T^3}\right)\left[\overline{v} + B_0\left(1 - \frac{b}{\overline{v}}\right)\right] - \frac{A_0}{\overline{v}^2}\left(1 - \frac{a}{\overline{v}}\right) \tag{2.19}$$

where $\overline{v}$ is the *molar specific volume* defined by $\overline{v} = vM$. (A bar over a variable signifies a per mole basis.) The gas constants A_0, B_0, a, b, and c are included in Table 2.4 for four gases.

In general, the more gas constants in the model, the better the fit between the state equation and the actual relationship of the properties.

Another real gas equation of state that should be mentioned is the virial equation of state. It is expressed as the series

$$P = \frac{RT}{v} + \frac{a(T)}{v^2} + \frac{b(T)}{v^3} + \cdots \tag{2.20}$$

The constants $a(T)$, $b(T)$, and so on must be given if this virial state equation is used. It is not used in this text and is presented for information purposes only.

The following example compares the pressure determined by the van der Waals and the Beattie–Bridgeman models to that determined by the ideal-gas law.

Observe: The reduced properties are nondimensional properties.

Van der Waals equation: An equation of state that accounts for real gas behavior.

Units: The units on a are Pa·m^6/kg^2 and on b the units are m^3/kg.

Molar specific volume: It is defined by the relationship $\overline{v} = vM$, where M is the molar mass (see Table B-3).

Table 2.4 The Constants in the Beattie–Bridgeman Equation

Gas	A_0	B_0	a	b	c
Air	131.84	0.04611	0.01931	0.00110	4.34×10^4
Nitrogen	136.23	0.05046	0.02617	0.00691	4.20×10^4
Oxygen	151.09	0.04624	0.02562	0.00421	4.80×10^4
Carbon Dioxide	507.28	0.10476	0.07132	0.07235	6.60×10^5

Note: The constants in Table 2.4 have units requiring P to be in kPa, $\overline{v}$ in m^3/kmol, T in K, and R_u in kJ/kmol·K.

Example 2.19 The pressure using three equations of state

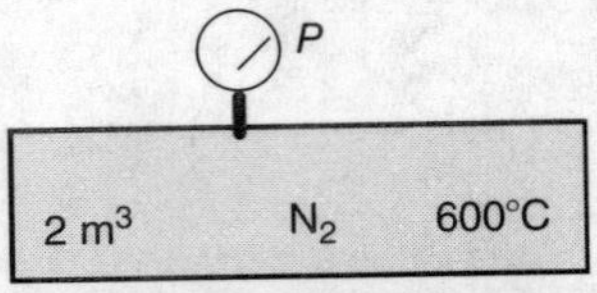

Figure 2.24

Forty kilograms of nitrogen exist at 600°C in the 2-m³ container of Fig. 2.24. What is the pressure in the container? Use i) the ideal-gas law, ii) the van der Waals equation, and iii) the Beattie–Bridgeman equation.

Solution

i) The gas constant for nitrogen is found in Table B-2 to be 0.297 kJ/kg·K. The ideal-gas law of Eq. 2.14 is used to find the pressure:

$$P = \frac{mRT}{V} = \frac{40\text{ kg} \times 0.297\text{ kJ/kg}\cdot\text{K} \times (600 + 273)\text{ °R}}{2\text{ m}^3} = \underline{5190\text{ kPa}}$$

ii) The two constants in the van der Waals equation are calculated (or look them up in Table B-10) to be

$$a = \frac{27R^2T_{\text{cr}}^2}{64P_{\text{cr}}} = \frac{27 \times 297^2 \times 126.2^2}{64 \times 3.39 \times 10^6} = 174.8\text{ Pa}\cdot\text{m}^6/\text{kg}^2$$

$$b = \frac{RT_{\text{cr}}}{8P_{\text{cr}}} = \frac{297 \times 126.2}{8 \times 3.39 \times 10^6} = 0.001382\text{ m}^3/\text{kg}$$

The van der Waals equation, with $v = 2/40 = 0.05\text{ m}^3/\text{kg}$, gives

$$P = \frac{RT}{v - b} - \frac{a}{v^2}$$

$$= \frac{297 \times 873}{0.05 - 0.001382} - \frac{174.8}{0.05^2} = 5.26 \times 10^6\text{ Pa or }\underline{5260\text{ kPa}}$$

iii) The Beattie–Bridgeman equation constants from Table 2.4 are

$$A_0 = 136.23,\ B_0 = 0.05046,\ a = 0.2617,\ b = -0.00691,\ c = 4.2 \times 10^4$$

Observe: The *Z*-factor of Fig. H-2 in the Appendix would be greater than 1. It is difficult to read an accurate value of *Z* from the chart.

The Beattie–Bridgeman equation, with $\bar{v} = Mv = 28 \times 0.05 = 1.4\text{ m}^3/\text{kmol}$, gives

$$P = \frac{R_uT}{\bar{v}^2}\left(1 - \frac{c}{\bar{v}T^3}\right)\left[\bar{v} + B_0\left(1 - \frac{b}{\bar{v}}\right)\right] - \frac{A_0}{\bar{v}^2}\left(1 - \frac{a}{\bar{v}}\right)$$

$$= \frac{8.314 \times 873}{1.4^2}\left(1 - \frac{4.2 \times 10^4}{1.4 \times 873^3}\right)\left[1.4 + 0.05046\left(1 + \frac{0.00691}{1.4}\right)\right]$$

$$- \frac{136.23}{1.4^2}\left(1 - \frac{0.2617}{1.4}\right) = \underline{5316\text{ kPa}}$$

Results:
Ideal-gas equation: $P = 5.19$ MPa
Van der Waals: $P = 5.26$ MPa
Beattie–Bridgeman: $P = 5.32$ MPa

Compared to the Beattie–Bridgeman equation, which is considered the most accurate, the ideal-gas law has an error of −2.4%. An ideal-gas assumption is good at relatively low pressures; 5 MPa is relatively low. Because pressures exceeding 5 MPa are not encountered that often in our study, the ideal-gas law is most often used.

☑ **You have completed Learning Outcome (7)**

2.6 Internal Energy and Enthalpy of Ideal Gases

The properties of internal energy and enthalpy for an ideal gas are calculated using properties called specific heats. The *specific heat* of a substance is the amount of energy needed to raise a unit mass of that substance one degree. Textbooks in other fields may use the term *heat capacity* rather than specific heats.

Specific heat: The amount of energy needed to raise a unit mass of a substance one degree.

For a simple compressible system, only two independent variables are necessary to establish the state of the system. Consequently, we can consider specific internal energy to be a function of temperature and specific volume; that is, $u = u(T,v)$. Using the chain rule from calculus, we express the differential in terms of partial derivatives as

$$du = \left(\frac{\partial u}{\partial T}\right)_v dT + \left(\frac{\partial u}{\partial v}\right)_T dv \qquad (2.21)$$

Constant-volume specific heat: $C_v = (\partial u/\partial T)_v$

Since u, v, and T are all properties, each partial derivative is also a property; the first one on the right-hand side is defined to be the *constant-volume specific heat* C_v; that is,

$$C_v = \left(\frac{\partial u}{\partial T}\right)_v \qquad (2.22)$$

One of the classical experiments of thermodynamics, performed by Joule in 1843, is illustrated in Fig. 2.25. A high-pressure volume with an ideal gas is connected to an evacuated volume, as shown. After equilibrium is attained, the valve is opened. Even though the pressure and volume of the ideal gas have changed markedly, the temperature of the water does not change. If there was no change in the water temperature, there was no energy flow to the water, so the internal energy of the air must not have changed. Because the pressure and specific volume of the air certainly changed, Joule concluded that the internal energy of an ideal gas does not depend on pressure or volume so that, for an ideal gas,

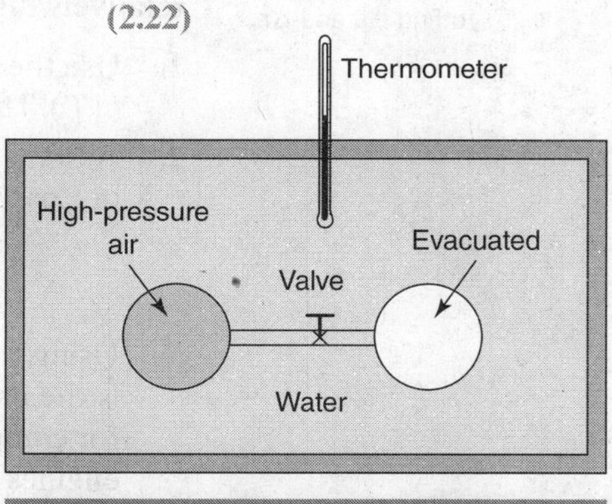

Figure 2.25
A sketch of the experimental setup used by Joule.

$$\left(\frac{\partial u}{\partial v}\right)_T = 0 \qquad (2.23)$$

For a gas that behaves as an ideal gas, Eqs. 2.21 and 2.22 allow us to write

$$du = C_v dT \qquad (2.24)$$

Similarly, if we consider enthalpy to depend on the two properties T and P, that is, $h = h(T, P)$, we can express an infinitesimal change in enthalpy as

$$dh = \left(\frac{\partial h}{\partial T}\right)_P dT + \left(\frac{\partial h}{\partial P}\right)_T dP \qquad (2.25)$$

We define the *constant-pressure specific heat* to be

Constant-pressure specific heat: $C_p = (\partial h/\partial T)_p$

$$C_p = \left(\frac{\partial h}{\partial T}\right)_P \qquad (2.26)$$

and, since $h = u + Pv = u + RT$, assuming an ideal gas, we observe that h is also only a function of temperature; hence, $\partial h/\partial P|_T = 0$. With the definition of C_p, we see that

$$dh = C_p dT \tag{2.27}$$

Comment

The specific heats are properties and do not require constant volume or constant-pressure processes when used. That is simply what they have been named.

The changes in internal energy and enthalpy needed to change from state 1 to state 2 can now be found for an ideal gas by integrating Eqs. 2.24 and 2.27:

$$\Delta u = u_2 - u_1 = \int_{T_1}^{T_2} C_v dT$$
$$\Delta h = h_2 - h_1 = \int_{T_1}^{T_2} C_p dT \tag{2.28}$$

Like most physical properties, C_v and C_p vary with the temperature of the substance. However, if the temperature change for a process is not large, they can be assumed to be constants. Constant values for C_v and C_p for temperatures around 300 K are shown in Table B-2 in both sets of units.

Comment

There are five methods used to find Δu and Δh.

If a large change in temperatures is encountered, or if the temperatures are relatively high, there are five methods to find Δu and Δh for an ideal gas.

1. Use the ideal-gas tables included as Appendix F, each of which lists $u(T)$ and $h(T)$. This method is relatively simple, and it is the most accurate.
2. Assume constant specific heats from Table B-2 and perform the integrations in Eqs. 2.28 to obtain

$$u_2 - u_1 = C_v(T_2 - T_1)$$
$$h_2 - h_1 = C_p(T_2 - T_1) \tag{2.29}$$

   Using this method, the temperature should not exceed about 500°C, which is the case for many of the problems we will encounter. It is used, however, for comparison purposes when analyzing the performance of power cycles in engines with much higher temperatures.
3. Table B-6 lists the values of C_v and C_p over a wide temperature range. The procedure for using them is to calculate the average temperature for the process and use the specific heat at that average temperature. Then use Eqs. 2.29 to find the desired quantity.
4. Use an equation for the specific heat C_p that is a direct function of the temperature. Table 2.5 lists the coefficients of an equation for C_p that is a cubic function of temperature:

$$C_p = a + bT + cT^2 + dT^3 \tag{2.30}$$

Comment

The units on C_p are sometimes listed as kJ/kg·K and sometimes as kJ/kg·°C. These are equivalent since the size of 1°C is the same size as 1 K.

Table 2.5 C_p (in kJ/kg·K) as a Function of Absolute Temperature

Gas	a	b	c	d
Air	0.9703	6.8×10^{-5}	1.66×10^{-7}	-6.8×10^{-11}
Oxygen	0.7965	4.74×10^{-4}	-2.24×10^{-7}	4.1×10^{-11}
Nitrogen	1.0317	-5.6×10^{-5}	2.88×10^{-7}	-1.02×10^{-10}

where C_p is in kJ/kg·K and T is in kelvins. The constants $a, b, c,$ and d have dimensions in order to make Eq. 2.30 dimensionally homogeneous. Molar values to calculate $\overline{C}_p$ for numerous gases are included in Table B-5.

5. Use the IRC Calculator for air as described in Example 2.7.

Let's find the relationship among C_v, C_p, and the gas constant R that exists for an ideal gas. Return to the definition $h = u + Pv$ in Eq. 2.11 and write the differential form as

$$dh = du + RdT \tag{2.31}$$

using $d(Pv) = d(RT) = RdT$, since R is a constant. Now, substitute the expressions for dh from Eq. 2.27 and du from Eq. 2.24 and divide by dT and obtain

$$C_p = C_v + R \tag{2.32}$$

Now we know why only C_p is listed in many tables. If C_p is known, C_v can be easily determined using Eq. 2.32. Since C_p is most often used, it is the one listed.

If the definition of $\overline{h}$ on a per mole basis was used, we would find

$$\overline{C}_p = \overline{C}_v + R_u \tag{2.33}$$

Often the functional form similar to Eq. 2.30 is given on a per mole basis so that the units on $\overline{C}_p$ are kJ/kmol·K.

Before we leave this section, let's define another property that is related to the specific heats, the *ratio of specific heats* k:

$$k = \frac{C_p}{C_v} \tag{2.34}$$

The ratio of specific heats: It is defined as $k = C_p/C_v$.

It occurs quite often when working with ideal gases, so it is usually listed in tables. It also allows, with Eq. 2.32, to express C_p and C_v as

$$C_v = \frac{R}{k - 1} \quad \text{and} \quad C_p = \frac{kR}{k - 1} \tag{2.35}$$

These relationships will be used to simplify expressions.

Estimate C_p of steam — Example 2.20

Estimate C_p of steam at 6 MPa and 600°C. Use i) a forward-difference method, ii) a backward-difference method, and iii) a central-difference method.

Solution

i) A forward-difference method is based on the entries at a forward state and the desired state—in this case, at the same pressure as required by Eq. 2.26. We use the definition

$$C_p \cong \left.\frac{\Delta h}{\Delta T}\right|_{P = 6\text{ MPa}} = \frac{(3894.2 - 3658.4)\text{ kJ/kg}}{(700 - 600)\text{ °C}} = 2.358\text{ kJ/kg·°C}$$

(Continued)

Example 2.20 *(Continued)*

ii) A backward-difference method is based on the entries at the desired state and a backward state. We have

$$C_p \cong \left.\frac{\Delta h}{\Delta T}\right|_{P=6\ \text{MPa}} = \frac{3658.4 - 3540.6}{600 - 550} = \underline{2.356\ \text{kJ/kg}\cdot{}^\circ\text{C}}$$

iii) A central-difference method is based on the entries at states on each side of the desired state at an equal temperature difference. The method gives

$$C_p \cong \left.\frac{\Delta h}{\Delta T}\right|_{P=6\ \text{MPa}} = \frac{3894.2 - 3422.2}{700 - 500} = \underline{2.36\ \text{kJ/kg}\cdot{}^\circ\text{C}}$$

To three significant digits, each method gives $C_p = 2.36$ kJ/kg·°C. If C_p were more sensitive to temperature change, as in Fig. 2.26, the difference would be significantly larger and the central-difference method would obviously be the most accurate. The desired C_p is the slope of the curve at 600°C.

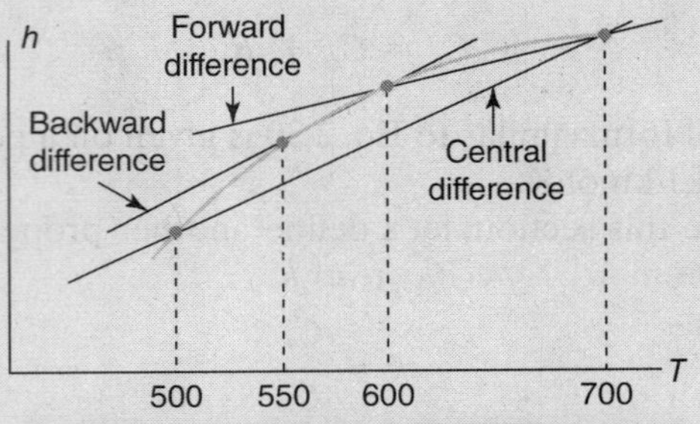

Figure 2.26

Example 2.21 Find ΔH of air using five methods

Estimate the change in the enthalpy of 5 kg of air if the temperature changes from 27°C to 400°C.

i) Use the air table, Table F-1.
ii) Assume the constant specific heat from Table B-2.
iii) Use an average specific heat found in Table B-6.
iv) Use Eq. 2.30.
v) Use the IRC Calculator.

Suggestion: To interpolate, observe that from 660 K to 680 K, $\Delta h = 21.4$ kJ/kg so that's a little more than 1 kJ/kg per degree. So add 14 kJ/kg to 670.47 and $h_2 = 684$ kJ/kg. Extreme accuracy is not required. Three significant digits are fine. When you interpolate, think a little first.

Solution

The temperature 27°C is equal to 300 K, and 400°C is 673 K. Remember: Enthalpy does not depend on pressure for an ideal gas.

i) In Table F-1 we find $h_1 = 300$ kJ/kg as a direct entry and interpolate to find $h_2 = 684$ kJ/kg. There results

$$\Delta H = m(h_2 - h_1) = 5\ \text{kg} \times (684 - 300)\ \text{kJ/kg} = \underline{1920\ \text{kJ}}$$

ii) In Table B-2, we find $C_p = 1.003$ kJ/kg·K. This provides

$$\Delta H = mC_p(T_2 - T_1)$$
$$= 5 \text{ kg} \times 1.003 \frac{\text{kJ}}{\text{kg·K}} \times (673 - 300) \text{ K} \cong \underline{1870 \text{ kJ}}$$

iii) The specific heat for an average temperature of (673 + 300)/2 = 486 K, found in Table B-6, is 1.026 kJ/ kg·K (this is estimated since the variation is slight, and we are not interested in extreme accuracy). We find

$$\Delta H = mC_p(T_2 - T_1)$$
$$= 5 \text{ kg} \times 1.026 \frac{\text{kJ}}{\text{kg·K}} \times (673 - 300) \text{ K} \cong \underline{1910 \text{ kJ}}$$

iv) The coefficients in the cubic equation used to estimate C_p are found in Table 2.4. They are used in Eq. 2.30 and integrated using Eq. 2.28 to give

$$\Delta H = m\int_{T_1}^{T_2} C_p dT = m\int_{T_1}^{T_2} (a + bT + cT^2 + dT^3)dT$$
$$= 5 \text{ kg}\left[0.9703(673 - 300) + 6.8 \times 10^{-5}\frac{673^2 - 300^2}{2}\right.$$
$$\left. + 1.66 \times 10^{-7}\frac{673^3 - 300^3}{3} - 6.8 \times 10^{-11}\frac{673^4 - 300^4}{4}\right]\frac{\text{kJ}}{\text{kg}}$$
$$\cong \underline{1950 \text{ kJ}}$$

v) Using the IRC Calculator selecting "Dry air" at 27°C (select P_1 = 100 kPa since it doesn't matter), h_1 = 300 kJ/kg, and at 400°C we find h_2 = 685 kJ/kg. Hence,

$$\Delta H = m(h_2 - h_1) = 5 \text{ kg} \times (685 - 300) \text{ kJ/kg} = \underline{1925 \text{ kJ}}$$

The answer from the air table in (i) is considered the most accurate. The errors in the other parts are found to be (ii) −2.6%, (iii) −0.5%, (iv) 1.6%, and (v) 0.26%. All answers are within engineering accuracy. The answer in (ii) is probably the one most often used, providing $\Delta T < \approx$ 500 K and T_1 or T_2 is near standard temperature.

2.7 Specific Heats of Liquids and Solids

The volume of a liquid or a solid changes very little as the temperature and pressure change, even though a very small change can lead to large stresses in a solid under certain conditions and to significant movement in a liquid, such as the expansion of a liquid in a thermometer. In a later chapter, we will show that $C_p = C_v$ for an incompressible substance—that is, one for which the specific volume does not change during a process. In general, we make the assumption

Liquids and solids: We usually make the assumption that $C_p \cong C_v$.

that $C_p \cong C_v$ for a *liquid* and a *solid*. Often the subscript is simply dropped, and C represents the specific heat for liquids and solids. In most thermodynamic books, C_p will be used. For water we use $C_p = 4.18$ kJ/kg·K. For ice $C_p = 2.1 + 0.0069T$ kJ/kg·°C, with T measured in °C. The specific heat does not vary significantly with pressure, except for very special situations, and will be considered independent of pressure in this introductory text. The specific heats of some common liquids and solids are presented in Table B-4 in the Appendix.

Comment

We most often assume the constant specific heats found in Table B-4 as C_p.

For most applications to solids and liquids, Eq. 2.29 provides the internal energy change as

$$u_2 - u_1 = h_2 - h_1 = C(T_2 - T_1) \tag{2.36}$$

unless there's a large pressure change with liquids, in which case the Pv term must be included.

Example 2.22 Find ΔU of ice

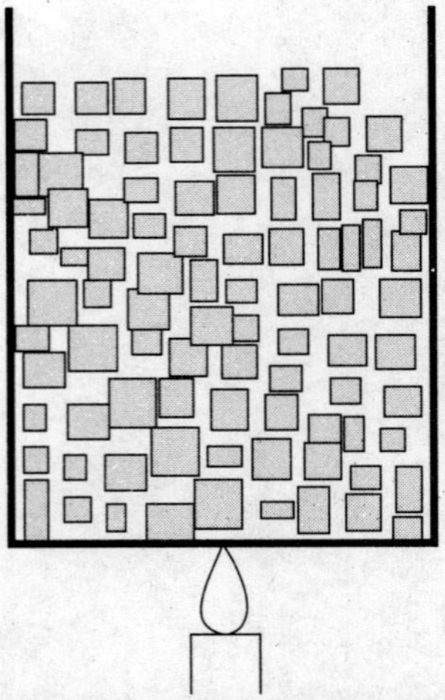

Figure 2.27

Two kilograms of ice exist at −18°C in the open container of Fig. 2.27. Estimate the change in the internal energy of the ice if the temperature changes from −18°C to 82°C.

Solution

Integrate $C_p = 2.1 + 0.0069T$ to find the internal energy change of the ice between −18°C and 0°C. Assuming $C_p \cong C_v$,

$$\Delta U = m\Delta u$$

$$= 2\int_{-18}^{0} (2.1 + 0.0069T)dT = 2\left(2.1 \times 18 - \frac{0.0069}{2} \times 18^2\right) = 73.36 \text{ kJ}$$

The ice is now at 0°C, at which state it melts, requiring the heat of fusion, 330 kJ/kg:

$$\Delta U_{\text{melt}} = m \times (\text{heat of fusion}) = 2 \text{ kg} \times 330 \frac{\text{kJ}}{\text{kg}} = 660 \text{ kJ}$$

If the temperature of the water is raised from 0°C to 82°C, the internal energy change is

$$\Delta U = m \times C_{\text{water}}\Delta T = 2 \text{ kg} \times 4.18 \frac{\text{kJ}}{\text{kg·K}} \times (82 - 0) \text{ K} = 685.5 \text{ kJ}$$

The overall change $\Delta U = 73.36 + 660 + 685.5 = \underline{1419 \text{ kJ}}$.

$\Delta H = \Delta U$ for this problem since pressure was held constant and the change in volume was considered insignificant. In general, $\Delta(Pv) \cong 0$ for a solid or a liquid.

☑ **You have completed Learning Outcome (8)**

2.8 Summary

In this chapter we considered the different phases in which a pure substance exists and introduced properties of fluids used in power generation and refrigeration systems. The equations of state for an ideal gas and state equations for real gases were presented. Enthalpy and internal energy for an ideal gas, a solid, and a liquid were determined with the use of tables and specific heats. Numerous additional terms were defined:

Compressed liquid: *The temperature is below the boiling point.*

Compressibility factor: *A factor Z that accounts for real gas behavior.*

Dependent property: *One that cannot vary if the other property is held constant.*

Fluid: *A substance that flows when subjected to a shear stress, no matter how small that stress may be.*

Fusion: *The process of a liquid becoming a solid.*

Heat of fusion: *The energy required to melt a unit mass of a substance.*

Heat of sublimation: *The energy required to completely vaporize a unit mass of a solid.*

Heat of vaporization: *The energy required to completely vaporize a unit mass of saturated liquid.*

Law: *The summary of the results of repeated observation.*

Quality: *The mass of the vapor divided by the total mass of a mixture.*

Saturated liquid: *The state in which a substance is just starting to vaporize (boil).*

Saturated mixture: *A mixture consisting of both liquid and vapor.*

Saturated vapor: *The state in which a mixture is just completely vaporized.*

Specific heat: *The amount of energy needed to raise a unit mass of a substance one degree.*

State postulate: *If two intensive, independent properties of a simple pure substance are given, all others can be determined.*

Sublimation: *The phase change from a solid directly into a gas.*

Vapor: *The gas phase of a substance, usually when it is near condensation.*

Vaporization: *The phase change from a liquid to a gas.*

Working fluid: *A fluid to which we add or subtract energy while it is most often undergoing a cycle.*

Several important equations were presented:

Quality: $x = \frac{m_g}{m}$

Specific volume of a mixture: $v = v_f + x(v_g - v_f)$

Enthalpy: $h = u + Pv$

Enthalpy change: $h_2 - h_1 = C_p(T_2 - T_1)$

Ideal-gas law: $Pv = RT$

Internal energy change: $u_2 - u_1 = C_v(T_2 - T_1)$

Gas constant: $R = \frac{R_u}{M}$

Compressibility factor: $Z = \frac{Pv}{RT}$

Reduced properties: $P_R = \frac{P}{P_{cr}}, \quad T_R = \frac{T}{T_{cr}}, \quad v_R = \frac{v}{RT_{cr}/P_{cr}}$

van der Waals equation: $P = \frac{RT}{v - b} - \frac{a}{v^2}$

Constant-volume specific heat: $C_v = \left(\frac{\partial u}{\partial T}\right)_v$

Constant-pressure specific heat: $C_p = \left(\frac{\partial h}{\partial T}\right)_p$

Specific heat relationship: $C_p = C_v + R$

Ratio of specific heats: $k = \frac{C_p}{C_v}$

Problems

FE Exam Practice Questions

2.1 What two properties are dependent during the phase-change process?

(A) Pressure and specific volume
(B) Temperature and specific volume
(C) Specific volume and enthalpy
(D) Temperature and pressure

2.2 When molecules break away from a solid and become a vapor, the process is referred to as:

(A) Melting
(B) Sublimation
(C) Vaporization
(D) Evaporation

2.3 The height of Mount Everest is nearly 9000 m. Water would boil at that height at a temperature nearest:

(A) 65°C
(B) 70°C
(C) 75°C
(D) 80°C

2.4 When you take a very hot shower, the water is:

(A) Subcooled
(B) A mixture
(C) Saturated water
(D) Superheated

2.5 Water is maintained at a constant temperature of 200°C, while the specific volume changes from 0.002 m^3/kg to 0.2 m^3/kg. Select the diagram that best represents this process.

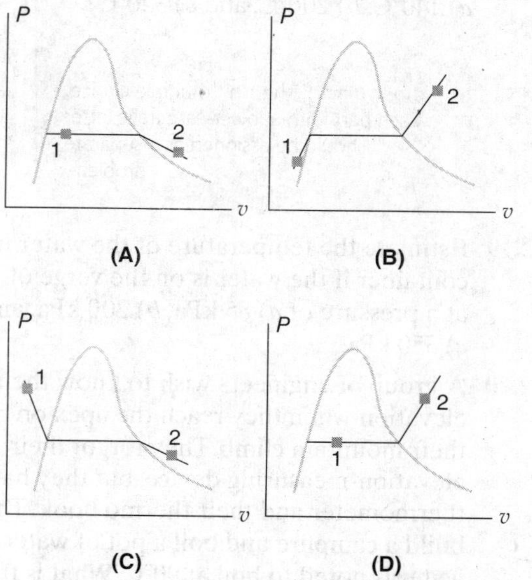

2.6 Two kilograms of water occupy 80 000 cm^3 at 200°C. Its quality is nearest:

(A) 0.175

(B) 0.211

(C) 0.287

(D) 0.314

2.7 The volume occupied by 20 kg of water at 300°C and 10 MPa is nearest:

(A) 0.028 m^3

(B) 0.088 m^3

(C) 0.22 m^3

(D) 0.36 m^3

2.8 Superheated steam is contained in a rigid volume at 400°C and 4 MPa. Energy is released from the steam, resulting in a decreased pressure. The pressure that will cause the steam to just begin to condense is nearest:

(A) 2.4 MPa

(B) 2.6 MPa

(C) 2.8 MPa

(D) 3.0 MPa

2.9 A 3-L cylinder holds 2 L of steam and 1 L of liquid at 200°C. The quality is nearest:

(A) 1.8%

(B) 2.2%

(C) 2.6%

(D) 3.2%

2.10 Superheated steam exists at 6.5 MPa and 675°C. The enthalpy is nearest:

(A) 3828 kJ/kg

(B) 3832 kJ/kg

(C) 3836 kJ/kg

(D) 3840 kJ/kg

2.11 A 1-m-diameter bike tire is pumped up to 500 kPa. It is assumed to have a circular cross section with a diameter of 2 cm. The mass of 20°C air in the tire is nearest:

(A) 0.0059 kg

(B) 0.0082 kg

(C) 0.018 kg

(D) 0.028 kg

2.12 The cylinder shown in Fig. 2.28 contains 0.02 kg of air. The weight of the frictionless piston is nearest:

(A) 8000 N

(B) 6000 N

(C) 4000 N

(D) 3000 N

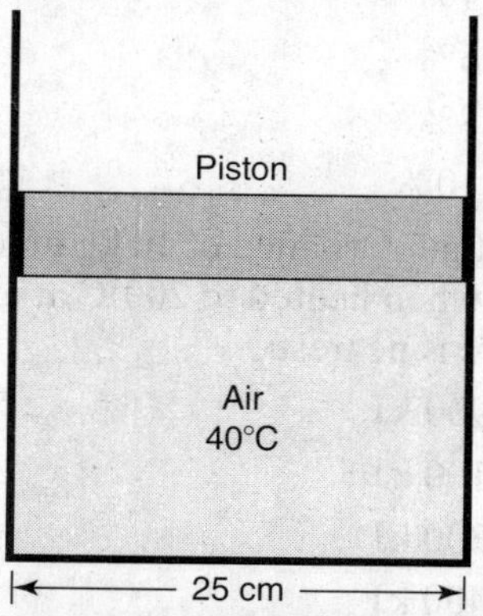

Figure 2.28

2.13 The mass of air contained in a rigid 20-L volume at 3 MPa with a temperature of 130 K is nearest:

(A) 2.7 kg
(B) 2.1 kg
(C) 1.8 kg
(D) 1.5 kg

2.14 Infiltration in a house in the winter where $T = -25°C$ occurs even when there is no wind due to the difference in the air density between the inside and outside of a house. For an inside temperature of 22°C, the density difference is nearest:

(A) 0.207 kg/m^3
(B) 0.224 kg/m^3
(C) 0.241 kg/m^3
(D) 0.288 kg/m^3

2.15 Estimate the value of the C_p of steam at 400°C and 4 MPa:

(A) 2.6 kg
(B) 2.4 kg
(C) 2.2 kg
(D) 2.0 kg

2.16 Estimate the percent error if the C_p value is used from Table B-2 to find the change in enthalpy of steam that has a temperature change from 200°C to 600°C at 200 kPa:

(A) −4%
(B) −6%
(C) −8%
(D) −10%

2.17 The enthalpy change of 10 kg of ice initially at −20°C when heated to 200°C at atmospheric pressure is nearest:

(A) 9900 kJ
(B) 9100 kJ
(C) 8300 kJ
(D) 6400 kJ

Phases of a Substance

2.18 What is the pressure in kPa inside a container if it holds water that is just beginning to boil at *a*) 140°C, *b*) 200°C, and *c*) 320°C?

> Instructor: This isn't multiple choice. Each part with a lower-case italic letter should be assigned as a separate problem.

2.19 Estimate the temperature of the water in a container if the water is on the verge of boiling at a pressure of *a*) 85 kPa, *b*) 200 kPa, and *c*) 350 kPa.

2.20 A group of engineers wish to know their elevation when they reach the apex on their mountain climb. They forgot their elevation-measuring device, but they have a thermometer and their thermo book. They build a campfire and boil a pot of water. The water is noted to boil at 90°C. What is their elevation?

2.21 A mountain climber reaches an elevation of *a*) 3600 m, *b*) 6000 m, and *c*) 9600 m on the mountain of Fig. 2.29. At what temperature will water just boil at the selected elevation?

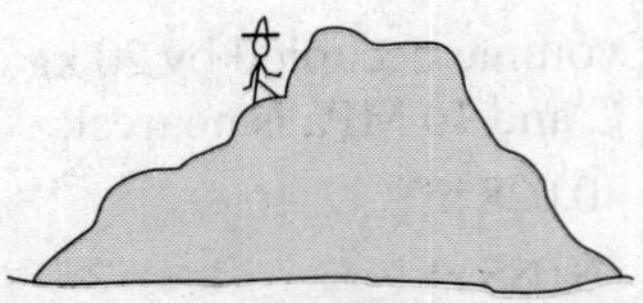

Figure 2.29

2.22 A container holds 10 kg of liquid water and 7 kg of water vapor at 140°C. Calculate the quality of this saturated mixture.

2.23 A container holds 0.03 m^3 of liquid water and 0.15 m^3 of water vapor at 90°C. Calculate the quality of this saturated mixture.

2.24 The 4 kg of saturated liquid water in Fig. 2.30 are completely vaporized at a constant pressure of 200 kPa. Determine the volume of the water at state 1 and at state 2.

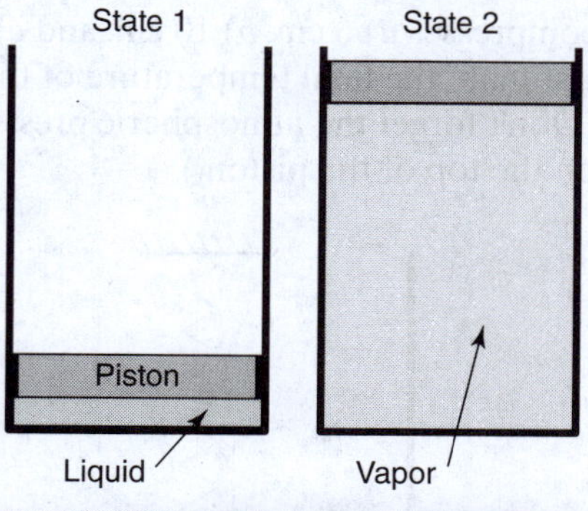

Figure 2.30

2.25 Saturated liquid water is contained in a volume of 400 cm^3. Heat is added at a constant pressure of 400 kPa until the water is completely vaporized. What is the final volume?

2.26 A vessel contains 10 kg of water with a quality of 0.85 at a pressure of *a*) 140 kPa, *b*) 200 kPa, and *c*) 2 MPa. Determine the volume of liquid and the volume of vapor. What is the temperature of the mixture?

2.27 Ten kilograms of saturated mixture are held in the container of Fig. 2.31 at 40°C and a quality of 0.6. What is the volume of liquid, the volume of vapor, and the pressure of the mixture?

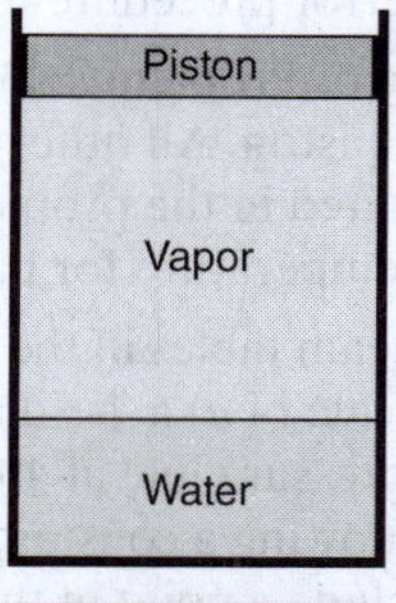

Figure 2.31

2.28 We want to boil water in a cylinder that maintains a constant pressure of *a*) 500 kPa, *b*) 1 MPa, and *c*) 3 MPa. To what temperature do we need to heat the water to make this happen?

2.29 The atmospheric pressure on the surface of Mars is 0.0003 atmospheres. At what temperature would water boil on Mars?

Vorotylin Roman/Shutterstock.com

Pressure cooker

2.30 Use the Internet to find three brands of kitchen pressure cookers. What is the maximum rated pressure for each of these? What would be the temperature at which water would boil at these pressures?

2.31 A 5-m^3 container holds water that has a quality of 0.7 at *a*) 120°C, *b*) 240°C, and *c*) 370°C. What are the pressure and the mass of water in the container?

2.32 A constant-pressure, piston-cylinder device contains 0.5 kg of saturated liquid water at 50°C. Heat is added to the water until it becomes a saturated vapor. What is the initial volume of the water? What is the final volume of the water?

2.33 A rigid container holds water at the critical point. The container is cooled until the pressure reaches *a*) 2 MPa, *b*) 1200 kPa, and *c*) 400 kPa. Determine the final quality of the water.

2.34 Water with a quality of 0.1 is contained in the rigid volume of Fig. 2.32 at 200 kPa. It is heated until the temperature reaches *a*) 140°C, *b*) 180°C, and *c*) 210°C. Calculate the quality of the water and the pressure at the final state.

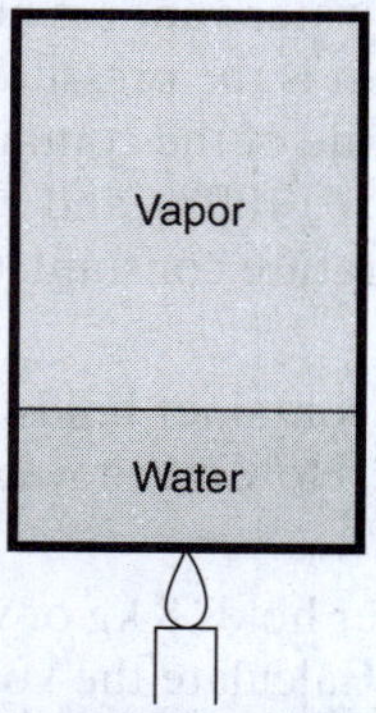

Figure 2.32

2.35 Four kilograms of water at 25°C are heated at constant pressure, as in Fig. 2.31, until it is saturated liquid at 120°C. Approximate its change in volume.

2.36 Two kilograms of water are contained in a piston-cylinder arrangement at 20°C and 5 MPa. It is heated at constant pressure until the temperature reaches 200°C. Calculate its change in volume using i) Table C-4 and ii) Table C-1. What do you conclude?

Instructor: All parts with choices (i, ii, etc.) are intended to be solved.

2.37 Two kilograms of saturated liquid water at 200 kPa are heated until the pressure and temperature reach 2 MPa and 400°C, respectively. Calculate the change in volume.

2.38 A container holds 10 kg of water at 130°C with a quality of 0.4. What is the pressure in the container? What is the volume of the container? What is the volume of the liquid phase?

2.39 A 1.9 m^3 container holds 4.6 kg of water at i) 350 kPa, ii) 450 kPa, and iii) 700 kPa. What is the state of the water?

Note: It is assumed that the IRC Calculator has been introduced and is now available to be used for property determination. Your instructor may, however, require that the tables in the Appendix be used.

2.40 Saturated water vapor is held in a container at 200°C. What is the pressure in the container? If the volume of the container is increased by *a*) 50%, *b*) 100%, and *c*) 200%, holding the temperature constant, estimate the final pressure.

2.41 A 0.27 m^3 container holds water at 300 kPa and 200°C. How much water is in the container?

2.42 A container holds 8 kg of water at a pressure of 5 MPa. Calculate the volume of the container if the temperature is *a*) −5°C, *b*) 20°C, *c*) 400°C, and *d*) 800°C.

2.43 A mass of 0.02 kg of saturated water vapor is contained in the cylinder of Fig. 2.33. The spring with spring constant $K = 60$ kN/m just touches the top of the frictionless 160-kg piston. Heat is added until the spring compresses *a*) 6 cm, *b*) 10 cm, and *c*) 15 cm. Estimate the final temperature of the steam. (Don't forget the atmospheric pressure acting on the top of the piston.)

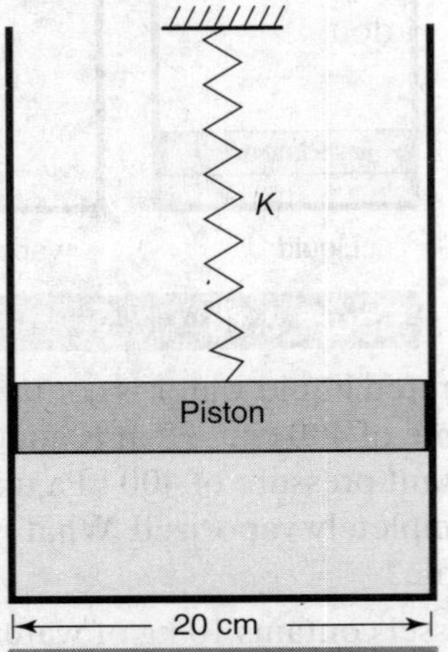

Figure 2.33

2.44 A mass of 0.02 kg of steam with $x_1 = 0.8$ is contained in the cylinder of Fig. 2.33. The spring with spring constant $K = 60$ kN/m just touches the top of the frictionless 160-kg piston. Heat is added until the temperature reaches 800°C. Estimate the final pressure of the steam. A trial-and-error procedure may be necessary.

2.45 The spring of Problem 2.43 is 2 cm above the top of the piston. All other quantities are the same as stated in the problem. Calculate the final temperature for part (*a*).

2.46 Use the steam tables in the Appendix and make a plot to scale of *a*) a *T-v* diagram showing a constant-pressure line at 400 kPa, and *b*) a *P-v* diagram showing a constant-temperature line at 200°C. Include a point in the subcooled region and a point in the superheated region. Observe the distortions of the various diagrams in this text. The distortions are necessary to show the various regions, as the diagrams of parts (*a*) and (*b*) should show.

Internal Energy and Enthalpy

2.47 Water at a pressure of 1600 kPa has a quality of 0.4. Determine the temperature, the average internal energy, and the average enthalpy for this mixture. Use the steam tables.

2.48 Superheated water exists at 1.86 MPa and 420°C. Determine its specific enthalpy using:

i) The steam tables in the appendix (a double interpolation is needed)
ii) The IRC Calculator
iii) The equation $h = u + Pv$ by using u and v from the IRC Calculator

Are all the answers acceptable?

Comment

The word "specific" is often omitted when it is obvious that the specific property is desired since the mass is not given. In fact, in the tables, the specific internal energy may be referred to simply as "energy" and the specific enthalpy as "enthalpy."

2.49 5 kg of saturated water vapor at 150°C are cooled at constant temperature to a saturated liquid. What is the change in pressure? The change in the total internal energy? The change in total enthalpy? Does $\Delta H = \Delta U + \Delta(PV)$? *a*) Use the steam tables and *b*) the IRC Calculator.

2.50 Saturated water vapor is compressed from 200°C to 2 MPa and 600°C. Determine the change in specific volume and enthalpy. *a*) Use the steam tables and *b*) the IRC Calculator.

2.51 Four kilograms of water at 20°C are heated at constant pressure of 200 kPa to 400°C. Determine the change in density, internal energy, and enthalpy. *a*) Use the steam tables and *b*) the IRC Calculator.

2.52 Twenty kilograms of water are heated at constant pressure from 300°C and 0.6 MPa to 400°C. Calculate the change in enthalpy. *a*) Use the steam tables and *b*) the IRC Calculator.

2.53 Determine the change in internal energy and enthalpy of water at 40°C and 3500 kPa if the temperature increases to 260°C while the pressure in Fig. 2.34 is held constant. Use *a*) the steam tables and *b*) the IRC Calculator.

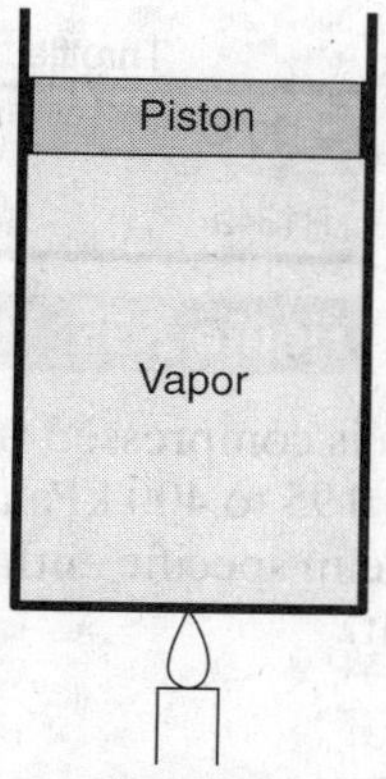

Figure 2.34

2.54 How much energy is needed to increase the temperature of 10 kg of ice from −20°C to 20°C at atmospheric pressure?

2.55 Estimate the amount of energy required to sublimate the 4 kg of water contained in the frozen clothes in a northern state which are at −20°C. See Fig. 2.35.

Figure 2.35

Refrigerants

2.56 A 0.05-m^3 container contains 50 kg of R134a. If the pressure is *a*) 2.5 MPa and *b*) 240 kPa, what is the state of the refrigerant?

2.57 A 0.27 m^3 container contains 45 kg of R134a at −30°C. What is the state of the refrigerant?

2.58 A constant-pressure, piston-cylinder device contains 100 kg R134a at −12°C and 200 kPa. If the refrigerant is heated at constant pressure until the temperature reaches 60°C, determine:

i) The change in volume
ii) The change in total internal energy
iii) The change in total enthalpy

2.59 A flow of saturated liquid R134a at 30°C passes through a throttle where the pressure drops to 70 kPa. If the enthalpy remains constant as the refrigerant passes through the throttle of Fig. 2.36, determine the temperature and the quality of the flow exiting the valve.

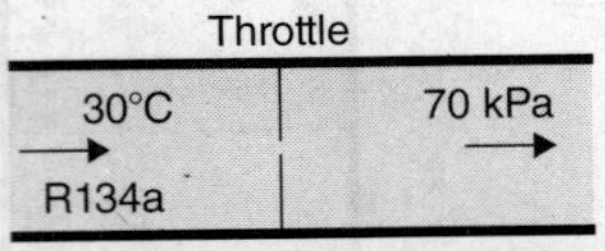

Figure 2.36

2.60 Ammonia is compressed from −40°C and a quality of 0.95 to 400 kPa and 40°C. Calculate the change in specific enthalpy of the refrigerant.

Ideal-Gas Law

2.61 A spherical balloon contains 0.1 kg of air at 120 kPa and 20°C. If the temperature of the air is increased to 35°C and the pressure remains the same, determine the initial and final volume of the sphere.

2.62 A spherical 0.6 m-diameter balloon contains helium at 22°C. What is the mass of helium in the balloon assuming the pressure is approximately 100 kPa? Approximate the diameter of the balloon at an elevation of 1500 m, as shown in Fig. 2.37. (Assume the pressure difference across the balloon surface is negligible.)

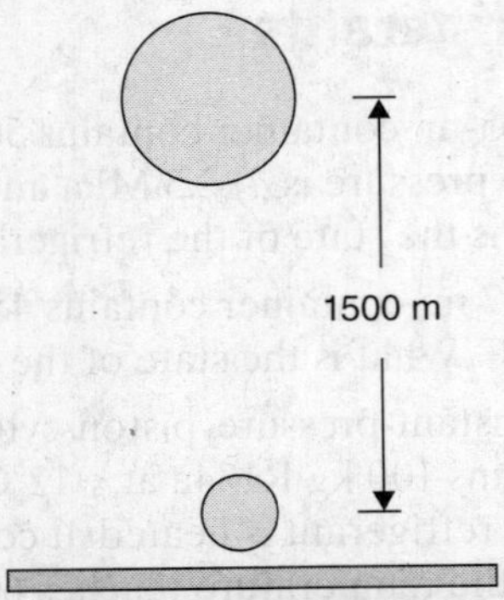

Figure 2.37

2.63 The atmosphere of Mars is composed of carbon dioxide at 1 kPa and −60°C. Determine the specific volume of the atmosphere.

2.64 A 5-m^3-rigid container holds 0.2 kg of air at 300 K. How much air at 300 K must be added to the container to bring the pressure to 500 kPa?

2.65 5 kg of air at 100 kPa and 27°C are compressed so that the final volume is one-fourth the initial volume and the pressure is 280 kPa. Determine the final temperature of the air.

2.66 A 4-m^3 volume contains 2 kg of an unknown gas at 400 kPa and 112°C. What gas do you think occupies the volume?

2.67 An enclosed football stadium is maintained at 100 kPa and 20°C. If its volume is 2×10^6 m^3, estimate the weight of the air.

2.68 Fill in the missing properties in the following table if the air occupies a 10-m^3 volume.

	P (kPa)	T (°C)	v (m^3/kg)	ρ (kg/m^3)	W (N)
i)	200	200			
ii)		400	0.08		
iii)	600				40
iv)		−40		20	

2.69 Fill in the missing properties in the following table if the air occupies a 0.8 m^3 volume.

	P (kPa)	T (°C)	v (m^3/kg)	ρ (kg/m^3)	W (N)
i)	140	207			
ii)		317	0.07		
iii)	700				24
iv)		−50		26	

2.70 The air in a house is maintained at 22°C. The air outside the house is at −28°C. What is the difference in the density of the air between the inside and the outside of the house? There are always small passages where air can infiltrate into or from a house even if extreme care is taken to seal all those passages. Explain what the air particles do even if the outside air is perfectly motionless. See Fig. 2.38.

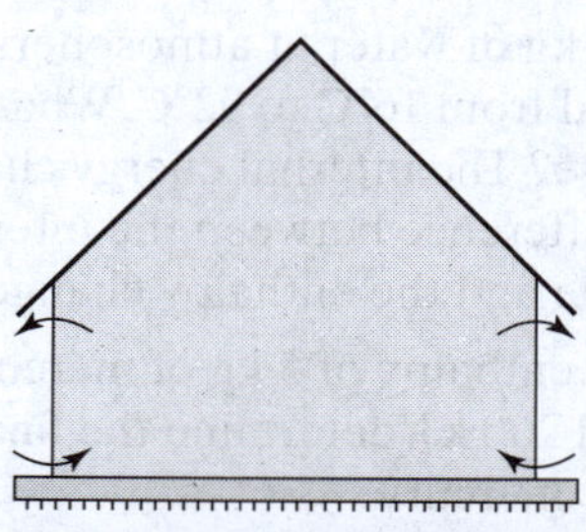

Figure 2.38

2.71 The tires on an automobile are inflated to 250 kPa gage in New York, where the temperature is −23°C. The auto is driven to Arizona, where the temperature on the blacktop is 70°C. Estimate the gage pressure in the tires assuming the volume does not change.

2.72 The golfer in Fig. 2.39 on a very humid day states that the ball doesn't travel as far because of the heavy air. If the air were less humid at the same pressure and location, would the ball have traveled further? Explain, referring to the ideal-gas law and the fact that the drag F_D on the golf ball is proportional to the density of the air: $F_D = \rho A V^2/2$.

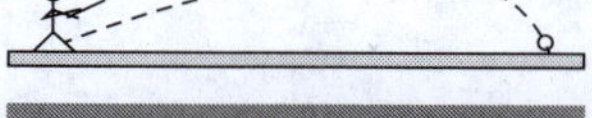

Figure 2.39

Real-Gas Equations of State

2.73 Nitrogen exists at 4 MPa and 130 K. Approximate the specific volume using i) the ideal-gas law and ii) the compressibility factor from Fig. 2.23.

2.74 Ten kilograms of air occupies 1 m^3 of space at −40°C. Calculate and compare the pressure using the:

i) Ideal-gas law
ii) Van der Waals equation
iii) Beattie–Bridgeman equation
iv) *Z*-factor (see Appendix H)
v) IRC Calculator

2.75 Estimate the pressure of nitrogen at −60°C and v = 0.036 m^3/kg using the:

i) Ideal-gas law
ii) Van der Waals equation
iii) *Z*-factor (see Appendix H)

2.76 The pressure and temperature are measured in a tank containing air to be 60 kPa and −80°C, respectively. Determine the specific volume using the:

i) Ideal-gas law
ii) Van der Waals equation
iii) *Z*-factor (see Appendix H)
iv) IRC Calculator

Internal Energy and Enthalpy of Ideal Gasses

2.77 Calculate the internal energy change and the enthalpy change if the temperature changes from 20°C to 450°C for each of the following, assumed to an ideal gas with constant specific heats:

a) Air
b) Nitrogen
c) Hydrogen
d) Propane
e) Steam

Note: Use specific heat values from Table B-2 unless otherwise stated.

2.78 Steam is assumed to be an ideal gas. Determine the enthalpy change if state 1 is saturated vapor at 30°C and state 2 is at 150°C and 200 kPa.

i) Assume constant specific heats.

ii) Use the steam tables.

iii) Use the IRC Calculator.

2.79 Nitrogen is heated from 20°C to 500°C. Calculate the change in the internal energy and enthalpy. Use:

i) Constant specific heats from Table B-2

ii) The ideal-gas table in Appendix F

iii) The average specific heat from Table B-5

iv) Equation 2.30

2.80 Ten kg of air at −4°C is heated to 536°C. Determine the change in internal energy and enthalpy. Use:

i) Constant specific heats from Table B-2

ii) The ideal-gas table in Appendix F

iii) The specific heat at the average temperature from Table B-6

iv) Equation 2.30

v) The IRC Calculator

2.81 Estimate the specific heat C_p of steam at 1.6 MPa and 400°C using i) backward differences, ii) forward differences, iii) central differences, and iv) Table B-2. Which is the most accurate?

2.82 Estimate the specific heat C_p of steam at:

a) 0.2 MPa and 400°C

b) 1.8 MPa and 400°C

c) 7 MPa and 400°C

2.83 Estimate the specific heat C_p of steam at:

a) 300 kPa and 300°C

b) 1 MPa and 300°C

c) 4 MPa and 300°C

Specific Heats of Liquids and Solids

2.84 Forty kg of water at atmospheric pressure are heated from 16°C to 92°C. What is the enthalpy change? The internal energy change? Why is the difference between the internal energy change and the enthalpy change so small?

2.85 If the enthalpy of 5 kg of mercury at 25°C is raised 200 kJ, determine the final temperature of the mercury.

2.86 The temperature of 10 kg of ice at −20°C is raised to 50°C, displayed by Fig. 2.40. Determine the enthalpy change.

a) Assume $C_{p,\text{ice}} = 2.1$ kJ/kg·°C.

b) Use $C_{p,\text{ice}} = 2.1 + 0.0069T$ kJ/kg·°C.

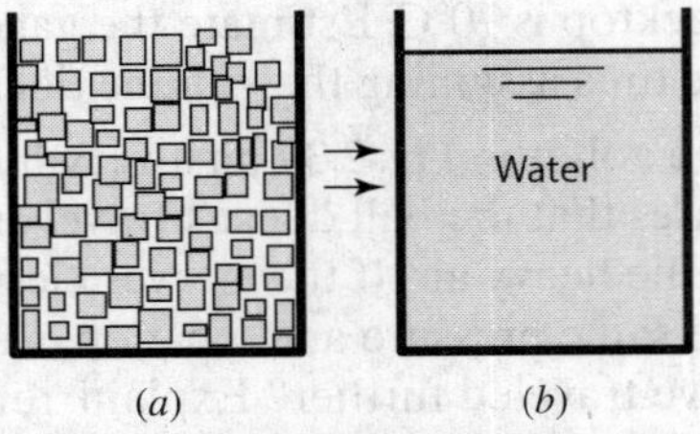

Figure 2.40

CHAPTER

3

The First Law for Systems

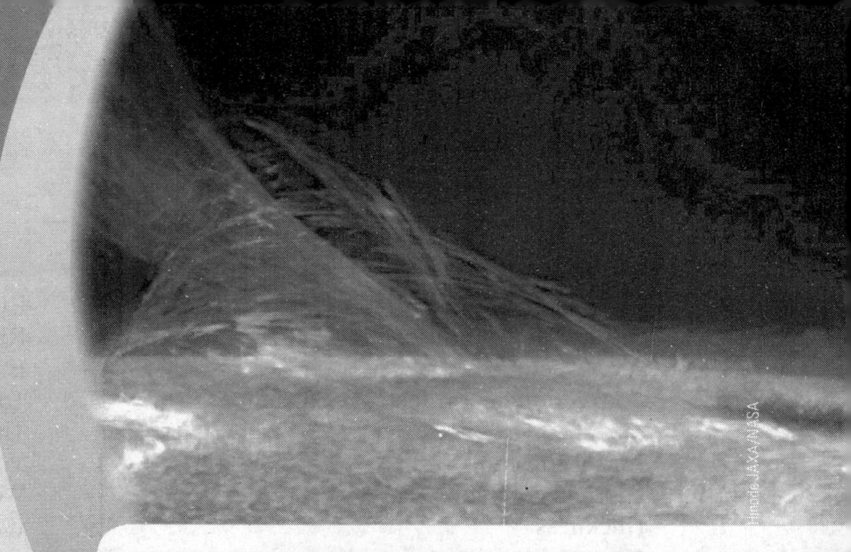

Outline

Nomenclature

The following variables are introduced in this chapter:

C	A constant	R	Electrical resistivity, R-factor
ds	Differential length segment	r	Radius
$\mathbf{F}$	Force vector	s	Entropy
h_c	Convection heat transfer coefficient	T	Torque
i	Current	t	Time
K	Spring constant	V	Voltage
k	Thermal conductivity	W	Work
L	Thickness	$\dot{W}$	Work rate
n	A constant	w	Work per unit mass
Q	Heat transfer	ε	Emissivity
$\dot{Q}$	Rate of heat transfer	σ	Stefan–Boltzmann constant
q	Specific heat transfer	ψ	A special property
		τ	Shear stress

Learning Outcomes

- ❑ **Solve work integrals**
- ❑ **Identify heat transfer mechanisms**
- ❑ **Apply the first law to systems**
- ❑ **Identify and analyze common thermodynamic processes**

Comment

We have interest primarily in the conservation of mass and the 1st law.

First law of thermodynamics: An accounting of the flow of energy into and out of a system and its accumulation.

System: A volume with fixed mass. The volume can change, as it does in a cylinder when the air is compressed by a piston.

Many of the problems we solve in engineering involve the application of one of the three laws: the conservation of mass, Newton's second law,[1] and the first law of thermodynamics. This chapter focuses on the conservation of mass and the first law of thermodynamics, with minor emphasis on Newton's second law. The *first law of thermodynamics* is simply an accounting of the energy that flows into and out of a system and its accumulation. The general approach to solving problems applying the two laws of interest can be summed up with

$$\text{Input} - \text{Output} = \text{Accumulation} \quad (3.1)$$

The system that is being analyzed must be carefully defined. A *system* consists of a fixed mass of substance that is contained in a volume. Mass is not allowed to enter or leave the volume. Even though the mass of the system is fixed, the shape of the volume that contains this mass may vary; the volume in a piston-cylinder arrangement decreases as the piston compresses the air. The boundary of the system we are analyzing will be defined by the boundary of the mass of the system. In Chapter 4, the substance will enter and/or leave the volume, referred to as a control volume.

[1] A fourth law, the conservation of angular momentum, is a result of Newton's second law.

Motivational Example—The piston-cylinder processes in an automobile engine

A good example of a thermodynamic system is the air in a cylinder being compressed by a piston in an automobile engine. The engine cycle consists of five distinct processes. First, air is drawn into the cylinder through the inlet valve by moving the piston down, as shown in Fig. 3.1. Next, the inlet valve closes and the piston moves upward, compressing the air during the compression stroke thereby increasing its pressure and temperature. When the piston nearly reaches its maximum extension into the cylinder, a spark plug ignites the injected fuel, with the result that the pressure and temperature of the air are greatly increased. Next, this high-pressure air exerts a large force on the

The five strokes in a piston engine:
Intake
Compression
Ignition/Combustion
Power
Exhaust

TDC—Top dead center

BDC—Bottom dead center

Comment

The piston-engine process can be seen in motion on the Internet on the site at: howstuffworks.com /engine1.htm

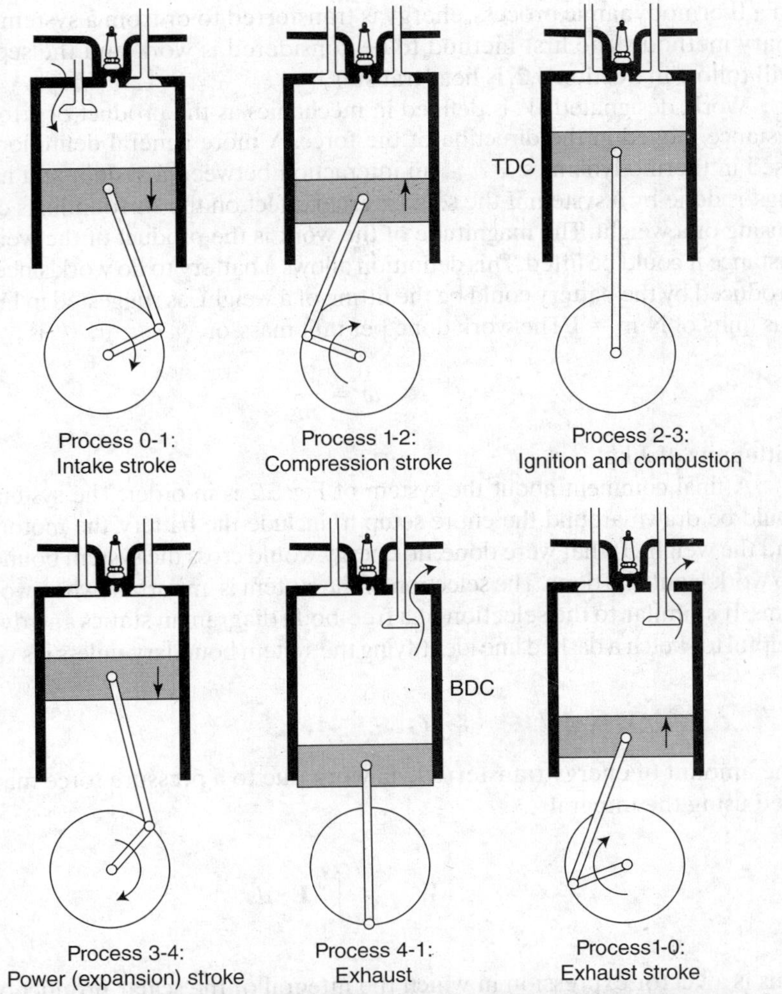

Figure 3.1

Piston-engine processes.

(Continued)

Motivational Example (*Continued*)

piston during the power stroke. As the cycle continues, the exit valve is opened and the piston moves upward, forcing the products of combustion from the cylinder. The system is now ready to repeat the series of processes. The compression stroke, ignition, and power stroke are processes that involve a fixed amount of mass, a system. Once the valves are closed, the mass of air in the cylinder is fixed. During the intake and exhaust strokes, the system is not identified.

Note: The mass of fuel injected is negligible compared to the mass of air.

3.1 Work

3.1.1 Definition and units

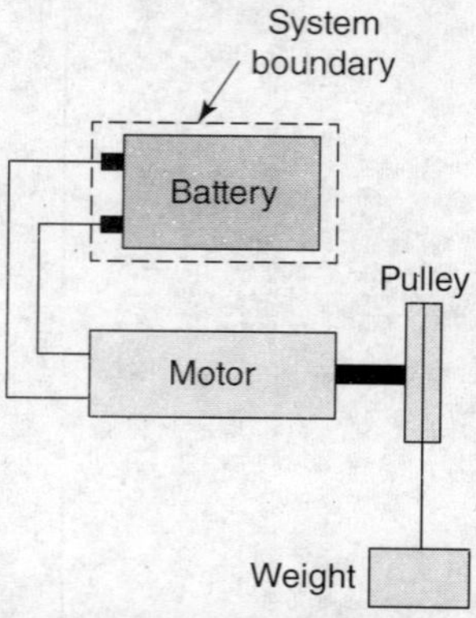

Figure 3.2
Work being done by a battery.

In a thermodynamic process, energy is transferred to or from a system by two primary methods. The first method to be considered is work and the second, which will follow in Section 3.2, is heat transfer.

Work, designated W, is defined in mechanics as the product of a force and the distance moved in the direction of the force. A more general definition of work is used in thermodynamics: *Work*, an interaction between a system and its surroundings, is done by a system if the sole external effect on the surroundings could be the raising of a weight. The magnitude of the work is the product of the weight and the distance it could be lifted. This definition allows a battery to do work since the energy produced by the battery could be the lifting of a weight, as suggested in Fig. 3.2. Work has units of N·m = J. The work done per unit mass, or *specific work*, is

Work: Is done by a system if the sole external effect could be the raising of a weight.

$$w = \frac{W}{m} \tag{3.2}$$

with units of J/kg.

A final comment about the system of Fig. 3.2 is in order. The system boundary could be drawn around the entire setup to include the battery, the motor, the pulley, and the weight. If that were done, no energy would cross the system boundary and so no work would be done. The selection of the system is important when working problems. It's similar to the selection of a free-body diagram in statics and dynamics. It's helpful to sketch a dashed line identifying the system boundary, unless it's very obvious.

3.1.2 Work due to pressure

The amount of energy transferred via work due to a pressure force may be evaluated using the integral

$$W_{1\text{-}2} = \int_{s_1}^{s_2} \mathbf{F} \cdot d\mathbf{s} \tag{3.3}$$

This is a vector expression in which the integral of the scalar product of the force vector $\mathbf{F}$ and the differential path element $d\mathbf{s}$, a vector tangential to the path, is integrated over the path from the initial state to the final state. The work done is a path function; hence the subscripts on W indicate a process from state 1 to state 2. Since it depends on the path, its differential is not exact, so it is represented by

Comment

Work is not a property and is never written as W_1 or W_2.

δW. Work is not a property and is never written as W_1 or W_2 but always as $W_{1\text{-}2}$, or simply as W.

In thermodynamics, the primary force used to do work is a pressure force. Pressure forces always act normal to the surface in contact with the fluid; for example, in Fig. 3.3 the pressure force is moving the piston a small displacement ds. The resulting displacement is often perpendicular to the surface in which case the dot product, using $F = PA$, takes the form

$$W_{1\text{-}2} = \int_{s_1}^{s_2} PA\,ds \tag{3.4}$$

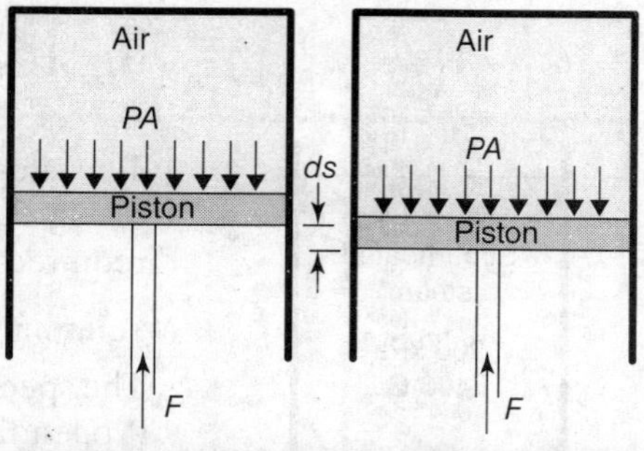

Figure 3.3
A pressure force doing work on a moving boundary.

Recognize that $A\,ds = dV$, the infinitesimal increase in volume due to the displacement ds. The expression for the work due to pressure acting on a boundary is an important and oft-used relationship, in both differential and integral form,

Comment

The force $F = PA$ is in the direction of ds, so the differential work is simply $F\,ds = PA\,ds = P\,dV$.

$$\delta W = P\,dV \qquad W_{1\text{-}2} = \int_{V_1}^{V_2} P\,dV \qquad w_{1\text{-}2} = \int_{v_1}^{v_2} P\,dv \tag{3.5}$$

If the pressure is a known function for each position of the boundary in Fig. 3.3, the integration could be performed and the work determined. Notice how the path can be very different between state 1 and state 2, resulting in less work (the area under the curve) in Fig. 3.4*a* than in Fig. 3.4*b*.

If the pressure is constant between states 1 and 2, the expression for the work would be

$$w_{1\text{-}2} = P(v_2 - v_1) \qquad \text{if } P = \text{const} \tag{3.6}$$

The sign convention: Work done by the system on the surroundings is considered positive, and work done by the surroundings on the system is negative.

Remember: Work done on a system is negative.

One additional observation should be made concerning the work defined by the integral of Eq. 3.5. It is assumed that the pressure and volume are known at each state between the initial and final states; that is, a quasi-equilibrium process is assumed between state 1 and state 2 of Fig. 3.4. So, Eq. 3.5 represents a quasi-equilibrium work mode. Nonequilibrium work modes involve, for example, friction or electrical resistance heaters. If a nonequilibrium work mode is present, the work from state 1 to state 2 is not given by Eq. 3.5. If the information given does not allow one to determine the type of process, a quasi-equilibrium process will be assumed.

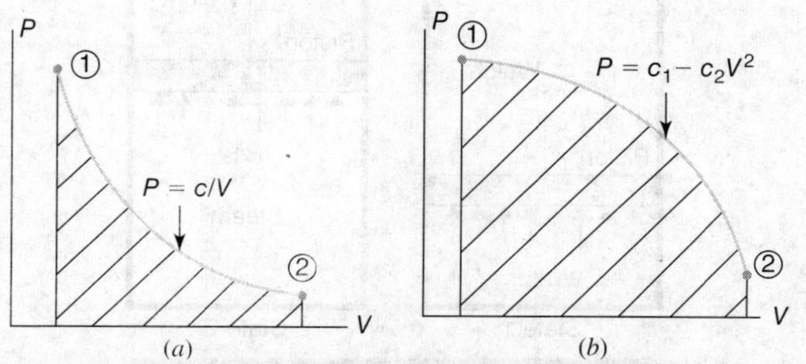

Figure 3.4
Work is the cross-hatched area from state 1 to state 2.

Example 3.1 Work needed to condense steam

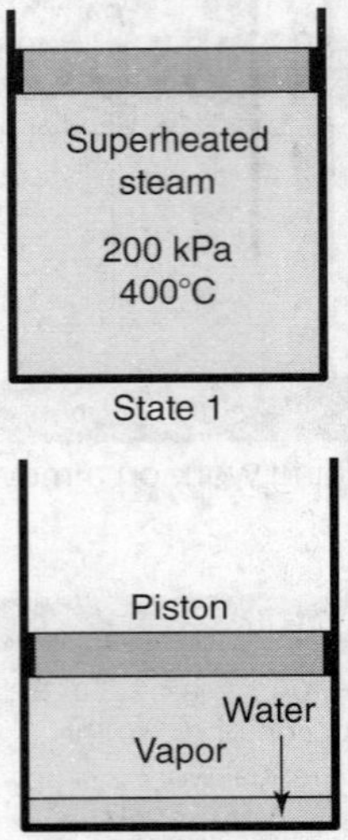

Figure 3.5

Ten kilograms of steam at 200 kPa and 400°C are condensed in the cylinder of Fig. 3.5 at constant pressure until the quality is 50%. Determine the work required between the two states.

Solution

The specific volumes at states 1 and 2 are found using Tables C-2 and C-3 in the Appendix. They are

$$v_1 = 1.5493 \text{ m}^3/\text{kg}$$
$$v_2 = v_f + x(v_g - v_f)$$
$$= 0.00106 + 0.5 \times (0.8857 - 0.00106) = 0.4434 \text{ m}^3/\text{kg}$$

The work during this condensation process, assumed to be a quasi-equilibrium process, is then, using Eq. 3.6,

$$W_{1\text{-}2} = Pm(v_2 - v_1)$$
$$= 200 \text{ kN/m}^2 \times 10 \text{ kg} \times (0.4434 - 1.5493) \text{ m}^3/\text{kg} = \underline{-2212 \text{ kJ}}$$

recognizing that kN·m = kJ. Note that the work is negative since the volume is decreasing. A negative work means that work is being done on the system, the steam.

Example 3.2 The work needed to raise a piston

One hundred grams of water at 50°C are contained in a 220-mm-diameter cylinder at state 1, as shown in Fig. 3.6. Energy is added until the temperature reaches 150°C in the steam at state 2. If the frictionless piston has a mass of 387 kg, find the work done by the steam on the piston.

Solution

The pressure in the cylinder, due to the piston resting on top of the water is

$$P = \frac{mg}{A} = \frac{387 \text{ kg} \times 9.81 \text{ m/s}^2}{\pi \times 0.11^2 \text{ m}^2} = 99\,870 \text{ Pa gage} \quad \text{or} \approx 200 \text{ kPa abs}$$

Pressure: The absolute pressure is 99.9 + 100 = 199.9 ≈ 200 kPa. Assume P_{atm} = 100 kPa, unless otherwise stated.

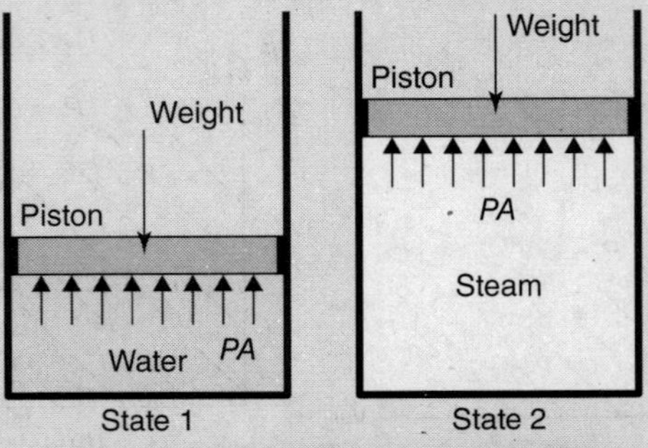

Figure 3.6

Comment

One can visualize atmospheric pressure acting on top of the piston or simply add 100 kPa to give absolute pressure, as required.

The work done by the steam to lift the piston is due to the force PA and is given by $mP\Delta v$ (see Eq. 3.6). State 1 is compressed liquid; at 50°C, Table C-1 gives $v_1 = 0.0102\ \text{m}^3/\text{kg}$. State 2 is superheat; from Table C-3, $v_2 = 0.9596\ \text{m}^3/\text{kg}$ at 0.20 MPa and 150°C. The work is

$$
\begin{aligned}
W_{1\text{-}2} &= Pm(v_2 - v_1) \\
&= 200\,\frac{\text{kN}}{\text{m}^2} \times \left(\frac{100}{1000}\right)\text{kg} \times (0.9596 - 0.01012)\,\frac{\text{m}^3}{\text{kg}} \\
&= \underline{18.99\ \text{kJ}}
\end{aligned}
$$

Because we used kPa for pressure, the work is in kJ.

Observation: This is the work done by the steam on the piston. The air above the piston does negative work on the piston due to the atmospheric pressure.

Work needed to increase a volume — Example 3.3

Air at 5 MPa is contained in a cylinder at state 1, as shown in Figure 3.7. A piston is moved such that the volume changes according to $PV = 10^5$, with P in Pa and V in m^3. The piston moves until the volume doubles. Find the work done by the air on the piston.

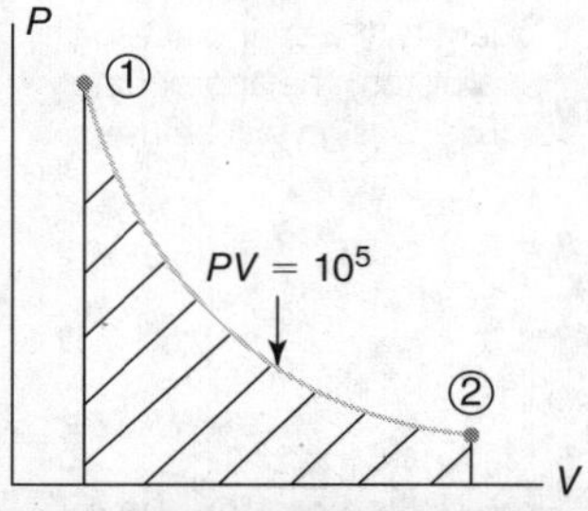

Figure 3.7

Solution

The work done due to the motion of the piston is given by the integral of Eq. 3.5, so the volumes must be known since they are the limits of integration. They are

$$
V_1 = \frac{10^5}{P_1} = \frac{10^5\ \text{N}\cdot\text{m}}{50 \times 10^5\ \text{N/m}^2} = 0.02\ \text{m}^3 \quad \text{and} \quad V_2 = 0.04\ \text{m}^3
$$

Note that 10^5 is a constant that has units of N·m, the same units as on PV. The work is then

$$
W_{1\text{-}2} = \int_{V_1}^{V_2} P\,dV = \int_{0.02}^{0.04} \frac{10^5}{V}\,dV = 10^5 \ln 2 = 69\,300\ \text{J} = \underline{69.3\ \text{kJ}}
$$

The work is positive, so it is doing work on the surroundings, that is, the piston, as expected since the volume increases.

3.1.3 Other forms of work

There are four other forms of work that we will occasionally encounter in our study. They are the work required to stretch a linear spring, to rotate a shaft, to turn a paddle wheel, and to cause an electrical current to flow through a resistor. Other forms of work exist, such as work due to electrical and magnetic fields and the work needed to stretch a liquid film, but they will not be included. Let's analyze each of the four.

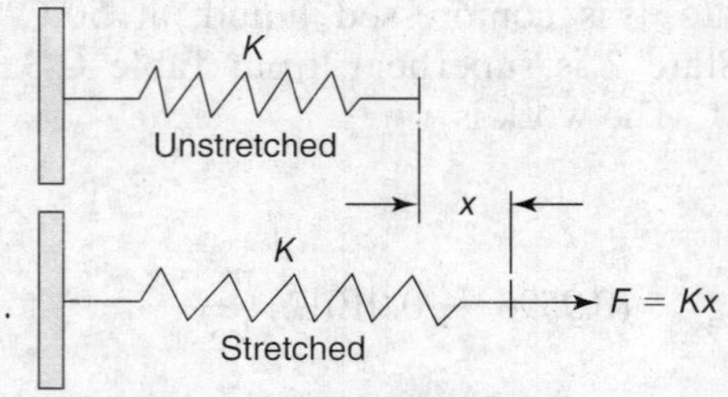

Figure 3.8

A linear spring being stretched.

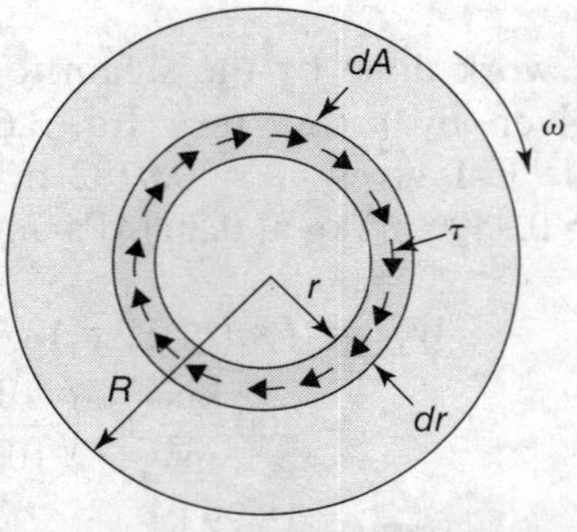

Figure 3.9

The stresses acting on an area element of a rotating shaft.

Comment

We have used x for displacement. The work here is written as a positive quantity. If a spring acts on a piston, the appropriate sign will be used.

A linear spring: The force needed to stretch a linear spring, shown in Fig. 3.8, from its unstretched position is

$$F = Kx \tag{3.7}$$

where K is the *spring constant* with units of N/m. The work required to stretch the spring from the stretched position x_1 to position x_2 is

$$W = \int_{x_1}^{x_2} F dx = K \int_{x_1}^{x_2} x dx = \frac{1}{2}K\left(x_2^2 - x_1^2\right) \tag{3.8}$$

where we have written W rather than $W_{1\text{-}2}$, as is often done. If the spring is the system, the work done on the spring has crossed the boundary of the spring and energy is stored in the spring.

Power: The rate of doing work.

Shear force: Shear stress times the area upon which it acts.

A rotating shaft: Next, consider the rate of doing work, called *power*, that is transmitted by a rotating shaft. The differential *shear force* due to the shearing stress τ that acts on the infinitesimal area dA, displayed in Fig. 3.9, is $dF = \tau dA$. Recognizing that the velocity with which this differential force τdA moves is $r\omega$ and that force times velocity is the rate of doing work, symbolized by $\dot{W}$, we see that

$$\dot{W} = \int_A r\omega\tau dA = \int_0^R r\omega\tau(2\pi r dr) = 2\pi\omega \int_0^R \tau r^2 dr \tag{3.9}$$

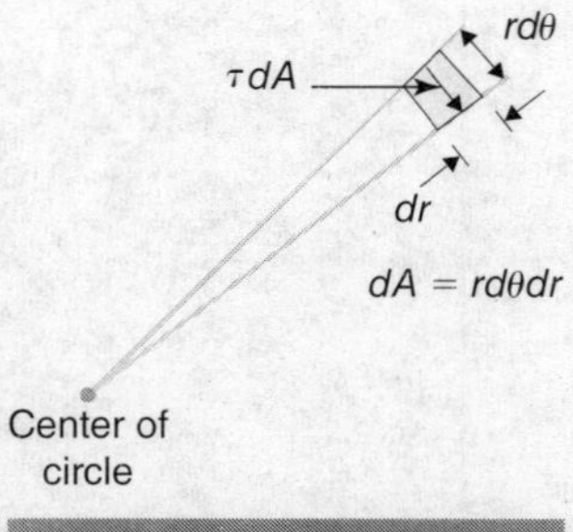

Figure 3.10

Actually, a double integration could have been performed using the area element shown in Fig. 3.10. First, the integration would be from $\theta = 0$ to $\theta = 2\pi$, then from $r = 0$ to $r = R$.

The torque T transmitted by the shaft is found by integrating the infinitesimal torque $rdF = r\tau dA$ over the area:

$$T = \int_A r\tau dA = \int_0^R r\tau(2\pi r dr) = 2\pi \int_0^R \tau r^2 dr \tag{3.10}$$

Comparing Eq. 3.9 with Eq. 3.10, we observe that the power transmitted by a shaft is

$$\dot{W} = \omega T \tag{3.11}$$

Units: $[\omega T]$ = (rad/s)(N·m) = N·m/s = J/s = W

If the work over a time increment Δt is desired, the power is multiplied by the time, giving

$$W = \omega T \Delta t \tag{3.12}$$

The power or work delivered by a shaft can now be calculated.

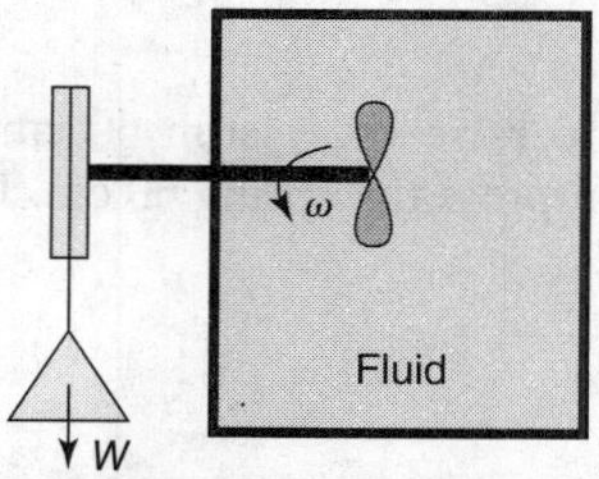

Figure 3.11
A paddle wheel doing work on a fluid.

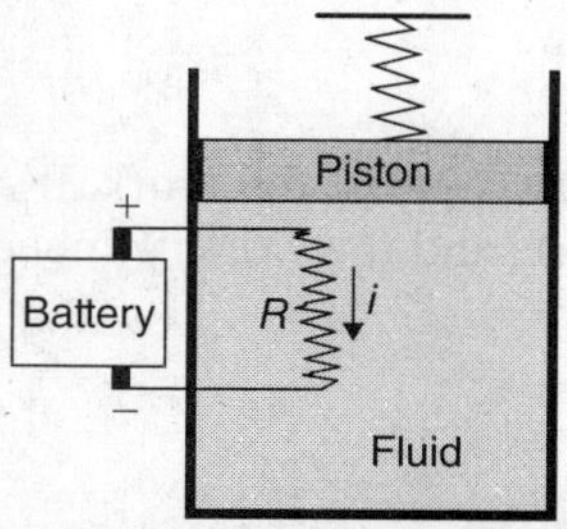

Figure 3.12
An electrical resistance heating a fluid.

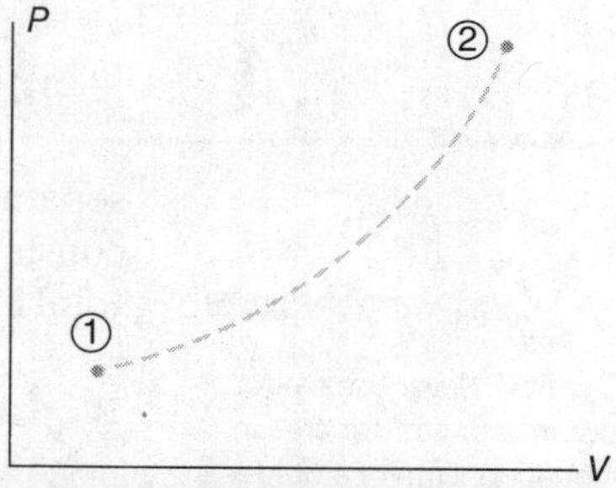

Figure 3.13
A nonequilibrium process.

A paddle wheel: One situation where a rotating shaft contributes to a thermodynamic system is when a paddle wheel is stirring a fluid, as in Fig. 3.11. The work can be found either by using Eq. 3.12 or by using the weight times the distance the weight falls. This is an example of a nonequilibrium[2] work mode. The viscous effects in the fluid dissipate this negative work into stored energy by raising the temperature of the fluid.

An electrical resistance heater: Another example of a nonequilibrium work mode is the electrical resistance heater in a volume of fluid, as shown by Fig. 3.12. This provides the same effect to the fluid as does that of the paddle wheel. It is considered negative work since it is energy that crosses the boundary of the fluid, resulting in an increase in the temperature of the fluid. The power transferred to the fluid by the resistance R is given by

$$\text{Power} = Vi = \frac{V^2}{R} \tag{3.13}$$

Note: Don't confuse voltage V with volume V. The context will make the meaning obvious.

where the usual variable for the electric potential (voltage) is V, i is the current, and R is the resistance. The power Vi is measured in watts. The work transferred by the resistor is power multiplied by the time, that is, $i^2R\Delta t$.

If a P-V diagram were sketched after a period of time elapsed of the setup of Fig. 3.12, the actual process from state 1 to state 2 would not be known, so a dashed line connecting the states could be used, as in Fig. 3.13.

The work from a spring — Example 3.4

Air at 150 kPa is contained in an 80-cm-diameter cylinder by a frictionless piston, initially 60 cm from the bottom of the cylinder. A spring with $K = 80$ kN/m is brought into contact with the piston so that no force exists in the spring, as shown in Fig. 3.14. Energy is then added to the air until the volume doubles. Calculate the work done by the air on the piston.

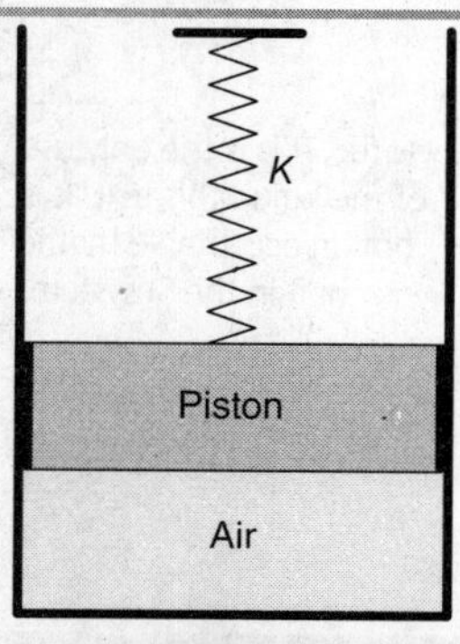

Figure 3.14

Solution

Since V_2 is double V_1, the height of air in the cylinder must double so the piston is raised 60 cm, thereby compressing the spring 60 cm. The work can be

(*Continued*)

[2]An equilibrium work mode can be reversed and the energy recovered. A paddle wheel does not provide an equilibrium work mode, since if it could be reversed, it would not result in a raising of the weight and a lowering of the temperature of the fluid.

Example 3.4 (Continued)

separated into two parts: the work W_P needed to raise the piston 60 cm at constant pressure, and the work W_S needed to compress the spring 60 cm. The total is

Units: Never use cm for distance or cm² for area in an equation. Either Pa or kPa can be used for pressure, but make sure the same units are used on each term in an equation.

$$W = W_P + W_S = PAh + \frac{1}{2}Kx^2$$

$$= 150\frac{\text{kN}}{\text{m}^2} \times (\pi \times 0.4^2)\ \text{m}^2 \times 0.6\ \text{m} + \frac{1}{2} \times 80\frac{\text{kN}}{\text{m}} \times 0.6^2\ \text{m}^2 = \underline{59.6\ \text{kJ}}$$

Rather than using Eq. 3.6 to find the work to raise the piston, we used the relationship $W = F \times \textit{distance}$. Because both the pressure and the spring constant were expressed using kPa, the work is in kJ.

Example 3.5 Work provided by a paddle wheel

Comment

On Earth, gravity varies between 9.80 m/s² (on the highest mountain) and 9.79 m/s² (in the lowest ocean trench) so assuming sea-level gravity is acceptable for all problems unless you're in a space station!

A two-hundred-kg mass attached to a paddle wheel via a pulley, as in Fig. 3.15, drops 2 m. Estimate the energy added to the water in the tank.

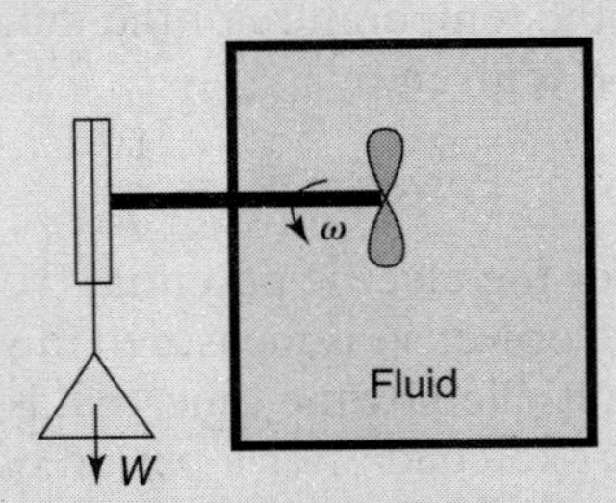

Figure 3.15

Solution

The energy added to the water by the rotation of the paddle wheel is

$$W = F \times h = 200\ \text{kg} \times 9.81\frac{\text{N}}{\text{kg}} \times 2\ \text{m} = 3920\ \text{N}\cdot\text{m}$$

Note: This is the only use of the kgf in this text. It is not an acceptable metric unit in the SI system.

Work is usually measured in J or kJ. The energy added to the water by the paddle wheel is given as

$$W = 3920\ \text{J} = \underline{3.92\ \text{kJ}}$$

This would be a negative nonequilibrium work mode relative to the fluid since it represents energy added to the fluid.

☑ **You have completed Learning Outcome (1)**

3.2 Heat Transfer

Heat transfer is another mechanism used to transfer energy to or from a system. *Heat transfer*, indicated by Q, is the transfer of energy due to a temperature difference between a system and its surroundings. The rate of heat transfer is designated $\dot{Q}$. If the heat transfer for a process is zero, it is referred to as an *adiabatic process*. An insulated container would provide for zero heat transfer.

Heat transfer is accomplished by one or more of three physical mechanisms. These mechanisms will be briefly described but will not be analyzed in detail in this text. This would require a full semester course called Heat Transfer.

Conduction heat transfer is the transfer of thermal energy by intermolecular action. If the molecules on the boundary of the surroundings are more active, and have a higher kinetic energy than the molecules that make up the system boundary, a transfer of energy takes place from the faster molecules to the slower ones. We know that temperature is related to the speed of the molecules; consequently, in order for energy to be transferred from the surroundings to a system, the temperature of the boundary of the surroundings must be higher than the temperature of the system boundary. Fourier's law of heat conduction, for the one-dimensional plane wall of Fig. 3.16, takes the form

$$\dot{Q} = kA\frac{\Delta T}{L} = A\frac{\Delta T}{R} \quad (3.14)$$

where k is thermal conductivity with units of W/m·K, A is the wall area through which the heat flows, ΔT is the temperature difference across the wall, and L is the thickness of the wall. Sometimes the R-factor in Eq. 3.14, the resitivity, is provided; it is $R = L/k$. It should be noted that the R-factors can be added for multilayered surfaces.

Convection heat transfer is the transfer of thermal energy by conduction and by the bulk flow of a fluid, called *advection*. This means that a fluid that contains thermal energy will transfer energy primarily by conduction when the fluid motion is very low, and then primarily by advection when the fluid motion is relatively high. Consider a hot plate sitting in air. Initially, the air will absorb heat from the plate by conduction from the hot plate to the cooler air. As the air absorbs thermal energy, it expands and becomes buoyant. The air then rises from the plate, carrying thermal energy away from the plate, while cooler air takes its place, and the process continues with the air speeds increasing as the plate heats up. If air is blown over the plate in some manner, it is *forced convection*; otherwise; with no forced air, it is referred to as free convection. *Newton's law of cooling* provides an equation that allows convective heat transfer to be calculated from an area A; it is

$$\dot{Q} = h_c A(T_s - T_\infty) \quad (3.15)$$

where h_c is the convective heat transfer coefficient with units of W/m^2·K, T_s is the surface temperature, and T_∞ is the bulk fluid temperature. Convection is a complex combination of heat transfer and fluid motion and can be difficult to model. It is studied in detail as part of a heat transfer course. For convection, the R-factor would be $R = 1/h_c$.

Heat transfer: The transfer of energy due to a temperature difference between a system and its surroundings.

Adiabatic process: A process for which the heat transfer is zero.

Conduction: The transfer of thermal energy by intermolecular action.

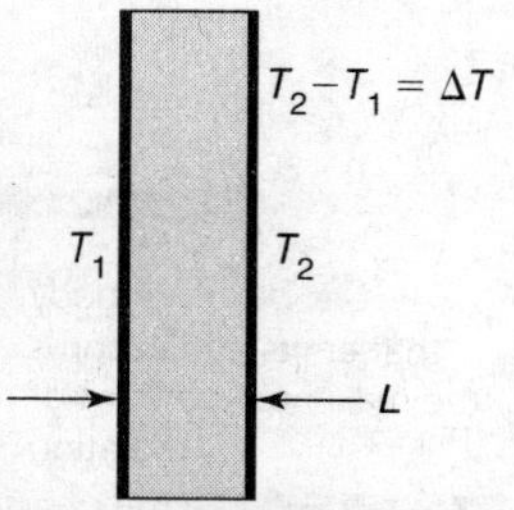

Figure 3.16
A plane wall

Note: R-factors can be added for mulilayered surfaces: $R = R_1 + R_2 + \cdots + R_n$

Convection: The transfer of thermal energy primarily due to the motion of a fluid.

Note: First, convection occurs primarily due to conduction and then primarily due to advection as the air begins to rise.

Forced convection: When air is forced over a surface.

Newton's law of cooling: The equation that allows convective heat transfer to be calculated.

Radiation: The transfer of thermal energy by electromagnetic radiation.

Stefan–Boltzmann law: The equation that allows radiation heat transfer to be calculated.

Radiation heat transfer is the transfer of thermal energy by electromagnetic radiation. Any substance with a temperature above absolute zero emits electromagnetic energy. The frequency of the radiation emitted is a function of the temperature of the substance. Radiation is unique in that it can be transferred through a perfect vacuum as well as through gases. The human body has a normal temperature of 310 K. At this temperature, the frequency of radiation is primarily in the infrared zone, which is why infrared night vision goggles are so effective in locating people. Radiation is very unique in that the energy output varies with absolute temperature to the fourth power. This type of relationship does not occur elsewhere in nature. To calculate the heat transfer from a solid surface to the surroundings, the *Stefan–Boltzmann law* provides

$$\dot{Q} = \varepsilon\sigma A\left(T_s^4 - T_{\text{surr}}^4\right) \tag{3.16}$$

where ε is the emissivity and σ is the Stefan–Boltzmann constant (5.67×10^{-8} J/s·m^2·K^4. The temperature T_s of the surface and T_{surr} of the surroundings must be absolute temperatures. The emissivity is between 0 and 1 where $\varepsilon = 1$ is for a *black body* that emits and absorbs the maximum amount of radiation. Radiation does not transfer significant amounts of heat until high temperatures are reached. Before the development of rockets and jet engines in the 1940s, radiation was mainly a subject reserved for the astronomical modeling of stars.

Black body: A body that emits and absorbs the maximum amount of radiation

In this text we will not analyze the details of the heat transfer process. A quantity of heat will simply be transferred to or from a system without specifying how it is done. We will, however, provide several examples and problems illustrating the simplified heat transfer equations listed above. They are the type of problems included on national exams such as the Fundamentals of Engineering Exam and the GRE/Engineering Exam.

The sign conventions are very important to the analyses of processes. Heat transferred to a system is positive, and heat removed from a system is negative. The unit of heat transfer in the SI system is the joule (J). In this text we will use the kilojoule (kJ) since the joule is quite small.

Comment

Heat transferred to a system is positive, and heat removed from a system is negative.

Heat transfer and work have several traits in common.

1. Both work and heat transfer are boundary phenomena. Work and heat represent energy that enters or leaves a system through the system boundary. A system does not contain work or heat.
2. Both heat transfer and work are path functions. The amount of energy transferred is not just a function of the initial and final states but of the path the process takes to get from one state to the other. Hence, the differential of heat transfer is inexact and is written δQ.
3. Both heat transfer and work are time-dependent phenomena. The processes take place over a period of time and are not instantaneous.
4. Both heat transfer and work are processes, not state functions. As processes, heat transfer and work cause a change of state. Heat transfer is written as $Q_{1\text{-}2}$, never Q_1 or Q_2. Often it is simply Q when the process is obvious.

The amount of energy transferred by heat transfer can be defined in specific terms by dividing the amount of energy transferred by the mass of the system. The symbol for specific heat transfer is q, defined by

$$q = \frac{Q}{m} \tag{3.17}$$

In the SI system it is common to assign units of kJ/kg to q and kN·m/kg to w, even though they are equivalent units.

Conduction heat transfer — Example 3.6

Estimate the rate of heat transfer through a wall that measures 2.4 m by 3.6 m composed of two identical wood layers with $R = 0.1$ m·K/W and an insulation layer with $R = 0.8$ m·K/W, as displayed in Fig. 3.17. The inside temperature is 22°C, and the outside temperature is −16°C. Neglect the convective heat transfer due to the air layers on the inside and outside. (The convective layers on a glass pane cannot be neglected.)

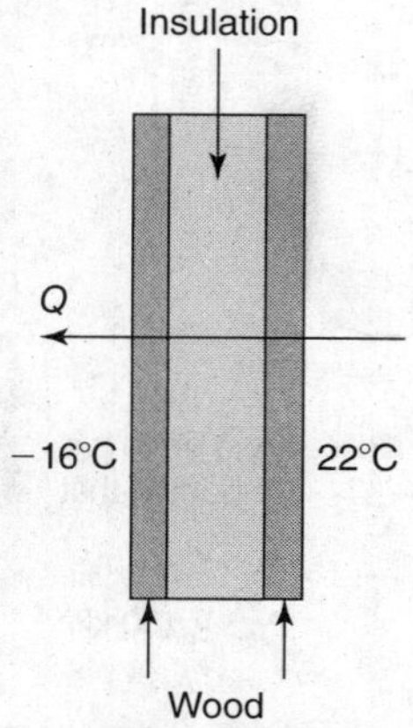

Figure 3.17

Solution

Equation 3.14 provides the heat transfer through the wall. Using R rather than k (we can add R-factors) gives

$$\dot{Q} = \frac{A\Delta T}{R_{\text{total}}} = \frac{A\Delta T}{2R_{\text{wood}} + R_{\text{ins}}}$$

$$= \frac{(2.4 \times 3.6)\ \text{m}^2 \times (22 + 16)\ ^\circ\text{C}}{(2 \times 0.1 + 0.8)\ \text{m}\cdot\text{K/W}} = \underline{328\ \text{W}}$$

Radiation heat transfer — Example 3.7

The walls of the electric oven of Fig. 3.18 are maintained at 2000°C. An artist inserts a 10-cm-diameter spherical mass of glass at 20°C into the center of the oven. Estimate the initial maximum rate of heat transfer to the glass sphere.

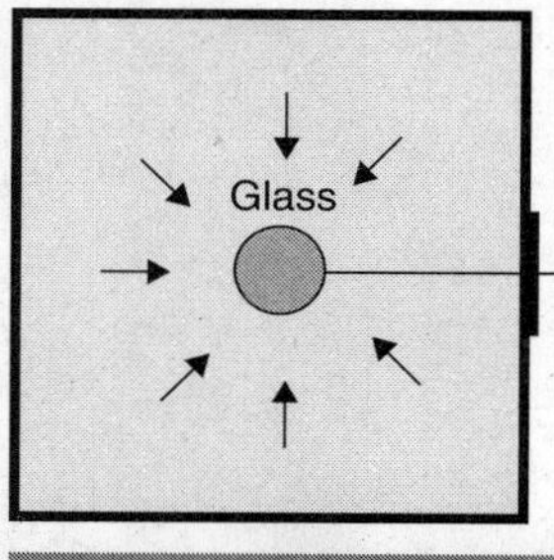

Figure 3.18

Solution

The heat transfer rate due to radiation is given by Eq. 3.16. The maximum rate of heat transfer occurs with a black body for which $\varepsilon = 1$. That rate is

$$Q = \varepsilon\sigma A\left(T_s^4 - T_{\text{surr}}^4\right)$$

$$= 1 \times 5.67 \times 10^{-8}\ \frac{\text{J/s}}{\text{m}^2\cdot\text{K}^4} \times (4\pi \times 0.05^2)\ \text{m}^2 \times (2273^4 - 293^4)\ \text{K}^4$$

$$= 47\ 500\ \text{J/s or } \underline{47.5\ \text{kJ/s}}$$

We used the surface area of a sphere as $4\pi r^2$. Temperature must be in kelvins and the area in m^2.

☑ **You have completed Learning Outcome (2)**

3.3 Problem-Solving Method

A course in thermodynamics is by nature a problem-solving course. You are being taught the skills necessary to solve practical problems in thermodynamics. In high school this type of problem was called a "word" problem. A story is told about the physical system and what happens to it. Usually, but not always, a specific answer is requested at the end of the story. It is up to the reader to extract the necessary information and develop a solution that will yield the answer. You will be taking many courses like this in your engineering career, and it would serve your career well to develop sound, systematic tools for setting up and solving problems.

In this text, emphasis will be on solving problems. The language of thermodynamics is unique, and special attention must be paid to the wording and interpretation of problems. Read the problem carefully. All problems will contain exactly enough information to solve the problem. Some of the necessary information will be given directly or can be found in the Appendix. Other information must be obtained from the description of the processes used. A systematic procedure for setting up and solving problems is highly recommended. An eight-step, problem-solving method is listed below and will be demonstrated in solving example problems. If an example is solved without adhering to each of these steps, the reader can insert the desired information in the margin next to the example's solution. The eight steps are as follows.

Note: Most often it is assumed that

$g = 9.81 \text{ m/s}^2$

$P_{atm} = 100 \text{ kPa}$

1. Read the problem carefully and identify all the information given in the problem. Treat the problem statement as though every word holds a clue to solving the problem.
2. Sketch a figure showing the system or the control volume. Engineers often think in visual terms. Try and picture a problem. Drawing a sketch of the physical problem will often give insight into understanding and solving the problem.
3. List the given information for all states. Know what type of substance you are analyzing. Convert all data to useful units.
4. Check the problem for special processes.
5. State any assumptions you make about the problem. This is important if you are making simplifying assumptions. If you simplify a problem too much, the problem you finally solve may not be the problem originally stated.
6. Determine the required states and/or properties.
7. Apply the appropriate equations. In this course it will most often include the first law of thermodynamics, the energy equation. Be sure to use the correct sign conventions on your terms, and always be aware that the units on every term in an equation must be the same.
8. Check your answer thoroughly. Does it appear to be too large or too small? Are the units on your answer the proper units? If the units are incorrect, you definitely made a mistake somewhere.

These steps are not followed literally, maybe occasionally, but they are in one's mind while solving all problems. They become part of the thought process as a problem is being solved.

3.4 The First Law Applied to Systems

The first law of thermodynamics, usually simply called the first law or the energy equation, focuses primarily on thermodynamics. We use it to analyze thermodynamic processes and thermodynamic cycles, which are made up of several sequential processes. In this chapter the first law will be applied to a fixed mass of a substance, a system. Change in the energy level of a system can be caused by heat transfer or work. The standard form of the first law applied to a system undergoing a process is

$$Q - W = \Delta E \tag{3.18}$$

The terms on the left-hand side of the equation represent the energy transferred across the boundary of a system. They are the heat Q transferred to the system and the work W done by the system. They represent the two ways in which energy is transferred into and out of a system. Energy can also be transported into and out of a volume by a flowing fluid, but that will be considered in the next chapter. In this chapter, no fluid will cross the boundary of a system.

The work and heat transfer across the boundary of a system results in a change in the energy of the system, represented by Eq. 3.18. The change in the system's energy ΔE is the sum of three terms:

$$Q - W = \Delta U + \Delta KE + \Delta PE \tag{3.19}$$

where ΔU is the change in the internal energy of the system, ΔKE is the change in the kinetic energy of the system, and ΔPE is the change in the potential energy of the system. The energy Q and W transferred across the system's boundary represents the amount of energy stored or released from the system. The systems of interest in the remainder of this chapter are stationary, so the changes in potential and kinetic energy are zero. This reduces Eq. 3.19 to

$$Q - W = \Delta U \tag{3.20}$$

If a system undergoes a cycle so that the system returns to its initial state, such as the cycle sketched in Fig. 3.19, which consists of four processes, the internal energy change is zero for the cycle. Then Eq. 3.20 takes the form

$$Q_{net} = W_{net} \tag{3.21}$$

where Q_{net} and W_{net} are the net heat transfer and the net work for the entire cycle. This form of the first law will be used quite often when considering cycles.

We can write the first law in terms of specific properties by dividing Eq. 3.20 by the mass of the system to give, for a system undergoing a process,

$$q - w = \Delta u \tag{3.22}$$

Remember: The sign convention is that heat transferred from the surroundings to the system is positive, whereas work done by the system on the surroundings is positive, which is why the negative sign exists in Eq. 3.19. In some areas of study, the same convention is used for both, resulting in a plus sign, although the negative sign used in thermodynamics is more common.

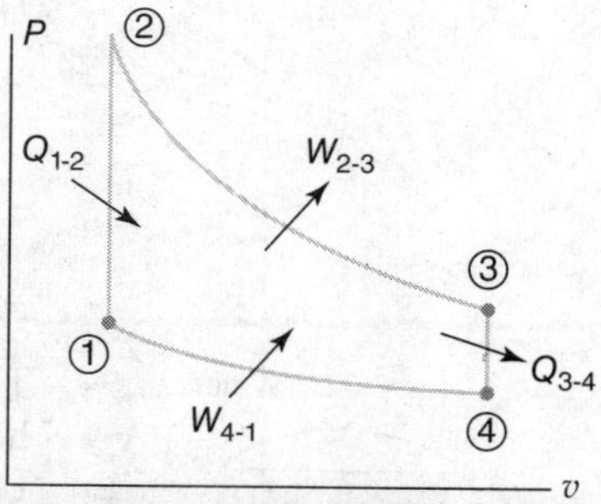

Figure 3.19
A cycle composed of four processes.

1st law for a cycle: $Q_{net} = W_{net}$

In this equation, q is the heat transfer per unit mass and w is the work done per unit mass. All terms in this equation have units of kJ/kg.

The first law is also used in differential form as

$$\delta q - \delta w = du \tag{3.23}$$

where the differentials of q and w are inexact but the differential of u is exact since u is a property; q and w are not properties.

Example 3.8 An application of the first law

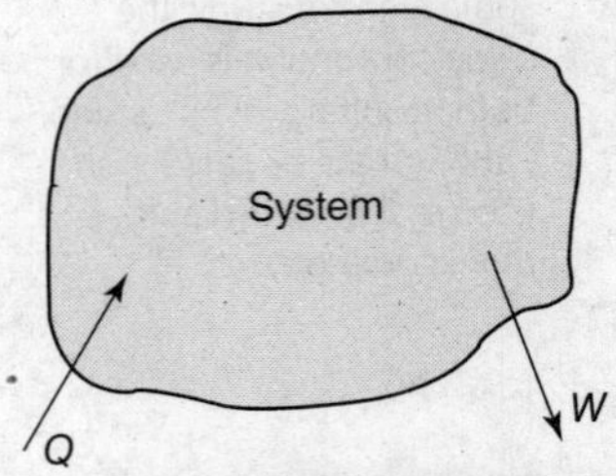

Figure 3.20

One hundred kilojoules of heat is transferred to a system while the system does 200 kJ of work on the surroundings. Calculate the change in the energy possessed by the system for this process. A system is any fixed quantity of mass, like that sketched in Fig. 3.20.

Solution

Since heat is added to the system, $Q = 100$ kJ. The work is done by the system on the surroundings so $W = 200$ kJ. When we apply Eq. 3.20, the change in the system's energy is

$$Q - W = \Delta E$$

$$100 \text{ kJ} - (+200 \text{ kJ}) = \Delta E \qquad \therefore \Delta E = \underline{-100 \text{ kJ}}$$

The system loses 100 kJ of energy.

Example 3.9 A first-law problem

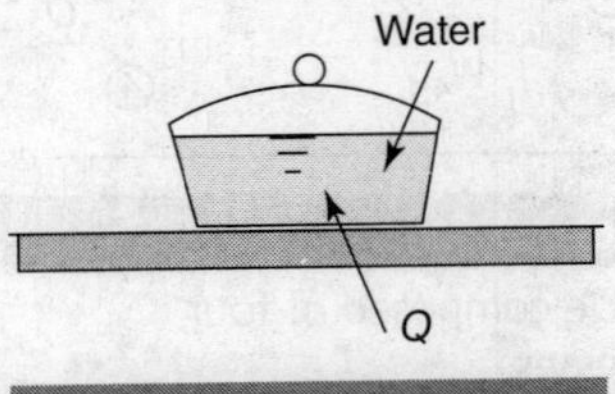

Figure 3.21

The covered pot of Fig. 3.21 contains 2 kg of liquid water at 20°C. How much heat will the stovetop have to transfer to the water to heat the water to 50°C? Use more than one method to find Q.

Solution

At 20°C and 50°C and atmospheric pressure, the water is a subcooled liquid, so we ignore the pressure and use the temperature. Table C-1 provides the following properties:

$$\text{At } T_1 = 20°\text{C: } u_1 = 83.9 \text{ kJ/kg}$$

$$\text{At } T_2 = 50°\text{C: } u_2 = 209.3 \text{ kJ/kg}$$

The top of the pot is covered, so the heat loss from the top of the water will be neglected. With $W = 0$, the first law reduces to the following:

$$Q = \Delta U = m(u_2 - u_1)$$

$$= 2 \text{ kg} \times (209.3 - 83.9) \text{ kJ/kg} = \underline{250.8 \text{ kJ}}$$

Comment

For many problems, in fact most of the problems in a first course in thermodynamics, the kinetic and potential energy changes are either zero or negligible, as in this example.

A positive Q indicates that heat was indeed transferred to the water.

Or, the IRC Calculator at 20°C provides $u_1 = 83.9$ kJ/kg, and at 50°C it provides $u_2 = 209$ kJ/kg, which gives

$$Q = \Delta U = m(u_2 - u_1)$$
$$= 2 \text{ kg} \times (209 - 83.9) \text{ kJ/kg} = \underline{250 \text{ kJ}}$$

Or, we could have used the specific heat C_p of water from Table B-4 to be 4.18 kJ/kg and found, using Eq. 2.36,

$$Q = \Delta U = mC_p(T_2 - T_1)$$
$$= 2 \text{ kg} \times 4.18 \frac{\text{kJ}}{\text{kg} \cdot {}^\circ\text{C}} \times (50 - 20)\,^\circ\text{C} = \underline{251 \text{ kJ}}$$

All three methods resulted in the same answer. The above rather simple estimate $\Delta U = mC_p\Delta T$ is applicable to either a solid or a liquid and is often the method of choice.

Remember: $C_p \cong C_v$ for a solid or a liquid.

☑ You have completed Learning Outcome (3)

3.5 The First Law Applied to Various Processes

We will now apply the first law to several common thermodynamic processes involving systems. In order to evaluate the work done during a process, the type of process must be known. We will encounter five primary processes in our study. In all the processes presented in this chapter, it is assumed that the mass in the volume of interest remains constant; that is, it is a system. No mass will be added to or removed from the volume, but, in general, the volume can change its shape and size, such as a piston moving in a cylinder. The first is the constant-volume process.

3.5.1 The constant-volume process

The constant-volume process is a process where by the volume of the system remains constant and the system boundary does not move. A balloon filled with water would not qualify even though the volume would not change. If the boundary is fixed, there is no $P\Delta V$ work done, so Eq. 3.20 would become

Comment

The boundary is assumed fixed in a constant-volume process.

$$Q = \Delta U \tag{3.24}$$

If a table of properties is available for a substance, the specific internal energy can be found. For a gas, we can integrate Eq. 2.24 to provide ΔU; the heat transfer is then

$$Q = m\int_{T_1}^{T_2} C_v dT \tag{3.25}$$

Note: There could be nonequilibrium work due to a paddle wheel or an electrical resistance heater, which must be added to the right-hand side.

Comment

Specific heats will be assumed constant unless otherwise stated. For high temperatures, this may lead to significant error.

If C_v can be assumed to be constant, then

$$Q = mC_v(T_2 - T_1) \tag{3.26}$$

If nonequilibrium work is present in a problem, such as a paddle wheel, it must be included in the above equations.

Isometric process: A constant-volume process.

A constant-volume process is also called an *isometric process*. This type of process is often identified by the description of the problem. Terms such as *rigid container* or *fixed volume* indicate that the process is a constant-volume process. If the volume of a system stays constant, the specific volume will also remain constant.

Problems must be read very carefully. All the information needed to solve a problem will be given in the problem statement. It remains for the reader to be able to extract the necessary information.

Example 3.10 A constant-volume process

The rigid container of Fig. 3.22 holds 7 kg of nitrogen at −30°C and a pressure of 430 kPa. Determine the final pressure and the heat transfer required to heat the nitrogen to 90°C.

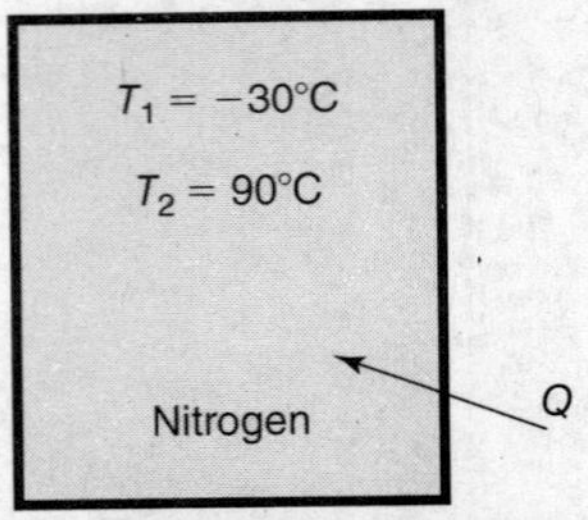

Figure 3.22

Solution

Nitrogen is assumed to be an ideal gas, so we can use the ideal-gas law with $R = 297$ J/kg·K to solve for properties. Given for state 1: $m = 7$ kg, $T_1 = -30°C = 243$ K, $P_1 = 430$ kPa. From the ideal-gas law

$$v_1 = \frac{RT_1}{P_1} = \frac{297 \text{ J/kg·K} \times 243 \text{ K}}{430 \times 1000 \text{ Pa}} = 0.168 \text{ m}^3/\text{kg}$$

Since this device is a rigid container, the volume will stay constant, so $v_2 = 0.168$ m³/kg. With $T_2 = 90°C = 363$ K, the final pressure is

$$P_2 = \frac{RT_2}{v_2} = \frac{297 \times 363}{0.168} = 642000 \text{ Pa} = \underline{642 \text{ kPa}}$$

The heat transfer is found using Eq. 3.26 to be

$$Q = mC_v(T_2 - T_1)$$
$$= 7 \text{ kg} \times 0.742 \text{ kJ/kg·K} \times (363 - 243)\text{K} = \underline{623 \text{ kJ}}$$

3.5.2 The constant-pressure process

The constant-pressure process is one in which the pressure of the system remains constant for the duration of the process. If we assume a constant-pressure process, Eq. 3.6 expresses the quasi-equilibrium work as

$$W = P(V_2 - V_1) \tag{3.27}$$

A constant-pressure process is called an *isobaric process*. A device that characterizes the constant-pressure process is the frictionless piston moving inside a cylinder. Figure 3.23 shows a simple vertical piston-cylinder device. The piston is allowed to slide vertically up and down inside the piston without friction.

The pressure beneath the piston is maintained by the weight of the piston acting downward. If the substance under the piston is heated, the substance will expand. The piston will move as necessary to maintain a constant pressure in the cylinder. No substance is allowed to escape around the edge of the piston.

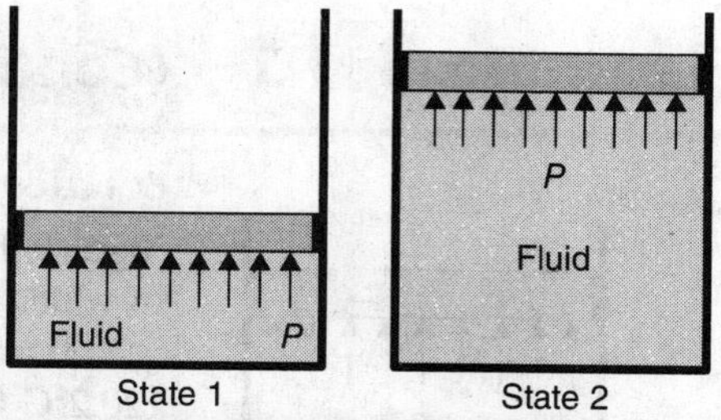

Figure 3.23
A constant-pressure process

The first law for this quasi-equilibrium constant-pressure process, neglecting kinetic and potential energy changes, using Eq. 3.27 for the work, takes the form

$$Q - P(V_2 - V_1) = U_2 - U_1 \tag{3.28}$$

which can be written as

$$Q = (PV + U)_2 - (PV + U)_1 \tag{3.29}$$

This is expressed in terms of enthalpies as

$$Q = H_2 - H_1 \tag{3.30}$$

or

$$q = h_2 - h_1 \tag{3.31}$$

Comment
Enthalpy is very handy when considering a constant-pressure process.

For an ideal gas, we can use

$$q = \Delta h = \int_{T_1}^{T_2} C_p dT \tag{3.32}$$

$$= C_p(T_2 - T_1) \qquad (\text{if } C_p = \text{const}) \tag{3.33}$$

For large temperatures, the ideal-gas tables in the Appendix F or the average C_p value from Table B-6 could be used, especially if accuracy is important rather than simple comparisons.

If any nonequilibrium work mode is present, such as an electricresistance heater, that work must be included as an additional term in the above energy equations.

Remember: Any of the five methods presented in Section 2.6 can be used to find Δh. If the temperatures are relatively large, say above 500°C, it's best to use the ideal-gas tables, the average C_p from Table B-6 (or the IRC Calculator for air only).

A constant-pressure process — Example 3.11

A vertical piston-cylinder device contains 20 kg of saturated liquid water. The weight of the piston is such that a constant pressure of 200 kPa is maintained in the cylinder. Heat is added to the water by a resistance heater until the quality is 0.3, as shown at state 2 in Fig. 3.24. A 12-V battery supplies a current of 2 A for 200 minutes. Determine the initial and final volumes of the water in the cylinder, the work done, and the heat transfer required for this process.

(Continued)

Example 3.11 (Continued)

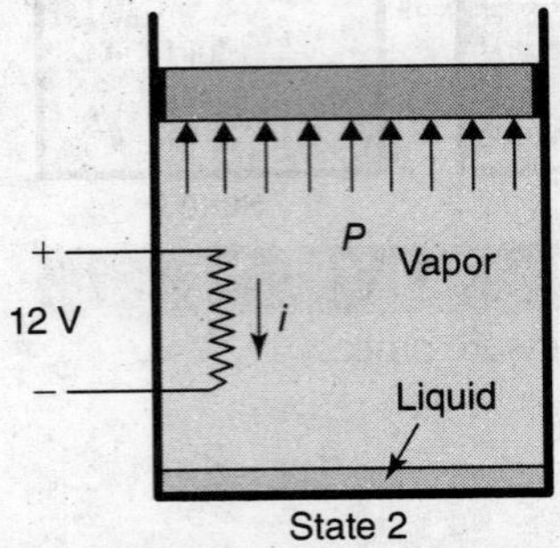

Figure 3.24

Solution

State 1: The known properties at state 1 are $m = 20$ kg and $P_1 = 200$ kPa, both of which will remain constant. Since the water is a saturated liquid at 200 kPa, $x_1 = 0$ and the temperature is the saturation temperature at 200 kPa, which is 120.2°C from Table C-2. The specific volume at state 1 is v_f at 200 kPa, which from Table C-2 is $v_f = 0.001061$ m³/kg. Knowing the mass of the system, we can solve for the initial volume:

$$V_1 = mv_1 = 20 \text{ kg} \times 0.001061 \text{ m}^3/\text{kg} = \underline{0.02122 \text{ m}^3}$$

State 2: The mass and the pressure remain constant from state 1 to state 2. Since state 2 is a saturated mixture ($x_2 = 0.3$), the temperature also remains constant at the saturation temperature of 120.2°C. To find v_2, the quality is used as follows:

$$\begin{aligned} v_2 &= v_f + x_2(v_g - v_f) \\ &= 0.001061 + 0.3(0.8857 - 0.001061) = 0.2665 \text{ m}^3/\text{kg} \end{aligned}$$

The volume at state 2 is:

$$V_2 = mv_2 = 20 \times 0.2665 = \underline{5.329 \text{ m}^3}$$

Now we can calculate the work done during this process. It is

$$\begin{aligned} W_{1\text{-}2} &= P(V_2 - V_1) - (Vi)_{\text{elect}}\Delta t \\ &= 200\,000 \text{ Pa} \times (5.329 - 0.2122) \text{ m}^3 - (12 \times 2)\frac{\text{J}}{\text{s}} \times (200 \times 60) \text{ s} \\ &= 735\,400 \text{ J} \quad \text{or} \quad \underline{735 \text{ kJ}} \end{aligned}$$

Note: The units on $(Vi)_{\text{elect}}$ are watts, so it is necessary to multiply by the time to obtain joules.

The negative electrical nonequilibrium work was included as $Vi\Delta t$ which provides joules (see Eq. 3.13). There is a positive net work done by the system on the surroundings. The boiling mixture is pushing the piston upward.

The necessary heat transfer is found from the first law, Eq. 3.29, to be

$$Q - W_{\text{elect}} = m(h_2 - h_1)$$

Using $h_2 = h_f + x\,h_{fg} = 504.7 + 0.3 \times 2201.9 = 1165$ kJ/kg and $h_1 = 504.7$ kJ/kg, we find the necessary heat transfer to be as follows:

$$\begin{aligned} Q &= m(h_2 - h_1) + (-Vi)_{\text{elect}}\Delta t \\ &= 20 \text{ kg} \times (1165 - 504.7) \text{ kJ/kg} \times 1000 \text{ J/kJ} - 12 \times 2 \times (200 \times 60) \text{ J} \\ &= 12.92 \times 10^6 \text{ J} \quad \text{or} \quad \underline{12.92 \text{ MJ}} \end{aligned}$$

Note: Because of the electrical resistance heater, the required heat transfer is less than would otherwise be required.

The factor "1000" converts kJ to J since the enthalpies are in kJ/kg.

3.5.3 The constant-temperature process

For the *isothermal process*, during which the temperature remains constant, various equations and tables can be used to find the properties required when solving a problem. Properties are fairly insensitive to pressure when temperature is held constant, but there are situations for which the variation of properties with pressure must be taken into account. The first law, neglecting kinetic and potential energy changes, for an isothermal process does not simplify and is used for a system in the form

Isothermal process: A constant-temperature process.

$$Q - W = m(u_2 - u_1) \tag{3.34}$$

Note: For a substance such as steam, tables are used to find Δu.

If the substance can be approximated as an ideal gas, the first law can be simplified since internal energy depends only on temperature so that $u_2 - u_1 = 0$. The first law is then

$$Q = W \qquad \text{(for an ideal gas)} \tag{3.35}$$

Using the ideal-gas law $PV = mRT$, the work can be expressed, recognizing that mRT is constant, as

$$W = \int_{V_1}^{V_2} P\,dV = mRT\int_{V_1}^{V_2} \frac{dV}{V} = mRT\ln\frac{V_2}{V_1} \tag{3.36}$$

Note: For an ideal gas only.

Note it can also be written in terms of the pressure ratio since for an isothermal process $P_1V_1 = P_2V_2$, showing that $V_2/V_1 = P_1/P_2$. The work can then be expressed as

$$W = mRT\ln\frac{P_1}{P_2} \tag{3.37}$$

Note: For an ideal gas only.

This accounts for only the quasi-equilibrium work. Any nonequilibrium work must be included appropriately.

A constant-temperature process — Example 3.12

Propane at 100°C and 120 kPa with a volume of 2 m³ is compressed using an isothermal process until the pressure is 600 kPa, as shown in Fig. 3.25. Determine the work done during this process and the heat required.

Solution

Propane is an ideal gas with a gas constant $R = 0.1886$ kJ/kg·K.

Given: **State 1:** $T_1 = 100°\text{C} = 373\ \text{K},\ P_1 = 120\ \text{kPa},\ V_1 = 2\ \text{m}^3$

State 2: $T_2 = T_1 = 373$ K (isothermal process), $P_2 = 600$ kPa

The mass of the system is constant and can be determined from state 1:

$$m = \frac{PV}{RT} = \frac{120 \times 2}{0.1886 \times 373} = 3.412\ \text{kg}$$

(Continued)

Example 3.12 (Continued)

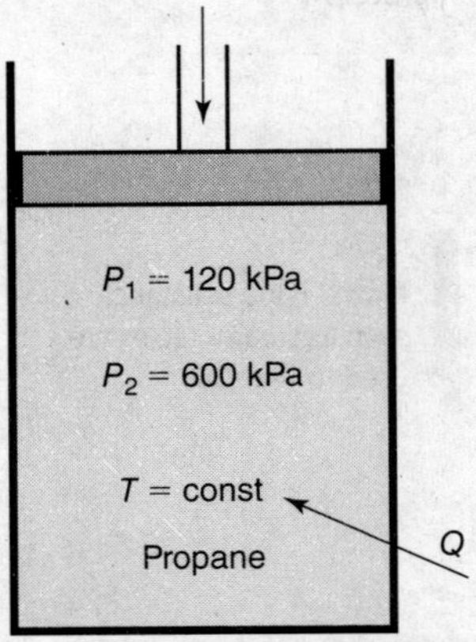

Figure 3.25

Since we know the change in pressure rather than the change in volume, we will use Eq. 3.37 to find the work:

$$W = mRT \ln \frac{P_1}{P_2}$$

$$= 3.412 \text{ kg} \times 0.1886 \frac{\text{kJ}}{\text{kg·K}} \times 373 \text{ K} \times \ln \frac{120}{600} = \underline{-386.3 \text{ kJ}}$$

In this example the work is negative. This indicates that work was done by the surroundings on the propane.

The heat transfer is found by applying the first law and is

$$Q = W = \underline{-386.3 \text{ kJ}}$$

since no nonequilibrium work is present and $\Delta u = 0$. The minus sign means that heat leaves the system.

3.5.4 The adiabatic process

Adiabatic process: No heat transfer is allowed.

The *adiabatic process*, a process with no heat transfer, is applicable in a number of situations. Many texts do not present this process until after the second law of thermodynamics is discussed, but that delay is unnecessary and prevents a very important process from logically being introduced at this point in thermodynamics. The compression and expansion of the gases in the cylinder of an engine are perhaps the most common applications of this process. They require an insulated volume, or a process that takes place so quickly that there is no time for a significant amount of heat transfer, as when an engine is operating at, say, 3000 rpm. They also require a quasi-equilibrium process, of which the idealized processes of compression and expansion in an engine are examples. Molecular speeds are very fast and outpace the 3000 rpm speed of an engine.

Comment

The adiabatic, quasi-equilibrium process is applicable in compression and expansion processes in engines. There is heat transfer from the engine cylinders, but it is negligibly small for each cycle.

The differential form of the first law (see Eq. 3.23) for an adiabatic process, during which $\delta q = 0$, takes the form

$$-\delta w = du \tag{3.38}$$

For a quasi-equilibrium process during which $\delta w = Pdv$ (no nonequilibrium work modes, such as a paddle wheel or an electrical resistance, can be present), the differential form of the first law becomes

$$du + Pdv = 0 \tag{3.39}$$

We now restrict our system to be an ideal gas. The ideal-gas law $P = RT/v$ and the relationship $du = C_v dT$ (see Eq. 2.24) allow Eq. 3.39 to be written as

$$C_v dT + \frac{RT}{v} dv = 0 \tag{3.40}$$

or, rewritten as

$$\frac{C_v}{R}\frac{dT}{T} = -\frac{dv}{v} \tag{3.41}$$

For many processes, C_v can be assumed constant, and even in the expansion process in the cylinder of an engine, it is assumed constant for comparison purposes or for a quick approximation. So, assuming a constant C_v, Eq. 3.41 can be integrated between states 1 and 2:

$$\frac{C_v}{R}\int_{T_1}^{T_2}\frac{dT}{T} = -\int_{v_1}^{v_2}\frac{dv}{v} \tag{3.42}$$

$$\frac{C_v}{R}\ln\frac{T_2}{T_1} = -\ln\frac{v_2}{v_1} \tag{3.43}$$

Comment

C_v is often assumed constant, especially for comparison purposes, even if it introduces significant error. If more accuracy is desired, the ideal-gas tables, Table B-5 or Table B-6 can be utilized; that will be considered in detail in Chapter 6. Here, only the constant specific-heat process will be considered.

Note: Values for R, C_v, and k are found in Table B-2 in the Appendix.

With a little algebra, this equation takes the form

$$\frac{T_2}{T_1} = \left(\frac{v_1}{v_2}\right)^{k-1} \quad \text{or} \quad Tv^{k-1} = \text{const} \tag{3.44}$$

Note: For an ideal gas with constant specific heats.

where Eq. 2.35 was used in the form $R/C_v = k - 1$.

Using the ideal-gas law, Eq. 3.44 can take on the following forms:

$$\frac{T_2}{T_1} = \left(\frac{P_2}{P_1}\right)^{(k-1)/k} \quad \text{or} \quad T/P^{(k-1)/k} = \text{const} \tag{3.45}$$

$$\frac{P_2}{P_1} = \left(\frac{v_1}{v_2}\right)^{k} \quad \text{or} \quad Pv^{k} = \text{const} \tag{3.46}$$

Equations 3.44, 3.45, and 3.46 are for an ideal gas undergoing an adiabatic ($q = 0$), quasi-equilibrium process. No nonequilibrium work modes are allowed; they cannot simply be added.

Most often an engine is designed with a predetermined compression ratio r, represented in Fig. 3.26, which is defined to be the maximum volume divided by the minimum volume of the gas in the cylinder:

$$r = \frac{v_1}{v_2} \tag{3.47}$$

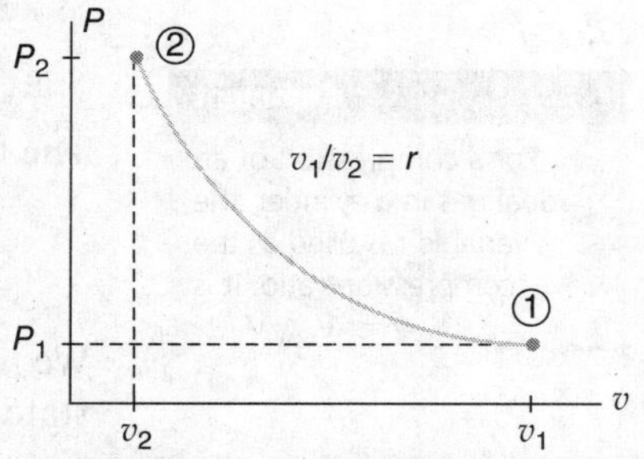

Figure 3.26

The pressure-volume diagram for a quasi-equilibrium adiabatic compression process.

If the working fluid is not an ideal gas, we return to Eq. 3.39 and observe that the sum of the two differential properties represents a third differential property, which we shall call $d\psi$. This is analogous to our definition of enthalpy in Eq. 2.11. This allows us to write

$$d\psi = du + Pdv \tag{3.48}$$

Since $d\psi = 0$, $\psi_2 = \psi_1$. We do not give the property ψ (psi) a formal name, but we do recognize that it is constant whenever the property,

Note: ψ = const for a quasi-equilibrium adiabatic process.

denoted by s, the entropy, is constant. Hence, it is not necessary to create a new function and list it in the tables since s is listed in the various tables, as we shall discover in Chapter 6 (it is not necessary or helpful to define entropy at this point in our study). Consequently, an adiabatic, quasi-equilibrium process requires $\psi_2 = \psi_1$ or $s_2 = s_1$.

Comment: Entropy will be defined in Chapter 6.

Example 3.13 An adiabatic ($Q = 0$) process for air

Air at 20°C and 100 kPa is compressed in an insulated cylinder from a volume of 400 cm^3, as shown in Fig. 3.27. For a compression ratio of 8 to 1, find the required work. Assume constant specific heats and a quasi-equilibrium process.

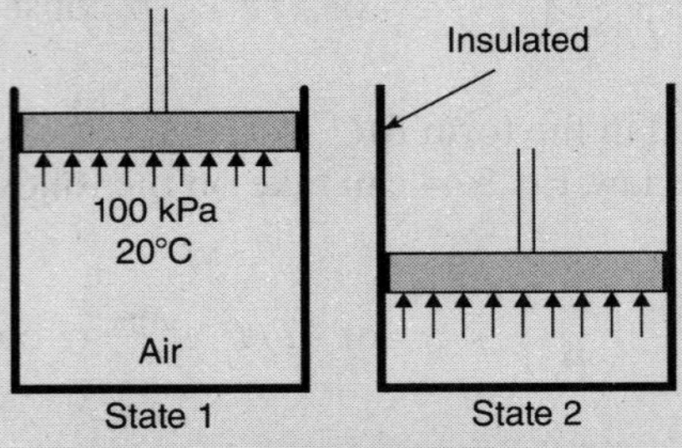

Figure 3.27

Solution

Air is an ideal gas with a gas constant $R = 0.287$ kJ/kg·K.

Given:

State 1: $T_1 = 20°\text{C} = 293\,\text{K},\ \ P_1 = 100\,\text{kPa},\ \ V_1 = 400\,\text{cm}^3$

State 2: $V_2 = V_1/8$ (a compression ratio is a volume ratio)

The first law states that

$$-W = m(u_2 - u_1) = mC_v(T_2 - T_1)$$

Comment

For a compression of an ideal gas in a cylinder, the variable r is used as the compression ratio. It is $r = V_{max}/V_{min}$.

We must determine m and T_2. The mass of the system can be determined from state 1:

$$m = \frac{P_1 V_1}{RT_1} = \frac{100 \times (400 \times 10^{-6})}{0.287 \times 293} = 4.757 \times 10^{-4}\ \text{kg}$$

The final temperature is found using Eq. 3.44 to be

$$T_2 = T_1\left(\frac{v_1}{v_2}\right)^{k-1} = 293 \times 8^{1.4-1} = 673\text{ K}$$

Absolute temperature must be used, and $k = 1.4$ was found in Table B-2 in the Appendix. The work is now calculated to be

$$W = mC_v(T_2 - T_1)$$
$$= 4.757 \times 10^{-4}\text{ kg} \times 717\frac{\text{J}}{\text{kg·K}}(673 - 293)\text{ K} = \underline{129.6\text{ J}}$$

The specific heat C_v was found in Table B-2 to be 717 J/kg·K. If 0.717 kJ/kg·K was used for C_v, the answer would be 0.1296 kJ.

An adiabatic process for steam Example 3.14

Steam at 400°C and 4 MPa expands in an insulated cylinder to 600 kPa, as sketched in Fig. 3.28. Estimate the work output if the cylinder's initial volume is 8000 cm³. Assume a quasi-equilibrium process, as was done in the preceding four examples.

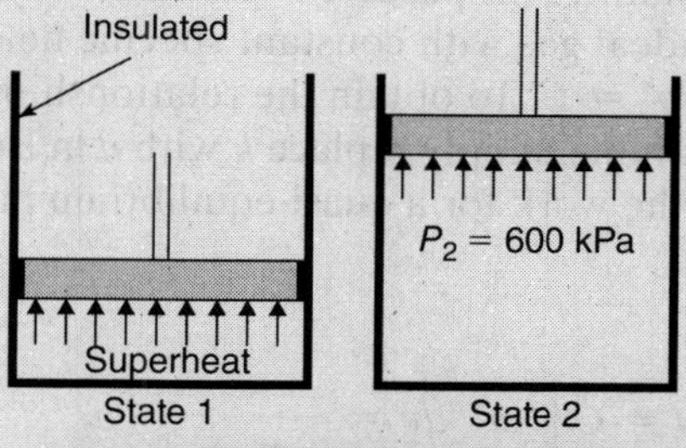

Figure 3.28

Solution

State 1 is in the superheat region, so Table C-3 is used.

Given: **State 1:** $T_1 = 400°\text{C}$, $P_1 = 4\text{ MPa}$, $V_1 = 8000\text{ cm}^3$

State 2: $P_2 = 600\text{ kPa}$

The process is adiabatic since the cylinder is insulated. The work output for the quasi-equilibrium process, with $Q = 0$, is given by

$$W = m(u_1 - u_2)$$

(Continued)

Example 3.14 (Continued)

To find the work, we must know m, u_1, and u_2. The mass is found by using $v_1 = 0.07341\ m^3/kg$ from Table C-3:

$$m = \frac{V_1}{v_1} = \frac{8000 \times 10^{-6}\ m^3}{0.07341\ m^3/kg} = 0.109\ kg$$

Comment

If an entry is close, an interpolation is not needed. An answer to three significant digits is acceptable.

Also, from Table C-3, we find $u_1 = 2919.9$ kJ/kg. To find u_2 we must use $\psi_2 = \psi_1$ (see the discussion below Eq. 3.48) or $s_2 = s_1 = 6.769$ kJ/kg·K (we aren't concerned about the units since we are equating the numerical values). At $P_2 = 600$ kPa $= 0.6$ MPa, we search for the state where $s_2 = 6.769$. It is very close to the first entry in Table C-3 where $P_2 = 0.6$ MPa (600 kPa), so we use $u_2 = 2567.4$ kJ/kg. The work is estimated to be

$$W = m(u_1 - u_2)$$
$$= 0.109\ kg(2919.9 - 2567.4)\ kJ/kg = \underline{38.4\ kJ}$$

3.5.5 The polytropic process

Polytropic process: A process in which the pressure and volume vary according to $Pv^n = C$, where C is a constant.

In a *polytropic process*, the pressure and volume vary according to the relation

$$Pv^n = C \tag{3.49}$$

where n and C are constants. This process is motivated by the adiabatic process equation (3.46) for an ideal gas with constant specific heats, which can be written as $P_1v_1^k = P_2v_2^k$ or $Pv^k = C$. To obtain the relationships among temperature, pressure, and specific volume, simply replace k with n in Eqs. 3.44, 3.45, and 3.46.

An expression for the work for a quasi-equilibrium process is found by performing the integration:

$$w = \int_{v_1}^{v_2} P dv = C \int_{v_1}^{v_2} v^{-n} dv = \frac{C}{-n+1}\left(v_2^{-n+1} - v_1^{-n+1}\right) \tag{3.50}$$

Note: For an ideal gas with constant specific heats.

But $C = Pv^n = P_1v_1^n = P_2v_2^n$. Inserting these into Eq. 3.49, we get:

$$w = \frac{P_2v_2 - P_1v_1}{1 - n} \quad \text{or} \quad W = \frac{P_2V_2 - P_1V_1}{1 - n} \tag{3.51}$$

For an ideal gas, the ideal-gas law allows this to be expressed as

$$w = \frac{R(T_2 - T_1)}{1 - n} \tag{3.52}$$

The work integral for a polytropic process can be evaluated if you know either the pressure and volume changes for the process or the temperature change for an ideal gas. But we're reminded that these equations for the work are for quasi-equilibrium processes only.

Assume that an ideal gas is the working fluid. Then, we observe for a quasi-equilibrium polytropic process:

$n = 0$	constant-pressure process
$n = \infty$	constant-volume process
$n = 1$	isothermal process
$n = k$	adiabatic process

If n is not one of these numbers, it is simply a polytropic process. Also, observe that the expressions for the work are not valid if $n = 1$, that is, for an isothermal process. For an isothermal process, the work is found using Eq. 3.36 or 3.37.

A polytropic process with $n = 1.5$ Example 3.15

Air at a pressure of 200 kPa and a volume of 2 m³ is allowed to expand in a polytropic process until the volume of the gas is 5 m³, as shown in Fig. 3.29. Determine the work done and the heat transfer during this process if $Pv^{1.5} = C$.

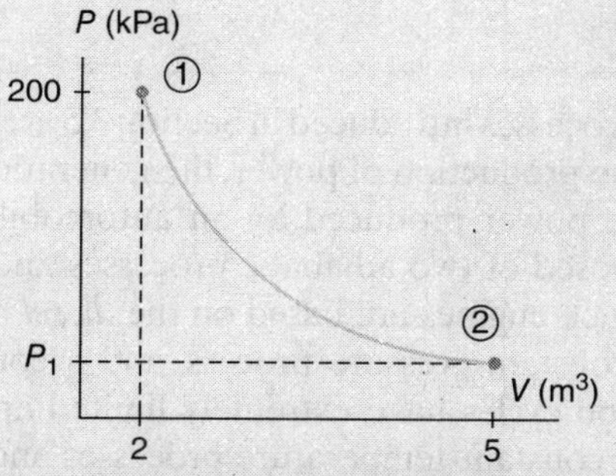

Figure 3.29

Note: If $Pv^{1.5} = C$, then $PV^{1.5} = C$ since m is constant for a process and $V = mv$.

Solution

Given:

State 1: $P_1 = 200$ kPa, $V_1 = 2$ m³, $C = P_1V_1^{1.5} = 200 \times 2^{1.5} = 565.7$

State 2: $V_2 = 5$ m³

The pressure at state 2, using the polytropic equation, is

$$P_2 = \frac{565.7}{V_2^{1.5}} = \frac{565.7}{5^{1.5}} = 50.6 \text{ kPa}$$

We can calculate the work using Eq. 3.51 to be

$$W = \frac{P_2V_2 - P_1V_1}{1 - n} = \frac{50.6 \times 5 - 200 \times 2}{1 - 1.5} = \underline{294 \text{ kJ}}$$

Positive work means that work was done by the gas on the surroundings.

(Continued)

Example 3.15 (Continued)

The heat transfer is found from the first law. It is

$$Q = \Delta U + W$$

$$= mC_v(T_2 - T_1) + W = \frac{C_v}{R}(P_2V_2 - P_1V_1) + W$$

$$= \frac{0.717}{0.287}(50.6 \times 5 - 200 \times 2) + 294$$

$$= \underline{-73.2 \text{ kJ}}$$

Note: For an ideal gas with constant specific heats.

where the ideal-gas law allowed us to equate $mT = PV/R$. If we had used $n = 1.4$, the heat transfer would be zero.

☑ You have completed Learning Outcome (4)

3.6 Cycles

The quasi-equilibrium processes introduced in Section 3.5 are used to form a number of important cycles in the production of power, the generation of refrigeration, and in gas compressors. The power produced by an automobile engine uses primarily the *Otto cycle*, composed of two adiabatic processes and two constant-volume processes. Most large truck engines are based on the *diesel cycle*, composed of two adiabatic processes, a constant-pressure process, and a constant-volume process. The Stirling and Ericsson cycles have extremely limited applications; the *Stirling cycle* operates with two constant-temperature processes and two constant-volume processes, and the *Ericsson cycle* operates with two constant-temperature processes and two constant-pressure processes. There is, however, a fictitious cycle that is used to establish the most efficient cycle possible; it is the *Carnot cycle*, composed of four reversible processes: two adiabatic processes and two constant-temperature processes. No cycle can have an efficiency greater than that of the Carnot cycle, so it provides an upper limit that cannot be exceeded. Each of these cycles will be studied in detail in Chapter 9. There are several unidentified cycles in the problems.

Comment

The Carnot cycle is fictitious because it has two heat transfer processes at constant temperature. Quasi-equilibrium heat transfer must occur across an infinitesimal temperature difference, a process that is not reproducible in the laboratory. A large finite temperature difference is necessary for heat transfer to occur in a reasonable time period.

3.7 Summary

The first law is simply the application of conservation of energy to a system. Many of the problems of interest in engineering involve the application of one of the three conservation laws: the conservation of mass, the conservation of energy, and Newton's second law. The conservation of energy, the first law, was applied to a system in this chapter. No inflow or outflow was considered; that will be the focus in Chapter 4. To apply the first law, it was necessary to define work and heat transfer. This was done and applied to a variety of processes, including isometric ($\Delta V = 0$), isobaric ($\Delta P = 0$), isothermal ($\Delta T = 0$), adiabatic ($Q = 0$), and polytropic processes. While applying the first law to the various processes, the following terms were defined:

Adiabatic process: *A process for which the heat transfer is zero.*
Conduction: *The transfer of thermal energy by intermolecular action.*
Convection: *The transfer of thermal energy by the motion of a fluid.*
Heat transfer: *The transfer of energy due to a temperature difference.*
Isobaric process: *A constant-pressure process.*
Isometric process: *A constant-volume process.*
Isothermal process: *A constant-temperature process.*
Polytropic process: *A process in which $Pv^n = C$.*
Power: *The rate at which work is transmitted.*
Radiation: *The transfer of thermal energy by electromagnetic radiation.*
Work: *The product of a force and the distance moved in the direction of the force.*

Several important equations were presented:

Work:	$W_{1\text{-}2} = \int_{V_1}^{V_2} P dV$
Work, due to a spring:	$W_{1\text{-}2} = \frac{1}{2}K(x_2^2 - x_1^2)$
Power transmitted by a shaft:	$\dot{W} = \omega T$
Power due to a resistance:	$\dot{W} = Vi = i^2R$
Work due to a resistance:	$W = i^2R\Delta t$
Conduction:	$\dot{Q} = kA\frac{\Delta T}{L}$
Convection:	$\dot{Q} = h_c A(T_s - T_\infty)$
Radiation:	$\dot{Q} = \varepsilon\sigma A(T_s^4 - T_{surr}^4)$
Constant-volume process:	$Q = mC_v(T_2 - T_1),\ W = 0$
First law for a cycle:	$Q_{net} = W_{net}$
Constant-pressure process:	$Q = m(h_2 - h_1),\ W = P(V_2 - V_1)$
Constant-temperature process:	$Q = W = mRT\ln\frac{V_2}{V_1}$
Adiabatic process:	$w = C_v(T_1 - T_2), \quad \frac{T_2}{T_1} = \left(\frac{v_1}{v_2}\right)^{k-1} = \left(\frac{P_2}{P_1}\right)^{(k-1)/k}$
Polytropic process:	$w = \frac{P_2v_2 - P_1v_1}{1-n}, \quad Pv^n = \text{Const}$

Problems

FE Exam Practice Questions

3.1 A person carries a 50-N weight moving 30 m vertically up a hill and down the other side dropping a distance of 30 m. The work done is nearest:

(A) 1500 N·m

(B) 3000 N·m

(C) 0 N·m

(D) −1500 N·m

3.2 Select the only correct statement about a quasi-equilibrium work mode:

(A) It can always be written as PdV.

(B) The differential work is written as dw.

(C) The work at state 1 can be identified as W_1.

(D) It can be reversed with no losses.

3.3 Which work mode is not a quasi-equilibrium work mode?

(A) A rotating drive shaft in an automobile

(B) A rotating stirrer in a container of paint

(C) A parabolic spring being stretched

(D) A piston compressing air in a cylinder of a car

3.4 Which is not an acceptable SI unit of work?

(A) N·m

(B) N·m/s

(C) J

(D) MJ

3.5 Air is expanding in a piston/cylinder arrangement such that $PV^2 = 2$, with P in kPa and V in m^3. The work when the volume changes from 0.04 m^3 to 0.2 m^3 is nearest:

(A) 20 kJ

(B) 25 kJ

(C) 40 kJ

(D) 55 kJ

3.6 Two kilograms of saturated liquid water are heated at constant pressure until the temperature and pressure are 400°C and 2 MPa. The work is nearest:

(A) 700 kJ

(B) 650 kJ

(C) 600 kJ

(D) 550 kJ

3.7 The paddle wheel in Fig. 3.30 does work on a contained fluid with a 40 N·m torque rotating a shaft at 400 rpm for 10 minutes. The work transferred by the paddle wheel to the fluid is nearest:

(A) 800 kJ

(B) 1000 kJ

(C) 1670 kJ

(D) 16 700 kJ

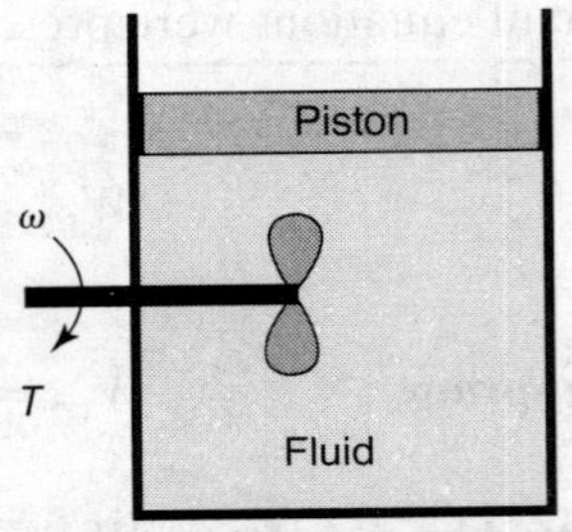

Figure 3.30

3.8 The electric heater in Fig. 3.31 operates on 12 V and requires 10 A. After 10 minutes, the work transferred to the fluid is about:

(A) 72 kJ

(B) 1200 kJ

(C) 1200 kJ

(D) 72 000 kJ

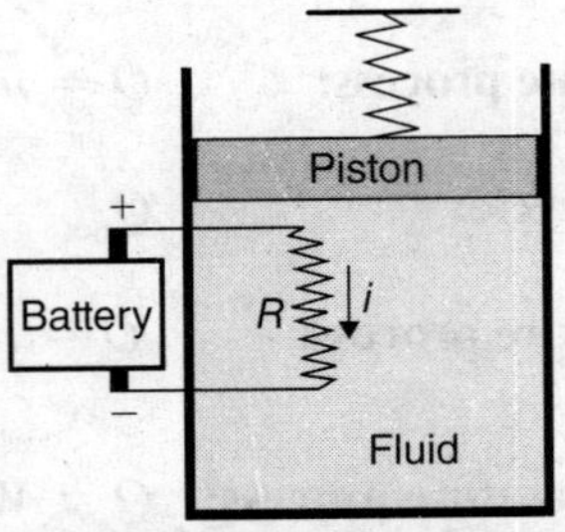

Figure 3.31

3.9 Saturated steam holds the 80-kg, 10-cm-diameter piston such that it just touches the spring in Fig. 3.32. The spring constant is 8000 N/m. The work required to raise the frictionless piston 20 cm is nearest:

(A) 510 J
(B) 470 J
(C) 320 J
(D) 250 J

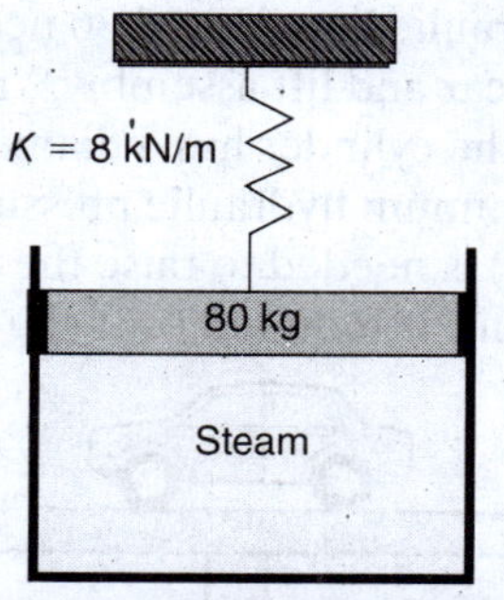

Figure 3.32

3.10 You are standing near the bonfire in Fig. 3.33 on a cool evening and feel very warm. You are heated primarily by:

(A) Radiation
(B) Conduction
(C) Convection
(D) A combination of the three

Figure 3.33

3.11 Approximate the heat transfer rate through a 0.8-m by 2-m double-pane window if a convective layer on a vertical pane of glass can be approximated as having an R-factor of 1 $m^2 \cdot °C/W$. The resistivity of the air layer between the panes can be approximated as $R = 0.4\ m^2 \cdot °C/W$. The temperature difference across the window is 30°C.

(A) 60 J/s
(B) 40 J/s
(C) 20 J/s
(D) 10 J/s

3.12 Two kilograms of water at 20°C are heated at a constant pressure of 100 kPa until the water is saturated steam. The work is nearest:

(A) 190 kJ
(B) 280 kJ
(C) 320 kJ
(D) 340 kJ

3.13 A conference room with dimensions 4 m by 10 m by 20 m is filled with 100 conference attendees. The air-conditioning system fails, resulting in an increasing temperature. There are 50 100-watt light bulbs, and each person emits approximately 0.117 kJ/s. Assuming a constant pressure, in 15 minutes the temperature will rise approximately:

(A) 10°C
(B) 14°C
(C) 18°C
(D) 22°C

3.14 The air in a rigid 8-m^3 container is at 200 kPa and 60°C. It is heated to 400°C. The heat transfer is nearest:

(A) 4100 kJ
(B) 2410 kJ
(C) 1660 kJ
(D) 260 kJ

3.15 The 4 kg of air, shown in Fig. 3.34, are heated at constant temperature from 200°C and 600 kPa until the pressure is 4 MPa. The heat transfer demanded by this process is nearest:

(A) −830 kJ
(B) −930 kJ
(C) −1030 kJ
(D) −1130 kJ

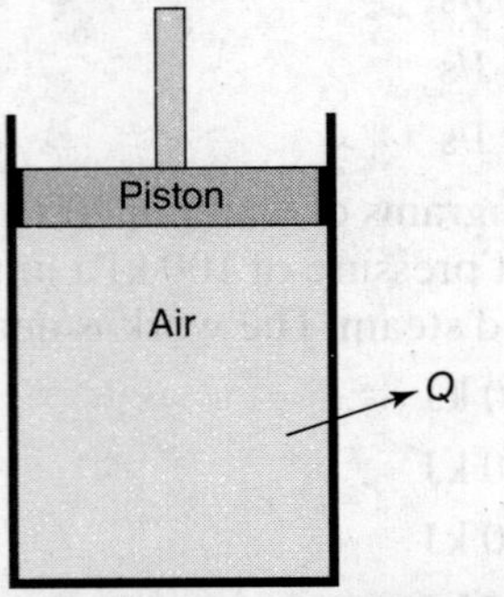

Figure 3.34

3.16 A piston compresses air in the automobile cylinder of Fig. 3.35 from 100 kPa and 20°C such that $v_2/v_1 = 0.1$. The temperature at state 2 is nearest:

(A) 360°C
(B) 460°C
(C) 560°C
(D) 660°C

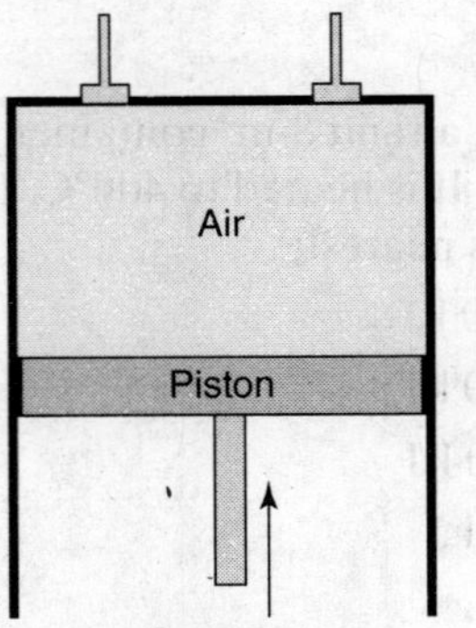

Figure 3.35

Work

3.17 If 200 kJ of work is done on a surface moving it 2 m by a force acting at an angle of 30° to a normal to the area, what is the applied force? The force moves in the direction of the normal.

3.18 A force of 450 N is applied normal to a surface that is 3 m by 3 m. If the surface is moved 0.9 m in the direction of the force, how much work is done?

3.19 The hydraulic lift of Fig. 3.36 needs to raise a 2000 kg car and lift assembly 2 m above the floor. If the cylinder has a diameter of 15 cm, what minimum hydraulic pressure (a gage pressure) is needed to raise the car? How much work is required over the 2 m distance?

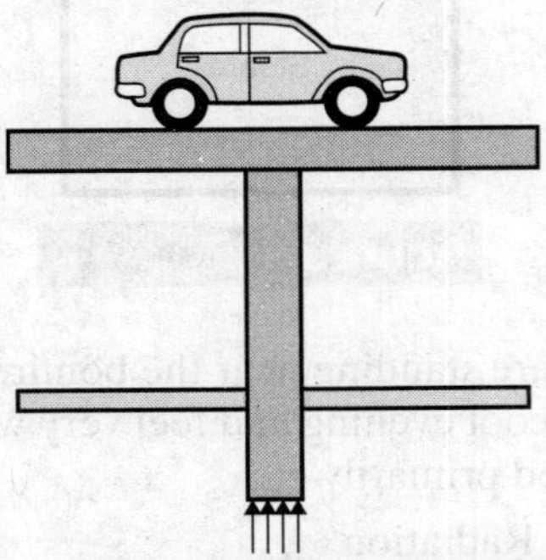
Figure 3.36

3.20 A constant pressure of 200 kPa is applied to a 1-m-diameter vertical cylinder in a frictionless piston-cylinder device that is open at the top. Calculate the mass of the piston and the work needed to move the piston 5 m in the vertical direction.

3.21 A force varies linearly with position according to $F = 2500x$ N, where x is the displacement in m. Calculate the work done as this force moves an object horizontally from $x = 0$ m to $x = 15$ m.

3.22 A force varies according to the equation $F = 10e^{x/5}$ N. Calculate the work done as this force pushes an object horizontally from $x = 0$ m to $x = 8$ m.

3.23 Pressure is related to volume by the relationship $PV^{1.2} =$ Const, where P is in kPa and V is in m^3. Determine the work required to compress an ideal gas from 0.8 m^3 to 4000 cm^3 if the initial pressure is 140 kPa.

3.24 Five kilograms of oxygen are compressed from 140 kP and 10°C in a process governed by the equation PV^n = const, where the pressure is in kPa and the volume is in m^3. The final temperature in this process is 37°C. Calculate the work done if *a*) $n = 1.4$, *b*) $n = 1.8$, and *c*) $n = 2$.

Instructor: This isn't multiple choice. Each part with a lower-case italic letter should be assigned as a separate problem.

3.25 Air is compressed in the cylinder of Fig. 3.37 such that PV = const, where the pressure is in kPa and the volume is in m^3. The initial pressure and temperature are 280 kPa and 90°C, respectively. If the volume changes from 0.015 m^3 to 0.0015 m^3, calculate the work required.

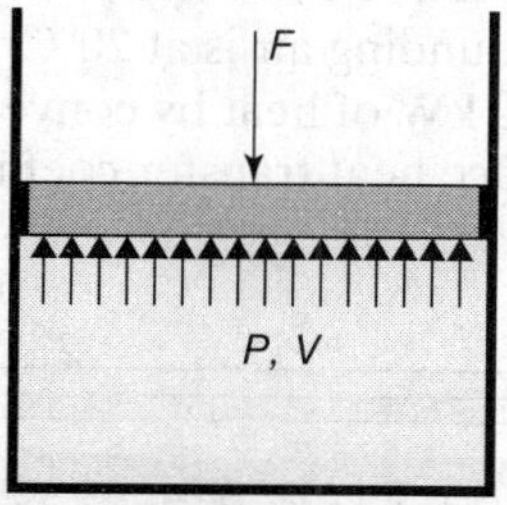

Figure 3.37

3.26 A balloon is filled with air from an initial radius of 1.2 m until it reaches a radius of 4 m. The pressure is related to the radius by $P(r) = 120 - 2(r - 1.2)^2$ kPa. Calculate the work. (The surface area of a sphere is $4\pi r^2$.)

3.27 Argon is contained in a piston-cylinder device with a volume of 1 m^3. The argon is compressed according to the equation $P = 5V^{-2}$ (P is in kPa and V is in m^3 until the volume is 0.2 m^3.) Calculate the work done during this process.

3.28 A gas is allowed to expand in an experimental process where the pressure is measured and the volume is determined. The data is listed below. Estimate the work done during this process.

P (kPa)	50	100	150	150	150	100
V (m^3)	10	15	20	25	30	35

3.29 Pressure is measured at several positions in an experiment. The volume at each position is determined. The results of the experiment are presented in the list that follows. Estimate the work performed by the process.

P (kPa)	700	870	950	980	980	950	900	830
V (m^3)	0.003	0.006	0.009	0.012	0.015	0.018	0.021	0.024

Other Forms of Work

3.30 A linear spring has a spring constant of 25 N/m. From its unstretched length it is compressed 10 cm. Calculate the work done on the spring.

3.31 The linear spring of Fig. 3.38 has a spring constant of 4 kN/m. Initially, the spring has been stretched 0.1 m. from its unstretched length. The spring is stretched another 0.125 m. Calculate the work done on the spring during this process.

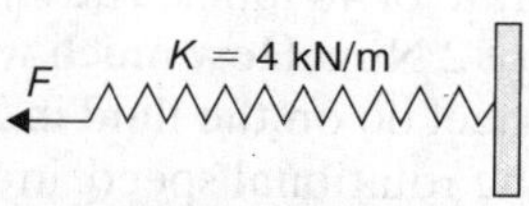

Figure 3.38

3.32 A nonlinear spring creates a force that varies with the amount of stretching according to the equation $F = Kx^2$, where $K = 100$ N/m^2, F is in newtons, and x is in meters. If the spring is stretched 15 cm from its unstretched length, calculate the work done on the spring.

3.33 A torsional spring creates a moment when it is twisted from its undeformed state according to the equation $T = K\theta$. T is in N·m and θ is in radians. If the spring constant $K = 10$ N·m/rad and the moment arm is 20 cm, calculate the work done twisting this spring from its undeformed position of 0° to an angle of 180°.

3.34 The air in the cylinder of Fig. 3.39 is initially at 300 kPa. Heat is added until the pressure increases to 900 kPa. Determine the work done by the air on the frictionless piston.

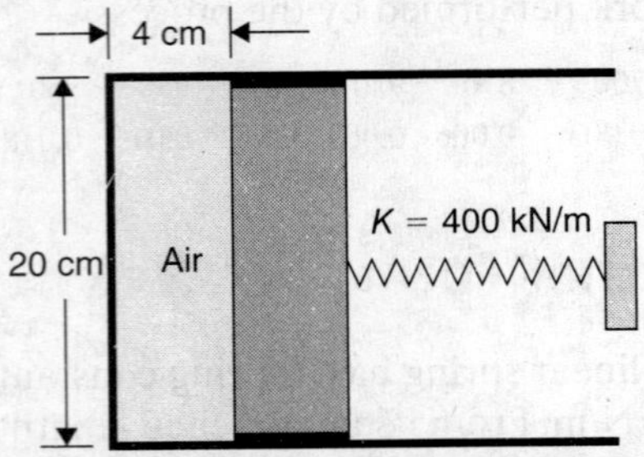

Figure 3.39

3.35 A paddle wheel rotates at 200 rpm in a substance. Determine the power transmitted by the shaft and the work it inputs into the substance in 20 minutes if it requires 4 N·m to rotate the wheel.

3.36 The 400-N weight in Fig. 3.40 is falling at the constant rate of 40 mm/s. The shaft transmits a torque of 2 N·m. How much work does the paddle wheel do on the fluid in 20 seconds? What is the rotational speed, in rpm, of the shaft during that time period?

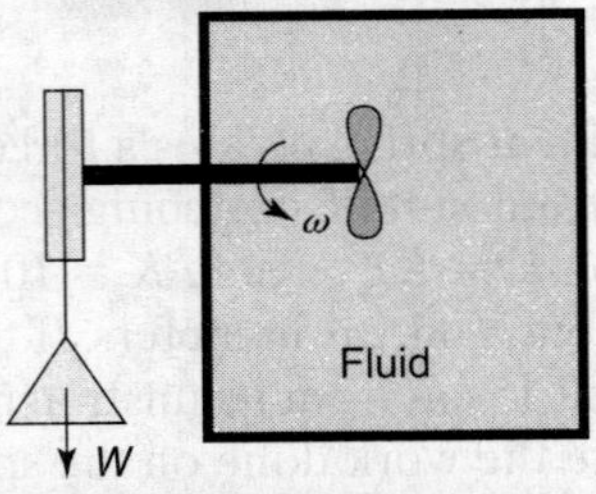

Figure 3.40

3.37 The 12-V battery of Fig. 3.41 powers the 10-ohm resistance for 3 minutes. The work from the battery inputs energy into the air. How much erengy is added to the air?

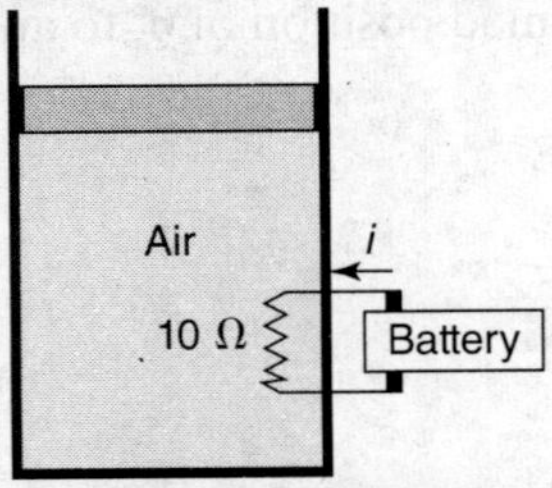

Figure 3.41

Heat Transfer

3.38 A steel wall is 2 cm thick. One side of the wall is at 80°C, and the other side is at 20°C. If the steel has a thermal conductivity $k = 45$ W/m·°C, calculate the amount of heat transferred through the wall per unit area.

3.39 A 1.2 m by 2.4 m window has an R-factor of 2 m^2·K/W. If the average outside and inside temperatures are −20°C and 22°C, respectively, how much heat is lost in one day? Estimate how much energy would be lost if there were 20 such windows in a house.

3.40 A pan with a diameter of 0.2 m. holds water at 77°C in a room at 27°C. If the heat transfer coefficient at the surface of the water is 633 kW/m^2-K, calculate the rate of heat transfer from the surface.

3.41 The horizontal flat plate of Fig. 3.42 has a surface area of 1 m^2. The plate is at 60°C, and the surrounding air is at 20°C. If the plate is losing 10 kW of heat by convection, calculate the surface heat transfer coefficient.

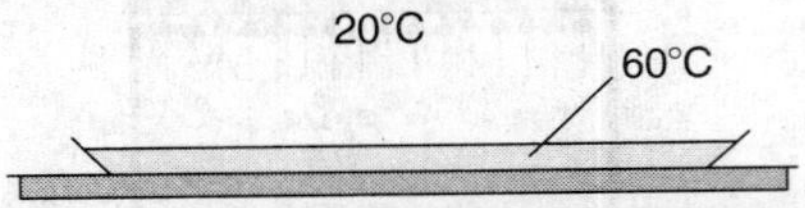

Figure 3.42

3.42 The average temperature of the human body is 37°C. Calculate the maximum rate of heat transfer per unit area by radiation from the human body in a room that is at 10°C. What is the percent increase in the heat transfer rate if you are running a fever with a temperature of 39°C?

3.43 A 1-m-diameter spherical satellite is traveling in intergalactic space where the surrounding temperature is 0 K. If the satellite has a surface temperature of −50°C, estimate the rate of heat loss by radiation from the satellite.

Problem-Solving Method

3.44 As a junior engineer, you have been assigned the task of finding a way to mount a 0.6 m by 0.6 m ceramic tile mosaic on a vertical concrete wall. The mosaic weighs 130 N. Using the steps listed in the problem-solving method, devise a way to mount this mosaic.

3.45 Your boss has asked you to devise a method to develop a safety device that will prevent a kerosene space heater from starting fires in the home. Use the problem-solving method to develop this "safety device."

The First Law Applied to Systems

3.46 A rigid container holds 4 kg of water at 20°C. If 200 kJ of heat is added to this system, what will be the final temperature? How much work is done in this process?

3.47 The radiator on a car engine removes 20 kW of heat from the engine. If the temperature of the radiator remains constant and no work is done on the coolant, how much heat is being developed by the engine?

3.48 A power plant generates 100 MW of heat in the boiler and rejects 75 MW of heat to the surroundings, as sketched in Fig. 3.43. How much work is being produced?

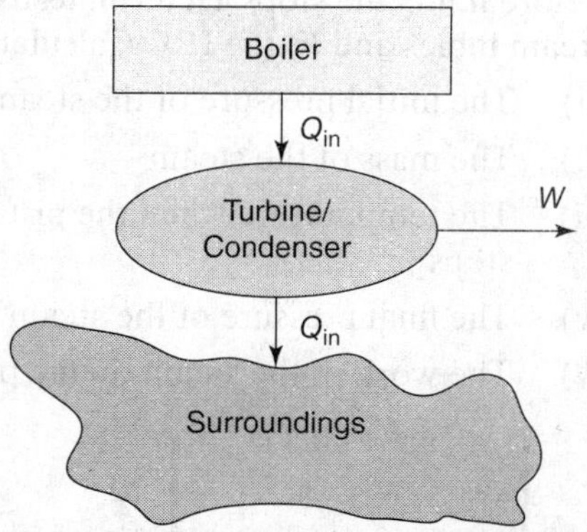

Figure 3.43

Note: It is assumed that kinetic and potential energy changes are insignificant when information to include such changes is not given.

3.49 A system produces 5000 kJ of work while at the same time 8000 kJ of heat is added to the system and 4000 kJ of heat leaves the system. What is the net change of total internal energy for this system?

3.50 A system undergoes a four-process cycle. Several quantities for the various processes are listed in the following table. Determine the missing entries. All quantities are in kJ.

Process	Q	W	ΔU
$1 \rightarrow 2$	$Q_{1\text{-}2}$	−200	0
$2 \rightarrow 3$	$Q_{2\text{-}3}$	600	$U_3 - U_2$
$3 \rightarrow 4$	600	$W_{1\text{-}2}$	200
$4 \rightarrow 1$	$Q_{4\text{-}1}$	0	−600

3.51 A system undergoes a four-process cycle. Several quantities for the various processes are listed in the following table. Determine the missing entries. All quantities are in kJ.

Process	Q	W	ΔU
$1 \rightarrow 2$	400	$W_{1\text{-}2}$	200
$2 \rightarrow 3$	$Q_{2\text{-}3}$	600	400
$3 \rightarrow 4$	$Q_{3\text{-}4}$	100	$U_4 - U_3$
$4 \rightarrow 1$	−800	$W_{4\text{-}1}$	0

3.52 Ice cubes at −20°C, each occupying a volume of 10 mL, are added to the insulated container of Fig. 3.44 that contains one liter of water at 20°C. Estimate the final temperature, or the mass of ice that melts, after equilibrium is reached if the number of ice cubes is *a*) 10, *b*) 25, and *c*) 100.

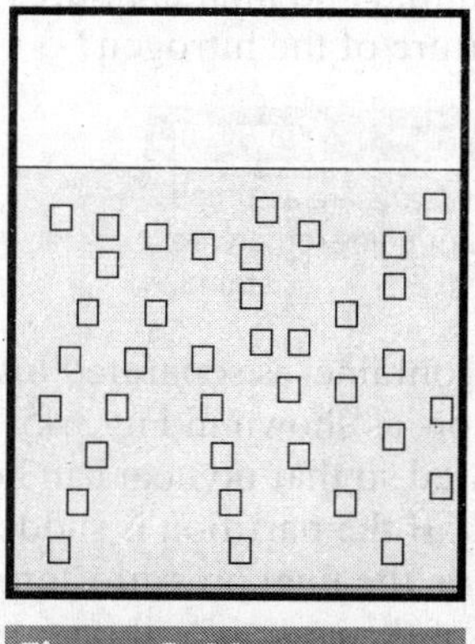

Figure 3.44

Note: The parts of problems with lower-case italic letters are intended to be assigned as separate problems.

3.53 Ice cubes at −7°C, each occupying a volume of 7 cm^3, are added to an insulated container holding 1.0 kg of water at 21°C. Estimate the equilibrium temperature, or the mass of ice that melts, if the number of ice cubes is *a*) 10, *b*) 40, and *c*) 200.

3.54 A block of aluminum at 38°C is brought into contact with a 40 kg block of copper at 93°C in an insulated container. What is the final equilibrium temperature if the mass of the aluminum block is *a*) 20 kg, *b*) 40 kg, and *c*) 60 kg?

The First Law Applied to Various Processes

3.55 One hundred kilograms of water are held in a rigid tank at 15°C. Determine the final temperature of the water if *a*) 3000 kJ, *b*) 5000 kJ, or *c*) 8000 kJ of heat are added to the tank.

3.56 A 10-m^3-rigid tank contains R134a at 200 kPa and 20°C. Heat is added to this tank until the pressure reaches 300 kPa. How much heat has been added? What is the final temperature of the refrigerant?

3.57 A 10-m^3-rigid tank contains 50 kg of nitrogen at 20°C. A fan inside the tank does 2500 kJ of work on the nitrogen. During this process, 1500 kJ of heat are lost from the system. What are the initial and final pressures and the final temperature of the nitrogen?

Comment

Specific heats are assumed constant unless otherwise stated.

3.58 A rigid container is separated into two parts by a partition, as shown in Fig. 3.45. The container is insulated so that no heat can be transferred in or out. If the partition is suddenly removed, determine the final pressure, temperature, and volume of the container if one part contains 10 kg of air at 560 kPa and 93°C, while the other part holds 15 kg of air at 770 kPa and 67°C. Assume constant specific heats.

3.59 The container of Fig. 3.45 is separated into two parts by a partition. The container is insulated so that no heat can be transferred in or out. If the partition is suddenly removed, determine the final pressure, temperature, and volume of the container if one part contains 10 kg of air at 500 kPa and 100°C, while the other part holds 15 kg of air at 800 kPa and 80°C. Assume constant specific heats.

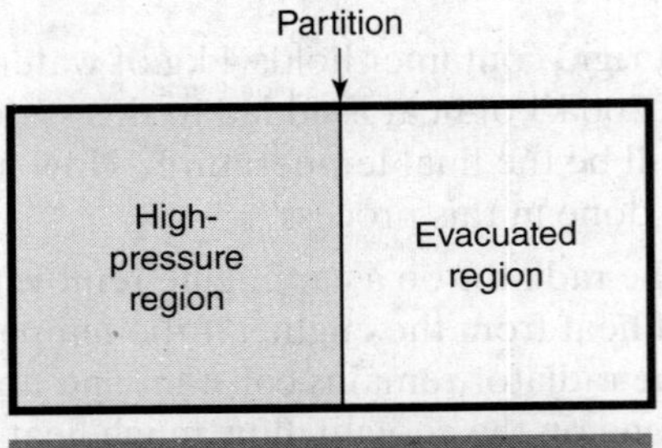

Figure 3.45

3.60 The steam in the circular cylinder of Fig. 3.46 has an initial quality of 10%. Heat is added until the temperature of the steam reaches 500°C. The frictionless piston rises 50 mm before it hits the stops. Determine, using *a*) the steam tables and *b*) the IRC Calculator:

i) The initial pressure of the steam
ii) The mass of the steam
iii) The temperature when the piston hits the stops
iv) The final pressure of the steam
v) The work of the steam on the piston

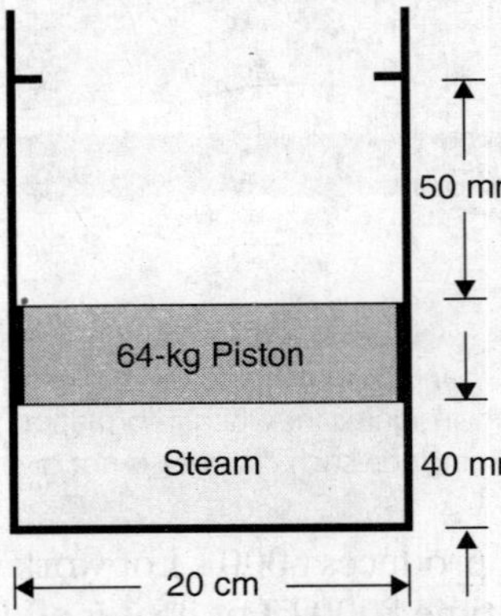

Figure 3.46

3.61 A spring with constant $K = 500$ kN/m just touches the top of the circular piston of Fig. 3.46 (see Fig. 3.14 also). The cylinder contains steam initially at a quality of 40%. Estimate the distance the spring will be compressed when the temperature reaches *a*) 300°C, *b*) 400°C, and *c*) 500°C. The stops have been removed.

3.62 A 0.04-m^3 open pan of water at 20°C is sitting on the burner of a stove. How much heat is needed to completely vaporize the water?

3.63 A vertical frictionless piston-cylinder device contains 50 kg of steam at 200 kPa and 300°C. Determine the heat transfer and the work done during the process if the steam is cooled at constant pressure until the temperature is *a*) 180°C, *b*) 125°C, and *c*) 80°C.

3.64 A frictionless piston maintains a constant pressure of 120 kPa as 0.6 kg of air is being heated in a cylinder. Estimate the work and the heat transfer required if the initial temperature of the air is 25°C and *a*) the volume is doubled, and *b*) the volume is quadrupled. Assume constant specific heats.

3.65 Two kilograms of water with a quality of 50% and a pressure of 1000 kPa are contained in the cylinder of Fig. 3.47 by a frictionless piston. The water is heated at constant pressure until the temperature reaches 600°C. Determine the heat transfer and the work done by the water using i) the steam tables and ii) the IRC Calculator. Show the area representing the work on a *P*-*v* diagram.

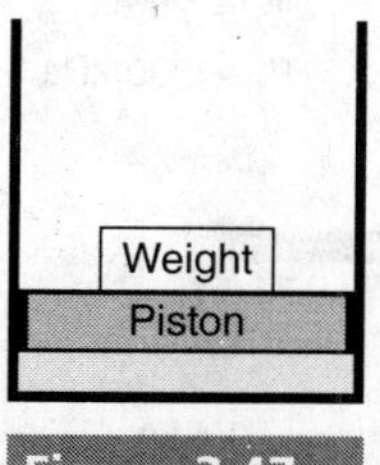

Figure 3.47

3.66 Three kilograms of superheated steam at 4 MPa and 600°C are compressed in a cylinder at constant pressure. Determine the heat transfer required and the work required by this process if the final volume is *a*) 0.08 m^3, *b*) 0.04 m^3, and *c*) 0.02 m^3.

3.67 Air is contained in the cylinder of Fig. 3.48. The air is heated until the frictionless piston is raised 40 cm above the stops. At what temperature does the piston just leave the stops? Determine the final temperature, the heat transfer, and the work done by the air on the piston. Assume constant specific heats.

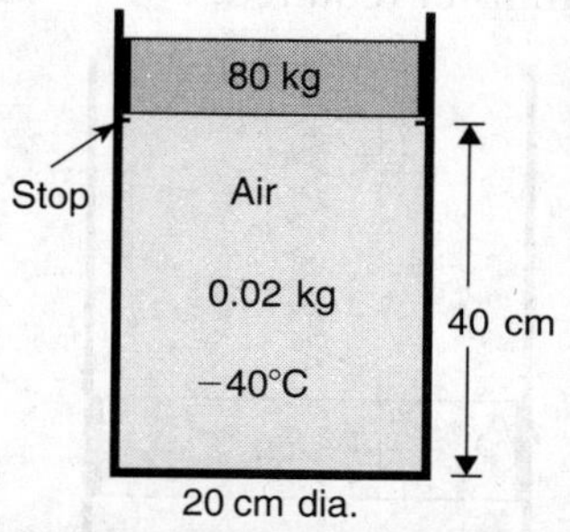

Figure 3.48

3.68 Four kilograms of solid copper at 200°C are immersed with 20 kg of water at 40°C in an insulated container. What is the final equilibrium temperature?

3.69 Calculate the heat transfer needed to heat 20 kg of argon from 10°C to 100°C i) at constant pressure and ii) at constant volume. Assume constant specific heats.

3.70 Ten kilograms of nitrogen are compressed from 500 kPa in an isothermal process at 240°C. Calculate the heat transfer and the work if the final pressure is *a*) 1500 kPa, *b*) 2500 kPa, and *c*) 4000 kPa. Assume constant specific heats.

3.71 Steam is compressed in a cylinder such that the temperature remains constant. The initial pressure, temperature, and volume are 400 kPa, 200°C, and 0.08 m^3, respectively. If the final state has a quality of 0.5, estimate the work required. (This requires a graphical solution since there is no *P*-*V* relationship.)

3.72 A mass of 0.02 kg of saturated water vapor is contained in the cylinder of Fig. 3.49. The spring with spring constant $K = 60$ kN/m just touches the top of the frictionless 160-kg piston. Heat is added until the spring compresses 10 cm. Estimate the final temperature of the steam, the work done, and the heat transfer required.

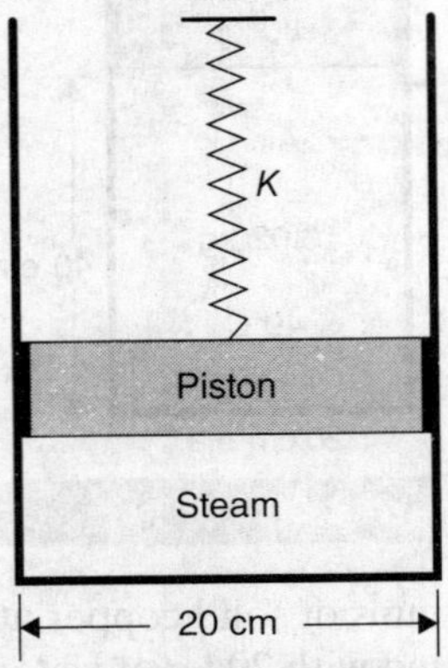

Figure 3.49

3.73 The paddle wheel in Fig. 3.50 requires 0.8 N·m of torque to rotate at 200 rpm. If it rotates for 40 minutes, determine the final temperature of the 10 kg of nitrogen contained in the insulated rigid volume, initially at 25°C and 200 kPa. Assume constant specific heats.

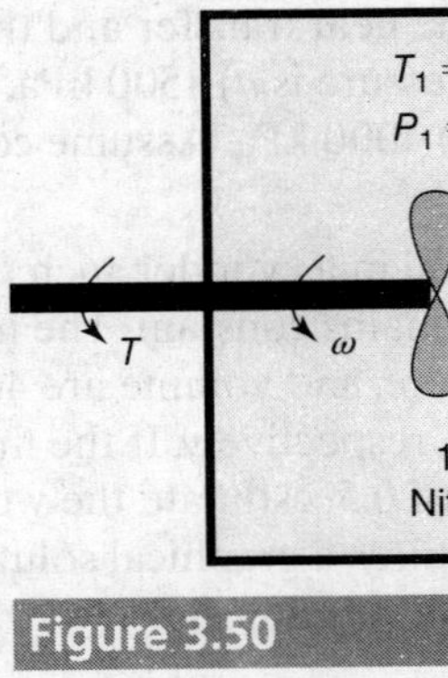

Figure 3.50

3.74 If the paddle wheel is removed from the volume of Problem 3.73 and 4800 kJ of heat is added to the volume with a resistance heater, determine the final temperature of the nitrogen.

3.75 Determine the heat transfer needed to increase the pressure of 2 kg of 50%-quality steam confined in a rigid container from 140 kPa to 1200 kPa. Also state the final temperature. *a*) Use the steam tables, and *b*) use the IRC Calculator. Sketch the process on a *T*-*v* diagram.

3.76 Water at 400 kPa with a quality of 0.2 is heated in a rigid container until the temperature is 200°C. Calculate the final quality, the heat transfer, and the work done using *a*) the steam tables and *b*) the IRC Calculator. Show the process on a *P*-*v* diagram.

3.77 Determine the heat transfer needed to increase the temperature of 2 kg of 50%-quality steam confined in a rigid container from 140°C to 1000°C. Also state the final pressure. *a*) Use the steam tables, and *b*) use the IRC Calculator. Sketch the process on a *T*-*v* diagram.

3.78 A frictionless piston allows a constant pressure in a cylinder containing 4 kg of saturated water vapor originally at 180°C. If 4000 kJ of heat is added to the vapor, estimate the final temperature i) without the use of enthalpy and ii) with the use of enthalpy.

3.79 The paddle wheel in Fig. 3.51 requires 0.8 N·m of torque to rotate at 200 rpm. If it rotates for 40 minutes, determine the final temperature of the 0.4 kg of nitrogen contained in the insulated volume, initially at 25°C and 200 kPa. The frictionless piston maintains a constant pressure. Assume constant specific heats.

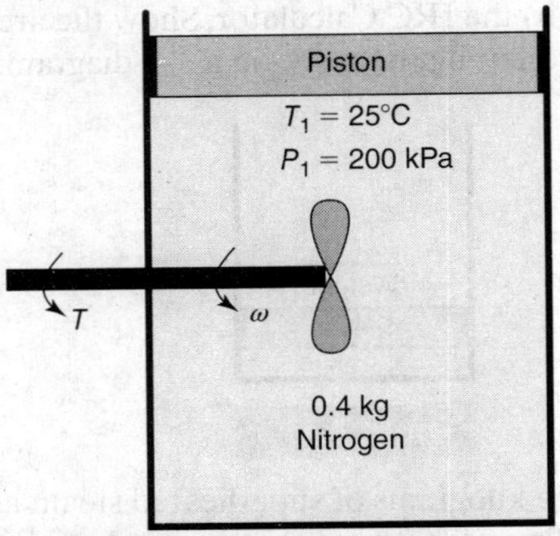

Figure 3.51

Comment

Constant specific heats are assumed unless otherwise stated.

3.80 Nitrogen at 100°C is compressed in the cylinder of Fig. 3.52 such that the temperature remains constant. If the pressure changes from 600 kPa to 1200 kPa, determine how much heat must be transferred from the 0.4 kg of nitrogen. Assume constant specific heats.

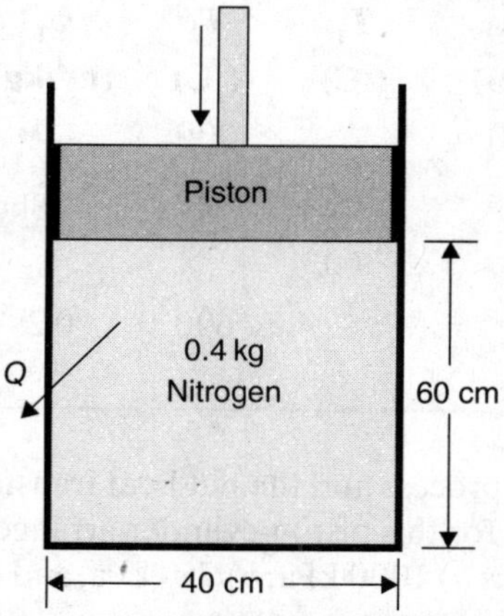

Figure 3.52

3.81 Air at 107°C and 5 L expands in a cylinder such that the temperature remains constant. If the pressure changes from 2.8 MPa to 300 kPa, determine how much heat must be transferred to the air. Assume constant specific heats.

3.82 Air enters an insulated cylinder at 20°C and 100 kPa. It is compressed so that its volume decreases by a factor of 8. What is the final temperature? How much work is required to compress the 0.2 kg of air? Assume constant specific heats.

3.83 Air at 200 kPa is compressed in the insulated cylinder of Fig. 3.53 with an initial volume of 4000 cm^3. What is the work required if the final volume is *a*) 1000 cm^3, *b*) 600 cm^3, and *c*) 400 cm^3? Assume constant specific heats.

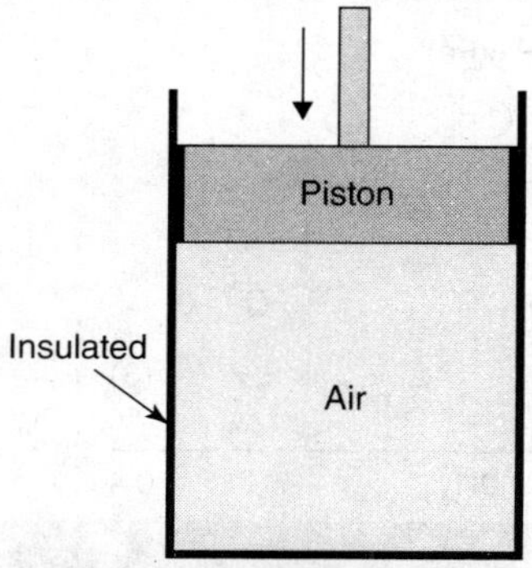

Figure 3.53

3.84 Air is allowed to expand in a cylinder from 700 kPa and 10°C to 100 kPa. Calculate the change in specific volume and specific internal energy for this process if the process is *a*) isothermal, *b*) adiabatic quasi-equilibrium, and *c*) polytropic with $n = 2.5$. Assume constant specific heats.

3.85 The air–fuel mixture enters an automobile engine cylinder at 120°C and 100 kPa. The engine has a compression ratio of 8, which means the air–fuel mixture is compressed to one-eighth the original volume. This process is a polytropic process that follows the equation Pv^n = const. Calculate the specific work and the specific heat transfer for this process if *a*) $n = 1.3$, *b*) $n = 1.4$, and *c*) $n = 1.5$. Assume constant specific heats.

Comment

Variable specific heats will be considered in a later chapter.

3.86 Air at 700°C is expanded in an insulated cylinder such that the volume increases by a factor of 8. Estimate the final temperature, assuming a quasi-equilibrium process. Also, calculate the work provided by the 0.2 kg of air. Assume constant specific heats.

3.87 Twenty kilograms of nitrogen are allowed to expand in a polytropic process according to the equation PV^n = const. The nitrogen is initially at 500 kPa with a volume of 10 m^3. The final volume is 40 m^3. Calculate the work done during this process and the amount of heat transferred from the system if *a*) $n = 2.5$, *b*) $n = 2.0$, and *c*) $n = 1.4$.

Comment

Constant specific heats are assumed unless otherwise stated.

3.88 One kilogram of air is compressed in a cylinder for each of the quasi-equilibrium processes listed in the following table. Fill in the missing quantities for the selected process.

Process	Q (kJ)	W (kJ)	ΔU (kJ)	ΔH (kJ)	P_1 (kPa)	P_2 (kPa)	T_1 (°C)	T_2 (°C)	v_1 (m^3/kg)	v_2 (m^3/kg)
a) $V = C$			200			200		200		
b) $P = C$					400		200	400		
c) $T = C$		80			200		60			
d) $Q = 0$								600	0.2	0.02

3.89 Nitrogen at 420 kPa and 150°C with a volume of 0.27 m^3 is allowed to expand in a polytropic process such that $Pv^n = C$, where C is a constant. Determine the work done and the heat transfer during this process if the final volume is 1.35 m^3 and *a*) $n = 1.2$, *b*) $n = 1.4$, and *c*) $n = 1.6$.

3.90 Air undergoes a cycle that is composed of three processes, as shown in Fig. 3.54, with $T_{high} = 2400$ K and $v_{max}/v_{min} = 6$. Assuming quasi-equilibrium processes with constant specific heats, calculate the work for each process and the net heat transfer for the cycle for this piston-cylinder arrangement if P_{high} is *a*) 1000 kPa, *b*) 750 kPa, and *c*) 500 kPa.

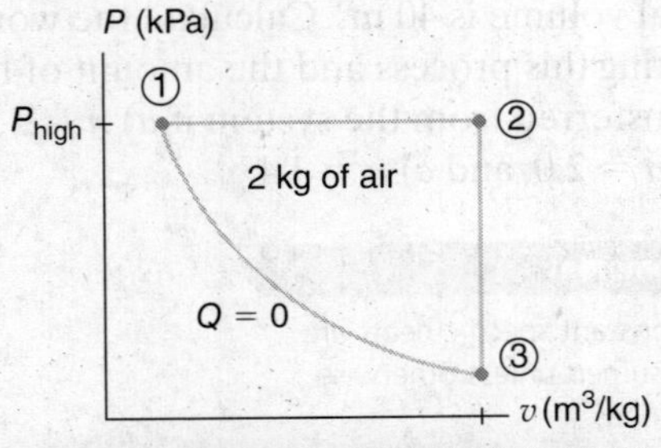

Figure 3.54

3.91 Air undergoes a cycle that is composed of three processes, as shown in Fig. 3.55. Assuming quasi-equilibrium processes and constant specific heats with $T_{high} = 427$°C, calculate the work for each process and the net heat transfer for the cycle for this piston-cylinder arrangement if P_{high} is *a*) 1000 kPa, *b*) 750 kPa, and *c*) 600 kPa.

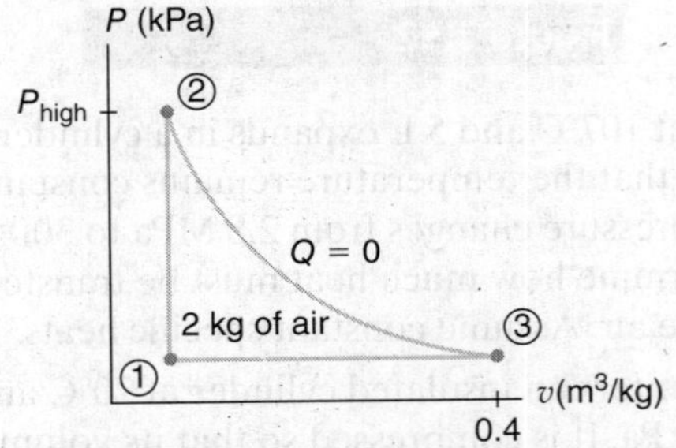

Figure 3.55

3.92 Two kilograms of air undergo a cycle that is composed of three processes, as shown in Fig. 3.56. Assuming quasi-equilibrium processes and constant specific heats, calculate the work for each process and the net heat transfer for the cycle for this piston-cylinder arrangement if P_{high} is *a*) 1000 kPa, *b*) 750 kPa, and *c*) 500 kPa.

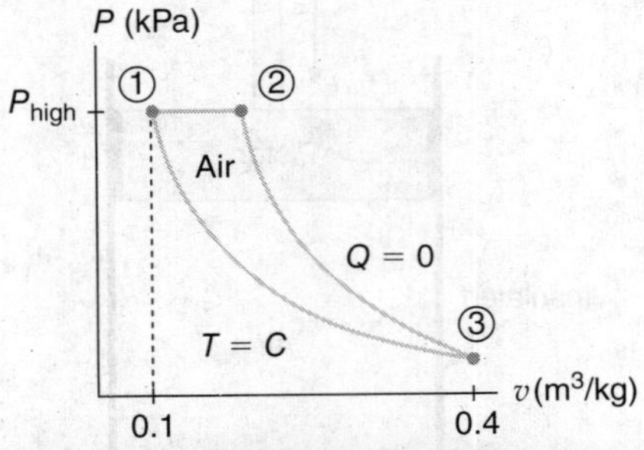

Figure 3.56

3.93 States 2 and 3 are saturated states as the steam undergoes the cycle shown in Fig. 3.57. Assuming quasi-equilibrium processes, determine the net work produced and the heat transfer required if 4 kg of steam make up the working fluid and P_{low} is *a*) 14 kPa, *b*) 40 kPa, and *c*) 80 kPa.

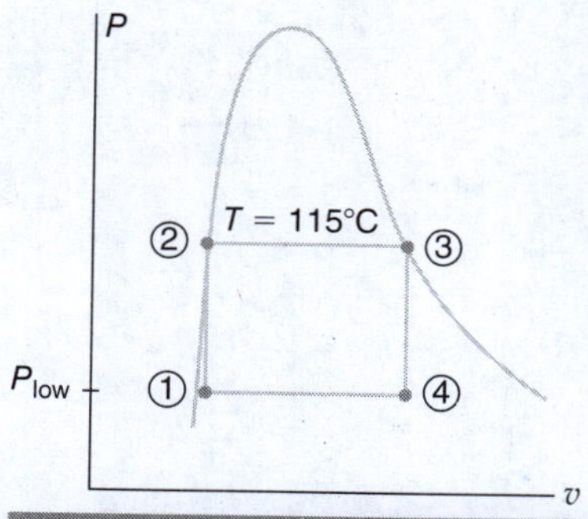

Figure 3.57

CHAPTER

4

The First Law Applied to Control Volumes

Arogant/Shutterstock.com

Outline

Nomenclature

The following variables are introduced in this chapter:

BWR	Back-work ratio
COP_{HP}	Coefficient of performance of a heat pump
COP_R	Coefficient of performance of a refrigerator
$dE_{c.v.}/dt$	Change in energy per unit time
m_f	The final mass
m_i	The initial mass
$\dot{m}$	Mass flow rate (or mass flux)
Q	Heat transfer
$\dot{Q}$	Rate of heat transfer
$\dot{Q}_B$	Boiler rate of heat transfer
$\dot{Q}_C$	Condenser rate of heat transfer
$\dot{Q}_{Cond}$	Condenser rate of heat transfer
$\dot{Q}_{Evap}$	Evaporator rate of heat transfer
P_f	The final pressure
P_i	The initial pressure
T_f	The final temperature
T_i	The initial temperature
V	Velocity
$\dot{V}$	Volumetric flow rate (or, simply, flow rate)
$\dot{W}_C$	Compressor work rate
$\dot{W}_{Comp}$	Compressor work rate
$\dot{W}_S$	Shaft work rate
w_S	Shaft work per unit mass
$\dot{W}_T$	Turbine work rate
η	Efficiency

Learning Outcomes

- ❑ **Select a proper control volume that allows quantities of interest to be calculated**
- ❑ **Apply the conservation of mass to control volumes**
- ❑ **Apply the first law of thermodynamics to control volumes**
- ❑ **Calculate the performance of common engineeri ıg devices**
- ❑ **Understand the filling and discharge of tanks**
- ❑ **Combine engineering devices to form power and refrigeration cycles**

Control volume: The volume into which and/or from which a fluid flows.

Control surface: The surface that surrounds a control volume.

In the last chapter, we learned how to apply the first law of thermodynamics to systems where the mass was completely contained within the system boundaries. In this chapter the first law will be applied to devices that have the fluid transported into and out of volumes. Such a volume will be referred to as a *control volume*. A system is an identified set of mass elements that would be very difficult to follow as the fluid flows through, for example, a turbine. So, the volume into which and/or from which a fluid flows is called a control volume.[1] The surface that surrounds the control volume is the *control surface*. Control volumes include a wide variety of engineering devices, such as pumps, turbines, boilers, condensers, and evaporators that will be used in the thermodynamic cycles of later chapters. We will apply the conservation of mass and the first law of thermodynamics to ensure that both mass and energy are accounted for as the fluid flows through a device, such as the general-shaped control volume of Fig. 4.1. There may be more than two locations where the fluid flows in and out of a device, such as the furnace ductwork in a house. Identifying a control volume is analogous to identifying a free-body diagram in mechanics courses.

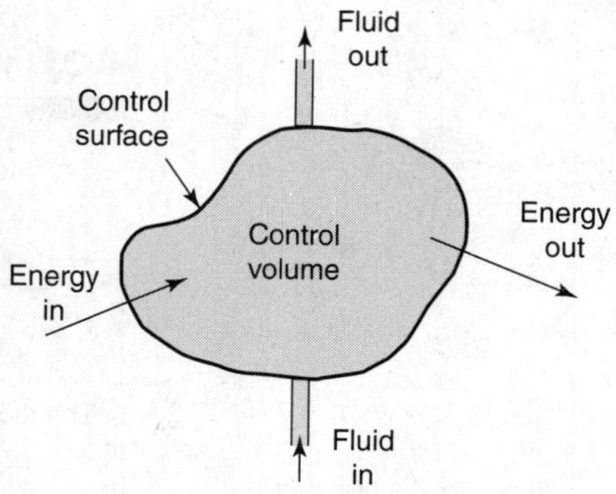

Figure 4.1

Fluid and energy may cross the control surface of a control volume, such as a turbine, a nozzle, or a radiator.

Motivational Example—The Air Conditioner

An air conditioner is a device that forces heat to be transferred from a cool region inside to a warm region outside. Normal flow of heat is from a high-temperature region to a low-temperature region. A picture of a typical window air conditioner along with a schematic of how it works is shown in Figs. 4.2*a* and 4.2*b*. The air conditioner transfers heat from inside a house that is being kept cool to the warmer air outside of the house. To accomplish this, we use several devices to be presented in this chapter: a compressor, a throttle, an evaporator, and a condenser, as displayed in the cycle of Fig. 4.2*c*.

Figure 4.2*a*

A window air conditioner.

The compressor is a device that is powered by an electric motor that increases the pressure and temperature of the refrigerant being pumped through this device. A compressor is similar in function to a pump except that, in general, a compressor pressurizes a vapor and a pump pressurizes a liquid. After leaving the compressor, the refrigerant travels through a condenser where heat is transferred from the hot refrigerant to the outside atmosphere, which is at a lower temperature than the refrigerant. During this constant-pressure process, the refrigerant is condensed from a superheated vapor to a saturated liquid, hence the name "condenser." The refrigerant then travels through a throttle that is simply a constriction in the pipe. The throttle causes a large pressure drop that causes the refrigerant to partially evaporate and become much colder. This process is similar to the spraying of an aerosol can. As the liquid expands to a lower temperature, it vaporizes and becomes cold because of the latent heat during the phase change.

The cold refrigerant, which is a saturated mixture at low temperature, then passes through the evaporator. Here heat is transferred from the inside of

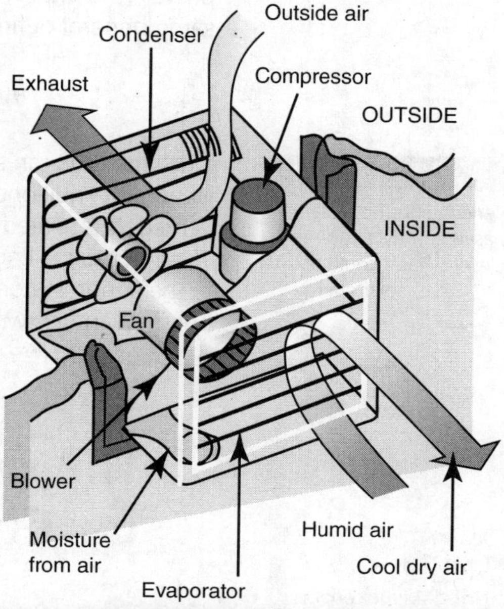

Figure 4.2*b*

The schematic of a window air conditioner.

(*Continued*)

[1] Control volumes are also called open systems, particularly in older textbooks.

Motivational Example—(*Continued*)

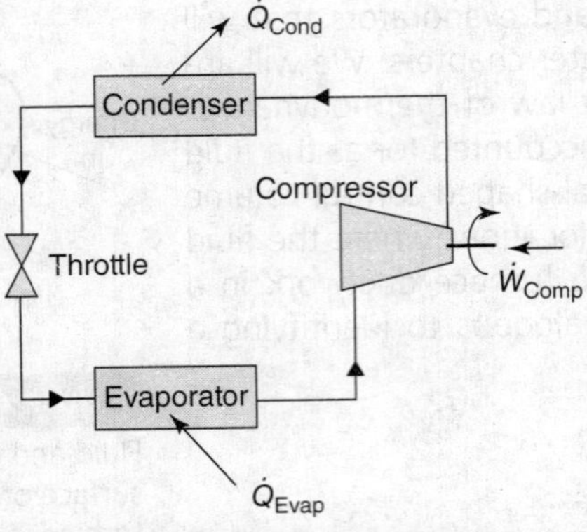

Figure 4.2c

The air conditioner cycle showing its components.

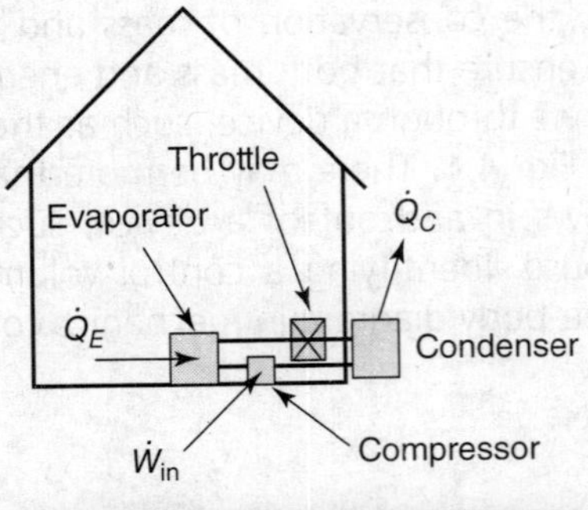

Figure 4.3

An air conditioner in a house.

the house (see Fig. 4.3) to the cold refrigerant. As heat is absorbed from the house, the remaining liquid in the refrigerant is evaporated, hence the name "evaporator." The heat removed from the air-conditioned house by the evaporator is the cooling load. The refrigerant then enters the compressor to begin the cycle again.

The performance parameter that quantifies the effectiveness of the air conditioner is the *coefficient of performance* (COP). This parameter is equal to the ratio of the cooling load (the desired effect) to the input mechanical power (the energy input) to the compressor. Efficiency η and COP have the same general definition, the *desired effect* divided by *purchased energy*:

$$\eta_{\text{engine}} = \frac{\text{desired effect}}{\text{energy input}} \qquad \text{COP} = \frac{\text{desired effect}}{\text{energy input}}$$

Comment

The energy input is usually purchased energy, although a clever person could use solar energy.

Since the COP can exceed 1, and usually does, we do not use the term *efficiency* of an air conditioner. By definition, efficiency cannot exceed 1, so COP was invented. It is related to the term used in stores, the *energy efficiency ratio* (EER). The EER has the English units of Btu/hr in the numerator and the SI units of watts in the denominator. The relationship between COP and EER is EER = 3.412 COP. These quantities will be considered in detail later in this chapter in Section 4.4.2.

4.1 The Conservation of Mass for Control Volumes

Note: All the control volumes of interest are fixed, that is, a volume does not change shape, as does a balloon or a heart.

4.1.1 Basic information

For the problems analyzed in this chapter, there will be one or more flows of a fluid into the fixed control volume (c.v.) to be analyzed and one or more flows out of the volume. Most devices of interest have one inlet and one outlet, as

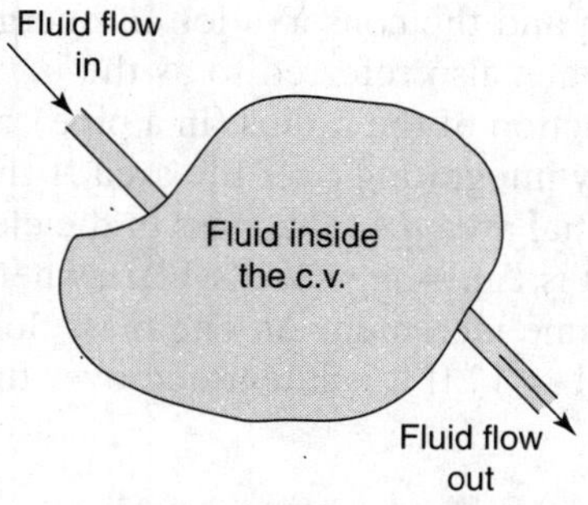

Figure 4.4

Mass flows into and out of a general fixed control volume. Each flow and the mass inside the control volume could be time dependent.

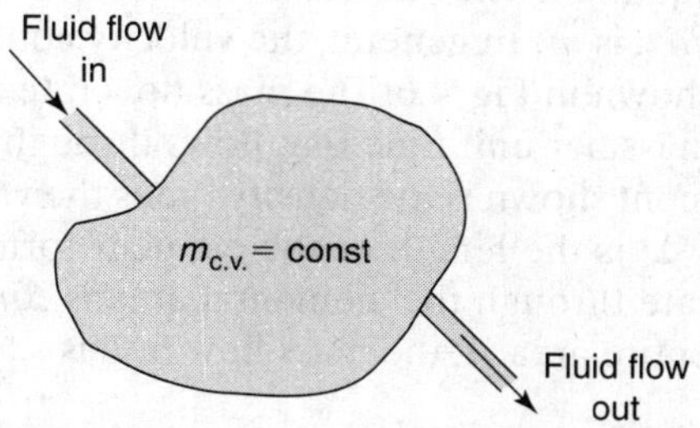

Figure 4.5

In steady flow, all flows are independent of time.

displayed in Fig. 4.4. The conservation of mass as applied to a control volume can be summed up as follows:

$$\begin{array}{c}\text{Mass of fluid flow}\\ \text{into the volume}\end{array} - \begin{array}{c}\text{Mass of fluid flow}\\ \text{out of the volume}\end{array} = \begin{array}{c}\text{Change of mass of}\\ \text{fluid in the volume}\end{array} \quad (4.1)$$

If more fluid is entering than is leaving, it means that either the fixed control volume is filling up and/or the density of the fluid in the fixed control volume is increasing. If either of these situations exists, it is an unsteady flow. For the problems we will solve in this chapter, with the exception of those in Section 4.3, a *steady flow* will be assumed, as shown in Fig. 4.5; this means that the physical properties of the fluid do not change with time. For such a flow, there is no accumulation of fluid within the control volume, which means that

Steady flow: The physical properties of the fluid do not change with time.

$$\text{Mass of fluid flow in} = \text{Mass of fluid flow out} \quad (4.2)$$

As a result of this approach, physical properties are defined at the boundary of the control volume and do not change with time. This approach greatly simplifies the analysis of devices but will be applicable only to devices that are operating in the steady-state mode. This includes the many devices in the thermodynamic systems of interest. It is important to identify the control volume and show inlets and outlets when solving problems.

☑ You have completed Learning Outcome (1)

4.1.2 The continuity equation

Mass flows across the surface area of every device of interest in this chapter. The rate at which it flows across an area is measured by the mass flow rate or the volumetric flow rate. *Mass flow rate* is defined as the amount of mass that goes through a cross-sectional area of a conduit, usually a pipe, per unit time. In the SI system the units are kg/s, and in the English system the units are lbm/s. Mass flow rate is

Mass flow rate: The amount of mass that crosses the cross section of a conduit per unit time. It is also called the *mass flux*.

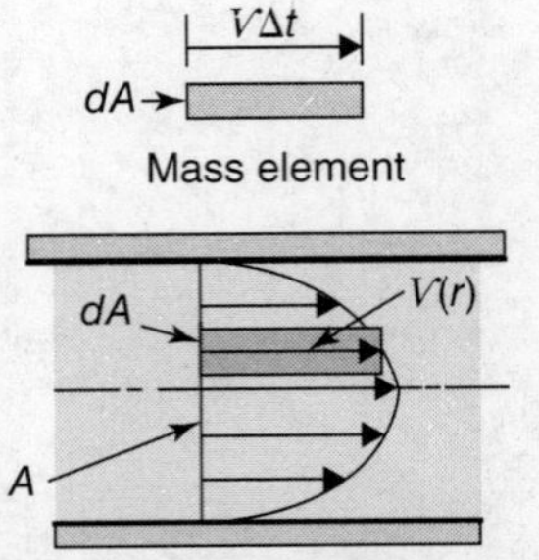

Figure 4.6
A general velocity profile in a pipe.

Comment
Most flows of interest in conduits are turbulent flows, which have essentially uniform velocity profiles.

Mass flux: $\dot{m} = \rho AV$

used in both the conservation of mass equation and the conservation of energy equation. The standard variable for mass flow rate, also referred to as the *mass flux*, is $\dot{m}$. In general, the velocity could be a function of the radius (in a pipe), as shown in Fig. 4.6. The mass flow rate is found by integrating over the area A the mass per unit time that flows through a differential area dA. The mass in the element shown is the density times the volume; that is, $\Delta m = \rho \times (dA \times V\Delta t)$, where $V\Delta t$ is the length of the element formed over time increment Δt. The mass flow rate through the elemental area is $\Delta m/\Delta t = \rho dA \times V$. If it is integrated over the entire area A, the mass flow rate is

$$\dot{m} = \int_A \rho V dA \tag{4.3}$$

where the velocity V is the velocity component perpendicular to the area element dA. If the velocity vector is not perpendicular to the area element, the tangential component would not move any mass through the area; it is only the normal component that moves the fluid through the stationary area. If the velocity vector is normal to the area and uniform across the area, as is the case for the problems of interest in an introductory course in thermodynamics, the above integral would provide

$$\dot{m} = \rho AV = \frac{AV}{v} \tag{4.4}$$

where ρ is the density, assumed to be uniform over the area A, v is the specific volume, and V is the uniform velocity in the conduit normal to the area. In this text, the velocity will always be assumed normal to and uniform across the area. A uniform profile is an acceptable approximation for the velocity in a turbulent[1] flow, which represents the flows of interest.

The masses of fluid entering and leaving the control volume of Fig. 4.5 in the time increment Δt are written by referring to Fig. 4.7:

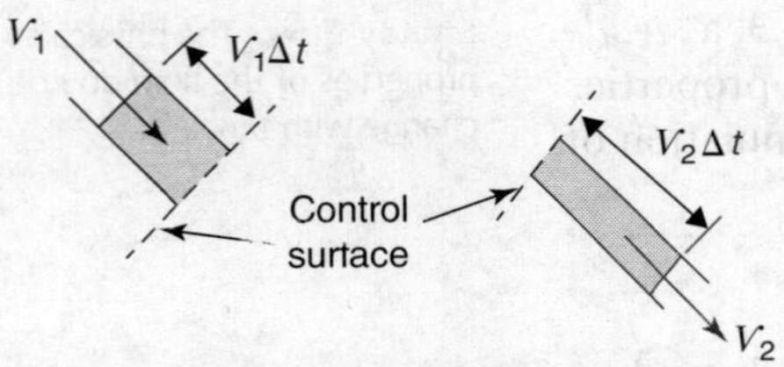

Figure 4.7
The fluid that enters and leaves the control volume of Fig. 4.5. The velocity components are assumed normal to their respective area.

$$(\Delta m)_1 = [density \times volume]_1 = [density \times Area \times length]_1$$
$$= \rho_1 A_1 V_1 \Delta t \tag{4.5}$$

$$(\Delta m)_2 = [density \times volume]_2 = [density \times Area \times length]_2$$
$$= \rho_2 A_2 V_2 \Delta t \tag{4.6}$$

If the flow is unsteady, the mass inside the control volume can change. If that's the case, the conservation of mass expressed by Eq. 4.2 would require

$$\rho_1 A_1 V_1 \Delta t - \rho_2 A_2 V_2 \Delta t = \Delta m_{\text{c.v.}} \tag{4.7}$$

[1] Most flows of engineering importance are turbulent flows. If a pipe is 2.5 cm in diameter and water at 20°C is flowing in the pipe, the flow will be turbulent if the velocity exceeds 0.1 m/s. If air is flowing, a turbulent flow exists when the velocity exceeds 1 m/s. Whenever you get a drink of water from a fountain, the flow is turbulent.

Divide both sides by Δt and let $\Delta t \to 0$; the above equation can then be written in the form

$$\rho_1 A_1 V_1 - \rho_2 A_2 V_2 = \frac{dm_{c.v.}}{dt} \quad (4.8)$$

Remember:

$$\lim_{\Delta t \to 0} \frac{\Delta m_{c.v.}}{\Delta t} = \frac{dm_{c.v.}}{dt}$$

The devices of interest in our study invariably assume a steady flow; that is, the amount of mass that enters must equal the amount that leaves, so $dm_{c.v.}/dt = 0$. Then Eq. 4.8 takes the simplified form:

$$\rho_1 A_1 V_1 = \rho_2 A_2 V_2 \quad (4.9)$$

Assumptions for the continuity equation

1. Steady flow in a conduit.
2. Uniform flow (velocity and density are constant across each area).
3. The velocities are perpendicular to the respective areas.
4. One inlet and one exit.

This is often referred to as the *continuity equation*. It is the equation that will most often be used when analyzing devices of interest.

If a volume had two inlets and two outlets, as sketched in Fig. 4.8, the continuity equation would take the form

$$\rho_1 A_1 V_1 + \rho_2 A_2 V_2 = \rho_3 A_3 V_3 + \rho_4 A_4 V_4 \quad (4.10)$$

Or, to be more general, with multiple inlets and outlets, the conservation of mass for an unsteady flow would be

$$\sum \rho_i A_i V_i - \sum \rho_e A_e V_e = \frac{dm_{c.v.}}{dt} \quad (4.11)$$

where i represents an inlet and e an exit.

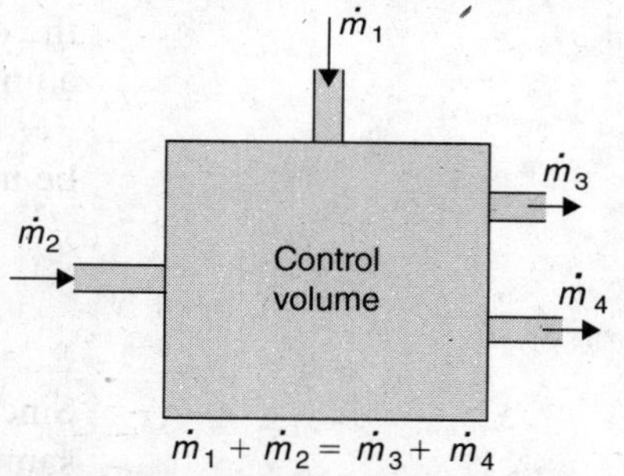

Figure 4.8 A control volume with two inlets and two outlets.

Occasionally, the volumetric flow rate is used, which is the amount of volume of fluid that passes a cross-sectional area of a conduit per unit time. It has dimensions of volume per unit time with units of m^3/s. An accepted symbol for volumetric flow rate is $\dot{V}$. Often, $\dot{V}$ is simply called the *flow rate*. The flow rate is equal to the product of the flow velocity and the cross-sectional area of the flow perpendicular to the velocity:

$$\dot{V} = AV \quad (4.12)$$

Flow rate: The amount of volume of fluid that passes a cross-sectional area per unit time, indicated by $\dot{V}$.

If the fluid is incompressible, that is, $\rho_1 = \rho_2$, the density divides out of Eq. 4.9, so that the continuity equation for an incompressible flow with one inlet and one outlet, a rather common flow, is

$$A_1 V_1 = A_2 V_2 \quad (4.13)$$

The relationship between mass flux and flow rate is

$$\dot{m} = \rho \dot{V} \quad (4.14)$$

Examples will illustrate the use of the conservation of mass represented by the continuity equation.

Example 4.1 Water flowing through a device

Water enters a device through a 2-cm-diameter pipe and exits from a 4-cm-diameter pipe. Assume steady, incompressible flow and calculate the exiting velocity if the entering velocity is 10 m/s. See Fig. 4.9.

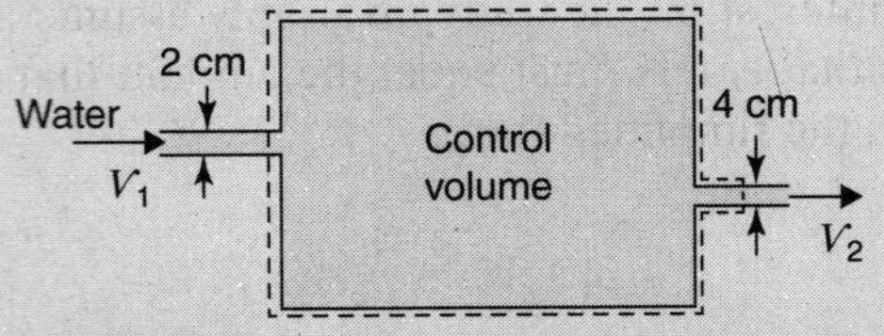

Figure 4.9

Solution

The control volume is selected so that it encompasses the device and cuts the inlet pipe anywhere upstream of where it enters the device but some distance downstream of the exit where the pipe leaves the device. (The streamlines in the downstream pipe take some distance to become parallel to the wall so that a uniform velocity V_2 exists.)

Since water is essentially incompressible (constant density), Eq. 4.13 can be used. It provides

$$A_1V_1 = A_2V_2$$

$$(\pi \times 0.01^2) \times 10\text{ m/s} = (\pi \times 0.02^2) \times V_2 \qquad \therefore V_2 = \underline{2.5\text{ m/s}}$$

Since the units are the same on both sides of the equation, V_2 will have the same units as V_1.

Example 4.2 Water with variable density flowing through a device

Water enters a device through a 0.1m-diameter pipe at 10°C and 3.5 MPa and exits from an 0.2m-diameter pipe at 200°C and 3.5 MPa. Assume steady flow and calculate the exiting velocity if the entering velocity is 3 m/s (see Fig. 4.10). Do not assume the water to be incompressible.

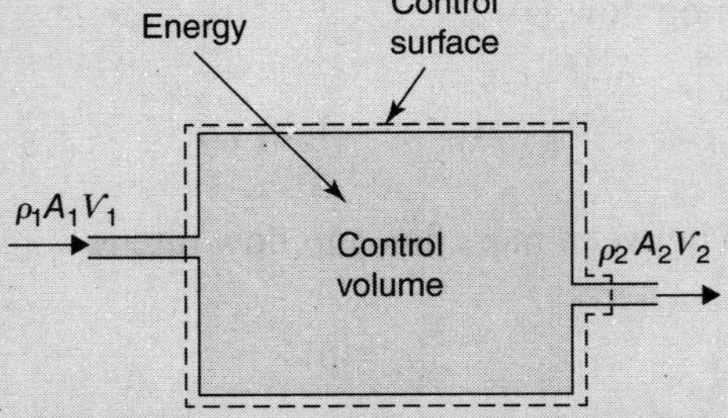

Figure 4.10

Solution

Given:

	State 1	State 2
	$P_1 = 3.5$ MPa	$P_2 = 3.5$ MPa
	$T_1 = 10°C$	$T_2 = 200°C$
	$D_1 = 0.1$ m	$D_2 = 0.2$ m
	$V_1 = 3$ m/s	

(Both states are subcooled liquid water.)

The control volume is again selected so that it encompasses the device and cuts the inlet pipe anywhere upstream of where it enters the device but some distance downstream of where the exit pipe leaves the device. (The pressure P_2 is uniform across the downstream section if the streamlines in the pipe are parallel to the walls, so section 2 must be downstream of the exit.)

Steam Table C-4 is used to find the specific volumes (although Table C-1 using the temperature only could be used with sufficient accuracy):

$$v_1 = 0.001 \text{ m}^3/\text{kg} \qquad v_2 = 0.00115 \text{ m}^3/\text{kg}$$

The continuity equation (4.6) provides

$$\rho_1 A_1 V_1 = \rho_2 A_2 V_2$$

$$\frac{1}{0.0010 \text{ m}^3/\text{kg}} \times \pi(0.05)^2 \text{ m}^2 \times 3 \text{ m/s} = \frac{1}{0.00115 \text{ m}^3/\text{kg}} \times \pi(0.1) \text{ m} \times V_2$$

$$\therefore V_2 = \underline{0.86 \text{ m/s}}$$

where the density was used as $\rho = 1/v$. The units must be the same on either side of the equation, so the velocity must be in m/s. If the density was considered constant, $V_2 = 0.75$ m/s. Higher temperature differences in liquids can have a significant influence on the density in the continuity equation.

Note: Subcooled liquids, compressed liquids, or just plain liquids are all synonymous.

Flow of air through a room — Example 4.3

Air enters a room at two inlets and leaves at one outlet, as shown in Fig. 4.11. Determine V_3 for a steady-flow situation. The temperature and pressure of the air at all locations are approximately 20°C and 100 kPa, respectively.

Solution

The mass flow rate into the control volume must equal the mass flow rate out of the control volume if the flow is steady, so that the mass of air in the control

(Continued)

Example 4.3 (*Continued*)

volume is not increasing or decreasing. The continuity equation (refer to Eq. 4.7) would take the form

$$\rho_1 A_1 V_1 + \rho_2 A_2 V_2 = \rho_3 A_3 V_3$$

This can be written as (see Eqs. 4.3 and 4.8)

$$\dot{m}_1 + \rho_2 \dot{V}_2 = \rho_3 A_3 V_3$$

Using the values provided in Fig. 4.10 with $\rho_2 = \rho_3 = P/RT$, this equation provides

$$0.002\,\frac{\text{kg}}{\text{s}} + \frac{100}{0.287 \times 293}\,\frac{\text{kg}}{\text{m}^3} \times 0.002\,\frac{\text{m}^3}{s} = \frac{100}{0.287 \times 293} \times \pi \times 0.1^2 \times V_3$$

$$\therefore V_3 = \underline{0.1172 \text{ m/s}}$$

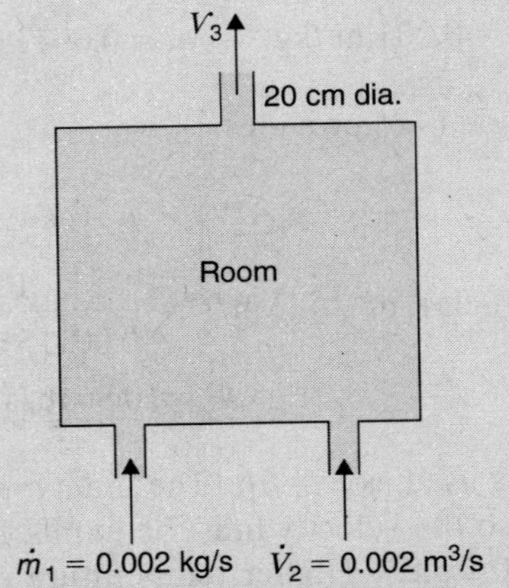

Figure 4.11

In all three examples above, the flows were steady flows. In Example 4.1 the density was constant, in Example 4.2 the conditions were specified and determined by the steam tables, and the inlet density and exit density were different. Example 4.3 involved air that was assumed to be incompressible. If the flow involves a compressible substance, as in flow through a nozzle, a process equation may be needed. In addition, the first law of thermodynamics, developed for a control volume in Section 4.2, may be required.

Note: A process could be isothermal, isobaric, adiabatic, or polytropic.

☑ **You have completed Learning Outcome** **(2)**

4.2 The First Law for Control Volumes

The major difference between the first law applied to control volumes and the first law applied to systems is that the equation is now a rate equation. It still represents conservation of energy, except now we consider rate of heat transfer and work rate across the boundary of the control volume. All terms in the equation will now have units of energy per unit time: kJ/s or kW. Consider the control volume sketched in Fig. 4.12.

Comment

For a control volume, we consider rate of heat transfer and work rate across the boundary of the control volume.

To arrive at the desired energy equation, consider the rate of heat transfer $\dot{Q}$, work rate $\dot{W}$, and the rate of change of the kinetic energy (review Eq. 1.22), the internal energy, and the potential energy. The work rate term $\dot{W}$ is composed of two parts: the work rate required to move the fluid into and out of the control volume due to the pressure forces, and the work rate $\dot{W}_S$ that crosses the surface usually due to a shaft crossing the boundary. For a steady flow with one inlet and one outlet, the conservation of energy demands that

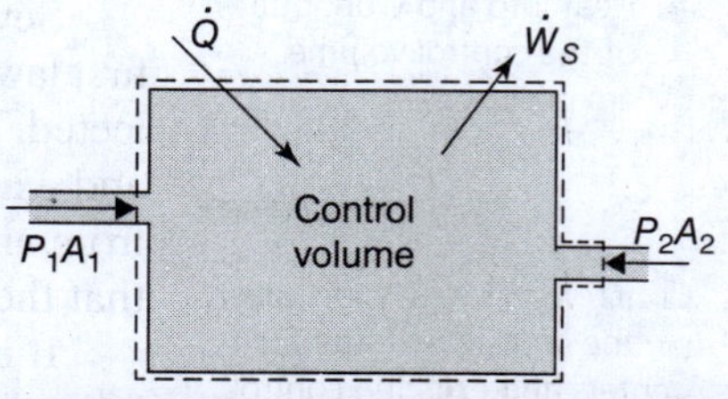

Figure 4.12

The control volume used to derive the energy equation.

$$\begin{pmatrix}\text{Energy crossing}\\ \text{control surface}\end{pmatrix} + \begin{pmatrix}\text{Energy entering}\\ \text{the c.v.}\end{pmatrix} = \begin{pmatrix}\text{Energy exiting}\\ \text{the c.v.}\end{pmatrix}$$

$$\dot{Q} - \dot{W} + \dot{m}_1\left(\frac{1}{2}V_1^2 + u_1 + gz_1\right) = \dot{m}_2\left(\frac{1}{2}V_2^2 + u_2 + gz_2\right) \tag{4.15}$$

The mass flux into, the mass flux out, and the work rate (a force times velocity) are, respectively,

$$\dot{m}_1 = \rho_1 A_1 V_1$$
$$\dot{m}_2 = \rho_2 A_2 V_2 \tag{4.16}$$
$$\dot{W} = P_2A_2V_2 - P_1A_1V_1 + \dot{W}_S$$

Work rate $= \text{Force} \times \text{Velocity}$ $= PA \times V = PAV$

This work due to a pressure force is occasionally called *flow work*.

where $\dot{W}_S$ is the shaft work that crosses the boundary primarily through a rotating shaft. Substitute Eqs. 4.16 into Eq. 4.15 and arrive at

$$\dot{Q} - P_2A_2V_2 + P_1A_1V_1 - \dot{W}_S + \rho_1A_1V_1\left(\frac{1}{2}V_1^2 + u_1 + gz_1\right)$$
$$= \rho_2A_2V_2\left(\frac{1}{2}V_2^2 + u_2 + gz_2\right) \tag{4.17}$$

Additional terms can be added to this equation to account for additional inlets and/or exits. Since there is one inlet and one exit in this steady flow, continuity provides $\dot{m} = \rho_1A_1V_1 = \rho_2A_2V_2$ and Eq. 4.17 takes the form

$$\dot{Q} - \dot{W}_S = \dot{m}\left[\frac{V_2^2 - V_1^2}{2} + u_2 - u_1 + \frac{P_2}{\rho_2} - \frac{P_1}{\rho_1} + g(z_2 - z_1)\right] \tag{4.18}$$

Comment

If the flow is not steady, the term $dE_{c.v.}/dt$ is added to the right-hand side of Eq. 4.17. It accounts for the change in energy in the control volume. The energy $E_{c.v.}$ would include internal energy U, kinetic energy KE, and potential energy PE.

Assumptions for the energy equation:

1. Steady flow.
2. One inlet and one outlet.
3. Uniform flow at the inlet and outlet (the velocity, pressure, and enthalpy are uniform across the areas).
4. Heat into and work out of the control volume are positive.

Control surface: It includes the areas where the fluid enters and exits the control volume.

Note: The shaft work term could include a resistance heater or a paddle wheel in the control volume, in which case the term would be negative.

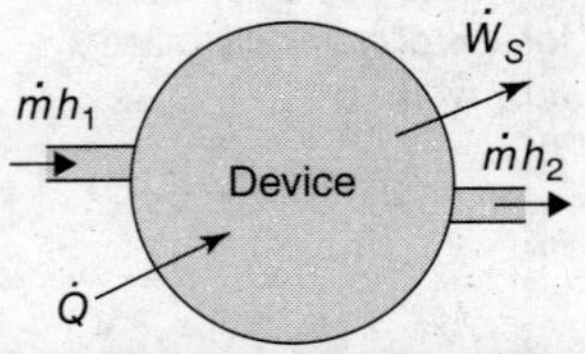

Figure 4.13

The steady flow of energy into and out of a device.

Enthalpy ($h = u + P/\rho$) can be introduced to obtain

$$\dot{Q} - \dot{W}_S = \dot{m}\left[h_2 - h_1 + \frac{1}{2}\left(V_2^2 - V_1^2\right) + g(z_2 - z_1)\right] \tag{4.19}$$

We refer to this equation as the *energy equation* or the *first law*.

The left-hand side of Eq. 4.19 is the sum of the net rate of heat transfer into the control volume minus the net rate at which work is done by the control volume, most often by a rotating shaft. Similar to the energy consideration in Chapter 3, the boundary phenomena of the heat transfer rate minus the work rate represents a net rate of energy transfered to the control volume across the *control surface*.

Several points must be considered in the analysis of most problems in which the first law is used. As a first step, it is very important to identify the control volume selected. Those parts of the control surface should be identified where the fluid enters and exits the control volume. The control surface should be chosen sufficiently far from an abrupt area change (an entrance, an exit, a valve, or a sudden contraction) that the velocity and pressure can be approximated by uniform distributions.

It is also necessary to specify the process by which the flow variables change from the inlet to the outlet. If the fluid can be treated as an ideal gas with constant specific heats, then simplified equations may be used; if not, one of the methods listed in Section 2.6 must be applied. If the fluid is a low-speed gas or a liquid, a constant density process is often assumed.

In most devices of interest in thermodynamics, the kinetic and potential energy changes are negligible. For such devices, illustrated in Fig. 4.13, the steady-flow energy equation takes the simplified form

$$\dot{Q} - \dot{W}_S = \dot{m}(h_2 - h_1) \tag{4.20}$$

Devices for which Eq. 4.20 provides acceptable results include those for which the fluid is compressible such as turbines, compressors, and heat exchangers. The kinetic energy term must be retained in nozzles and diffusers. The gravity term is most often, if not always, neglected; it is very important in the analysis of a dam used to generate power, but that subject is covered in a fluid mechanics course.

If a device utilizes an incompressible fluid (constant density) with negligible kinetic and potential energy changes, with $u_1 = u_2$ (negligible temperature change is assumed), return to Eq. 4.18 and a simplified form of the energy equation results:

$$\dot{Q} - \dot{W}_S = \dot{m}\left(\frac{P_2 - P_1}{\rho}\right) \tag{4.21}$$

This equation is used to find the minimum power required to drive, for example, a water pump, a device that exists in every electrical power plant, or to determine the maximum power generated by a hydroturbine. The heat transfer term is typically neglected.

The next several subsections present the common devices encountered in thermodynamics that are modeled as control volumes.

☑ You have completed Learning Outcome (3)

4.2.1 Turbines, compressors, and pumps

A *pump* is a device that transfers energy to a liquid with the result that the pressure of the liquid is increased. Power is supplied to the pump by a motor. *Compressors* and *blowers* also fall into this category but have the primary purpose of increasing the pressure in a gas. A *turbine*, on the other hand, is a device in which work is done by the fluid on a set of rotating blades. As a result, there is a pressure drop from the inlet to the outlet and mechanical power leaves the turbine. In some situations, heat may be transferred from any of these devices to the surroundings, but often the heat transfer is negligible. In addition, the kinetic and potential energy changes are negligible. For adiabatic devices ($\dot{Q} = 0$), for which the kinetic and potential energy changes can be neglected, operating in a steady-state mode, the energy equation (see Eq. 4.20) takes the form

Pumps, compressors, and blowers: Devices that increase the pressure in a fluid.

Turbine: A device that produces work by using a set of rotating blades to extract energy from a fluid.

$$-\dot{W}_S = \dot{m}(h_2 - h_1) \qquad \text{or} \qquad -w_S = h_2 - h_1 \tag{4.22}$$

Note: Kinetic and potential energy changes are typically neglected in pumps, turbines, and compressors.

The power $\dot{W}_S$ and specific work w_S are positive for a turbine and negative for a compressor.

If an ideal gas, such as air, is the fluid in a compressor or a turbine, as in the engine of a commercial aircraft, the process from the inlet to the outlet can be assumed to be an adiabatic quasi-equilibrium process (no heat transfer and no losses). Thus, assuming constant specific heats, an approximation to the unknown temperature is

$$T_2 = T_1\left(\frac{P_2}{P_1}\right)^{(k-1)/k} \tag{4.23}$$

This is a very important equation since the compression or expansion of a gas is a common process in several devices.

The specific heats of choice in this chapter will be found in Table B-2 in the Appendix, which is common practice, although a more accurate answer would be found using those found in Table B-6 (an average temperature would be assumed for problems where the temperatures are not given). Variable specific heats for ideal gases, which will provide for more accurate calculations, will be considered in Chapter 6. The constant specific heats used in this chapter allow quick comparisons of various design changes; they are also used on national exams such as the Fundamentals of Engineering Examination.

For liquids that are incompressible, such as water flowing through a pump, neglecting heat transfer and losses, the power required by the pump is $\dot{W}_P = -\dot{W}_S$ so Eq. 4.21 becomes

$$\dot{W}_P = \dot{m}\left(\frac{P_2 - P_1}{\rho}\right) \tag{4.24}$$

For a hydroturbine with the same assumptions, the power produced is $\dot{W}_T = \dot{W}_S$, so

$$\dot{W}_T = \dot{m}\left(\frac{P_1 - P_2}{\rho}\right) \tag{4.25}$$

Example 4.4 The compression of air

Air is compressed from 100 kPa to 500 kPa. If the inlet temperature is 20°C, as displayed in Fig. 4.14, determine the compressor power required for a quasi-equilibrium adiabatic process from inlet to outlet if the mass flux is 0.4 kg/s.

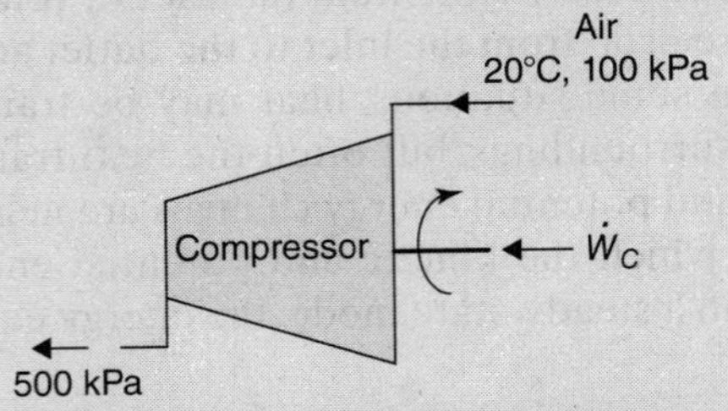

Figure 4.14

Recall: Assume an ideal gas with constant specific heats taken from Table B-2.

Solution

Given:

State 1	State 2
$P_1 = 100$ kPa	$P_2 = 500$ kPa
$T_1 = 20°C$	

Note: $\dot{W}_C = -\dot{W}_S$ since compressor work is negative shaft work. This allows the power required to drive the compressor to be a positive number.

The process across the compressor is an adiabatic quasi-equilibrium process so that Eq. 4.23 provides

$$T_2 = T_1\left(\frac{P_2}{P_1}\right)^{(k-1)/k} = (20 + 273) \times 5^{0.4/1.4} = 464 \text{ K}$$

The power required, using $\dot{W}_C = -\dot{W}_S$, is then

$$\dot{W}_C = -\dot{W}_S = \dot{m}(h_2 - h_1) = \dot{m}C_p(T_2 - T_1)$$

$$= 0.4\frac{\text{kg}}{\text{s}} \times 1.0\frac{\text{kJ}}{\text{kg}\cdot\text{K}} \times (464 - 293)\text{ K} = \underline{68.4 \text{ kW}}$$

This represents the least power required since an ideal process was assumed. Losses in the form of heat transfer from the compressor, and friction and separation as the air moves over the blades could be accounted for by using an efficiency for the device. The losses would increase the required power. Efficiency will be introduced in Chapter 6.

Example 4.5 A turbine generating power using steam

Steam enters the turbine of Fig. 4.15 at 4000 kPa and 500°C and leaves as a saturated vapor at 80 kPa. For an inlet velocity of 200 m/s, calculate the turbine power output. Neglect any heat transfer and kinetic energy change. (This process does not allow Eq. 4.23 to be applied since it is not an ideal gas. Also, there may be significant losses around the turbine blades.)

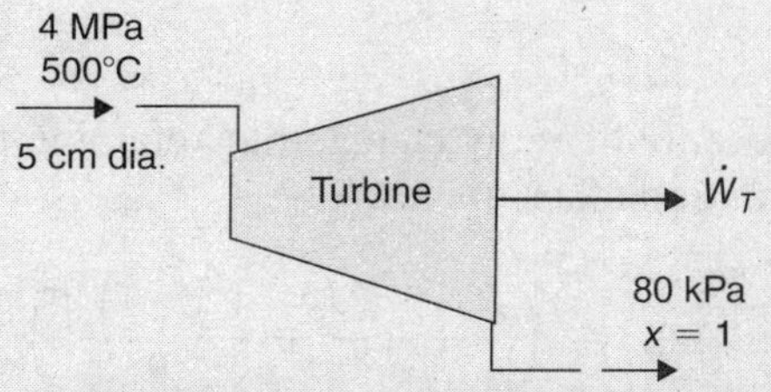

Figure 4.15

Solution

Given:

	State 1	State 2
	$P_1 = 4000$ kPa	$P_2 = 80$ kPa
	$T_1 = 500°C$	$x = 1$
Steam tables:	$h_1 = 3445$ kJ/kg	$h_2 = 2666$ kJ/kg

The energy equation, in the form of Eq. 4.22, is

$$-\dot{W}_T = \dot{m}(h_2 - h_1)$$

The mass flux $\dot{m}$ is found as follows:

$$\dot{m} = \rho_1 A_1 V_1 = \frac{1}{v_1} A_1 V_1 = \frac{1}{0.08643} \times \pi \times 0.025^2 \times 200 = 4.544 \text{ kg/s}$$

The maximum power output is then

$$\dot{W}_T = -4.544 \frac{\text{kg}}{\text{s}} \times (2666 - 3445) \frac{\text{kJ}}{\text{kg}} = 3540 \text{ kJ/s} \quad \text{or} \quad \underline{3540 \text{ kW}}$$

Units: Always use kg, m, and s and the units will work out. It is not necessary to check the units in every problem, just the ones that include terms with unusual units. But check them occasionally to be sure.

A pump increasing the pressure of water — Example 4.6

Determine the maximum pressure increase provided by the pump that requires 10 hp. Figure 4.16 shows a velocity of 4 m/s at the 6-cm-diameter inlet; there is a 10-cm-diameter exit. Do not neglect the kinetic energy change.

Note: If the kinetic energy change is to be ignored, the velocity V_2 would not be of interest so continuity would not be used.

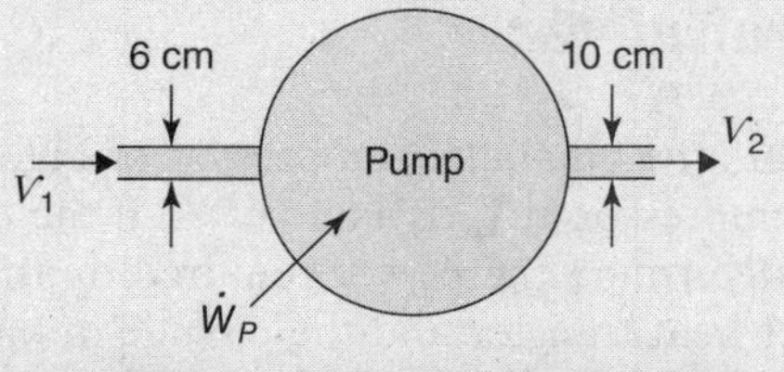

Figure 4.16

Solution

The energy equation (4.18) is used. By neglecting the heat transfer and assuming no increase in internal energy (100% efficiency), we establish the maximum pressure

(*Continued*)

Note: Using the density of water as 1000 kg/m^3 is acceptable unless the temperature exceeds about 100°C, as in Example 4.2.

Example 4.6 (Continued)

rise. Neglecting the potential energy change but retaining the kinetic energy change, the energy equation takes the form

$$-\dot{W}_S = \dot{m}\left(\frac{P_2 - P_1}{\rho} + \frac{V_2^2 - V_1^2}{2}\right)$$

The velocity V_1 is given, and V_2 is found from the continuity equation as follows:

$$A_1V_1 = A_2V_2 \quad \pi \times 0.03^2 \times 4 = \pi \times 0.05^2 \times V_2 \quad \therefore V_2 = 1.44 \text{ m/s}$$

The mass flow rate needed in the energy equation, using $\rho = 1000 \text{ kg/m}^3$, is

$$\dot{m} = \rho A_1 V_1 = 1000 \frac{\text{kg}}{\text{m}^3} \times (\pi \times 0.03^2) \text{ m}^3 \times 4 \frac{\text{m}}{\text{s}} = 11.31 \text{ kg/s}$$

Recognizing that the pump work is negative, we find that the energy equation is

$$-(-10) \text{ hp} \times 746 \frac{\text{W}}{\text{hp}} = 11.31 \frac{\text{kg}}{\text{s}}\left[\frac{\Delta P}{1000} + \frac{1.44^2 - 4^2}{2}\right]\frac{\text{J}}{\text{kg}}.$$

$$\therefore \Delta P = 667\,000 \text{ Pa}$$

or a maximum pressure rise of 667 kPa. The pump efficiency (to be introduced in Chapter 6) would decrease this pressure. Check the units in this equation. If ΔP is in kPa, then $\dot{W}_P$ would be in kW, and a factor of 1000 would be needed in the denominator of the kinetic energy term to convert from J/kg to kJ/kg.

Note: In this example the kinetic energy terms account for about 1% of the energy transfer. This is very typical, and in most applications the inlet and exit areas are not given. But even if they are, as in this example, kinetic energy changes can usually be ignored in a pump or turbine.

Units on $\Delta P/\rho$: This is encountered quite often so make sure it is understood:

$$\frac{\text{N/m}^2}{\text{kg/m}^3} = \frac{\text{N}\cdot\text{m}}{\text{kg}} = \frac{\text{J}}{\text{kg}}$$

Units on $V^2/2$:

$$\frac{\text{m}^2}{\text{s}^2} = \frac{\text{N}\cdot\text{m}}{\text{N}}\cdot\frac{\text{m}}{\text{s}^2}\frac{\text{J}}{\text{N}\cdot\text{s}^2/\text{m}} = \frac{\text{J}}{\text{kg}}$$

Comment

Engineers, in general, accept accuracy within 3%, or so. Most problems have dimensions or material properties that are not known to four significant digits, and probably known to just two or three, at best. The numbers in the tables are not measured values but are simply curve fits. So, a kinetic energy change of under 3% is usually ignored.

Example 4.7 Compressing a refrigerant

Refrigerant 134a enters the adiabatic compressor of Fig. 4.17 at 100 kPa and −7°C. It leaves the compressor at 1 MPa and 65°C. If the mass flow rate of the refrigerant is 1 kg/s, determine the power required by the compressor to accomplish this. Neglect heat transfer and the change in kinetic energy. (There could be losses, so we do not assume a quasi-equilibrium process.)

R134a:	State 1 (entrance)	State 2 (exit)
	$P_1 = 100$ kPa	$P_2 = 1$ MPa
	$T_1 = -7$°C	$T_2 = 65$°C
	superheated vapor	superheated vapor

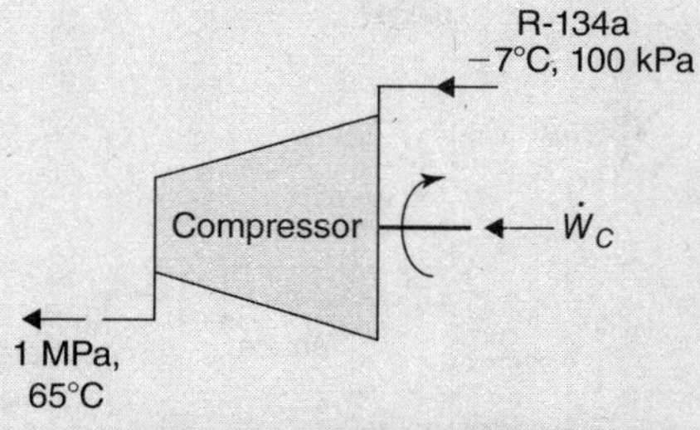

Figure 4.17

Using the IRC Calculator,

$$h_1 = 251 \text{ kJ/kg} \qquad h_2 = 300 \text{ kJ/kg}$$

Solution

Use the first law to calculate the work rate required by the compressor:

$$\dot{W}_C = -\dot{W}_S = \dot{m}(h_2 - h_1) = 1\frac{\text{kg}}{\text{s}} \times (300 - 251)\frac{\text{kJ}}{\text{kg}} = 49 \text{ kW}$$

$$\text{or} \quad \frac{49 \text{ kW}}{0.746 \text{ kW/hp}} = \underline{65.7 \text{ hp}}$$

Digits: Only two significant digits remain after the subtraction.

As usual, the kinetic and potential energy changes were neglected. Information to include those changes was not given, as is usually the case.

4.2.2 Throttling devices

A *throttling device*, or simply a *throttle*, involves a steady-flow process that provides a relatively large pressure drop with no heat transfer or work and no significant change in potential energy and kinetic energy. Two such devices are sketched in Fig. 4.18. For this process (see Eq. 4.20),

Throttling device: An adiabatic device that provides a sudden pressure drop.

$$h_2 = h_1 \tag{4.26}$$

where section 1 is upstream and section 2 is downstream. Most valves are throttling devices, for which the energy equation takes the form of Eq. 4.26. They are also used in many refrigeration units in which the sudden drop in pressure causes a change in phase of the refrigerant, resulting in the required temperature drop. Example 4.9 shows that, as a refrigerant passes through a throttle, the temperature may drop by over 25°C.

A throttling calorimeter, sketched in Fig. 4.19, is a device used to determine the quality of the steam, for which $x < 1$, flowing in a pipe. A valve is placed in a length of pipe so that it can vent steam from the pipe. An orifice is attached to the outlet of the wide-open valve, and a thermometer measures the temperature of the steam downstream of the orifice, as shown. Knowing the pressure and temperature of the steam venting into the atmosphere, the enthalpy of the exiting steam can be determined. Since the enthalpy doesn't change across an orifice, the

(*a*) An orifice plate

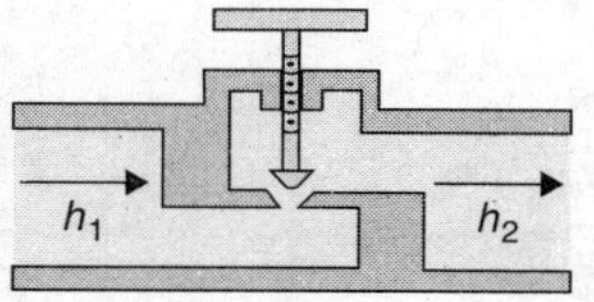

(*b*) A valve

Figure 4.18

Throttling devices.

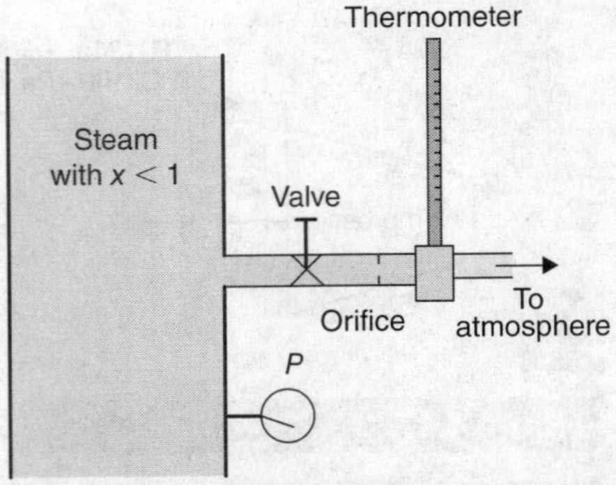

Figure 4.19

A throttling calorimeter.

enthalpy of the saturated steam in the pipe will be the same as the exhaust steam; it's an isenthalpic process. Knowing the enthalpy and pressure of the steam inside the pipe allows us to determine other properties of the steam which is in the wet zone.

A throttling calorimeter is quite complicated, and care must be taken in determining the quality of the steam. The involved process is much more complicated than illustrated in the following example, but the example includes the basic concepts.

Example 4.8 A throttling calorimeter

Wet steam ($x < 1$) at 600 kPa flows in a pipe. A throttling calorimeter similar to the one shown in Fig. 4.19 has been mounted on a pipe, and the thermometer on the calorimeter reads 120°C. Determine the quality of the steam in the pipe.

Solution

The steam coming from the calorimeter has a temperature of 120°C and is being vented to the atmosphere. At 100 kPa, the IRC Calculator shows this steam to be a superheated vapor with a specific enthalpy of 2 720 000 J/kg. Thus, the steam in the pipe also has a specific enthalpy of 2 720 000 J/kg with a pressure of 600 kPa. The IRC Calculator for 600 kPa with $h = 2\,720\,000$ J/kg provides the quality of the steam in the pipe:

$$x = \underline{0.983}$$

Note: If the atmospheric pressure is not given, it is assumed to be the standard pressure of 100 kPa.

Example 4.9 R134a passing through a throttling plate

Refrigerant 134a leaves the condenser of an air-conditioning system at 900 kPa and 34°C and passes through the throttle of Fig. 4.20, where the pressure is reduced to 140 kPa. Determine the temperature and quality of the refrigerant as it leaves the throttle.

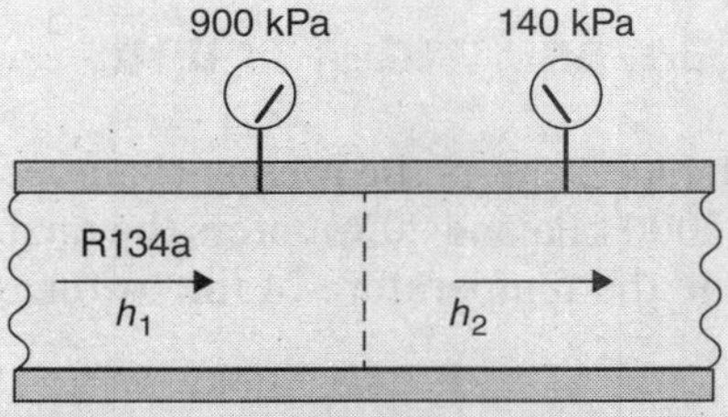

Figure 4.20

Solution

The refrigerant enters the throttle as a subcooled liquid at 900 kPa and 34°C. The specific enthalpy of the refrigerant is, from Table D-1,

$$h_1 = h_{f@34°C} = 97.31 \text{ kJ/kg}$$

Since the enthalpy is constant across a throttle, the exit-specific enthalpy is also 97.31 kJ/kg and the exit pressure is 140 kPa.

For a pressure of 140 kPa, $h_f = 25.77$ kJ/kg and $h_g = 236$ kJ/kg. Thus, the exit state is in the quality region since $h_2 = h_1 = 97.31$ kJ/kg. The corresponding saturation temperature is

$$T_{\text{exit}} = T_{\text{sat}} = \underline{-18.8°\text{C}}$$

The exit quality is calculated to be

$$x_{\text{exit}} = \frac{97.31 - 25.77}{236 - 25.77} = \underline{0.340}$$

The IRC Calculator could have been used. With $h = 97300$ J/kg and $P = 140$ kPa, we find $T_{\text{exit}} = \underline{-18.8°\text{C}}$ and $x_{\text{exit}} = \underline{0.283}$.

Observe: A refrigerant must experience a relatively low temperature after it passes through a throttle in order to be classified as a refrigerant.

Comment

The IRC Calculator is more accurate than the straight-line interpolation.

4.2.3 Mixing chambers

A *mixing chamber* is a device that combines at least two inlet flows to form a single outlet flow. There is no work done on or by the mixing chamber and negligible heat transfer to or from the chamber. Since mass must be conserved for steady flow, the sum of the mass flow rates into the mixing chamber will equal the mass flow rate out of the chamber. In order to ensure a steady flow, the pressures at the inlets and the outlet will be the same. The only energy transfer is done by the physical mixing of the inlet flows to form a mixed exit flow, as displayed in Fig. 4.21. The energy equation, similar to Eq. 4.19, with no work or heat transfer and negligible kinetic and potential energy changes, would require the inlet enthalpy to be equal to the exit enthalpy, that is,

$$\dot{m}_1 h_1 + \dot{m}_2 h_2 = \dot{m}_3 h_3 \tag{4.27}$$

A chamber with two inlets and one exit is encountered quite often when a mixing chamber is used in a cycle. As illustrations, see the next two examples.

Mixing chamber: A device that combines at least two inlet flows to form a single outlet flow.

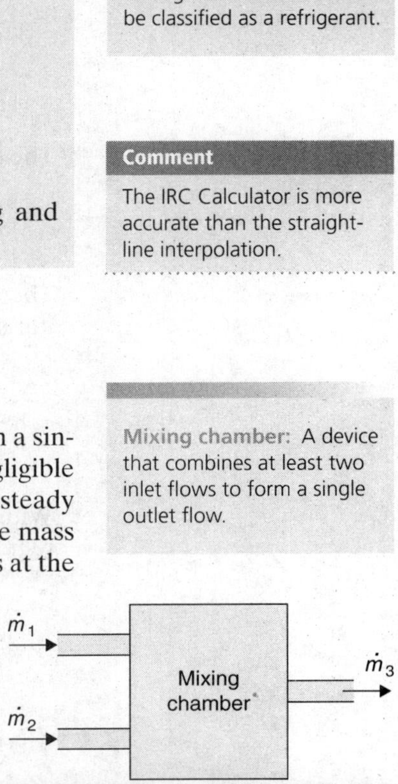

Figure 4.21
A mixing chamber where two fluids are physically mixed.

Example 4.10 The mixing of steam and water, SI units

Steam at 2000 kPa and 400°C enters the mixing chamber of Fig. 4.22 at a flow rate of 5 kg/s. Water at 2000 kPa and 20°C enters the mixing chamber at a flow rate of 2 kg/s. Determine the temperature of the exiting mixture, or quality if in the wet region.

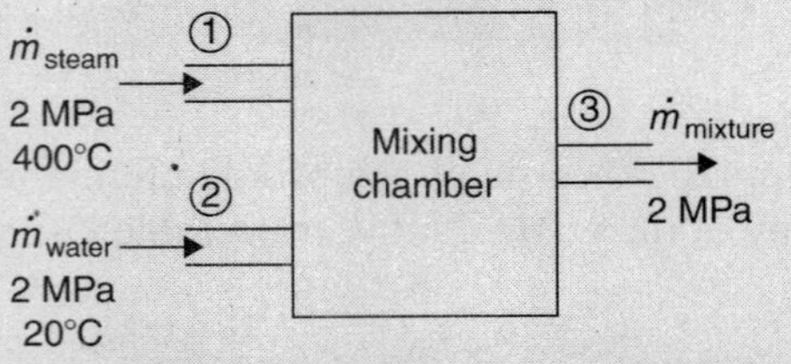

Figure 4.22

Solution

State 1 (steam)	State 2 (water)	State 3 (mixture)
$P_1 = 2$ MPa	$P_2 = 2$ MPa	$P_3 = 2$ MPa
$T_1 = 400°C$	$T_2 = 20°C$	
Superheated	Subcooled	
$h_1 = 3247.6$ kJ/kg	$h_2 = 83.96$ kJ/kg	

The conservation of mass demands that

$$\dot{m}_1 + \dot{m}_2 = \dot{m}_3 = 7\text{ kg/s}$$

There is no heat transfer or work that crosses the boundary of the chamber, nor are there any kinetic or potential energy changes, so Eq. 4.27 is used:

$$\dot{m}_1 h_1 + \dot{m}_2 h_2 = \dot{m}_3 h_3$$

$$5 \times 3247.6 + 2 \times 83.96 = 7 \times h_3. \qquad \therefore h_3 = 2344\text{ kJ/kg}$$

With $h_3 = 2344$ kJ/kg and $P_3 = 2$ MPa, it is observed that state 3 is in the quality region with

$$T_3 = T_{sat} = \underline{212.4°C} \quad \text{and} \quad x_3 = \frac{2344 - h_f}{h_{fg}} = \frac{2344 - 908.8}{1890.7} = \underline{0.759}$$

If the IRC Calculator is used, input $h_3 = 234\,4000$ J/kg and $P_3 = 2000$ kPa:

$$T_3 = \underline{212°C} \text{ and } x_3 = \underline{0.76}.$$

The mixing of water and water, English units — Example 4.11

A flow of water at 1.4 MPa and 30°C enters the mixing chamber of Fig. 4.23 at a flow rate of 2.7 m³/min. A second flow of water at 1.4 MPa and 5°C enters at a flow rate of 2 m³/min. Determine the mass flow rate and temperature of the exit flow.

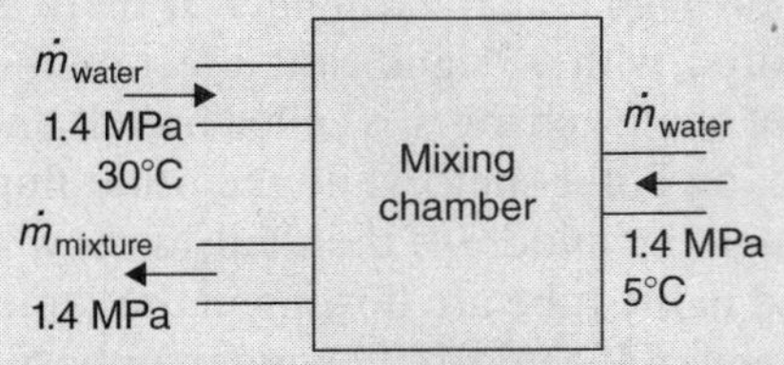

Figure 4.23

Solution

State 1	State 2	State 3
$P_1 = 1.4$ MPa	$P_2 = 1.4$ MPa	$P_3 = 1.4$ MPa
$T_1 = 30°C$	$T_2 = 5°C$	
$v_1 = 9.541 \times 10^{-4}$ m³/kg	$v_2 = 9.548 \times 10^{-4}$ m³/kg	
$h_1 = 135.2$ kJ/kg	$h_2 = 18.7$ kJ/kg	

Use the steam tables to find the density of water, which is $\rho = 1/v$. For the flow at state 1 (see Eq. 4.14):

$$\dot{m}_1 = \frac{\dot{V}_1}{v_1} = \frac{(2.7/60)\ \text{m}^3/\text{s}}{9.541 \times 10^{-4}\ \text{m}^3/\text{kg}} = 47.16\ \text{kg/s}$$

For the inflow at state 2:

$$\dot{m}_2 = \frac{\dot{V}_2}{v_2} = \frac{(2/60)\ \text{m}^3/\text{s}}{9.548 \times 10^{-4}\ \text{m}^3/\text{kg}} = 34.91\ \text{kg/s}$$

Use the conservation of mass and the energy equation (4.27) to establish state 3:

$$\dot{m}_3 = \dot{m}_1 + \dot{m}_2 = 47.16 + 34.91 = \underline{82.07\ \text{kg/s}}$$

$$\dot{m}_1 h_1 + \dot{m}_2 h_2 = \dot{m}_3 h_3 \qquad 47.16 \times 135.2 + 34.91 \times 18.7 = 82.07 \times h_3$$

The exiting enthalpy is calculated to be $h_3 = 85.64$ kJ/kg. At this enthalpy and $P_3 = 1.4$ MPa, the temperature is interpolated to be

$$T_3 = 18 + \frac{85.64 - 77.6}{86.8 - 77.6}(20 - 18) = \underline{19.7°\text{C}}$$

If the IRC Calculator is used to find the exiting temperature, input $P_3 =$ 1400 kPa and $h_3 = 85\,640$ J/kg:

$$T_3 = \underline{20.1°\text{C}}$$

4.2.4 Heat exchangers

Heat exchanger: A device used to transfer thermal energy from a hotter fluid to a colder fluid.

Comment

Pressure is assumed constant as the fluid flows through a heat exchanger.

Heat exchangers are used to transfer thermal energy from a hotter fluid to a colder fluid or from a hot fluid to the cool surroundings. In a power plant, energy is transferred from the hot gases after combustion to the water in the pipes of the boiler. Hot coolant that leaves an automobile engine transfers heat to the surrounding atmosphere by use of a radiator. In the analysis, the pressure is always assumed constant as the fluid flows through a heat exchanger. If there is a small pressure drop due to friction, it is ignored, with no significant effect on the quantities calculated.

The simplest type of heat exchanger is called a shell-and-tube heat exchanger. One fluid flows through an outer shell, while the other fluid flows through a pipe (tube) that passes through the middle of the shell, as shown in Fig. 4.24. If the fluid in the shell and the fluid in the tube are flowing in the same direction, it is called a parallel-flow heat exchanger. If they are flowing in opposite directions, it is called a counter-flow heat exchanger; it is more effective than the parallel-flow exchanger. Certain characteristics may result in one or the other being selected. The larger the temperature difference between the two fluids at any point along the heat exchanger, the more heat per unit surface area is exchanged between the two fluids.

Many design options are available to engineers in selecting the best type of heat exchanger for a particular application. It is possible to have the tube turn around and pass again through the shell. This is called a one shell–two tube heat exchanger. Other options might be a one shell–four tube heat exchanger or a two shell–six tube heat exchanger. Mechanical and chemical engineers are expected to know how to design and analyze heat exchangers for use in industry. A heat transfer course includes a detailed analysis of heat exchangers.

Another type of heat exchanger is the cross-flow heat exchanger. This type is usually used when one fluid is a liquid and the other is a gas, the most common example being a car radiator. Hot coolant from the engine flows through a criss-crossing pattern of tubes. The tubes have plate-type fins attached to them to move heat away from the coolant, as sketched in Fig. 4.25. Air flows perpendicular to the coolant flow across the tubes and fins, picking up heat in its brief contact with the hot tubes. Another good example is the heat exchanger used in air conditioners. These heat exchangers transfer heat between the refrigerant used in the air conditioner and the surrounding air.

In this section we will learn how to perform a basic heat exchanger analysis and determine the effects on the two fluids. Unlike the mixing chamber, the two fluids in a heat exchanger never come in contact; hence, the flow velocity of either

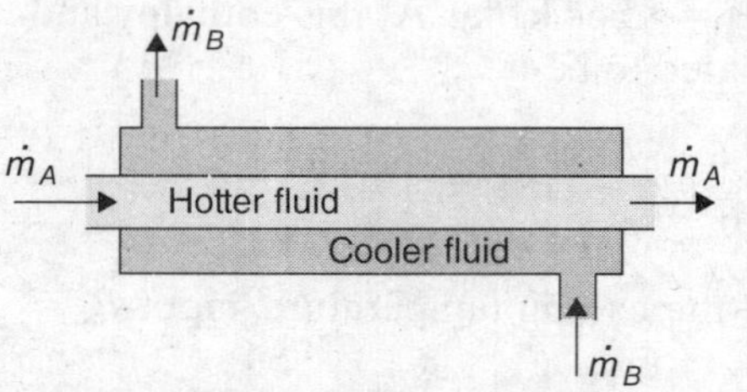

Figure 4.24

A shell-and-tube heat exchanger.

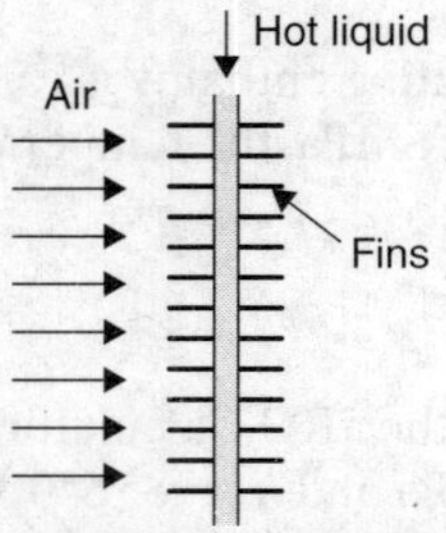

Figure 4.25

One of the tubes, through which a hot liquid flows, with fins attached.

fluid does not normally change. The pressure drop through the passage is usually neglected. Anything that causes a drop in pressure or a change in velocity will require additional pump energy, which is typically quite small, to maintain the flow rate. For this reason, shells and tubing systems in heat exchangers are designed assuming minimal flow resistance and pressure drops. Since heat exchangers are usually small and compact, the potential energy change is assumed zero.

Comment

The pressure in each passage is constant, no work is done, and kinetic and potential energy changes are negligible.

The first law of thermodynamics applied to the two fluids flowing through a heat exchanger will show that the rate of heat flow from the hotter fluid in the exchanger will equal the rate at which the cooler fluid absorbs heat. So, in order to perform a useful analysis, the first law is applied to each fluid and analyzed separately. What relates the two is the fact that the amount of heat removed from the hotter fluid equals the heat added to the cooler fluid. The first law applied to either of the two fluids provides

Key Concept: The rate of heat flow from the hotter fluid in the exchanger will equal the rate at which the cooler fluid absorbs heat.

$$\dot{Q} = \dot{m}(h_2 - h_1) \tag{4.28}$$

The outside of the heat exchanger of Fig. 4.24 is assumed to be insulated, so that all the heat transfer from fluid A is transferred to fluid B; this is expressed by

$$0 = \dot{m}_A(h_{A2} - h_{A1}) + \dot{m}_B(h_{B2} - h_{B1}) \tag{4.29}$$

For the control volumes shown in Figs. 4.26*a* and *b*, the rate of heat transfer would be

$$\dot{Q} = \dot{m}_A(h_{A2} - h_{A1}) \qquad -\dot{Q} = \dot{m}_B(h_{B2} - h_{B1}) \tag{4.30}$$

Each fluid will have an inlet state and an outlet state. In solving heat exchanger problems, begin with the fluid for which you have the most information. Apply the first law to that fluid to find the rate of heat transfer between the two fluids. Then use that information to determine the properties of the other fluid.

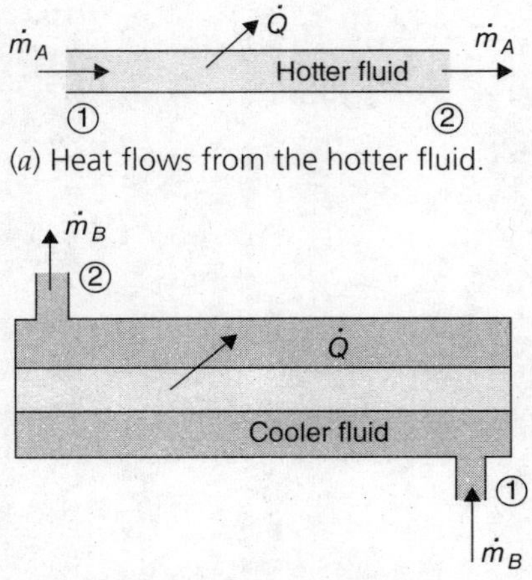

(*a*) Heat flows from the hotter fluid.

(*b*) Heat flows to the cooler fluid.

Figure 4.26
The two flows of Fig. 4.24.

The flow of antifreeze through a radiator — Example 4.12

A radiator uses air to cool ethylene glycol (antifreeze) in a cross-flow heat exchanger. The ethylene glycol (C_p = 2.85 kJ/kg·K) enters the radiator, sketched in Fig. 4.27, at 65°C with a flow rate of 2.2 kg/s. It leaves the radiator at 26°C. Air at 20°C blows over the radiator at a flow rate of 270 m^3/min. Determine the average exit temperature T_4 of the air.

Solution

The antifreeze is analyzed first.

Ethylene Glycol:	<u>**State 1** (entrance)</u>	<u>**State 2** (exit)</u>
	$T_1 = 65°C$	$T_2 = 26°C$
	$\dot{m}_1 = 2.2$ kg/s	$\dot{m}_2 = 2.2$ kg/s

(Continued)

Example 4.12 (*Continued*)

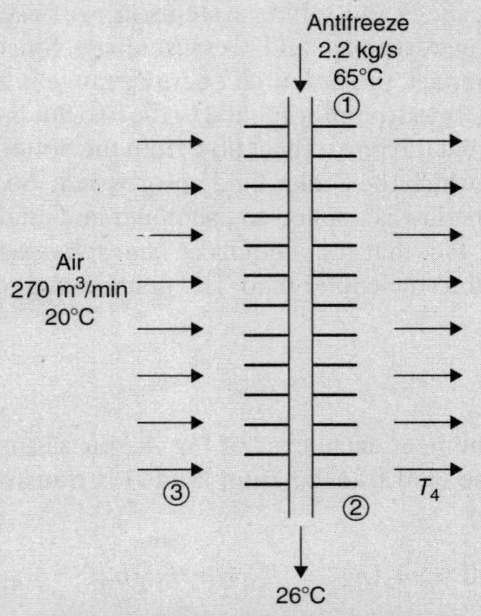

Figure 4.27

The rate of heat transfer from the antifreeze is

$$\dot{Q} = \dot{m}_{eg}C_p(T_2 - T_1)$$

$$= 2.2\frac{\text{kg}}{\text{s}}\left(2.85\frac{\text{kJ}}{\text{kg}\cdot\text{K}}\right)(26 - 65)°\text{K} = -244.5\text{ kW}$$

The rate of heat transfer to the air will be +244.5 kW.

We can now solve for the outlet conditions for the air using the inlet temperature as $T_3 = 20°\text{C}$. The specific volume is

$$v_3 = \frac{RT_3}{P_3} = \frac{\left(0.287\frac{\text{kJ}}{\text{kg}\cdot\text{K}}\right)(293\text{ K})}{101.3\text{ kPa}} = 0.829\text{ m}^3/\text{kg}$$

The mass flux and heat transfer rate to the air are

$$\dot{m} = \frac{\dot{V}}{v} = \frac{(270/60)\text{ m}^3/\text{s}}{0.829\text{ m}^3/\text{kg}} = 5.43\text{ kg/s}$$

$$\dot{Q} = \dot{m}C_p\Delta T$$

$$244.5\text{ kW} = 5.43\frac{\text{kg}}{\text{s}} \times 1.004\frac{\text{kJ}}{\text{kg}\cdot\text{K}} \times (\text{T}_4 - 20)°\text{K}$$

Solving for the exit temperature of the air results in

$$T_4 = \underline{64.8°\text{C}}$$

A heat exchanger in a fast breeder reactor — Example 4.13

Fast breeder nuclear reactors may use liquid sodium to transfer heat from the reactor to the steam used to power the turbine. Liquid sodium, flowing at 100 kg/s, enters the heat exchanger of Fig. 4.28 at 450°C and exits at 350°C. The C_p of liquid sodium is 1.25 kJ/kg·°C. Water enters the heat exchanger at 5000 kPa and 20°C. Calculate the rate of heat transfer and the minimum mass flow rate of the water so that the water exits at 400°C.

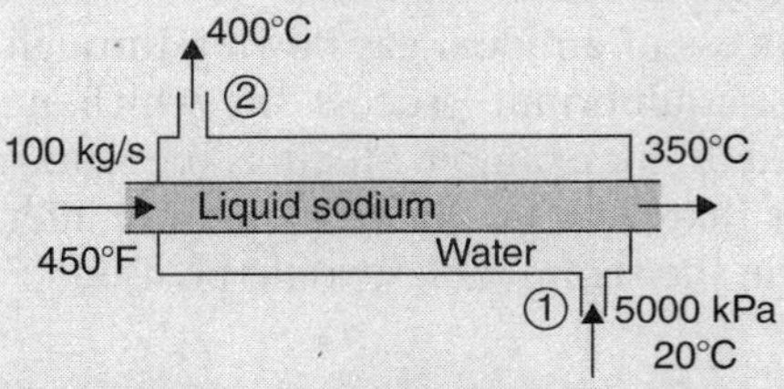

Figure 4.28

Remember: The units on C_v can be written as kJ/kg·K or kJ/kg·°C when it is multiplied by a temperature difference since the temperature difference is the same in kelvins or °C.

Solution

The heat transfer from the liquid sodium can be immediately calculated for this constant-pressure process:

$$\dot{Q} = \dot{m}C_p\Delta T = 100\frac{\text{kg}}{\text{s}} \times 1.25\frac{\text{kJ}}{\text{kg}\cdot{}^\circ\text{C}} \times (350 - 450)\,{}^\circ\text{C} = \underline{-12\,500 \text{ kJ/s}}$$

The negative sign indicates that heat leaves the sodium.

The water enters as a subcooled liquid at 5 MPa and 20°C. The water is expected to leave the heat exchanger as superheated vapor at 5 MPa and 400°C. Properties are found in Tables C in the Appendix.

Water:	**State 1**	**State 2**
	$P_1 = 5$ MPa	$P_2 = 5$ MPa
	$T_1 = 20$°C	$T_1 = 400$°C
	$h_1 = h_{f\text{ at }20^\circ\text{C}} = 83.92$ kJ/kg	$h_2 = 3196$ kJ/kg

Note: Always assume constant pressure as the fluid flows through a heat exchanger.

The heat transfer to the water allows the mass flux to be calculated:

$$\dot{Q} = \dot{m}\Delta h$$

$$12\,500 = \dot{m} \times (3196 - 83.92) \qquad \therefore \dot{m} = \underline{4.02 \text{ kg/s}}$$

4.2.5 Nozzles and diffusers

A *nozzle* is a device that is used to increase the velocity of a flowing fluid. It does this by reducing the flow area in a subsonic or incompressible flow, or by increasing the flow area in a supersonic flow. A *diffuser* is a device that decreases the velocity in a flowing fluid. There is no work input into the devices, and there is usually negligible heat transfer so that they are most often adiabatic devices.

Nozzle: A device that increases the velocity of a flowing fluid.

Diffuser: A device that decreases the velocity in a flowing fluid.

Neglecting potential energy changes, we see that the energy equation (4.15) takes the form

$$0 = \frac{V_2^2 - V_1^2}{2} + h_2 - h_1 \tag{4.31}$$

We must be careful with units when using Eq. 4.31. If SI units are being used, specific enthalpy has units of kJ/kg. The specific kinetic energy $V^2/2$ has units of J/kg. This means that the specific kinetic energy term must be divided by 1000 J/kJ in order to make the term consistent in units.

An important process of an ideal gas flowing through a nozzle or a diffuser is the adiabatic quasi-equilibrium process for which it can be assumed that Eq. 3.45 can be applied. If, for example, steam is the working fluid, the discussion after Eq. 3.47 suggests that $\psi_2 = \psi_1$ (or $s_2 = s_1$) from inlet to outlet, so the property called "entropy" in the steam tables would be used.

Example 4.14 Air flow through a diffuser

Air flows through the diffuser shown in Fig. 4.29. The inlet conditions are 10 kPa and 200°C. The diffuser exit diameter is adjusted such that the exiting velocity is 20 m/s when the inlet velocity is 160 m/s. Calculate: i) the exit temperature, ii) the mass flow rate, and iii) the exit diameter. Assume an adiabatic quasi-equilibrium flow. Unless stated otherwise, the air will be treated as an ideal gas with constant specific heats.

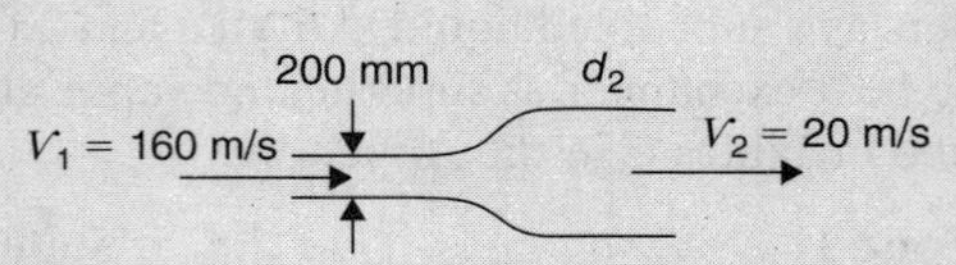

Figure 4.29

Solution

Given:

	State 1	**State 2**
	$P_1 = 10$ kPa	$V_2 = 20$ m/s
	$T_1 = 200$°C	
	$V_1 = 160$ m/s	
	$D_1 = 200$ mm	

i) To find the exit temperature, the energy equation (4.18) is used. Using $\Delta h = C_p \Delta T$, it is

$$0 = \frac{V_2^2 - V_1^2}{2} + C_p(T_2 - T_1)$$

We then have, using $C_p = 1000$ J/kg·K (or 1000 J/kg·°C),

$$T_2 = \frac{V_1^2 - V_2^2}{2C_p} + T_1 = \frac{(160^2 - 20^2)\ \text{m}^2/\text{s}^2}{2 \times 1000\ \text{J/kg}\cdot°\text{C}} + 200°\text{C} = \underline{213°\text{C}}$$

Note: If T_1 is expressed in kelvins, then the answer will be in kelvins.

ii) To find the mass flow rate, we must find the density at the entrance. From the inlet conditions the ideal-gas law provides

$$\rho_1 = \frac{P_1}{RT_1} = \frac{10}{0.287 \times 473} = 0.0737\ \text{kg/m}^3$$

The mass flow rate is then

$$\dot{m} = \rho_1 A_1 V_1 = 0.0737\frac{\text{kg}}{\text{m}^3} \times (\pi \times 0.1^2)\ \text{m}^2 \times 160\frac{\text{m}}{\text{s}} = \underline{0.370\ \text{kg/s}}$$

iii) To find the exit diameter, we use the continuity equation,

$$\rho_1 A_1 V_1 = \rho_2 A_2 V_2$$

This requires the density at the exit. It is found by assuming adiabatic quasi-equilibrium flow of an ideal gas. Referring to Eq. 3.44 and using $\rho = 1/v$, we have

$$\rho_2 = \rho_1\left(\frac{T_2}{T_1}\right)^{1/(k-1)} = 0.0737\left(\frac{485.6}{473}\right)^{2.5} = 0.0787\ \text{kg/m}^3$$

Hence, with $A = \pi d^2/4$,

$$d_2^2 = \frac{\rho_1 d_1^2 V_1}{\rho_2 V_2} = \frac{0.0737 \times 0.2^2 \times 160}{0.0787 \times 20} = 0.299\ \text{m}^2$$

$$\therefore d_2 = 0.547\ \text{m} \quad \text{or} \quad \underline{547\ \text{mm}}$$

Comment

C_p invariably multiplies a temperature difference, such as $\Delta T = 10°\text{C}$ or $\Delta T = 10$ K, so it doesn't matter if K or °C is written in the denominator.

Also, we used $\frac{\text{J}}{\text{kg}} = \frac{\text{N}\cdot\text{m}}{\text{N}\cdot\text{s}^2/\text{m}} = \text{m}^2/\text{s}^2$

Units: The units on the ideal-gas law should be very familiar so that writing them is probably unnecessary.

Problem-solving hints
The following steps will help you set up and solve problems more effectively.

1) Identify the type of device that is being analyzed.
2) Identify the fluid. Can the ideal-gas law be used, or are property tables required?
3) Identify the information given in the problem.

The flow of steam through a nozzle — Example 4.15

Steam enters the nozzle of Fig. 4.30 at 3 MPa and 300°C with a velocity of 3 m/s and leaves at 2 MPa and 300°C. The inlet area is 375 cm^2, and heat is being added to the nozzle at a specific rate of 40 kJ/kg of steam. Determine: i) the mass flow rate, ii) the exit velocity, and iii) the exit area of the nozzle.

(*Continued*)

Example 4.15 (*Continued*)

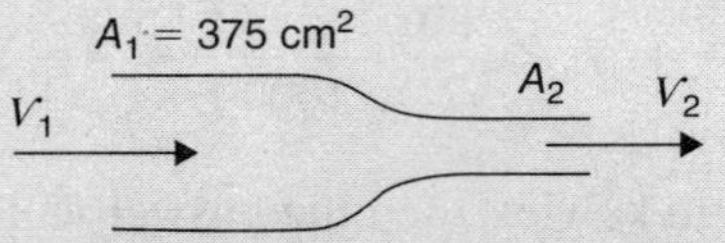

Figure 4.30

Solution

> **Comment**
>
> The steam tables will be used when steam is the fluid, even though there are regions of the steam table where the steam can be considered an ideal gas with acceptable results. The IRC Calculator can also be used.

Given:

	State 1	State 2
	$P_1 = 3$ MPa	$P_2 = 2$ MPa
	$T_1 = 300°$C	$T_2 = 300°$C
	$V_1 = 3$ m/s	
	$A_1 = 375$ cm$^2 = 3.75 \times 10^{-2}$m^2	

From Table C-3:

$v_1 = 0.081$ m^3/kg $\quad v_2 = 0.125$ m^3/kg

$h_1 = 2994$ kJ/kg $\quad h_2 = 3024$ kJ/kg

i) The mass flow rate through the nozzle is

$$\dot{m} = \frac{A_1 V_1}{v_1} = \frac{3.75 \times 10^{-2}\,\text{m}^2 \times 3\text{ m/s}}{0.081\text{ m}^3/\text{kg}} = \underline{1.39\text{ kg/s}}$$

ii) Write the first law (4.15) by dividing by $\dot{m}$ to obtain

$$q = \frac{\dot{Q}}{\dot{m}} = \frac{V_2^2 - V_1^2}{2} + h_2 - h_1$$

$$40\text{ kJ/kg} = \frac{(V_2^2 - 3^2)\text{ m}^2/\text{s}^2}{2000} + (3024 - 2994)\text{ kJ/kg}$$

Solve for the exit velocity V_2 to obtain $V_2 = \underline{141\text{ m/s.}}$

iii) The mass flow rate out of the nozzle equals the mass flow rate into the nozzle:

$$\dot{m}_2 = \rho_2 A_2 V_2 = \frac{A_2 V_2}{v_2} = \frac{A_2 \times 141\text{ m/s}}{0.125\text{ m}^3/\text{kg}} = 1.39\text{ kg/s}$$

$$\therefore A_2 = 1.23 \times 10^{-3}\text{ m}^2 = \underline{12.3\text{ cm}^2}$$

☑ **You have completed Learning Outcome (4)**

4.3 Unsteady Flow

This section will present the unsteady flow of the filling and discharge of a tank with an ideal gas. A steady flow, a flow independent of time, has been assumed in the problems considered in the previous sections of this chapter. In this section, the time-dependent terms in the conservation of mass and the energy equation must be retained when analyzing a flow that is changing with time. Filling a balloon or a tire with air, the depletion of an underground cavern of natural gas where it has been stored during the summer, are examples of unsteady-flow problems.

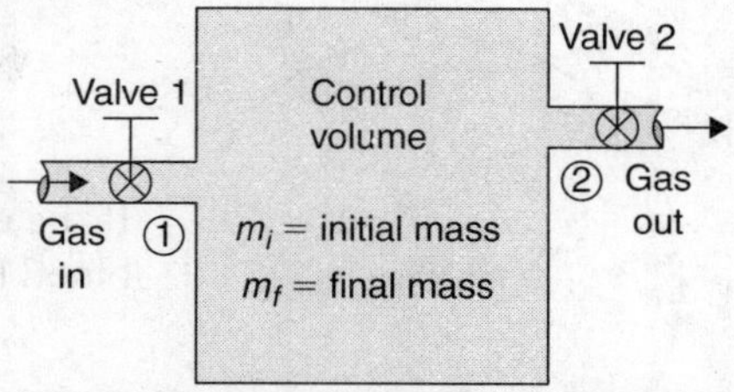

Figure 4.31
The unsteady flow of a gas into or out of a volume.

The flow entering and leaving a volume through pipes, as in Fig. 4.31, will be assumed to be uniform flow so that the energy equation (4.15), assuming $\dot{m}_2 \neq \dot{m}_1$, takes the form

$$\dot{Q} - \dot{W}_S = \dot{m}_2\left(h_2 + \frac{V_2^2}{2} + gz_2\right) - \dot{m}_1\left(h_1 + \frac{V_1^2}{2} + gz_1\right) + \frac{dE_{\text{c.v.}}}{dt} \quad (4.32)$$

Recall:
$$E = m\left(\frac{V^2}{2} + gz + u\right)$$

where $dE_{\text{c.v.}}/dt$ represents the time-rate-of-change of the energy contained in the control volume. If the kinetic and potential energy changes are insignificant and no shaft work exists, the energy equation reduces to

$$\dot{Q} = \dot{m}_2h_2 - \dot{m}_1h_1 + \frac{d(mu)_{\text{c.v.}}}{dt} \quad (4.33)$$

Consider the situation where a tank is being filled—that is, valve 2 of Fig. 4.31 is closed and valve 1 is open, equivalent to that shown in Fig. 4.32. The energy equation may then be written as

$$\dot{Q} = \frac{d(mu)_{\text{c.v.}}}{dt} - \dot{m}_1h_1 \qquad \text{(filling a volume)} \quad (4.34)$$

If the initial mass in the volume is m_i and the final mass is m_f, an integration (multiply by dt) from time t_i to t_f provides

$$Q = m_fu_f - m_iu_i - m_1h_1 \quad (4.35)$$

assuming h_1 is constant over the time interval of integration, as it would be if the working fluid in the pipe were maintained at constant conditions; enthalpy remains constant across a valve.

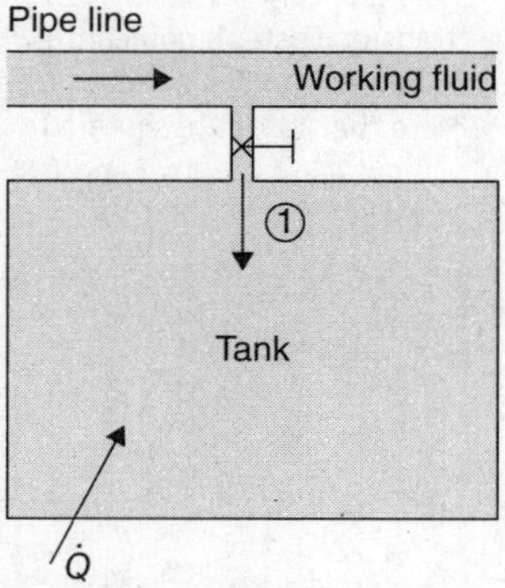

Figure 4.32
The filling of a tank.

The conservation of mass requires that the final mass m_f in the volume is equal to the initial mass m_i plus the mass m_1 that entered, that is,

$$m_f = m_i + m_1 \qquad \text{(filling a volume)} \quad (4.36)$$

Integration of $\dot{m}_1h_1dt$:
$$\int\left(\frac{dm_1}{dt}h_1\right)dt = h_1\int dm_1 = h_1m_1$$

If gas is being discharged from a volume with valve 1 of Fig. 4.31 closed and valve 2 opened, equivalent to Fig. 4.33, the enthalpy of the exiting gas cannot be assumed to be constant. The energy equation, neglecting kinetic and potential energy changes, assuming negligible heat transfer and no shaft work, takes the form

$$0 = \dot{m}_2h_2 + \frac{d(mu)_{\text{c.v.}}}{dt} \qquad \text{(discharging a volume)} \quad (4.37)$$

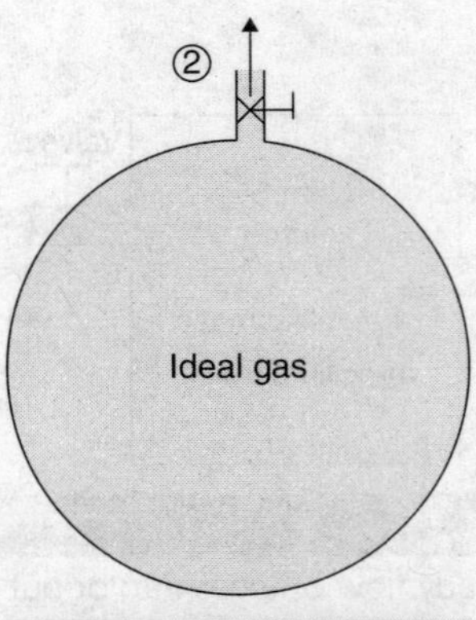

Figure 4.33
A volume being discharged.

The conservation of mass demands that the mass in the control volume decreases as mass leaves the control volume. This is expressed as

$$\frac{dm}{dt} = -\dot{m}_2 \tag{4.38}$$

If the enthalpy is expressed as $h = u + Pv$, the above two equations can be combined to give

$$dm(u_2 + P_2 v_2) = d(mu)_{\text{c.v.}} \tag{4.39}$$

If the velocity is relatively small, as it is assumed to be, just before valve 2 at area A_2, the properties u_2, v_2, and P_2 can be assumed to be approximately equal to u, v, and P in the volume. This allows Eq. 4.39 to be simplified to

$$(u + Pv)dm = udm + mdu \quad \text{or} \quad Pvdm = mdu \tag{4.40}$$

Let's restrict the analysis to an ideal gas with constant specific heats. Then $Pv = RT$ and $du = C_v dT$ so that Eq. 4.40 becomes

$$\frac{dm}{m} = \frac{C_v}{R}\frac{dT}{T} \tag{4.41}$$

Comment

Equation (3.45) for an adiabatic quasi-equilibrium process of an ideal gas with constant specific heats is applicable to a system or in a flow from state 1 to state 2, providing there are no significant losses or heat transfer. Losses significantly affect molecular activity. Here the states are signified by *i* and *f*.

Integrate from an initial state to a final state to obtain

$$\ln\frac{m_f}{m_i} = \frac{1}{k-1}\ln\frac{T_f}{T_i} \tag{4.42}$$

where Eq. 2.35 was used to relate C_v and R. Equation (4.42) can be put in the following forms for this adiabatic quasi-equilibrium process (see Eq. 3.45):

$$\frac{m_f}{m_i} = \left(\frac{T_f}{T_i}\right)^{1/(k-1)} \quad \text{or} \quad \frac{m_f}{m_i} = \left(\frac{P_f}{P_i}\right)^{1/k} \tag{4.43}$$

Several assumptions have been made, such as constant specific heats and low exit velocity which that may make quantities calculated using the above equations only estimates, but such estimates can often aid in decision making.

Example 4.16 The filling of a tank with steam

A 30-m^3 rigid, insulated tank is attached to a steam line maintained at 2 MPa and 400°C, as shown in Fig. 4.34. If it is evacuated before the valve is opened, determine the temperature of the steam in the tank when the pressure just reaches 2 MPa, and the mass of steam that flows into the tank.

Solution

The energy equation (4.35) for an evacuated ($m_i = 0$) insulated ($Q = 0$) tank simplifies to

$$m_f u_f = m_1 h_1$$

But conservation of mass demands that the final mass in the tank must equal the mass that entered, that is, $m_f = m_1$. The above energy equation then provides the surprising result

$$u_f = h_1 = 3250 \text{ kJ/kg}$$

After sufficient time, the final pressure will be the same as the pressure in the steam line, which is 2 MPa. So, at $u_f = 3250$ kJ/kg and $P_f = 2$ MPa, Table C-3 gives

$$T_f = \underline{576°\text{C}}$$

The temperature is much higher in the tank than in the steam line. The "energy" due to the Pv term in $h = u + Pv$ is converted to thermal energy, resulting in the higher temperature.

The mass of steam in the tank can be approximated by interpolating in Table C-3 (or use the IRC Calculator) at 2000 kPa and 576°C,

$$m_f = \frac{V}{v} = \frac{30 \text{ m}^3}{0.194 \text{ m}^3/\text{kg}} = \underline{155 \text{ kg}}$$

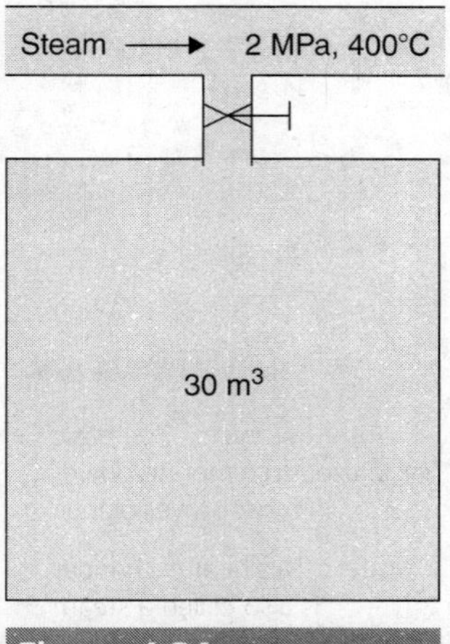

Figure 4.34

The discharge of air from a tank — Example 4.17

A 30-m³ rigid tank containing air is pressurized to 2 MPa and allowed to eventually reach a temperature of 30°C. In Fig. 4.31 valve 2 is then opened with valve 1 closed. Assuming a low velocity of escape and negligible heat transfer, estimate the mass of air in the tank and its temperature when the pressure reaches 200 kPa.

Solution

To use Eq. 4.43, the final mass of air in the tank can be found if the initial mass m_i is known. The initial mass is

$$m_i = \frac{P_i V}{RT_i} = \frac{2000 \times 30}{0.287 \times (30 + 273)} = 690 \text{ kg}$$

Equation (4.43) then is used to find

$$m_f = m_i \left(\frac{P_f}{P_i}\right)^{1/k} = 690\left(\frac{200}{2000}\right)^{1/1.4} = \underline{133 \text{ kg}}$$

The final temperature can now be found also using Eq. 4.43:

$$\frac{m_f}{m_i} = \left(\frac{T_f}{T_i}\right)^{1/(k-1)} \quad \frac{133}{690} = \left(\frac{T_f}{303}\right)^{2.5} \quad \therefore T_f = 157 \text{ K or } \underline{-116°\text{C}}$$

Pressurized nitrogen in large tanks, used as fertilizer by farmers, can be very dangerous if a hose breaks and whips across a farmer's body. The low temperature immediately freezes any exposed body part.

Units: Use kPa when writing pressure and kJ when writing *R*, and the units will check. Of course, temperature is in kelvins. Writing the units whenever you work a problem is like looking at the keys on a piano as you play. It will be pretty slow, but it does avoid errors when you're not sure of yourself!

☑ You have completed Learning Outcome (5)

4.4 Devices Combined into Cycles

All of the devices introduced in Section 4.2 are used in a number of cycles that are commonplace in today's everyday world. Refrigerators, power plants, aircraft engines, large truck engines, all utilize cycles that are based on the devices of this chapter. These cycles are the primary focus of thermodynamics taught in Mechanical Engineering departments, as opposed to thermodynamics taught in Physics or Chemical Engineering departments. In fact, several subsequent chapters will be devoted to a detailed analysis of the three basic cycles introduced in this section. Modifications to these cycles are often made to increase efficiencies so they use less energy. Only the basic cycles will be presented in this section.

Rankine cycle: The basic cycle used to generate power in most power plants.

Boiler: This heat exchanger is also called a steam generator.

Comment

The energy content of trash is about 50% of the energy content of coal, per unit weight. It also reduces the amount of landfill required by over 90%. Visit http://www.ecomaine.org/electricgen/index.shtm.

4.4.1 The Rankine power cycle

The devices in Section 4.2 can be combined into several cycles that produce power. The basic cycle used in nuclear, coal, or trash-burning power plants is the *Rankine cycle*, shown in Fig. 4.35. The cycle is composed of four ideal devices:

- The *pump* compresses saturated liquid water to a high pressure.
- The *boiler* (a heat exchanger) adds heat, often from burning coal, to the high-pressure water to create superheated steam.
- High-pressure, high-temperature steam passes through the *turbine* and produces power.
- The steam that exits the turbine is condensed in a *condenser* (a heat exchanger) to saturated liquid water, which again enters the pump.

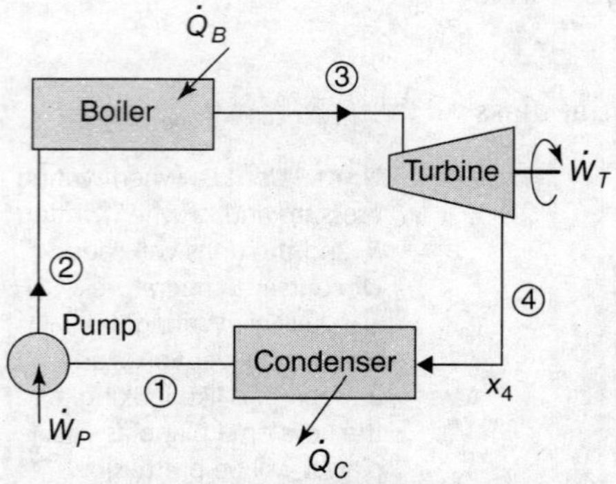

Figure 4.35
The Rankine cycle.

Since the Rankine cycle is used to generate power, we define an efficiency of the cycle to be

$$\eta = \frac{\text{net power output}}{\text{input power}} = \frac{\dot{W}_T - \dot{W}_P}{\dot{Q}_B} \qquad (4.44)$$

The efficiency is usually in the vicinity of 30%; that is, about 30% of the energy needed to heat the water in the boiler is actually converted to useful power from the turbine. The other 70% is discarded by the condenser by discharging the heat transfer $\dot{Q}_C$ since, considering the cycle as a system, the first law requires

$$\dot{Q}_B = \dot{W}_T + \dot{Q}_C \qquad (4.45)$$

Comment

Discarded heat from a power plant can be used to heat (and cool) buildings in a city and a university campus, thereby substantially increasing the effective efficiency of the power plant to perhaps 60%.

Often, some of the discarded heat $\dot{Q}_C$ can be utilized to heat and cool (absorption refrigeration is used; see Chapter 10) campus buildings, a downtown area, or large industrial plants. If this is done, the cycle efficiency remains as defined by Eq. 4.44, even though the desired effect is increased. Half of the discarded heat may be used, so the effectiveness is essentially doubled. That is why many universities have their own power plants, even though their power cycles with smaller devices are less efficient than those used by the large power companies.

The Rankine cycle will be studied in detail in Chapter 8. Several modifications, also introduced in Chapter 8, are made to the cycle in actual power plants to improve the efficiency.

A simple Rankine cycle — Example 4.18

The pump in a Rankine cycle raises the pressure of 8 kg/s of water from 20 kPa to 4 MPa. The boiler raises the temperature to 500°C, and the steam exits the turbine as saturated vapor, as sketched in Fig. 4.36. Calculate:

i) The power required by the pump
ii) The energy input in the boiler
iii) The power produced by the turbine
iv) The cycle efficiency

Neglect any losses by the various components.

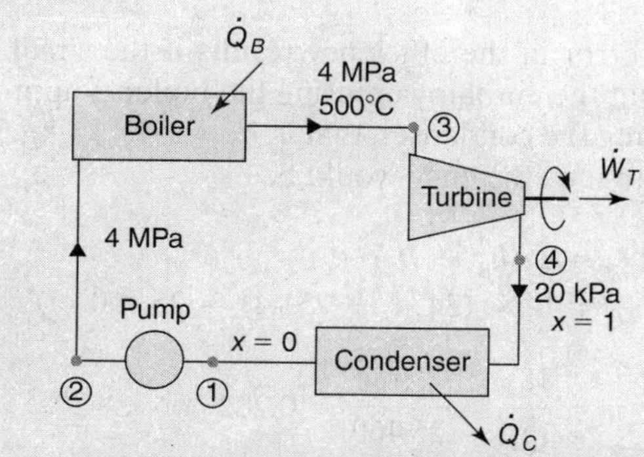

Figure 4.36

Solution

i) The power required to operate the pump is found from Eq. 4.21 to be

$$\dot{W}_P = \dot{m}\frac{P_2 - P_1}{\rho}$$
$$= 8\,\frac{\text{kg}}{\text{s}} \times \frac{(5000 - 20)\ \text{kN/m}^2}{1000\ \text{kg/m}^3} = \underline{40\ \text{kW}}$$

> **Remember:** The pressure through the condenser is assumed to be constant, so the inlet pressure to the pump is 20 kPa.

> **Units:** kN·m/s = kW

ii) The enthalpy h_2 is required to analyze the boiler. It is found by writing the energy equation across the pump (see Eq. 4.19) as

$$\dot{W}_P = \dot{m}(h_2 - h_1)$$
$$40 = 8 \times (h_2 - 251.4) \qquad \therefore h_2 = 256.4\ \text{kJ/kg}$$

> **Observe:** The enthalpy change across the pump is only 5 kJ/kg, less than 2% of h_2. So, we can let $h_2 = h_1$ with no significant effect on the cycle calculations.

Now, the energy equation across the boiler (a heat exchanger), expressed by Eq. 4.28, yields

$$\dot{Q}_B = \dot{m}(h_3 - h_2)$$
$$= 8 \times (3434 - 256.4) = \underline{25\,420\ \text{kJ/s}}.$$

> **Note:** The rate of heat transfer is often written using kJ/s, whereas power is written using kW. The two units are equivalent, so either can be used.

(*Continued*)

Example 4.18 (Continued)

Units: Are these units obvious yet? If not, keep including them until they are. The units should be automatic after sufficient problems are worked.

iii) The turbine produces the power

$$\dot{W}_T = -\dot{m}(h_4 - h_3)$$
$$= -8 \times (2538 - 3434) = \underline{7168 \text{ kW}}$$

iv) The cycle efficiency is found to be

$$\eta = \frac{\text{output}}{\text{input}} = \frac{\dot{W}_T - \dot{W}_P}{\dot{Q}_B} = \frac{7168 - 40}{25\,420} = 0.280 \text{ or } \underline{28.0\%}$$

Comment

The cycle efficiency does not include the energy lost in the combustion of the coal or the energy lost converting the mechanical energy that leaves the turbine to electrical energy that leaves the generator. Those losses would be included in "plant efficiency."

Actually, negligible error in the efficiency results if the small pump power is simply ignored so that the enthalpy entering the boiler is approximately equal to the enthalpy leaving the condenser, that is, $h_2 \cong 251.4$ kJ/kg. Then, the boiler heat transfer rate and the efficiency would be

$$\dot{Q}_B = \dot{m}(h_3 - h_2)$$
$$= 8 \times (3434 - 251.4) = 25\,460 \text{ kJ/s}$$

$$\eta = \frac{\dot{W}_T}{\dot{Q}_B} = \frac{7168}{25\,460} = 0.282 \text{ or } \underline{28.2\%}$$

These are essentially the same numbers (within 1%) as when the pump was included. The energy required to compress a liquid is negligible when compared to the power produced by the turbine.

4.4.2 The refrigeration cycle

Refrigeration cycle: A cycle used to refrigerate or heat a space.

Several of the devices can be combined into cycles that are used to create low temperatures for airconditioning, refrigeration, and heating. These three modes of operation utilize the *refrigeration cycle*, as sketched in Fig. 4.37. It is composed of four devices:

- A *compressor* provides the purchased power $\dot{W}_{Comp}$ needed to convert the fluid, called the refrigerant, from a saturated vapor to a superheated state at a higher pressure and temperature. Heat transfer from a compressor is ignored unless otherwise stated.

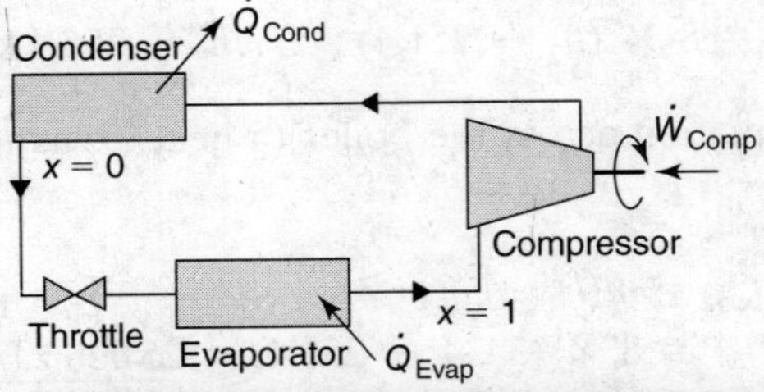

Figure 4.37

A refrigeration cycle.

- A *condenser* (a heat exchanger) condenses the superheated refrigerant to a saturated liquid.
- A *throttle* significantly reduces the pressure of the refrigerant, resulting in a significantly reduced temperature.
- An *evaporator* (a heat exchanger) takes on heat from the surroundings so that the refrigerant leaves as a saturated vapor, as shown in Fig. 4.37.

If the desired effect is to cool a space by transferring heat $\dot{Q}_{Evap}$ from the space to the refrigerant, the grouping of devices is called a *refrigerator*, which includes a freezer, an airconditioner, or a refrigerator. The performance of the cycle is measured by the coefficient of performance, the COP, which is the ratio of the desired effect divided by the purchased power. The COP of a refrigerator that is used to cool a space, with the evaporator providing the desired effect, is

$$\text{COP}_R = \frac{\text{desired effect}}{\text{input power}} = \frac{\dot{Q}_{Evap}}{\dot{W}_{Comp}} \tag{4.46}$$

Comment

Efficiencies are, by definition, less than 100%, but the COP can, and usually is, greater than 1. So, a different term for the performance of a refrigeration cycle is used.

If a control surface is placed around the entire cycle of Fig. 4.37, the arrangement would appear as in Fig. 4.38, with all four devices inside the "black" box. The conservation of energy would then require

$$\dot{W}_{Comp} = \dot{Q}_{Cond} - \dot{Q}_{Evap} \tag{4.47}$$

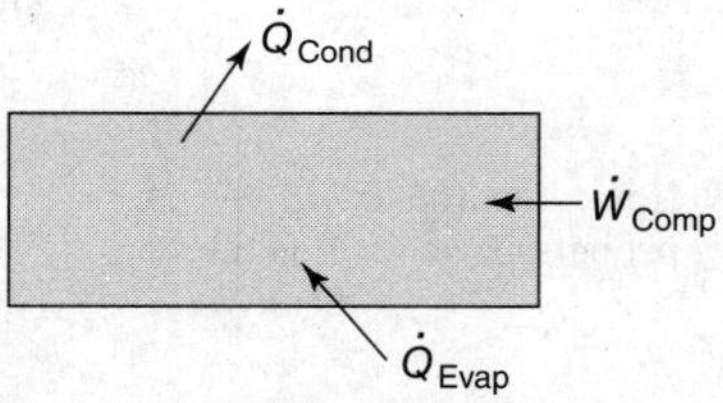

Figure 4.38

The energy entering and leaving the system composed of the devices of Fig. 4.37.

The COP can then be written as

$$\text{COP}_R = \frac{\dot{Q}_{Evap}}{\dot{Q}_{Cond} - \dot{Q}_{Evap}} = \frac{1}{\dot{Q}_{Cond}/\dot{Q}_{Evap} - 1} \tag{4.48}$$

If the desired effect is to heat a space using $\dot{Q}_{Cond}$ from the cycle of Fig. 4.37, the grouping of devices is referred to as a *heat pump*. Its coefficient of performance, with the condenser providing the desired effect, becomes

$$\text{COP}_{HP} = \frac{\text{desired effect}}{\text{purchased power}} = \frac{\dot{Q}_{Cond}}{\dot{W}_{Comp}} \tag{4.49}$$

Comment

A refrigerator cools a space, and a heat pump heats a space. A cycle that does both, as often used in a motel, is usually called a heat pump. Both a heat pump and a refrigerator utilize a refrigeration cycle.

Or, using Eq. 4.47, it is

$$\text{COP}_{HP} = \frac{\dot{Q}_{Cond}}{\dot{Q}_{Cond} - \dot{Q}_{Evap}} = \frac{1}{1 - \dot{Q}_{Evap}/\dot{Q}_{Cond}} \tag{4.50}$$

A comparison of Eqs. 4.48 and 4.50 shows that

$$\text{COP}_{HP} = \text{COP}_R + 1 \tag{4.51}$$

Heat pumps are often used in motels to provide cooling in the summer and heating in the winter. They are often the choice for homes in regions where the cooling load and the heating load are relatively the same, such as through the midsection of the United States.

Example 4.19 A refrigeration cycle

The refrigeration cycle shown cools a space by compressing R134a from 320 kPa to 1000 kPa. Other information is shown in Fig. 4.39. The mass flux of R134a through the cycle is 0.2 kg/s. Determine i) the adiabatic compressor power required, ii) the tons of refrigeration, and iii) the COP. Neglect any losses.

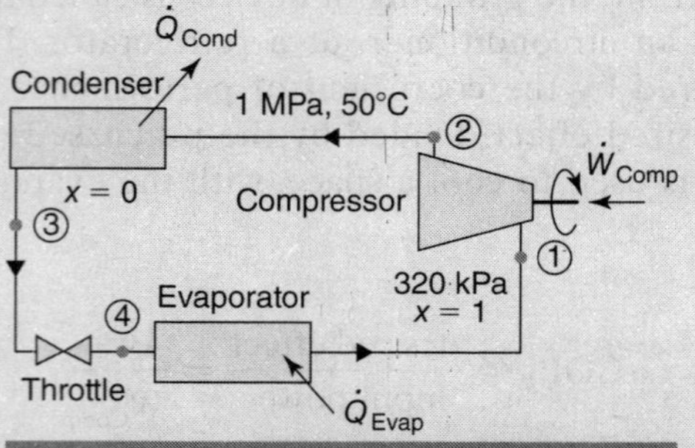

Figure 4.39

Solution

i) The compressor power is found using Eq. 4.22 to be

> **Note:** There are 0.746 kW per horsepower or 1.34 hp per kW.

$$\dot{W}_{Comp} = \dot{m}(h_2 - h_1)$$
$$= 0.2 \times (280.2 - 248.7) = 6.30 \text{ kW} \quad \text{or} \quad \underline{8.44 \text{ hp}}$$

where we used $\dot{W}_{Comp} = -\dot{W}_S$.

ii) The evaporator provides the refrigeration $\dot{Q}_{Evap}$. There is no change in enthalpy across the throttle, so the enthalpy entering the evaporator is the same as the enthalpy leaving the condenser. Since the pressure does not change across the condenser, the enthalpy at 1 MPa and $x = 0$ is $h_3 = h_4 = 105.3$ kJ/kg. So, the energy equation across the evaporator is

> **Enthalpy:** The enthalpy h_3 is h_f found in Table D-2 at 1000 kPa.

$$\dot{Q}_{Evap} = \dot{m}(h_1 - h_4)$$
$$= 0.2 \times (248.7 - 105.3) = 28.68 \text{ kJ/s} \quad \text{or} \quad \underline{8.15 \text{ tons}}$$

> **Note:** One ton of refrigeration is 3.52 kJ/s. A ton of refrigeration is a common measure used to size a refrigeration unit.

iii) The COP is defined by Eq. 4.46. It is

$$\text{COP} = \frac{\dot{Q}_{Evap}}{\dot{W}_{Comp}} = \frac{28.68}{6.30} = \underline{4.55}$$

This means that the cooling energy is 4.55 times the power input. Losses through the various components, especially the compressor, would decrease this COP.

4.4.3 The Brayton cycle

The devices in Section 4.2 can also be combined into a power cycle that operates on a gas, usually air, called the *Brayton cycle*. It is shown in Fig. 4.40. When operating in a large truck, an army tank, or a jet engine, the heat exchanger at the bottom of the cycle is omitted and the air is simply exhausted from the turbine into the atmosphere. The heat exchanger does not affect the analysis of the cycle since the same amount of heat is simply exhausted in an open cycle, and no work is done in either the open or the closed cycle. The energy supplied $\dot{Q}_{in}$ is usually supplied by natural gas in the combustor. The cycle is composed of four devices:

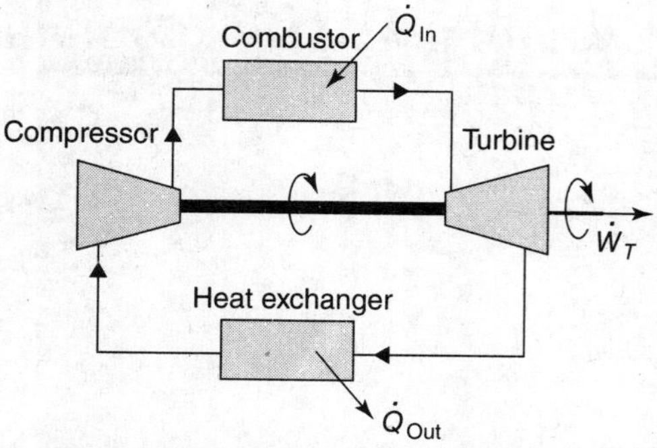

Figure 4.40
The Brayton cycle.

- Air is compressed by the *compressor* to an outlet pressure up to about 30 times the inlet pressure in a jet engine. An adiabatic quasi-equilibrium process is assumed across the compressor.
- Heat is added at constant pressure in the *combustor* (a heat exchanger).
- High-pressure, high-temperature air passes through the *turbine* producing power. An adiabatic quasi-equilibrium process is assumed across the turbine.
- The air that exits the turbine is cooled in a second constant-pressure *heat exchanger* to again enter the compressor.

Since the Brayton cycle generates power, we define an efficiency of the cycle to be

$$\eta = \frac{\text{net power output}}{\text{input power}} = \frac{\dot{W}_{Out}}{\dot{Q}_{In}} \tag{4.52}$$

Note that the power output is

$$\dot{W}_{Out} = \dot{W}_T - \dot{W}_C \tag{4.53}$$

since the turbine drives the compressor, as shown in Fig. 4.40, which requires a considerable amount of power. In fact, it wasn't until more recent times that the turbine and compressor of a Brayton cycle were sufficiently efficient[2] to allow the cycle to be a competitive producer of power.

A simple Brayton cycle — Example 4.20

Air at 2 kg/s enters the compressor of the Brayton cycle of Fig. 4.41 at 100 kPa and 20°C and leaves the compressor at 600 kPa. It leaves the combustor at 900°C. Calculate i) the combustor heat input; ii) the turbine exit temperature; iii) the back-work ratio (BWR), the ratio of the power required by the compressor to the power produced by the turbine; iv) the power output; and v) the efficiency of the cycle. Assume ideal components. *(Continued)*

Back-work ratio: The ratio of the power required by the compressor to the power produced by the turbine.

[2] Inefficiencies in the turbine and compressor are due primarily to the losses associated with the flow of gases around the blades of turbomachinery. The blades are very complex and limit the high temperature in the turbine.

Example 4.20 (Continued)

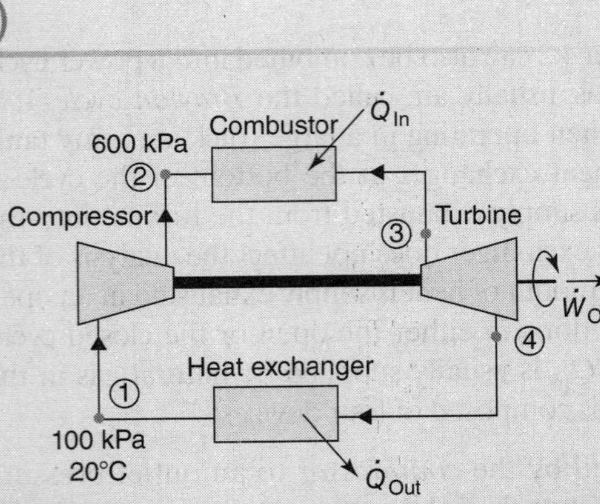

Figure 4.41

Comment

The quasi-equilibrium process in a compressor or a turbine is an ideal process that provides the minimum exit temperature for an actual compressor. An efficiency, to be introduced in Chapter 6, will account for the losses in a real device.

Solution

i) The temperature of the air leaving the compressor is found assuming a quasi-equilibrium adiabatic process with no losses (see Eq. 3.45) to be

$$T_2 = T_1\left(\frac{P_2}{P_1}\right)^{(k-1)/k} = (20 + 273) \times 6^{0.2857} = 489 \text{ K}$$

The heat input by the combustor, a heat exchanger, is

$$\dot{Q}_{In} = \dot{m}(h_3 - h_2) = \dot{m}C_p(T_3 - T_2)$$
$$= 2\frac{\text{kg}}{\text{s}} \times 1.0\frac{\text{kJ}}{\text{kg}\cdot\text{K}} \times (1173 - 489)\text{ K} = \underline{1368 \text{ kJ/s}}$$

ii) The pressure is assumed constant across the combustor (it's a heat exchanger), so the pressure entering the turbine is 600 kPa and the pressure leaving the turbine is 100 kPa. The air passing through the turbine undergoes a quasi-equilibrium adiabatic process (no losses), so the exiting temperature is

$$T_4 = T_3\left(\frac{P_4}{P_3}\right)^{k-1/k} = 1173 \times \left(\frac{100}{600}\right)^{0.2857} = 703 \text{ K} \quad \text{or} \quad \underline{430°\text{C}}$$

iii) The power required by the compressor and produced by the turbine are, respectively,

$$\dot{W}_C = \dot{m}(h_2 - h_1) = \dot{m}C_p(T_2 - T_1)$$
$$= 2 \times 1.0 \times (489 - 293) = 392 \text{ kW}$$
$$\dot{W}_T = \dot{m}(h_3 - h_4) = \dot{m}C_p(T_3 - T_4)$$
$$= 2 \times 1.0 \times (1173 - 703) = 940 \text{ kW}$$

Note: The compressor requires 41.7% of the power generated by the turbine.

The back-work ratio is then

$$\text{BWR} = \frac{\dot{W}_C}{\dot{W}_T} = \frac{392}{940} = \underline{0.417} \quad \text{or} \quad 41.7\%$$

iv) The power produced by the cycle is

$$\dot{W}_{Out} = \dot{W}_T - \dot{W}_C = 940 - 392 = \underline{548 \text{ kW}}$$

v) The cycle efficiency is found to be

$$\eta = \frac{\dot{W}_{Out}}{\dot{Q}_{In}} = \frac{548}{1368} = 0.400 \text{ or } \underline{40\%}$$

The processes across the compressor and turbine are not without losses, as assumed in this example. When losses are included, using realistic efficiencies, the cycle efficiency is reduced to about 25% to 30%.

Observe: The compressor power acts as though it is a loss in the cycle efficiency.

☑ You have completed Learning Outcome (6)

4.5 Summary

In this chapter, the first law was applied to a control volume Most of the problems of interest involved the application of the conservation of mass and the conservation of energy to various devices, which included pumps, turbines, heat exchangers, mixing chambers, throttles, and nozzles. The filling and discharge of rigid volumes were the unsteady flows that were analyzed. An introduction to cycles that utilized the devices listed above closed out the chapter. Included were the very important Rankine, refrigeration, and Brayton cycles. The conservation of momentum is not of interest in our study of thermodynamics; it is of particular interest in fluid mechanics.

The following additional terms were defined in this chapter:

Brayton cycle: *A cycle that utilizes a compressor and a turbine to generate power.*

Coefficient of performance: *The COP, which is the ratio of the desired effect divided by the purchased power for a refrigeration cycle.*

Continuity equation: *The equation that represents the conservation of mass applied to a control volume.*

Control surface: *The surface that surrounds the control volume.*

Control volume: *The volume into which and/or from which a fluid flows.*

Diffuser: *A device that decreases the velocity in a flowing fluid.*

Efficiency: *The ratio of power output to power input of a cycle*

Heat exchanger: *A device used to transfer energy from a hotter fluid to a colder fluid.*

Nozzle: *A device that increases the velocity of a flowing fluid.*

Mass flow rate (mass flux): *The product of the density, the area, and the velocity normal to the area.*

Mixing chamber: *A device that combines at least two inlet flows to form a single-outlet flow.*

Pumps, compressors, and blowers: *Devices that increase the pressure in a fluid.*

Rankine cycle: *A cycle used to generate power in most power plants.*

Refrigeration cycle: *A cycle used to refrigerate or heat a space.*

Shaft work: *The work that crosses the control-volume boundary via a rotating shaft.*

Throttling device: *An adiabatic device that provides a sudden pressure drop.*

Unsteady flow: *The time-dependent flow into and/or from a volume.*

Turbine: *A device that produces work by using a set of rotating blades to extract energy from a fluid.*

Volumetric flow rate (flow rate): *The product of the area and the average velocity normal to the area.*

Several important equations were presented:

Performance: $$\eta_{\text{engine}} \text{ or COP} = \frac{\text{desired effect}}{\text{purchased energy}}$$

Mass flow rate (mass flux): $$\dot{m} = \rho A V$$

Continuity equation: $$\rho_1 A_1 V_1 = \rho_2 A_2 V_2$$

Volumetric flow rate (flow rate): $$\dot{V} = AV$$

Energy equation ($\Delta PE = 0$): $$\dot{Q} - \dot{W}_S = \dot{m}\left[h_2 - h_1 + \frac{1}{2}\left(V_2^2 - V_1^2\right)\right]$$

First law for a pump: $$\dot{Q} - \dot{W}_S = \dot{m}\left(\frac{P_2 - P_1}{\rho}\right)$$

First law for a nozzle or diffuser: $$0 = \frac{V_2^2 - V_1^2}{2} + h_2 - h_1$$

First law for a throttle: $$h_2 = h_1$$

Filling of a tank: $$Q = m_f u_f - m_i u_i - m_1 h_1$$

Discharging a tank: $$\frac{m_f}{m_i} = \left(\frac{T_f}{T_i}\right)^{1/(k-1)} = \left(\frac{P_f}{P_i}\right)^{1/k}$$

Performance of a refrigerator: $$\text{COP}_R = \frac{\dot{Q}_{\text{Evap}}}{\dot{W}_{\text{Comp}}}$$

Performance of a heat pump: $\text{COP}_{HP} = \dfrac{\dot{Q}_{\text{Cond}}}{\dot{W}_{\text{Comp}}}$

Back-work ratio: $\text{BWR} = \dfrac{\dot{W}_C}{\dot{W}_T}$

Problems

FE Exam Practice Questions

4.1 Which of the following would not be analyzed using a control volume?

(A) Air being compressed and expanded in a piston-cylinder arrangement
(B) Air being compressed in a jet engine
(C) Air being released from a balloon
(D) Air being compressed by a bicycle pump

4.2 Select the control volume that could best be analyzed assuming a steady flow.

(A) Air being compressed by a bicycle pump
(B) Air being breathed into the lungs
(C) Air being released from a balloon
(D) Air flowing through the fans on a dirigible airship

4.3 When using the continuity equation as $A_1V_1 = A_2V_2$, select the assumption that is not appropriate for flow in a conduit:

(A) Steady flow
(B) Constant-diameter flow
(C) Uniform flow
(D) Each velocity component must be perpendicular to its respective area

4.4 Steam at 4 MPa and 400°C flows at 40 m/s in a constant-diameter pipe. Downstream where $P_2 = 2$ MPa and $T_2 = 260$°C, the velocity is nearest:

(A) 29 m/s
(B) 37 m/s
(C) 51 m/s
(D) 62 m/s

4.5 Water enters the 2-cm-diameter pipe of Fig. 4.42 at 20 m/s and exits between two parallel disks that are 2 mm apart. Determine the velocity V_2 of the water when $r = 10$ cm.

(A) 5 m/s
(B) 10 m/s
(C) 15 m/s
(D) 20 m/s

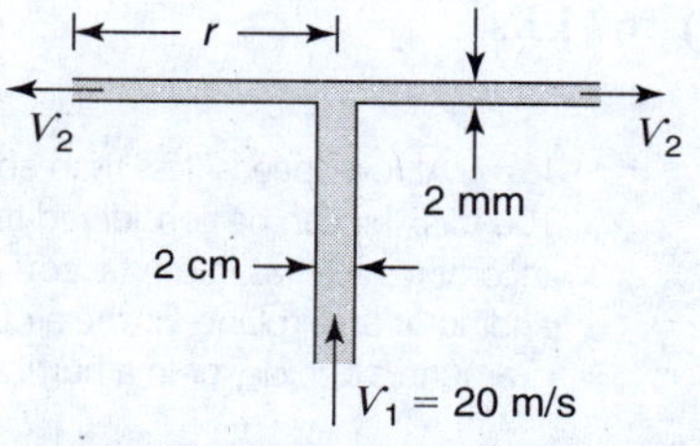

Figure 4.42

4.6 The first law applied to a steady flow of water through an insulated pump, neglecting kinetic and potential energy changes and losses, is represented by which equation?

(A) $\dot{W}_P = \dot{m}(h_2 - h_1)$
(B) $-\dot{W}_P = \dot{m}(h_2 - h_1)$
(C) $\dot{W}_P = \dot{m}\Delta P/\rho$
(D) $-\dot{W}_P = \dot{m}\Delta P/\rho$

4.7 The diffuser of Fig. 4.43 decreases the velocity of air from 180 m/s to 20 m/s. Estimate the temperature change of the air in the insulated section.

(A) 10°C

(B) 16°C

(C) 24°C

(D) 32°C

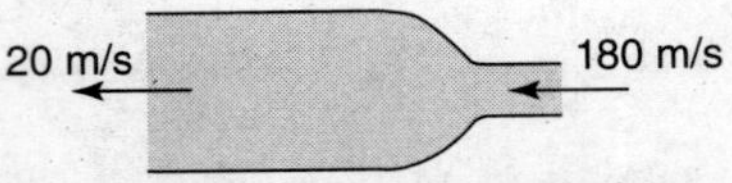

Figure 4.43

4.8 Air, with a velocity of 40 m/s, enters a relatively short constant-diameter pipe section at 40°C and 120 kPa and leaves at 200°C with a velocity of 44 m/s. The heat transferred to the air in the 10-cm-diameter pipe is nearest:

(A) 242 kJ/s

(B) 187 kJ/s

(C) 92 kJ/s

(D) 67 kJ/s

Note: At low speeds (less than about 100 m/s), air can be considered to be incompressible, as in the takeoff and landing of an airplane, or the air flow around a vehicle, or in a hurricane.

4.9 Refrigerant 134a enters a valve at 1.4 MPa as a saturated liquid and exits the valve at 120 kPa. The temperature immediately after the valve is nearest:

(A) −18°C

(B) −20°C

(C) −22°C

(D) −24°C

4.10 An insulated pump increases the pressure of water in a power plant from 10 kPa to 2 MPa. The minimum horsepower required for a mass flux of 2 kg/s is nearest:

(A) 5.3 hp

(B) 8.2 hp

(C) 12.6 hp

(D) 18.3 hp

4.11 Air flowing at 20 kg/min enters the insulated compressor of Fig. 4.44 at 100 kPa and 20°C and exits at 800 kPa. The minimum work rate required to compress the air, assuming constant specific heats, is nearest:

(A) 106 hp

(B) 92 hp

(C) 81 hp

(D) 67 hp

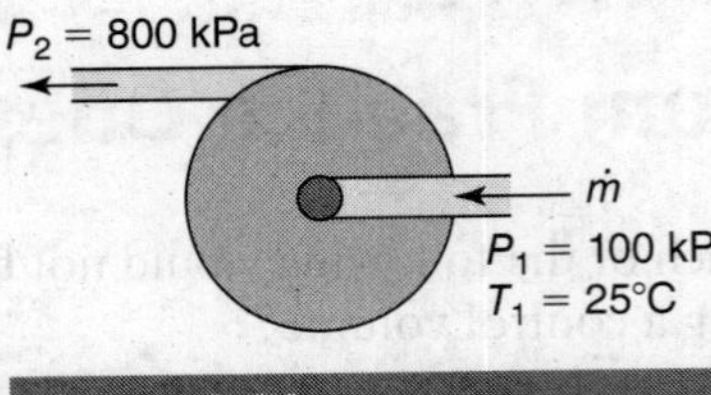

Figure 4.44

4.12 Air at 600°C and 2 MPa flows into the turbine of Fig. 4.45 and exits at 100 kPa. For a mass flux of 1.2 kg/s, the maximum power output (assume an adiabatic, quasi-equilibrium process with constant specific heats) is nearest:

(A) 342 kW

(B) 398 kW

(C) 464 kW

(D) 602 kW

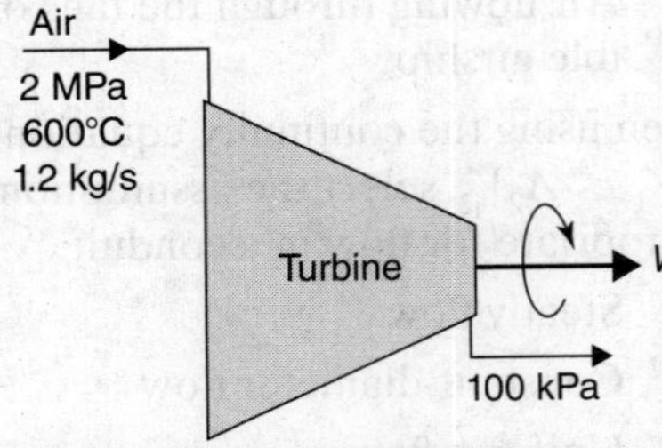

Figure 4.45

4.13 Water at 90°C flows through a radiator with a mass flux of 0.2 kg/s and exits at 87°C. It heats 10 m^3/min of standard atmospheric air. If the heat that leaves the water enters the air, estimate the temperature increase of the air.

(A) 26.9°C

(B) 22.4°C

(C) 16.8°C

(D) 12.3°C

4.14 A Rankine cycle operates with a mass flux of 2 kg/s, as shown in Fig. 4.46. Determine the pump power requirement.

(A) 7.8 kW

(B) 5.6 kW

(C) 4.2 kW

(D) 3.9 kW

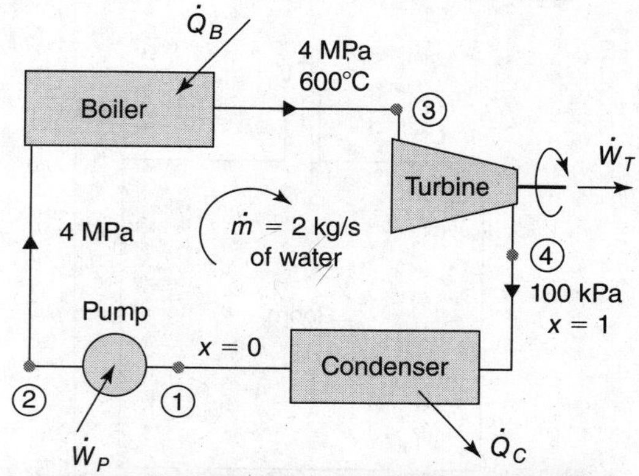

Figure 4.46

Problems 4.14–4.18

4.15 The energy requirement of the boiler of Fig. 4.46 is nearest:

(A) 6.5 MJ/s

(B) 5.2 MJ/s

(C) 4.8 MJ/s

(D) 3.6 MJ/s

4.16 The velocity of the steam in the 40-cm-diameter pipe exiting the boiler of Fig. 4.46 is nearest:

(A) 3.2 m/s

(B) 2.4 m/s

(C) 1.6 m/s

(D) 1.2 m/s

4.17 The power produced by the turbine of Fig. 4.46 is nearest:

(A) 4 MW

(B) 3 MW

(C) 2 MW

(D) 1 MW

4.18 The efficiency of the cycle of Fig. 4.46 is nearest:

(A) 41%

(B) 39%

(C) 35%

(D) 31%

4.19 A refrigeration cycle operates as shown in Fig. 4.47. If the refrigerant R134a mass flux is 0.5 kg/s, the adiabatic compressor power requirement is nearest:

(A) 19 kW

(B) 17 kW

(C) 14 kW

(D) 9 kW

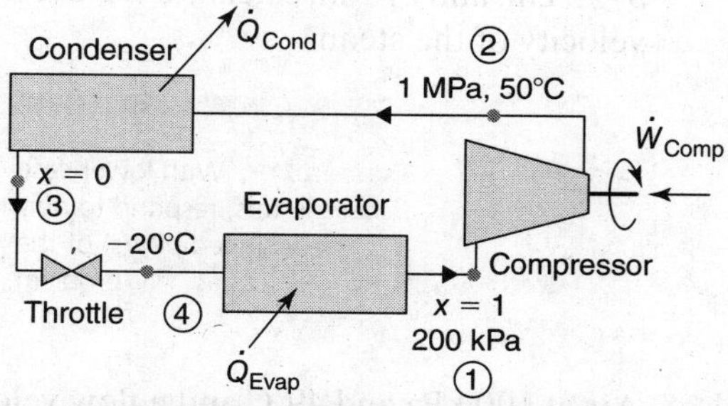

Figure 4.47

Problems 4.19–4.21

4.20 The cooling capacity of the cycle of Fig. 4.47 with $\dot{m}$ = 0.5 kg/s, is nearest:

(A) 99 kJ/s

(B) 82 kJ/s

(C) 68 kJ/s

(D) 56 kJ/s

4.21 The temperature drop across the throttle of Fig. 4.47 is nearest:

(A) 55°C

(B) 50°C

(C) 45°C

(D) 40°C

Conservation of Mass

4.22 Liquid water is supplied to a nozzle at a velocity of 2 m/s. The nozzle has an entrance diameter of 3 cm and an exit diameter of 1 cm. Determine i) the volumetric flow rate of the water, ii) the mass flow rate, and iii) the exit velocity of the water.

> **Remember:** Respond to all parts with i), ii), iii), etc.

4.23 A steam pipe must deliver 20 kg/s of steam at 1 MPa and 400°C to a processing facility. If the pipe has an inside diameter of *a*) 40 cm, *b*) 75 cm, and *c*) 1 m, calculate the average flow velocity of the steam.

> **Remember:** With lower-case italic letters, respond to only the assigned part(s) of the problem

4.24 Air at 100 kPa and 49°C and a flow velocity of 4.5 m/s enters a nozzle whose inlet area is 75 cm^2. Determine the mass flow rate and volume flow rate of the air.

4.25 Water enters a nozzle at 8 m/s and exits to the atmosphere, as shown in Fig. 4.48. Calculate the velocity V_2 of the water at the exit if *a*) $d_1 = 4$ cm, *b*) $d_1 = 6$ cm, and *c*) $d_1 = 8$ cm. Assume the water to be incompressible.

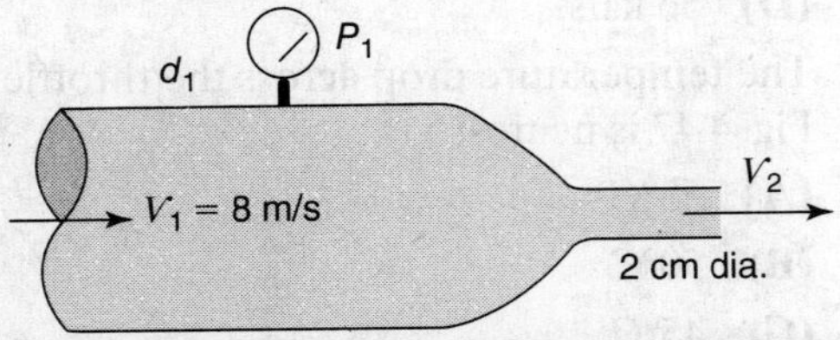

Figure 4.48

4.26 Air enters a room at two inlets and leaves at one outlet, as shown in Fig. 4.49. Determine $\dot{V}_3$ for a steady-flow situation if the velocity V_2 is *a*) 10 m/s, *b*) 20 m/s, and *c*) 30 m/s. The temperature and pressure of the air at all locations are approximately 25°C and 90 kPa, respectively. The air can be assumed to be incompressible for low velocities (below about 100 m/s).

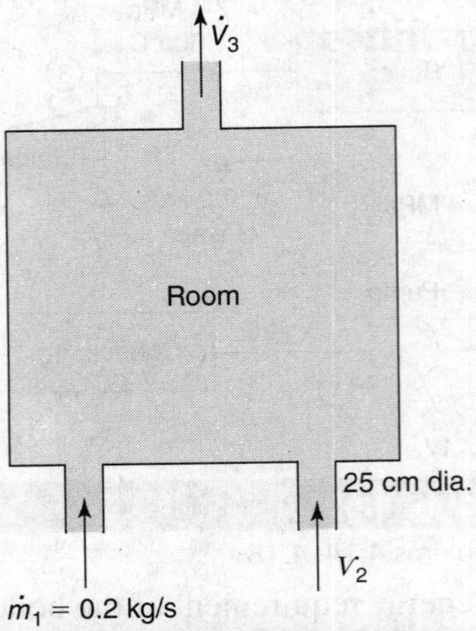

Figure 4.49

The First Law

4.27 Air is flowing in a 10-cm-constant diameter pipe at 50 m/s. At section 1 the temperature and pressure are 60°C and 400 kPa, respectively. Heat is added to the air, and at section 2 downstream the temperature and pressure are measured to be 300°C and 380 kPa, respectively. Calculate the mass flux and the downstream velocity. Also, calculate the added heat transfer by i) neglecting the kinetic energy change, and ii) by including the kinetic energy change. Comment as to the importance of including the kinetic energy change. (Potential energy changes are ignored unless elevation changes are included.)

4.28 If water leaves the nozzle of

a) Problem 4.25*a*

b) Problem 4.25*b*

c) Problem 4.25*c*

and exits to the atmosphere at 0 kPa gage, estimate the gage pressure of the water upstream where $V_1 = 8$ m/s. Use the information given in Problem 4.25.

Turbines, Compressors, and Pumps

4.29 Steam enters the adiabatic turbine of Fig. 4.50 at 10 MPa and 600°C. If the mass flow rate of the steam is 2 kg/s, determine the power output of the turbine if the steam leaves at *a)* 20 kPa with $x = 0.9$, *b)* 10 kPa with $x = 1$, and *c)* at 20 kPa with $s_2 = s_1$ (use s, which is entropy, as given in the tables).

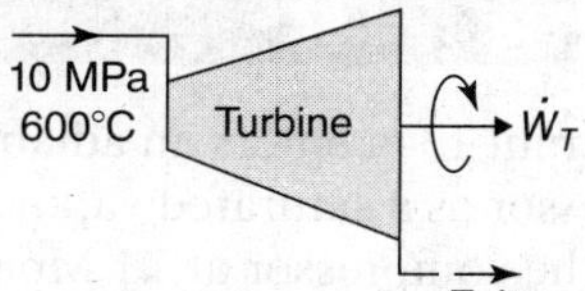

Figure 4.50

4.30 Air is expanded in an adiabatic turbine from 1.5 MPa and 500°C to 120 kPa and 110°C. The volumetric flow rate of the air is 10 m^3/min at the inlet. Determine the power output of the turbine. Assume no losses (quasi-equilibrium) and constant specific heats.

Note: Kinetic and potential energy changes are ignored when information to include them is not given. Also, assume an ideal gas with constant specific heats unless otherwise stated.

4.31 Steam enters an adiabatic turbine at 10 MPa, 500°C, and a flow velocity of 100 m/s through four 2-cm-diameter jets. It leaves the turbine at 30 kPa with a velocity of 20 m/s and a quality of 0.94. Determine the output power of the turbine. Calculate the error in the output power if the kinetic energy change is neglected. Should the kinetic energy change be included?

4.32 Steam enters an adiabatic turbine at 14 MPa, 550°C, with a flow velocity of 90 m/s through four 25 mm-diameter jets. It leaves at 70 kPa with a velocity of 12 m/s and a quality of 0.90. Determine the output power of the turbine. Calculate the error in the output power if the kinetic energy change is neglected. Should the kinetic energy change be included?

4.33 Air enters the adiabatic turbine of Fig. 4.51 at 300 kPa and 500°C and leaves the turbine at 100 kPa. If the power output of the turbine is 600 hp, determine the mass flow rate of the air through the turbine. Assume a quasi-equilibrium process with constant specific heats.

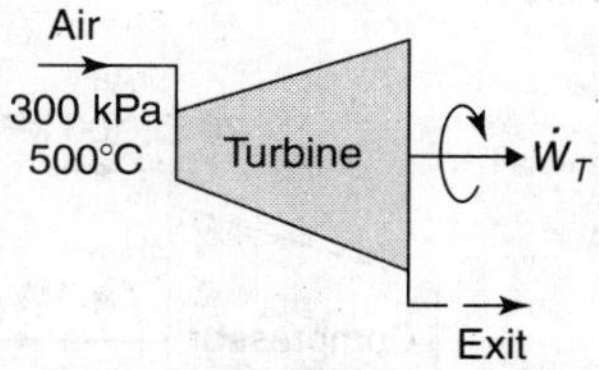

Figure 4.51

4.34 Air with a velocity of 40 m/s enters an adiabatic turbine through a 4-cm-diameter pipe at 2 MPa and 400°C and expands to the exit maintained at 100 kPa. Determine the power produced if *a)* the air exits at 30°C and *b)* an adiabatic, quasi-equilibrium process is experienced by the air.

Note: Assume an ideal gas with constant specific heats unless otherwise stated.

4.35 A small stream near a mountain cabin is dammed up to produce a head of 2 m of water at the inlet to a hydroturbine, as sketched in Fig. 4.52. A distance upstream the flow is estimated to be 4 m/s in the 1.2-m-wide, 4-cm-deep stream. Estimate the maximum power that could be delivered by the hydroturbine. (This is one of the few times, maybe the only time, in this text that the potential energy must be retained.)

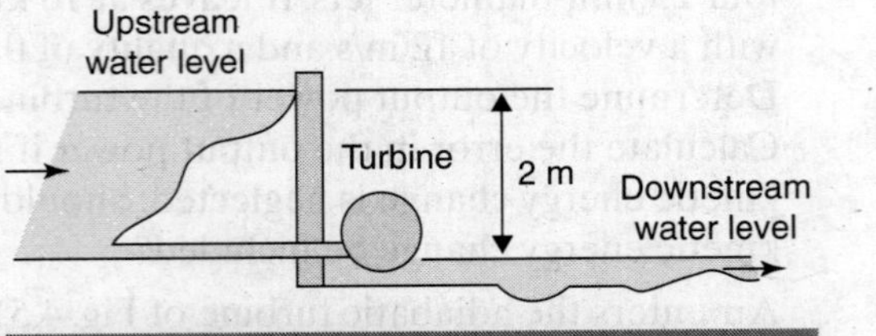

Figure 4.52

4.36 Air is compressed from 100 kPa and 20°C to 800 kPa and 260°C. If the input power to the adiabatic compressor of Fig. 4.53 is 20 kW, determine the mass flow rate of air through the compressor.

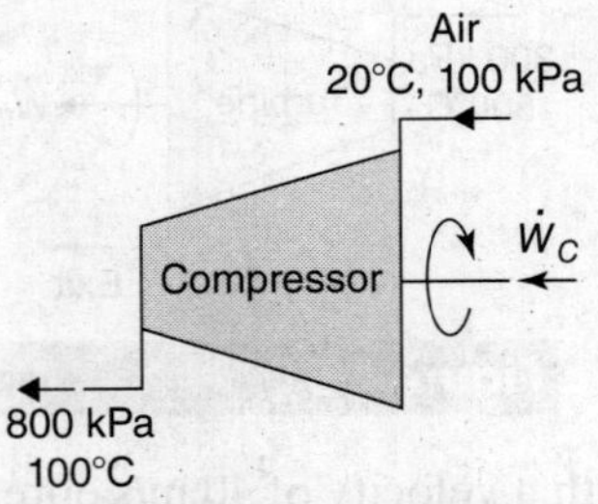

Figure 4.53

4.37 Air is compressed from 100 kPa and 40°C. The input power to the adiabatic compressor is 20 kW. Determine the mass flow rate of air through the compressor assuming the air undergoes a quasi-equilibrium process to *a*) 400 kPa, *b*) 600 kPa, and *c*) 800 kPa.

Note: Continue to assume constant specific heats.

4.38 Air with a velocity of 40 m/s enters an adiabatic compressor through a 4-cm-diameter pipe at 100 kPa and 30°C and is compressed to 2000 kPa. Determine the power required if *a*) the air exits at 450°C and *b*) an adiabatic, quasi-equilibrium process is experienced by the air.

4.39 Refrigerant 134a enters the adiabatic compressor of Fig. 4.54 as a saturated vapor at 20°C. It leaves the compressor at 1 MPa and 50°C. If the mass flow rate of the refrigerant is 4 kg/s, determine the power input to the compressor.

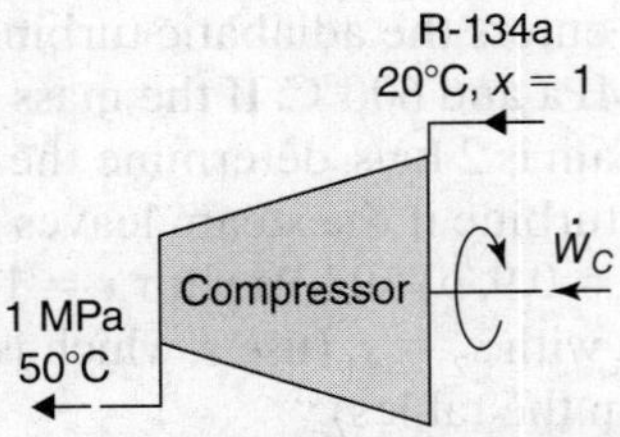

Figure 4.54

4.40 Refrigerant 134a enters an adiabatic compressor as a saturated vapor at 20°C. It leaves the compressor at 1.1 Mpa and 50°C. If the mass flow rate of the refrigerant is 4.5 kg/s, determine the horsepower input to the compressor.

4.41 Ammonia is compressed from 120 kPa with $x = 1$ to a pressure of 1.2 MPa and a temperature of 100°C. For a mass flux of 3 kg/s, determine the power required to drive the adiabatic compressor.

4.42 A compressor requires 200 hp to compress 0.02 kg/s of steam from saturated vapor at 150°C to 2 MPa and 400°C. Determine the heat transfer rate from the compressor.

4.43 Liquid water is pumped from 100 kPa to 600 kPa at a flow rate of 1.2 m^3/s. Calculate the necessary input power to the pump.

4.44 The pump of Fig. 4.55 increases the pressure in the water from 0.1 MPa to 6 MPa. Estimate the horsepower required if the mass flux is 10 kg/min. The density of water is 1020 kg/m^3 (use $1/v = \rho$ at 16°C).

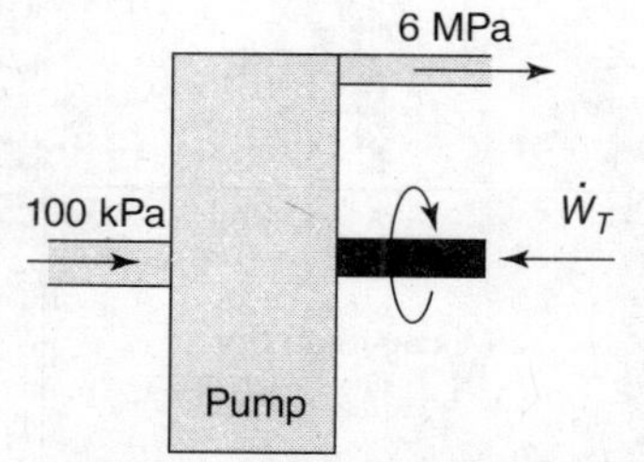

Figure 4.55

4.45 The feedwater pump in a power plant increases the pressure of the water exiting the condenser from 10 kPa to 6 MPa. Estimate the horsepower requirement if 10 kg/s of water is flowing.

Throttling Devices

4.46 Water travels in the pipe of Fig. 4.56 as a saturated liquid at 1 MPa. The water is throttled through a valve to a pressure of 100 kPa. Determine the quality and the temperature of the steam exiting the throttle.

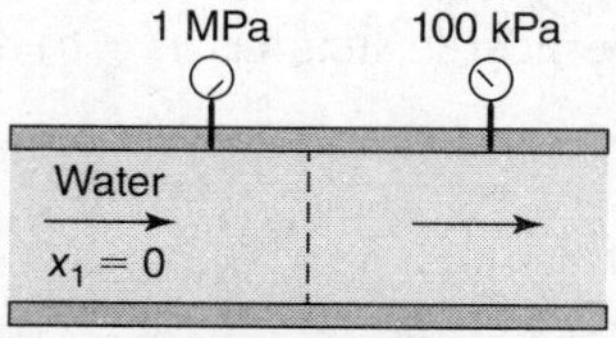

Figure 4.56

4.47 A fluid is throttled from 1 MPa and 38°C to a pressure of 100 kPa. Determine the temperature of the fluid exiting the throttle if it is *a*) R134a, *b*) ammonia, and *c*) air.

4.48 R134a at 350 kPa and 4°C is throttled to a pressure of *a*) 140 kPa, *b*) 100 kPa, and *c*) 60 kPa. Determine the exit temperature and quality.

4.49 The valve of Fig. 4.57 throttles 90°C water from 8 MPa to 40 kPa. What are the temperature and enthalpy of the water downstream of the valve? The upstream and downstream areas are the same.

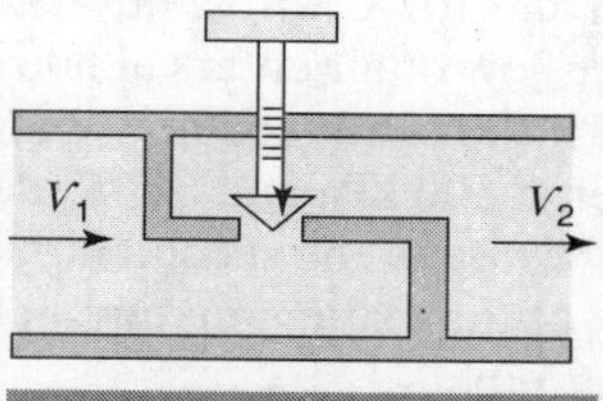

Figure 4.57

4.50 Ammonia flowing at 0.01 m^3/s is throttled from 900 kPa and 20°C to a pressure of 125 kPa by passing the refrigerant through the bank of small-diameter tubes shown in Fig. 4.58 that cause a sudden pressure drop. Determine the temperature and flow rate $\dot{V}_2$ of the ammonia exiting the throttle.

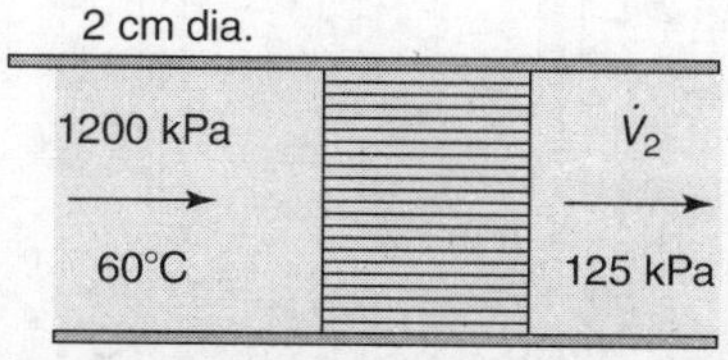

Figure 4.58

Mixing Chambers

4.51 Water enters a mixing chamber at 200 kPa and 40°C with a flow rate of 50 kg/s. Another flow of water enters at 200 kPa and 20°C with a flow rate of 100 kg/s. Determine the exit temperature of the combined flow. The exit pressure is also 200 kPa.

4.52 Air enters the mixing chamber of Fig. 4.59 at 500 kPa and 107°C with a flow rate of 5 m^3/s. Another flow of air enters at 500 kPa and 1027°C. If the combined flow exits the mixing chamber at 500 kPa and 627°C, determine the mass flow rate of the second input flow, and the velocity of the exiting mixture in the 80-cm-diameter pipe.

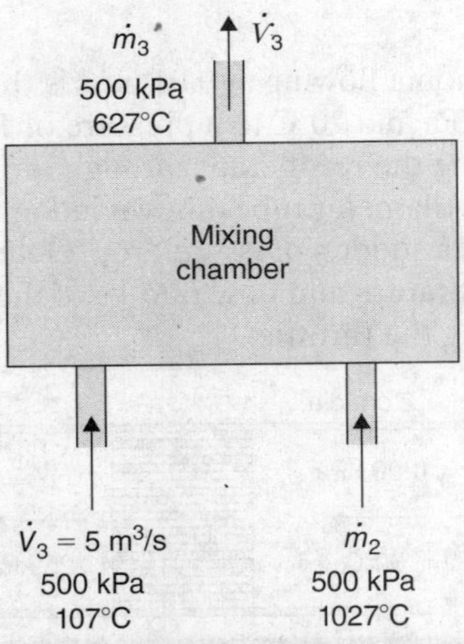

Figure 4.59

4.53 Steam enters a mixing chamber at 6 MPa and 400°C. Water enters the mixing chamber at 6 MPa and 80°C. Determine the ratio of the mass flow rate of the steam to the mass flow rate of the water if the exit flow leaves at a temperature of *a*) 200°C, *b*) 250°C, and *c*) 300°C.

4.54 Water at 24°C and 1.4 MPa is heated by mixing it with superheated steam at 300°C and 1.4 MPa, as shown in Fig. 4.60. If the mass flow rate of each entering flow is the same, calculate the temperature of the exiting flow.

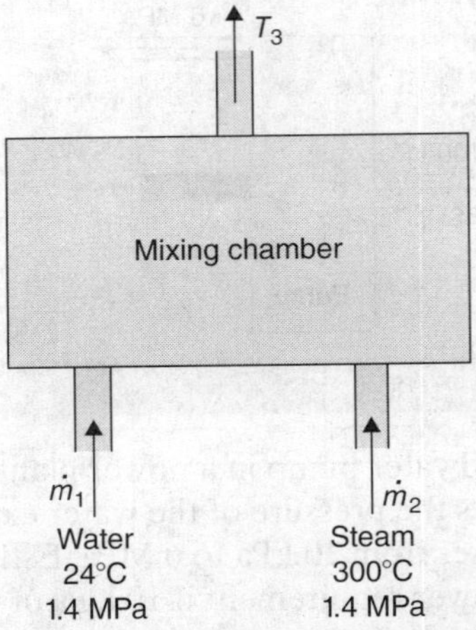

Figure 4.60

4.55 The water entering the boiler of a power plant is preheated by mixing the saturated water exiting the condenser pump at 400 kPa with superheated steam at 400 kPa and 160°C. Determine the temperature of the water entering the boiler if the ratio of the mass flow rates of the saturated water and the superheated steam is *a*) 6, *b*) 8, and *c*) 10.

Heat Exchangers

4.56 Water is used to cool R134a in the condenser of Fig. 4.61. The refrigerant enters the counter flow heat exchanger at 800 kPa, 80°C and a mass flow rate of 2 kg/s. The refrigerant exits as a saturated liquid. Cooling water enters the condenser at 500 kPa and 18°C and leaves the condenser at 30°C. Determine the necessary mass flow rate of the water. A reminder: Each fluid is assumed to flow at constant pressure.

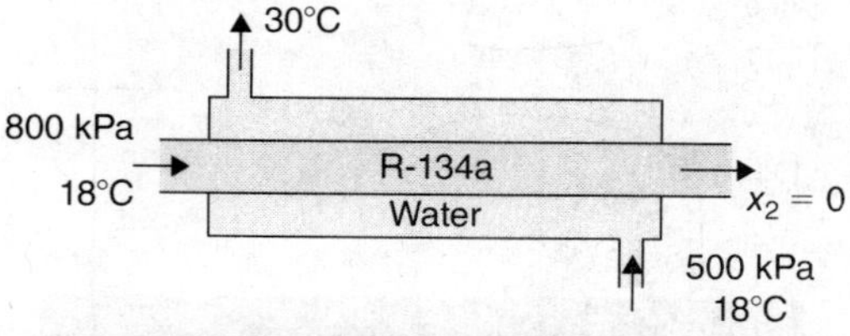

Figure 4.61

4.57 Water enters the condenser (a heat exchanger) shown in Fig. 4.62 of a power plant at 20°C and leaves at 70°C. If the mass flux of the steam is 4 kg/s, determine the mass flux required for the water flow. The conditions of the steam are displayed in the figure.

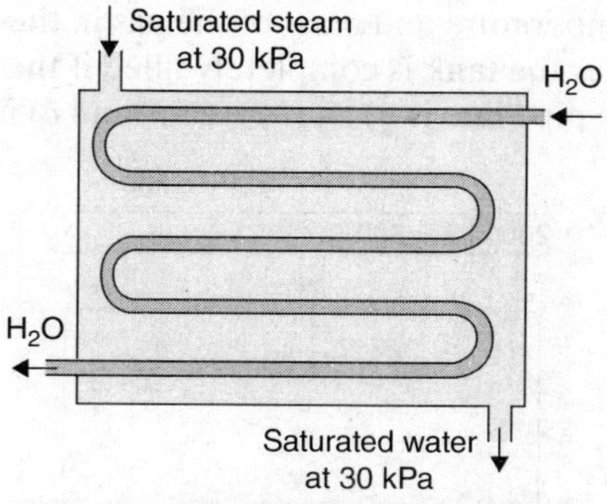

Figure 4.62

4.58 Steam enters a heat exchanger at 5 MPa, and 500°C and leaves at 250°C. Cooling water enters the heat exchanger at 500 kPa, and 25°C and leaves at 80°C. Determine the ratio of the mass flow rate of steam to the mass flow rate of the cooling water.

4.59 In the sketch of a car radiator in Fig. 4.63, air is used to cool ethylene glycol ($C_p = 2.5$ kJ/kg·°C). Air at a flow rate of 1 m^3/s enters the radiator at $T_1 = 20$°C and leaves at $T_2 = 100$°C. The ethylene glycol enters the radiator at $T_3 = 160$°C with a mass flux of 2 kg/s. Determine the exit temperature T_4 of the ethylene glycol.

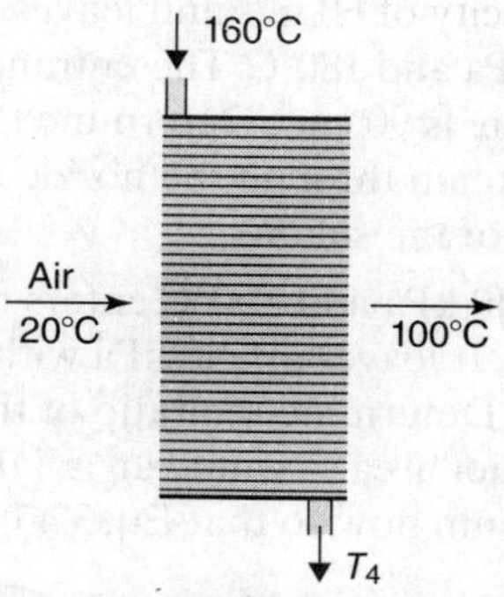

Figure 4.63

4.60 Superheated steam at 100 kPa and 110°C enters a condenser at 80 kg/s. The steam leaves the condenser as a saturated liquid at 100 kPa. Liquid water is used to cool the steam. The water enters the condenser heat exchanger at 20°C with a flow rate of 4 m^3/s (it does not mix with the steam). Determine the exit temperature of the cooling water.

Nozzles and Diffusers

4.61 Air enters the diffuser of Fig. 4.64 at 15°C with a flow velocity of 200 m/s. The inlet diameter is 4 cm. If the air leaves the diffuser at 100 kPa and 30°C, determine the exit velocity and the exit diameter. Assume no separation of the air from the walls of the diffuser (guide vanes are often used to ensure no separation) and an adiabatic quasi-equilibrium process (see Eq. 3.45).

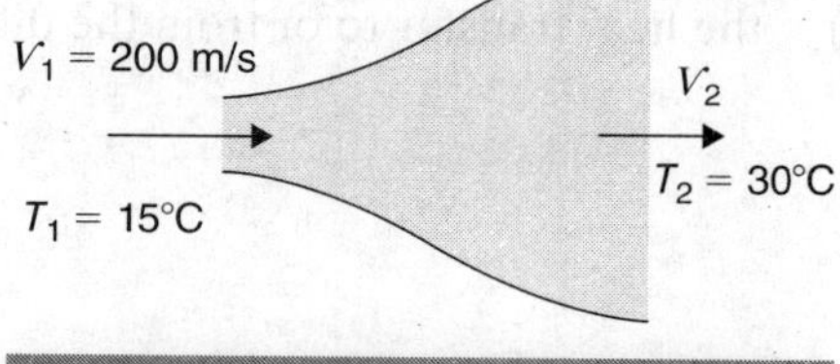

Figure 4.64

4.62 Steam enters a nozzle at 4 MPa and 500°C at a velocity of 100 m/s and leaves the nozzle at 1 MPa and 250°C. The entrance area of the nozzle is 0.01 m^2. Determine the mass flow rate of steam through the nozzle, the exit velocity of the steam, and the exit diameter.

4.63 Steam enters a nozzle at 140 kPa and 200°C at a velocity of 60 m/s and leaves the nozzle at 100 kPa and 180°C. The entrance area of the nozzle is 90 cm^2. Determine the mass flow rate of steam through the nozzle and the exit velocity of the steam.

4.64 Air at 140 kPa and 100°C enters the nozzle of Fig. 4.65. It leaves at 100 kPa with a velocity of 400 m/s. Determine the ratio of the exit area to the inlet area. Assume an adiabatic quasi-equilibrium flow so that Eq. 3.45 applies.

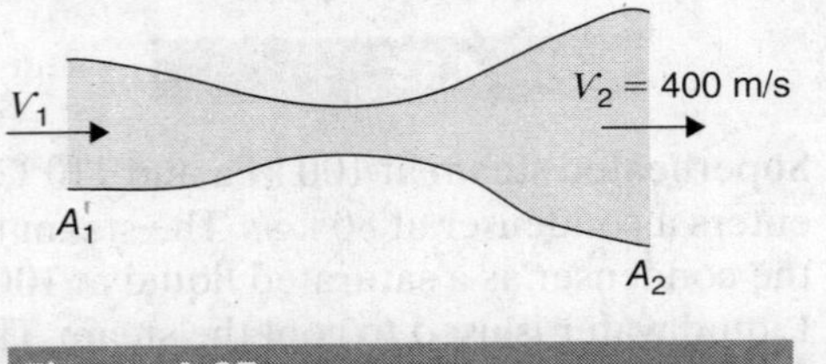

Figure 4.65

4.65 Refrigerant 134a enters a diffuser at 800 kPa, a temperature of 50°C, and a flow velocity of 160 m/s. It leaves the diffuser at 1 MPa and 60°C. The inlet area of the diffuser is 0.006 m^2. Determine the mass flow rate of the refrigerant and the exit area.

4.66 Air enters an adiabatic diffuser at 500 kPa and 30°C with a velocity of 20 m/s. The diffuser has an inlet diameter of 8 cm. Air leaves the diffuser at 800 kPa and 80°C. If the exit diameter of the diffuser is 20 cm, determine:

i) the mass flow rate
ii) the exit velocity
iii) the exit volumetric flow rate
iv) the heat transfer to or from the diffuser

4.67 The air in a laboratory reservoir is maintained at 20°C. It flows out through a converging-diverging nozzle into the laboratory maintained at 100 kPa, through a shape like that sketched in Fig. 4.66. Estimate the maximum velocity at the exit assuming an adiabatic, quasi-equilibrium flow with constant specific heats and a reservoir pressure of *a*) 600 kPa, *b*) 800 kPa, and *c*) 1200 kPa.

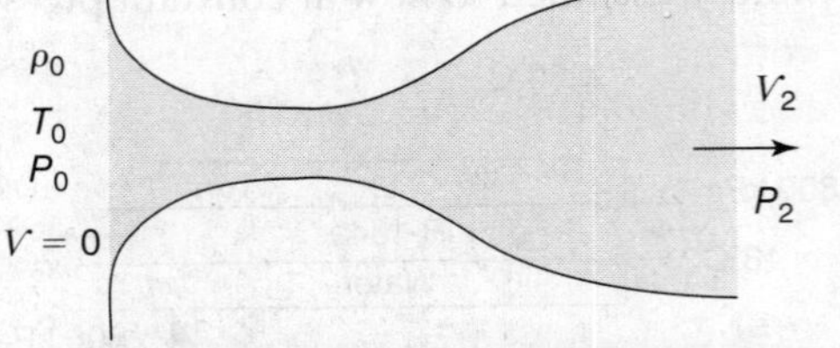

Figure 4.66

Unsteady Flow

4.68 A valve separates the 10-m^3 insulated, evacuated tank of Fig. 4.67 from a pipe line that contains a gas at 2000 kPa and 300°C. The gas in the pipe is held at constant conditions after the valve is opened. Determine the final temperature and mass of the gas in the tank after the tank is completely filled if the gas in the pipe line is *a*) air, *b*) steam, and *c*) nitrogen.

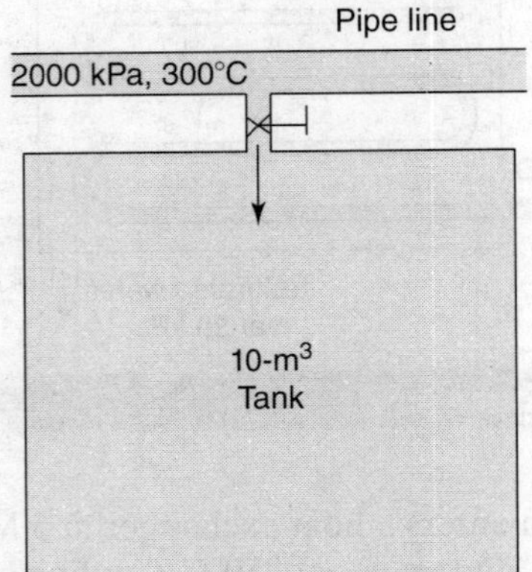

Figure 4.67

4.69 The evacuated tank in Problem 4.68 is not insulated. After a long period of time, the temperature is measured to be 347°C. Determine the heat transfer from the tank if the gas in the pipe line is *a*) air, *b*) steam, and *c*) nitrogen.

4.70 An insulated 4-m^3 volume contains a vacuum of air at 20°C. A valve is opened and atmospheric air at 100 kPa and 25°C rushes into the volume. After the air has stopped flowing in, estimate the temperature in the volume and the mass of air that entered if the initial pressure in the volume is *a*) 0 kPa, *b*) 10 kPa, and *c*) 25 kPa.

4.71 An air line at 700 kPa and 20°C is used to fill the 0.044 m^3 tire of a vehicle that contains air at only 35 kPa(g) and 20°C. Assuming negligible heat transfer from the rigid tire during filling, estimate the temperature of the air in the tire when the pressure reaches 245 kPa (g). Also, estimate the added mass of air. Assume that the conditions in the air line remain constant and that P_{atm} = 105 kPa.

4.72 The insulated 4-m^3 pressurized rigid volume of Fig. 4.68 contains air at 20°C. A valve is opened, and air discharges to the atmosphere at 100 kPa and 25°C. After the air has stopped flowing and the valve is closed, estimate the temperature in the volume if the initial pressure P_i in the volume is *a*) 2 MPa, *b*) 1200 kPa, and *c*) 800 kPa.

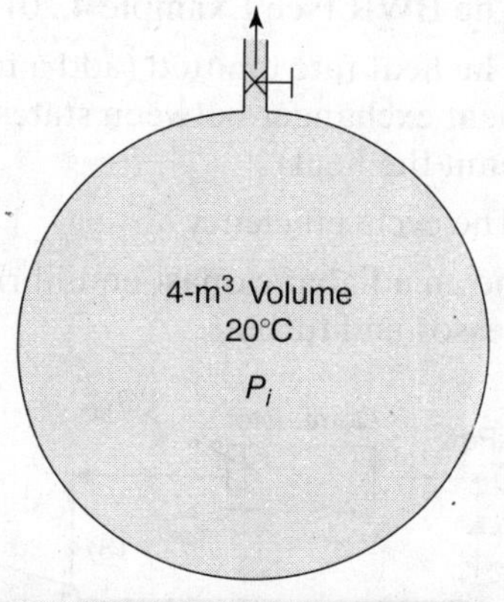

Figure 4.68

4.73 After a long period of time, the volume of Problem 4.72 *a* will reach the temperature of the atmosphere. Estimate the final pressure in the volume if the initial pressure was *a*) 2 MPa, *b*) 1200 kPa, and *c*) 800 kPa.

4.74 A 3-m-diameter insulated spherical balloon contains air at 1.2 MPa and 20°C. A special valve allows the air to leave at a constant rate of 0.18 kg/s. Calculate the temperature and pressure after *a*) 8 minutes and *b*) 15 minutes. Estimate the time needed for the temperature to reach −30°C.

Cycles

4.75 The Rankine cycle of Fig. 4.69 operates between 20 kPa and 4 MPa with a maximum temperature of 400°C. For a mass flow rate of 10 kg/s, determine the adiabatic turbine power output, the pump horsepower required, and the cycle efficiency (the pump power is negligible in the efficiency calculation) if the quality at the turbine exit is *a*) 1.0, *b*) 0.92, and *c*) 0.85. The boiler and condenser are treated as heat exchangers.

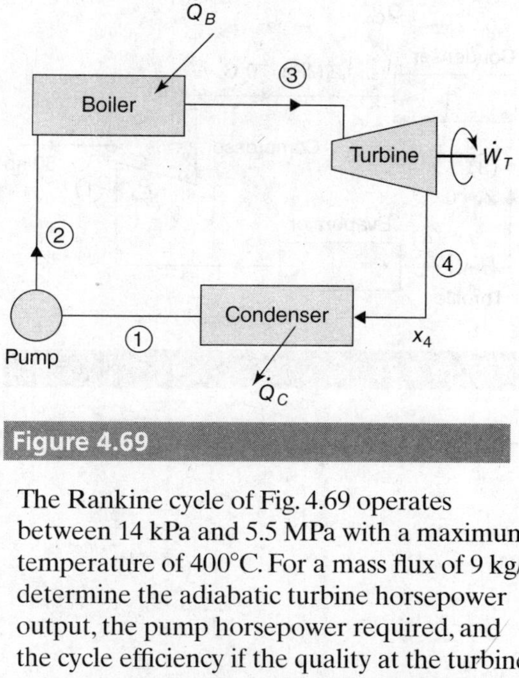

Figure 4.69

4.76 The Rankine cycle of Fig. 4.69 operates between 14 kPa and 5.5 MPa with a maximum temperature of 400°C. For a mass flux of 9 kg/s, determine the adiabatic turbine horsepower output, the pump horsepower required, and the cycle efficiency if the quality at the turbine exit is *a*) 1.0, *b*) 0.92, and *c*) 0.85. The boiler and condenser are treated as heat exchangers.

4.77 The refrigeration cycle of Fig. 4.70 operates with R134a. It is powered by an input of 30 kW to the compressor. Determine the cooling rate and the COP. The inlet pressure to the compressor is *a*) 120 kPa, *b*) 140 kPa, and *c*) 180 kPa.

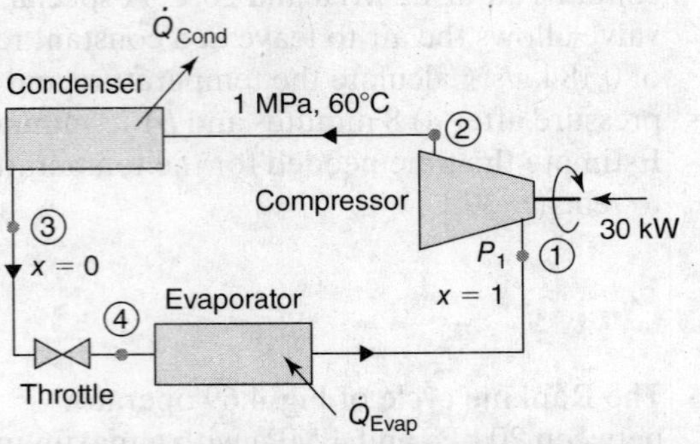

Figure 4.70

4.78 The refrigeration cycle of Fig. 4.71 operates with ammonia. If it is powered by an input of 50 hp to the compressor, determine the cooling rate and the COP. The inlet pressure to the adiabatic compressor is *a*) 100 kPa, *b*) 200 kPa, and *c*) 300 kPa.

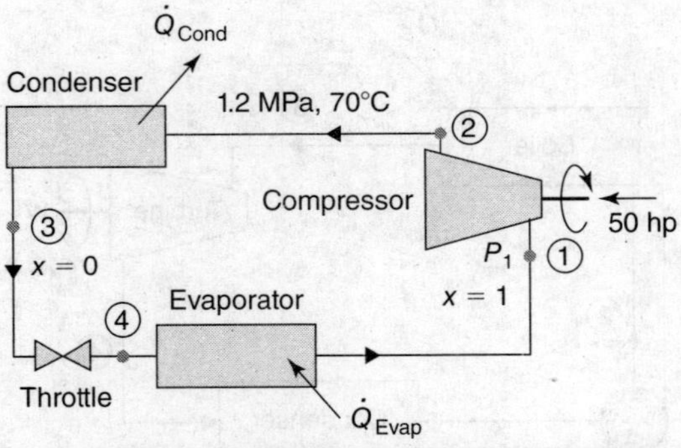

Figure 4.71

4.79 Air at 3.8 kg/s enters the Brayton cycle of Fig. 4.72 (such a cycle could be used in a large truck engine). For a combustor outlet temperature of *a*) 650°C, *b*) 850°C, and *c*) 960°C, determine:

i) The power output
ii) The BWR (see Example 4.20)
iii) The heat rate emitted (add a fictitious heat exchanger between states 4 and 1 to emit the heat)
iv) The cycle efficiency

Assume an adiabatic quasi-equilibrium compressor and turbine.

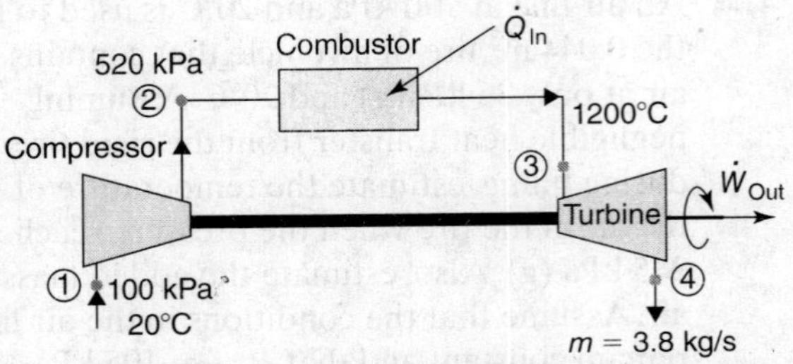

Figure 4.72

4.80 Air at 3.6 kg/s enters the Brayton cycle of Fig. 4.73 (such a cycle could be used in a large truck engine). For a combustor outlet temperature of *a*) 750°C, *b*) 850°C, and *c*) 1000°C, determine:

i) The power output
ii) The BWR (see Example 4.20)
iii) The heat rate emitted (add a fictitious heat exchanger between states 4 and 1 to emit the heat)
iv) The cycle efficiency

Assume an adiabatic quasi-equilibrium compressor and turbine.

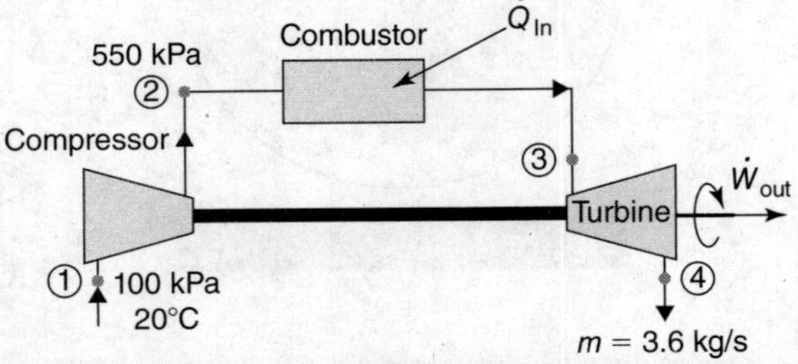

Figure 4.73

CHAPTER

5

The Second Law of Thermodynamics

Mariday/Shutterstock.com

Outline

Nomenclature

The following variables are introduced in this chapter:

Q_H	Heat transfer from a high-temperature reservoir
Q_L	Heat transfer from a low-temperature reservoir
T_H	Temperature of a high-temperature reservoir
T_L	Temperature of a low-temperature reservoir

Learning Outcomes

- **The concepts of reversibility and irreversibility**
- **Statements of the second law**
- **General cycle performance parameters**
- **The Carnot cycle performance parameters**

In this chapter, we will examine the second law of thermodynamics and study its implications. Unlike the first law of thermodynamics, which is a fundamental tool for problem solving, the second law informs us of not what can be done but of what cannot be done in thermodynamic processes and cycles. The second law of thermodynamics also puts very real limits on the efficiency of devices and cycles, limits that are often surprisingly low.

The second law has led to many discussions and statements. Here are two interesting statements:

1. "The tendency for entropy to increase in isolated systems is expressed by the second law of thermodynamics—perhaps the most pessimistic and amoral formulation in all human thought," by Gregory Hill and Kerry Thornley, *Principia Discordia* (1965).
2. "There are almost as many formulations of the second law as there have been discussions of it," by Philosopher/Physicist P. W. Bridgman (1941).

Unlike the philosophers, we will make a technical statement of the second law, one that can be used not only to make decisions on proposed designs but also to decide whether improvements should be made in existing devices and cycles.

5.1 Second-Law Concepts

Reservoir: A body that is able to provide heat or absorb heat without a change in its temperature.

In thermodynamics, a device is often in need of a source of energy, but it must also have a place for energy disposal. Such sources and disposals are termed *reservoirs.* A source of thermal energy is a *heat source,* and a thermal energy disposal site is a *heat sink*. They will usually be depicted as shown in Fig. 5.1. A reservoir is a sufficiently large body that it is able to provide heat or absorb heat without a change in its temperature. The ocean, a lake, a river, and the atmosphere are the most common heat sinks that serve as reservoirs where heat is dumped. A reservoir need not be large; a furnace that supplies heat at a constant temperature could be considered a reservoir that serves as a heat source, as sketched in the schematic of Fig. 5.1. An ice bath that stays at 0°C could be a heat sink. The important property of a reservoir is its constant temperature. In the case of an industrial furnace or the coal fire in a power plant, the reservoir's temperature could be very high.

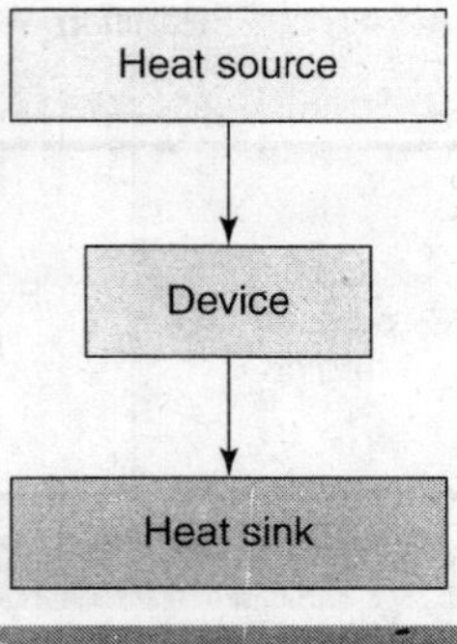

Figure 5.1
A heat source and a heat sink.

Between a heat source and a heat sink, a process takes place. A *reversible process* is one that, once completed, can be reversed and returned to its original state, leaving no change in either the system or the surroundings. Every system can be returned to its original state with some type of process, but every real process will require that the surroundings provide more energy than the surroundings received during the initial process. An example of a process that approaches a reversible process, which does not utilize heat reservoirs, is a spring that requires the surroundings to do work to stretch it; when released, it returns to its original shape, providing essentially the same amount of work to the surroundings. A frictionless piston compressing the air in the cylinder of an engine is essentially the same process as the spring and will be considered reversible; during its expansion, it gives back the energy required during the compression.

Reversible process: One that can be reversed and returned to its original state, leaving no change in either the system or the surroundings. The quasi-equilibrium process did not involve the surroundings.

All thermodynamic processes that utilize heat reservoirs are irreversible due to physical aspects of the processes that allow transformations only in one direction. Unlike the spring, heat flows in only one direction: from hot to cold. Several common effects that introduce irreversibilities follow.

1. **Heat transfer**—Heat transfer across a finite temperature difference is not reversible. Heat will naturally flow from a hotter region to a colder region, as in Fig. 5.2, but will never naturally flow from a colder region B to a hotter region A. The only way for heat transfer to flow from a colder region to a hotter region is to use a refrigeration cycle; it will not happen naturally. The surroundings must provide work to operate the cycle, as in the familiar refrigerator. The irreversibility of heat transfer across large temperature differences leads to the substantial losses in most thermodynamic processes, especially in vehicle engines and power plants.
2. **Friction**—When two surfaces rub against one another, some of the energy in one or both of the surfaces is turned into heat energy (see Fig. 5.3). A good example is when you rub the palms of your hands together to warm them. The friction produces heat. This process is not reversible. The two surfaces will not return to their original positions by returning the heat.
3. **Mixing**—If you mix two substances together, such as salt and pepper, or the red water and blue water in Fig. 5.4, they will not separate back into the original substances by themselves even with significant effort from the surroundings. The mixing process is not reversible.

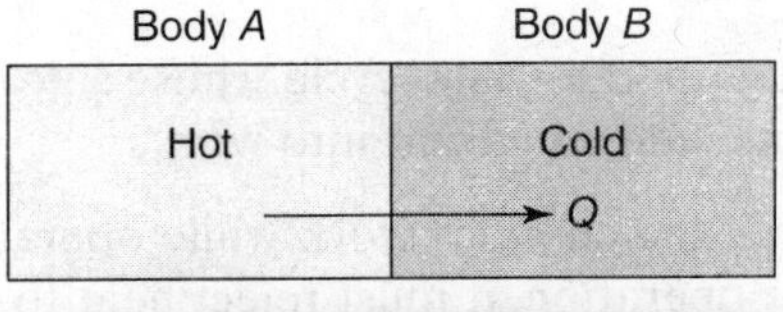

Figure 5.2
Heat flows from a hot to a cold body.

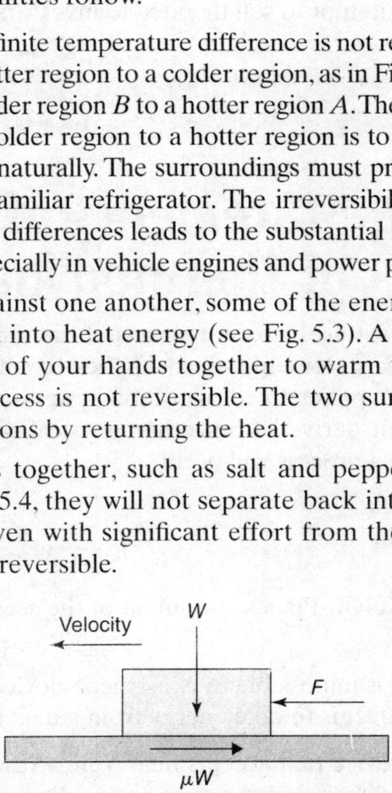

Figure 5.3
Friction between two surfaces.

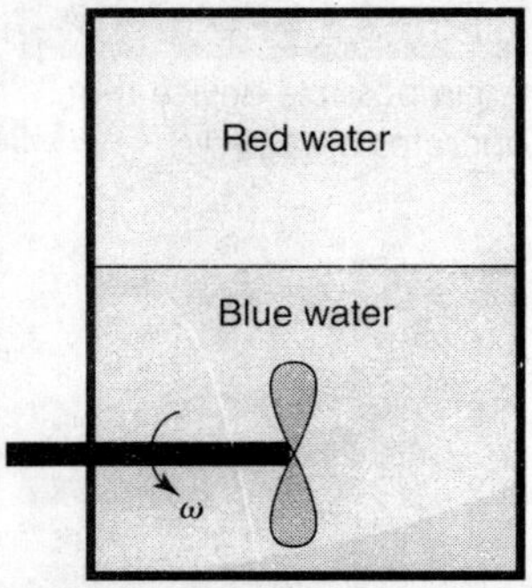

Figure 5.4
Mixing red water and blue water in a container.

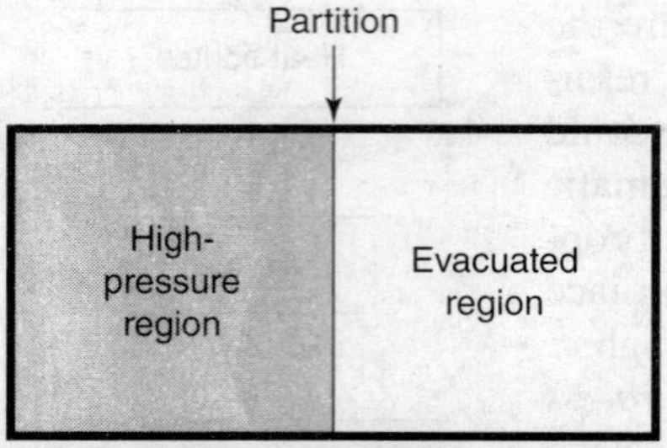

Figure 5.5

Gas suddenly flows from the high-pressure region to the evacuated region when the partition is ruptured.

4. **Unrestrained expansion**—The unrestrained expansion of a gas is not reversible. If you rupture or instantly remove the partition of Fig. 5.5, the gas will suddenly flow from the high-pressure region into the evacuated region. It will not, without energy supplied from the surroundings, return to its original state.

These four physical phenomena result in irreversible processes. No known process is free of all of these effects. All four of these phenomena demand that a process occurs in a particular "direction." For example, heat only flows with no outside effort, from hot to cold; friction does not help propel an object; rotating a blade in the opposite direction will not reorganize water that has been mixed; and the two gases will not return to their separate regions without outside help.

Comment

There is no known reversible process.

Some processes used in industry come close to being reversible—so close that we will consider the processes to be reversible in our solutions to problems. The compression and expansion processes in the cylinder of an engine are examples since friction can be essentially eliminated. The turbines described in Chapter 4 can also be designed to be nearly reversible. Note that most, if not all, of the processes that are nearly reversible do not involve heat reservoirs. All processes are inherently irreversible, especially those with heat transfer across large temperature differences. There is no such thing as a perpetual motion machine that operates with reversible processes, even though some clever entrepreneur may attempt to sell the idea to investors!

☑ You have completed Learning Outcome (1)

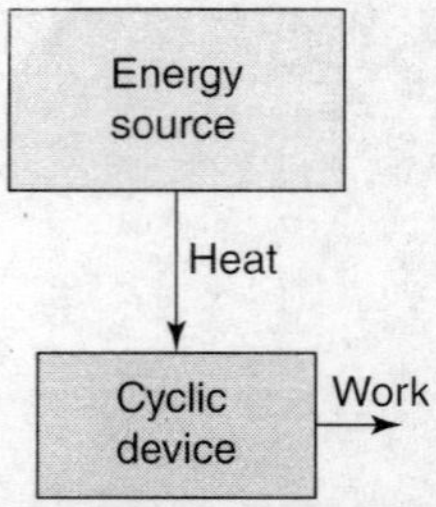

Figure 5.6

An impossible device that operates on a cycle.

5.2 Statements of the Second Law of Thermodynamics

This text uses two statements of the second law of thermodynamics: the Kelvin-Planck statement, which applies to heat engines, and the Clausius statement, which applies to refrigeration systems. Both are statements of observation and are not derived from other physical laws. Each also indicates the direction in which a process will occur.

5.2.1 Kelvin-Planck statement—heat engines

The Kelvin-Planck statement of the second law of thermodynamics (see Fig. 5.6) states:

> It is impossible to construct a device that operates on a cycle whose sole effect is to take energy from a heat source and convert it into work.

Heat engine: A device that accepts heat from a reservoir and supplies work while operating on a cycle.

The device that accepts heat from a reservoir and supplies work while operating on a cycle is called a *heat engine.* During its operation, it must reject heat to the surroundings.

Consider the operation of the simple heat engine in Fig. 5.7. Heat Q_H is transferred to the engine from the heat source at temperature T_H, work W is produced, and heat Q_L is transferred from the engine to the heat sink maintained at temperature T_L. The Kelvin–Planck statement of the second law states that the heat transfer Q_L to the heat sink cannot be zero. Some heat must be rejected by the engine. Essentially, W cannot equal Q_H. The first law of thermodynamics applied to a boundary around the engine would demand that

$$Q_H - Q_L = W$$

The second law will actually provide the minimum size of the discarded heat Q_L, as we shall determine in Section 5.4. Actually, Q_L must often be quite large, which is why inventors are encouraged to invent engines that violate the second law. Heat engines will be analyzed in detail in Chapters 8 and 9.

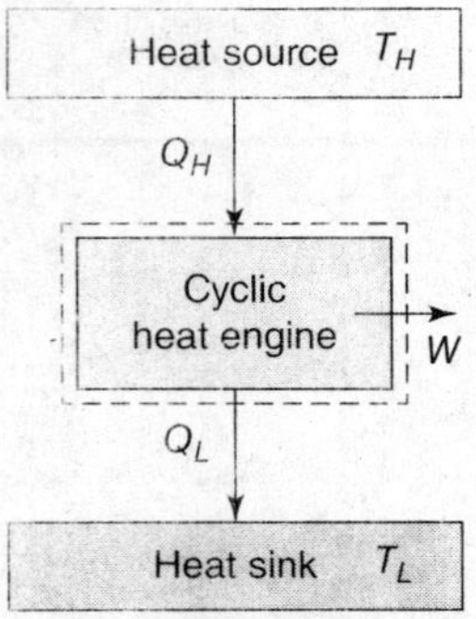

Figure 5.7
A heat engine accepts heat from a heat source and rejects heat to a heat sink.

5.2.2 Clausius statement—refrigerators

The Clausius statement of the second law of thermodynamics (see Fig. 5.8) states:

> It is impossible to construct a device that operates on a cycle whose sole effect is the transfer of heat from a low-temperature reservoir to a high-temperature reservoir.

Heat does not naturally flow from a cooler region to a hotter region; in fact, it does the reverse. In order to move heat from a cooler region to a hotter region, we must force it to do so by providing work from the surroundings. The cycle that does this is called a *refrigeration cycle* used by refrigerators, air conditioners, freezers, and heat pumps—generically referred to as *refrigerators*. The energy requirements of a refrigerator are shown in Fig. 5.9. The refrigerator removes heat Q_L from a low-temperature reservoir maintained at temperature T_L and rejects heat to the high-temperature reservoir maintained at temperature T_H. In order to operate the refrigerator, work W is required from the surroundings. The low-temperature region could be a freezer, and the

Refrigeration cycle: A cycle that moves heat from a cooler region to a hotter region.

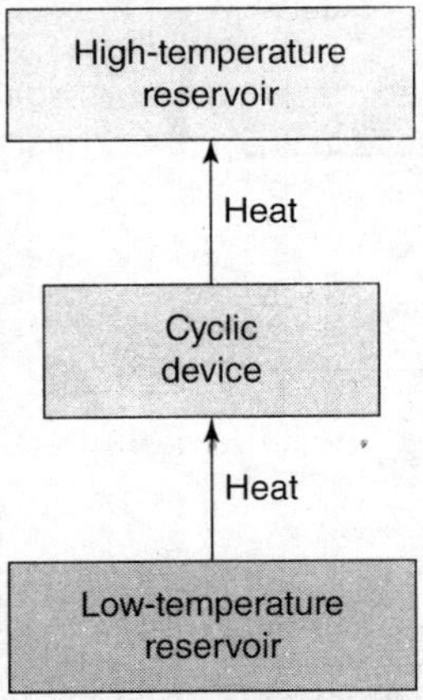

Figure 5.8
An impossible device.

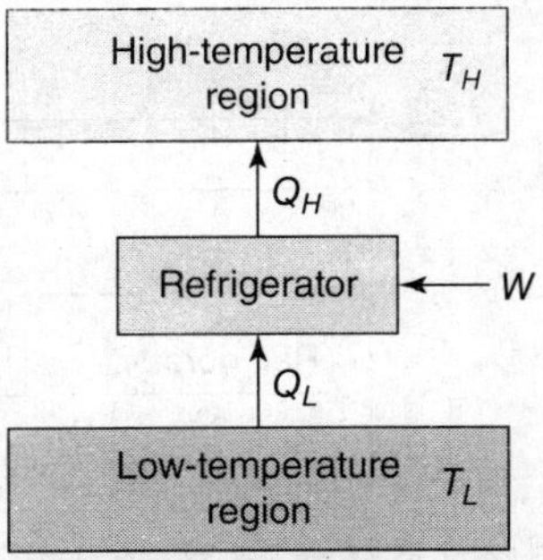

Figure 5.9
A refrigerator requires work to move heat from a cool region to a hotter region.

high-temperature region would be the hot coils on the back or bottom of the freezer.

The Clausius statement of the second law states that heat will not travel from the low-temperature reservoir to the high-temperature reservoir without the addition of work. In layperson's terms, you can't cool your house in the summer unless you spend money running an air conditioner. The first law applied to the refrigeration cycle is

$$W = Q_H - Q_L \tag{5.2}$$

Equation (5.2) is the same as Eq. 5.1 because the refrigeration cycle is simply the reverse of a heat engine cycle. The difference is that the heat engine creates work from purchased heat, whereas the refrigerator operates on purchased work. The refrigeration cycle will be studied in detail in Chapter 10.

Example 5.1 The Clausius and Kelvin-Planck statements

Show that the Clausius statement of the second law is equivalent to the Kelvin-Planck statement.

Solution

To show that the two statements are equivalent, consider the refrigerator on the left of Fig. 5.10*a*. This refrigerator violates the Clausius statement of the second law since it transfers heat from a low-temperature reservoir to a

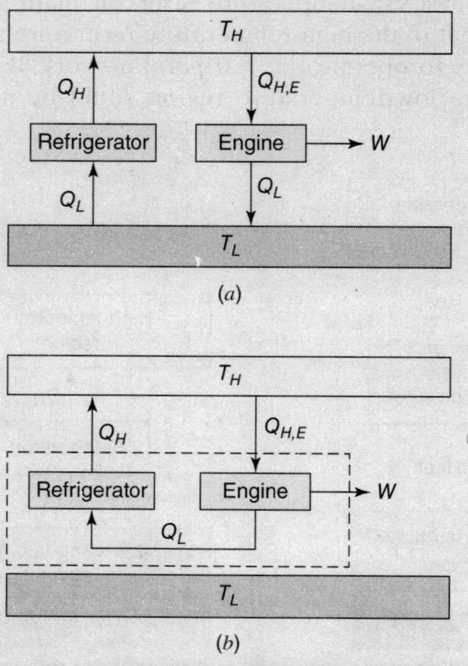

Figure 5.10

high-temperature reservoir with no work input. Assume that the engine on the right, which does not violate the Kelvin-Planck statement of the second law, rejects the same amount of heat Q_L used by the refrigerator.

Now, place a boundary around the engine and the refrigerator, as displayed in Fig. 5.10*b*, which shows that the heat rejected by the engine is used to supply the refrigerator. Since $Q_L = Q_H$, as required by the first law applied to the refrigerator, and $Q_{H,E} = Q_L + W$, as required by the first law applied to the engine ($Q_{H,E}$ is the heat from the high-temperature reservoir to the engine), the result is a net heat transfer of $(Q_{H,E} - Q_L)$ from the high-temperature reservoir directly into the "device." This device is composed of the refrigerator plus the engine inside the dotted boundary. This net heat transfer $(Q_{H,E} - Q_L)$ is converted directly into work W. Heat that is converted directly into work by a device operating on a cycle is a violation of the Kelvin-Planck statement of the second law. Hence, it is concluded that the Clausius statement and the Kelvin-Planck statement are equivalent.

☑ **You have completed Learning Outcome (2)**

5.3 Cycle Performance Parameters

The performance parameter for a thermodynamic cycle is defined as

$$\text{Performance parameter} = \frac{\text{Output}}{\text{Energy input}} \tag{5.3}$$

The required input is the energy (heat or work) that must be supplied to the cycle to make it operational. The output is the energy transferred by the cycle. In the case of a heat engine, it is the work transferred to the surroundings by the engine. In the case of an air conditioner, it is the heat accepted from the low-temperature region. In the case of a heat pump, it is the heat transferred to the high-temperature region. In the next sections, we will analyze performance parameters for the above cycles and determine how well these cycles function.

Comment

The required input is typically the energy that must be purchased, although work can be created using "green" energy such as wind, hydro, or solar. This sustainable energy is gaining considerable interest, but it still must be "purchased."

Efficiency:

$$\eta = \frac{\text{Output}}{\text{Required input}}$$

5.3.1 The heat engine

The performance parameter for a heat engine is the cycle efficiency η, defined by

$$\eta = \frac{\text{Output}}{\text{Required input}} = \frac{W}{Q_H} \tag{5.4}$$

The required input need to make a heat engine work is the heat transfer Q_H from a high-temperature reservoir. The output of an engine is the work produced W. The Kelvin-Planck statement of the second law states that the heat rejected to the

Comment

The heat rejected by an engine or a power plant cannot be zero. We shall see in Section 5.4 that the rejected heat must usually be quite large.

Comment

The maximum efficiency of most engines is much less than 100%. A power plant has an efficiency of around 40%, and that of an automobile engine is around 20%.

low-temperature reservoir cannot be zero. If we insert the left-hand side of Eq. 5.1 for the work, we obtain

$$\eta = \frac{W}{Q_H} = \frac{Q_H - Q_L}{Q_H} = 1 - \frac{Q_L}{Q_H} \quad (5.5)$$

Since Q_L cannot be zero by the second law, the engine efficiency must be less than unity ($\eta < 1$), so that a heat engine can never have an efficiency of 100%. In fact, we will discover that the efficiency of most heat engines is considerably less than 100%. Proposed engines with efficiencies close to 100% undoubtedly violate the second law.

It should be pointed out that rates of work $\dot{W}$ and heat transfer $\dot{Q}_L$ and $\dot{Q}_H$ can also be used in Eq. 5.5.

Example 5.2 The efficiency of a heat engine

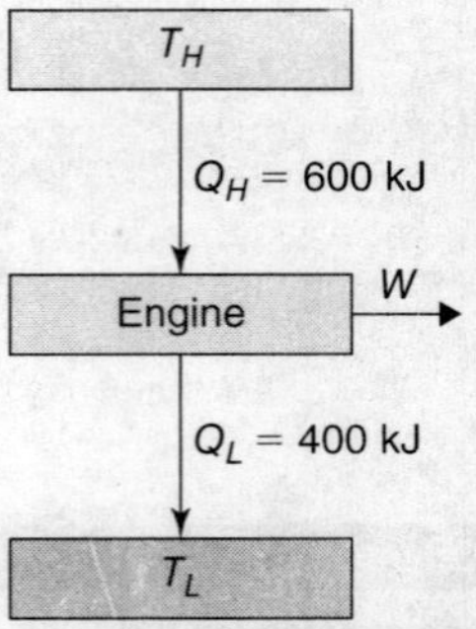

Figure 5.11

The heat engine of Fig. 5.11 receives 600 kJ of heat from a heat source. It rejects 400 kJ of heat to the environment. Calculate i) the work done by the heat engine and ii) the engine efficiency.

Solution

i) The work done by the engine is found by applying the first law:

$$W = Q_H - Q_L$$
$$= 600 \text{ kJ} - 400 \text{ kJ} = \underline{200 \text{ kJ}}$$

ii) Equation (5.5) allows the efficiency of the engine to be calculated. It is

$$\eta = \frac{W}{Q_H} = \frac{200 \text{ kJ}}{600 \text{ kJ}} = 0.333 \quad \text{or} \quad \underline{33.3\%}$$

This would represent a typical engine.

5.3.2 The refrigeration cycle

Refrigerator: A refrigeration cycle that cools a space.

The performance parameter for a refrigeration cycle is called the coefficient of performance, or simply the COP. A *refrigerator* operates on a cycle with the objective of cooling a space. It has a coefficient of performance of

$$COP_R = \frac{\text{Output}}{\text{Required input}} = \frac{Q_L}{W} \quad (5.6)$$

Note: An air conditioner and a freezer are included in the definition of a refrigerator.

The required input to make a refrigerator operate is the work input W to the compressor. The output is the heat removed Q_L from the low-temperature region, for example, the inside of a refrigerator. If we substitute Eq. 5.2 for the compressor work, we obtain

$$COP_R = \frac{Q_L}{W} = \frac{Q_L}{Q_H - Q_L} = \frac{1}{\dfrac{Q_H}{Q_L} - 1} \quad (5.7)$$

The coefficient of performance has no defined upper limit as does the efficiency of a heat engine; that's why it is not referred to as "efficiency." If the work required by a refrigerator approaches zero, then the COP_R for the system could theoretically approach infinity. When the second law is applied to the refrigeration cycle, we will discover that it has a limit that is closer to 10 than to infinity. A $COP_R < 1$ for a refrigeration cycle is an indicator of poor performance, although in certain situations, a $COP_R < 1$ is acceptable, as in a commercial airplane or on a university campus, where absorption refrigeration (to be presented in Chapter 10) may be used.

Comment

The coefficient of performance of a refrigerator has no defined upper limit.

Heat transfer rates $\dot{Q}_L$ and $\dot{Q}_H$, rather than Q_L and Q_H, are often used in the equations for COP.

Figure 5.12 shows the basic components of a central air-conditioning system for a building. The compressor takes in refrigerant, the fluid used in a refrigeration cycle, at low pressure and low temperature and compresses it to a higher-temperature superheated vapor. As the pressure of the refrigerant increases, so does the temperature. The hot refrigerant exits the compressor at about 50°C. The hot refrigerant is then piped outside, where it passes through a heat exchanger, the condenser. The high-pressure refrigerant exchanges heat with the outside air where it is condensed to a saturated liquid. The refrigerant then passes through a valve where the sudden drop in pressure creates a sudden drop in the refrigerant temperature. The mixture is now at about −30°C. This cold refrigerant then passes through an evaporator where it exchanges heat with the air in the house. The refrigerant then returns to the compressor. The processes in this cycle will be studied in detail in Chapter 10.

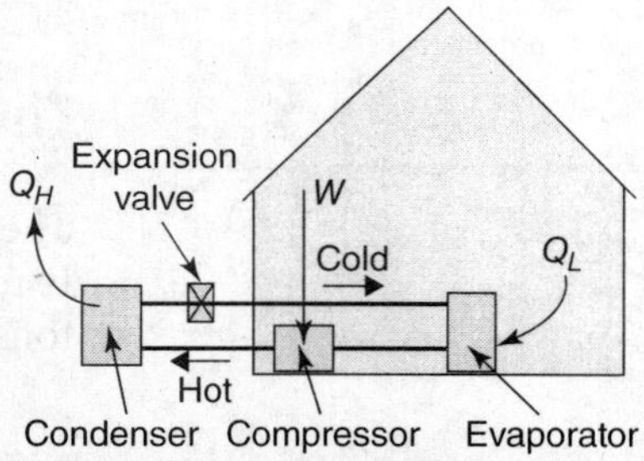

Figure 5.12

Basic components of an air conditioner.

Comment

A simple valve suddenly reduces the pressure in the refrigerant with a simultaneous reduction in temperature.

A *heat pump* is a device that operates on a refrigeration cycle in order to heat a building, for example, the house of Fig. 5.13. The hot refrigerant from the compressor is diverted into the heat exchanger used in the house for air conditioning. Heat is exchanged between the hot refrigerant and the air in the house. The refrigerant then flows through the valve, where the pressure drops and the refrigerant becomes very cold. Next the cold refrigerant passes through the heat exchanger outside the house, where it absorbs heat from the outside atmosphere. This happens because the refrigerant must be much colder than the outside air, so heat pumps are not common in locations where the outside temperature is less than about −7°C. The refrigerant is then sent back to the compressor. This is actually accomplished using clever valving and piping with the same system used for the air conditioner.

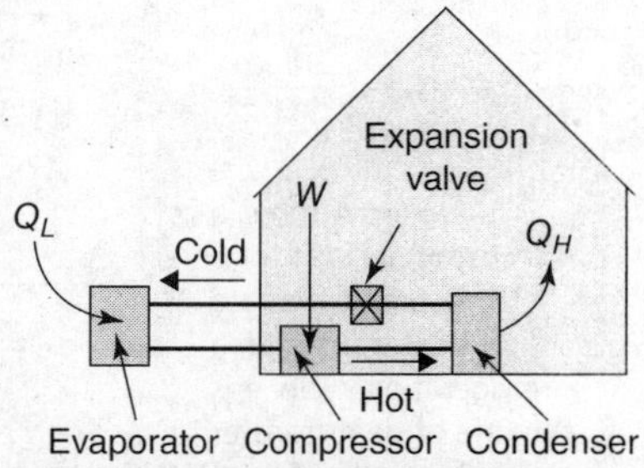

Figure 5.13

A heat pump uses the same refrigerant and components used by the air conditioner of Fig. 5.12.

The performance parameter for a heat pump is also called the coefficient of performance. The output is now the heat delivered to the house Q_H, so the coefficient of performance of a heat pump is

$$COP_{HP} = \frac{Q_H}{W} = \frac{Q_H}{Q_H - Q_L} = \frac{1}{1 - \dfrac{Q_L}{Q_H}} \qquad (5.8)$$

Heat pump: A device that heats a space by operating on a refrigeration cycle.

A comparison of Eq. 5.7 with Eq. 5.8 would show that

$$COP_{HP} = COP_R + 1 \qquad (5.9)$$

There is also no limit on COP_{HP}. In Section 5.4 we will find that the COP is not very large—usually less than 10 for an ideal heat pump.

It should be pointed out that the COP of an air conditioner is seldom, if ever, listed. Manufacturers list the EER (energy efficiency ratio), which is Q_L/W, often with different units for both Q_L and W.

Example 5.3 The performance of an air conditioner

The compressor in the schematic of an air conditioner in Fig. 5.14 requires 5 hp. It rejects 12 kJ/s of heat to the environment. Calculate i) the cooling rate, in tons, and ii) the performance of the air conditioner.

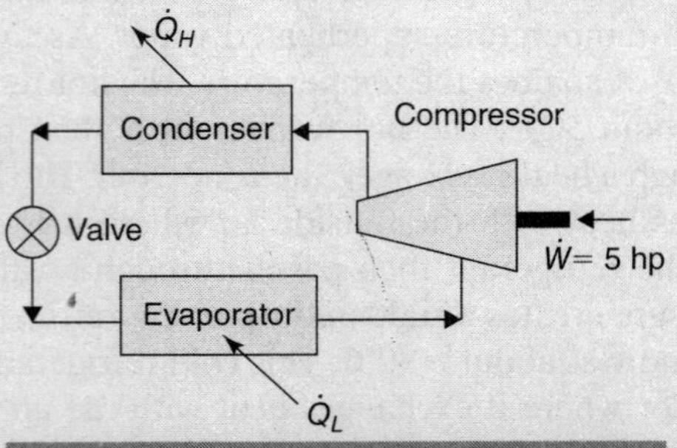

Figure 5.14

Solution

i) The cooling rate is found by applying the first law to the air conditioner as the "refrigerator" box in Fig. 5.9:

$$\dot{Q}_L = \dot{Q}_H - \dot{W}$$

$$= \left(12 \text{ kW} - 5 \text{ hp} \times 0.746 \frac{\text{kW}}{\text{hp}}\right) \times \frac{1}{3.517} \frac{\text{ton}}{\text{kW}} = \underline{2.35 \text{ tons}}$$

ii) Equation (5.6) allows the coefficient of performance to be calculated. It is

$$\text{COP}_R = \frac{\dot{Q}_L}{\dot{W}} = \frac{8.27 \text{ kW}}{5 \times 0.746 \text{ kW}} = \underline{2.22}$$

Note: A ton is a customary unit for a measure of the rate of heat transfer in a refrigeration cycle. It is the rate of heat transfer needed to melt a ton of ice at 0°C in one day.

Unit: 1 ton = 3.517 kW

Example 5.4 The heat pump

A refrigeration cycle operates as a heat pump requiring 5 hp by the compressor, as shown in Fig. 5.14. It supplies 12 kJ/s of heat to a space. Calculate i) the rate at which heat is absorbed from the environment and ii) the performance of the heat pump.

Solution

i) The rate at which heat is absorbed from the environment is found by applying the first law to the heat pump as a unit (the "refrigerator" box in Fig. 5.9):

$$\dot{Q}_L = \dot{Q}_H - \dot{W}$$

$$= 12\text{ kW} - 5\text{ hp} \times 0.746\frac{\text{kW}}{\text{hp}} = \underline{8.27\text{ kJ/s}}$$

Observe that the energy equation is identical to that used for the air conditioner of Example 5.3.

ii) Equation (5.8) allows the coefficient of performance to be calculated. It is

$$\text{COP}_{HP} = \frac{\dot{Q}_H}{\dot{W}} = \frac{12\text{ kW}}{5 \times 0.746\text{ kW}} = \underline{3.22}$$

The definition of the coefficient of performance changed compared to that of the air conditioner of Example 5.3 since the desired output changed from $\dot{Q}_L$ to $\dot{Q}_H$.

Observe: The COPs from Examples 5.3 and 5.4 satisfy Eq. 5.9.

☑ **You have completed Learning Outcome (3)**

5.4 The Carnot Cycle

Now that the efficiency of a heat engine has been defined, we consider an engine that operates with quasi-equilibrium processes—processes that have no friction, no sudden expansion, and no heat transfer across finite temperature differences. Such an engine does not exist, but it does define the most efficient engine possible since it has no irreversibilities. It is referred to as a *Carnot engine.* Consider our fictitious engine to operate using a piston-cylinder arrangement with the four quasi-equilibrium processes of Fig. 5.15; an ideal gas with constant specific heats will be assumed. The processes are as follows.

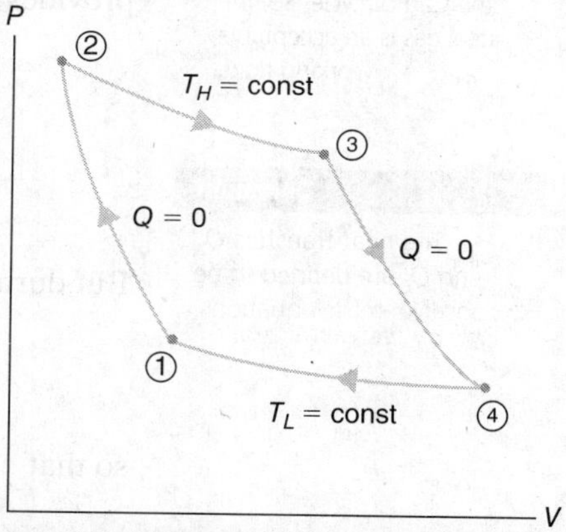

Figure 5.15

The processes that make up the Carnot gas cycle.

1→2: *An adiabatic compression.* The piston compresses a gas from a low temperature T_1 to a high temperature T_2 with the cylinder completely insulated.

2→3: *An isothermal expansion.* Heat is transferred from a high-temperature reservoir to the gas at the high temperature $T_2 = T_3 = T_H$. The reservoir temperature is a fraction of a degree above the gas temperature, so both are approximated to be at T_H.

3→4: *An adiabatic expansion.* The gas is expanded from a high temperature T_3 to a low temperature T_4, with the cylinder completely insulated.

Carnot engine: An ideal engine that operates in a cycle with no irreversibilities. It can operate with any working fluid, but an ideal gas will be assumed for which simplified equations are available.

Note: It is the two heat transfer processes that are assumed to occur across an infinitesimal temperature difference that make the Carnot cycle a fictitious cycle. For heat transfer to occur quickly, a large temperature difference is necessary.

Comment

The two processes used in the Carnot gas cycle are the adiabatic process and the isothermal heat transfer process.

Note: We are assuming an ideal gas with constant specific heats. The objective of Problem 5.32 is to show that the working fluid does not affect the efficiency of the Carnot cycle. So, the ideal gas is an acceptable working fluid.

Comment

The heat transfers Q_H and Q_L are defined to be positive in the equations.

Comment

The ratio Q_L/Q_H is replaced with T_L/T_H in this ideal Carnot cycle.

4→1: *An isothermal compression.* Heat is transferred from the gas to a low-temperature reservoir at the low temperature $T_4 = T_1 = T_L$. The gas temperature is a fraction of a degree above the reservoir temperature, so both are approximated to be at T_L.

This Carnot engine operates with ideal processes that are reversible with no friction, so they cannot be improved upon; the cycle must possess the maximum possible efficiency.

The two processes that make up the cycle have been analyzed in Chapter 3. The processes from state 1 to state 2 and from state 3 to state 4 are adiabatic processes for which $Q_{1\text{-}2} = Q_{3\text{-}4} = 0$. The isothermal process from state 2 to state 3 was analyzed in Section 3.5.3, with the result that the heat transfer $Q_{2\text{-}3} = Q_H$ is given for the ideal gas by Eq. 3.35 and Eq. 3.36 to be

$$Q_H = mRT_H \ln \frac{V_3}{V_2} \tag{5.10}$$

The heat transfer $Q_L = -Q_{4\text{-}1}$ for the isothermal process from state 4 to state 1 is given by

$$Q_L = mRT_L \ln \frac{V_4}{V_1} \tag{5.11}$$

The heat transfers Q_H and Q_L are expressed to be positive so that Eq. 5.5 for the cycle efficiency is acceptable (we recognize that $Q_{4\text{-}1}$ from state 4 to state 1 is negative since heat is leaving the gas). Equation (5.5) for the cycle efficiency then provides

$$\eta = 1 - \frac{Q_L}{Q_H} = 1 - \frac{mRT_L \ln V_4/V_1}{mRT_H \ln V_3/V_2}$$

$$= 1 - \frac{T_L}{T_H}\frac{\ln V_4/V_1}{\ln V_3/V_2} \tag{5.12}$$

But during the adiabatic quasi-equilibrium processes, Eq. 3.44 is used to give

$$\frac{T_2}{T_1} = \frac{T_H}{T_L} = \left(\frac{V_1}{V_2}\right)^{k-1} \qquad \frac{T_3}{T_4} = \frac{T_H}{T_L} = \left(\frac{V_4}{V_3}\right)^{k-1} \tag{5.13}$$

so that

$$\frac{V_1}{V_2} = \frac{V_4}{V_3} \quad \text{or} \quad \frac{V_4}{V_1} = \frac{V_3}{V_2} \tag{5.14}$$

Consequently, the two ln terms in Eq. 5.12 cancel out, providing us with the surprisingly simple expression for the efficiency of the Carnot cycle:

$$\eta = 1 - \frac{T_L}{T_H} \tag{5.15}$$

Compare Eq. 5.5 with Eq. 5.15. We have replaced Q_L/Q_H with T_L/T_H; that is,

$$\frac{Q_H}{Q_L} = \frac{T_H}{T_L} \tag{5.16}$$

The efficiency of this ideal reversible cycle, the Carnot cycle, is dependent only on the absolute temperatures of the two heat reservoirs. This is, in fact, true of all reversible cycles; they all have the same efficiency. The efficiency does not depend on the working fluid (the conclusion of Problem 5.32), and it is impossible for any cycle operating between the same two reservoirs to have an efficiency greater than the Carnot efficiency.

The Carnot cycle can be reversed resulting in a refrigerator or a heat pump, depending on the objective of the cycle. If the objective is to cool, we use the COP_R, defined by Eq. 5.7. We substitute T_H/T_L for Q_H/Q_L, and we obtain the maximum possible COP for a refrigerator:

Comment

If the Carnot power cycle of Fig. 5.15 is reversed, the Carnot refrigeration cycle results.

$$COP_R = \frac{1}{\dfrac{T_H}{T_L} - 1} = \frac{T_L}{T_H - T_L} \tag{5.17}$$

If the objective is to heat, we substitute T_L/T_H for Q_L/Q_H in Eq. 5.8 to obtain the maximum possible COP for a heat pump:

$$COP_{HP} = \frac{1}{1 - \dfrac{T_L}{T_H}} = \frac{T_H}{T_H - T_L} \tag{5.18}$$

It is true that the Carnot cycles considered above are fictitious cycles that cannot be built, let alone tested in the laboratory. Why then are they defined? It is important to know the maximum possible performance that an engine or a refrigerator can have when operating between two temperature reservoirs. If an engine has an efficiency of 30% and the Carnot efficiency operating between the same temperatures is 33%, it would not be advisable to devote substantial effort to increase the performance of that engine. Also, if an inventor proposes an engine that operates with an efficiency of 60% but the Carnot engine, operating between the same high and low temperatures, has an efficiency of 55%, the engine will not operate as proposed.

Comment

All Carnot cycles are fictitious. They serve as limits for real cycles.

Can the Carnot efficiency be exceeded? Example 5.5

Show that the efficiency of a Carnot engine cannot be exceeded.

Solution

A very clever engine is proposed that has an efficiency that exceeds the Carnot efficiency. The engine accepts Q_H from the high-temperature reservoir, and a Carnot refrigerator rejects the same amount of heat to that reservoir, as shown in Fig. 5.16. The work W_E produced by the engine is greater than the work W required by the Carnot refrigerator since the proposed engine is more efficient than the refrigerator.

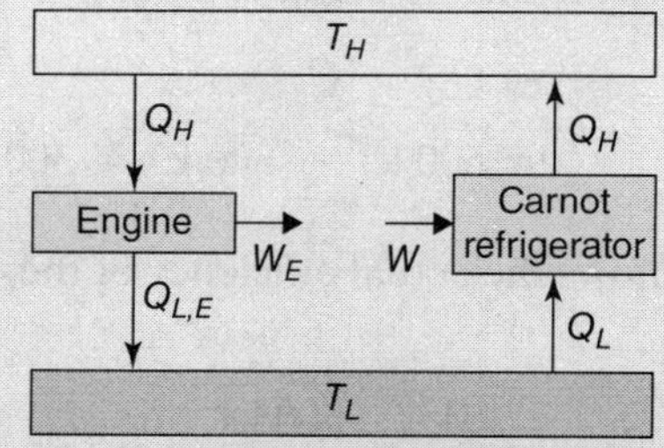

Figure 5.16

(*Continued*)

Example 5.5 (Continued)

Now, consider the engine and the refrigerator to operate as a single device, as identified inside the dotted boundary of Fig. 5.17. The heat rejected by the refrigerator is used to drive the engine. The heat Q_L from the low-temperature reservoir is greater than the heat rejected $Q_{L,E}$ by the engine since $W_E > W$. Consequently, the single device inside the dotted boundary accepts heat $(Q_L - Q_{L,E})$ from a single reservoir and produces net work $(W_E - W)$; this is a violation of the Kelvin-Planck statement of the second law. It is concluded that no engine operating on a cycle can have an efficiency greater than the Carnot efficiency.

Suggestion: Assume that $W > W_E$ and show that the proposed engine is possible.

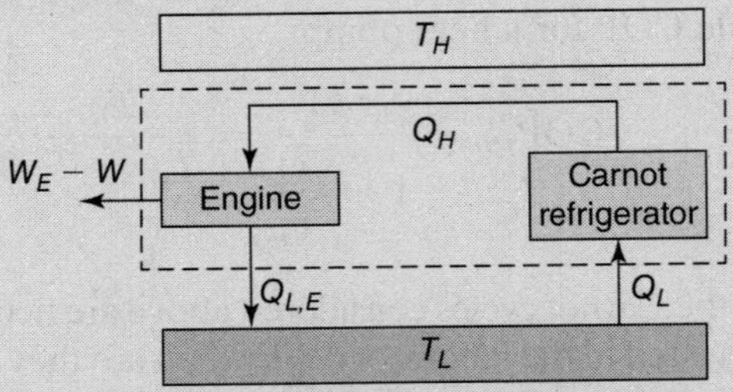

Figure 5.17

Example 5.6 The efficiency of an engine

The heat engine of Fig. 5.18 receives 600 kJ of heat from a heat source at 600°C. It rejects 300 kJ of heat to an environment at 10°C. Calculate i) the work done by the heat engine, ii) the actual engine efficiency, and iii) the maximum possible engine efficiency.

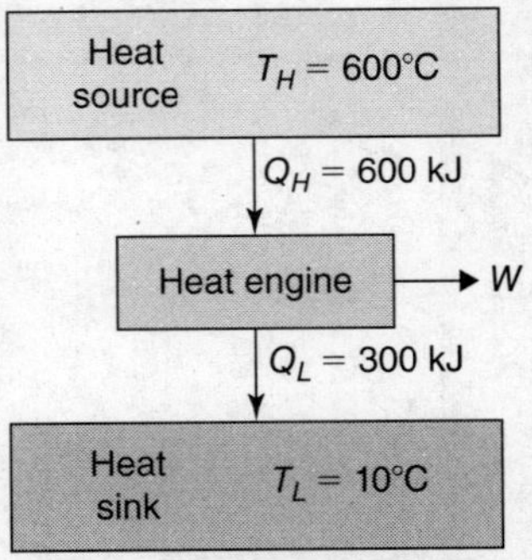

Figure 5.18

Solution

For the two reservoirs:

$$Q_H = 600 \text{ kJ} \qquad T_H = 600°\text{C} = 873 \text{ K}$$

$$Q_L = 300 \text{ kJ} \qquad T_L = 10°\text{C} = 283 \text{ K}$$

i) The work done by the engine is found by applying the first law:

$$W = Q_H - Q_L$$
$$= 600 \text{ kJ} - 300 \text{ kJ} = \underline{300 \text{ kJ}}$$

ii) Equation (5.5) allows the actual efficiency of the engine to be calculated. It is

$$\eta = \frac{W}{Q_H} = \frac{300 \text{ kJ}}{600 \text{ kJ}} = 0.5 \text{ or } \underline{50\%}$$

iii) The maximum possible efficiency, the Carnot efficiency, is found, using Eq. 5.15, to be

$$\eta_{\text{Carnot}} = 1 - \frac{T_L}{T_H}$$

$$= 1 - \frac{283\text{ K}}{873\text{ K}} = 0.676 \quad \text{or} \quad \underline{67.6\%}$$

The actual engine efficiency can never be greater than the Carnot efficiency. If it is greater, it is considered a violation of the 2nd law.

Observation: When a difference of temperatures is desired, it doesn't matter if absolute or relative temperatures are used, but when a temperature appears alone, as in the numerator of Eq. 5.17, or as a ratio, invariably absolute units must be used.

Is this invention possible? Example 5.7

An inventor claims to have developed an automobile engine that burns biofuels at 317°C with an engine efficiency of 50%, as demonstrated by Fig. 5.19. If the local air temperature is 17°C, can this claim be believed?

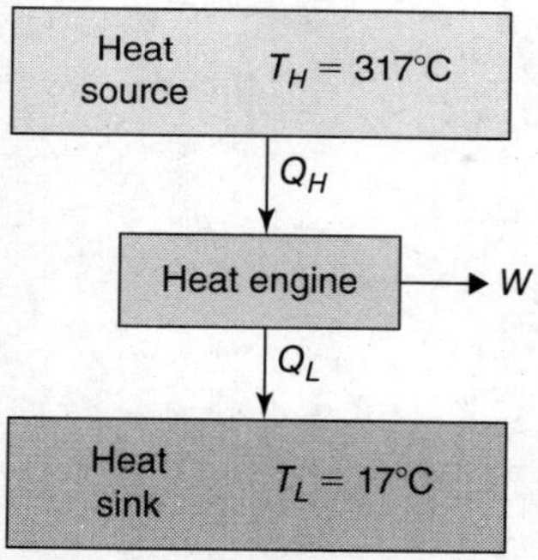

Figure 5.19

Solution

The upper and lower absolute temperatures for this engine are

$$T_H = 317°\text{C} = 590\text{ K} \qquad T_L = 17°\text{C} = 290\text{ K}$$

The Carnot efficiency of the engine, using Eq. 5.15, is

$$\eta_{\text{Carnot}} = 1 - \frac{T_L}{T_H}$$

$$= 1 - \frac{290\text{ K}}{590\text{ K}} = 0.51 \quad \text{or} \quad 51\%$$

An engine efficiency of 50% is possible according to the second law but is not very likely. Real effects, especially heat transfer across a finite temperature difference, would lower the actual efficiency by a considerable amount.

The COP of an air conditioner Example 5.8

An air conditioner maintains a house temperature of 23°C while the outside temperature is 40°C. If the refrigeration cycle removes 8 kJ/s from the house and the compressor does 2 hp of work on the refrigerant, as shown in Fig. 5.20, determine i) the heat rejected to the outside environment, ii) the actual coefficient of performance for the air conditioner, and iii) the maximum possible coefficient of performance for this air conditioner.

(*Continued*)

Example 5.8 (Continued)

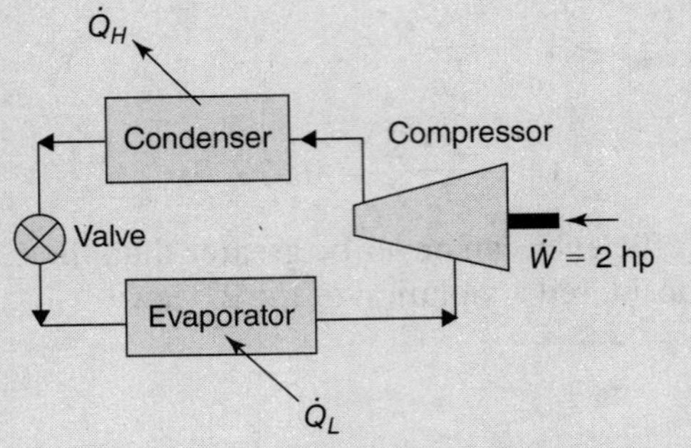

Figure 5.20

Solution

The given information is

$$T_L = 23°C = 296 \text{ K} \qquad T_H = 40°C = 313 \text{ K}$$

$$\dot{Q}_L = 8 \text{ kJ/s} \qquad \dot{W} = 2 \text{ hp} = 1.492 \text{ kW}$$

i) The first law provides the heat rejected to the high-temperature reservoir:

$$\dot{Q}_H = \dot{W} + \dot{Q}_L$$

$$= 1.492 + 8 = \underline{9.49 \text{ kJ/s}}$$

ii) The actual COP for this air conditioner, using Eq. 5.6, is

$$\text{COP}_R = \frac{\dot{Q}_L}{\dot{W}} = \frac{8}{1.492} = \underline{5.36}$$

iii) To obtain the maximum possible COP, the Carnot COP, we use Eq. 5.17:

$$\text{COP}_{\text{max}} = \frac{T_L}{T_H - T_L} = \frac{296}{313 - 296} = \underline{17.41}$$

This must be greater than the actual COP, which it is.

☑ **You have completed Learning Outcome** **(4)**

5.5 Summary

In this chapter, the second law was introduced along with the concepts of reversibility and irreversibility. The primary contributors to irreversibility were identified: friction, heat transfer across a finite temperature difference, mixing, and unrestrained expansion. The Kelvin-Planck and Clausius statements of the second law were then presented. The performance of cycles was investigated with the most efficient cycle possible, the Carnot cycle, given special attention. It was analyzed for a four-process, ideal-gas cycle with its maximum efficiency expressed in terms of temperatures. The following additional terms were defined in this chapter:

Carnot cycle: *An ideal cycle that operates with no losses. It can be an engine, a refrigerator, or a heat pump.*

Clausius statement: *It is impossible to construct a device that operates on a cycle whose sole effect is the transfer of heat from a low-temperature reservoir to a high-temperature reservoir.*

EER (energy efficiency ratio): Q_L/W *with units of Btu's for* Q_L *and watt-hours for W.*

Heat engine: *A cycle that takes energy from a heat source and converts a portion of it to work.*

Heat pump: *A device operating on a refrigeration cycle that heats a space.*

Kelvin-Planck statement: *It is impossible to construct a device that operates on a cycle whose sole effect is to take energy from a heat source and convert it to work.*

Performance parameter: *The ratio of the output of a device to the required input.*

Refrigeration cycle: *A cycle that moves heat from a cool region to a hotter region.*

Refrigerator: *A device operating on a refrigeration cycle to cool a space.*

Reservoir: *A body that it is able to provide heat or absorb heat without a change in its temperature.*

Reversible process: *One that can be reversed and returned to its original state leaving no change in either the system or the surroundings. No process is reversible.*

SEER (the seasonal EER): *The Btu's divided by watt-hours for the entire cooling season.*

Several important equations were presented:

Performance: $$\eta_{\text{engine}} \text{ or COP} = \frac{\text{desired effect}}{\text{purchased energy}}$$

Relation between COP_R and COP_{HP}: $$COP_{HP} = COP_R + 1$$

Heat to temperature ratio for Carnot cycle: $$\frac{Q_H}{Q_L} = \frac{T_H}{T_L}$$

Efficiency of a Carnot engine: $$\eta = 1 - \frac{T_L}{T_H}$$

COP of a Carnot refrigerator: $$COP_R = \frac{1}{T_H/T_L - 1}$$

COP of a Carnot heat pump: $$COP_{HP} = \frac{1}{1 - T_L/T_H}$$

Comment

"Carnot" can be replaced by "reversible."

Problems

FE Exam Practice Questions

5.1 Which one of the following could not be considered a reservoir?

(A) The fire in a coal furnace in a house in the country
(B) A bathtub filled with ice water
(C) The combustion process in an automobile engine
(D) The water in a river

5.2 Which of the following processes could not be considered a reversible process?

(A) The process of a piston compressing the air in a cylinder
(B) Inflating a balloon
(C) The stretching of a rubber band
(D) The freezing of water into ice in a refrigerator

5.3 Which statement comes closest to being a Kelvin-Planck second-law statement?

(A) You can't get something for nothing.
(B) You can't convert heat completely into work transmitted by a rotating shaft.
(C) You can't use heat to cool a space.
(D) You can't move heat from a cold space to a hot space.

5.4 The heat engine of Fig. 5.21 operates between the two reservoirs shown, producing 100 kW of power and 50 kJ/s of rejected heat. The engine's efficiency is nearest:

(A) 33%
(B) 50%
(C) 67%
(D) 80%

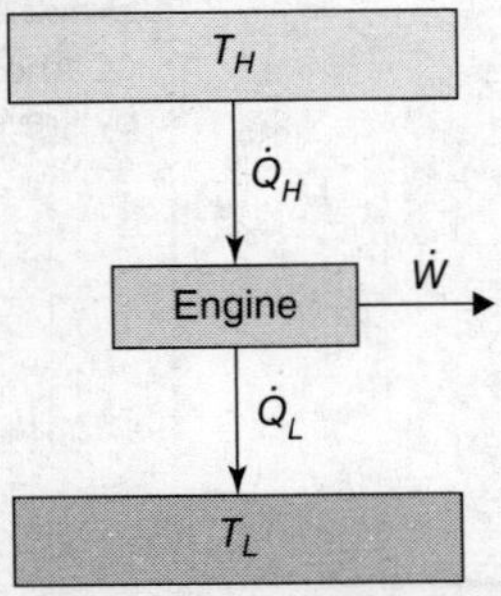

Figure 5.21

5.5 A refrigerator requires 5 hp to provide 24 000 kJ/h of cooling. The COP of the refrigerator is nearest:

(A) 1.8
(B) 2.0
(C) 2.4
(D) 2.8

5.6 If the ideal heat engine of Fig. 5.22 rejects 50 kJ of heat, estimate the energy required to operate the engine.

(A) 110 kJ
(B) 100 kJ
(C) 90 kJ
(D) 80 kJ

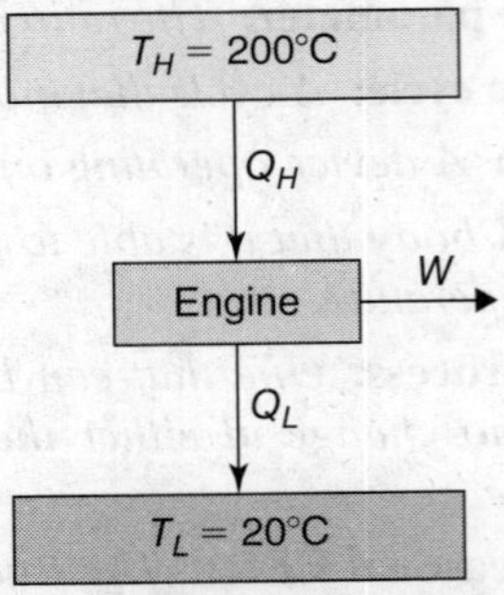

Figure 5.22

5.7 An inventor proposes an engine that utilizes the boiling water of a hot spring and suggests it can operate with an efficiency of 22% in an area where a local stream has an average temperature of 12°C. Such a proposal is:

(A) Not possible
(B) Very likely possible
(C) Possible but improbable
(D) Possible many days of the year

5.8 A Carnot engine expands 0.2 kg of air at 40°C from 30 cm^3 to 300 cm^3 during the isothermal heat addition process in the gas cycle of Fig. 5.15. The heat required is nearest:

(A) 24 kJ
(B) 31 kJ
(C) 36 kJ
(D) 41 kJ

5.9 The heat pump of Fig. 5.23 requires 5 hp and rejects 17 kJ/s of heat to the high-temperature reservoir. The COP is nearest:

(A) 4.98
(B) 4.56
(C) 3.82
(D) 3.56

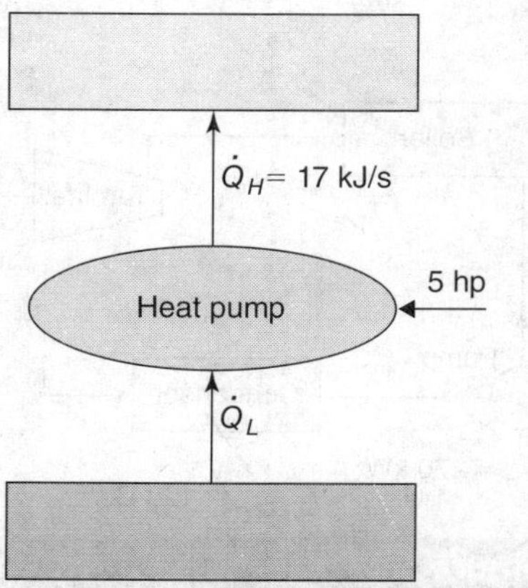

Figure 5.23

5.10 If the high-temperature reservoir in Problem 5.9 is at 160°C, the low-temperature reservoir for a Carnot heat pump is at:

(A) 65°C
(B) 55°C
(C) 45°C
(D) 35°C

Statements of the Second Law

5.11 Show that a violation of the Kelvin-Planck statement of the second law implies a violation of the Clausius statement of the second law. This requires an argument similar to that used in Example 5.1.

The Heat Engine

5.12 A telemarketer advertises an emergency electric generator powered by a gasoline engine. The gasoline-powered engine supposedly produces 0.9 hp of mechanical power while the electric generator produces i) 600 W, ii) 800 W, and iii) 1000 W of electric power. Which of these generators are possible?

5.13 The heat engine of Fig. 5.24 accepts 30 kJ/s of heat from a high-temperature reservoir and produces *a*) 10 hp, *b*) 15 hp, and *c*) 20 hp. Determine the rejected heat and the efficiency of the engine.

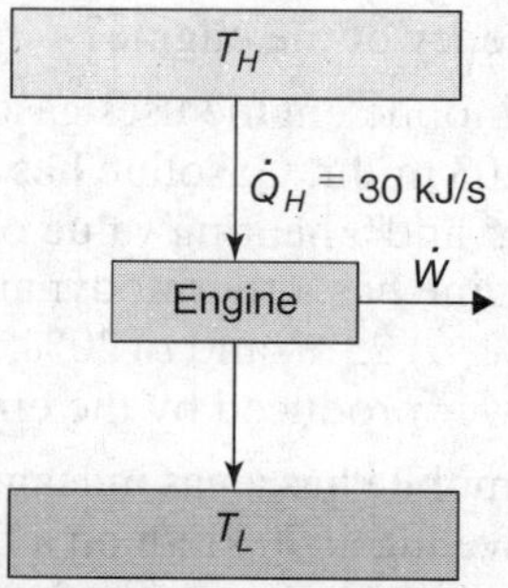

Figure 5.24

5.14 A power plant produces 20 MW of power by burning trash and rejects 14×10^7 kJ each hour from a cooling tower. Calculate the efficiency of the power plant and the amount of heat supplied by the trash burner every hour.

5.15 A gas turbine engine is used to propel a ship. It has a thermodynamic efficiency of 0.75. How much heat must be supplied to this engine for it to produce *a*) 4000 hp, *b*) 3500 kW, and *c*) 260 MJ/min?

Remember: Parts with lower-case italic letters are separate problems and may be assigned separately.

5.16 An automobile engine produces 200 hp at an efficiency of 28%. What rate of heat transfer is required from the gasoline? How many kilograms of gasoline would be needed per hour if there are 50 MJ/kg of gasoline?

5.17 A power plant burns coal to deliver 80 MJ/s of heat to the steam. The plant turbines produce 15 000 hp. What is the rate of heat rejection from this plant, and what is its thermodynamic efficiency?

5.18 A heat engine is required to produce *a*) 120 hp, *b*) 70 kW, and *c*) 110 kW by accepting 150 kW from a heat source. How much heat is dumped to the low-temperature reservoir, and what is the efficiency of the engine?

5.19 An automobile engine uses gasoline at the rate of 0.03 m^3/hr. Gasoline has a density of 755 kg/m^3 and a heating value of 48 800 kJ/kg. If the engine has a thermodynamic efficiency of *a*) 30%, *b*) 25%, and *c*) 20%, what is the horsepower produced by the engine?

5.20 An automobile has a gas mileage of 12 km/liter when traveling at 90 km/h on a flat road with no head wind. Gas contains about 48 800 kJ/kg with a density of 755 kg/m^3. Estimate the efficiency of the engine if it produces *a*) 17 hp, *b*) 14 hp, and *c*) 12 hp under those conditions.

5.21 The engine of an automobile requires 20 hp to travel at 100 km/hr. If the engine's efficiency is 18%, determine the rate of fuel consumption, in L/km, assuming an energy content of 45 MJ/kg and a density of 720 kg/m^3.

5.22 A power plant burns 50 000 kg of coal each hour. Coal contains about 26 MJ/kg of energy. If the efficiency of the plant is 34%, estimate the heat transfer each day to the nearby river used to receive the rejected energy.

5.23 The power plant of Fig. 5.25 provides *a*) 200 GJ/hr, *b*) 3000 MJ/min, and *c*) 60 000 kJ/s from the boiler to the steam that flows through the plant. It rejects 120 GJ/hr to the environment, and the pump requires the power indicated. Calculate the horsepower output and the cycle efficiency. Is it necessary to include the pump power?

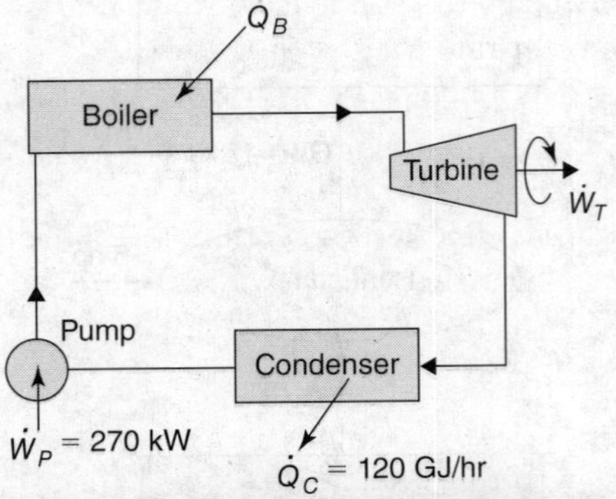

Figure 5.25

The Refrigeration Cycle

5.24 A refrigerator is advertised that will remove 10 kW from the refrigerator while dumping 15 kW of heat into the kitchen. How much horsepower is the compressor using to run this system?

5.25 The refrigeration cycle of Fig. 5.26 is intended to cool a space by supplying *a*) 10 kJ/s, *b*) 15 kJ/s, and *c*) 20 kJ/s of cooling. Determine the rate of heat transfer that is rejected from the condenser and the performance of the cycle.

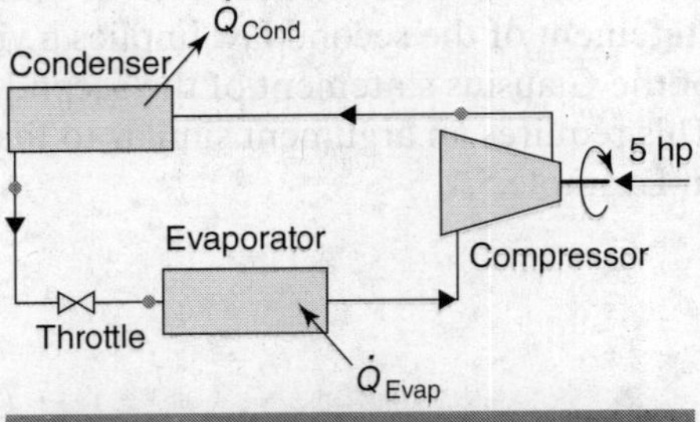

Figure 5.26

5.26 The refrigeration cycle of Fig. 5.26, which requires 5 hp by the compressor, as shown, is intended to heat a space. It has a COP of 5 when operating in the heating mode. Determine the rate of heat transfer that is provided to the space being heated and the rate of heat transfer accepted by the evaporator.

5.27 The refrigeration cycle of Fig. 5.26 requires 5 hp by the compressor, as shown, and rejects 21 kJ/s by the condenser. Calculate the heat transfer rate to the evaporator and the COP if used as a heater.

5.28 An air conditioner removes 26 kW from a room. The compressor of Fig. 5.27 uses 7 hp to run the system. What is the rate of heat rejection to the outside environment? What is the coefficient of performance for this air conditioner?

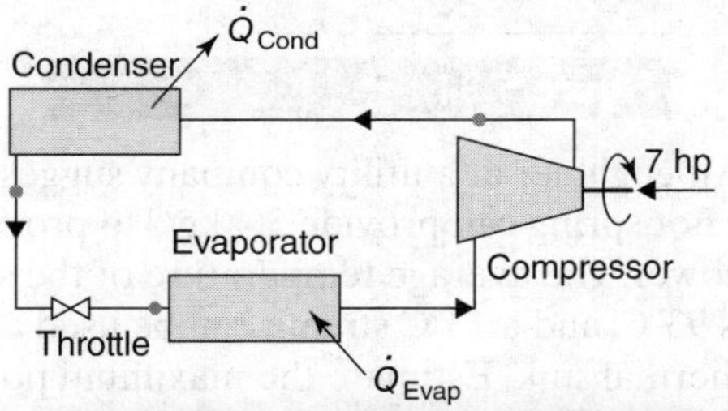

Figure 5.27

5.29 Seven thousand kJ/hr of heat is removed from the inside of a refrigerator while transferring 9000 kJ/hr of heat into a kitchen. What is the coefficient of performance of this refrigerator?

5.30 An air-conditioning system transfers 100 kJ/s of heat from the condenser. How much horsepower must the compressor supply if the COP is *a*) 8, *b*) 6, and *c*) 4?

5.31 The compressor of the heat pump, shown in Fig. 5.28, with a COP of 8 requires 2 kW of power. How much heat is being supplied to the house each hour?

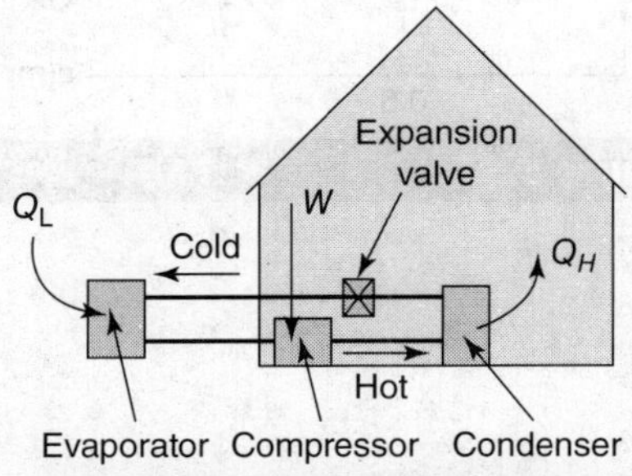

Figure 5.28

The Carnot Cycle

5.32 Show that the working fluid used in a Carnot cycle does not influence the efficiency of the cycle. Use an argument similar to that used in Example 5.5.

5.33 A Carnot engine operates between a heat source at 300°C and a heat sink at i) 100°C, ii) 20°C, and iii) −30°C. What is the maximum possible efficiency of this engine for each heat sink?

5.34 Ocean Thermal Energy Conversion (OTEC) is an alternative energy system that uses the warm water at the surface of the ocean to operate a power plant, as sketched in Fig. 5.29. If the surface water at 27°C is used to supply energy to the plant and if water at 4°C deep in the ocean is used as the low-temperature reservoir, what will be the maximum possible efficiency of such a power plant?

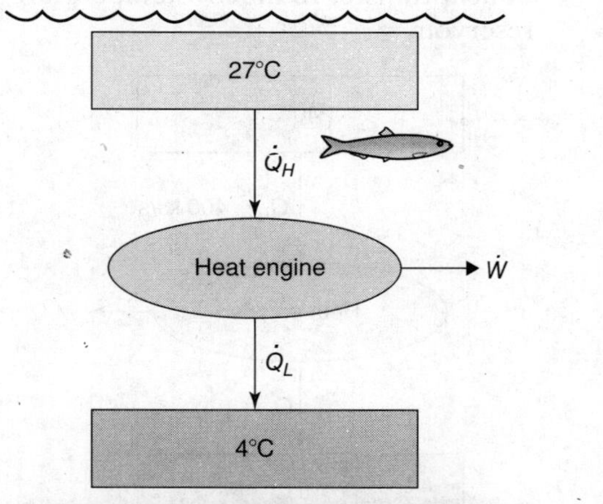

Figure 5.29

5.35 Geothermal heat sources can be used to operate power plants to produce electricity. If a source of steam at 50°C is located near a river that has a water temperature of 5°C, what is the maximum efficiency possible for a power plant using these as the heat source and heat sink?

5.36 The Eddystone Power plant described at the start of Chapter 2 uses supercritical steam at 600°C. If the condenser of this power plant uses Delaware River water at 4°C as the heat sink for the plant, what is the maximum possible efficiency for this power plant?

5.37 An ideal heat engine produces 25 hp by transferring 60 kJ/s from a reservoir maintained at 380°C. What is the temperature of the heat sink?

5.38 A Carnot heat engine receives 1 MJ/s of heat from a source at 1000°C and rejects 0.4 MJ/s of heat to a sink of unknown temperature. Calculate the temperature of the heat sink and the efficiency of the engine.

5.39 An entrepreneur proposes a heat engine that presumably receives 100 kJ/s from a heat source at 147°C and produces 30 kW of power while rejecting heat to a heat sink at 37°C. Determine whether or not this engine can possibly work.

5.40 The heat engine of Fig. 5.30 operates between two reservoirs maintained at 240°C and 40°C, respectively. If 400 kJ/s is transferred from the high-temperature reservoir, determine the maximum horsepower that can be produced by the engine and the accompanying rate of heat transfer to the low-temperature reservoir.

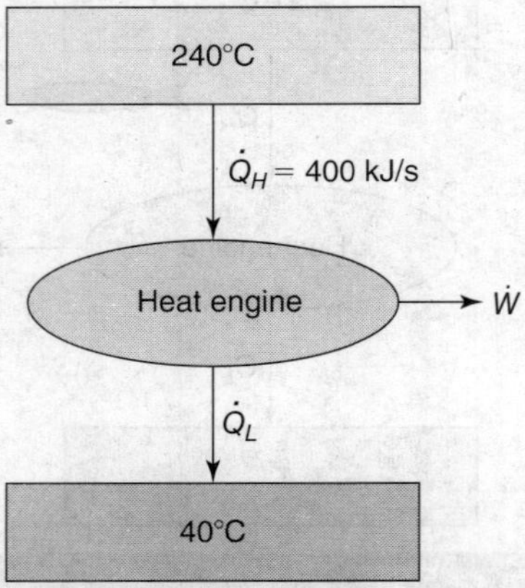

Figure 5.30

5.41 A heat engine is proposed to operate between the two reservoirs shown in Fig. 5.31. It is proposed to produce 30 hp by transferring 4000 kJ/min from the high-temperature reservoir. Make an engineering analysis of the proposal.

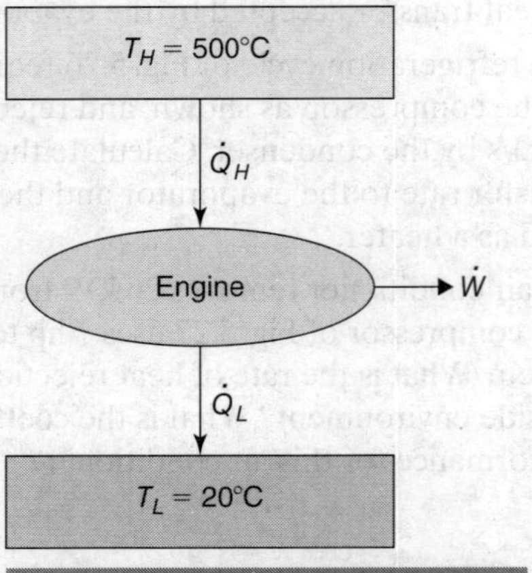

Figure 5.31

5.42 An engineer at a utility company suggests that a hot spring can provide 3.4 kg/s to produce power. The average temperature of the spring is 93°C, and a 13°C stream can be used as a thermal sink. Estimate the maximum power that can be produced.

5.43 A Carnot engine undergoes the cycle of Fig. 5.32. The net work output is 200 kJ. Determine the required heat addition and T_H if the cycle efficiency is *a*) 50%, *b*) 60%, and *c*) 70%. Air is the working fluid.

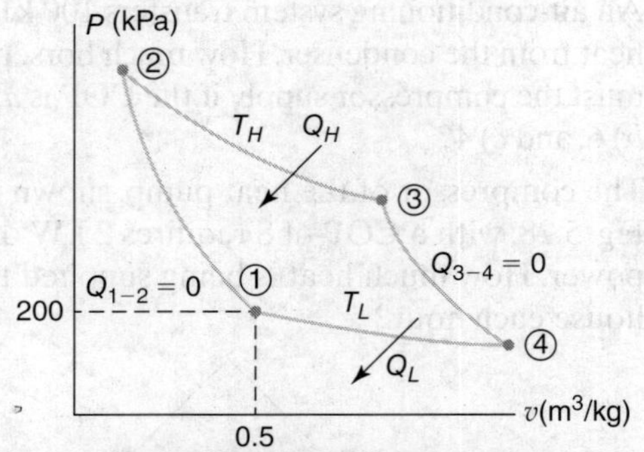

Figure 5.32

5.44 An inventor proposes to extract 100°C water from a hot spring and use it to power a heat engine. The atmosphere is at 20°C. The proposed engine would operate with a flow rate of 150 kg/min and produce 50 hp, according to the inventor. Comment as to the feasibility of the proposal using calculated numbers.

5.45 The Carnot engine of Fig. 5.33 operates with an efficiency of 60% by rejecting energy to a heat sink maintained at *a*) 30°C, *b*) 20°C, and *c*) 10°C. If the rejected heat rate is 50 kJ/s, determine the power output, the heat rate from the heat source, and the temperature of the heat source.

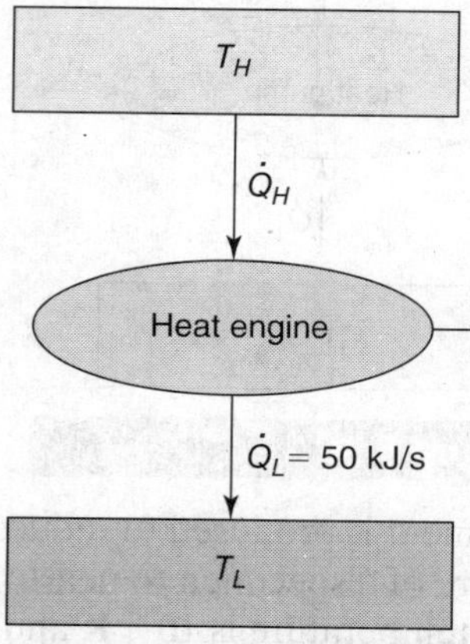

Figure 5.33

5.46 A Carnot engine operates between two heat reservoirs with an efficiency of 75%. If a Carnot refrigerator operates between the same two reservoirs, calculate its COP.

5.47 Three reservoirs are at temperatures T_1, T_2, and T_3. A Carnot engine is positioned between T_1 and T_2, a second Carnot engine between T_2 and T_3, and a third Carnot engine between T_1 and T_3. Express the efficiency η_3 of the third Carnot engine in terms of the efficiencies η_1 and η_2 of the other two engines.

5.48 Two Carnot engines operate between the three reservoirs displayed in Fig. 5.34. The heat rejected by Engine 1 is the heat input to Engine 2. Determine the temperature T_2 that will make the efficiency of each engine the same.

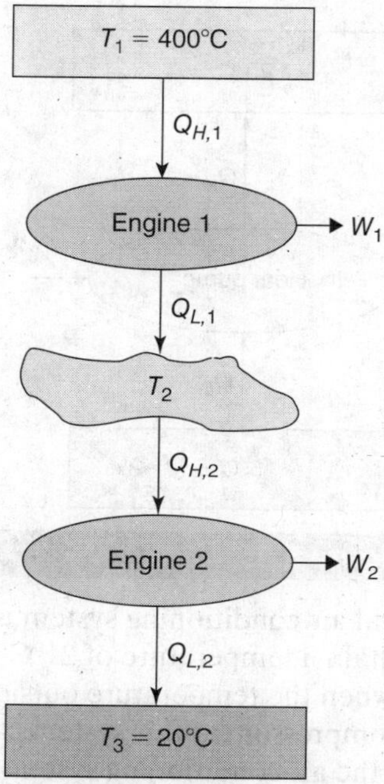

Figure 5.34

5.49 The Carnot refrigeration cycle of Fig. 5.35 operates between two reservoirs maintained at 5°C and 23°C, respectively. If 400 kJ/s is transferred from the low-temperature reservoir, determine the horsepower required by the compressor and the rate of heat transfer to the high-temperature reservoir.

5.50 The Carnot refrigeration cycle of Fig. 5.35 is used as a heat pump to supply heat to the high-temperature reservoir. Determine the horsepower required by the compressor and the rate of heat transfer from the low-temperature reservoir if the heat supply rate is *a*) 200 kJ/s, *b*) 700 kJ/s, and *c*) 2000 kJ/s.

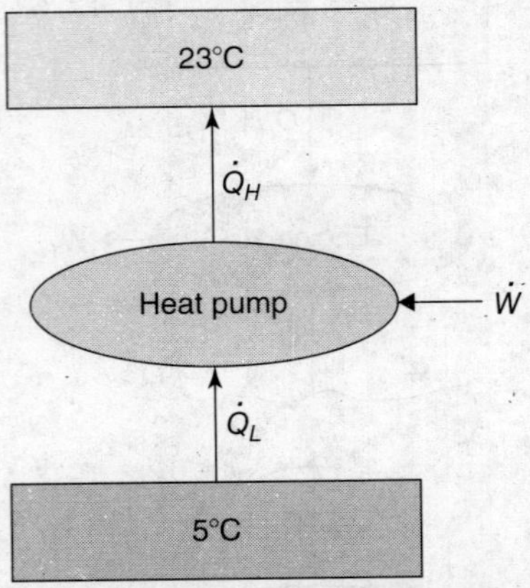

Figure 5.35

5.51 A central air-conditioning system is used to maintain a temperature of 21°C inside a house when the temperature outside is 35°C. If the compressor for this system can deliver 8 hp to the air-conditioning system, what is the maximum rate in kJ/hr that heat can be removed from the house?

5.52 A central air-conditioning system is reversed to operate as a heat pump in the winter. If the inside temperature is maintained at 20°C when the outside temperature is −12°C and the house loses heat at a rate of 72,000 kJ/hr, what is the minimum power required to operate the compressor?

5.53 If the air outside a house has a temperature of 31°C and the indoor temperature is 19°C, what is the maximum cooling load, in kJ/hr, that the system can produce if the compressor can deliver 5 hp?

5.54 The heat pump of Fig. 5.36 is proposed to heat an oil rig positioned in the ocean where the surface water is at 28°C and the water at a reasonable depth is at 11°C. If 230 kW are desired, what minimum compressor horsepower would be required?

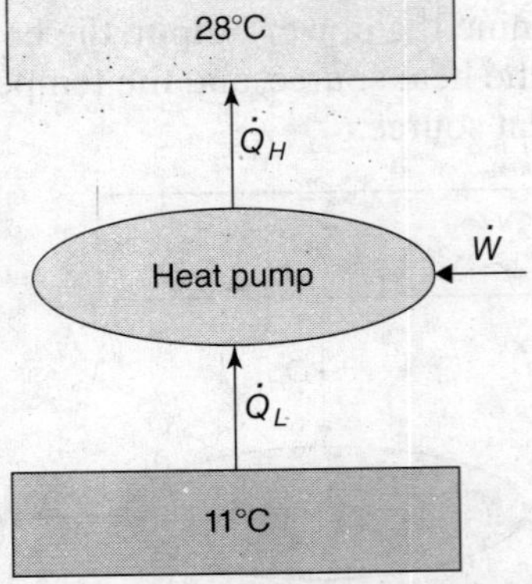

Figure 5.36

5.55 An experiment is proposed to reduce the temperature of a specimen to nearly absolute zero. If its temperature is 10^{-5} K and it is desired to remove 0.02 J of energy from the specimen, what is the minimum work requirement if the laboratory at 18°C is the high-temperature reservoir?

CHAPTER

6

Entropy

Outline

Nomenclature

The following variables are introduced in this chapter:

I	Irreversibility
i	Irreversibility per unit mass
P_r	Relative pressure
S	Entropy
S_{gen}	Entropy generated
S_{surr}	Entropy of surroundings
$\dot{S}_{prod}$	Entropy production
s	Specific entropy
$\bar{s}$	Specific entropy per mol
s°	An entropy function
T_0	Dead state temperature
v_r	Relative specific volume
W_a	Actual work
η_{II}	Second-law efficiency
Ψ	Exergy (or availability)
ψ	Exergy per unit mass
ΔS_{gen}	Generated entropy
ΔS_{net}	Net entropy change
ΔS_{univ}	Entropy change in the universe

Learning Outcomes

- **Understand the inequality of Clausius for engineering cycles**
- **Understand the concept of entropy and how it is used in thermodynamics**
- **Evaluate changes in entropy for thermodynamic processes**
- **Use the concept of entropy to develop performance parameters for common engineering devices**
- **Use availability and exergy to solve second-law problems**

In Chapter 5 the second law of thermodynamics was observed to limit the performance of heat engines and refrigerators. In this chapter we will quantify the second law and make it a tool for the analysis of cycles using the property of entropy. Changes in entropy for a system will be evaluated, and cycle performance will be analyzed.

Entropy is a difficult property to define. Like energy, the effects of entropy can be observed, but it is not a directly measurable property such as pressure or temperature. In this chapter we will define entropy and calculate the value of this property for various substances and the change in entropy for several processes of interest. This property will help us determine if the second law of thermodynamics is satisfied or if it is violated. Entropy will also be used in the definition of performance parameters for some common engineering devices.

Motivational Example—High-Efficiency Electric Motor

A research group from Tokai University under the direction of Professor Hideki Kimura has developed a 24-volt DC motor that has an energy efficiency in excess of 96%. The motor, shown in Fig. 6.1, was developed to be used in electric vehicles such as the one shown in Fig. 6.2. The motor was developed by studying the causes of energy loss in motors and then making improvements to reduce these losses. The two major areas of energy loss were friction in the bearings and wind friction over the rotating core. Electric motors are very efficient; the inefficiency in electrical energy production is in the processing of the steam in power plants where efficiencies of about 33% have been realized since the 1950s. New technologies have raised efficiencies, but added regulations have managed to decrease efficiencies with the result that power-plant efficiencies have remained relatively constant. Google "electricity-generating efficiency" to read more about this subject.

© Lyroky/Alamy

Figure 6.1

Core of a high-efficiency DC motor.

Tokai University

Figure 6.2

A streamlined electric car.

6.1 Inequality of Clausius

The study of entropy begins by developing the *Clausius inequality*, expressed as

$$\oint \frac{\delta Q}{T} \le 0 \tag{6.1}$$

Clausius inequality:

$$\oint \frac{\delta Q}{T} \le 0$$

where the circle through the integral symbol indicates integration around a cycle. The German physicist J. Clausius (1822–1888) postulated, based on experiments, that the ratio of the differential heat transfer δQ to the absolute temperature T at which that heat transfer took place integrated around a cycle is conserved for reversible cycles and negative for irreversible cycles (for all real cycles). Let us interrupt this analysis for a short discussion of a reversible process.

The Clausius inequality is applicable to a system that undergoes a *reversible* or an irreversible cycle. For example, the air in a cylinder could be compressed with a piston with negligible friction between the piston and the cylinder wall, or it could be compressed the same distance, with the piston being forced to move due to extreme friction between the piston and the cylinder wall. The air, the system, would undergo the same process in both cases; the process with friction could be considered a quasi-equilibrium process[1] for the system, the air, but *irreversible* when both the system and surroundings are considered. The friction would make the process, as a whole, irreversible. Generally speaking, both the system and surroundings are included when a process is declared reversible. However, in this discussion of entropy, it is the system as it passes through a cycle that is being considered in the Clausius inequality. If all processes in the cycle are quasi-equilibrium processes, the "equal" sign is used. If even one process involves heat transfer with the system boundary at a different temperature than the system, a sudden expansion, internal heating (a resistor), or mixing, the process is a nonequilibrium process and the "less than" sign is used for the cycle.

Reversible process: The system and surroundings possess no irreversibilities. A reversible cycle possesses all reversible processes.

Irreversible process: The system or surroundings experience irreversibilities.

[1]Some authors refer to this process as being internally reversible.

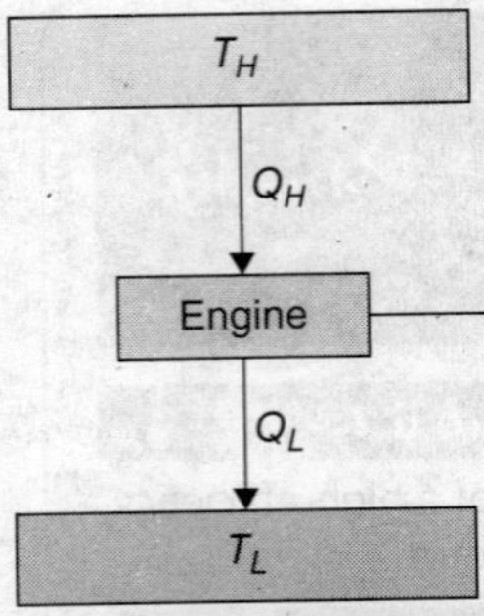

Figure 6.3

A heat engine.

To show that the Clausius inequality is true, consider $\oint \delta Q/T$ for the heat engine of Fig. 6.3 that operates on a cycle. Since T_H and T_L are constant during each respective heat transfer process, the integral becomes

$$\oint \frac{\delta Q}{T} = \frac{Q_H}{T_H} - \frac{Q_L}{T_L} \tag{6.2}$$

For a Carnot heat engine, the integral in Eq. 6.2 will equal zero since it has been shown (see Eq. 5.16) that $Q_H/Q_L = T_H/T_L$ or, rewritten, $Q_H/T_H = Q_L/T_L$. Consequently, for the reversible cycle,

$$\oint \frac{\delta Q}{T} = 0 \quad \text{(reversible cycle)} \tag{6.3}$$

Now, consider a heat engine operating on an irreversible cycle between the same two reservoirs with an equal amount of energy supplied to each, that is, $(Q_H)_{\text{rev}} = (Q_H)_{\text{irr}}$. Since $W_{\text{rev}} > W_{\text{irr}}$, the first law demands that $(Q_L)_{\text{rev}} < (Q_L)_{\text{irr}}$. Hence, for the irreversible cycle,

Remember: The first law for the heat engine requires $Q_H - Q_L = W$.

$$\oint \frac{\delta Q}{T} = \frac{Q_H}{T_H} - \frac{Q_L}{T_L} < 0 \quad \text{(irreversible cycle)} \tag{6.4}$$

and the Clausius inequality is shown to be true.

Equation (6.3) implies that $\delta Q/T$ for a reversible process is an exact differential since the cyclic integral is zero, an observation to be used in the next section. To show that this is true, consider the two-process cycle of Fig. 6.4. A perfect differential df integrated between states 1 and 2 would give $f_2 - f_1$. Then, integrated between states 2 and 1, to complete the cycle, the result would be $f_1 - f_2$. The cyclic integral would be the sum, which is zero. If an inexact differential, such as the work δW (or heat δQ), were integrated around a complete cycle, the result would not be zero but the net work (or net heat transfer) for the cycle, that is, the area under the P-v diagram of Fig. 6.4. It is concluded that $\delta Q/T$ is an exact differential.

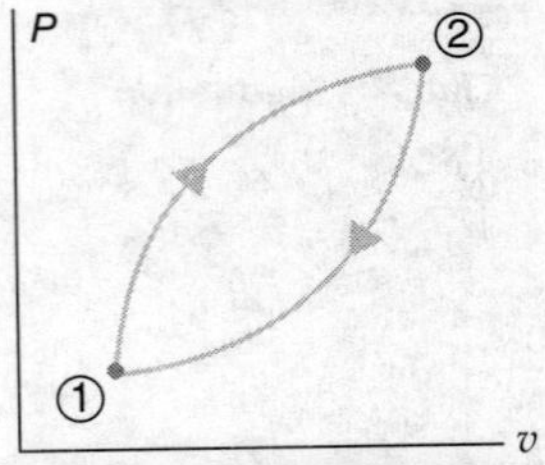

Figure 6.4

A two-process cycle.

The above analysis holds true for both heat engine cycles and refrigeration cycles.

Example 6.1 Check the Clausius inequality

The schematic of a steam power plant operates on the cycle shown in Fig. 6.5 with its P-v diagram in Fig. 6.6. Heat transfer takes place in the boiler and condenser only. (The boiler and condenser are heat exchangers, so it is assumed that they operate at constant pressure.) Check the Clausius inequality for this cycle if the properties of the steam at the four states are as follows:

State 1: $P_1 = 10$ kPa, $x_1 = 0.01$ **State 2:** $P_2 = 800$ kPa, $x_2 = 0$

State 3: $P_3 = 800$ kPa, $x_3 = 1$ **State 4:** $P_4 = 10$ kPa, $x_4 = 0.9$

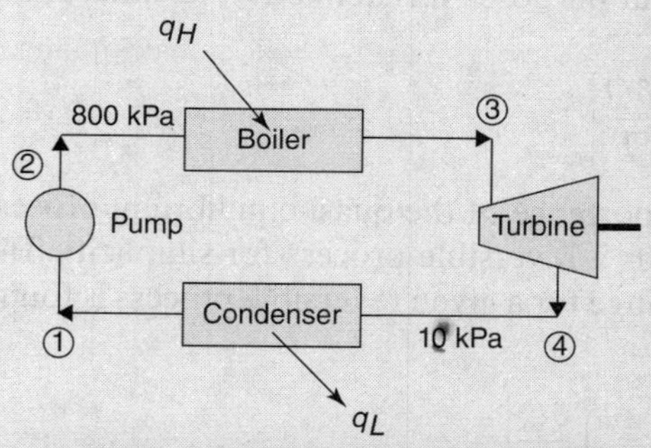

Figure 6.5

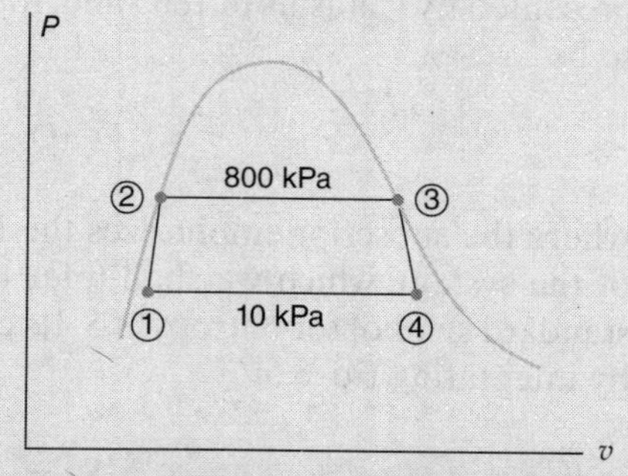

Figure 6.6

Solution

The enthalpies and temperatures are found in the steam tables (the IRC Calculator can also be used) and are listed or calculated as follows:

$$h_1 = 216 \text{ kJ/kg} \qquad h_2 = 721 \text{ kJ/kg},$$

$$h_3 = 2770 \text{ kJ/kg}, T_3 = 170°\text{C} \qquad h_4 = 2340 \text{ kJ/kg}, T_4 = 45.8°\text{C}$$

Using the energy equation (4.16), the specific heat transfer in the boiler is

$$q_H = h_3 - h_2 = 2770 - 721 = 2049 \text{ kJ/kg}$$

The heat transfer emitted by the condenser is

$$q_L = h_4 - h_1 = 2340 - 216 = 2124 \text{ kJ/kg}$$

The boiler temperature is $T_H = 170°\text{C} = 443$ K, and the low temperature is $T_L = 45.8°\text{C} = 318.8$ K. Because the high and low temperatures are constant, we can apply Eq. 6.3 to verify the Clausius inequality:

$$\oint \frac{\delta q}{T} = \frac{q_H}{T_H} - \frac{q_L}{T_L} = \frac{2049}{443} - \frac{2124}{318.8} = \underline{-2.04 \text{ kJ/kg}\cdot\text{K}}$$

The value of the cyclic integral is less than zero, which it must be. If the integral was positive, the cycle would not be possible since it would violate the second law, as expressed by Eq. 6.1.

☑ **You have completed Learning Outcome (1)**

6.2 Entropy

The Clausius inequality has identified $\delta Q/T$ as an exact differential for a reversible process (refer to the discussion referencing Fig. 6.4). Hence, it represents the differential of a property of the system. This property, called *entropy*, will be used as an expression of the second law as it applies to physical systems. It was first

Entropy: A property defined by $dS = \delta Q/T|_{rev}$. Other definitions exist in other disciplines, but in thermodynamics this one suffices.

presented by Clausius in 1865 and, for our purposes, it is defined by its differential to be

$$dS = \left.\frac{\delta Q}{T}\right|_{\text{rev}} \tag{6.5}$$

Isentropic process: A process for which the entropy change is zero. Not all adiabatic processes are isentropic, and not all isentropic processes are adiabatic. Adiabatic processes that are assumed to be internally reversible are isentropic.

where the subscript emphasizes the importance of the quasi-equilibrium process of the system, which we shall refer to as a reversible process for simplicity. The standard symbol for entropy is S. Its change for a given reversible process is found by integrating Eq. 6.5:

$$S_2 - S_1 = \int_1^2 \left.\frac{\delta Q}{T}\right|_{\text{rev}} \tag{6.6}$$

The entropy change $\Delta S = S_2 - S_1$ can be positive or negative depending on whether heat is added to or extracted from a system. For an adiabatic reversible process (review Section 3.5.4), the entropy change is zero. If the entropy change of a process is zero, it is an *isentropic process*. Not all adiabatic processes are isentropic processes, only the quasi-equilibrium ones. There would be an entropy change for an irreversible adiabatic process. Heat could leave an irreversible process such that the entropy change would be zero, resulting in an isentropic process.

Consider a system undergoing a cycle consisting of a process from state 1 to state 2 that could be reversible along path A or irreversible along path C. The return process is along the reversible path B, as shown in Fig. 6.7. Equation (6.1) for the cycle can be expressed as

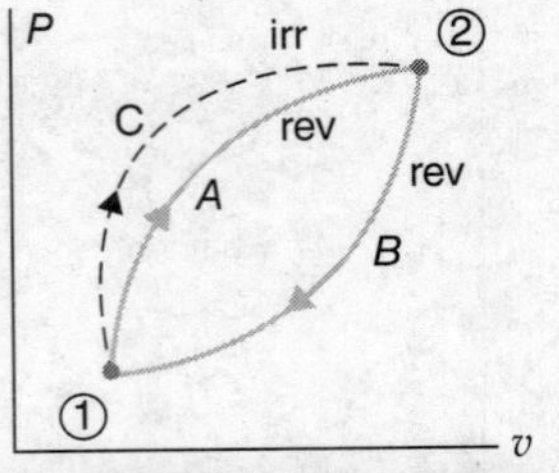

Figure 6.7

A cycle consisting of a reversible process A or an irreversible process C from state 1 to state 2 and a reversible process B back to state 1.

$$\underset{\substack{\text{along } A\\ \text{or } C}}{\int_1^2} \frac{\delta Q}{T} + \underset{\text{along } B}{\int_2^1} \frac{\delta Q}{T} \leq 0 \tag{6.7}$$

The integral along the reversible path B is defined by Eq. 6.6 as $S_1 - S_2$, so Eq. 6.7 can be written as

$$\underset{\substack{\text{along } A\\ \text{or } C}}{\int_1^2} \frac{\delta Q}{T} + S_1 - S_2 \leq 0 \tag{6.8}$$

Note: The limits have been switched, so a minus sign is introduced resulting in $S_1 - S_2$. When the limits are switched on an integral, a minus sign must be introduced, as was learned in calculus.

This equation can be written in the equivalent form,

$$S_2 - S_1 \geq \int_1^2 \frac{\delta Q}{T} \tag{6.9}$$

where the equality holds for a quasi-equilibrium process and the inequality for an irreversible process, that is, any real process. Equation (6.9) can be considered a mathematical statement of the second law for a system undergoing a process.

Several conclusions can be reached considering the relationship between entropy and heat transfer expressed by Eq. 6.9.

Comment

The effect of an irreversibility is to increase the change in entropy for a process.

1. Consider two processes each with the same infinitesimal heat transfer δQ and temperature T, one a reversible process for which the infinitesimal entropy change $dS = \delta Q/T$ and the other an irreversible process for which $dS > \delta Q/T$. The effect of the irreversibility (e.g., heat transfer across a finite temperature difference) is to increase the change in entropy for a process.

2. Consider an isolated system, one that does not interact with the surroundings so that there is no heat transfer. Equation (6.9) would then be expressed as

$$\Delta S \geq 0 \tag{6.10}$$

The second law demands that the entropy of an isolated system remains constant for a reversible process or increases for a real process.

3. Since the universe is an isolated system, its entropy must always increase for every real process, as demanded by Eq. 6.10. This can be stated for any process as

$$\Delta S_{univ} = \Delta S_{sys} + \Delta S_{surr} \geq 0 \tag{6.11}$$

The entropy change of the universe, ΔS_{univ}, is also referred to as ΔS_{net}, the net entropy increase, or ΔS_{gen}, the entropy generated. This has led to various statements of the second law, all of which refer to a generation of entropy as the result of every real process.

Specific entropy, often referred to simply as entropy, is defined as the amount of entropy per unit mass; the standard symbol is a lower-case *s*, defined by $s = S/m$. This can often be confusing since upper-case *S* and lower-case *s* look quite similar, especially when hand written during the solution to a problem. Entropy *S* has units of kJ/K, and specific entropy *s* has units of kJ/kg·K.

A more qualitative definition of entropy is that it is the measure of the amount of disorder in a substance. In a solid, the molecules are held in a tight crystalline structure. Solids have a very low level of entropy. The molecules in a solid move very little from their set position. A solid existing at absolute zero temperature has zero entropy. In a liquid, the molecules are able to move freely, with intermolecular forces keeping them together. Liquids have a higher level of entropy compared to a solid at the same temperature. In a gas, the molecules move independently and randomly with a relatively high level of entropy.

Comment

The entropy of every real process increases in an isolated system.

A result of the second law: Every real process results in a net entropy increase, represented by $\Delta S_{univ} = \Delta S_{net} = \Delta S_{gen}$, in the universe.

Suggestion: Be careful when writing *S* and *s*. Make the *S* quite large and the *s* quite small so that they are obviously different.

Comment

Entropy is also considered to be a measure of the amount of disorder in a substance.

Observation: Since the actual value of entropy is not of interest in a liquid or a gas, a zero value for entropy is assigned at an arbitrary temperature.

☑ **You have completed Learning Outcome (2)**

6.3 Entropy Change in Substances for Systems

6.3.1 Basic relationships

The incremental heat transfer for a reversible process (no friction, sudden expansion, electrical resistance heaters, or paddle wheels) can be related to the differential entropy change by Eq. 6.4 as

$$\delta Q = TdS \tag{6.12}$$

or

$$\delta q = Tds \tag{6.13}$$

When integrated, there results

$$Q_{1-2} = \int_1^2 TdS \quad \text{or} \quad q_{1-2} = \int_1^2 Tds \tag{6.14}$$

Comment

The area under a *T-S* diagram between two states represents the heat transfer between those two states for a reversible process. The area under a *T-S* diagram for a complete cycle represents $Q_{net} = W_{net}$ for a cycle composed of reversible processes.

which shows that the area under a *T-S* diagram between two states represents the heat transfer between those two states for the reversible process. If the process is irreversible, the area does not represent the heat transfer, much like work is not represented by the area under a *P-v* diagram for an irreversible process. Consequently, for a closed cycle of reversible processes, the area under a *T-S* diagram represents Q_{net} for the cycle, which, according to the first law (see Eq. 3.21), is also equal to W_{net}.

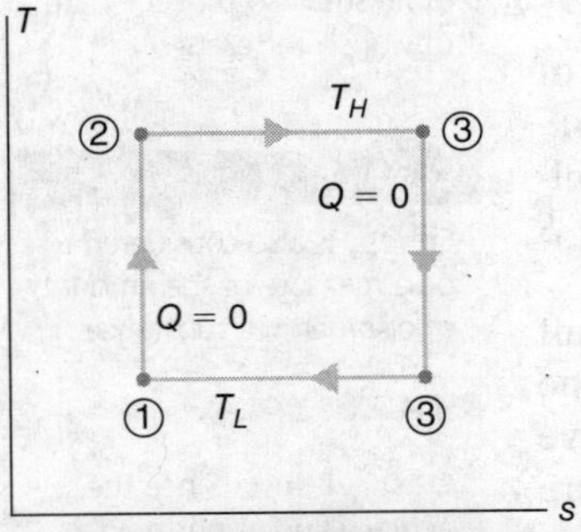

Figure 6.8

The Carnot cycle.

If the Carnot cycle of Fig. 5.15 were to be sketched on a *T-S* diagram, it would appear as shown in Fig. 6.8. The rectangular area represents $Q_{net} = W_{net}$; hence, the *T-S* diagram is often the diagram of choice for cycles since the adiabatic process, a vertical line on the *T-S* diagram, is so common.

To find an expression for the entropy change for a process, we return to the differential form of the first law, expressed by Eq. 3.23,

$$Tds = du + Pdv \tag{6.15}$$

where we used $\delta q = Tds$ (from Eq. 6.13) and $\delta w = Pdv$ (from Eq. 3.5). In terms of enthalpy, $h = u + Pv$, Eq. 6.15 can be written as

$$\begin{aligned} Tds &= du + d(Pv) - vdP \\ &= dh - vdP \end{aligned} \tag{6.16}$$

Calculus: $d(Pv) = Pdv + vdP$ so that $Pdv = d(Pv) - vdP$

It should be emphasized that entropy is a property that, for a simple substance, is dependent on two other properties of the substance, usually temperature and pressure, the two measureable properties. The differential relationships of Eqs. 6.15 and 6.16 can be used for reversible or irreversible processes since they simply relate properties.

Comment

Entropy is dependent on two other properties for a simple substance.

The entropy change over a process is often of interest since it is so closely related to the second law. It will be of particular interest for the adiabatic reversible process, since the entropy change for such a process is zero; it is an isentropic process, as defined by Eq. 6.6. In the next several subsections, entropy changes will be related to various processes for systems.

Comment

The differential relationships of Eqs. 6.15 and 6.16 can be used for reversible or irreversible processes since they relate properties.

6.3.2 Entropy change of an ideal gas with constant C_p and C_v

To relate the entropy change for an ideal gas with constant specific heats to temperature and specific volume, divide Eq. 6.15 by the absolute temperature and use Eq. 2.22 and the ideal-gas law to find

$$ds = \frac{du}{T} + \frac{Pdv}{T} = \frac{C_v dT}{T} + \frac{Rdv}{v} \tag{6.17}$$

Assuming constant specific heat C_v, this equation can be integrated for a process to obtain the change in entropy:

$$s_2 - s_1 = \int_{T_1}^{T_2} C_v \frac{dT}{T} + \int_{v_1}^{v_2} R \frac{dv}{v}$$

$$= C_v \ln \frac{T_2}{T_1} + R \ln \frac{v_2}{v_1} \quad (6.18)$$

If the first law were used in terms of enthalpy as Eq. 6.16, the entropy change, after dividing by T and integrating, assuming constant C_p, would be

$$s_2 - s_1 = \int_{T_1}^{T_2} C_p \frac{dT}{T} - \int_{P_1}^{P_2} R \frac{dP}{P}$$

$$= C_p \ln \frac{T_2}{T_1} - R \ln \frac{P_2}{P_1} \quad (6.19)$$

Either Eq. 6.18 or Eq. 6.19 can be used to find the entropy change in an ideal gas with constant specific heats. Which form is used depends on the information given.

Comment

The specific heats of the ideal gas must be constant for these equations to be used.

It should be emphasized that the above relationships can be used for either reversible (quasi-equilibrium) of irreversible processes since they are simply equations relating properties for an ideal gas. For an isentropic process, that is, one for which $\Delta s = 0$, Eq. 6.18 and Eq. 6.19 and the ideal-gas law result in the three relationships

$$\frac{T_2}{T_1} = \left(\frac{v_1}{v_2}\right)^{k-1} \qquad \frac{T_2}{T_1} = \left(\frac{P_2}{P_1}\right)^{(k-1)/k} \qquad \frac{P_2}{P_1} = \left(\frac{v_1}{v_2}\right)^{k} \quad (6.20)$$

Note: We also used $R/C_v = k - 1$.

These are the same equations as we found for the quasi-equilibrium adiabatic process of Section 3.5.4, which they should be since the same assumptions apply. These equations will be used extensively in the remainder of our study because it is most common to assume constant specific heats for an ideal gas. It does introduce significant error if the temperatures are relatively high, but the results are of interest in comparisons and do give quick estimates of unknown properties. The following subsection presents the equations for ideal gases with variable specific heats.

The entropy change in an ideal gas — Example 6.2

Ten kilograms of air at 600 kPa and 40°C are heated in the rigid container of Fig. 6.9 to 250°C. Assuming constant specific heats, calculate i) the heat transfer and ii) the change in entropy for this process.

Solution

From the example statement, we know that the air has the following properties:

$$P_1 = 600 \text{ kPa},\ T_1 = 40°\text{C} = 313 \text{ K},\ T_2 = 250°\text{C} = 523 \text{ K}$$

(Continued)

Example 6.2 (Continued)

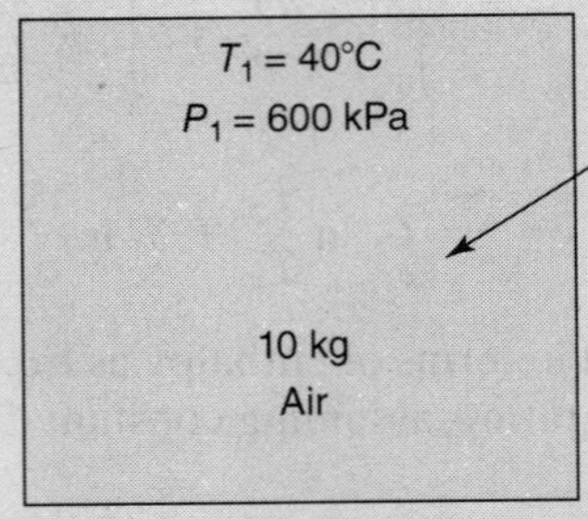

Figure 6.9

i) The heat transfer can be calculated from the given information. The first law provides

$$Q = m\Delta u = mC_v(T_2 - T_1)$$
$$= 10 \text{ kg} \times 0.717 \text{ kJ/kg·K} \times (250 - 40) = \underline{1506 \text{ kJ}}$$

ii) To find the entropy change, Eq. 6.18 is used since $v_2 = v_1$ (the container is rigid). The entropy change is

$$\Delta S = mC_v \ln\frac{T_2}{T_1} + mR\ln\frac{v_2}{v_1}$$

$$= 10 \text{ kg}\left[(0.717 \text{ kJ/kg·K}) \ln\frac{523}{313} + 0.287 \cancel{\ln 1}\right] = \underline{3.68 \text{ kJ/K}}$$

Comment

Since the mass is given, we have found ΔS. If the mass had not been given, we would have found Δs.

We have used $C_v = 0.717$ kJ/kg·K and $R = 0.287$ kJ/kg·K from Table B-2 in the Appendix and recognized that $\ln 1 = 0$.

Example 6.3 An isentropic process of an ideal gas

Air at 100 kPa and 20°C is compressed in a cylinder to a pressure of 1800 kPa. Assuming an isentropic process with constant specific heats, calculate i) the final temperature, ii) the initial and final specific volumes, and iii) the work required. The process is shown on *T-s* and *P-v* diagrams in Fig. 6.10.

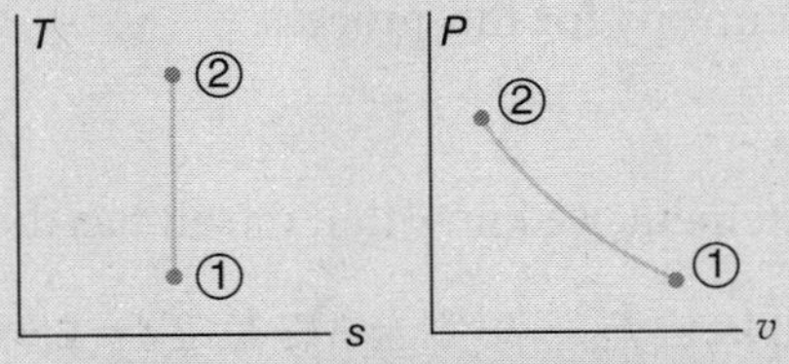

Figure 6.10

Solution

From the example statement, we know that the air has the following properties:

$$P_1 = 100 \text{ kPa}, \quad T_1 = 20°\text{C} = 293 \text{ K}, \quad P_2 = 1800 \text{ kPa}$$

i) The temperature at the end of the compression process is found, using Eq. 6.20 with $k = 1.4$, to be

$$T_2 = T_1\left(\frac{P_2}{P_1}\right)^{(k-1)/k} = 293\left(\frac{1800}{100}\right)^{0.4/1.4} = 669 \text{ K} \quad \text{or} \quad \underline{396°\text{C}}$$

ii) The initial and final specific volumes are found to be

$$v_1 = \frac{RT_1}{P_1} = \frac{0.287 \times 293}{100} = \underline{0.841 \text{ m}^3/\text{kg}}$$

$$v_2 = \frac{RT_2}{P_2} = \frac{0.287 \times 669}{1800} = \underline{0.1067 \text{ m}^3/\text{kg}}$$

iii) To find the work, the first law is used. Since $Q = 0$, it is

$$-w = u_2 - u_1 = C_v(T_2 - T_1) = 0.717(396 - 20) = \underline{270 \text{ kJ/kg}}$$

We have used $C_v = 0.717$ kJ/kg·K from Table B-2. Observe that the work is negative, which was expected since work must be input in order to compress a system. The work per unit mass was calculated since neither the mass nor the volume of the air was specified.

6.3.3 Entropy change of a solid, a liquid, and a reservoir

For a solid or a liquid for which the specific heats can be assumed constant, as is usually done, the first law, Eq. 6.15, is put in the form

$$ds = \frac{du}{T} \tag{6.21}$$

since solids and liquids are assumed to be incompressible so that $dv = 0$. Letting $du = CdT$, Eq. 6.21 can be integrated to provide

$$s_2 - s_1 = \int_{T_1}^{T_2} C\frac{dT}{T} = C\ln\frac{T_2}{T_1} \tag{6.22}$$

For most solids and liquids, the specific heat $C = C_p = C_v =$ const, so the integration was easily performed. For some solids or liquids, it may be necessary to integrate using a given $C(T)$. In tables, the specific heat C is most often listed as C_p.

The entropy change of a substance, whether a solid, a liquid, or a gas, which composes a reservoir (a source of energy at constant temperature) is often of interest. Since the temperature of a reservoir is constant, its entropy change is

$$\Delta S = \int \frac{\delta Q}{T} = \frac{1}{T}\int \delta Q = \frac{Q}{T} \tag{6.23}$$

Several examples now illustrate the above equations for entropy change of a system.

Example 6.4 Entropy change of the universe

Air in the 2-m^3-rigid container of Fig. 6.11 cools from 200°C and 400 kPa to 100°C. The surroundings are at 20°C. Determine the entropy change of the universe.

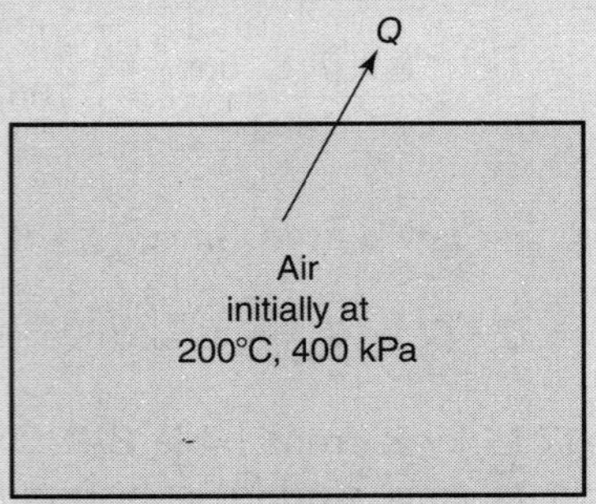

Figure 6.11

Solution

The mass of air in the container is

$$m = \frac{PV}{RT} = \frac{400 \times 2}{0.287 \times 473} = 5.893 \text{ kg}$$

Using Eq. 6.18, we find the entropy change in the air to be

$$\Delta S_{air} = mC_v \ln\frac{T_2}{T_1} + mR\ln\frac{v_2}{v_1}$$

$$= 5.893 \text{ kg} \times 0.717 \frac{\text{kJ}}{\text{kg}\cdot\text{K}} \times \ln\frac{373}{473} = -1.004 \text{ kJ/K}$$

where we have used $v_2 = v_1$ since the volume is rigid. The heat transfer from the air is found using the first law:

$$Q_{air} - W = \Delta U = mC_v(T_2 - T_1)$$

$$= 5.893 \text{ kg} \times 0.717 \frac{\text{kJ}}{\text{kg}\cdot\text{K}}(373 - 473) \text{ K} = -422.5 \text{ kJ}$$

The heat transfer is negative since it is leaving the system.

The entropy change of the surroundings (a large reservoir), which are at 20°C, is positive since heat is added to the surroundings using Eq. 6.23; it is equal to

$$\Delta S_{\text{surr}} = \frac{Q}{T} = \frac{422.5}{293} = 1.442 \text{ kJ/K}$$

The entropy change of the universe is found to be

$$\Delta S_{\text{surr}} = \Delta S_{\text{air}} + \Delta S_{\text{surr}}$$
$$= -1.004 + 1.442 = \underline{0.438 \text{ kJ/K}}$$

The net entropy change is positive as demanded by the second law.

Entropy change of ice and water — Example 6.5

Two kg of ice at −7°C and 12 kg of water at 16°C are mixed in an insulated container, as shown in Fig. 6.12. Estimate i) the final temperature of the mixture and ii) the entropy change.

Solution

i) The first law is applied to the mixture to determine the final temperature. If all the ice is assumed to melt so that the final temperature is above 0°C, we have

$$m_i\left[C_{p,i}(0 + 7) + \Delta h_i + C_{p,w}(T - 0)\right] = m_w C_{p,w}(16 - T)$$

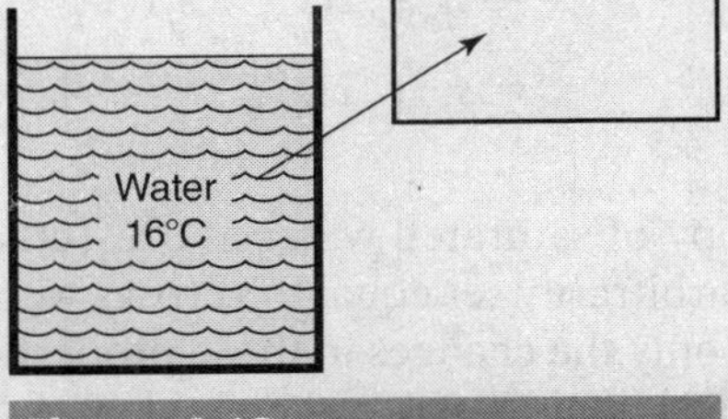

Figure 6.12

(Continued)

Example 6.5 (*Continued*)

where T is the final temperature of the mixture. The subscript 'i' refers to ice and 'w' to water. Substitute the known quantities into the above equation, using units on C_p of kJ/kg·K, and obtain

Comment

It takes 330 kJ to melt one pound of ice.

$$2[2.033 \times 7 + 330 + 4.177(T - 0)] = 12 \times 4.177 \times (16 - T)$$

$$\therefore T = \underline{1.93°\text{C}}$$

C_p for water and ice was found in Table B-4. The latent heat Δh_i of ice was listed in Section 2.2.4 as 330 kJ/kg.

ii) The entropy change of the ice and the water is found to be

Units: The units on ΔS are the same as the units on (mC_p), which are kg(kJ/kg·K) = kJ/K.

$$\Delta S_{\text{net}} = m_i\left[C_{p,i}\ln\frac{273}{266} + (s_w - s_i) + C_{p,w}\ln\frac{274.93}{273}\right] + m_w C_{p,w}\ln\frac{274.93}{289}$$

The entropy of saturated water at 0°C is found in Table C-1, and that of ice at 0°C is found in Table C-5. This provides

$$\Delta S_{\text{net}} = 2\left[2.033\ln\frac{273}{266} + [0-(-1.221)] + 4.177\ln\frac{274.93}{273}\right] + 12\times 4.177\ln\frac{274.93}{289}$$

$$= \underline{0.1051\ \text{kJ/kg·K}}$$

It is positive, as it must be.

6.3.4 Entropy change of a phase-change substance

The entropy of a pure substance is calculated in the same manner as internal energy and enthalpy. Values of specific entropy are listed in tables in the Appendix for steam, R134a, and ammonia. Follow the same steps taken when calculating the properties of steam in Chapter 2. The specific entropy for a saturated mixture is calculated using

Comment

The entropy is arbitrarily set equal to zero at 0°C for saturated water.

$$s = s_f + x(s_g - s_f)$$
$$= s_f + xs_{fg} \qquad (6.24)$$

Comment

The saturated temperature table can be used for the entropy of a compressed liquid.

Observe that the entropy of saturated water at 0°C (other substances use a different temperature) is arbitrarily set equal to zero, as were u_f and h_f. This is of no consequence since it is only the changes in those properties that are of interest in our study. For compressed liquid water, either Table C-4 or the saturated value at the given temperature can be used. The superheated values for steam are listed in Table C-3. Similar tables for R134a and ammonia are included in the Appendix. The IRC Calculator can also be used.

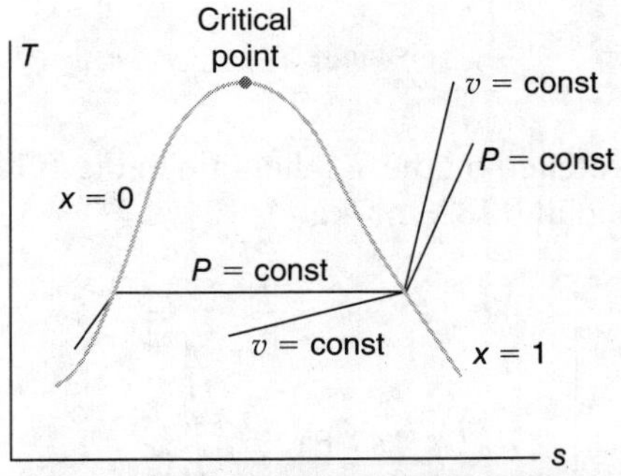

Figure 6.13
A T-s diagram for steam.

A *T-s* diagram is often sketched, especially when the working fluid is one of the three substances listed above. For water, it is relatively symmetric and resembles Fig. 6.13 when drawn to scale.

The entropy change in steam — Example 6.6

Steam is contained in a rigid volume at 1400 kPa and 300°C. Heat is transferred from the volume until the pressure is 60 kPa. Determine i) the entropy change and ii) the heat transfer required. iii) Also, sketch the process on a *T-s* diagram.

Solution

From the example statement, we know that the steam has the following properties:

$$P_1 = 1400\,\text{kPa},\quad T_1 = 300°\text{C},\quad P_2 = 60\,\text{kPa}$$

i) For a rigid volume, $v_1 = v_2$. The steam table C-3 provides $v_1 = 0.1823\ \text{m}^3/\text{kg}$. Table C-2 then allows the quality to be calculated:

$$v_2 = v_f + x_2(v_g - v_f)$$

$$0.1823 = 0.00103 + x_2(2.732 - 0.00103) \qquad \therefore x_2 = 0.0664$$

The entropy at state 2 is then

$$s_2 = s_f + x_2 s_{fg} = 1.1455 + 0.0664 \times 6.387 = \underline{1.57\ \text{kJ/kg·K}}$$

ii) The heat transfer can be calculated from

$$q = u_2 - u_1$$

$$= [359.8 + 0.0664 \times (2489.6 - 359.8)] - 2785.2 = \underline{-2284\ \text{kJ/kg}}$$

The minus sign means heat leaves the volume.

(Continued)

Example 6.6 (*Continued*)

iii) The process is sketched on the *T-s* diagram in Fig. 6.14. The specific volume must be held fixed at 0.1823 m^3/kg.

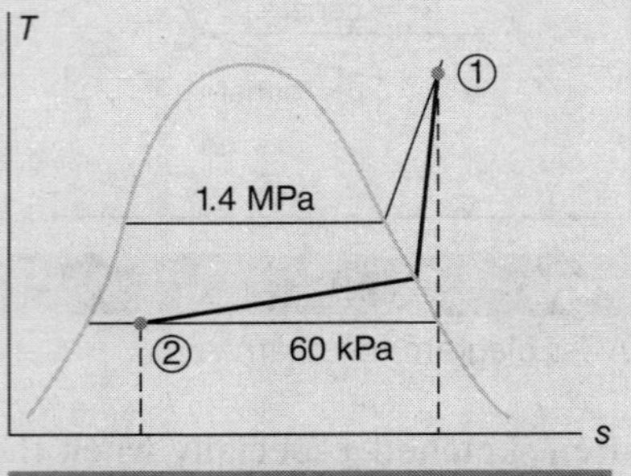

Figure 6.14

6.3.5 Entropy change of an ideal gas with variable specific heats

If the temperatures are sufficiently high, the assumption of constant specific heats results in significant error. To account for variable specific heats, return to Eq. 6.16 using $dh = C_p dT$ and $v/T = R/P$:

$$ds = \frac{dh}{T} - \frac{vdP}{T} = \frac{C_p}{T}dT - \frac{R}{P}dP \tag{6.25}$$

Integrate this equation, recognizing that R is a constant and $C_p = C_p(T)$:

Note: C_p depends only on temperature.

$$s_2 - s_1 = \int_{T_1}^{T_2} C_p \frac{dT}{T} - R \ln \frac{P_2}{P_1} \tag{6.26}$$

The integral in this equation is a function of the temperature only. It can be evaluated from the gas tables in Appendix F using a function s° (some authors prefer ϕ rather than s°) that is defined so that

$$s_2^\circ - s_1^\circ = \int_{T_1}^{T_2} C_p \frac{dT}{T} \tag{6.27}$$

Using this equivalence, we see that Eq. 6.26 becomes

Comment

This relationship is most often used when considering variable specific heats.

$$s_2 - s_1 = s_2^\circ - s_1^\circ - R \ln \frac{P_2}{P_1} \tag{6.28}$$

This more exact expression for entropy change is used only when requested. Equations (6.18) and (6.19) are often used for comparison purposes, even though they are not accurate when either of the temperatures is relatively high.

For some gases listed in Appendix F, the entropy function is presented on a per mol basis, so the change in entropy takes the form

$$\bar{s}_2 - \bar{s}_1 = \bar{s}_2^\circ - \bar{s}_1^\circ - R_u \ln \frac{P_2}{P_1} \tag{6.29}$$

For an isentropic process, Eqs. 6.20 cannot be used if the specific heats are not constant. We can, however, use Eq. 6.28 to obtain, for $\Delta s = 0$,

$$\frac{P_2}{P_1} = e^{(s^\circ_2 - s^\circ_1)/R} = \frac{e^{s^\circ_2/R}}{e^{s^\circ_1/R}} = \frac{f(T_2)}{f(T_1)} \tag{6.30}$$

This leads to the definition of the *relative pressure* P_r, defined by

$$P_r = e^{s^\circ/R} \tag{6.31}$$

which is included as an entry in the air Table F-1. In terms of P_r, Eq. 6.30 takes the form

$$\frac{P_2}{P_1} = \frac{P_{r2}}{P_{r1}} \tag{6.32}$$

Observe: The relative pressure P_r is non-dimensional, as was the reduced pressure P_R of Chapter 2.

The ideal-gas law allows

$$\frac{v_2}{v_1} = \frac{P_1}{P_2}\frac{T_2}{T_1} = \frac{T_2/P_{r2}}{T_1/P_{r1}} \tag{6.33}$$

This leads to the *relative specific volume* $v_r = \text{const} \times T/P_r$, which allows the use of

$$\frac{v_2}{v_1} = \frac{v_{r2}}{v_{r1}} \tag{6.34}$$

The relative specific volume is also an entry in the air Table F-1.

Comment

The entries P_r and v_r in Appendix F-1 are used for isentropic processes with variable specific heats.

Entropy change with variable specific heats — Example 6.7

Ten kilograms of air at 600 kPa and 40°C are heated in the rigid container of Fig. 6.15 to 250°C. Calculate i) the heat transfer and ii) the change in entropy for this process. Do not assume constant specific heats as was done in Example 6.2.

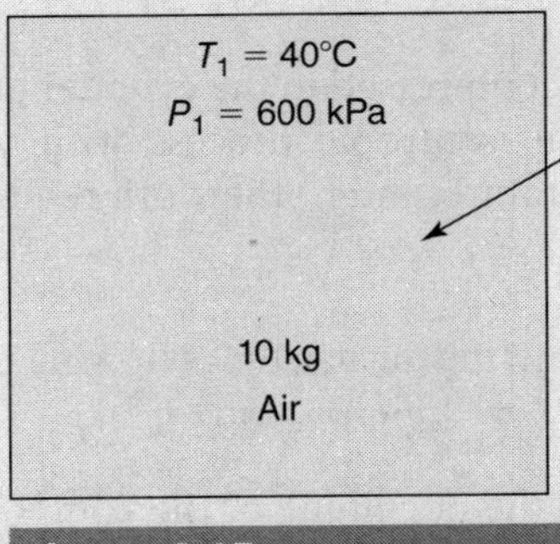

Figure 6.15

(Continued)

Example 6.7 (Continued)

Solution

From the example statement, we know that the air has the following properties:

$$P_1 = 600 \text{ kPa}, \quad T_1 = 40°\text{C} = 313 \text{ K}, \quad T_2 = 250°\text{C} = 523 \text{ K}$$

i) The heat transfer can be calculated directly from the given information. The first law provides, using the ideal gas Table F-1 (or the IRC Calculator),

$$Q = m(u_2 - u_1) = 10 \text{ kg} \times (377 - 223) \text{ kJ/kg} = \underline{1540 \text{ kJ}}$$

ii) To find the entropy change, Eq. 6.28 is used since the specific heats are not assumed to be constant. The entropy change is

$$S_2 - S_1 = m\left(s_2^\circ - s_1^\circ - R\ln\frac{P_2}{P_1}\right)$$

$$= 10 \text{ kg}\left(2.266 - 1.745 - 0.287\ln\frac{1003}{600}\right)\frac{\text{kJ}}{\text{kg}\cdot\text{K}} = \underline{3.74 \text{ kJ/K}}$$

Note: We used the ideal-gas law to find $P_2 = P_1T_2/T_1 = 600 \times 523/313 = 1003$ kPa.

Using the IRC Calculator, we would find

$$S_2 - S_1 = m(s_2 - s_1) = 10(6.77 - 6.40) = \underline{3.70 \text{ kJ/K}}$$

Because the temperatures of states 1 and 2 are not very high, the results of Example 6.2 and this example are not significantly different. If T_2 had been, say, 1000°C, the error would be much larger.

Example 6.8 An isentropic process with variable specific heats

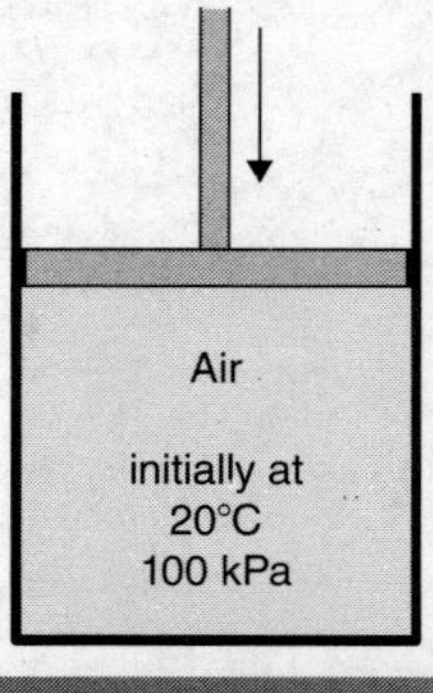

Figure 6.16

Air at 100 kPa and 20°C is compressed in the cylinder of Fig. 6.16 to a pressure of 1800 kPa. Assuming an isentropic process with variable specific heats, calculate i) the final temperature, and ii) the work required.

Solution

From the example statement, the air has the following properties:

$$P_1 = 100 \text{ kPa}, \quad T_1 = 20°\text{C} = 293 \text{ K}, \quad P_2 = 1800 \text{ kPa}$$

i) The relative pressure at the beginning of the compression process is found in Table F-1 at 293 K to be

$$P_{r1} = 0.3 \times (1.386 - 1.2311) + 1.2311 = 1.2776$$

The relative pressure P_{r2} (use Eq. 6.28) is found to be

$$P_{r2} = P_{r1}\frac{P_2}{P_1} = 1.2776 \times \frac{1800}{100} = 23.00$$

Table F-1 is used to locate the temperature at which $P_{r2} = 23.00$. It is interpolated to be

$$\frac{T_2 - 640}{660 - 640} = \frac{23.00 - 20.65}{23.13 - 20.65} \quad \therefore T_2 = 659 \text{ K or } 386°\text{C}$$

ii) To find the work, the first law is used. With $Q = 0$ and the specific internal energies found from Table F-1, the work is

$$-w = u_2 - u_1 = 480 - 209 = \underline{271 \text{ kJ/kg}}$$

The work is essentially the same as that of Example 6.2, but the temperature at the end of the compression process assuming constant specific heats is 10°C too low. The higher the pressure ratio, the larger the difference would be. If either of the temperatures was in excess of 500°C, the error would be significant.

6.4 Entropy Changes for a Control Volume

Thus far in this chapter we have considered entropy changes for a system, a specified mass of a substance. In this section the entropy change between the outlet and the inlet of a control volume, which represents several devices of interest, will be considered. Following the derivation of the first law (see Eq. 4.11), the second law expression by Eq. 6.11, which represents the entropy change of the control volume plus that of the surroundings, can be represented by

$$\begin{pmatrix}\text{Entropy change}\\ \text{in c.v.}\end{pmatrix} + \begin{pmatrix}\text{Entropy}\\ \text{exiting}\end{pmatrix} - \begin{pmatrix}\text{Entropy}\\ \text{entering}\end{pmatrix} + \begin{pmatrix}\text{Entropy change}\\ \text{of surroundings}\end{pmatrix} \geq 0 \quad (6.35)$$

Consider a control volume that has one inlet and one outlet through which a fluid flows, as sketched in Fig. 6.17. The mass in the control volume may change with time. Allow a small mass m_1 of substance to enter and a small mass m_2 to leave the control volume over a small time increment Δt. The expression of the second law represented by Eq. 6.35 can be written as

$$\Delta S_{\text{c.v.}} + m_2 s_2 - m_1 s_1 + \frac{Q_{\text{surr}}}{T_0} \geq 0 \quad (6.36)$$

where the temperature of the surroundings is at T_0. Heat is assumed to flow from the control volume to the surroundings since we assume the surroundings to be at a temperature lower than the temperature of the control volume, as is usually the case. Added inlets and outlets are included by adding

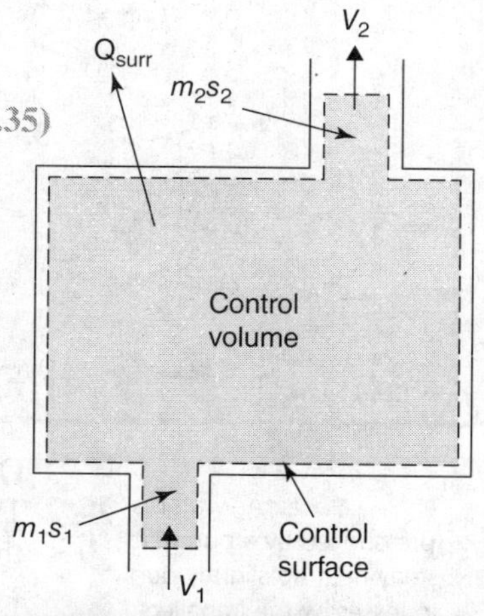

Figure 6.17
A general control volume with an inlet and an outlet.

Comment

If heat is added to the surroundings, the entropy of the surroundings increases, as stated by Eq. 6.23.

appropriate terms to Eq. 6.36. Divide each term in this equation by Δt and, in the limit as $\Delta t \to 0$, the equation takes the form

$$\frac{dS_{\text{c.v.}}}{dt} + \dot{m}_2 s_2 - \dot{m}_1 s_1 + \frac{\dot{Q}_{\text{surr}}}{T_0} \geq 0 \tag{6.37}$$

The equality is used for an ideal flow: no losses due to viscous effects, mixing, separated flow, or shock waves, in addition to no heat transfer across the control surface over a finite temperature difference. If any losses, that is, irreversible effects, exist as the fluid flows through the control volume or if heat transfer exists from the control surface to the surroundings over a finite temperature difference, the inequality is used.

Recall: It was assumed in Fig. 6.17 that $\dot{Q}_{\text{surr}}$ is positive since it represents the heat transfer to the surroundings, which, from the surroundings point of view, is positive.

Let us write Eq. 6.37 in terms of entropy production. Rather than use the symbol "≥," we define the rate of entropy production to be

$$\dot{S}_{\text{prod}} = \frac{dS_{\text{c.v.}}}{dt} + \dot{m}_2 s_2 - \dot{m}_1 s_1 + \frac{\dot{Q}_{\text{surr}}}{T_0} \tag{6.38}$$

For an ideal flow with no losses or heat transfer to the surroundings, the entropy production would be zero. Any losses or heat transfer to the surroundings across a finite temperature difference would result in entropy production. Consequently, any flow through a control volume with losses results in entropy production.

In most devices of interest, a steady flow is assumed, so that changes do not occur within the control volume. With one inlet and one outlet, a steady flow requires $\dot{m}_1 = \dot{m}_2 = \dot{m}$ so that Eq. 6.38 becomes

$$\dot{m}(s_2 - s_1) + \frac{\dot{Q}_{\text{surr}}}{T_0} \geq 0 \quad \text{(steady flow)} \tag{6.39}$$

Several observations can be made from this simplified equation.

1. If the control volume is insulated and there are negligible losses, $s_2 = s_1$, resulting in an isentropic flow.
2. If the control volume is insulated and there are losses between the inlet and outlet, then $s_2 - s_1 > 0$ so that $s_2 > s_1$. The entropy would increase, as expected.
3. If heat transfer crosses the control surface into the control volume, in which case $\dot{Q}_{\text{surr}}$ would be negative, the entropy increases from the inlet to the outlet.

Note: $\dot{Q} = -\dot{Q}_{\text{surr}}$

Example 6.9 Entropy calculations for steam in a power cycle

Note: As water is pressurized by a pump, it may heat up slightly due to losses by the impellers, but the temperature change can be neglected.

Determine the entropy change across each device of the cycle shown in Fig. 6.18. The boiler and condenser are simple constant-pressure heat exchangers. Use the following properties:

State 1: $P_1 = 10$ kPa, $x_1 = 0$ **State 2:** $P_2 = 5$ MPa, $T_2 = 46$°C

State 3: $P_3 = 5$ MPa, $T_3 = 500$°C **State 4:** $P_4 = 10$ kPa, $x_4 = 0.9$

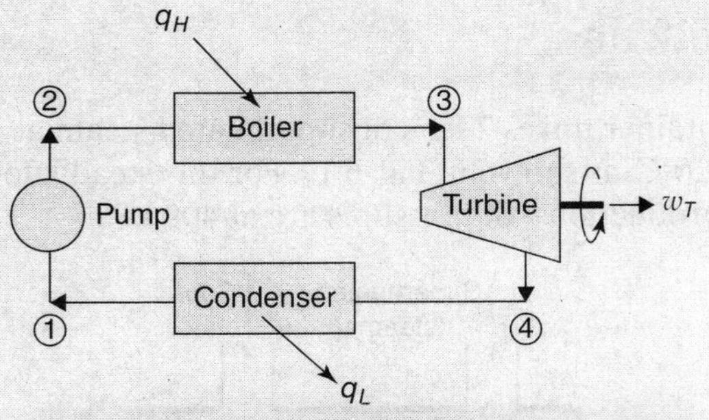

Figure 6.18

Solution

State 1 is a saturated liquid, so from Table C-2 at 0.010 MPa, s_1 = 0.649 kJ/kg·K and T_1 = 45.8°C. At state 2 the IRC Calculator at 46°C and 5 MPa gives s_2 = 0.650 kJ/kg·K. The entropy change across the pump is then

$$s_2 - s_1 = 0.650 - 0.649 = \underline{0.001 \text{ kJ/kg·K}}$$

> **Comment**
> This entropy change across the pump is essentially zero since it has negligible heat transfer and losses.

At state 3 the entropy is found as a direct entry in Table C-3 to be s_3 = 6.976 kJ/kg·K. Consequently, the entropy change across the boiler is

$$s_3 - s_2 = 6.976 - 0.650 = \underline{6.326 \text{ kJ/kg·K}}$$

> **Observe:** Substantial heat is added to the water, resulting in a large entropy increase.

State 4 is in the quality region at 10 kPa. The specific entropy at this state is obtained from a quality calculation to be

$$s_4 = s_f + x_4(s_g - s_f)$$
$$= 0.6491 + 0.90(8.151 - 0.6491) = 7.401 \text{ kJ/kg·K}$$

The entropy change across the turbine is then

$$s_4 - s_3 = 7.401 - 6.976 = \underline{0.425 \text{ kJ/kg·K}}$$

This entropy change suggests that heat is lost from the turbine or there are losses through the turbine. If the turbine were adiabatic, that is, no heat transfer, the entropy increase would be due only to the losses around the blades.

Finally, the entropy change across the condenser is

$$s_1 - s_4 = 0.649 - 7.401 = \underline{-6.752 \text{ kJ/kg·K}}$$

It is expected that the entropy change around the cycle is zero since the steam returns to its initial state. Adding the entropy changes of the four devices together results in

$$\Delta s_{cycle} = 0.001 + 6.326 + 0.425 - 6.752 = 0$$

The result could be a small number, either positive or negative, due to round-off error.

The entropy of the universe would be increasing due to the heat transfer from the condenser and to the boiler. The entropy of the surroundings would decrease due to the boiler but increase due to the condenser with a net increase.

> **Comment**
> Entropy is produced for all flows with losses, through any device. The entropy in the universe can never decrease due to a process of any kind.

Example 6.10 Entropy production

An insulated container mixes 2 kg/s of superheated steam at 400°C and 6 kg/s of cool water at 20°C, as shown in Fig. 6.19. For this steady flow, determine the rate of entropy production. All pressures are at 400 kPa.

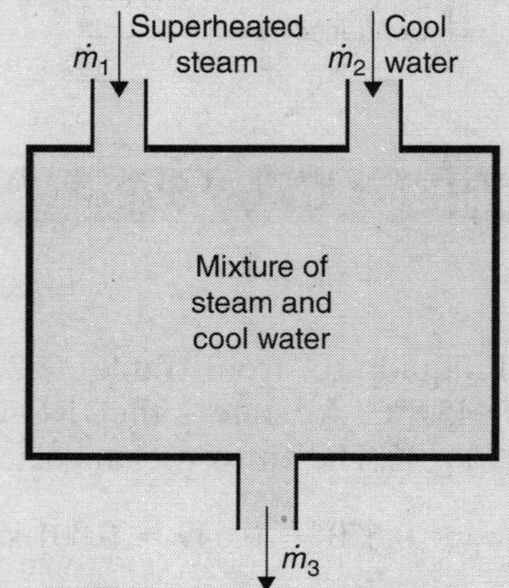

Figure 6.19

Solution

The exiting temperature is found by applying the first law. First, however, the conservation of mass demands that the mass flux exiting the container is $\dot m_1 + \dot m_2 = \dot m_3$ or $2 + 6 = 8$ kg/s. The first law for the insulated container with $\dot Q = 0$ and $\dot W_S = 0$ takes the form (refer to Example 4.10)

$$\dot m_1 h_1 + \dot m_2 h_2 = \dot m_3 h_3$$

$$2 \times 3273 + 6 \times 84.8 = 8h_3 \qquad \therefore h_3 = 882 \text{ kJ/kg}$$

where

Units: kN·m = kJ

$$h_2 = u_2 + P_2 v_2 = 83.9 \frac{\text{kJ}}{\text{kg}} + 400 \frac{\text{kN}}{\text{m}^2} \times 0.00234 \frac{\text{m}^3}{\text{kg}} = 84.8 \text{ kJ/kg}$$

is the enthalpy of compressed water at 20°C. The enthalpy of 882 kJ/kg is greater than h_f at 0.4 MPa (see Table C.2), so the quality of the exiting mixture is found as follows:

$$882 = 604.7 + x_3(2133.8) \qquad \therefore x_3 = 0.130$$

The entropy production of Eq. 6.38 requires the entropy at each inlet and outlet. They are $s_1 = 7.8985$ kJ/kg·K, $s_2 = 0.2965$ kJ/kg·K, and $s_3 = 1.777 + 0.13 \times 5.1197 = 2.443$ kJ/kg·K. Since $\dot Q_{\text{surr}} = 0$ and $dS_{\text{c.v.}}/dt = 0$ for the steady flow, the entropy production is the entropy that exits minus the entropy that enters:

Units: The units on the term $\dot m\, s$ are (kg/s)(kJ/kg·K) = kJ/K·s

$$\dot S_{\text{prod}} = \dot m_3 s_3 - \dot m_1 s_1 - \dot m_2 s_2$$

$$= 8 \times 2.443 - 2 \times 7.8985 - 6 \times 0.2965 = \underline{1.97 \text{ kJ/K·s}}$$

It is positive as it must be as required by the second law. Any real process always generates entropy. The more irreversible a process, the more entropy it produces. A mixing process is highly irreversible.

☑ **You have completed Learning Outcome (3)**

6.5 Isentropic Efficiency

In Chapter 5 the performance parameters for a variety of cycles were defined. They included the efficiency of a heat engine and the coefficient of performance for refrigerators, air conditioners, and heat pumps. We will now define performance parameters for individual devices.

6.5.1 Isentropic turbine efficiency

The actual work produced by a turbine, using Eq. 4.20, is found to be

$$-w_T = h_2 - h_1 \tag{6.40}$$

where $w_T = w_S$, state 1 is the inlet, and state 2 is the outlet. The maximum work that the turbine could produce would be associated with an adiabatic quasi-equilibrium process between the inlet and the outlet, that is, an isentropic process. No losses would exist in the turbine, and no energy would escape from the turbine in the form of heat transfer. For such a process, Eq. 6.40 would represent the maximum work output. To emphasize that state 2 is found by using an isentropic process from state 1 to state 2, we write Eq. 6.40 as

Comment

The shaft work is the turbine work.

$$w_{T,s} = h_1 - h_{2s} \tag{6.41}$$

where h_{2s} is the exit enthalpy found by assuming an isentropic process. The *isentropic efficiency of a turbine* is defined as the ratio of the actual work output to the isentropic or maximum work output:

$$\eta_T = \frac{w_T}{w_{T,s}} = \frac{h_1 - h_2}{h_1 - h_{2s}} \tag{6.42}$$

This is also referred to as the *adiabatic efficiency*. The actual work would be less than the maximum work due to heat transfer from the turbine, losses due to separated flow from the blades, and friction.

Comment

Isentropic efficiency is also called adiabatic efficiency. Most often, the terms "isentropic" and "adiabatic" are dropped, and it is simply referred to as the turbine efficiency.

Efficiency of a turbine — **Example 6.11**

Steam enters the turbine of Fig. 6.20 at 5 MPa and 500°C. It exits as a saturated vapor at 10 kPa. Determine the isentropic efficiency of the turbine.

(Continued)

Example 6.11 (*Continued*)

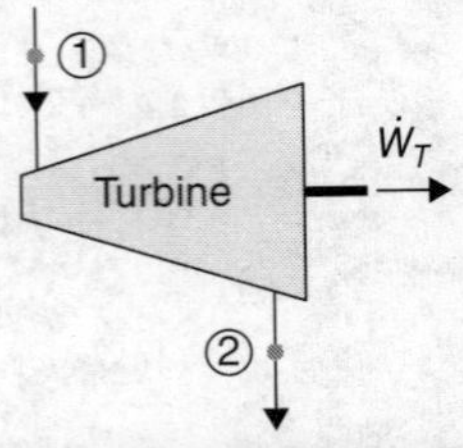

Figure 6.20

Solution

Steam enters the turbine as a superheated vapor at state 1 and leaves as a saturated vapor at state 2, as shown on the *T-s* diagram of Fig. 6.21. The enthalpies at the inlet and exit are found from the steam tables to be

$$h_1 = 3433.8 \text{ kJ/kg} \quad \text{and} \quad h_2 = 2584.6 \text{ kJ/kg}$$

To determine the isentropic efficiency of this turbine, we must find the exit condition assuming an isentropic process from state 1 to state 2, shown by the solid vertical line in the *T-s* diagram of Fig. 6.21. The entropy at state 1 from Table C-3 is $s_1 = 6.976$ kJ/kg·K. This must equal the entropy at state 2*s*. Using values from Table C-2, we find the quality of state 2*s*, and then h_{2s} can be found:

$$s_{2s} = 6.976 = s_f + x_{2s}s_{fg}$$

$$= 0.6491 + x_{2s} \times 7.5019 \quad \therefore x_{2s} = 0.8433$$

Note: The IRC Calculator also gives $h_{2s} = 2210$ kJ/kg.

$$h_{2s} = h_f + x_{2s}h_{fg} = 191.8 + 0.8433 \times 2392.8 = 2210 \text{ kJ/kg}$$

This means that if the turbine were isentropic, the steam would leave with a quality of 0.8433 instead of 1.0. An isentropic turbine would take more energy out of the steam, thereby producing more work. The isentropic efficiency of this turbine is calculated to be

Note: A turbine would not be designed with a quality less than 1.0 near its exit since that means water droplets could form on the blades. This will be considered in Chapter 8.

$$\eta_T = \frac{w_{\text{actual}}}{w_s} = \frac{h_1 - h_2}{h_1 - h_{2s}} = \frac{3434 - 2585}{3434 - 2210} = 0.694 \quad \text{or} \quad \underline{69.4\%}$$

where w_s is the isentropic work, that is, the maximum turbine work.

Comment

For this turbine, about 31% of the potential work that the turbine could develop is lost to irreversibilities.

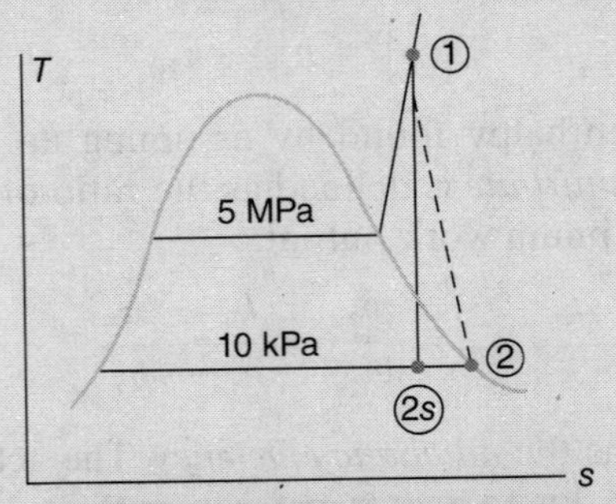

Figure 6.21

Comment

As with a turbine, we often drop the word "isentropic." The efficiency of a compressor is also assumed to be based on an isentropic process.

6.5.2 Isentropic compressor efficiency

The performance parameter for a device that uses power like a pump or compressor is defined differently than the performance parameter for a device that produces power like the turbine. A compressor uses input power to increase the pressure of the working fluid, often a vapor, as in a refrigeration cycle. As the compressor becomes more efficient, less power is needed to produce the same pressure increase. The *isentropic efficiency of a compressor* is defined as

$$\eta_C = \frac{w_{C,s}}{w_C} = \frac{h_{2s} - h_1}{h_2 - h_1} \quad (6.43)$$

The actual work would be greater than the minimum work due to heat transfer from the compressor, losses due to separated flow from the blades, and friction.

Efficiency at state 1 of a compressor — Example 6.12

Refrigerant 134a flows into the compressor of Fig. 6.22 as saturated vapor at 0.2 MPa and exits the compressor at State 2, superheated vapour at 0.6 MPa. If the isentropic efficiency of the compressor is 0.85, determine the work needed to run the compressor.

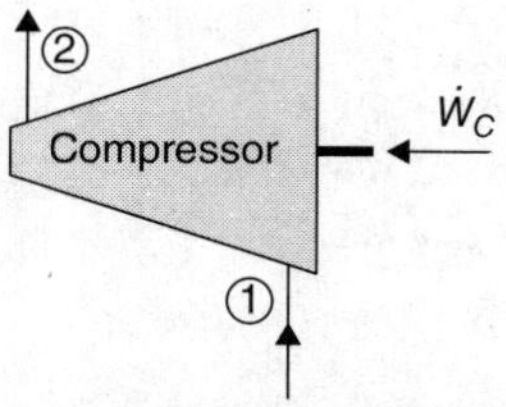

Figure 6.22

Solution

The *T-s* diagram of this process is shown in Fig. 6.23. Properties are found in 0.9253 kJ/kg·K, or the IRC Calculator can be used.

State 1, Saturated vapor: $P_1 = 0.2$ MPa, $T_1 = -10.09$°C, $h_1 = 241.3$ kJ/kg, $s_1 = 0.9253$ kJ/kg·K.

State 2, Superheated vapor: $P_2 = 0.6$ MPa, $s_{2s} = s_1 = 0.9253$ kJ/kg·K, $h_{2s} =$ 263.5 kJ/kg.

The minimum work is associated with the isentropic process. It is

$$w_s = h_{2s} - h_1 = 263.50 - 241.30 = 22.2 \text{ kJ/kg}$$

To calculate the actual work required by the compressor, we use Eq. 6.43:

$$\eta_C = \frac{w_s}{w_C} \qquad 0.85 = \frac{22.2}{w_C} \qquad \therefore w_C = \underline{26.1 \text{ kJ/kg}}$$

The specific work multiplied by the mass flow rate, should it be specified, would give the power required by the compressor.

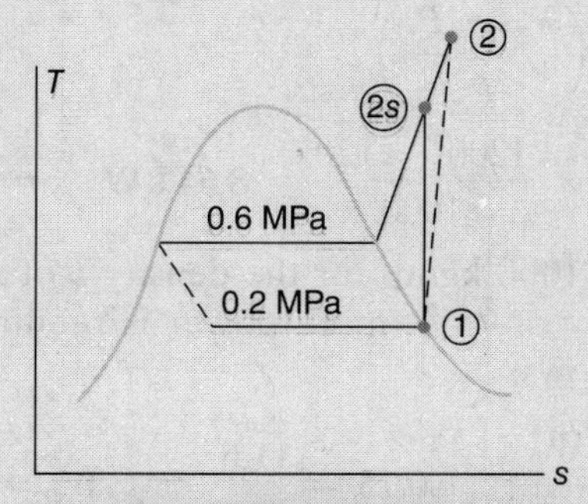

Figure 6.23

Note: Occasionally, a pump is used to increase the pressure in a gas, such as an air pump, but not in our study.

6.5.3 Pump efficiency

A *pump* is a device used to increase the pressure in a working fluid, primarily a liquid. The efficiency of a pump that pumps a liquid is defined using the same basic relationship as that used for a compressor. It is

$$\eta_P = \frac{\dot{W}_{P,s}}{\dot{W}_P} \tag{6.44}$$

The minimum work input assumes no heat transfer, no losses, and no changes in kinetic and potential energy.

Example 6.13 Efficiency of a pump

The pump of Fig. 6.24 has an efficiency of 0.75. If the pressure of 8 kg/s of water is increased from 120 kPa to 1200 kPa, determine the horsepower required by the pump.

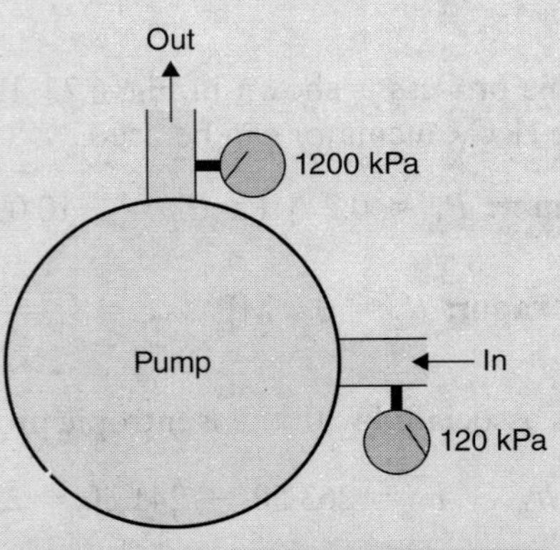

Figure 6.24

Solution

The minimum power required by the pump to increase the pressure from 120 kPa to 1200 kPa is found by neglecting any losses in addition to kinetic and potential energy changes. The first law (see Eq. 4.21) provides

$$\dot{W}_{P,s} = \dot{m}\left(\frac{P_2 - P_1}{\rho}\right)$$

$$= 8 \times \frac{1200 - 120}{1000} = 8.64 \text{ kW} \quad \text{or} \quad 11.6 \text{ hp}$$

Units:

$$\frac{\text{kg}}{\text{s}}\frac{\text{kN/m}^2}{\text{kg/m}^3} = \frac{\text{kN}\cdot\text{m}}{\text{s}} = \text{kJ/s} = \text{kW}$$

The standard value of 1000 kg/m³ for the density ρ of water was used. If the pump has an efficiency of 75%, more power is required. The actual power needed is found as follows:

$$\eta_P = \frac{\dot{W}_{P,s}}{\dot{W}_P} \qquad 0.75 = \frac{11.6}{\dot{W}_P} \qquad \therefore \dot{W}_P = \underline{15.5 \text{ hp}}$$

6.5.4 Isentropic efficiency of a nozzle

A nozzle is used to convert energy, due primarily to a high pressure, into an increased kinetic energy of the working fluid, be it a converging nozzle with a liquid or a subsonic gas flow, or a diverging nozzle with a supersonic gas flow, as displayed in Fig. 6.25. The isentropic efficiency of a nozzle is defined by the ratio of the actual kinetic energy at the exit to the kinetic energy at the exit that would exist for an isentropic process between the inlet and the exit. Its efficiency is

$$\eta_N = \frac{V_2^2/2}{V_{2s}^2/2} = \frac{h_1 - h_2}{h_1 - h_{2s}} \tag{6.45}$$

where the first law for a nozzle was expressed by Eq. 4.31. Uniform velocity profiles are assumed at the inlet and the exit; this is always done unless otherwise stated since it is a good approximation for a turbulent flow, which invariably exists.

A diffuser, on the other hand, attempts to recover the energy contained in the kinetic energy entering the diffuser. The performance of a diffuser is measured by how effective the diffuser is at pressure recovery. Hence, a pressure-recovery coefficient is the measure of its effectiveness. Diffusers are sketched in Fig. 6.26. A supersonic diffuser actually has a decreasing area, as displayed in Fig. 6.26*b*.

The performance of the diffuser of Fig. 6.26*a* is relatively poor because the viscous effects in the wall boundary layers are quite pronounced and the fluid tends to separate from the boundary if the divergence angle is very large, greater than about 10°. To shorten the length of a diffuser, vanes can be installed with the divergence angle between vanes less than 10°.

The flow through nozzles and diffusers is studied in detail in fluid mechanics and is not of significant interest in thermodynamics.

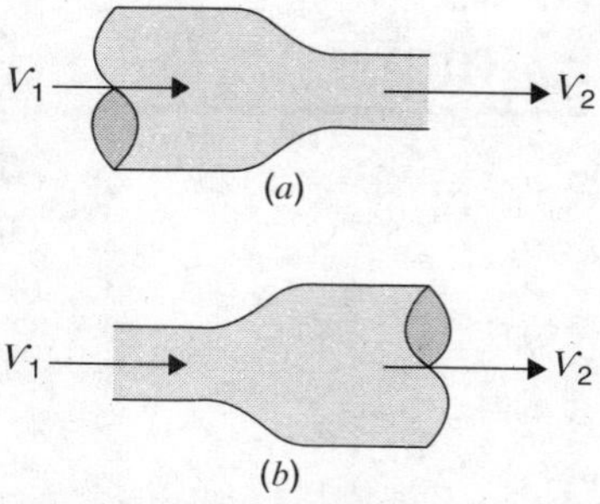

Figure 6.25
A nozzle for *a*) a liquid or a subsonic flow and *b*) a supersonic flow.

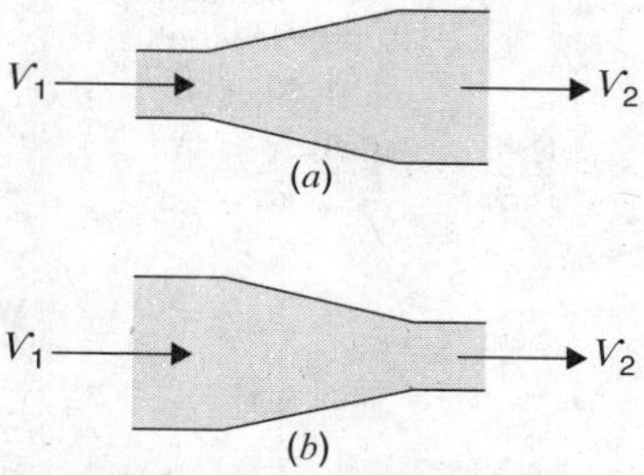

Figure 6.26
A diffuser for *a*) a liquid or a subsonic flow and *b*) for a supersonic flow.

Efficiency of a nozzle — Example 6.14

Nitrogen enters a nozzle at 240 kPa and 400 K with a velocity of 10 m/s. It exits at 100 kPa. If the nozzle adiabatic efficiency is 90%, determine the exiting velocity. Assume constant specific heats and uniform velocity profiles.

Solution

Nitrogen can be modeled as an ideal gas. The isentropic process for nitrogen traveling through the nozzle, using Eq. 6.20 with $k = 1.4$, provides the exiting temperature:

$$T_{2s} = T_1\left(\frac{P_2}{P_1}\right)^{\frac{k-1}{k}} = 400\left(\frac{100}{240}\right)^{\frac{1.4-1}{1.4}} = 400 \times \frac{1}{2.4^{0.2857}} = 311.5\text{ K}$$

(Continued)

Example 6.14 (*Continued*)

The nozzle's efficiency is given so that the exiting temperature can be found using Eq. 6.45 as follows:

$$\eta_N = \frac{h_1 - h_2}{h_1 - h_{2s}} = \frac{\mathscr{C}_p(T_1 - T_2)}{\mathscr{C}_p(T_1 - T_{2s})}$$

$$0.9 = \frac{400 - T_2}{400 - 311.5} \qquad \therefore T_2 = 320 \text{ K}$$

The first law in the form of Eq. 4.31 is used to determine the exiting velocity assuming uniform velocity profiles at the inlet and exit:

$$0 = \frac{V_2^2 - V_1^2}{2} + C_p(T_2 - T_1)$$

$$0 = \frac{V_2^2 - 10^2}{2} \frac{\text{m}^2}{\text{s}^2} + 1042 \frac{\text{N·m}}{\text{kg·K}} \times (320 - 400) \text{ K}$$

$$\therefore V_2 = \underline{408 \text{ m/s}}$$

Units: To understand the units, use Newton's second law:

$$\frac{\text{N·m}}{\text{kg}} = \frac{\text{kg·m}}{\text{s}^2}\frac{\text{m}}{\text{kg}} = \frac{\text{m}^2}{\text{s}^2}$$

where C_p for nitrogen was found in Table B-2.

This is a supersonic flow ($V > 300$ m/s), so the nozzles of Fig. 6.26 must be combined to form a converging diverging nozzle similar to the one shown in Fig. 6.27 that would provide a supersonic flow. This type of nozzle is observed on rockets to put satellites into orbit.

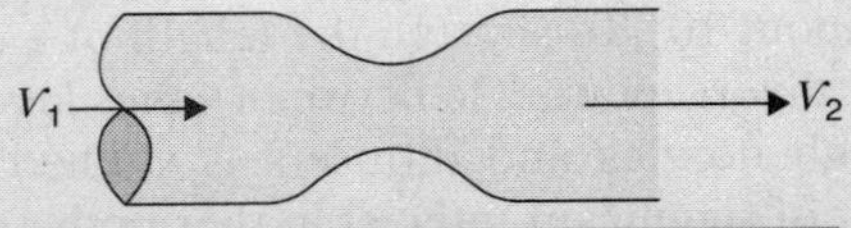

Figure 6.27

☑ **You have completed Learning Outcome (4)**

6.6 Exergy (Availability) and Irreversibility

Exergy: The maximum reversible work possible during a process that brings a system into equilibrium with the surroundings, that is, to a dead state.

Dead state: The system is at equilibrium with the surroundings, taken as 25°C and 100 kPa, unless otherwise stated.

The *exergy*, or *availability*, of a system is the maximum work possible during a process that brings the system into equilibrium with the surroundings. After a system reaches equilibrium with the surroundings, the exergy is zero and the system is in a *dead state*; that is, it has no additional thermal energy, kinetic energy, or potential energy to do work. The dead state is taken as 25°C and 100 kPa with no kinetic or potential energy, unless otherwise specified. Chemical, electrical, magnetic, and nuclear energies could also be included, but we are not concerned with those energies in our study. If a body has only kinetic and/or potential energy, the exergy is equal to the kinetic and/or potential energy since it represents the maximum work that can be delivered using a process, a real or a fictional process, that converts all the kinetic and/or potential energy to useful work.

A process must be a reversible process if the maximum work is delivered by a device, such as a turbine, or if the minimum work is required by a device, such as a compressor. Such work is referred to as *reversible work* W_{rev}. To provide reversible work, recall that a process must be subject to the following restrictions:

Reversible work: The maximum work that can be delivered or the minimum work that is needed between two given states. Keep in mind the restrictions.

- No heat transfer across a finite temperature difference
- No friction
- No unrestrained expansion
- No mixing
- No turbulence
- No combustion

If the final state of a process is the environment, that is, the dead state, the reversible work and exergy are equal. The symbol used to denote exergy[1] or availability is the Greek letter psi (pronounced "sigh"). It is the maximum possible reversible work:

Comment

Exergy, availability, and maximum reversible work are all synonymous.

$$\Psi = W_{rev, max} \tag{6.46}$$

or, per unit mass,

$$\psi = w_{rev, max} \tag{6.47}$$

Exergy must have the same units as work and energy.

Second-law efficiency is now introduced. It is based on the maximum possible reversible work that can be accomplished between an initial state and the environment. The reversible work introduced earlier is based on an isentropic process between states 1 and 2, where state 2 may be at any temperature and pressure; the second-law efficiency is based on an isentropic process between state 1 and the environment at temperature T_0 and pressure P_0, that is, the dead state. It represents the maximum efficiency possible for a process. The *second-law efficiency* η_{II} is the actual work W_a divided by the maximum work possible by an energy-producing device or cycle:

Second-law efficiency: The ratio of the actual work W_a to the maximum work possible for an isentropic process between an initial state and the dead state.

$$\eta_{II} = \frac{W_a}{W_{rev, max}} \quad \text{(turbine)} \tag{6.48}$$

For a device (a compressor) or cycle (a refrigerator) that consumes work, the second-law efficiency would be defined by

$$\eta_{II} = \frac{W_{rev, min}}{W_a} \quad \text{(compressor)} \tag{6.49}$$

or

$$\eta_{II} = \frac{COP_a}{COP_{rev}} \quad \text{(refrigerator)} \tag{6.50}$$

[1] We use Ψ (psi) for exergy since exergy and availability are synonymous and that is the letter most often used for availability. Some textbooks suggest a slight difference between exergy and availability.

If the actual work involves an adiabatic reversible process associated with an inlet (compressor) or outlet (turbine) that is at the dead state, the second-law efficiency would be 100%.

Irreversibility: A measure of how efficient a process is.

A quantity called *irreversibility* is also a measure of how efficient a process is. It should be relatively small compared to the reversible work produced by the process. It is defined as

$$I = W_{\text{rev, max}} - W_a \tag{6.51}$$

or, on a per-unit-mass basis,

$$i = w_{\text{rev, max}} - w_a \tag{6.52}$$

It may be easier to assess a process based on the second-law efficiency compared to an assessment based on the irreversibility since irreversibility does not possess an upper limit but second-law efficiency can never exceed 100%.

Let us find expressions for the maximum reversible work and the associated irreversibility for a control volume. Consider the control volume of Fig. 6.28. The heat transfer is shown leaving the control volume since the environment is at the dead state temperature T_0, which must be lower than the temperature of the control volume. The first law (see Eq. 4.15), assuming a nonsteady flow, is written as

$$-\dot{Q}_{\text{surr}} - \dot{W}_S = \dot{m}_2\left(h_2 + \frac{V_2^2}{2} + gz_2\right) - \dot{m}_1\left(h_1 + \frac{V_1^2}{2} + gz_1\right) + \frac{dE_{\text{c.v.}}}{dt} \tag{6.53}$$

where a negative sign is used on the heat transfer term since heat is assumed to be leaving the control volume and Q_{surr} is considered positive. We can express Eq. 6.38 as

$$\dot{Q}_{\text{surr}} = T_0\dot{m}_1 s_1 - T_0\dot{m}_2 s_2 + T_0\dot{S}_{\text{prod}} - T_0\frac{dS_{\text{c.v.}}}{dt} \tag{6.54}$$

Substitute this expression into Eq. 6.53 and the first law takes the form

$$\dot{W}_S = T_0\frac{dS_{\text{c.v.}}}{dt} - \dot{m}_2\left(h_2 + \frac{V_2^2}{2} + gz_2 - T_0 s_2\right) + \dot{m}_1\left(h_1 + \frac{V_1^2}{2} + gz_1 - T_0 s_1\right) - \frac{dE_{\text{c.v.}}}{dt} - T_0\dot{S}_{\text{prod}} \tag{6.55}$$

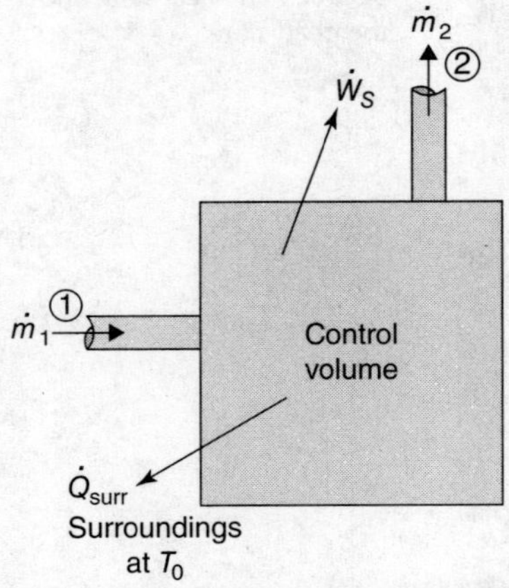

Figure 6.28

The control volume used for the maximum reversible work and the irreversibility.

In most problems of interest in an introductory course in thermodynamics a steady flow is assumed; also, we neglect changes in kinetic and potential energy (a nozzle or a diffuser are exceptions). For such a flow with one inlet and one outlet (see Fig. 6.29) so that $\dot{m}_1 = \dot{m}_2 = \dot{m}$, Eq. 6.55 takes the form

$$\dot{W}_S = \dot{m}\left[(h_1 - h_2) - T_0(s_1 - s_2)\right] - T_0\dot{S}_{\text{prod}} \tag{6.56}$$

where $\dot{W}_S$ represents the shaft work.

The maximum reversible work for a process results if the entropy production is zero since no irreversibilities would exist and the surroundings are at the dead state. With no irreversibilities and the surroundings at T_0 and P_0, Eq. 6.56 provides the maximum reversible work. Assuming $\dot{m}_1 = \dot{m}_2 = \dot{m}$ with no kinetic and potential energy changes, the maximum reversible work rate is

$$\dot{W}_{\text{rev, max}} = \dot{m}[h_1 - h_2 - T_0(s_1 - s_2)] \quad (6.57)$$

or, in terms of exergy, letting $\dot{W}_{\text{rev, max}}/\dot{m} = w_{\text{rev, max}}$,

$$\psi_1 - \psi_2 = w_{\text{rev, max}} = h_1 - h_2 - T_0(s_1 - s_2) \quad (6.58)$$

The actual work rate with the same restrictions, using Eq. 6.53, is

$$\dot{W}_a = (h_1 - h_2) + \dot{Q} \quad (6.59)$$

in which $\dot{Q} = -\dot{Q}_{\text{surr}}$ in Fig. 6.28. The irreversibility is then

$$\dot{I} = \dot{m}T_0(s_2 - s_1) - \dot{Q} \quad (6.60)$$

The exergy at a given location in a flow, say state 1, is given with reference to Eq. 6.57. It is

$$\psi_1 = h_1 - h_0 - T_0(s_1 - s_0) \quad (6.61)$$

If kinetic energy and potential energy are significant, simply add $V_1^2/2 + gz_1$ to the right-hand side of Eq. 6.61. The dead state is assumed to have no kinetic or potential energy, so $V_0^2/2 + gz_0 = 0$.

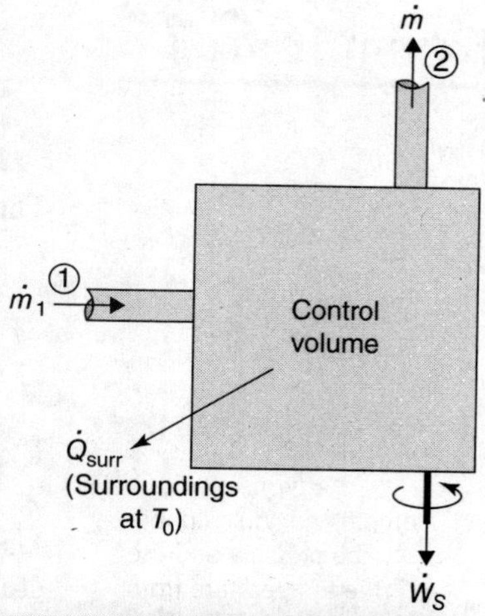

Figure 6.29
A steady-flow control volume with one inlet and one outlet.

Note: If a nozzle or diffuser is analyzed, add $V_1^2/2 + gz_1$ to the right-hand side.

Exergy of steam — Example 6.15

Steam leaves the boiler of Fig. 6.30 at 6 MPa and 600°C with a mass flow rate of 2 kg/s. Determine the maximum possible work rate that can be performed by the steam; that is, determine its exergy rate. Then, calculate the power that an isentropic turbine will produce if its outlet pressure is 20 kPa.

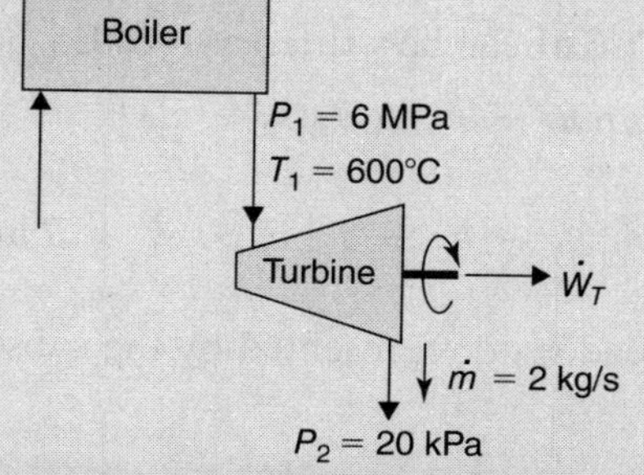

Figure 6.30

(Continued)

Example 6.15 (*Continued*)

Solution

The properties at the exit from the boiler and the dead state, assumed to be at 25°C and 100 kPa, are:

State 1: $P_1 = 6$ MPa, $T_1 = 600$°C
$h_1 = 3658$ kJ/kg, $s_1 = 7.168$ kJ/kg·K

Dead state 0: $T_0 = 25$°C, $P_0 = 100$ kPa (compressed liquid)
$h_0 = 104.9$ kJ/kg, $s_0 = 0.3672$ kJ/kg·K

Remember: For a compressed liquid, ignore the pressure and use the temperature from Table C-1.

Use Eq. 6.61 (multiply by the mass flux) to determine the exergy rate of the steam at the boiler exit, just before it enters the turbine:

$$\dot{\Psi} = \dot{W}_{\text{rev, max}} = \dot{m}[h_1 - h_0 - T_0(s_1 - s_0)]$$

$$= 2\frac{\text{kg}}{\text{s}}\left[(3658 - 104.9)\frac{\text{kJ}}{\text{kg}} - 298\text{ K} \times (7.168 - 0.3672)\frac{\text{kJ}}{\text{kg}\cdot\text{K}}\right]$$

$$= \underline{3050\text{ kW}}$$

State 2 will have the same entropy as state 1, that is, $s_2 = 7.17$ kJ/kg·K. With $P_2 = 20$ kPa, the IRC Calculator gives $h_2 = 2360$ kJ/kg. The isentropic turbine would then produce

$$\dot{W}_T = \dot{m}(h_1 - h_2) = 2\text{ kg/s} \times (3658 - 2360)\text{ kJ/kg} = \underline{2600\text{ kW}}$$

which is less than the exergy rate at the turbine inlet.

Example 6.16 Exergy of substances

Which substance can do more useful work, 2 kg of N_2 or 2 kg of O_2, both at 500 K and 400 kPa? Assume that kinetic and potential energy changes are negligible with a dead state at 25°C and 100 kPa.

Solution

The exergy given in Eq. 6.61 can be written as (see Eq. 6.28)

$$\Psi = W_{\text{rev, max}} = m[h_1 - h_0 + T_0(s_0 - s_1)]$$

$$= m\left[h_1 - h_0 + T_0\left(s_0^\circ - s_1^\circ - R\ln\frac{P_0}{P_1}\right)\right]$$

where state 2 is the dead state represented by the subscript "0." For nitrogen N_2, there results

$$\Psi_{N_2} = 2\text{ kg}\left[\frac{14\,581 - 8669}{28} + 298\left(\frac{191.5 - 206.6}{28} - 0.297\ln\frac{100}{400}\right)\right]\frac{\text{kJ}}{\text{kg}}$$

$$= \underline{346\text{ kJ}}$$

For oxygen O_2, the maximum possible reversible work, the exergy, is

$$\Psi_{O_2} = m\left[h_1 - h_0 + T_0\left(s_0^o - s_1^o - R\ln\frac{P_0}{P_1}\right)\right]$$

$$= 2 \times \left[\frac{14770 - 8682}{32} + 298\left(\frac{205 - 221}{32} - 0.26\ln\frac{100}{400}\right)\right] = \underline{297\text{ kJ}}$$

The nitrogen is capable of producing more work than the oxygen if both are between state 1 at 500 K and 400 kPa, and the dead state.

Note: The ideal-gas tables provide enthalpy and entropy on a molar basis. So, $h = \bar{h}/M$ and $s^o = \bar{s}^o/M$.

Units: $\text{kg}\left[\frac{\text{kJ/kmol}}{\text{kg/kmol}} + \text{K}\frac{\text{kJ/kmol}\cdot\text{K}}{\text{kg/kmol}}\right] = \text{kJ}$

Second-law efficiency of a turbine — Example 6.17

Steam enters the adiabatic turbine of Fig. 6.31 at 6 MPa and 550°C and leaves at 70 kPa, with a mass flow rate of 4.5 kg/s. It has an isentropic efficiency of 85%. Determine i) the turbine power output, ii) the maximum reversible work rate, iii) the rate of irreversibility, and iv) the second-law efficiency. Assume a dead-state temperature T_0 = 20°C and 100 kPa.

Solution

The properties at the inlet and exit to the turbine are, using the IRC Calculator,

State 1: $P_1 = 6$ MPa, $T_1 = 550$°C
$h_1 = 3540$ kJ/kg, $s_1 = 7.03$ kJ/kg·K

State 2s: $P_2 = 70$ kPa, $s_{2s} = 7.03$ kJ/kg·K
$h_{2s} = 2490$ kJ/kg, $T_{2s} = T_{sat} = 89$°C, $x_{2s} = 0.935$

i) The turbine efficiency is based on an isentropic process (see Eq. 6.42). Its output is

$$\dot{W}_T = \dot{m}\eta_T(h_1 - h_{2s})$$

$$= 4.5\frac{\text{kg}}{\text{s}} \times 0.85 \times (3540 - 2490)\frac{\text{kJ}}{\text{kg}}$$

$$= \underline{4017\text{ kW}} \quad \text{or} \quad \underline{5380\text{ hp}}$$

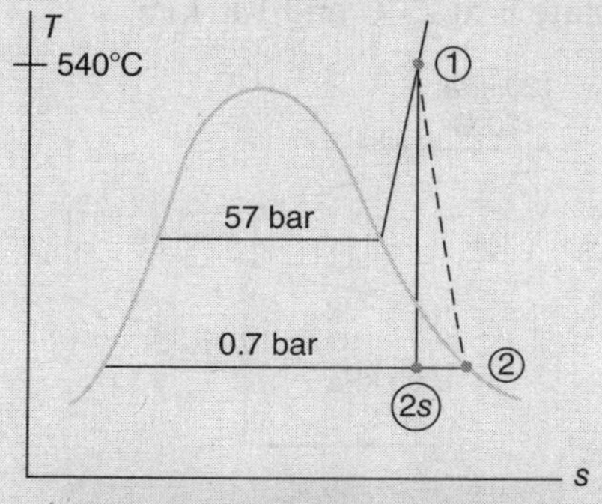

Figure 6.31

(Continued)

Example 6.17 (Continued)

ii) First, let's find the exit state of the turbine. Using Eq. 6.42, we have

$$3540 - h_2 = 0.85 \times (3540 - 2490) \qquad \therefore h_2 = 2647 \text{ kJ/kg}$$

Using the IRC Calculator, we find $s_2 = 7.45$ kJ/kg·K. The maximum reversible power is then

$$\dot{W}_{\text{rev, max}} = \dot{m}[h_1 - h_2 - T_0(s_1 - s_2)]$$

$$= 4.5 \times [3540 - 2647 - 293(7.03 - 7.45)]$$

$$= \underline{4571 \text{ kW}} \quad \text{or} \quad \underline{6130 \text{ hp}}$$

Units: 1 hp = 746 W = 0.746 kW

iii) The rate of irreversibility is

$$\dot{I} = \dot{m}T_0(s_2 - s_1) - \cancel{\dot{Q}}$$

$$= 4.5 \frac{\text{kg}}{\text{s}} \times 293 \text{ K}(7.45 - 7.03)\frac{\text{kJ}}{\text{kg·K}} = \underline{554 \text{ kW}}$$

Or, we could use Eq. 6.51:

$$\dot{I} = \dot{W}_{\text{rev, max}} - \dot{W}_T = 4571 - 4017 = \underline{554 \text{ kW}}$$

iv) The second-law efficiency is

$$\eta_{\text{II}} = \frac{W_T}{W_{\text{rev, max}}} = \frac{4017}{4571} = 0.879 \quad \text{or} \quad \underline{87.9\%}$$

The second-law efficiency is slightly higher than the isentropic efficiency.

Example 6.18 Exergy change across a compressor

R134a is compressed from −20°C and 120 kPa to 80°C and 1.2 MPa. Determine the exergy in, exergy out, and the minimum work required by the compressor of Fig. 6.32. The dead state is at 25°C and 100 kPa.

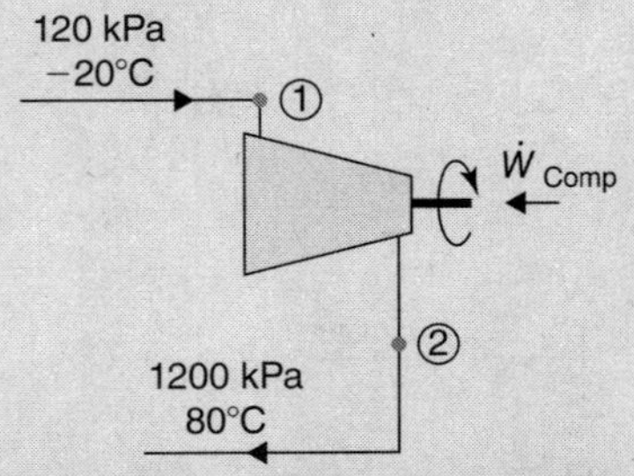

Figure 6.32

Solution

The properties at inlet 1 and exit 2 to the compressor and the dead state are:

State 1: $P_1 = 120$ kPa, $T_1 = -20°C$
$h_1 = 239$ kJ/kg, $s_1 = 0.955$ kJ/kg·K

State 2: $P_2 = 1.2$ MPa, $T_2 = 60°C$
$h_2 = 311$ kJ/kg, $s_2 = 1.02$ kJ/kg·K

Dead state: $P_0 = 100$ kPa, $T_0 = 25°C$
$h_0 = 276$ kJ/kg, $s_0 = 1.11$ kJ/kg·K

The exergy at the inlet and at the exit are

$$\psi_1 = h_1 - h_0 - T_0(s_1 - s_0)$$

$$= (239 - 276)\frac{\text{kJ}}{\text{kg}} - 298\text{ K} \times (0.955 - 1.11)\frac{\text{kJ}}{\text{kg·K}} = \underline{9.2\text{ kJ/kg}}$$

$$\psi_2 = h_2 - h_0 - T_0(s_2 - s_0)$$

$$= (311 - 276)\frac{\text{kJ}}{\text{kg}} - 298\text{ K} \times (1.02 - 1.11)\frac{\text{kJ}}{\text{kg·K}} = \underline{61.8\text{ kJ/kg}}$$

The minimum work required to compress the R134a is

$$w_{\text{min}} = \psi_2 - \psi_1$$

$$= 61.8 - 9.2 = \underline{52.6\text{ kJ/kg}}$$

Note: As usual, the kinetic and potential energy changes are assumed to be negligible since no information is given otherwise.

☑ **You have completed Learning Outcome (5)**

6.7 Summary

This chapter began with the inequality of Clausius, which was utilized to define entropy. Entropy was then used as a measure of whether the second law was satisfied or whether it was violated. To make such a determination, entropy was calculated for the various substances of interest: ideal gases, steam, solids, and liquids. It was applied to both systems and control volumes. It was also used to identify the most ideal processes for the various devices and then to define their efficiencies based on those ideal processes. Finally, the property exergy was introduced and used to establish the second-law efficiency. The following terms were defined in this chapter:

Area under a *T-S* diagram: *For a complete cycle with reversible processes, it represents* $Q_{\text{net}} = W_{\text{net}}$.

Clausius inequality: *It is the mathematical relationship* $\oint \delta Q/T \leq 0$.

Dead state: *The state of a system that is at equilibrium with the surroundings.*

Entropy: *A property defined by its differential:* $dS = \delta Q/T|_{\text{rev}}$. *It is a measure of the amount of disorder in a substance.*

Exergy (or availability): *The maximum reversible work possible during a process that brings a system into equilibrium with the surroundings.*

Irreversibility: *A measure of the efficiency of a process.*

Isentropic process: *An adiabatic reversible process for which* $\Delta s = 0$.

Isentropic efficiency: *The ratio of the actual work to the isentropic or maximum work.*

Reversible work: *The maximum work that can be delivered or the minimum work that is required between two given states.*

Second-law efficiency: *The ratio of the actual work* W_a *to the isentropic work between an initial state and the dead state.*

The second law: *The entropy of every real process increases in an isolated system. Or every real process results in a net entropy increase in the universe.*

Several important equations were presented:

Entropy: $$dS = \left.\frac{\delta Q}{T}\right|_{\text{rev}} \qquad S_2 - S_1 = \int_1^2 \left.\frac{\delta Q}{T}\right|_{\text{rev}}$$

Entropy change for a process: $$\Delta S_{\text{univ}} = \Delta S_{\text{sys}} + \Delta S_{\text{surr}} \geq 0$$

Entropy change for an ideal gas C_p = const:

$$s_2 - s_1 = C_v \ln\frac{T_2}{T_1} + R \ln\frac{v_2}{v_1} = C_p \ln\frac{T_2}{T_1} - R \ln\frac{P_2}{P_1}$$

Isentropic process of an ideal gas C_p = const: $$\frac{T_2}{T_1} = \left(\frac{v_1}{v_2}\right)^{k-1} = \left(\frac{P_2}{P_1}\right)^{(k-1)/k}$$

Entropy change for an ideal gas $C_p \neq$ const (use gas tables):

$$s_2 - s_1 = s_2^\circ - s_1^\circ - R \ln\frac{P_2}{P_1}$$

Isentropic process of an ideal gas $C_p \neq$ const (use gas tables): $$\frac{P_2}{P_1} = \frac{P_{r2}}{P_{r1}}$$

Entropy change for solids or liquids: $$s_2 - s_1 = C \ln\frac{T_2}{T_1}$$

Entropy change for a reservoir: $$\Delta S = \frac{Q}{T}$$

Entropy production through a device: $$\dot{S}_{\text{prod}} = \frac{dS_{\text{c.v.}}}{dt} + \dot{m}_2 s_2 - \dot{m}_1 s_1 + \frac{\dot{Q}_{\text{surr}}}{T_0}$$

Isentropic efficiency of a turbine:	$\eta_T = \dfrac{w_T}{w_{T,\max}} = \dfrac{h_1 - h_2}{h_1 - h_{2s}}$
Second-law efficiency of a turbine:	$\eta_{II} = \dfrac{W_a}{W_{\text{rev, max}}}$
Exergy:	$\Psi = W_{\text{rev, max}} \qquad \psi = w_{\text{rev, max}}$
Irreversibility:	$I = W_{\text{rev}} - W_a \qquad i = w_{\text{rev}} - w_a$
Exergy in a flow:	$\psi = h - h_0 - T_0(s - s_0)$

Problems

FE Exam Practice Questions

6.1 The Clausius inequality places which restriction on an actual cycle?
- **(A)** The net heat transfer must be positive.
- **(B)** The net heat transfer must be negative.
- **(C)** The differential quantity $\delta Q/T$ integrated around a cycle must be negative.
- **(D)** The differential quantity $\delta Q/T$ integrated around a cycle must be positive.

6.2 The Clausius inequality is applicable for only:
- **(A)** Heat engines around a cycle
- **(B)** All processes in the cycle of a heat engine
- **(C)** All processes in heat engines and refrigerators
- **(D)** All cycles involving heat engines and refrigerators

6.3 Assume ideal conditions for each component of the refrigeration cycle of Fig. 6.33 (with *T-s* diagram in Fig. 6.34) and find $\oint \delta q/T$. R134a is the refrigerant. (Remember, q_H is negative and q_L is positive.)
- **(A)** −0.062 kJ/kg·K
- **(B)** −0.085 kJ/kg·K
- **(C)** −0.095 kJ/kg·K
- **(D)** −0.16 kJ/kg·K

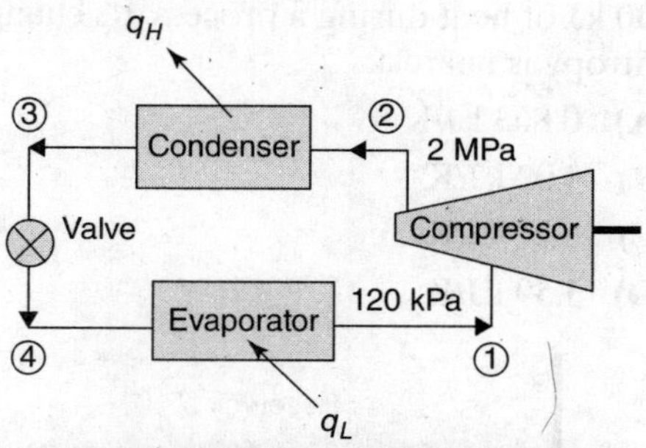

Figure 6.33

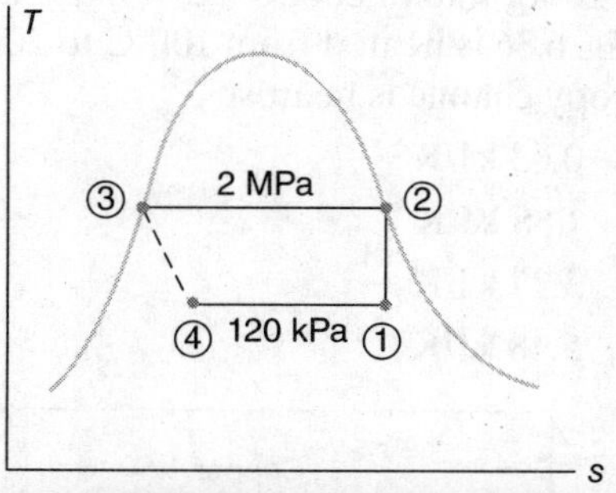

Figure 6.34

6.4 Air is heated from 20°C to 800°C at constant pressure of 200 kPa in a cylinder with an initial volume of 4000 cm^3. The entropy change, assuming an ideal gas with constant specific heats, is nearest:

(A) 0.0123 kJ/K
(B) 0.0972 kJ/K
(C) 0.563 kJ/K
(D) 3.11 kJ/K

6.5 A piston compresses air in a cylinder from 20°C and 100 kPa to 2 MPa. The final temperature, assuming an ideal gas with constant specific heats, is approximately:

(A) 690°C
(B) 540°C
(C) 490°C
(D) 420°C

6.6 The 200°C heat reservoir of Fig. 6.35 gains 600 kJ of heat during a process. Its change in entropy is nearest:

(A) 0.863 kJ/K
(B) 1.02 kJ/K
(C) 1.27 kJ/K
(D) 1.59 kJ/K

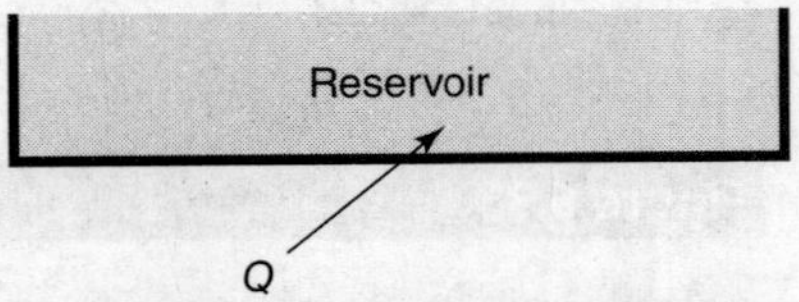

Figure 6.35

6.7 The 20-kg slab of copper ($C = 0.395$ kJ/kg·K) in Fig. 6.36 is heated from 100°C to 200°C. Its entropy change is nearest

(A) 0.63 kJ/K
(B) 1.88 kJ/K
(C) 3.27 kJ/K
(D) 5.48 kJ/K

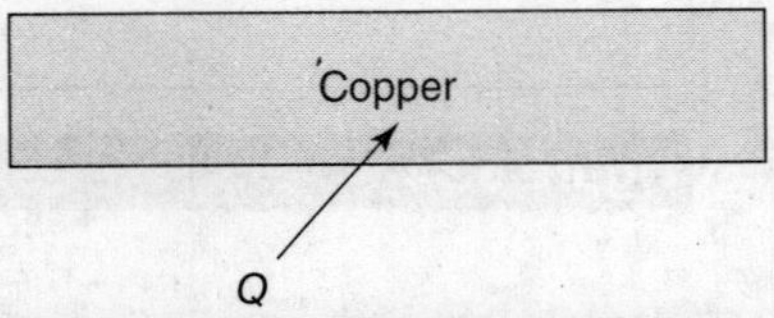

Figure 6.36

6.8 The air in the constant-pressure cylinder of Fig. 6.37 is initially at 400°C and 120 kPa. It is cooled until its temperature is 40°C. The surroundings are at 20°C. The entropy generated during this process is nearest:

(A) 0.463 kJ/kg·K
(B) 0.591 kJ/kg·K
(C) 0.766 kJ/kg·K
(D) 1.229 kJ/kg·K

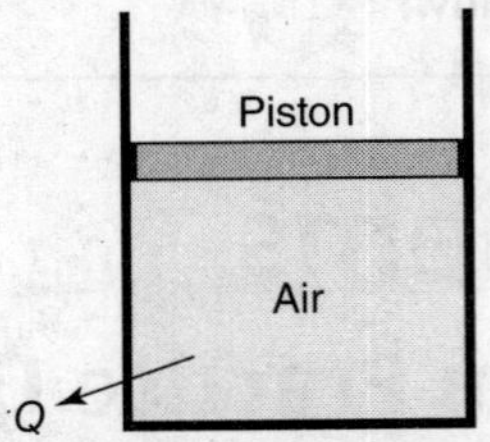

Figure 6.37

6.9 Two kilograms of steam enter a turbine each second at 600°C and 4 MPa and leave as saturated vapor at 20 kPa while 100 kJ/s of heat leave the turbine. The surroundings are at 25°C. The entropy production is nearest:

(A) 1.42 kJ/K·s
(B) 1.08 kJ/K·s
(C) 0.63 kJ/K·s
(D) 0.34 kJ/K·s

6.10 Steam at 500°C and 4 MPa enters a turbine with an isentropic efficiency of 86%. The exit pressure is 100 kPa. The exiting temperature of the steam is nearest:

(A) 95°C
(B) 107°C
(C) 118°C
(D) 126°C

6.11 Saturated R134a at 120 kPa enters an adiabatic compressor with an efficiency of 80%. The exit pressure is 1.6 MPa, as shown in the *T-s* diagram of Fig. 6.38. The exiting temperature of the refrigerant from the compressor is nearest:

(A) 104°C
(B) 99°C
(C) 95°C
(D) 89°C

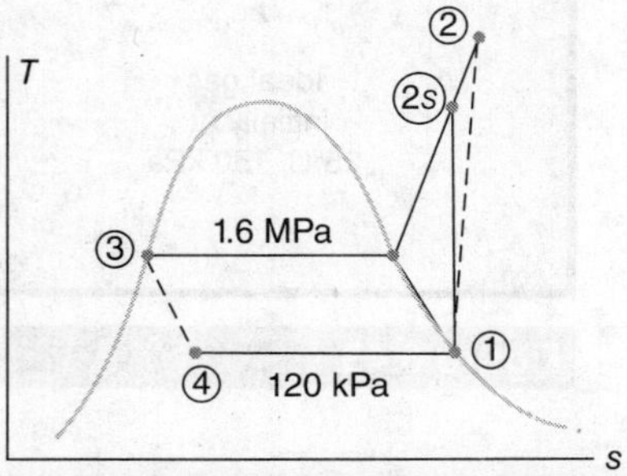

Figure 6.38

Questions 6.12–6.15

Steam enters a turbine (see Figures 6.39 and 6.40) at 4 MPa and 500°C and leaves as saturated steam at 40 kPa. Assume standard conditions for the dead state.

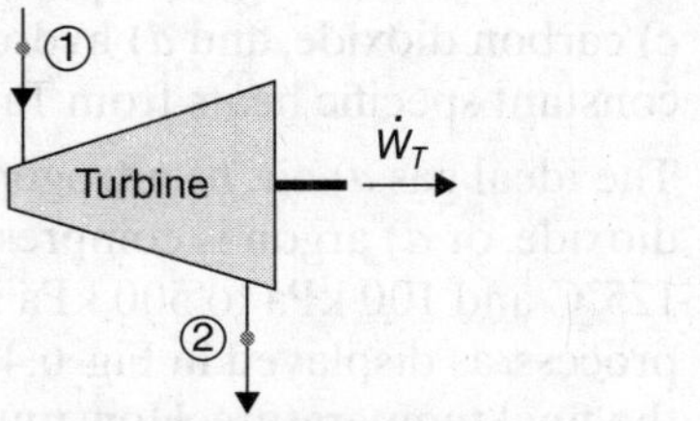

Figure 6.39

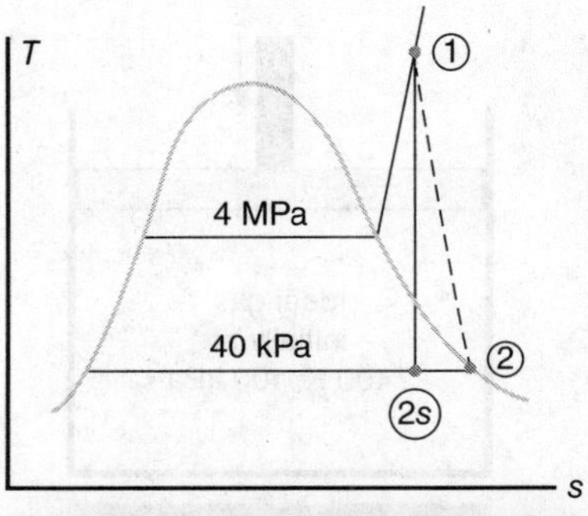

Figure 6.40

6.12 The exergy of the steam at state 1 is nearest:

(A) 1340 kJ/kg
(B) 17900 kJ/kg
(C) 3030 kJ/kg
(D) 4400 kJ/kg

6.13 The second-law efficiency of the turbine is nearest:

(A) 68%
(B) 72%
(C) 78%
(D) 82%

6.14 The isentropic efficiency of the turbine is nearest:

(A) 80%
(B) 75%
(C) 70%
(D) 67%

6.15 The irreversibility of the process is nearest:

(A) 529 kJ/kg
(B) 389 kJ/kg
(C) 319 kJ/kg
(D) 229 kJ/kg

Inequality of Clausius

6.16 The *T-s* diagram of a Carnot refrigeration cycle utilizing R134a is shown in Fig. 6.41. Heat transfer takes place in the condenser and evaporator at constant temperature. Verify the Clausius inequality using $x_2 = 1$ and $x_3 = 0$.

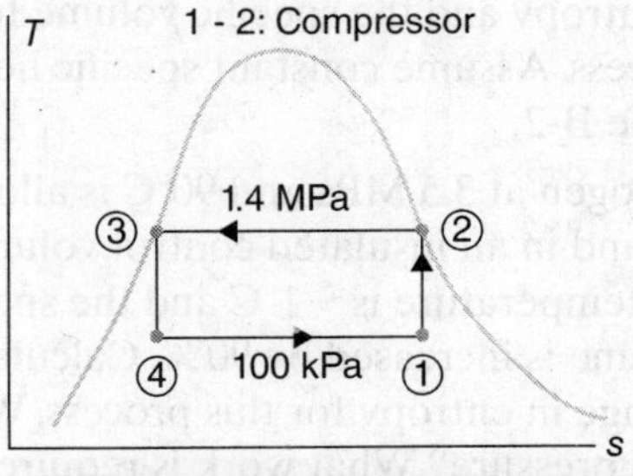

Figure 6.41

6.17 The *T*-s diagram of a Carnot steam power cycle is shown in Fig. 6.42. Heat transfer takes place in the boiler and condenser only. Assume an isentropic turbine and pump. Determine the Clausius inequality numerical result for this cycle if $x_2 = 0$ and $x_3 = 1$ and P_B is *a*) 2 MPa, *b*) 4 MPa, and *c*) 6 MPa.

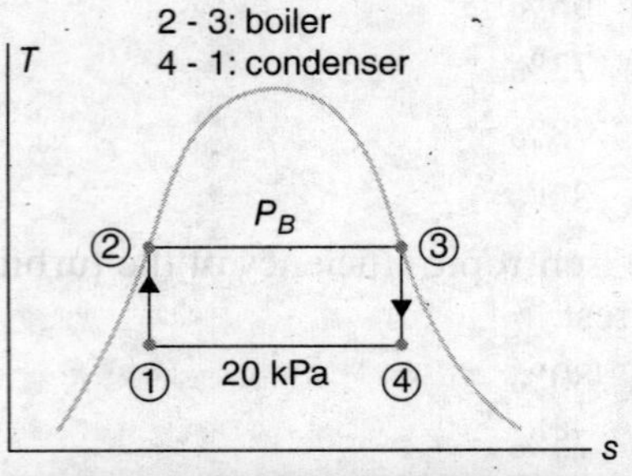

Figure 6.42

Entropy Change in an Ideal Gas with Constant C_p and C_v

6.18 A piston moves in a cylinder at constant temperature while heat transfer occurs. If the pressure doubles, determine entropy change in the air assuming constant specific heats from Table B-2.

6.19 A piston moves in a cylinder at constant pressure while heat is added. If the volume doubles, determine entropy change in the air assuming constant specific heats from Table B-2.

6.20 Air at 200°C and 300 kPa is compressed to 400°C and 800 kPa. Calculate the change in entropy and the specific volume for this process. Assume constant specific heats from Table B-2.

6.21 Nitrogen at 3.5 MPa and 90°C is allowed to expand in an insulated control volume until the temperature is −1°C and the specific volume is increased by 90%. Calculate the change in entropy for this process. What is the final pressure? What work is required? Assume constant specific heats from Table B-2.

6.22 Four kilograms of an ideal gas are initially at 25°C and 150 kPa in the rigid container of Fig. 6.43. Heat is added until the pressure reaches 900 kPa. Determine the entropy change of gas if it is *a*) air, *b*) nitrogen, *c*) carbon dioxide, and *d*) hydrogen. Assume constant specific heats from Table B-6.

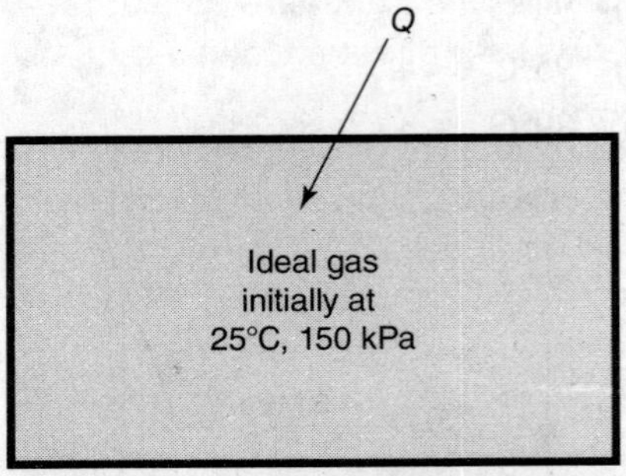

Figure 6.43

Remember: Each part with a lower-case italic letter is a separate problem.

6.23 An ideal gas is initially at 120 kPa and 20°C when contained in a 5-m^3 volume. If it experiences an entropy increase of 4 kJ/K as the temperature is raised to 100°C, what is its final volume if the gas is *a*) air, *b*) nitrogen, *c*) carbon dioxide, and *d*) hydrogen? Assume constant specific heats from Table B-2.

6.24 The ideal gas *a*) air, *b*) nitrogen, *c*) carbon dioxide, or *d*) argon is compressed from 125°C and 100 kPa to 500 kPa in an isentropic process, as displayed in Fig. 6.44. Calculate the final temperature. How much work occurs during this process? Assume constant specific heats from Table B-2.

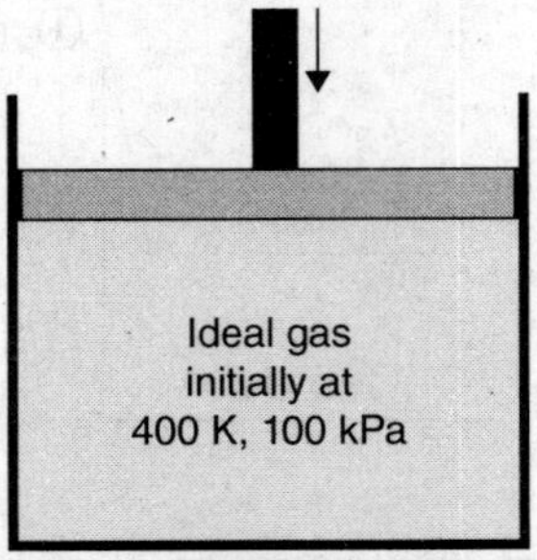

Figure 6.44

6.25 The ideal gas *a*) air, *b*) nitrogen, *c*) carbon dioxide, or *d*) argon is allowed to triple its volume in an isentropic process. If the final temperature is −7°C, what is the initial temperature and what work is produced? Assume constant specific heats from Table B-2.

6.26 Two thousand kilojoules of heat are added to 10 kg of air contained in a rigid container. Calculate the final temperature and the entropy change of the air if the initial temperature and pressure are, respectively, *a*) 200°C and 200 kPa, *b*) 100°C and 400 kPa, and *c*) 400°C and 100 kPa. Assume constant specific heats from Table B-2.

6.27 The frictionless piston/cylinder arrangement of Fig. 6.45 contains 2 kg of nitrogen at 40°C and 200 kPa. If the pressure is held constant, calculate the final temperature and the entropy change if a paddle wheel inserted in the insulated cylinder adds *a*) 100 kJ of work, *b*) 200 kJ of work, and *c*) 400 kJ of work. Assume constant specific heats from Table B-2.

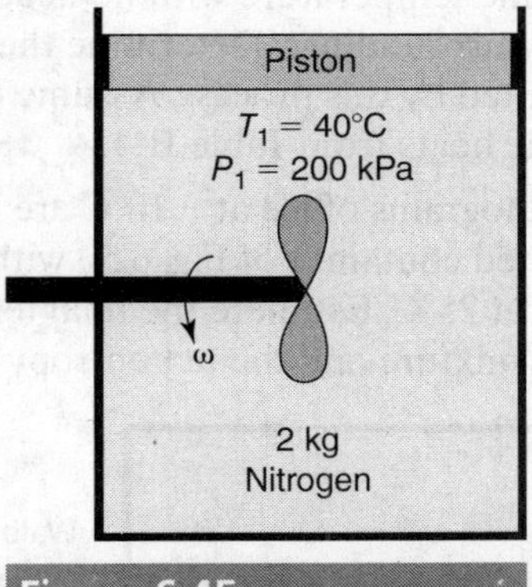

Figure 6.45

6.28 Heat is added to a volume that contains 2 kg of air, initially at 80°C and 200 kPa. Determine the work, the heat transfer, and the entropy change if the temperature is held constant while the volume is expanded to *a*) 2 m^3, *b*) 3 m^3, and *c*) 4 m^3. Assume constant specific heats from Table B-2.

6.29 Air is compressed in an isentropic process from 20ºC and 100 kPa until the final volume is *a*) six times the initial volume, *b*) eight times the initial volume, and *c*) ten times the initial volume,. Calculate the final pressure and temperature assuming constant specific heats from Table B-2.

6.30 Four hundred kilojoules of heat are transferred to 2 kg of air in a piston/cylinder arrangement that is maintained at constant pressure. Calculate the final temperature and the entropy change if the initial temperature is *a*) 100°C, *b*) 200°C, and *c*) 400°C. Assume constant specific heats from Table B-2.

6.31 An electric heater provides 800 kJ of heat to nitrogen in a rigid 2-m^3 volume. If the initial temperature is 100°C, determine the entropy increase if the initial pressure is *a*) 100 kPa, *b*) 400 kPa, and *c*) 1000 kPa. Assume constant specific heats from Table B-2.

6.32 A frictionless piston compresses 0.2 kg of air in the insulated cylinder of Fig. 6.46 from the initial conditions shown. Estimate the final temperature and the work required if the final pressure is *a*) 800 kPa, *b*) 1400 kPa, and *c*) 2 MPa. Assume constant specific heats from Table B-2.

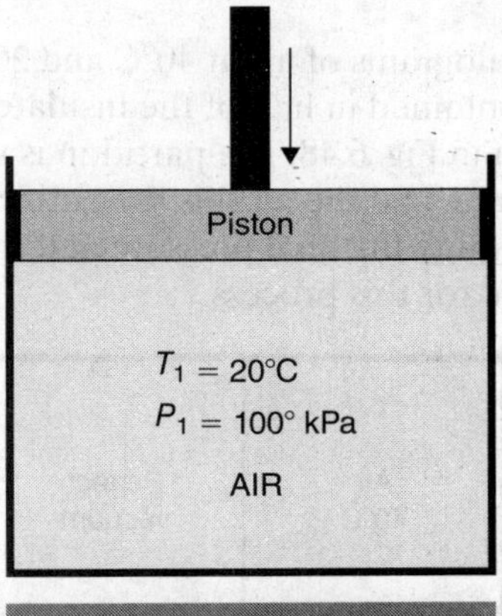

Figure 6.46

6.33 The gases that closely resemble air after combustion in an automobile cylinder are expanded such that the volume increases by a factor of 8. If the initial pressure is 2 MPa, estimate the final temperature and the work produced if the initial temperature is *a*) 427°C, *b*) 590°C, and *c*) 760°C. Assume constant specific heats from Table B-2. Assume a frictionless process.

6.34 A torque of 50 N·m is required to rotate the paddle wheel shown in Fig. 6.47 at 60 rad/s. If it rotates for 2 minutes during which time 200 kJ of heat is transferred to the air from a reservoir, calculate the final temperature in this rigid volume and the entropy increase in the air. Assume constant specific heats from Table B-2.

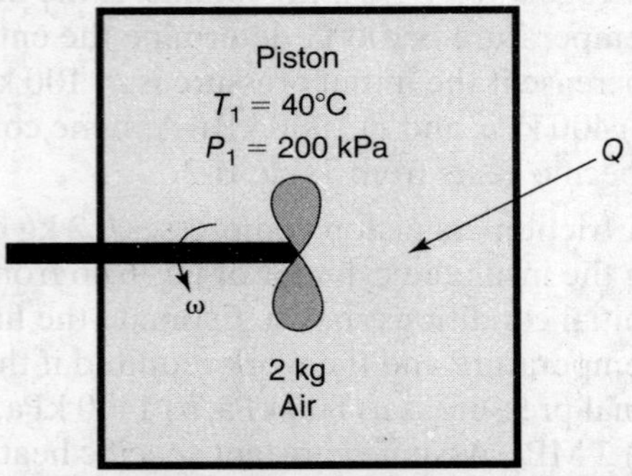

Figure 6.47

6.35 Two kilograms of air at 40°C and 200 kPa are contained in half of the insulated volume shown in Fig. 6.48. The partition is suddenly ruptured, and the air fills the entire volume. Determine the final pressure and the entropy change for this process.

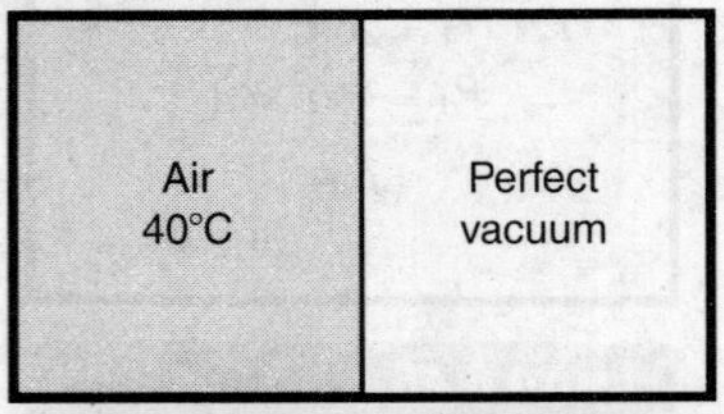

Figure 6.48

Entropy Change of a Solid, a Liquid, and a Reservoir

6.36 The engine in Fig. 6.49 delivers 50 kJ by operating between two reservoirs at 900°C and 40°C. Its efficiency is 40%. Determine the entropy change of each reservoir.

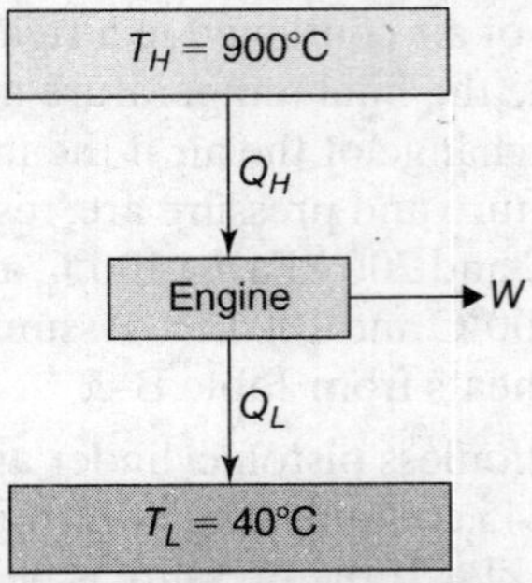

Figure 6.49

6.37 A 4 kg copper slab at 93°C is brought into contact with a 10 kg aluminum slab at 16°C. Insulation allows the two slabs to arrive at the same temperature with no heat transfer to the surroundings. Determine the entropy generated by this process. Assume constant specific heats from Table B-4.

6.38 Two kilograms of ice at −10°C are mixed in the insulated container of Fig. 6.50 with 12 kg of water at 25°C. Estimate the final temperature of the mixture and the net entropy change.

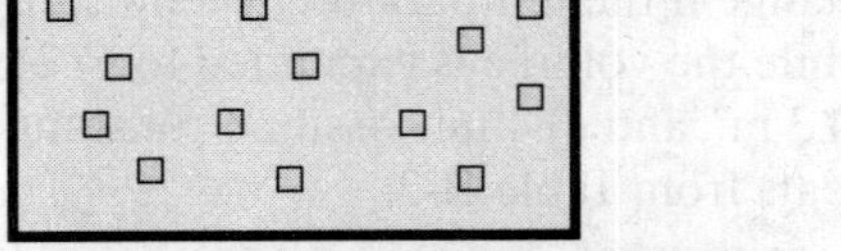

Figure 6.50

6.39 Twenty kilograms of copper at 80°C are submerged in a container that holds water at 24°C. Two hundred kilojoules of heat is transferred from the water to the 25°C surroundings. Calculate the eventual temperature of the copper, the entropy change of the water plus the copper, and the entropy change of the universe if the mass of water is *a*) 6 kg, *b*) 10 kg, and *c*) 15 kg. Use average C_p values from Table B-4.

Entropy Change of a Phase-Change Substance

6.40 Steam at 4000 kPa and 400°C is allowed to cool to a saturated vapor at 4000 kPa. Calculate the ratio of the initial specific volume to the final specific volume and the change in specific entropy for this process.

6.41 Refrigerant 134a undergoes a process from 400 kPa and 400°C to a pressure of *a*) 400 kPa, *b*) 800 kPa, and *c*) 1200 kPa in a rigid container. Calculate the heat transfer and the change in specific entropy.

6.42 Ten kg of water at 100 kPa and 5°C are heated to *a*) 137°C, *b*) 160°C, and *c*) 260°C in a constant-pressure process. Calculate the change in entropy for this process.

6.43 Four kilograms of saturated steam are being condensed at a constant pressure of 120 kPa in the cylinder of Fig. 6.51 by transferring heat to the 25°C surroundings. Calculate the entropy generated by this process if *a*) $x_2 = 0.6$, *b*) $x_2 = 0.2$, and *c*) $T_2 = 40°C$.

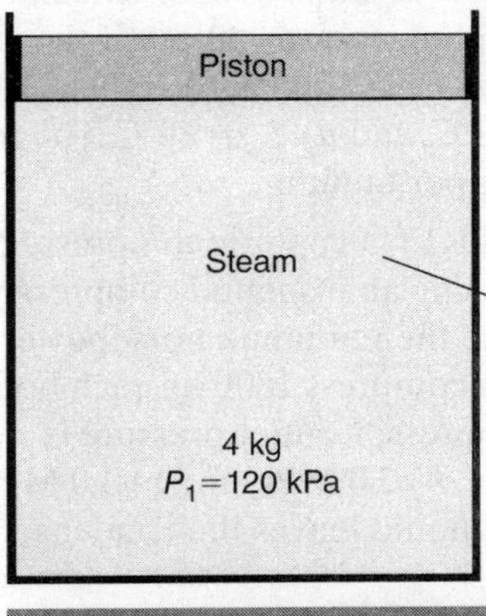

Figure 6.51

6.44 Two kilograms of steam are contained in a 40-L rigid container originally at 120 kPa. If 2 MJ of heat are transferred from a 400°C reservoir, estimate the entropy generated.

6.45 Superheated steam is in a 0.3 m³ rigid container at 1.4 MPa and 450°C. The container is cooled until the pressure reaches 140 kPa. Determine the necessary heat transfer and the entropy change of the universe if the heat is transferred to a 21°C atmosphere.

6.46 Steam at 200°C is compressed reversibly in an insulated cylinder from 100 kPa to *a*) 500 kPa, *b*) 800 kPa, and *c*) 1200 kPa. Calculate the final temperature and the work requirement.

6.47 Determine the work output when 4 kg of steam, contained in a 0.4-m³ cylinder at 600°C and 8000 kPa, are expanded isentropically until the pressure is *a*) 1200 kPa, *b*) 800 kPa, and *c*) 400 kPa.

6.48 The temperature of 5 kg of water changes from 40°C to 400°C, while the pressure remains constant at 200 kPa. Calculate the required heat transfer and the entropy increase of the universe if the heat comes from a 600°C reservoir.

Entropy Change of an Ideal Gas with Variable Specific Heats

6.49 Air is compressed from 400 K and 100 kPa in an isentropic process. Assuming constant specific heats from Table B-2 and variable specific heats, calculate the final temperature if the final pressure is *a*) 500 kPa, *b*) 1000 kPa, and *c*) 2000 kPa.

6.50 Air is allowed to quadruple its volume in an isentropic process. If the final temperature of the air is 17°C and we assume variable specific heats, what is the initial temperature? What would it be if constant specific heats (from Table B-2) were assumed?

6.51 The farmer in Fig. 6.52 inserts nitrogen from a pressurized tank to fertilize a field. An isentropic process can be assumed as the gas leaves the tank. Estimate the temperature of the nitrogen at the point where it enters the soil if the pressure in the tank is *a*) 8 MPa and *b*) 14 MPa. Make any necessary assumptions and assume constant specific heats and then variable specific heats. (This would be the same temperature should a hose burst and the nitrogen spray across the farmer.)

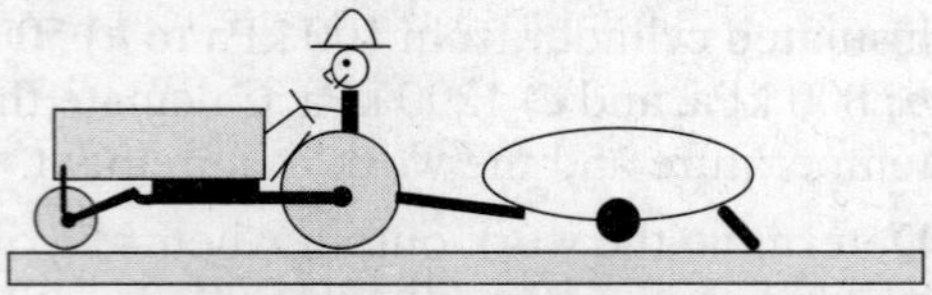

Figure 6.52

6.52 Rework the following problems assuming variable specific heats:

a) Prob. 6.22(*a*) *b*) Prob. 6.22(*b*)
c) Prob. 6.23(*a*) *d*) Prob. 6.24(*a*)
e) Prob. 6.25(*a*) *f*) Prob. 6.26(*b*)
g) Prob. 6.27(*a*)

Entropy Changes for a Control Volume

6.53 R134a enters the valve of Fig. 6.53 at 1.2 MPa and $x = 0$. It leaves the valve at 120 kPa. Determine the entropy change.

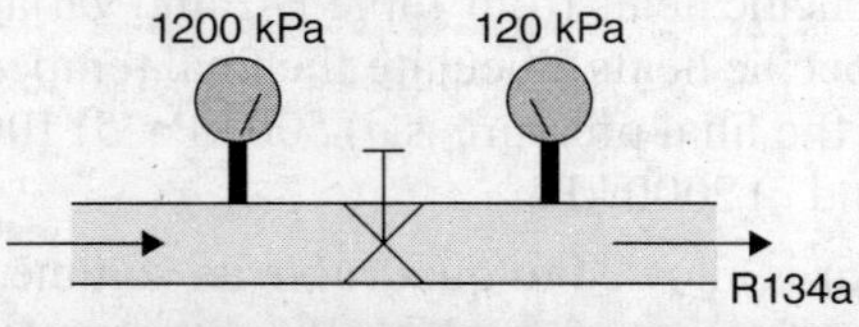

Figure 6.53

6.54 Ammonia enters a throttle as a saturated liquid at 600 kPa. It exits the throttle to a pressure of 120 kPa. Calculate the exit temperature of the refrigerant and the change in specific entropy for this process. Use the tables.

6.55 Refrigerant 134a enters a heat exchanger at 0.9 MPa and a temperature of 70°C and leaves at the same pressure with a quality of 0.6. The mass flow rate is 10 kg/s. What are the heat transfer and rate of change of entropy for the refrigerant?

6.56 Water enters a boiler at 6 MPa and 60°C and leaves at 400°C. Calculate the rate of heat transfer if 2000 kg of water pass through the boiler each minute. What is the entropy generated if the heat that stokes the boiler comes from a flame maintained at 2000°C?

6.57 Steam at 20 kPa and a quality of 0.9 enters the condenser of Fig. 6.54 and leaves as saturated liquid. Calculate the heat transfer and entropy change. If the heat is transferred to a 10°C lake, what entropy is generated due to this process?

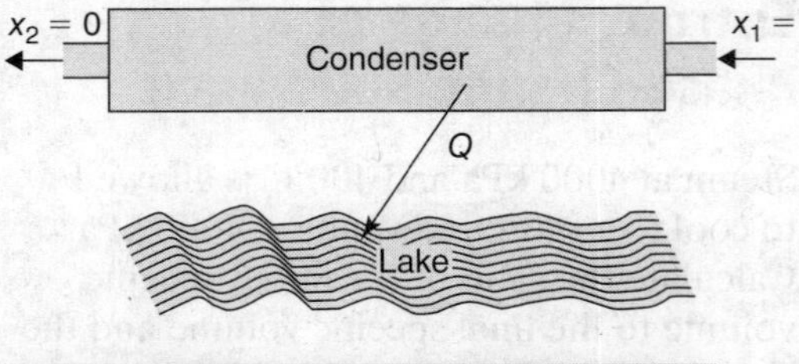

Figure 6.54

6.58 Two kilograms of steam at 6 MPa and 600°C enter an insulated turbine each second. If the losses as the steam moves over the blades in the turbine are neglected, estimate the maximum horsepower output of the turbine if the exit pressure is 20 kPa.

6.59 Superheated steam enters an insulated turbine at 20 MPa and 600°C at a mass flow rate of 1000 kg/min. If the steam leaves the turbine at *a*) 10 kPa, *b*) 40 kPa, and *c*) 80 kPa as a saturated vapor, what are the power output and the rate of change of entropy of the steam?

6.60 Superheated steam flows into an insulated turbine at 4 MPa and 500°C. Calculate the maximum horsepower output if the exit pressure *a*) $P_2 = 10$ kPa, *b*) $P_2 = 60$ kPa, *c*) $T_2 = 50$°C, and *d*) $T_2 = 80$°C. The mass flux is 4000 kg per minute.

6.61 Saturated R134a vapor is compressed from 100 kPa using an insulated compressor. Determine the minimum horsepower needed to compress 1800 kg each hour if the compressor outlet pressure is *a*) 1.4 MPa, *b*) 2.0 MPa, and *c*) 3.0 MPa. Saturated liquid leaves the condenser.

6.62 Saturated ammonia at 120 kPa is compressed to 1200 kPa by the insulated compressor of Fig. 6.55. Neglect any losses and determine the horsepower requirement if 40 kg passes through the compressor each minute. What is the entropy change of the universe due to this compression process if the surroundings are at 20°C?

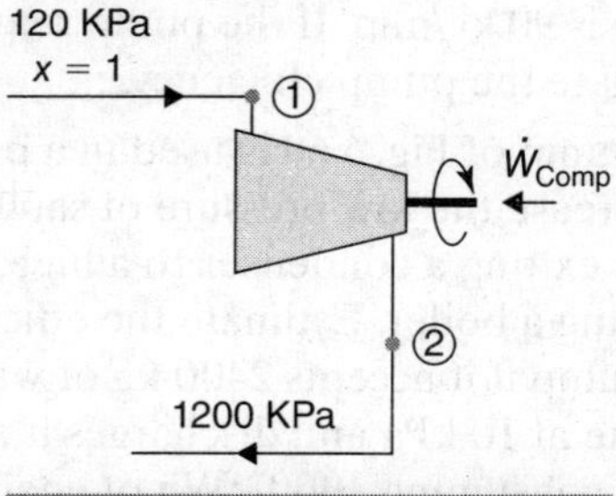

Figure 6.55

6.63 Two kilograms per second of superheated steam leave a steam generator and enter a turbine at 6 MPa and 500°C. The turbine produces 2500 hp by expanding the steam to a pressure of 20 kPa with $x = 0.9$. Calculate the rate of entropy production by the turbine if the surroundings are at 25°C.

6.64 Seven thousand kilograms of superheated steam enter a turbine each hour at 7 MPa and 550°C and leave the turbine at 35 kPa as saturated vapor producing 2000 hp. Calculate the rate of entropy change in the steam and in the universe if the surroundings are at 20°C.

6.65 The insulated turbine in the Rankine cycle in Fig. 6.56, operating with water, produces 3000 kW of power. Determine the entropy change across each device.

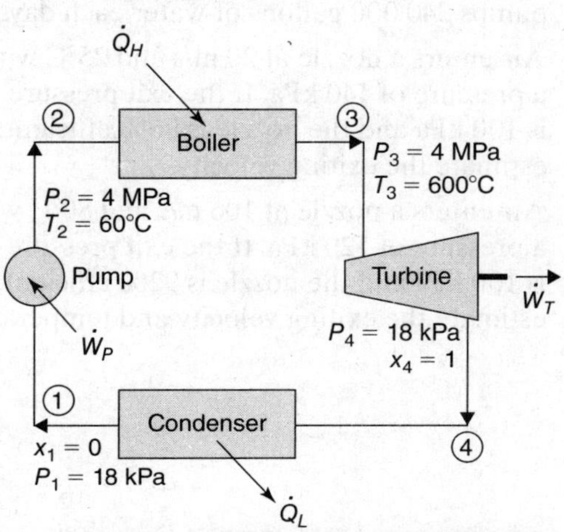

Figure 6.56

6.66 Water at atmospheric pressure and 82°C enters a mixing chamber at a flow rate of 5 L/s. Cold water enters the mixing chamber at 4°C at a flow rate of 7.5 L/s. Calculate the temperature of the water leaving the mixing chamber and the rate of change of entropy for this insulated chamber.

6.67 The boiler in a power plant heats water to a temperature near 600°C before it enters the turbine. The water is preheated by mixing it with superheated steam extracted from the turbine before it enters the boiler, resulting in less fuel used by the boiler. The insulated preheater is shown in Fig. 6.57. Determine the entropy production if all pressures are at 600 kPa and

a) $\dot{m}_1 = 2$ kg/s of steam at 250°C
$\dot{m}_2 = 20$ kg/s of water at 45°C

b) $\dot{m}_1 = 2.4$ kg/s of steam at 200°C
$\dot{m}_2 = 24$ kg/s of water at 40°C

c) $\dot{m}_1 = 2.8$ kg/s of steam at 180°C
$\dot{m}_2 = 30$ kg/s of water at 50°C

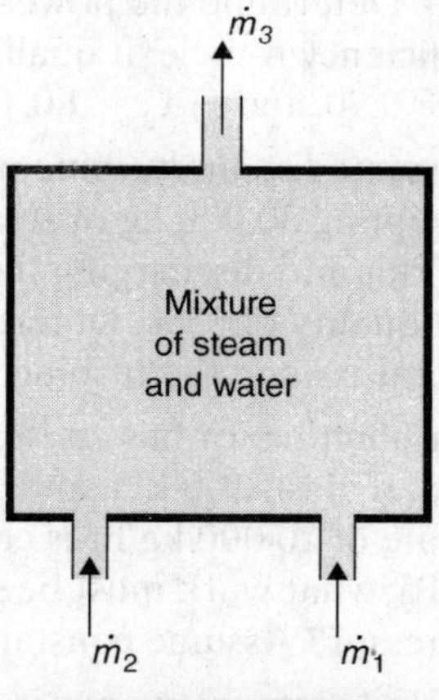

Figure 6.57

Isentropic Efficiency

6.68 Steam at 600 kPa and 350°C enters an adiabatic turbine at a flow rate of 20 kg/s. It leaves the turbine at a pressure of 40 kPa as a saturated vapor. Calculate the power produced by this turbine and the isentropic efficiency of the turbine.

6.69 Air enters an adiabatic gas turbine at 600 kPa and 400°C. It leaves the turbine at atmospheric pressure and a temperature of 140°C. Calculate the specific work output and the isentropic efficiency of this turbine. Assume constant specific heats.

6.70 For the adiabatic steam turbine of Fig. 6.58, calculate the isentropic efficiency if *a*) $x_2 = 0.9$, *b*) $x_2 = 1.0$, *c*) $T_2 = 50°C$, and *d*) $T_2 = 60°C$.

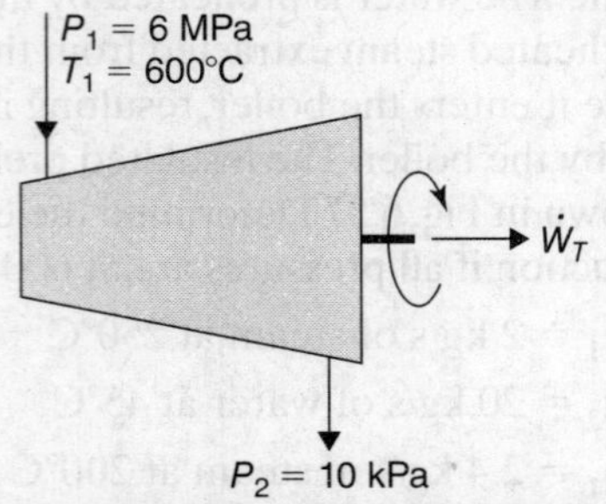

Figure 6.58

6.71 An insulated steam turbine accepts 16 kg/s at 14 MPa and 550°C and discharges it at 14 kPa. Determine the power produced and the efficiency if the exit quality is *a*) $x_2 = 0.82$, *b*) $x_2 = 0.90$, and *c*) $x_2 = 1.0$.

6.72 An insulated steam turbine produces 8000 hp by accepting 30 000 kg of steam at 10 MPa each hour and discharging the steam at 20 kPa with a quality of 85%. Determine the inlet temperature and the turbine efficiency.

6.73 An air compressor has an isentropic efficiency of 0.88. If air at 100 kPa and 25°C and a mass flow rate of 20 000 kg/hr is compressed to 500 kPa, what work must be supplied to the compressor? Assume constant specific heats.

6.74 Saturated ammonia vapor is compressed as shown in Fig. 6.59. Calculate the horsepower input if the compressor's adiabatic efficiency is *a*) 80%, *b*) 90%, and *c*) 100%.

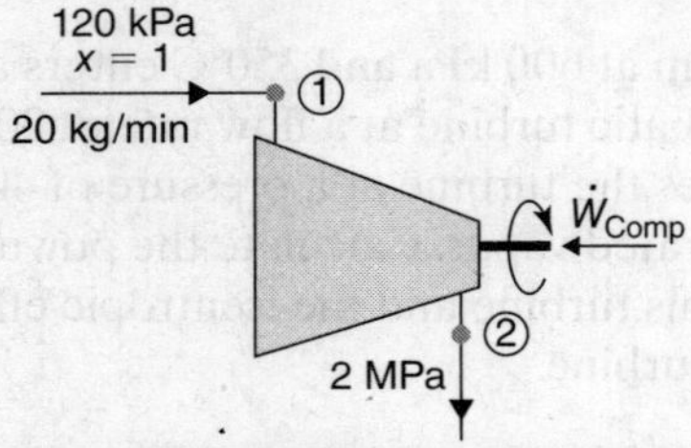

Figure 6.59

6.75 Saturated R134a vapor is to be compressed from 140 kPa to 1.6 MPa in a compressor rated at 86% efficiency. Determine the horsepower requirement if the mass flow rate is 180 kg each hour.

6.76 Water enters a pump at 80 kPa and leaves the pump at 6 MPa. The mass flow rate through the pump is 90 kg/min. If the pump requires 15 hp, calculate the pump efficiency.

6.77 The pump of Fig. 6.60 is used in a power plant to increase the low pressure of saturated liquid water exiting a condenser to a high pressure entering a boiler. Estimate the efficiency of the pump if it accepts 2400 kg of water each minute at 10 kPa and discharges it at 8 MPa while consuming 4000 kWh of energy over a 10-hour period.

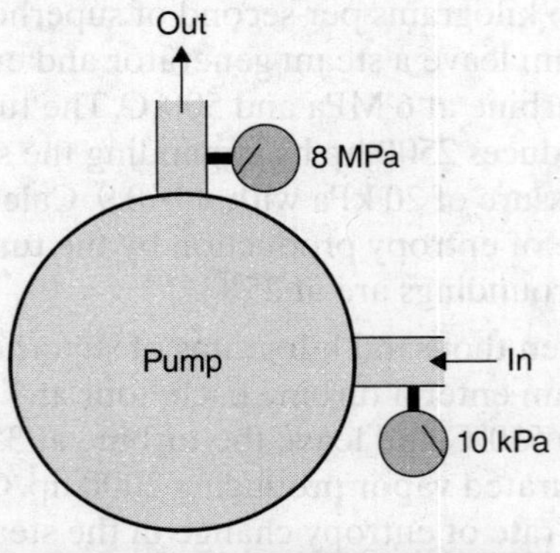

Figure 6.60

6.78 Determine the horsepower requirement of a pump that increases the pressure of water from 20 kPa to 5 MPa if it is 88% efficient and pumps 340 000 gallons of water each day.

6.79 Air enters a nozzle at 20 m/s and 25°C with a pressure of 140 kPa. If the exit pressure is 100 kPa and the nozzle is 96% efficient, estimate the exiting velocity.

6.80 Air enters a nozzle at 106 m/s and 80°C with a pressure of 120 kPa. If the exit pressure is 100 kPa and the nozzle is 92% efficient, estimate the exiting velocity and temperature.

Exergy (Availability) and Irreversibility

6.81 A steam generator produces superheated steam at a rate of 140 kg/min, a temperature of 500°C, and a pressure of 5 MPa. What is the maximum amount of power that could be obtained using this steam? Use the standard dead state.

6.82 Steam enters a turbine at 10 MPa and 600°C and exits at 20 kPa with a quality of 95%. Determine the irreversibility of this process.

6.83 What is the exergy of air at 300°C and 600 kPa if the dead state is 20°C and 101 kPa? *a*) Assume constant specific heats and *b*) use Table F-1.

6.84 R134a enters a compressor at 200 kPa and −10°C at a flow rate of 2.5 kg/s, and leaves with a temperature of 150°C and a pressure of 1.2 MPa. What is the rate of change of exergy for this device?

CHAPTER

7

$$ds = \left(\frac{\partial s}{\partial T}\right)_P dT + \left(\frac{\partial s}{\partial P}\right)_T dP$$

Thermodynamic Relations

Outline

Nomenclature

The following variables are introduced in this chapter:

a	Helmholtz function
B	Bulk modulus
g	Gibbs function
M	A general function
N	A general function
T^*	The star superscript indicates a state where the ideal-gas law $PV = mRT$ applies
β	Volume expansivity
μ_J	Joule–Thomson coefficient

Learning Outcomes

- **Use the Maxwell relations to develop relations among thermodynamic properties**
- **Apply the Clapeyron equation**
- **Find general relationships for internal energy, enthalpy, and entropy**
- **Develop and apply the Joule–Thomson coefficient**
- **Study the real-gas effects on thermodynamic properties**

In this chapter, we will relate thermodynamic properties that cannot be measured, such as enthalpy, internal energy, entropy, and specific heats, to properties that can be measured: temperature, pressure, mass, and volume. In previous chapters, tables have been used to determine properties of interest. How were all those properties determined? This chapter will provide those relationships that were used to determine properties over the wide range of temperatures and pressures encountered in the various devices analyzed in our study. In addition, the relationships developed in this chapter will allow us to perform more accurate calculations of properties than those used in previous chapters.

7.1 The Maxwell Relations

From Chapter 6 we developed the following differential relations for specific internal energy and specific enthalpy. Equations 6.15 and 6.16 are restated as

$$du = Tds - Pdv \tag{7.1}$$

$$dh = Tds + vdP \tag{7.2}$$

Two additional independent properties are now defined for a simple thermodynamic system. They are the *Helmholtz function a* and the *Gibbs function g*:

$$a = u - Ts \tag{7.3}$$

$$g = h - Ts \tag{7.4}$$

In differential form, these become, with help from Eqs. 7.1 and 7.2,

$$da = -Pdv - sdT \tag{7.5}$$

$$dg = vdP - sdT \tag{7.6}$$

Comment

The Helmholtz and Gibbs functions are independent properties defined in terms of other independent properties and are found to be useful in our study, similar to our use of enthalpy. They all have the units of energy and represent energy in certain situations.

From calculus the differential of a function z that depends on the two variables x and y is written as

$$dz = \left(\frac{\partial z}{\partial x}\right)_y dx + \left(\frac{\partial z}{\partial y}\right)_x dy \tag{7.7}$$

where the subscript y on the first partial derivative means that the variable y is held fixed while z is differentiated with respect to x; on the second partial derivative the variable x is held fixed while z is differentiated with respect to y. Equation (7.7) can be written in the form

$$dz = Mdx + Ndy \tag{7.8}$$

where $M = (\partial z/\partial x)_y$ and $N = (\partial z/\partial y)_x$. The partial derivatives of the functions M and N are

$$\begin{aligned}\left(\frac{\partial M}{\partial y}\right)_x &= \frac{\partial}{\partial y}\left(\frac{\partial z}{\partial x}\right)_y = \frac{\partial^2 z}{\partial y \partial x} = \frac{\partial^2 z}{\partial x \partial y}\\ \left(\frac{\partial N}{\partial x}\right)_y &= \frac{\partial}{\partial x}\left(\frac{\partial z}{\partial y}\right)_x = \frac{\partial^2 z}{\partial x \partial y} = \frac{\partial^2 z}{\partial y \partial x}\end{aligned} \tag{7.9}$$

If the differentiation can be interchanged, which is acceptable for well-behaved functions that do not have discontinuities or infinite slopes, observe that

$$\left(\frac{\partial M}{\partial y}\right)_x = \left(\frac{\partial N}{\partial x}\right)_y \tag{7.10}$$

Apply the above relations from calculus to Eqs. 7.1, 7.2, 7.5, and 7.6 and obtain the four *Maxwell relations*:

$$\left(\frac{\partial T}{\partial v}\right)_s = -\left(\frac{\partial P}{\partial s}\right)_v \tag{7.11}$$

$$\left(\frac{\partial T}{\partial P}\right)_s = \left(\frac{\partial v}{\partial s}\right)_P \tag{7.12}$$

$$\left(\frac{\partial P}{\partial T}\right)_v = \left(\frac{\partial s}{\partial v}\right)_T \tag{7.13}$$

$$\left(\frac{\partial v}{\partial T}\right)_P = -\left(\frac{\partial s}{\partial P}\right)_T \tag{7.14}$$

These four relations allow us to calculate changes in the property entropy that cannot be measured, knowing the changes in the properties T, P, and v that can be measured.[1] They are important relations in our study of thermodynamics; they allow the entropy to be listed in tables of properties such as the steam tables.

[1] The specific volume is considered measurable since we can measure volume and mass and then $v = V/m$.

Several other useful relations can be obtained from the differential forms of the four equations 7.1, 7.2, 7.5, and 7.6. Apply Eq. 7.7 directly to Eqs. 7.1, 7.2, 7.5, and 7.6 and observe that the following relations are true:

$$\left(\frac{\partial u}{\partial s}\right)_v = T \quad \text{and} \quad \left(\frac{\partial u}{\partial v}\right)_s = -P \tag{7.15}$$

$$\left(\frac{\partial h}{\partial s}\right)_P = T \quad \text{and} \quad \left(\frac{\partial h}{\partial P}\right)_s = v \tag{7.16}$$

$$\left(\frac{\partial a}{\partial v}\right)_T = -P \quad \text{and} \quad \left(\frac{\partial a}{\partial T}\right)_v = -s \tag{7.17}$$

$$\left(\frac{\partial g}{\partial P}\right)_T = v \quad \text{and} \quad \left(\frac{\partial g}{\partial T}\right)_P = -s \tag{7.18}$$

The first two sets, (7.15) and (7.16), can be used to relate certain properties in the tables. The Helmholtz and Gibbs functions are typically not listed in property tables.

Example 7.1 Verify a Maxwell relation

Superheated steam exists at 500 kPa and 500°C. Verify that Eq. 7.14 is indeed true using i) the values in Table C-3 and ii) the IRC Calculator.

Solution

i) The last of the relations is selected because temperature and pressure are held constant. In each of the other relations, either specific volume or entropy is held constant; this is a rather difficult task in tables where temperature and pressure are the entries. The derivatives are approximated (see Fig. 7.1), using the relatively large temperature and pressure differences in Steam Table C-3, to be

$$\left(\frac{\partial v}{\partial T}\right)_P \cong \left.\frac{\Delta v}{\Delta T}\right|_{P=0.5\text{ MPa}} = \frac{0.8041 - 0.6173}{600 - 400} = 0.000934\ \text{m}^3/\text{kg}\cdot\text{K}$$

$$-\left(\frac{\partial s}{\partial P}\right)_T \cong \left.\frac{\Delta s}{\Delta P}\right|_{T=500°\text{C}} = -\frac{8.0021 - 8.1913}{600 - 400} = 0.000946\ \text{m}^3/\text{kg}\cdot\text{K}$$

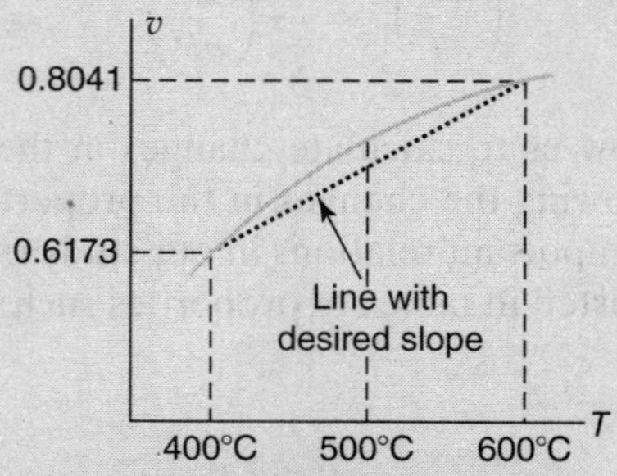

Figure 7.1

Note: The IRC Calculator could be used for the other relations to avoid complicated interpolations.

Comment

Central differences are used if at all possible, resulting in greater accuracy when compared to backward or forward differences:

Backward

$$\left.\frac{\Delta v}{\Delta T}\right|_{P=0.5\text{ MPa}} = \frac{0.711 - 0.617}{500 - 400} = 0.00094$$

Forward

$$\left.\frac{\Delta v}{\Delta T}\right|_{P=0.5\text{ MPa}} = \frac{0.804 - 0.711}{600 - 500} = 0.00093$$

If the curve had more curvature, the results using the three differences would be quite different from each other.

Observe that the accuracy is about 1%, showing the relation (7.14) to be satisfied, at least for engineers!

ii) Using the IRC Calculator, select 600°C and 400°C at 500 kPa and 600 kPa and 400 kPa at 500°C, which result in central differences. We find

$$\left(\frac{\partial v}{\partial T}\right)_P \cong \frac{\Delta v}{\Delta T}\bigg|_{P=0.5\text{ MPa}} = \frac{0.804 - 0.617}{600 - 400} = \underline{0.000935\ \text{m}^3/\text{kg}\cdot\text{K}}$$

$$-\left(\frac{\partial s}{\partial P}\right)_T \cong -\frac{\Delta s}{\Delta P}\bigg|_{T=500°\text{C}} = -\frac{8.00 - 8.19}{600 - 400} = \underline{0.00095\ \text{m}^3/\text{kg}\cdot\text{K}}$$

The approximations are essentially the same using the tables or the IRC Calculator. The values provide acceptable proof that $(\partial v/\partial T)_P = -(\partial s/\partial P)_T$.

Verify $(\partial h/\partial s)_P = T$ of Eq. 7.16 — Example 7.2

Verify, using steam at 500 kPa and 500°C, that $(\partial h/\partial s)_P = T$ is indeed true.

Solution

The derivative is approximated, using the values in Table C-3, to be

$$\left(\frac{\partial h}{\partial s}\right)_P \cong \frac{\Delta h}{\Delta s}\bigg|_{P=500\text{ kPa}} = \frac{3701.7 - 3271.9}{8.3522 - 7.7938} = 770\text{ K or }\underline{497°\text{C}}$$

where the temperatures were at 400°C and 600°C. It isn't closer because the properties result in relatively large differences, although 497°C is acceptably close to 500°C.

☑ You have completed Learning Outcome (1)

7.2 The Clapeyron Equation

The Clapeyron equation is developed from one of Maxwell's relations and allows us to determine the enthalpy of vaporization h_{fg} from the measurable properties P, T, and v. The Maxwell relation of interest is Eq. 7.13:

$$\left(\frac{\partial P}{\partial T}\right)_v = \left(\frac{\partial s}{\partial v}\right)_T \tag{7.19}$$

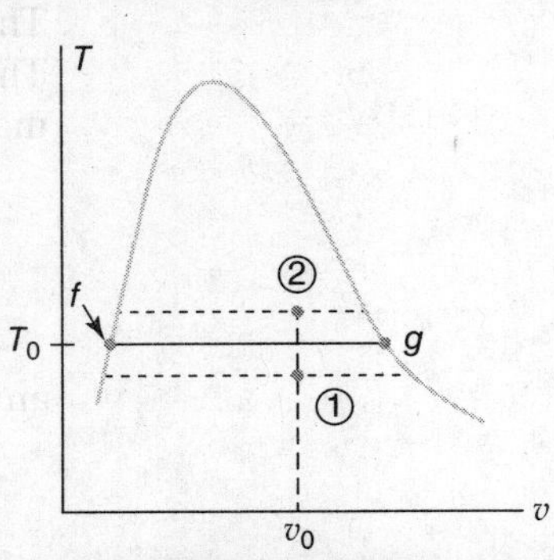

Figure 7.2
A T-v diagram showing a phase change.

Consider the constant-temperature process at $T = T_0$ (or constant pressure with $P = P_0$) of a saturated liquid vaporizing to a saturated vapor, as shown in Fig. 7.2. Since in the wet region, the pressure depends solely on the temperature, that is, $P = P(T)$, the partial derivative can be written as an ordinary derivative, so that Eq. 7.19 takes the form

$$\left.\frac{dP}{dT}\right|_{v=v_0} = \left(\frac{\partial s}{\partial v}\right)_T = \left.\frac{\Delta s}{\Delta v}\right|_{T=T_0} = \frac{s_g - s_f}{v_g - v_f} = \frac{s_{fg}}{v_{fg}} \tag{7.20}$$

If we integrate Eq. 7.2 from the saturated liquid state to the saturated vapor state, the equation can be written as

$$\int_f^g dh = \int_f^g T_0 ds + \int_f^g v\,d\!\!\!/P \tag{7.21}$$

The pressure integral dropped out since the pressure is constant from state f to state g at temperature T_0. Perform the integration and find

$$h_g - h_f = T_0(s_g - s_f) \quad \text{or} \quad h_{fg} = T_0 s_{fg} \tag{7.22}$$

Clapeyron equation: Allows the calculation of h_{fg} in terms of P, T, and v. See Example 7.3.

Substitute for s_{fg} in Eq. 7.20 and arrive at the *Clapeyron equation*:

$$\left.\frac{dP}{dT}\right|_{v=v_0} = \frac{h_{fg}}{T_0 v_{fg}} \tag{7.23}$$

which allows us to calculate h_{fg} by using only P, T, and v values. The derivative $dP/dT|_v$ can be approximated using values from tables that include the saturated liquid and vapor properties, similar to Table C-1 or C-2 in the Appendix.

For relatively low pressures, the tables show that $v_g \gg v_f$. Also, a good approximation is to treat the saturated vapor as an ideal gas, which is acceptable if the pressure is relatively low. Hence,

Calculation to show that the steam can be treated as an ideal gas: At 200 kPa we see that

$v = RT/P$
$= 0.908\ \text{m}^3/\text{kg}$

This is within 2.5% of $v_g = 0.886\ \text{m}^3/\text{kg}$ from Table C-2, which is acceptable accuracy.

$$v_{fg} \cong v_g = \frac{RT}{P} \tag{7.24}$$

When we use this approximation in the Clapeyron equation (7.23), the *Clapeyron–Clausius equation* results. It is

$$\left.\frac{dP}{dT}\right|_v = \frac{h_{fg}P}{RT^2} \tag{7.25}$$

The subscript "0" has been dropped on the temperature since it is not necessary. This equation can be used for any phase-change process, vaporization, or sublimation (replace h_{fg} with h_{ig}), providing the pressure is relatively low.

If Eq. 7.25 is written as

$$\left.\frac{dP}{P}\right|_{\text{sat}} = \left.\frac{h_{fg}dT}{R\,T^2}\right|_{\text{sat}} \tag{7.26}$$

an integration can be performed between states 1 and 2 (see Fig. 7.2), resulting in

$$\ln\frac{P_2}{P_1} = \frac{h_{fg}}{R}\left(\frac{1}{T_1} - \frac{1}{T_2}\right) \tag{7.27}$$

where h_{fg} was assumed constant at the average value between state 1 and state 2. Equation (7.27) can be used to provide the saturation temperature or pressure below the initial entry in a saturation table. See Example 7.4.

Find h_{fg} using the Clapeyron equation — Example 7.3

Assume that tables exist giving only properties involving P, T, and v. Using the Clapeyron equation, determine h_{fg} and then find s_{fg} for steam at 240°C.

Solution

The enthalpy h_{fg} is related to properties involving P, T, and v by Eq. 7.23. It provides, using properties at state 1 and state 2 as displayed in Fig. 7.2,

$$h_{fg} = T_0 v_{fg} \left.\frac{dP}{dT}\right|_v \cong T_0 v_{fg} \frac{P_2 - P_1}{T_2 - T_1}$$

$$= 513\text{ K} \times (0.05977 - 0.001229)\frac{\text{m}^3}{\text{kg}} \frac{(3973 - 2795)\text{ kN/m}^2}{(523 - 503)\text{ K}}$$

$$= \underline{1769\text{ kJ/kg}}$$

where central differences were used to determine the derivative. Equation (7.22) is then used to find

$$s_{fg} = \frac{h_{fg}}{T_0} = \frac{1769}{240 + 273} = \underline{3.45\text{ kJ/kg}\cdot\text{K}}$$

These compare favorably with $h_{fg} = 1766$ kJ/kg and $s_{fg} = 3.44$ kJ/kg·K from Table C-1 at 240°C. The values are within 0.3% of the tabulated values.

Find a saturation temperature — Example 7.4

Assume that the initial pressure in Table C-2 was 4 kPa but the saturation temperature at 1 kPa is desired. Use the integrated Clausius–Clapeyron equation (7.27) and approximate the saturation temperature of water at 1 kPa.

Solution

With the assumption that the saturated water vapor can be treated as an ideal gas, Eq. 7.27 requires that the average h_{fg} be known between 4 kPa and 1 kPa. It is estimated from Table C-2 at 0.0025 MPa to be $h_{fg} = 2452$ kJ/kg. Equation (7.27) then provides

$$\ln\frac{P_2}{P_1} = \frac{h_{fg}}{R}\left(\frac{1}{T_1} - \frac{1}{T_2}\right)$$

$$\ln\frac{1}{4} = \frac{2452\text{ kJ/kg}}{0.462\text{ kJ/kg}\cdot\text{K}}\left(\frac{1}{302\text{ K}} - \frac{1}{T_2}\right) \qquad \therefore T_2 = 280.1\text{ K or } \underline{7.1°\text{C}}$$

This value is quite close to the value of 7.0°C listed in Table C-2.

☑ You have completed Learning Outcome (2)

7.3 Relationships for Internal Energy, Enthalpy, Entropy, and Specific Heats

In this section thermodynamic relationships will be derived that will be used in the following two sections to calculate changes in internal energy, enthalpy, and entropy for nonideal gases.

Let the internal energy of a gas be selected as a function of temperature and specific volume, that is, $u = (T, v)$, as in Section 2.6. We can write the following expression for du:

$$du = \left(\frac{\partial u}{\partial T}\right)_v dT + \left(\frac{\partial u}{\partial v}\right)_T dv$$

$$= C_v dT + \left(\frac{\partial u}{\partial v}\right)_T dv \tag{7.28}$$

where the definition of C_v comes from Eq. 2.22. The differential form of the first law, expressed as Eq. 7.1, also relates du to thermodynamic properties. It is

$$du = Tds - Pdv \tag{7.29}$$

Consider $s = s(T, v)$, so that Eq. 7.29 can be written as

$$du = T\left[\left(\frac{\partial s}{\partial T}\right)_v dT + \left(\frac{\partial s}{\partial v}\right)_T dv\right] - Pdv$$

$$= T\left(\frac{\partial s}{\partial T}\right)_v dT + \left[T\left(\frac{\partial s}{\partial v}\right)_T - P\right] dv \tag{7.30}$$

We now have two expressions for du in Eqs. 7.28 and 7.30 which must be equal. Equating the coefficients of dT and of dv results in

$$C_v = T\left(\frac{\partial s}{\partial T}\right)_v \tag{7.31}$$

$$\left(\frac{\partial u}{\partial v}\right)_T = T\left(\frac{\partial s}{\partial v}\right)_T - P = T\left(\frac{\partial P}{\partial T}\right)_v - P \tag{7.32}$$

where the Maxwell relation (7.13) was used. Now, du can be related to P, T, v, and C_v by substituting Eq. 7.32 into Eq. 7.30:

$$du = C_v dT + \left[T\left(\frac{\partial P}{\partial T}\right)_v - P\right] dv \tag{7.33}$$

If an equation of state is used that relates P, T, and v and C_v is known, Eq. 7.33 can be integrated between two states to provide $u_2 - u_1$ for gases that do not obey the ideal gas law.

Let's find a similar relationship for enthalpy. Select $h = h(T, P)$, as was done in Section 2.6. The expression for dh is then written as

$$dh = \left(\frac{\partial h}{\partial T}\right)_P dT + \left(\frac{\partial h}{\partial P}\right)_T dP$$

$$= C_p dT + \left(\frac{\partial h}{\partial P}\right)_T dP \tag{7.34}$$

where the definition of C_p comes from Eq. 2.26. The differential form of the first law, expressed as Eq. 7.2, also relates dh to thermodynamic properties. It is

$$dh = Tds + vdP \tag{7.35}$$

Consider $s = s(T, P)$ so that Eq. 7.35 can be written as

$$dh = T\left[\left(\frac{\partial s}{\partial T}\right)_P dT + \left(\frac{\partial s}{\partial P}\right)_T dP\right] + vdP$$

$$= T\left(\frac{\partial s}{\partial T}\right)_P dT + \left[T\left(\frac{\partial s}{\partial P}\right)_T + v\right]dP \tag{7.36}$$

Note: If $s = s(T, P)$, the differential ds is written as:

$$ds = \left(\frac{\partial s}{\partial T}\right)_P dT + \left(\frac{\partial s}{\partial P}\right)_T dP$$

We now have two expressions for dh expressed in Eqs. 7.34 and 7.36. Equating the coefficients of dT and dP in those two equations results in

$$C_p = T\left(\frac{\partial s}{\partial T}\right)_P \tag{7.37}$$

$$\left(\frac{\partial h}{\partial P}\right)_T = T\left(\frac{\partial s}{\partial P}\right)_T + v = T\left(\frac{\partial v}{\partial T}\right)_P + v \tag{7.38}$$

where the Maxwell relation (7.14) was used. Now, dh can be related to P, T, v, and C_p by substituting Eq. 7.38 into Eq. 7.36:

$$dh = C_p dT + \left[v - T\left(\frac{\partial v}{\partial T}\right)_P\right]dP \tag{7.39}$$

If an equation of state is used that relates P, T, and v and C_p is known, Eq. 7.39 can be integrated between two states to provide $h_2 - h_1$ for gases that do not obey the ideal-gas law. That is quite a challenge and is the objective of Section 7.5.

Comment

The van der Waals equation of state, for example, could be used.

It is not necessary to integrate both Eq. 7.33 and Eq. 7.39 since if either Δu or Δh is found, the other can be determined using

$$\Delta h = \Delta u + P_2 v_2 - P_1 v_1 \tag{7.40}$$

An expression for ds can be found using $s = s(T, v)$ so that

$$ds = \left(\frac{\partial s}{\partial T}\right)_v dT + \left(\frac{\partial s}{\partial v}\right)_T dv$$

$$= \frac{C_v}{T}dT + \left(\frac{\partial P}{\partial T}\right)_v dv \tag{7.41}$$

where Eq. 7.31 was used for the first partial derivative and the Maxwell relation (7.13) was used for the second partial derivative. Alternatively, an expression for ds can be found considering $s = s(T, P)$, resulting in

$$ds = \left(\frac{\partial s}{\partial T}\right)_P dT + \left(\frac{\partial s}{\partial P}\right)_T dP$$

$$= \frac{C_p}{T}dT - \left(\frac{\partial v}{\partial T}\right)_P dP \tag{7.42}$$

where Eq. 7.37 was used for the first partial derivative and the Maxwell relation (7.14) was used for the second partial derivative. The two expressions for ds, Eqs. 7.41 and 7.42, will be used in Section 7.5 to find the entropy changes in real gases, gases that experience large changes in pressure and/or temperature for which neither the ideal-gas laws nor the ideal-gas tables in Appendix F give acceptable results.

Now, if C_p and C_v can be expressed in terms of P, v, and T, the expressions (7.39) and (7.42) for Δh and Δs will have been written in terms of only the measureable quantities P, v, and T. To do this, we start with the exact differential expressed by Eq. 7.42 and observe, referring to Eq. 7.8, that

$$M = \frac{C_p}{T} \quad \text{and} \quad N = -\left(\frac{\partial v}{\partial T}\right)_P \tag{7.43}$$

For an exact differential:

If $ds = MdT + NdP$

then $\left(\frac{\partial M}{\partial P}\right)_T = \left(\frac{\partial N}{\partial T}\right)_P$

There follows

$$\left[\frac{\partial}{\partial P}\left(\frac{C_p}{T}\right)\right]_T = -\left[\frac{\partial}{\partial T}\left(\frac{\partial v}{\partial T}\right)_P\right]_P \tag{7.44}$$

Or, equivalently, recognizing that $\partial T/\partial P = 0$,

$$\left(\frac{\partial C_p}{\partial P}\right)_T = -T\left(\frac{\partial^2 v}{\partial T^2}\right)_P \tag{7.45}$$

If the same steps that led to Eq. 7.45 are followed but beginning with Eq. 7.41, the relationship that would result would be

$$\left(\frac{\partial C_v}{\partial v}\right)_T = T\left(\frac{\partial^2 P}{\partial T^2}\right)_v \tag{7.46}$$

A third relationship is found by equating Eqs. 7.41 and 7.42:

$$\frac{C_p}{T}dT - \left(\frac{\partial v}{\partial T}\right)_P dP = \frac{C_v}{T}dT + \left(\frac{\partial P}{\partial T}\right)_v dv \tag{7.47}$$

or, rearranging,

$$dT = \frac{T}{C_p - C_v}\left(\frac{\partial v}{\partial T}\right)_P dP + \frac{T}{C_p - C_v}\left(\frac{\partial P}{\partial T}\right)_v dv \tag{7.48}$$

From calculus:

$\left(\frac{\partial p}{\partial T}\right)_v\left(\frac{\partial T}{\partial v}\right)_p\left(\frac{\partial v}{\partial P}\right)_T = -1$

and

$\left(\frac{\partial T}{\partial P}\right)_v = \frac{1}{(\partial P/\partial T)_v}$

Recognizing that $T = T(P, v)$, that is, $dT = (\partial T/\partial P)_v dP + (\partial T/\partial v)_P dv$, we find that

$$C_p - C_v = T\left(\frac{\partial P}{\partial T}\right)_v\left(\frac{\partial v}{\partial T}\right)_P$$

$$= -T\left(\frac{\partial P}{\partial v}\right)_T\left(\frac{\partial v}{\partial T}\right)_P^2 \tag{7.49}$$

Note: The volume of a solid or a liquid does not change significantly with temperature.

Note: Volume always decreases when pressure increases.

For a given relationship among T, P, and v, the expressions for du, dh, and ds can be integrated to provide $u_2 - u_1$, $h_2 - h_1$, and $s_2 - s_1$ for a nonideal gas; the expressions simplify for an ideal gas to the expressions found in Chapter 2.

Equation 7.49 can be used to make four important observations:

- $C_p = C_v$ for an incompressible substance (v = constant).
- $C_p \cong C_v$ for a liquid or a solid since $(\partial v/\partial T)_P$ is very small
- $C_p \rightarrow C_v$ as $T \rightarrow 0$ (absolute zero)
- $C_p \geq C_v$ since $(\partial v/\partial P)_T < 0$

Finally, the reason for the second expression for the difference in the specific heats in Eq. 7.49 was so it can be written in terms of the bulk modulus[2] B and the volume expansivity[3] β:

$$C_p - C_v = vTB\beta^2 \tag{7.50}$$

Handbooks of material properties contain B and β.

Definitions:

$$B = -v\left(\frac{\partial P}{\partial v}\right)_T$$

$$\beta = \frac{1}{v}\left(\frac{\partial v}{\partial T}\right)_P$$

Example 7.5 Find the enthalpy and entropy changes in air using the van der Waals equation of state

Air undergoes an isothermal process at 27°C from 1000 kPa to 8000 kPa. Calculate the enthalpy and entropy changes using the van der Waals equation of state (see Eq. 2.17). Assume that the specific heat C_v depends on temperature only.

Solution

It would be very difficult to integrate Eq. 7.39 directly for h since it involves an explicit form of v that is not available with the van der Waals equation of state $P = RT/(v - b) - a/v^2$. Consequently, we will integrate for u since the van der Waals equation involves an explicit form for P. We have, using Eq. 7.33,

$$u_2 - u_1 = \int_{T_1}^{T_2} C_v dT + \int_{v_1}^{v_2}\left[T\left(\frac{\partial P}{\partial T}\right)_v - P\right]dv$$

Suggestion: Try Eq. 7.39 and it will be quickly apparent that it is not wise to use that equation to find Δh directly!

The first integral is zero since $T_2 = T_1$. The second integral is simplified using the van der Waals equation (2.17), so that

$$u_2 - u_1 = \int_{v_1}^{v_2}\left[\frac{RT}{v-b} - \frac{RT}{v-b} + \frac{a}{v^2}\right]dv = 0.163\left(\frac{1}{v_1} - \frac{1}{v_2}\right)$$

Units: The units on a are kPa·m^6/kg^2. So, the units of $(a\,dv/v^2)$ are kJ/kg.

where $a = 0.163$ is found in Table B-10. The values of v_1 and v_2 are found using the van der Waals equation if the pressure and temperature are known. A trial-and-error procedure is used:

$$1000 = \frac{0.287 \times 300}{v_1 - 0.00127} - \frac{0.163}{v_1^2} \qquad \therefore v_1 = 0.0855 \text{ m}^3/\text{kg}$$

$$8000 = \frac{0.287 \times 300}{v_2 - 0.00127} - \frac{0.163}{v_2^2} \qquad \therefore v_2 = 0.01032 \text{ m}^3/\text{kg}$$

The internal energy change and consequently the enthalpy change are then

$$u_2 - u_1 = 0.163\left(\frac{1}{0.0855} - \frac{1}{0.01032}\right) = -13.9 \text{ kJ/kg}$$

$$h_2 - h_1 = u_2 - u_1 + P_2v_2 - P_1v_1$$

$$= -13.9 + 8000 \times 0.01032 - 1000 \times 0.0855 = \underline{-16.8 \text{ kJ/kg}}$$

(Continued)

[2] B is a measure of the change in specific volume with pressure at constant temperature.
[3] β is a measure of the change in specific volume with temperature at constant pressure.

Example 7.5 (*Continued*)

The entropy change is found using Eq. 7.41. Following the steps for Δu, we find

$$s_2 - s_1 = \cancel{\int_{T_1}^{T_2} \frac{C_v}{T} dT} + \int_{v_1}^{v_2} \left(\frac{\partial P}{\partial T}\right)_v dv = \int_{v_1}^{v_2} \frac{R}{v - b} dv = R \ln \frac{v_2 - b}{v_1 - b}$$

which results in

$$s_2 - s_1 = 0.287 \ln \frac{0.01032 - 0.00127}{0.0855 - 0.00127} = \underline{-0.640 \text{ kJ/kg·K}}$$

Units: The units on $s_2 - s_1$ are the same as on R.

For an ideal gas, $s_2^o = s_1^o$, so the entropy change would be

$$s_2 - s_1 = \cancel{s_2^o - s_1^o} - R \ln \frac{P_2}{P_1}$$

$$= -0.287 \ln \frac{8000}{1000} = -0.597 \text{ kJ/kg·K}$$

an error of about 6%.

If the air was treated as an ideal gas, the internal energy change and the enthalpy change would be zero for this isothermal process since for an ideal gas $u = u(T)$ and $h = h(T)$. The changes of $\Delta u = -13.9$ kJ/kg and $\Delta h = -16.8$ kJ/kg are relatively small for the large pressure change of 7000 kPa. We expected the changes to be negative since the superheat steam table C-3 shows negative changes for Δu, Δh, and Δs for a pressure increase if the temperature is held constant.

Comment

These changes would be caused by a small temperature change of about 6°C. The assumption of an ideal gas for air over large pressure and temperatures differences is quite acceptable.

Example 7.6 Internal energy of an ideal gas

Show that the internal energy of an ideal gas is a function only of the temperature, as has been assumed in previous chapters, given that the specific heats depend on temperature only.

Solution

The internal energy is related to T, P, and v by Eq. 7.33, which is

$$du = C_v dT + \left[T\left(\frac{\partial P}{\partial T}\right)_v - P\right] dv$$

For an ideal gas we know that $Pv = RT$, so, letting v = constant, we see that

$$T\left(\frac{\partial P}{\partial T}\right)_v = T\frac{R}{v} = P$$

making the quantity in brackets above go to zero. The specific heat $C_v = C_v(T)$ (see Eq. 2.30 and observe that $C_v = C_p - R$). Hence, we conclude that *the internal energy depends only on temperature.* The specific heats do depend somewhat on pressure for very high pressures.

Entropy change of a solid — Example 7.7

A 100-kg block of copper experiences a pressure change from 100 kPa to 100 MPa while the temperature remains constant. Determine its change in entropy. Use $\beta = 5 \times 10^{-5}\ \text{K}^{-1}$ and $\rho = 8770\ \text{kg/m}^3$.

Solution

The differential entropy change is given by Eq. 7.42 and the definition of the volume expansivity β to be

$$ds = \frac{C_P}{T}dT - \left(\frac{\partial v}{\partial T}\right)_P dP = -\beta v dP$$

An integration, assuming constant v and β, provides

$$\begin{aligned}\Delta S &= -m\beta v(P_2 - P_1)\\ &= -100\ \text{kg} \times (5 \times 10^{-5})\frac{1}{\text{K}} \times \frac{1}{8770\ \text{kg/m}^3}(100 - 0.1) \times 10^6\frac{\text{N}}{\text{m}^2}\\ &= \underline{-57.0\ \text{J/K}}\end{aligned}$$

If the copper was assumed to be incompressible ($dv = 0$ and then $\beta = 0$), Eq. 7.41 would indicate $\Delta S = 0$ since $dT = 0$ for this example. The entropy change is due to the small change in the volume of the copper.

Remember: N·m = J

☑ You have completed Learning Outcome (3)

7.4 The Joule–Thomson Coefficient

For a throttle, like an orifice or a valve, the enthalpy remains constant as the fluid flows through the throttle. Setting $dh = 0$ in Eq. 7.39 then provides

$$0 = C_p dT + \left[v - T\left(\frac{\partial v}{\partial T}\right)_P\right]dP \tag{7.51}$$

Divide this equation by dP and solve for dT/dP to obtain

$$\left.\frac{dT}{dP}\right|_h = \frac{1}{C_P}\left[T\left(\frac{\partial v}{\partial T}\right)_P - v\right] \tag{7.52}$$

We define the *Joule–Thomson coefficient* μ_J as the rate of change of temperature with respect to pressure for an *isenthalpic* (constant enthalpy) process, that is,

$$\mu_J = \left.\frac{dT}{dP}\right|_h = \frac{1}{C_P}\left[T\left(\frac{\partial v}{\partial T}\right)_P - v\right] \tag{7.53}$$

Isenthalpic: A constant enthalpy process. It's also called an isoenthalpic process.

This coefficient can be used to approximate the temperature change across a throttle for various refrigerants.

For an ideal gas, enthalpy is a function of temperature, and thus temperature is a function of enthalpy. For an isenthalpic process, as described in Eq. 7.53, temperature will also remain constant, giving a Joule–Thomson coefficient of zero for an ideal gas. For nonideal gases, Eq. 7.53 can also provide an expression for C_p in terms of measurable properties, T, P, and v.

Example 7.8 Joule–Thomson coefficients

Estimate the Joule–Thomson coefficient for R134a at 500 kPa and $x = 0$ (where $h_f = 73.4$ kJ/kg). Then estimate the exiting temperature from the throttle of Fig. 7.3 if the pressure drops from 1 MPa to 100 kPa and the entering temperature is 39°C. The IRC Calculator is used in this example (the properties are slightly different from those in Appendix D).

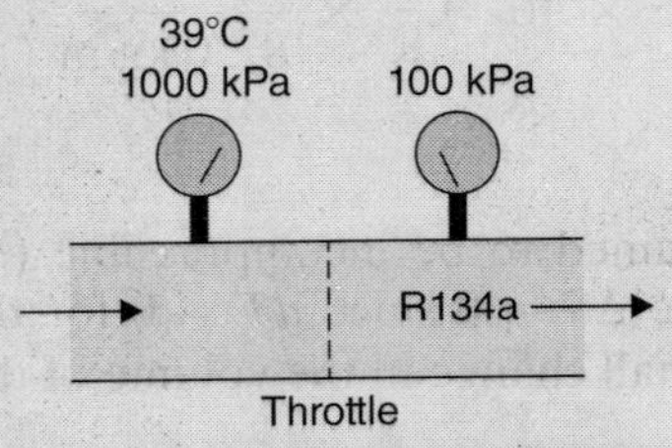

Figure 7.3

Comment

The IRC Calculator gives slightly different values from those in Appendix D.

Solution

Use forward differences by using a pressure at 500 kPa and 400 kPa (the refrigerant enters the throttle at $x = 0$, so a lower pressure is used in the quality region). The enthalpy is held constant at 73 400 J/kg as the refrigerant passes through the throttle. The IRC Calculator at 73 400 J/kg and 500 kPa gives $T =$ 15.7°C. At $h =$ 73 400 J/kg and 400 kPa, we find $T =$ 8.93°C. Using the definition of the Joule–Thomson coefficient results in

$$\mu_J \cong \left.\frac{\Delta T}{\Delta P}\right|_h = \left.\frac{T_2 - T_1}{P_2 - P_1}\right|_h = \frac{8.93 - 15.7}{400 - 500} = \underline{0.068°\text{C/kPa}}$$

The finite difference does not allow for more accuracy in the answer. More accuracy, though, is not demanded here; we are estimating the values.

Note that as the pressure decreases across a throttle for R134a, the temperature experiences a significant drop, which is the essential feature of a refrigerant. So, using the μ_J calculated, we find

$$\mu_J = 0.068 = \frac{T_2 - T_1}{P_2 - P_1} = \frac{T_2 - 39}{100 - 1000} \qquad \therefore T_2 = \underline{-22°\text{C}}$$

This exiting temperature is not accurate since μ_J varies significantly from 100 kPa to 1000 kPa while holding h fixed. But it does give a rough estimate since μ_J was calculated at an average pressure of 500 kPa. It is obvious why R134a is a good refrigerant. By simply passing it through a throttle, the temperature dropped about 60°C. If an ideal gas is passed through a throttle, the temperature does not change since enthalpy depends only on temperature for an ideal gas. So, if enthalpy remains constant, so does temperature for an ideal gas, but not for a refrigerant.

☑ **You have completed Learning Outcome (4)**

7.5 Real-Gas Effects

Gases at extreme conditions cannot always be assumed to be an ideal gas. Often, a low temperature combined with a high pressure results in nonideal-gas behavior (as Problems 7.40(b) and 7.43(c) would illustrate). Such temperature and pressure combinations are not usually encountered in applications of interest in an introductory course. This section provides the equations needed to make calculations given such extreme conditions.

The properties of T, P, and v were found in Chapter 2 using the Z-factor or an equation of state, such as the van der Waals equation. The differential changes in u, h, and s in nonideal gases for which an equation of state is not known can be determined using relations developed in Section 7.3. Expressions for Δh, Δu, and Δs will now be developed.

The relation for the differential enthalpy change in a gas was presented in Eq. 7.34. It is reproduced as

$$dh = C_p dT + \left[v - T\left(\frac{\partial v}{\partial T}\right)_P\right] dP \tag{7.54}$$

This equation can be integrated from state 1 to state 2 to give the change in enthalpy as

$$h_2 - h_1 = \int_{T_1}^{T_2} C_p dT - \int_{P_1}^{P_2} \left[v - T\left(\frac{\partial v}{\partial T}\right)_P\right] dP \tag{7.55}$$

To tackle these integrations directly is a difficult if not impossible task. So, we identify a path that results in more integrations but those that are possible to perform without too much trouble. The change in enthalpy is independent of the path between states 1 and 2 so we select a different path with three segments, as shown in Fig. 7.4 by the dashed lines. This alternate path involves two new states that exist at a very low pressure where the ideal-gas law can be applied with acceptable accuracy. The states will be designated as 1^* and 2^* so that $T_{1^*} = T_1$ and $T_{2^*} = T_2$. To move from 1 to 2, follow the path from 1 to 1^*, then from 1^* to 2^*, and finally from 2^* to 2. The enthalpy change can then be written as

$$h_2 - h_1 = (h_2 - h_2^*) + (h_2^* - h_1^*) + (h_1^* - h_1) \tag{7.56}$$

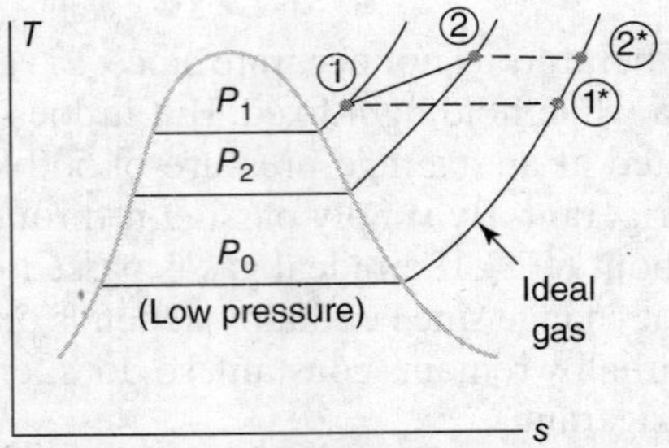

Figure 7.4

The path from state 1 to state 2 replaced by the alternate path $1 \rightarrow 1^* \rightarrow 2^* \rightarrow 2$.

The enthalpy change $(h_2^* - h_1^*)$ is for an ideal gas and is simply

$$h_2^* - h_1^* = \int_{T_1}^{T_2} C_p dT \tag{7.57}$$

since, for an ideal gas, the quantity in brackets in Eq. 7.55 is zero. The specific heat is assumed to be constant or the ideal-gas Tables F are used.

The other two enthalpy changes are at constant temperatures T_1 and T_2, respectively. An isothermal enthalpy change from some general state to a state of an ideal gas (for example, from state 2 to state 2* in Fig. 7.4) is given by Eq. 7.55 to be

Recall: The critical temperature T_{cr} and pressure P_{cr} can be found in Table B-3.

Hint: Recognize that, using $v = ZRT/P$,

$$T\frac{\partial v}{\partial T}\bigg|_P = \frac{RT^2}{P}\frac{\partial Z}{\partial T}\bigg|_P + \frac{TRZ}{P}$$

$$h - h^* = \int_{P^*}^{P} \left[v - T\left(\frac{\partial v}{\partial T}\right)_P \right] dP \tag{7.58}$$

since the first integral is zero if $T =$ const. To account for nonideal-gas behavior, the compressibility factor Z, defined by Eq. 2.15, is introduced along with the reduced pressure $P_R = P/P_{cr}$ and reduced temperature $T_R = T/T_{cr}$. The above enthalpy difference can, after some manipulation, be written as

$$\frac{h - h^*}{T_{cr}} = RT_R^2 \int_{P_R^*}^{P_R} \left(\frac{\partial Z}{\partial T_R}\right)_P \frac{dP_R}{P_R} \tag{7.59}$$

If both sides of this equation are divided by the molar mass M, Eq. 7.59 takes the form

$$\frac{\bar{h} - \bar{h}^*}{T_{cr}} = R_u T_R^2 \int_{P_R^*}^{P_R} \left(\frac{\partial Z}{\partial T_R}\right)_P \frac{dP_R}{P_R} \tag{7.60}$$

The difference $(\bar{h}^* - \bar{h})/T_{cr}$ is the *enthalpy departure* and can be obtained using Appendix I. Using Figure I-1 and Eq. 7.56, we can obtain the enthalpy difference in a real gas. An example will illustrate.

The internal energy change is found using the definition of enthalpy to give

$$u_2 - u_1 = h_2 - h_1 + P_2 v_2 - P_1 v_1$$
$$= h_2 - h_1 - R(Z_2 T_2 - Z_1 T_1) \tag{7.61}$$

The entropy change for a gas is found by integrating the expression provided in Eq. 7.42, giving

$$s_2 - s_1 = \int_{T_1}^{T_2} \frac{C_p}{T} dT - \int_{P_1}^{P_2} \left(\frac{\partial v}{\partial T}\right)_P dP \tag{7.62}$$

An alternate path is taken as was done in Fig. 7.4. The difference $\bar{s}^* - \bar{s}$ is the *entropy departure* and is presented in Appendix J. The entropy change between state 1 and state 2 is then found from

Entropy departure: The difference between the ideal-gas entropy at a sufficiently low pressure and the entropy at that same temperature.

$$s_2 - s_1 = (s_2 - s_2^*) + (s_2^* - s_1^*) - (s_1 - s_1^*) \tag{7.63}$$

There is one difference in this calculation when compared to the enthalpy difference calculation: The ideal-gas entropy change $s_2^* - s_1^*$ represents the ideal-gas difference between states 1^* and 2^*, so the pressure is not constant. The following example illustrates.

Property changes in a real gas — Example 7.9

Carbon dioxide undergoes a process from state 1 at 320 K and 200 kPa to state 2 at 600 K and 8000 kPa. Estimate the change in enthalpy, internal energy, and entropy, i) assuming the carbon dioxide is an ideal gas, using constant specific heats from Table B-2, ii) using the ideal-gas tables in Appendix F, and iii) using the equations of this section.

Solution

i) The ideal-gas equations assuming constant specific heats result in

$$\Delta h = h_2 - h_1 = C_p \Delta T = 0.842 \times (600 - 320) = \underline{236 \text{ kJ/kg}}$$

$$\Delta u = C_v \Delta T = 0.653 \times (600 - 320) = \underline{183 \text{ kJ/kg}}$$

$$\Delta s = s_2 - s_1 = C_p \ln \frac{T_2}{T_1} - R \ln \frac{P_2}{P_1}$$

$$= 0.842 \ln \frac{600}{320} - 0.1889 \ln \frac{8000}{200} = \underline{-0.168 \text{ kJ/kg·K}}$$

(Continued)

Example 7.9 (*Continued*)

ii) The ideal-gas Table F-4, which allows for variable specific heat, provides

$$\Delta h = \frac{\bar{h}_2 - \bar{h}_1}{M} = \frac{22\,280 - 10\,186}{44} = \underline{275\ \text{kJ/kg}}$$

$$\Delta u = \frac{\bar{u}_2 - \bar{u}_1}{M} = \frac{17\,291 - 7526}{44} = \underline{222\ \text{kJ/kg}}$$

$$\Delta s = \frac{\bar{s}_2^{\circ} - \bar{s}_1^{\circ}}{44} - R \ln \frac{P_2}{P_1}$$

$$= \frac{243.2 - 216.4}{44} - 0.1889 \ln \frac{8000}{200} = \underline{-0.0877\ \text{kJ/kg}\cdot\text{K}}$$

iii) Using the enthalpy departure chart of Fig. I-1 in the appendix, we need

Remember: $\bar{h} = Mh$

$$P_{R1} = \frac{P_1}{P_{cr}} = \frac{0.2}{7.39} = 0.027, \qquad T_{R1} = \frac{T_1}{T_{cr}} = \frac{320}{304.2} = 1.05$$

$$P_{R2} = \frac{P_2}{P_{cr}} = \frac{8}{7.39} = 1.08, \qquad T_{R2} = \frac{T_2}{T_{cr}} = \frac{600}{304.2} = 1.97$$

Accuracy: Two digits is all that can be obtained from the charts. Larger charts would allow for three digits.

Using values from the chart, the enthalpy differences are

$$\frac{\bar{h}_1^* - \bar{h}_1}{T_{cr}} = 0.05 \quad \therefore h_1^* - h_1 = \frac{0.05 \times 304.2}{44} = 0.35\ \text{kJ/kg}$$

$$\frac{\bar{h}_2^* - \bar{h}_2}{T_{cr}} = 2.4 \quad \therefore h_2^* - h_2 = \frac{2.4 \times 304.2}{44} = 17\ \text{kJ/kg}$$

Note: The terms $(h_2^* - h_1^*)$ and $(s_2^* - s_1^*)$ are for an ideal gas. Those from Part (ii) are more accurate than those in Part (i).

The property differences between state 1 and state 2 are found, using Eqs. 7.56, 7.61, and 7.63, to be

$$h_2 - h_1 = (h_2 - h_2^*) + (h_2^* - h_1^*) - (h_1 - h_1^*)$$

$$= -17 + 275 + 0.35 = \underline{258\ \text{kJ/kg}}$$

$$u_2 - u_1 = h_2 - h_1 - R(Z_2T_2 - Z_1T_1)$$

$$= 258 - 0.1889 \times (0.96 \times 600 - 0.99 \times 320) = \underline{209\ \text{kJ/kg}}$$

$$s_2 - s_1 = (s_2 - s_2^*) + (s_2^* - s_1^*) - (s_1 - s_1^*)$$

$$= -\frac{1}{44} - 0.0877 - \left(-\frac{1}{44}\right) = \underline{-0.0877\ \text{kJ/kg}\cdot\text{K}}$$

where the entropy departure $(\bar{s} - \bar{s}^*) = M(s - s^*)$ was found using Table J-1. The Z-values used to find $u_2 - u_1$ are found in Appendix H.

The high pressure (above about 10 MPa) at state 2 would make the ideal-gas calculations unacceptable as approximations. If the pressure at state 2 was less than about 4 MPa, which it usually is, the ideal-gas Table F would provide acceptable approximations. See Problem 7.40*a*.

Comment

It is in the combination of a high pressure and a low temperature (i.e., 8 MPa and 220 K) that real-gas effects are particularly significant (see Problem 7.40*b*). In our study, such a combination is seldom encountered.

☑ You have completed Learning Outcome (5)

7.6 Summary

In this chapter, relationships of thermodynamic properties were developed that could not be measured (enthalpy, internal energy, entropy, and specific heats) in terms of properties that can be measured: temperature, pressure, weight, and volume. In previous chapters, tables were used to determine properties of interest, but how were the entries in those tables determined if they were not measured? This chapter provided the relationships used to determine those properties over the wide range of temperatures and pressures that were encountered in the various devices analyzed in our study of thermodynamics. The following additional terms were defined in this chapter:

Clapeyron equation: *Allows the determination of the enthalpy of vaporization h_{fg} from measurable properties P, T, and v.*

Enthalpy departure: *The difference between the ideal-gas enthalpy at a sufficiently low pressure and the enthalpy at that same temperature divided by the critical-point temperature.*

Entropy departure: *The difference between the entropy at a sufficiently low pressure and the entropy at that same temperature.*

Helmholtz and Gibbs functions: *Properties that are defined in terms of other known properties.*

Joule–Thomson coefficient: *The rate of change of pressure with respect to temperature for a constant enthalpy process.*

Maxwell relations: *Relationships between entropy and measurable properties.*

Several important equations were presented:

Helmholtz function: $a = u - Ts$

Gibbs function: $g = h - Ts$

The differential: $dz = \left(\frac{\partial z}{\partial x}\right)_y dx + \left(\frac{\partial z}{\partial y}\right)_x dy = Mdx + Ndy$

Maxwell relations: $\left(\frac{\partial T}{\partial v}\right)_s = -\left(\frac{\partial P}{\partial s}\right)_v \quad \left(\frac{\partial T}{\partial P}\right)_s = \left(\frac{\partial v}{\partial s}\right)_P$

$$\left(\frac{\partial P}{\partial T}\right)_v = \left(\frac{\partial s}{\partial v}\right)_T \quad \left(\frac{\partial v}{\partial T}\right)_P = -\left(\frac{\partial s}{\partial P}\right)_T$$

Additional relations: $\left(\frac{\partial u}{\partial s}\right)_v = T\left(\frac{\partial u}{\partial v}\right)_s = -P$

$$\left(\frac{\partial h}{\partial s}\right)_P = T\left(\frac{\partial h}{\partial P}\right)_s = v$$

Clapeyron equation: $\left.\frac{dP}{dT}\right|_{v=v_0} = \frac{h_{fg}}{T_0 v_{fg}}$

Clapeyron–Clausius equation: $\left.\frac{dP}{dT}\right|_v = \frac{h_{fg}P}{RT^2}$

Differential internal energy: $du = C_v dT + \left[T\left(\frac{\partial P}{\partial T}\right)_v - P\right]dv$

Differential enthalpy: $dh = C_p dT + \left[v - T\left(\frac{\partial v}{\partial T}\right)_P\right]dP$

Differential entropy: $ds = \frac{C_p}{T}dT - \left(\frac{\partial v}{\partial T}\right)_P dP$

Joule–Thomson coefficient: $\mu_J = \left.\frac{dT}{dP}\right|_h = \frac{1}{C_P}\left[T\left(\frac{\partial v}{\partial T}\right)_P - v\right]$

Enthalpy departure: $\frac{h - h^*}{T_{cr}} = RT_R^2 \int_{P_R^*}^{P_R} \left(\frac{\partial Z}{\partial T_R}\right)_P \frac{dP_R}{P_R}$

Real-gas enthalpy change: $h_2 - h_1 = (h_2 - h_2^*) + (h_2^* - h_1^*) + (h_1^* - h_1)$

Problems

FE-Style Questions[4]

7.1 Estimate the partial derivative $(\partial s/\partial P)_T$ of steam at 400°C and 2.5 MPa:

(A) -0.0003 m³/kg·K

(B) -0.00025 m³/kg·K

(C) -0.0002 m³/kg·K

(D) -0.00015 m³/kg·K

7.2 The change in the Helmholtz function at 400°C in steam as the pressure changes from 2500 kPa to 3500 kPa is nearest:

(A) 224 kJ/kg

(B) 154 kJ/kg

(C) 122 kJ/kg

(D) 108 kJ/k

[4]The FE exam does not include questions on topics from this chapter.

7.3 Estimate the saturation temperature of R134a at 40 kPa using the Clapeyron equation and values from Table D-2:

(A) −44°C
(B) −46°C
(C) −48°C
(D) −50°C

7.4 Estimate the Joule–Thomson coefficients for ammonia at 1400 kPa and $h = 1570$ kJ/kg:

(A) 0.0190°C/kPa
(B) 0.0165°C/kPa
(C) 0.0145°C/kPa
(D) 0.0120°C/kPa

7.5 If the Joule–Thomson coefficient for a refrigerant at 1000 kPa is 0.08 °C/kPa, approximate the exiting temperature if the entering pressure and temperature are 2000 kPa and 60°C, respectively, and the exiting pressure is 120 kPa:

(A) −60°C
(B) −70°C
(C) −80°C
(D) −90°C

7.6 Estimate the change in the enthalpy of air if it is heated isothermally from 10°C and 100 kPa to 10 MPa:

(A) 21 kJ/kg
(B) 29 kJ/kg
(C) 41 kJ/kg
(D) 64 kJ/kg

7.7 The change in internal energy for the conditions of Problem 7.6 is nearest:

(A) 21 kJ/kg
(B) 19 kJ/kg
(C) 17 kJ/kg
(D) 15 kJ/kg

The Maxwell Relations Equations

7.8 Use the differential form for dv (see Eq. 7.7) assuming $v = RT/P$ and find the change in the specific volume Δv of air if the pressure and temperature change from 200 kPa and 100°C to 210 kPa and 120°C. Compare with Δv using the ideal-gas law directly.

7.9 Use the differential form for dP (see Eq. 7.7) assuming $P = RT/v$ and find the change in the pressure of air if the specific volume and temperature change from 0.5 m³/kg and 80°C to 0.52 m³/kg and 70°C. Compare with ΔP using the ideal-gas law directly.

7.10 Apply Eqs. 7.7 and 7.10 directly to Eqs. 7.1 and 7.2 and derive the relations of Eqs. 7.11 and 7.12.

7.11 Apply Eqs. 7.7 and 7.10 directly to Eqs. 7.5 and 7.6 and derive the relations of Eqs. 7.13 and 7.14.

7.12 Apply Eq. 7.7 directly to Eqs. 7.1 and 7.2 and derive the relations of Eqs. 7.15 and 7.16.

7.13 An ideal gas undergoes a process from state 1 to state 2. For the following data, approximate the change in specific entropy for this process using Eq. 7.13.

State	T (°C)	P (kPa)	v (m³/kg)
1	200	100	0.8
2	180	150	0.2

7.14 An ideal gas undergoes a process from state 1 to state 2. For the data shown below, approximate the change in specific entropy for this process using Eq. 7.11.

State	T (°C)	P (kPa)	v (m³/kg)
1	65	350	0.27
2	80	500	0.19

7.15 Verify Eq. 7.12, which is $(\partial T/\partial P)_s = (\partial v/\partial s)_P$, using the properties of steam at 1 MPa and 400°C.

7.16 R134a exists at 500 kPa and 100°C. Verify Eq. 7.14, which is $(\partial v/\partial T)_P = -(\partial s/\partial P)_T$, using *a*) the values in Table D-3 and *b*) the IRC Calculator.

7.17 Verify Eq. 7.14, which is $(\partial v/\partial T)_P = -(\partial s/\partial P)_T$, using the properties of R134a at 150°C and 1 MPa.

7.18 Verify, using ammonia at 200 kPa and 40°C, that $(\partial h/\partial s)_P = T$ is indeed true using the values in Table E-2.

7.19 Verify, using steam at 800 kPa and 400°C, that $(\partial h/\partial s)_P = T$ is indeed true using *a*) the values in Table C-3 and *b*) the IRC Calculator.

7.20 Verify, using steam at 6 MPa and 450°C, that $(\partial u/\partial s)_v = T$ is indeed true. Use the IRC Calculator.

The Clapeyron Equation

7.21 Show that saturated water vapor at 20 kPa can be treated as an ideal gas (compare $v = RT/P$ with v_g from Table C-2). Could saturated water vapor at 100 kPa be treated as an ideal gas?

7.22 Using the pressure, temperature, and specific volume found in Table C-1, determine h_{fg} for steam at 300°C using the Clapeyron equation. Also determine s_{fg}. Calculate the errors assuming the values obtained in Table C-1 are accurate.

7.23 Using the pressure, temperature, and specific volume found in Table D-1, determine h_{fg} for R134a at 16°C by applying the Clapeyron equation. Also determine s_{fg}. Calculate the errors assuming the values obtained in Table D-1 are accurate.

7.24 The saturation temperature of steam at a very low pressure is desired. Using the entries from Table C-2, estimate the saturation temperature at *a*) 400 Pa, *b*) 300 Pa, and *c*) 200 Pa.

Relationships for Δu, Δh, Δs, C_p and C_v

7.25 Determine an expression for the quantities in the brackets of Eqs. 7.33 and 7.39 for an ideal gas for which $Pv = RT$.

7.26 Assume an ideal gas with constant specific heats for which $Pv = RT$ and determine expressions for Δs using Eqs. 7.41 and 7.42.

7.27 Find the expression for the change in internal energy Δu for a gas obeying the equation of state $P = RT/(v - b) - a/v^2$ (the van der Waals equation) assuming constant specific heats.

7.28 Find the expression for the change in enthalpy Δh for a gas obeying the equation of state $P = RT/(v - b) - a/v^2$ (the van der Waals equation) assuming constant specific heats.

7.29 Find a relationship for $C_p - C_v$ for a gas for which the van der Waals equation of state $P = RT/(v - b) - a/v^2$ is applicable. Then let $a = b = 0$ and show that $C_p - C_v = R$.

7.30 Air undergoes an isothermal process at 100°C from 200 kPa to 6000 kPa. Calculate the enthalpy and entropy changes using the van der Waals equation of state $P = RT/(v - b) - a/v^2$. Assume that the specific heat C_v depends on temperature only. Then compare the results with those assuming constant specific heat with $C_p = 1.0$ kJ/kg·K.

7.31 A 50-kg block of aluminum experiences a pressure change from 100 kPa to 50 MPa, while the temperature increases from 50ºC to 100ºC. Estimate its change in entropy. Use $\beta = 7 \times 10^{-5}\ \mathrm{K}^{-1}$, $C_p = 0.9$ kJ/kg·°C, and $\rho = 2700\ \mathrm{kg/m^3}$. What is the contribution due to the pressure increase?

7.32 Estimate β and B for water at 10 MPa and 40°C and then find the difference $C_p - C_v$.

7.33 Estimate β and B for water at 7 MPa and 40°C and then find the difference $C_p - C_v$.

7.34 Estimate β and B for R134a at 800 kPa and 0°C and then find the difference $C_p - C_v$. Use the IRC Calculator.

The Joule–Thomson Coefficient

7.35 Estimate the Joule–Thomson coefficient for R134a at 350 kPa and $x = 0$. Then predict the exiting temperature from the valve of Fig. 7.5 if the entering pressure drops to 100 kPa from *a*) 400 kPa and 8°C, *b*) 600 kPa and 21°C, and *c*) 800 kPa and 31°C. Use the IRC Calculator.

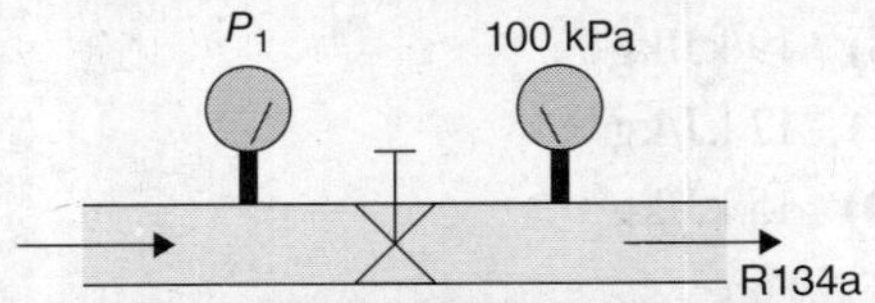

Figure 7.5

7.36 Estimate the Joule–Thomson coefficient for R134a at 1000 kPa and $x = 0$; then predict the exiting temperature from a valve if the pressure drops to 100 kPa from 2000 kPa and the entrance temperature of 67°C. Use the IRC Calculator.

7.37 Estimate the Joule–Thomson coefficient for air entering the valve of Fig. 7.6 at 1200 kPa and 400°C and show that the temperature does not significantly change when the pressure drops to 100 kPa.

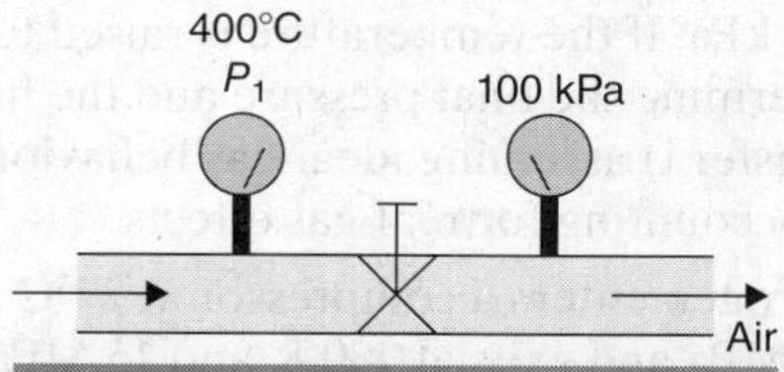

Figure 7.6

7.38 Estimate the Joule–Thomson coefficient of steam at 4 MPa and 400°C, and then estimate the value for C_p at that state using Eq. 7.53. Compare with the value found by using $C_p = (\partial h/\partial T)_P$.

7.39 Estimate the Joule–Thomson coefficient of steam at 6 MPa and 450°C, and then estimate the value for C_p at that state using Eq. 7.53. Compare with the value found by using $C_p = (\partial h/\partial T)_P$.

Real-Gas Effects

7.40 Calculate the change in enthalpy of air using both the ideal-gas table and the enthalpy departure chart if its state changes from:

a) 260 K and 900 kPa to 820 K and 4 MPa

b) 360 K and 900 kPa to 220 K and 8 MPa

c) 300 K and 200 kPa to 600 K and 10 MPa

7.41 Calculate the change in enthalpy of carbon dioxide using both the ideal-gas table and the enthalpy departure chart if its state changes from:

a) 320 K and 800 Pa to 300 K and 6 MPa

b) 400 K and 2 MPa to 400 K and 12 MPa

c) 500 K and 400 kPa to 900 K and 10 MPa

7.42 Calculate the change in entropy of air using both the ideal-gas table and the entropy departure chart if its state changes from:

a) 17°C and 1.3 MPa to 617°C and 7 MPa

b) 60°C and 1.8 MPa to 1000°K and 20 MPa

c) 220 K and 700 kPa to 250 K and 28 MPa

7.43 Calculate the change in the entropy of nitrogen using both the ideal-gas table and the entropy departure chart if its state changes from:

a) 260 K and 900 kPa to 800 K and 4 MPa

b) 350 K and 1200 kPa to 700 K and 8 MPa

c) 500 K and 1800 kPa to 220 K and 10 MPa

7.44 Air undergoes a process from state 1 at 400 K and 200 kPa to state 2 at 900 K and 12 MPa. Estimate the change in enthalpy and entropy:

i) Assuming the air is an ideal gas with constant specific heats from Table B-2

ii) Assuming the air is an ideal-gas with constant specific heats interpolated from Table B-6 at 650 K

iii) Using the ideal-gas tables in Appendix F-1

iv) Using the equations of Section 7.5

v) Using the IRC Calculator

7.45 Air undergoes a process from state 1 at 220 K and 200 kPa to state 2 at 300 K and 10 MPa. Estimate the change in enthalpy and entropy:

i) Assuming the air is an ideal-gas with constant specific heats from Table B-2

ii) Using the ideal-gas tables in the Appendix

iii) Using the equations of Section 7.5

iv) Using the IRC Calculator

7.46 Nitrogen undergoes a process from state 1 at 220 K and 2 MPa to state 2 at 280 K and 20 MPa. Estimate the change in enthalpy and entropy:

> **Note:** These extreme conditions are not indicative of typical problems but are meant to illustrate nonideal-gas behavior.

i) Assuming the nitrogen is an ideal gas with constant specific heats from Table B-2

ii) Assuming the nitrogen is an ideal gas with constant specific heats interpolated from Table B-6 at 250 K

iii) Using the ideal-gas tables in the Appendix

iv) Using the equations of Section 7.5

7.47 Carbon dioxide undergoes a process from state 1 at 360 K and 6 MPa to state 2 at 500 K and 20 MPa. Estimate the change in enthalpy, internal energy, and entropy:
i) Assuming the carbon dioxide is an ideal-gas with constant specific heats from Table B-2
ii) Using the ideal-gas tables in the Appendix
iii) Using the equations of Section 7.5

7.48 Determine the maximum power produced by the adiabatic turbine of Fig. 7.7 that accepts 4 kg/s of air at 10 MPa and 1100 K and exhausts the air to the atmosphere at 100 kPa. i) Use the ideal-gas table and ii) account for real-gas behavior.

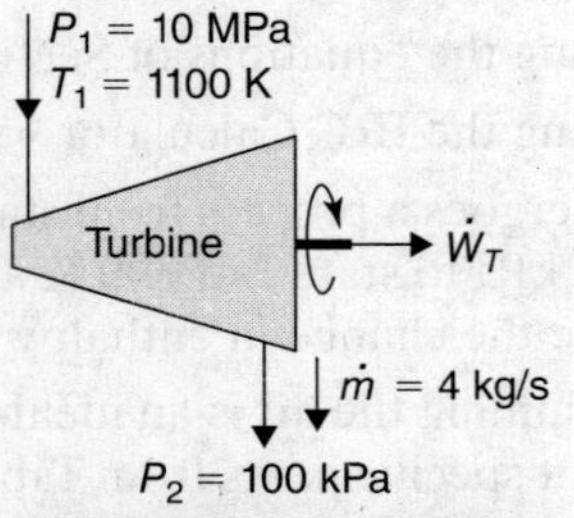

Figure 7.7

7.49 Calculate the minimum power required to compress 2 kg/s of nitrogen from 100 kPa and 300 K to 8 MPa assuming an adiabatic process. i) Use the ideal-gas table and ii) account for real-gas behavior.

7.50 A rigid volume contains air at 300 K and 400 kPa. If the temperature is raised to 1200 K, determine the final pressure and the heat transfer i) assuming ideal-gas behavior and ii) accounting for real-gas effects.

7.51 Nitrogen enters a compressor at 27°C and 1.5 MPa and exits at 480 K and 15 MPa. If the mass flux is 2 kg/s, determine the enthalpy change, the entropy change, and the power required if the heat loss is 20 kJ/s.

7.52 Air is compressed isothermally from 140 kPa and 37°C to 8.5 MPa, as shown in Fig. 7.8. Calculate the enthalpy change, the entropy change, and the work required if the heat loss is 186 kJ/kg.

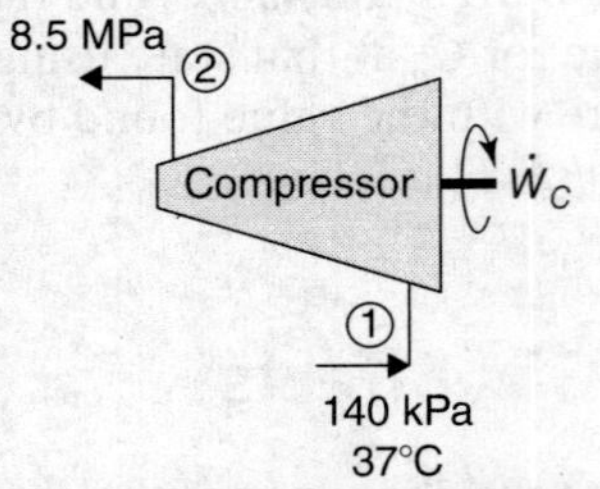

Figure 7.8

Part II

Applications

In Part I the fundamental concepts and problem-solving techniques used in thermodynamics were presented. In Part II these concepts and techniques are used to analyze power and refrigeration cycles, the analysis of which forms a primary objective in the introductory thermodynamics courses taught in many Mechanical Engineering departments. Chapters on mixtures and combustion are also included in this part since they are critical components in the production of power and the analysis of air-conditioning systems.

CHAPTER

8

The Rankine Power Cycle

Outline

Nomenclature

Only one new variable is introduced in this chapter:

ε Utilization factor

Learning Outcomes

- **Analyze a simple Rankine cycle**
- **Modify a Rankine cycle with reheat and preheat**
- **Understand the effectiveness of cogeneration**
- **Determine the overall efficiency of a power plant**

This chapter will analyze power plants. These plants use what is traditionally called a vapor Rankine cycle. The term *vapor* is used to indicate that the working fluid, water, undergoes phase changes between the liquid and gas phases. Phase change allows us to transfer large amounts of heat to or from the water. In Chapter 9, power cycles that use a gas as the working fluid will be considered. The equations and properties of the previous chapters will be utilized, so very few additional equations and quantities will be required. Basically, what has already been presented will be applied to the vapor cycle that produces power.

Motivational Example—The Cooling Tower

No power generating plant can operate at 100% efficiency. Some of the heat produced by the steam generator must be rejected to a low-temperature reservoir. In major power generating stations the device that accomplishes this job is a heat exchanger called a condenser. Ocean water, lake water, or river water is often used to cool the condensing steam. When an adequate source of water is not available, a cooling tower is a device that uses air to condense steam as a method of rejecting heat. Smaller cooling towers are also used as condensers in large air conditioning and refrigeration systems.

Note: A sketch of a cooling tower is shown in Fig. 11.21 on p. 411.

Cooling towers have been used to cool steam in major power plants since the early 1900's. In a dry tube cooling tower, the steam travels down pipes along the inside of the cooling tower. Cool air travels up the inside of the tower and transfers heat from the steam pipes. In a wet or evaporative cooling tower, the fluid to be cooled flows in the open, down the inside of the cooling tower. Again, air traveling up the inside of the cooling tower cools the fluid by convection and evaporation. The air moves up the cooling tower either by natural draft or by forced draft from fans. These cooling towers are so often used in nuclear power plants that they have become a symbol for nuclear plants even though they do not contain the reactor.

8.1 Energy Sustainability

The concept of sustainability applies to any system in which natural resources are used or impacted by the system. Any system that uses resources that cannot be replenished is ultimately unsustainable. At some point in time, the resources required in the operation of many of the systems in use today will disappear. Most of the systems that produce energy are unsustainable. They use fossil fuels like coal or oil for which there is a limited supply. Examples of energy supplies that are sustainable include solar energy, geothermal energy, hydropower, wind energy, and possibly the thermal energy and wave energy in the oceans.

Comment
Any system that uses resources that cannot be replenished is unsustainable.

Sustainability also influences the by-products developed in an engineering system. The by-products of an engineering process that damage the environment in which the system operates may also produce an unsustainable system. One example is the use of fossil fuels to produce electrical power. Even if there was an unlimited supply of coal and oil, a major product of combustion is carbon dioxide, which plants use to create oxygen in the atmosphere. Excessive carbon dioxide production, coupled with a decrease in the number of forests on our planet, has created a buildup of carbon dioxide in the atmosphere. This may be leading to a pattern of global warming that may cause significant damage to many ecosystems.

Comment
By-products of a process that damage the environment may also produce an unsustainable system.

Energy sustainability is essential to the future of humankind. Without energy, there is no economy. Problems with the sustainability of water resources and food production can be directly tied to the availability of energy. One of the major tasks for engineers in the next several decades will be to redirect our energy dependence from fossil fuels to alternative methods of energy conversion that are more sustainable. Chapter 13 will address several of these alternatives.

Conclusion: Energy sustainability is essential to the future of humankind.

8.2 The Rankine Cycle

8.2.1 Basic configuration and components

Boiler (steam generator): A heat exchanger used to transfer heat from the fuel to the working fluid.

The Rankine cycle is used to produce power; it has four major components, which will be described in some detail in the following paragraphs. A *boiler*, or *steam generator*, is a heat exchanger used to transfer large amounts of heat to the water, the working fluid. Figure 8.1 shows a sketch of a simple boiler. The heat for a boiler can be produced from coal, fuel oil, nuclear fission, geothermal heat sources, or even trash. The function of the boiler is to heat subcooled liquid water to superheated steam. In a coal-fired power plant, the coal is ground into a fine powder which is then fed down the center of the steam generator, where it burns at a very high temperature. The combustion temperature of coal increases with the amount of surface area available to burn. Pulverizing the coal creates this large surface area due to the small size of the coal particles. Water is pumped through pipes that cover the inside surface of the steam generator. As the water travels in these pipes, it absorbs heat from the burning coal in the center of the chamber.

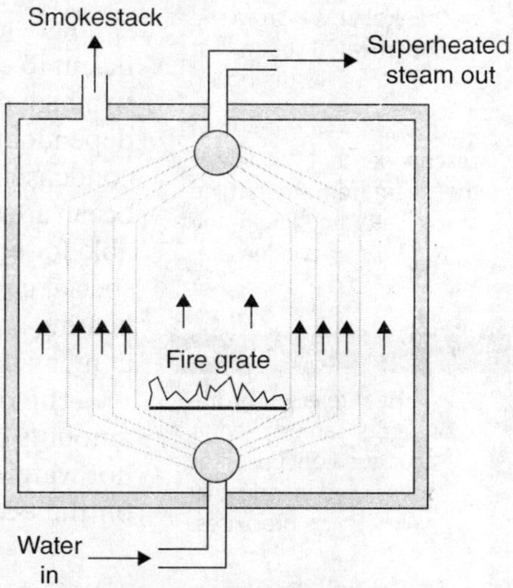

Figure 8.1
A sketch of a boiler.

In a nuclear power plant, the steam flows over hot rods of fissile uranium or plutonium. These rods do not get as hot as burning coal, so the steam temperature produced in a nuclear reactor is less than that developed in a coal-fired plant. As a result, the second law

Comment

Per pound, trash contains about 50% of the energy content of coal. However, serious emission problems accompany trash burning. Do a Google search on "waste energy" and discover what can be done.

indicates that the efficiency of a nuclear plant will be less than the efficiency of a coal-fired plant due to the lower boiler (reservoir) temperature. A well-designed coal-fired power plant will have an overall cycle efficiency of 40 to 42%. A nuclear power plant will have cycle efficiencies on the order of 32%. The heating process in the boiler, a heat exchanger, is assumed to be a constant-pressure process.

A major concern in the design of a power plant is the impact on the environment of operating the plant. Much of the pollution created by a power plant comes from the operation of the boiler. In a coal-fired power plant the main stack gas is carbon dioxide. Plants are now being designed which have facilities to remove the carbon dioxide and contain it for industrial use. Coal-fired boilers also emit sulfur into the atmosphere, the amount depending on the type of coal used. Nuclear reactors do not emit products of combustion; their main environmental problem is how to dispose of spent nuclear fuel rods. Currently, they are stored at the reactor site. Inventories of spent fuel rods are building up and pose a significant health risk if the storage facilities are not well maintained. The 2011 disaster at the Fukushima-Daichi Nuclear Plant in Japan is a clear example of the need to find a safe and sustainable solution to this problem.

Figure 8.2

The blades on a steam turbine rotor. [Permission from Tesla-Turbines, photo by Christian Kuhna.]

A *steam turbine* is used to create mechanical power from the superheated steam that flows from the boiler via a rotating shaft transmitting torque. The steam enters the turbine as a high-temperature, high-pressure superheated vapor and exits the turbine as a much lower energy steam, which can be superheated steam or a saturated mixture. If a mixture exits the turbine, it should not have a quality less than 0.90[1] in order to prevent water droplets from damaging the turbine blades. In a steam turbine, high-pressure steam impacts the blades of a rotor (see Fig. 8.2). The shaft of the turbine provides rotational mechanical power, which is the product of the torque produced by the turbine and the rotational speed of the turbine shaft. This mechanical power is used to drive an electric generator. The mechanical engineer loses control of the power after it leaves the turbine to the electrical engineer, who assumes control as the power enters the generator and passes to the electrical grid. So, our study of thermodynamics does not consider the power after it leaves the turbine, which operates ideally as an isentropic device.

Pollution: The boiler accounts for most of the pollution of CO_2 and sulfur in a coal-fired plant.

Steam turbine: The device that creates mechanical power from the superheated steam that leaves the boiler.

Condenser: A heat exchanger that condenses the steam from the turbine into a saturated liquid, the condensate.

The steam that exits the turbine travels through a heat exchanger, called a *condenser*. At the condenser, heat is removed from the steam from the turbine sufficient to condense it into a saturated liquid so that it can be efficiently pumped back to a high pressure. There are a variety of techniques used to accomplish this that are dependent on the cooling water resources available near the plant. A water-cooled condenser is sketched in Fig. 8.3. Ideally, large amounts of water from a river, lake, or ocean are used to condense the steam exiting the turbine. If water is not readily available for cooling, air can be used in a cooling tower facility to condense the steam. In special circumstances, much of the heat rejected by the condenser can be used for heating and cooling[2] buildings that are adjacent to a power plant, as has been done at many universities, cities, and industrial complexes.

Comment

Heat rejected by the condenser can be used to heat and cool buildings and/or used in industrial processes.

The operation of the condensers poses environmental concerns. When large amounts of heat are dumped into the environment, it is called *thermal pollution*. If hot water from a condenser is deposited into a river or lake, it will have an effect on the ecology of that body of water. Many types of fish cannot live at elevated

[1] A condensing turbine operates with an exiting quality less than 1.0. We will assume the exiting steam has a quality of 1.0 or is in the superheat region as a design criterion for an operating turbine.

[2] Cooling with heat uses a process called absorption refrigeration, which will be analyzed in Chapter 10.

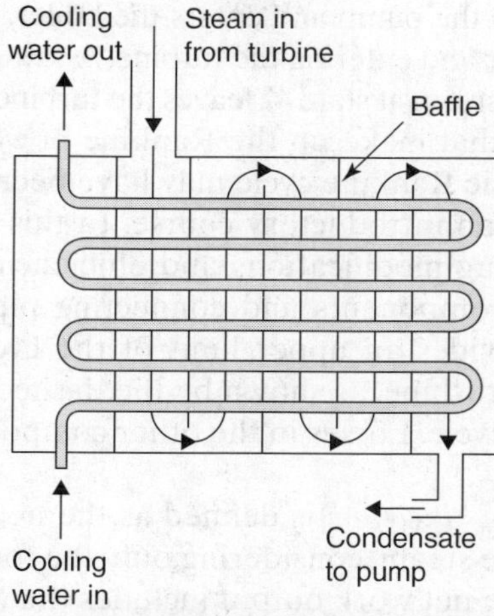

Figure 8.3

A water-cooled condenser in which steam enters and condensate leaves.

temperatures and will leave the area around a power plant, whereas other fish will be attracted to the warmer water. Some types of waterfowl will use the warm areas around a power plant as a winter sanctuary, an unnatural phenomenon that may or may not be welcomed. Also, vapors emitted by cooling towers, used to condense the steam, can cause fog and icing on local roads during cold weather, in addition to possible medical problems.

The liquid condensate leaving the condenser is at a low pressure, usually below atmospheric pressure. To obtain the high pressure required by the turbine, massive *feedwater pumps*, like the one shown in Fig. 8.4, are used to pump the water back to the boiler, thereby completing the cycle. Power produced by the plant must be used to run these pumps, although compared to the power produced by the turbine it is so small that it can actually be ignored when considering the efficiency of the power cycle.

A schematic of how the components of the simple Rankine cycle are arranged is shown in Fig. 8.5, and its *T-s* diagram is sketched in Fig. 8.6. State 1 is a saturated liquid that leaves the condenser and enters the pump. State 2 is the high-pressure

Suggestion: Let's just skip the condenser and use a compressor to raise the pressure back to the high pressure required by the turbine. Why dump all the heat into the environment? Answer this question at the end of this chapter.

Feedwater pump: The device used to produce the high pressure required by the turbine.

Observe: The boiler and condenser are constant-pressure heat exchangers.

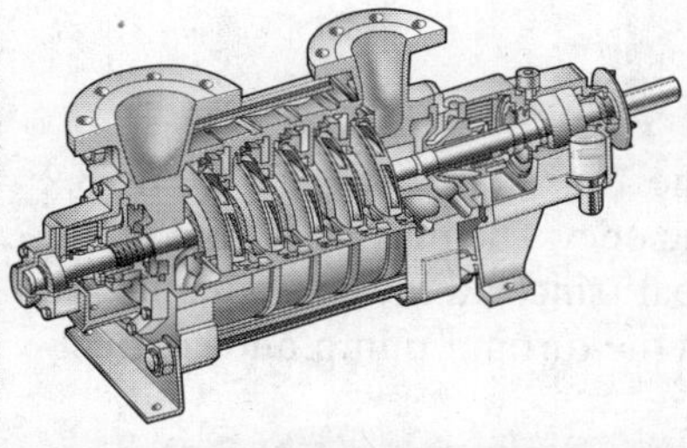

Based on of RAT Affiliates, rat.net.nz

Figure 8.4

Feedwater pump.

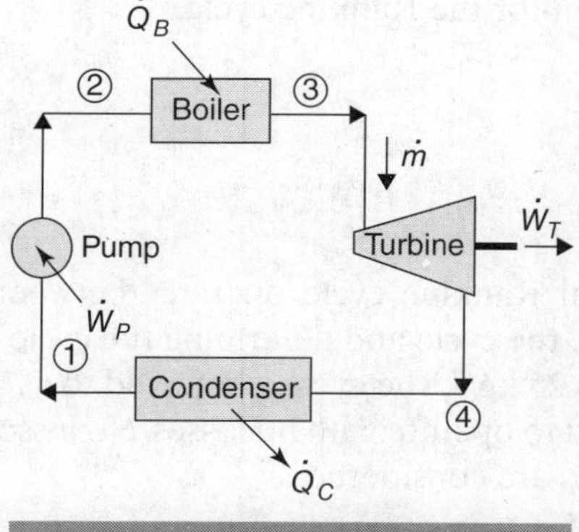

Figure 8.5

The simple Rankine cycle includes a pump, boiler, and turbine.

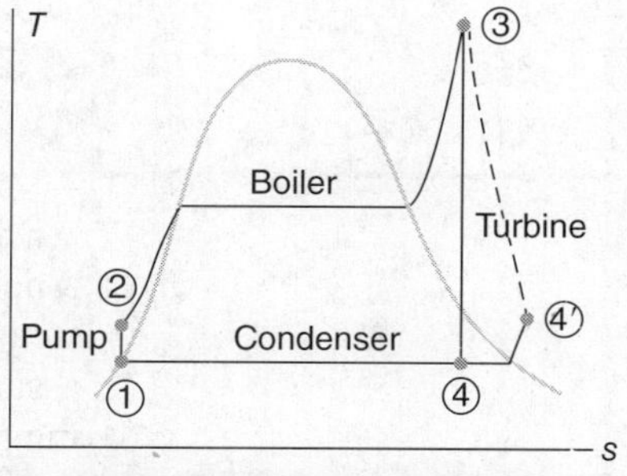

Figure 8.6

The *T-s* diagram for the simple Rankine cycle.

liquid water that leaves the pump and enters the boiler. State 3 is the superheated steam leaving the boiler and entering the turbine. A saturated mixture at state 4, or possibly a superheated steam at state 4′, leaves the turbine and enters the condenser.

The components that make up the Rankine power cycle were introduced in Section 4.4. The basic Rankine cycle may have been studied in Section 4.4.1; it is often skipped in an introductory course. In this chapter it is analyzed in greater detail, including modifications and efficiencies. If losses are assumed negligible in all the components and connecting pipes, it is called the *ideal Rankine cycle*. It provides an upper limit of the cycle efficiency. Losses are often included in the turbine, as shown by the dashed line in Fig. 8.6, in which case it is not an ideal cycle. Losses in the other components are relatively small and often ignored.

Ideal Rankine cycle: It provides an upper limit of the cycle efficiency since there are no losses in all the components, especially the turbine.

Rankine cycle efficiency: The net work output divided by the energy input, considering only the devices that constitute the Rankine cycle.

The *Rankine cycle efficiency* is defined as the net work output divided by the energy input to the steam, considering only the four devices that constitute the Rankine cycle. The net work output includes the work provided by the turbine minus the work required by the pump (the pump work is often neglected since it is relatively small, less than 1% of the turbine power). The energy input is the heat transfer received by the working fluid in the steam generator.[3] The cycle's *thermodynamic efficiency*, sometimes referred to simply as the *cycle efficiency*, is

$$\eta_{\text{cycle}} = \frac{w_T - w_P}{q_B} = \frac{W_T - W_P}{Q_B} = \frac{\dot{W}_T - \dot{W}_P}{\dot{Q}_B} \tag{8.1}$$

where W_T and W_P are the turbine and pump work, respectively, and Q_B is the heat input by the boiler.

Plant efficiency: The ratio of the energy output from the entire power plant divided by the energy in the fuel that enters the power plant. It is always less than the efficiency of the Rankine cycle.

The *plant efficiency* is the ratio of the energy output from the entire power plant divided by the energy in the fuel that enters the power plant. It would include the losses in the combustion chamber where the energy contained by a fuel is converted to the heat that enters the steam in the boiler pipes (losses of about 10%), and the losses that exist between the output of the turbine and the electrical energy that leaves the power plant (losses of about 8%). Also, about 6% of the electrical power is needed to operate the auxiliary equipment that is needed to operate the power plant. These losses do not enter the efficiency calculation for the Rankine cycle.

Example 8.1 An ideal Rankine cycle

An ideal Rankine cycle operates between the four states shown in Fig. 8.7. Analyze the cycle and determine the cycle efficiency. The mass flow rate of the steam is 25 kg/s. The cycle is referred to as "ideal" since all the components are assumed to operate with no losses. No losses in the turbine, pump, and connecting pipes are considered.

[3] We use both the terms *steam generator* and *boiler* for the same heat exchanger. This is done because both terms are in use and it's a good idea to become familiar with them. We tend to use *boiler* most often.

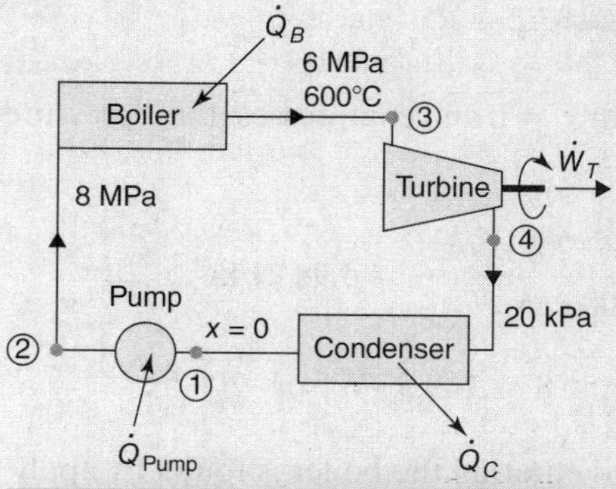

Figure 8.7

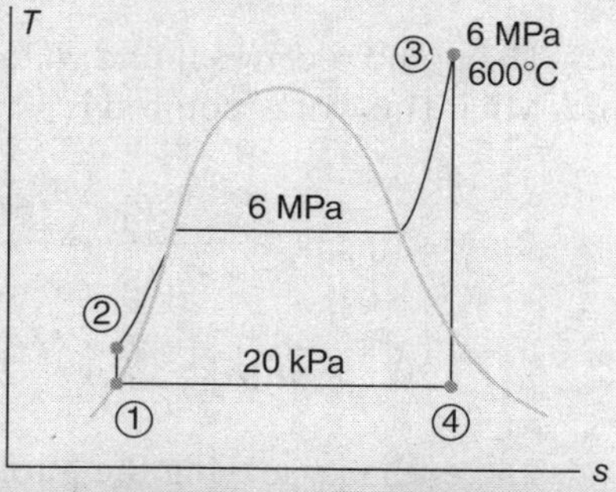

Figure 8.8

State 1: $P_1 = P_4,\ x_1 = 0$ **State 2:** $P_2 = P_3$

State 3: $T_3 = 600°C,\ P_3 = 6$ MPa **State 4:** $P_4 = 20$ kPa

Solution

First, a *T-s* diagram is often helpful. It should always be sketched, as in Fig. 8.8. Compressed liquid exits the pump, and superheated steam exits the boiler. Let's analyze the turbine first, for no particular reason. It accepts superheated steam at 6 MPa and 600°C, and, in the ideal cycle, the steam undergoes an isentropic process from state 3 to state 4; that is, $s_4 = s_3$. This allows state 4 to be identified since the pressure P_4 = 20 kPa. From Table C-3 at 6 MPa and 600°C, s_3 = 7.1677 kJ/kg·K, allowing us to locate state 4. Since $s_4 < s_g$ = 7.9093 kJ/kg·K at 20 kPa, we find x_4 as follows:

$$s_3 = s_4 = s_f + xs_{fg}$$
$$7.1677 = 0.8319 + x_4 \times 7.0774 \qquad \therefore x_4 = 0.8952$$

where s_f and s_{fg} at 20 kPa from Table C-2 were used. At state 4 the enthalpy is needed to calculate the isentropic turbine output. It is

$$h_4 = (h_f + xh_{fg})_4 = 251.4 + 0.8952 \times 2358.3 = 2363 \text{ kJ/kg}$$

At 6 MPa and 600°C, we find h_3 = 3658.4 kJ/kg from Table C-3, so the turbine output is

$$w_T = h_3 - h_4 = 3658.4 - 2363 = 1295 \text{ kJ/kg}$$
$$\therefore \dot{W}_T = w_T \Delta h = 25\,\frac{\text{kg}}{\text{s}} \times 1295\,\frac{\text{kJ}}{\text{kg}} = \underline{32\ 400 \text{ kW}}$$

The pressure remains constant through the condenser (it's simply a heat exchanger as described in Section 4.2.4). The enthalpy at state 1, where $x_1 = 0$ is h_1 = 251.4 kJ/kg from Table C-2 at 0.02 MPa, so the heat transfer from the condenser is

$$q_C = h_4 - h_1 = 2363 - 251.4 = 2112 \text{ kJ/kg}$$
$$\therefore \dot{Q}_C = \dot{m}(h_4 - h_1) = 25 \times 2112 = \underline{52\ 800 \text{ kJ/s}}$$

(Continued)

Suggestion: Always sketch a *T-s* diagram even if not requested. It helps visualize the processes.

Note: Only three or four significant digits are displayed. The numbers in the steam tables are from equations that result from curve fits on data that is known to, hopefully, three significant digits. The many digits included in tables implies accuracy that is simply not realistic.

Note: The calculations are done so that a positive number is found. The condenser heat transfer is negative, but the heat transfer from the condenser is considered positive.

Example 8.1 (*Continued*)

Note: The pump work is negative since it is an energy input. The work required by the pump is considered positive, that is, $\dot{W}_P = -\dot{W}_S$

Units: kN·m/kg = kJ/kg

The pump receives liquid water with $x_1 = 0$ and compresses it to a pressure of 6 MPa. The work required is

$$w_P = \frac{P_2 - P_1}{\rho} = \frac{(6000 - 20)\ \text{kN/m}^2}{1000\ \text{kg/m}^3} = 5.98\ \text{kJ/kg}$$

$$\therefore \dot{W}_P = \dot{m}(h_2 - h_1) = 25 \times 5.98 = 149.5\ \text{kW} \quad \text{or} \quad \underline{200\ \text{hp}}$$

The enthalpy leaving the pump and entering the boiler is found by applying the first law to the pump (see Eq. 4.19):

$$w_P = h_2 - h_1. \qquad \therefore h_2 = w_P + h_1 = 6 + 251 = 257\ \text{kJ/kg}$$

The heat transfer required by the steam generator is thus

$$q_B = h_3 - h_2 = 3658.4 - 257 = 3401\ \text{kJ/kg}$$

Units: Rate of heat transfer is often stated as kJ/s whereas work rate by kW. They are the same in that kJ/s = kW.

giving the rate of heat transfer as

$$\dot{Q}_B = \dot{m}(h_3 - h_2) = 25 \times 3401 = \underline{85\,000\ \text{kJ/s}}$$

The cycle efficiency is the cycle output divided by the input, or the turbine output minus the pump input divided by the energy required by the steam generator. It is

$$\eta_{\text{cycle}} = \frac{w_T - w_P}{q_B} = \frac{1295 - 5.98}{3401} = 0.379 \quad \text{or} \quad \underline{37.9\%}$$

Comment

A Carnot cycle operating between the high (600°C) and low (60.1°C) temperatures of this cycle would have an efficiency of 62%. The heat transfer process in the boiler is across a large temperature difference, so it is highly irreversible.

If, for the efficiency calculation, the pump work is set equal to zero, q_B would be 3658.4 – 251 = 3407 kJ/kg (we used $h_2 = h_1$). The efficiency would then be

$$\eta = \frac{w_T}{q_B} = \frac{1295}{3407} = 0.380 \quad \text{or} \quad \underline{38.0\%} \qquad \text{(without pump)}$$

The error is less than 0.3%, which is why the pump is often ignored in Rankine cycle efficiency calculations.

The remainder of the energy input ($q_B - w_T = q_C$) is rejected by the condenser. That is, 62% of the input energy leaves the cycle through the condenser and is lost to a river, a lake, or a cooling tower; or perhaps it is used to heat and cool university dorms and buildings, or buildings in a downtown area, or used by a large industrial complex. Use of the condenser waste heat does not change the cycle efficiency. The efficiency of the cycle, however, can be increased somewhat using various techniques, which are analyzed in subsequent sections.

Comment

If the waste heat is gainfully used, it does not change the Rankine cycle's efficiency. The increased utilization of the energy is measured by a "utilization factor," to be introduced in Section 8.4.

A Rankine cycle with some losses — Example 8.2

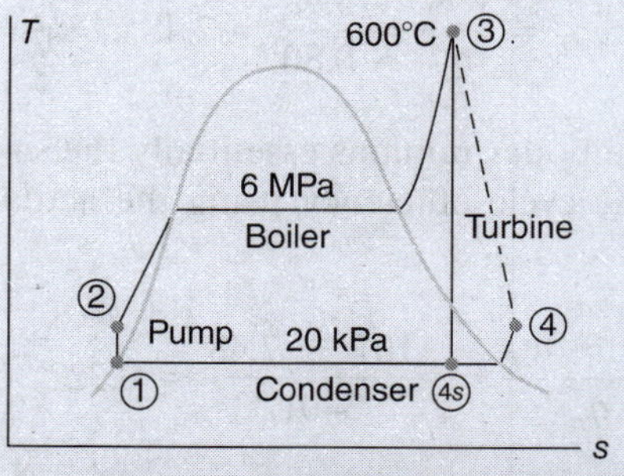

Figure 8.9

An actual Rankine cycle is formed using the same properties presented in Example 8.1 but with an 85%-efficient turbine and an 80%-efficient pump. The conditions on the boiler and condenser remain the same. Find the exiting state from the turbine and the efficiency of the actual Rankine cycle.

Solution

The steam entering the turbine at state 3 has a specific enthalpy of 3658 kJ/kg and a specific entropy of 7.1677 kJ/kg·K, found in Table C-3. An isentropic process was assumed in Example 8.1, shown from state 3 to state 4*s* in Fig. 8.9; the actual process is shown by the dotted line to state 4, assumed to be in the superheat region on the diagram. The isentropic process resulted in $h_{4s} = 2363$ kJ/kg, found in Example 8.1, providing the ideal turbine output of $w_{T,s} = 1295$ kJ/kg. The actual turbine output $w_{T,a}$ is only 85% of the ideal turbine output:

$$w_{T,a} = \eta_T \times w_{T,s} = 0.85 \times 1295 = 1101 \text{ kJ/kg}$$

The first law applied to the actual turbine provides

$$w_T = h_3 - h_4 = 3658.4 - h_4 = 1101 \qquad \therefore h_4 = 2557 \text{ kJ/kg}$$

The IRC Calculator, with $P_4 = 20$ kPa and $h_4 = 2\,557\,000$ J/kg, gives

$$T_4 = \underline{60.1°\text{C}} \text{ and } x_4 = \underline{0.978}$$

which is just in the wet region. The losses in the turbine actually have a positive effect by reducing the droplets that tend to form on the turbine blades since the steam that exits from the turbine is much dryer.[4]

The actual input work to the pump is the ideal pump work $w_{P,s} = 5.98$ kJ/kg, found in Example 8.1, divided by the pump efficiency. This results in

(Continued)

[4]A dry steam has a quality of 1.0 or is in the superheat region. Droplets can cause damage to the turbine blades, so they are avoided except in condensing turbines, in which qualities above 90% are acceptable.

Example 8.2 (Continued)

$$w_{P,a} = \frac{w_{P,s}}{\eta_P} = \frac{5.98}{0.80} = 7.48 \text{ kJ/kg}$$

The energy input to the boiler remains essentially the same as in Example 8.1 at $q_B = 3401$ kJ/kg. The cycle efficiency, using the actual turbine output and pump input, is

$$\eta_{\text{cycle}} = \frac{w_T - w_P}{q_B} = \frac{1101 - 7.48}{3401} = 0.322 \text{ or } \underline{32.2\%}$$

Comment

The pump work can be ignored when calculating the cycle efficiency.

which is 85% of the cycle efficiency of Example 8.1 since the pump doesn't significantly influence the cycle efficiency. The only significant effect on the cycle efficiency is the turbine efficiency.

8.2.2 Improving Rankine cycle efficiency

There are several ways in which design engineers increase the efficiency of a power plant that operates on the Rankine cycle. The efficiency of the ideal Rankine cycle can be increased in three primary ways:

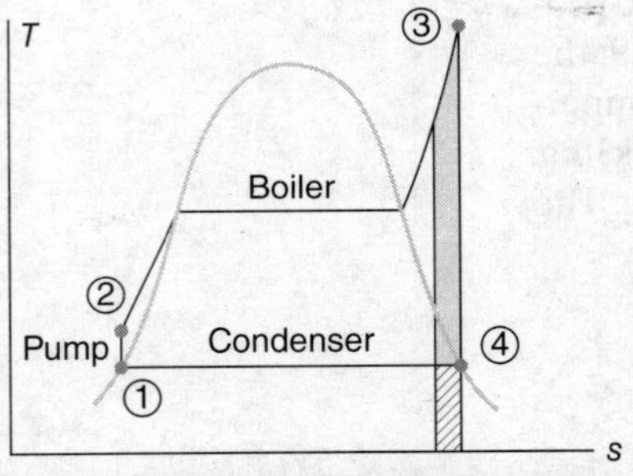

Figure 8.10

The effect of raising the boiler outlet temperature on the ideal Rankine cycle.

1. Increase the temperature of the vapor exiting the steam generator. This increases the turbine work output (the cross-hatched area[5] of Fig. 8.10) and also increases the boiler heat requirement (the sum of the cross-hatched area and the shaded area). If you recall the Carnot heat engine in Chapter 5, the efficiency increases as the temperature of the heat source increases, as required by Eq. 5.15. The use of high-temperature superheated steam will increase efficiency, but it is limited by the properties of the metals used to construct the turbine blades. A temperature of about 620°C is the maximum that present-day metals can tolerate. Higher temperatures are possible using ceramic blades or blade cooling.
2. Reduce the pressure at which the condenser operates. The lower the condenser pressure, the lower the enthalpy exiting the turbine; this results in increased cycle efficiency by increasing the turbine work output (the shaded area of Fig. 8.11). It should be noted that in Example 8.2, the outlet pressure of the turbine was well below atmospheric pressure. This is a common practice in power plant design.
3. Increase the pressure rise across the pump. Increase the steam generator pressure, which in turn increases the turbine inlet enthalpy. The turbine work increases by the difference of the two shaded areas of Fig. 8.12 (the turbine work is approximately the area under the T-s diagram, the area inside 1–2–3–4).

Comment

The failure of a brittle ceramic blade has serious consequences. Failure of a metal blade is much more forgiving.

[5]Remember: The area under the T-s diagram represents the heat transfer for the quasi-equilibrium processes. Since $w_T = q_B - q_C$ (ignoring the pump work), the area inside 1–2–3–4 represents the work.

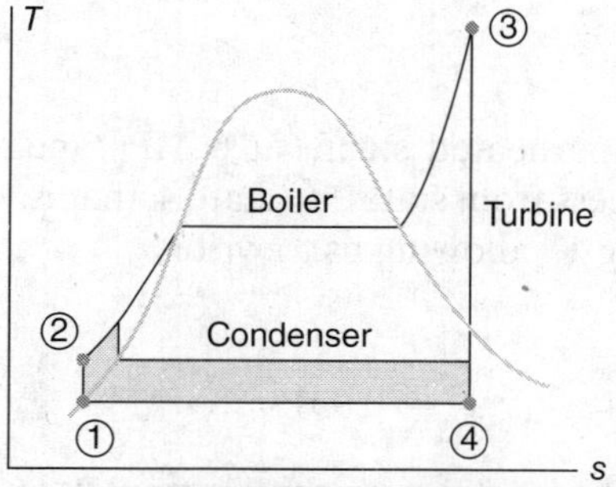

Figure 8.11

The effect of lowering the condenser pressure on the ideal Rankine cycle.

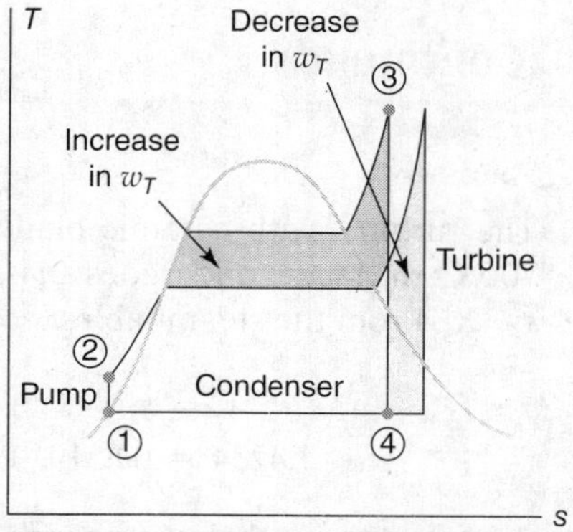

Figure 8.12

The effect of increasing the steam generator pressure on the ideal Rankine cycle.

The boiler heat transfer requirement also increases, but the net result is an increase in cycle efficiency. Supercritical pressures (pressures higher than the critical-point pressure) in excess of 22 MPa are not uncommon.

Modifications to the Rankine cycle can be made to further increase the cycle efficiency. These will be presented in the next section.

The effect of increased temperature — Example 8.3

The ideal Rankine power cycle of Example 8.1 has the temperature exiting the steam generator increased to 700°C, as sketched in Fig. 8.13. Calculate the cycle efficiency holding all other properties the same as in Example 8.1. Assume no losses.

State 1: $P_1 = P_4,\ x_1 = 0$ **State 2:** $P_2 = P_3$

State 3: $T_3 = 700°C,\ P_3 = 6$ MPa **State 4:** $P_4 = 20$ kPa

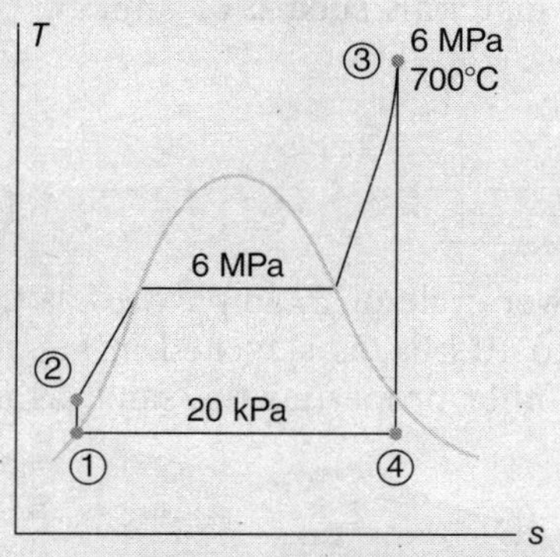

Figure 8.13

(Continued)

Example 8.3 (Continued)

Solution

The turbine, with ceramic blades, accepts superheated steam at 6 MPa and 700°C; the steam undergoes an isentropic process from state 3 to state 4; that is, $s_4 = s_3$. From the steam tables, $s_3 = 7.4234$ kJ/kg·K, allowing us to write

> **Note:** The higher temperature results in a slightly dryer mixture, a desirable effect, but still unacceptable. However, a turbine efficiency of 90% would move state 4 into the superheat region.

$$s_3 = s_4 = s_f + xs_{fg}$$
$$7.4234 = 0.8319 + x_4 \times 7.0774 \qquad \therefore x_4 = 0.9313$$

At state 4 the enthalpy is needed to calculate the isentropic turbine output. It is

$$h_4 = (h_f + xh_{fg})_4 = 251.4 + 0.9313 \times 2358.3 = 2448 \text{ kJ/kg}$$

At 6 MPa and 700°C, we find $h_3 = 3894.2$ kJ/kg from Table C-3, so the turbine output is

$$w_T = h_3 - h_4 = 3894.2 - 2448 = 1446 \text{ kJ/kg}$$

> **Note:** We ignore the pump work since it is so small compared to the turbine work.

Ignoring the pump work (it's too small to bother with) and using $h_2 \cong h_1 = 251$ kJ/kg from Table C-2 at 0.02 MPa, we find that the energy required by the boiler is

$$q_B = h_3 - h_2 = 3894 - 251 = 3643 \text{ kJ/kg}$$

The cycle efficiency is the turbine output divided by the energy required by the steam generator. It is

$$\eta_{\text{cycle}} = \frac{w_T}{q_B} = \frac{1446}{3643} = 0.397 \text{ or } \underline{39.7\%}$$

The increase of just 4% isn't much, but power companies invest millions of dollars to increase the efficiency even less than that. In fact, the metals from which turbine blades are crafted cannot withstand temperatures above about 620°C, so a turbine accepting steam at 700°C would have the first rows of blades made of special materials, such as ceramics.

Example 8.4 The effect of decreased condenser pressure

The ideal Rankine power cycle of Example 8.1 has the pressure entering the condenser decreased to 10 kPa, as sketched in Fig. 8.14. Calculate the cycle efficiency, holding all other properties the same as in Example 8.1. Assume no losses.

State 1: $P_1 = P_4,\ x_1 = 0$ **State 2:** $P_2 = P_3$

State 3: $T_3 = 600°\text{C},\ P_3 = 6$ MPa **State 4:** $P_4 = 10$ kPa

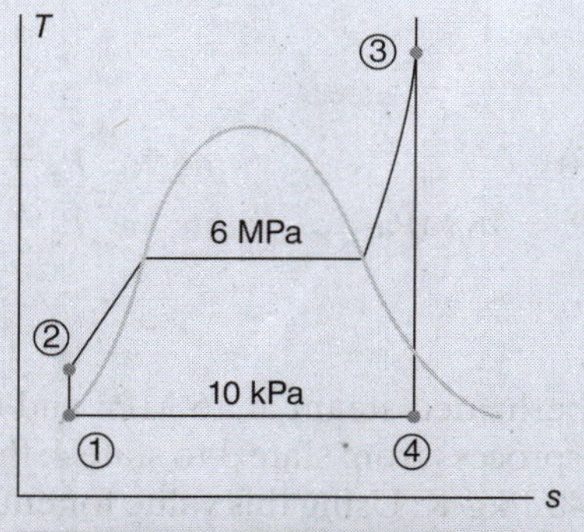

Figure 8.14

Solution

The turbine accepts superheated steam at 6 MPa and 600°C, and the steam undergoes an isentropic process from state 3 to state 4; that is, $s_4 = s_3$. From the IRC Calculator, $s_3 = 7170$ J/kg·K. Using this value for entropy and 10 kPa for the pressure, the IRC Calculator gives $h_4 = 2\,270\,000$ J/kg (2270 kJ/kg).

The enthalpy that enters the boiler is assumed to be $h_2 \cong h_1 = 192$ kJ/kg, which is found using $x_1 = 0$ and $P_1 = 10$ kPa. The enthalpy exiting the boiler at 6000 kPa and 600°C is $h_3 = 3660$ kJ/kg. Consequently, we find

$$\eta = \frac{w_T}{q_B} = \frac{h_3 - h_4}{h_3 - h_2} = \frac{3660 - 2270}{3660 - 192} = 0.401 \quad \text{or} \quad \underline{40.1\%}$$

The lowering of the condenser pressure by 10 kPa increased the efficiency by 5%, from 38.1% to 40.1%, but it had the undesirable effect of lowering the quality at the turbine exit to 0.869 (use the IRC Calculator at 10 kPa and $s_4 = s_3 = 7170$ kJ/kg·K).

Note: The reduced quality exiting the turbine is an undesirable feature.

The effect of increased boiler pressure — Example 8.5

The ideal Rankine power cycle of Example 8.1 has the pressure entering the boiler increased to 16 MPa, as sketched in Fig. 8.15. Calculate the cycle efficiency, holding all other properties the same as in Example 8.1. Assume no losses.

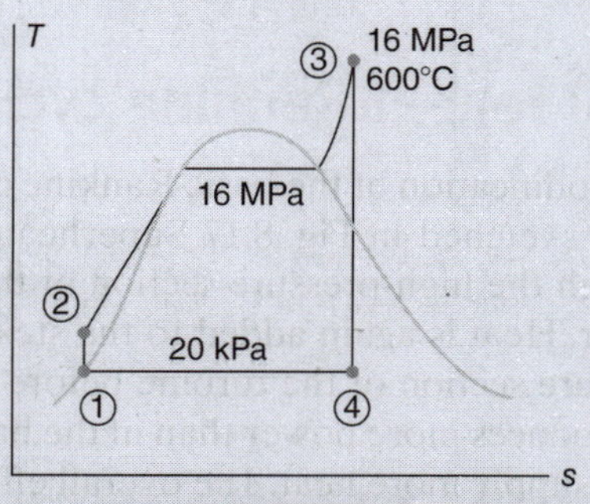

Figure 8.15

(Continued)

Example 8.5 (Continued)

State 1: $P_1 = P_4,\ x_1 = 0$ State 2: $P_2 = P_3$

State 3: $T_3 = 600°C,\ P_3 = 16$ MPa State 4: $P_4 = 20$ kPa

Solution

The turbine accepts superheated steam at 16 MPa and 600°C, and the steam undergoes an isentropic process from state 3 to state 4; that is, $s_4 = s_3$. From the IRC Calculator, $s_3 = 6640$ J/kg·K. Using this value for entropy s_4 and 20 kPa for the pressure, the IRC Calculator gives $h_4 = 2\,190\,000$ J/kg (2190 kJ/kg).

The enthalpy that enters the boiler is assumed to be $h_2 \cong h_1 = 251$ kJ/kg, which is found using $x_1 = 0$ and $P_1 = 20$ kPa. The enthalpy exiting the boiler at 16 000 kPa and 600°C is $h_3 = 3570$ kJ/kg. Consequently, we find

$$\eta = \frac{w_T}{q_B} = \frac{h_3 - h_4}{h_3 - h_2} = \frac{3570 - 2190}{3570 - 251} = 0.416 \text{ or } \underline{41.6\%}$$

Note: The quality exiting the turbine is very low, but cycle modifications, to be presented in the next section, can correct that situation.

The efficiency has increased by about 9%. Such high pressure in the boiler demands substantial increases in the strength of all the piping components, including the high-pressure inlet to the turbine, resulting in higher construction costs. Also, the quality exiting the turbine is quite low at 0.821. Modifications to the basic Rankine cycle (to be presented in the next section) can result in a steam quality at the turbine exit greater than 1.0.

☑ You have completed Learning Outcome (1)

8.3 Modified Rankine Cycles

This section presents two devices used to increase the efficiency of a power plant: the reheater and the feedwater heater. They are both commonly used in power-plant design. They result in higher cycle efficiencies, and the reheater reduces the excessive moisture exiting the turbine; with one or two reheaters, the steam at the turbine exit can be in the superheat region, a requirement of most turbines.

8.3.1 The ideal reheat Rankine cycle

The reheat cycle is a modification of the basic Rankine cycle, as shown in Fig. 8.16, and its *T-s* diagram is sketched in Fig. 8.17. Superheated steam from the steam generator flows through the high-pressure section of the turbine and is returned to the steam generator. Heat is again added to the steam, and it then flows back through the low-pressure section of the turbine before continuing on to the condenser. The turbine produces more power than in the basic Rankine cycle, but the steam generator must supply more heat. The overall effect is to increase the cycle efficiency a significant amount. The added costs are recoverable due to the large amounts of power produced by these plants over many years. The reheat process

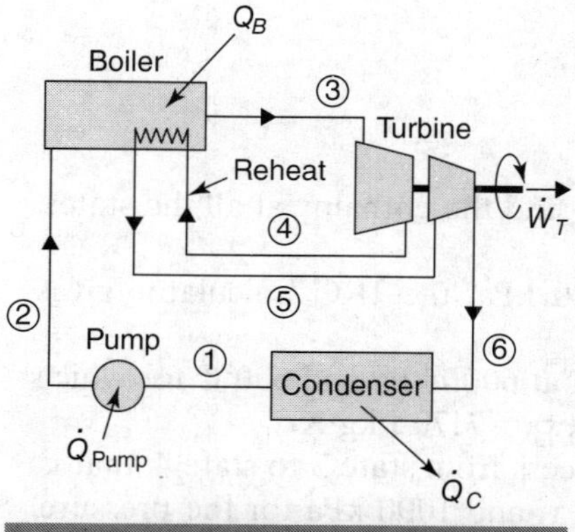

Figure 8.16
The reheat Rankine cycle.

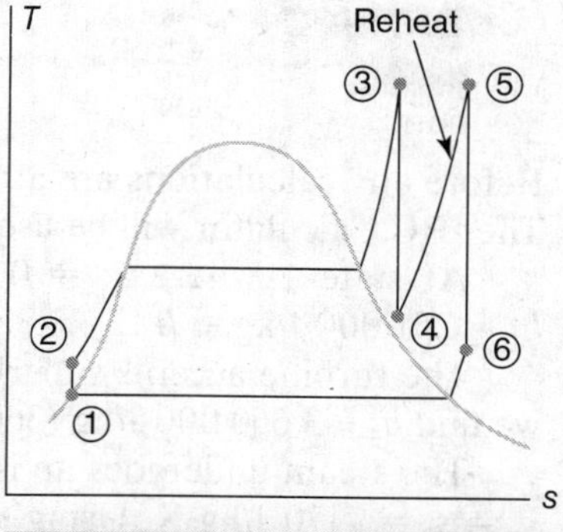

Figure 8.17
The *T-s* diagram for the ideal reheat Rankine cycle.

also results in reduced moisture near the turbine exit, a requirement for the low-pressure condensers in use. Some power plants will use two reheat cycles. The following example details the analysis of a reheat cycle. There are six states in this cycle, as displayed in Figs. 8.16 and 8.17.

The effect of a reheater — Example 8.6

An ideal Rankine power cycle of Example 8.1 has the steam in the high-pressure turbine returned to the boiler when it reaches a pressure of 1 MPa, as shown in Fig. 8.18. It is then returned to the low-pressure turbine at 600°C. Determine i) the heat transfer required by the steam generator, ii) the amount of heat transfer from the condenser, iii) the mass flow rate of the steam to produce 12 MW by the turbine, and iv) and the cycle efficiency. Neglect the pump work in the efficiency calculation. The state properties are listed as:

State 1: $P_1 = 20$ kPa, $x_1 = 0$ **State 2:** $P_2 = 6$ MPa
State 3: $P_3 = 6$ MPa, $T_3 = 600$°C **State 4:** $P_4 = 1$ MPa
State 5: $P_5 = 1$ MPa, $T_3 = 600$°C **State 6:** $P_6 = 20$ kPa

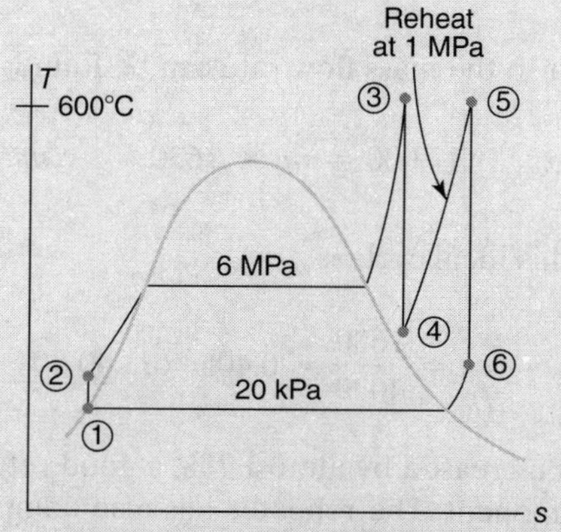

Figure 8.18

(Continued)

Example 8.6 (*Continued*)

Solution

Before any calculations are attempted, let's find the enthalpy at all the states. The IRC Calculator will be used.

At state 1 where $x_1 \doteq 0$ and $P_1 = 20$ kPa, the IRC Calculator gives $h_1 = 251\,000$ J/kg $\cong h_2$.

The turbine accepts superheated steam at 6000 kPa and 600°C for which we find $h_3 = 3\,660\,000$ J/kg (3660 kJ/kg) and $s_3 = 7170$ J/kg·K.

The steam undergoes an isentropic process from state 3 to state 4; that is, $s_4 = s_3 = 7170$ J/kg·K. Using this value for s_4 and 1000 kPa for the pressure, $h_4 = 3\,080\,000$ J/kg (3080 kJ/kg).

State 5 is at 1000 kPa and 600°C, so $h_5 = 3\,700\,000$ J/kg (3700 kJ/kg) and $s_5 = 8030$ J/kg·K.

At state 6, $s_6 = s_5$ and $P_5 = 20$ kPa, giving $h_6 = 2\,650\,000$ J/kg (2650 kJ/kg).

Note: State 6 is in the superheat region. No condensation would occur on the turbine blades.

We are now ready to respond to the four parts of this example.

i) The energy required by the boiler is the sum of the high-pressure section and the reheat section. It is

$$q_B = (h_3 - h_2) + (h_5 - h_4)$$
$$= (3660 - 251) + (3700 - 3080) = \underline{4029 \text{ kJ/kg}}$$

ii) The heat lost in the condenser is

$$q_C = h_6 - h_1 = 2650 - 251 = \underline{2400 \text{ kJ/kg}}$$

iii) The work provided by the turbine is the sum of both the high-pressure and low-pressure sections:

$$w_T = (h_3 - h_4) + (h_5 - h_6)$$
$$= (3660 - 3080) + (3700 - 2650) = 1630 \text{ kJ/kg}$$

The power is given so the mass flow rate can be found:

$$\dot{W}_T = \dot{m} \times w_T \qquad 12\,000 = \dot{m} \times 1630 \qquad \therefore \dot{m} = \underline{7.36 \text{ kg/s}}$$

iv) The efficiency of this ideal cycle is

$$\eta = \frac{w_T}{q_B} = \frac{1630}{4029} = 0.405 \text{ or } \underline{40.5\%}$$

The efficiency has increased by almost 7%, a good return for the addition of the reheat component. The reheater has also completely removed the excessive moisture near the exit of the turbine since state 6 is in the superheat region, a very important feature.

It should be noted that the efficiency could have been calculated after the boiler energy requirement and the condenser heat transfer were found since, considering a control surface surrounding the entire cycle, the first law requires

$$w_T = q_B - q_C = 4020 - 2399 = 1621 \text{ kJ/kg}$$

$$\therefore \eta_{\text{cycle}} = \frac{w_T}{q_B} = \frac{1621}{4020} = 0.403 \text{ or } 40.3\%$$

essentially the same as that calculated above.

Comment

The pump work is ignored in the efficiency calculation since it is small compared to the turbine work.

8.3.2 The ideal regenerative Rankine cycle

Another device that is used to improve the efficiency of a power plant is the feedwater heater. Significant energy is needed to heat the water from state 2 to the saturated temperature of the water in the boiler, as shown by the cross-hatched area of Fig. 8.19. This water can be preheated by extracting steam from the turbine at high temperature and pressure and mixing it with the condensate exiting the condenser in an *open feedwater heater,* thereby saving energy required by the boiler. Because the pressure of the condensate and the extracted steam must be at the same pressure, a condensate pump is needed to increase the pressure of the condensate to the pressure of the extracted steam. This is shown as an *ideal regenerative Rankine cycle* in Fig. 8.20. The heat saved in the boiler is somewhat offset by the fact that less steam is flowing through the turbine.

In the open feedwater heater, as sketched in Fig. 8.21, the steam from the turbine is directly mixed with the condensate from the condenser. The preheated feedwater then enters the feedwater pump at state 3 of Fig. 8.20. An open feedwater heater is a mixing chamber, as considered in Section 4.2.3. Mass and energy balances, assuming an insulated heater

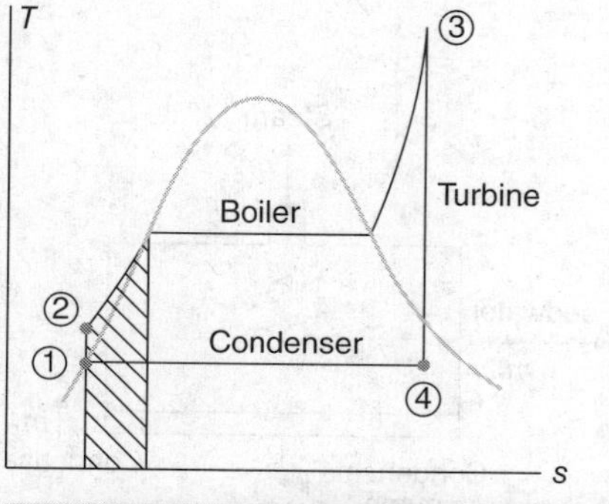

Figure 8.19

The heat transfer required to heat the water to the saturation state is shown by the cross-hatched area.

Open feedwater heater: A heat exchanger in which superheated steam and liquid water are mixed.

Regenerative Rankine cycle: A Rankine cycle that contains a feedwater heater.

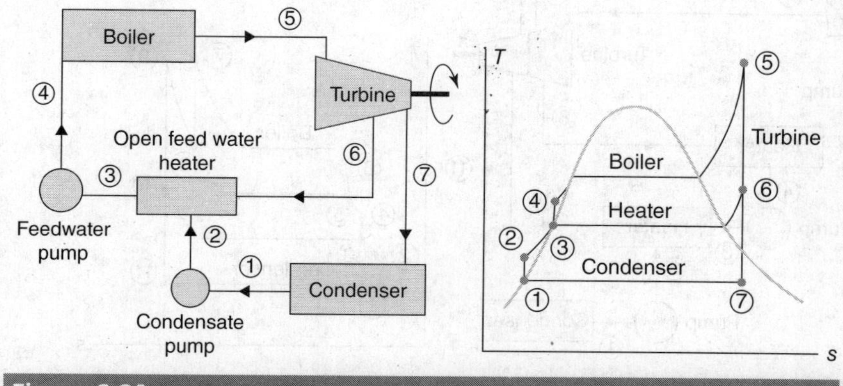

Figure 8.20

The ideal regenerative Rankine cycle with an open feedwater heater.

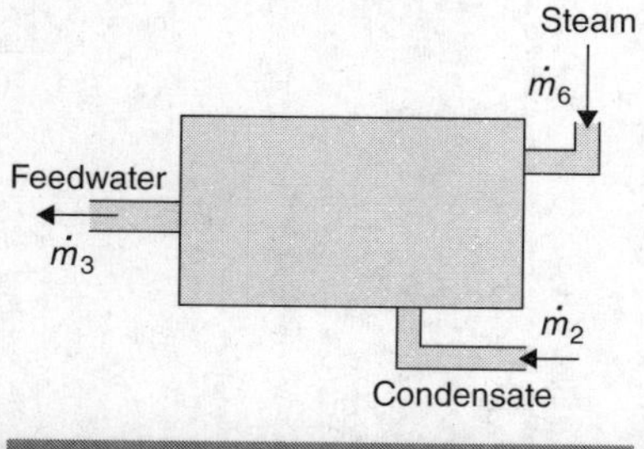

Figure 8.21

An open feedwater heater.

Comment

The extraction pressure for each heater is usually selected so that the saturation temperatures are equally spaced.

and negligible kinetic and potential energy changes, are written as (referring to Fig. 8.21 using the same state numbers as in Fig. 8.20),

$$\dot{m}_2 + \dot{m}_6 = \dot{m}_3 \qquad \text{(mass)} \tag{8.2}$$

$$\dot{m}_2 h_2 + \dot{m}_6 h_6 = \dot{m}_3 h_3 \qquad \text{(energy)} \tag{8.3}$$

providing the mass flux that must be extracted from the turbine:

$$\dot{m}_6 = \dot{m}_3 \frac{h_3 - h_2}{h_6 - h_2} \tag{8.4}$$

Knowing the properties of the extracted steam, the extracted mass flux $\dot{m}_6$ can be determined.

In the *closed feedwater heater* of Fig. 8.22, the two fluids are kept separate and do not mix, as considered in the heat exchanger of Section 4.2.4. The closed heater is more expensive but requires a small condensate pump, if one at all, and a throttle to reduce the pressure to the condenser pressure. If the small pump work is neglected, the same relation (Eq. 8.4) holds for a closed heater. Attention will be focused on the open heater in the examples and problems.

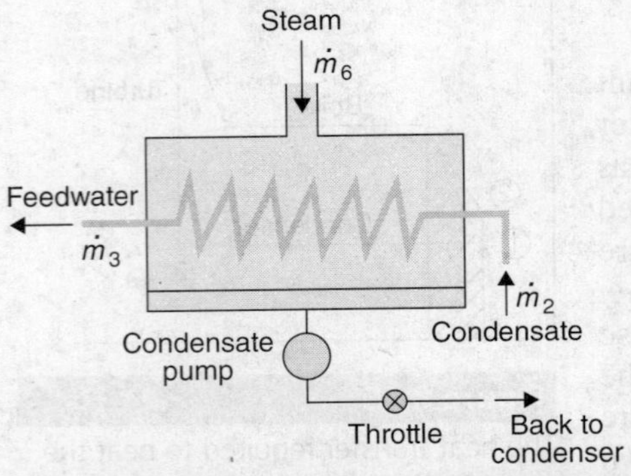

Figure 8.22

A closed feedwater heater.

Several feedwater heaters and one or two reheaters, needed to prevent condensation on the rear turbine blades, may be used in a power plant. The extraction pressure for each heater is usually selected so that the saturation temperatures are equally spaced. The reduction in mass flux, after the extracted steam is fed to a heater, results in less power produced by the turbine; however, the cycle efficiency is increased due to the reduction in energy required by the boiler. Millions of dollars are spent on even a small increase in efficiency since a power plant is in operation for such a long time.

An ideal Rankine cycle with a reheater and two feedwater heaters is sketched in Fig. 8.23. The feedwater heaters do not improve the condenser inlet quality, so a reheater is needed to move the condenser inlet quality near the

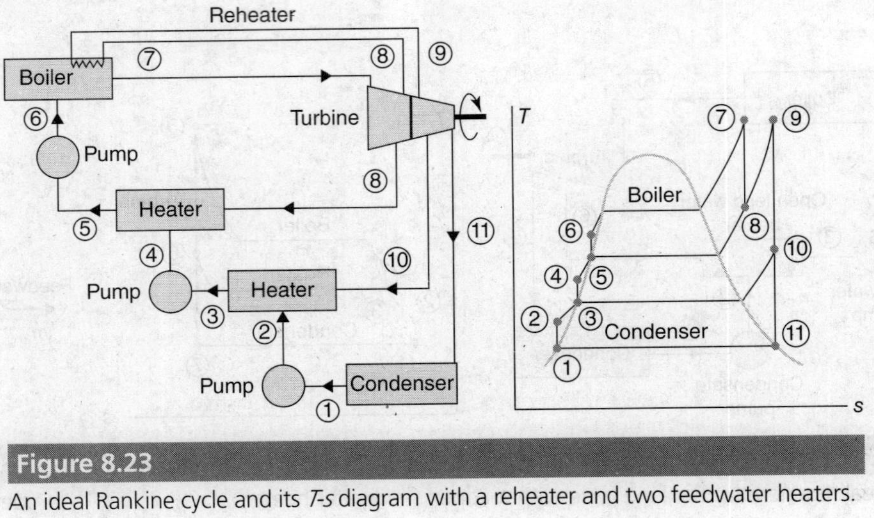

Figure 8.23

An ideal Rankine cycle and its *T-s* diagram with a reheater and two feedwater heaters.

saturated vapor state. The losses in the turbine will then allow the condenser inlet quality to be in the superheat region (see the dashed line of Fig. 8.9), thereby avoiding condensation on turbine blades, should that be a requirement of the turbine being used.

The effect of a feedwater heater — Example 8.7

The ideal Rankine cycle of Example 8.1 is modified by allowing an open feedwater heater to intercept steam at a pressure of 800 kPa (this about equally spaces the saturation temperatures for optimum performance), as sketched in the *T*-s diagram of Fig. 8.24. Calculate the cycle efficiency holding all other properties the same as in Example 8.1, as restated below. Assume no losses.

State 1: $P_1 = P_7$, $x_1 = 0$ — **State 2:** $P_2 = 800$ kPa

State 3: $P_3 = 800$ kPa, $x_3 = 0$ — **State 4:** $P_4 = 6$ MPa

State 5: $P_5 = 6$ MPa, $T_5 = 600°C$

State 6: $P_6 = 800$ kPa — **State 7:** $P_7 = 20$ kPa

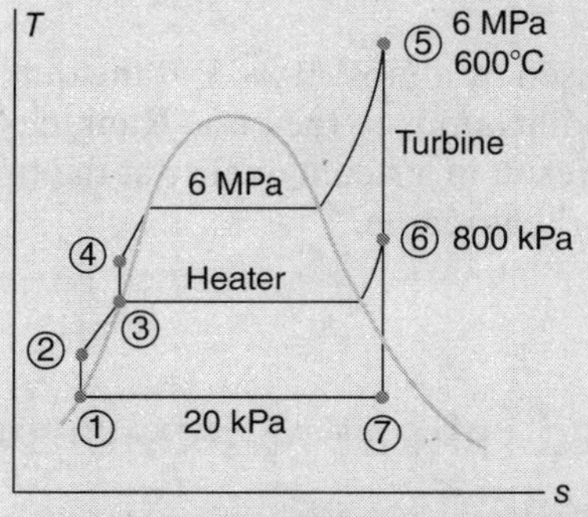

Figure 8.24

Solution

To begin, let's find the enthalpies at each state using the IRC Calculator. All enthalpies are needed when analyzing cycles.

State 1: $P_1 = 20$ kPa, $x_1 = 0$: $h_1 = 251$ kJ/kg

State 2: (neglect the pump work) $h_2 \cong h_1 = 251$ kJ/kg

State 3: $P_3 = 800$ kPa, $x_3 = 0$: $h_3 = 721$ kJ/kg

State 4: (neglect the pump work) $h_4 \cong h_3 = 721$ kJ/kg

State 5: $P_5 = 6$ MPa, $T_5 = 600°C$: $h_5 = 3660$ kJ/kg

State 6: $P_6 = 800$ kPa, $s_6 = s_5 = 7170$ J/kg·K: $h_6 = 3020$ kJ/kg

State 7: $P_7 = 20$ kPa, $s_7 = s_5 = 7170$ J/kg·K: $h_7 = 2360$ kJ/kg

Note: The units on entropy use J not kJ.

The mass flux of steam intercepted at state 6 is not given, so it must be calculated. Assuming $\dot{m}_3 = 1$ kg/s (since only the efficiency is of interest, the

(Continued)

Example 8.7 (*Continued*)

numerical value of the mass flux is not important, so select $\dot{m}_5 = 1$ kg/s for simplicity), Eqs. 8.2 and 8.3 provide $\dot{m}_2$ and $\dot{m}_6$ (see Fig. 8.21) as follows:

$$\left.\begin{aligned} \dot{m}_2 + \dot{m}_6 &= 1 \\ 251 \times \dot{m}_2 + 3020 \times \dot{m}_6 &= 721 \end{aligned}\right\} \quad \therefore \dot{m}_2 = 0.8303 \text{ kg/s}, \ \dot{m}_6 = 0.1697 \text{ kg/s}$$

The boiler heat requirement is, recognizing that $\dot{m}_3 = \dot{m}_4 = \dot{m}_5$ (see Fig. 8.20),

$$\dot{Q}_B = \dot{m}_5(h_5 - h_4) = 1 \times (3660 - 721) = 2940 \text{ kW}$$

The turbine output is

$$\begin{aligned} \dot{W}_T &= \dot{m}_5(h_5 - h_6) + (\dot{m}_5 - \dot{m}_6)(h_6 - h_7) \\ &= 1 \times (3660 - 3020) + (1 - 0.1697)(3020 - 2360) = 1188 \text{ kW} \end{aligned}$$

The cycle efficiency is now calculated to be

$$\eta = \frac{\dot{W}_T}{\dot{Q}_B} = \frac{1188}{2940} = 0.404 \quad \text{or} \quad \underline{40.4\%}$$

The efficiency has increased by almost 10%. But the quality exiting the turbine is quite low at 0.821. Modifications to the basic Rankine cycle (to be presented in the next section) can result in a steam quality at the turbine exit closer to 1.0 and possibly in the superheat region.

Note: An actual coal-fired power plant may have an efficiency in the neighborhood of 40%. Plants that combine the Rankine cycle with the Brayton gas cycle, considered in the next chapter in Section 9.7, can achieve efficiencies as high as 60%.

8.3.3 A combined reheat-regenerative ideal Rankine cycle

In today's power plants, both reheaters and feedwater heaters are used; in fact, several of each may be used. The reheaters guarantee that the steam exiting the turbine will be in the superheat region and the feedwater heaters keep the amount of fuel burned to a minimum. Together, they result in an ideal cycle efficiency that exceeds 50%. An example will illustrate two reheaters and two feedwater heaters. A power plant may have up to eight feedwater heaters.

Example 8.8 **A combined high-pressure cycle with two reheaters and two feedwater heaters**

A high-pressure (30 MPa) ideal Rankine cycle is modified by adding two reheaters and two open feedwater heaters, indicated by the *T-s* diagram of Fig. 8.25 with the properties listed below. The mass flux through the boiler at state 6 is 10 kg/s. Determine the turbine power output and the cycle efficiency, assuming no losses. The states are listed as follows:

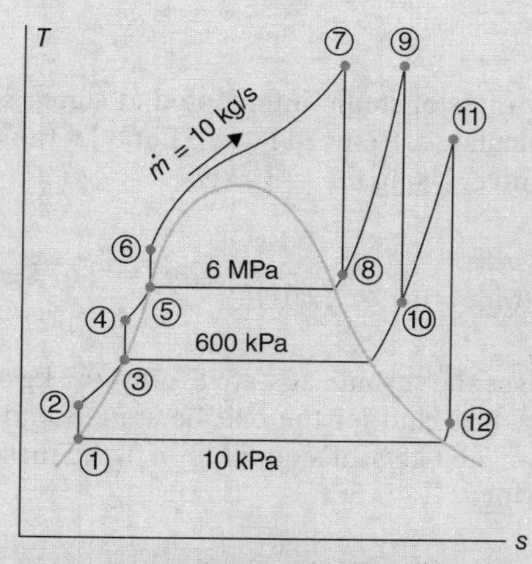

Figure 8.25

State 1: $P_1 = 10$ kPa, $x_1 = 0$ **State 2:** $P_2 = 600$ kPa
State 3: $P_3 = 600$ kPa, $x_3 = 0$ **State 4:** $P_4 = 6$ MPa
State 5: $P_5 = 6$ MPa, $x_5 = 0$ **State 6:** $P_6 = 30$ MPa
State 7: $P_7 = 30$ MPa, $T_7 = 600°C$ **State 8:** $P_8 = 6$ MPa
State 9: $P_9 = 6$ MPa, $T_9 = 600°C$ **State 10:** $P_{10} = 600$ kPa
State 11: $P_{11} = 600$ kPa, $T_{11} = 400°C$ **State 12:** $P_{12} = 10$ kPa

Solution

To begin, let's find the enthalpies, in kJ/kg, at each state using the IRC Calculator (all enthalpies have units of kJ/kg):

State 1: $P_1 = 10$ kPa, $x_1 = 0$ $\therefore h_1 = 192$
State 2: $P_2 = 600$ kPa $\therefore h_2 \cong h_1 = 192$
State 3: $P_3 = 600$ kPa, $x_3 = 0$ $\therefore h_3 = 670$
State 4: $P_4 = 6$ MPa $\therefore h_4 \cong h_3 = 670$
State 5: $P_5 = 6$ MPa, $x_5 = 0$ $\therefore h_5 = 1210$
State 6: $P_6 = 30$ MPa $\therefore h_6 \cong h_5 = 1210$
State 7: $P_7 = 30$ MPa, $T_7 = 600°C$ $\therefore h_7 = 3450$
State 8: $P_8 = 6$ MPa, $s_8 = s_7 = 6240$ J/kg·K $\therefore h_8 = 2990$
State 9: $P_9 = 6$ MPa, $T_9 = 600°C$ $\therefore h_9 = 3660$
State 10: $P_{10} = 600$ kPa, $s_{10} = s_9 = 7170$ J/kg·K $\therefore h_{10} = 2950$
State 11: $P_{11} = 600$ kPa, $T_{11} = 400°C$ $\therefore h_{11} = 3270$
State 12: $P_{12} = 10$ kPa, $s_{12} = s_{11} = 7710$ J/kg·K $\therefore h_{12} = 2440$

(Continued)

Example 8.8 (*Continued*)

The mass flow rates of steam intercepted at states 8 and 10 are not given, so they must be calculated. Refer to Fig. 8.21 and let the exit be state 5 and the inlets be states 4 and 8. Using $\dot{m}_5 = 10$ kg/s,

$$\left.\begin{aligned} \dot{m}_4 + \dot{m}_8 &= 10 \\ 670\dot{m}_4 + 2990\dot{m}_8 &= 10 \times 1210 \end{aligned}\right\} \quad \therefore \dot{m}_4 = 7.67 \text{ kg/s}, \ \dot{m}_8 = 2.33 \text{ kg/s}$$

Since 2.33 kg/s leaves the turbine at state 8, only 7.67 kg/s flows through state 9. Refer again to Fig. 8.21 and let the exit be state 3 and the inlets be states 2 and 10. Using $\dot{m}_3 = 7.67$ kg/s (it's equal to $\dot{m}_4$), the mass and energy balances provide the following:

$$\left.\begin{aligned} \dot{m}_2 + \dot{m}_{10} &= 7.67 \\ 192\dot{m}_2 + 2950\dot{m}_{10} &= 7.67 \times 670 \end{aligned}\right\} \quad \therefore \dot{m}_2 = 6.34 \text{ kg/s}, \ \dot{m}_{10} = 1.33 \text{ kg/s}$$

The turbine power output is, recognizing that

$$\begin{aligned} \dot{m}_{11} &= \dot{m}_9 - \dot{m}_{10} = 7.67 - 1.33 = 6.34 \text{ kg/s}, \\ \dot{W}_T &= \dot{m}_7(h_7 - h_8) + \dot{m}_9(h_9 - h_{10}) + \dot{m}_{11}(h_{11} - h_{12}) \\ &= 10(3450 - 2990) + 7.67(3660 - 2950) + 6.34(3270 - 2440) \\ &= \underline{15\,310 \text{ kW}} \end{aligned}$$

The boiler energy requirement is

$$\begin{aligned} \dot{Q}_B &= \dot{m}_7(h_7 - h_6) + \dot{m}_9(h_9 - h_8) + \dot{m}_{11}(h_{11} - h_{10}) \\ &= 10(3450 - 1210) + 7.67(3660 - 2990) + 6.34(3270 - 2950) \\ &= 29\,570 \text{ kW} \end{aligned}$$

The ideal cycle efficiency is now calculated to be

$$\eta = \frac{\dot{W}_T}{\dot{Q}_B} = \frac{15\,310}{29\,570} = 0.518 \quad \text{or} \quad \underline{51.8\%}$$

The quality of state 12 remains in the wet region at 94% but would move into the superheat region if the turbine efficiency was included in the analysis. Modern-day power plants will have perhaps eight feedwater heaters, with the result that higher ideal cycle efficiency is achieved.

☑ **You have completed Learning Outcome (2)**

8.4 Cogeneration Cycles

As observed in Example 8.2, up to two-thirds of the heat generated in a power plant is not converted to useful power by the turbine, but is often dumped into the environment from the condenser into lakes, rivers, oceans, or the atmosphere via cooling towers. The basic idea behind a cogeneration plant is to make use of this waste heat. Extracted superheated steam from the turbine can be used in industrial processes. Hot water from the condenser can be used to heat buildings or greenhouses, melt ice on roads or bridges, or even cool buildings using absorption refrigeration (see Section 10.3). To allow cogeneration, power plants are located near or in cities and next to universities. The relatively small power plant on a university campus cannot compete with the large utilities on the efficient conversion of energy from coal to electricity, but when the heat rejected from the condenser is utilized to heat and cool dorms and classroom buildings, the effective efficiency is sufficiently high that such power plants are cost effective. A *cogeneration cycle* is a power cycle that utilizes the output of the boiler to produce more than one useful form of energy. Perhaps the most common cogeneration cycle exists in every vehicle: Power is generated, and the waste heat can warm the occupants.

Cogeneration cycle: A power cycle that utilizes the output of the boiler to produce more than one useful form of energy.

A cogeneration cycle is sketched in Fig. 8.26 and an ideal *T-s* diagram in Fig. 8.27. The steam power plant produces mechanical power from the turbine, but steam from the turbine is also used in a process that requires heat. If large amounts of process steam are required, steam could also be extracted from the main steam line before it enters the turbine, shown by the dotted lines in Fig. 8.26; a throttle would then be required to reduce the pressure to the process heater pressure. The difference between the process heater and the feedwater heater (their *T-s* diagrams appear to be the same) is that heat leaves the cycle to be used in the process heater, whereas no heat leaves the cycle with a feedwater heater.

Comment

Waste heat is used to heat and cool university buildings, resulting in about a 70% utilization of the energy from coal, compared to possibly 40% utilization by commercial power plants.

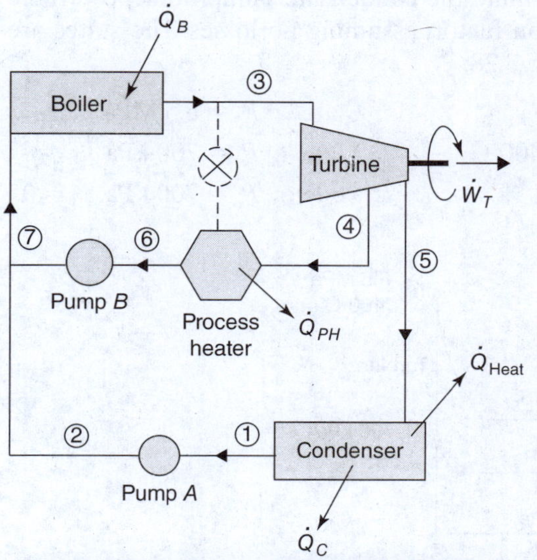

Figure 8.26

A cogeneration cycle.

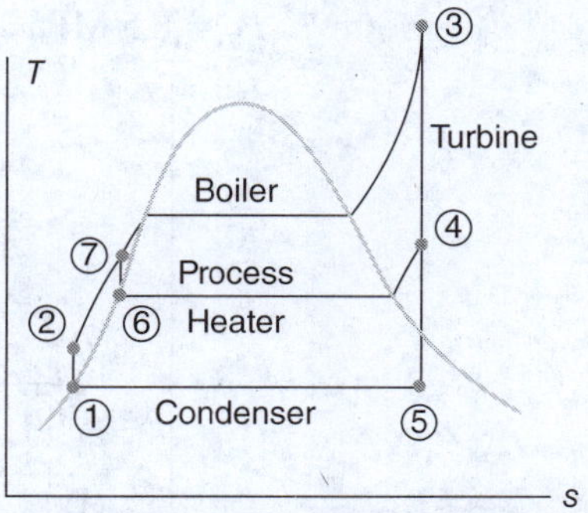

Figure 8.27

The *T-s* diagram of an ideal cogeneration Rankine cycle.

Utilization factor ε: The ratio of the energy utilized and the energy input to the boiler.

The performance parameter for a cogeneration cycle is the *utilization factor* ε, which is the ratio of the energy utilized and the energy input to the boiler. The total energy utilized is the power produced by the turbine plus the heat utilized by the process heater plus the heat used from the condenser. (We again ignore the small amount of energy to drive the pumps. It can certainly be included, if desired.) The utilization factor ε is

$$\varepsilon = \frac{\dot{W}_T + \dot{Q}_{PH} + \dot{Q}_{\text{Heat}}}{\dot{Q}_B} \tag{8.5}$$

Cogeneration is the environmentally friendly, economically sensible way to produce power, simultaneously saving significant amounts of money and also dramatically reducing total greenhouse gas emissions. However, to accomplish these desired effects, a power plant must be located near an industry, or a concentrated group of people requiring large amounts of heat and cooling, an objective not easily satisfied for the large power plants presently in operation. However, The U.S. Department of Energy has a goal of having cogeneration plants comprise 20% of the U.S. generation capacity by the year 2030.

Example 8.9 An ideal cogeneration cycle

Steam leaves the steam generator, as shown in the *T*-s diagram of Fig. 8.28, at 6.5 MPa and 400°C. The mass flux from the steam generator is 13.6 kg/s; 6.8 kg/s is removed at 700 kPa from the turbine at state 4 for process heating (not reheating) and reintroduced at zero quality. The steam exits the turbine at a pressure of 35 kPa. Fifty percent of the condenser heat transfer is used to heat and/or cool buildings. Determine the condensate pump horsepower, the cycle efficiency, and the utilization factor, assuming no losses. The states are summarized with reference to Fig. 8.28:

State 1: $P_1 = 35$ kPa, $x_1 = 0$ | **State 2:** $P_2 = 6.5$ MPa
State 3: $P_3 = 6.5$ MPa, $T_3 = 400$°C | **State 4:** $P_4 = 700$ kPa
State 5: $P_5 = 35$ kPa | **State 6:** $P_6 = 700$ kPa, $x_1 = 0$
State 7: $P_7 = 6.5$ MPa

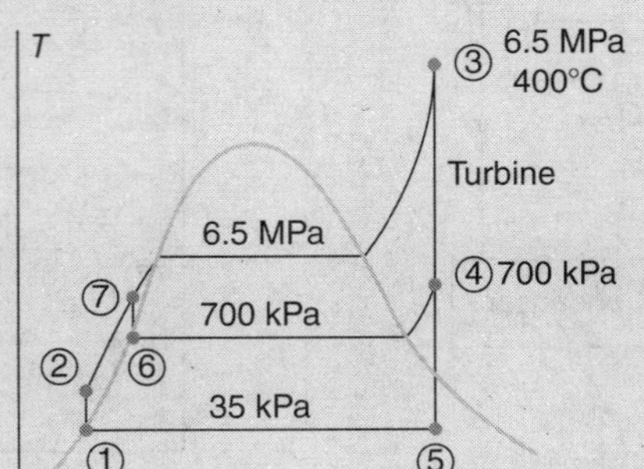

Figure 8.28

Solution

To begin, the enthalpies, in kJ/kg, at each state using the IRC Calculator are listed:

State 1: $P_1 = 35$ kPa, $x_1 = 0$	$\therefore h_1 = 300$
State 2: $P_2 = 6.5$ MPa	$\therefore h_2 \cong h_1 = 300$
State 3: $P_3 = 6.5$ MPa, $T_3 = 400°C$	$\therefore h_3 = 3237$
State 4: $P_4 = 700$ kPa, $s_4 = 6.62$ kJ/kg·K	$\therefore h_4 = 2725$
State 5: $P_5 = 35$ kPa, $s_5 = 6.62$ kJ/kg·K	$\therefore h_5 = 2252$
State 6: $P_6 = 700$ kPa, $x_1 = 0$	$\therefore h_6 = 696$
State 7: $P_7 = 6.5$ MPa	$\therefore h_7 \cong h_6 = 696$

The condensate pump horsepower requirement is

$$\dot{W}_{CP} = \dot{m}_5 \frac{P_2 - P_1}{\rho}$$

$$= 6.8 \frac{\text{kg}}{\text{s}} \times \frac{(6500 - 35)\text{ kPa}}{1000\text{ kg/m}^3} = 43.96\text{ kW or } \underline{58.9\text{ hp}}$$

The turbine power output is calculated to be

$$\dot{W}_T = \dot{m}_3(h_3 - h_4) + \dot{m}_5(h_4 - h_5)$$
$$= 13.6(3237 - 2725) + 6.8(2725 - 2252) = 10\,180\text{ kW}$$

The steam generator heat requirement is

$$\dot{Q}_B = \dot{m}_1(h_3 - h_1) + \dot{m}_6(h_3 - h_6)$$
$$= 6.8(3237 - 300) + 6.8(3237 - 696) = 37\,250\text{ kW}$$

The steam utilized by the process heater is

$$\dot{Q}_{PH} = \dot{m}_6(h_4 - h_6) = 6.8(2725 - 696) = 13\,800\text{ kW}$$

The condenser disposes of the following heat rate:

$$\dot{Q}_C = \dot{m}(h_5 - h_1) = 6.8(2252 - 300) = 13\,270\text{ kW}$$

Half of this wasted heat is used to heat and/or cool buildings, i.e.,

$$\dot{Q}_{\text{Heat}} = \frac{\dot{Q}_C}{2} = \frac{13\,270}{2} = 6635\text{ kW}$$

The cycle efficiency is calculated to be

$$\eta = \frac{\dot{W}_T}{\dot{Q}_B} = \frac{10\,180}{37\,250} = 0.273 \text{ or } 27.3\%$$

The utilization factor is a better measure of the performance of this cycle. It is

$$\varepsilon = \frac{\dot{W}_T + \dot{Q}_{PH} + \dot{Q}_{\text{Heat}}}{\dot{Q}_B} = \frac{10\,180 + 13\,800 + 6635}{37\,250} = 0.821 \text{ or } 82.1\%$$

Clearly, cogeneration allows much better utilization of the energy generated in the boiler of a power plant.

Comment

If drawn to scale, states 1 and 2 and states 6 and 7 would be on top of each other.

☑ You have completed Learning Outcome (3)

8.5 Losses in Power Plants

The actual vapor power cycle deviates from the Rankine cycle due to losses in all of the components that make up the cycle. The efficiency of a Carnot cycle operating between the high and low temperatures of Example 8.1 would be

$$\eta_{\text{Carnot}} = 1 - \frac{T_L}{T_H} = 1 - \frac{333.1}{873} = 0.618 \quad \text{or} \quad 61.8\% \tag{8.6}$$

Comment

The Rankine cycle efficiency is much less than the Carnot cycle efficiency due to the irreversible heat transfer over a large temperature difference in the boiler.

compared to 38% of the ideal Rankine cycle. Why is there such a difference in the efficiencies when each cycle is "ideal" and cannot exist in actual operation? The primary difference is due to the heat transfer in the boiler of the Rankine cycle; the heat transfer occurs over a very large temperature difference. The hot gases from the burning coal are at a temperature much, much higher than the steam in the pipes of the boiler. Hence, the heat transfer process is extremely irreversible, even though it is referred to as being "ideal." In the Carnot cycle the reversible heat transfer process is allowed to occur over only an infinitesimal temperature difference, a process that cannot be realized in a power plant, nor in any real-life situation. Thus, the efficiency of the Carnot cycle is much higher than that of the Rankine cycle.

Note: The efficiency of a modern-day turbine can approach 90%.

The only loss that makes the efficiency of an actual vapor cycle significantly lower than the efficiency of the ideal Rankine cycle is the loss in the turbine. The losses are primarily due to the flow of steam around the turbine blades. As observed in Example 8.2, the turbine losses result in a dryer steam at the turbine exit. With reheat and losses, the steam at the turbine exit will always be at a high quality or in the superheat region, a desirable objective in turbine operations. Modern-day turbines have an efficiency approaching 90%. New long blades have been developed to improve turbine efficiency. Computational fluid dynamics and advanced measurement techniques have resulted in high-performance blades.

Note: The combustion process is approximately 90% efficient. Conversion from mechanical to electrical energy is about 92% efficient. Plant auxiliaries use around 6% of the plant's electrical output.

Additional losses in every actual vapor power cycle occur in the pipes that connect all the components. These losses manifest themselves as pressure drops between the various components and in the piping system in the boiler. To compensate for the pressure drops, the outlet pressures from the various pumps are simply increased, requiring added power to operate each pump. But since the pump work can be neglected in efficiency calculations, the added power is not of significance in calculating cycle efficiency.

The overall plant efficiency is different from the cycle efficiency since it includes the transfer of energy in the fuel to energy in the steam, an efficiency of approximately 90%. And, the conversion of the mechanical power from the turbine via a rotating shaft to electrical power ready to be distributed to the electrical grid is about 92% efficient. Finally, numerous pieces of auxiliary equipment that use electrical power are required in the operation of the plant. Around 6% of the plant's power output is needed to operate this auxiliary equipment.

Comment

There are always irreversibilities in the boiler and the condenser since heat transfer occurs across large temperature differences, inherently irreversible processes.

For an actual vapor cycle efficiency of 50%, the added losses and power needs of equipment in the plant bring the overall efficiency of the power plant to around 39%. Just in excess of one-third of the energy contained in the fuel is converted to energy that is sent out over the electrical grid of a modern-day power plant.

The efficiency of a power plant Example 8.10

The turbine of the combined cycle of Example 8.8 has an efficiency of 87%. Assume that the pressures exiting the pumps are sufficiently high so as to provide the pressures shown in that example to the turbine and the heaters. Also assume that the inlet quality to the pumps is zero, with no cooling below the saturated liquid state. Estimate the overall plant efficiency using efficiencies suggested in this section.

Solution

The efficiency of the ideal combined cycle of Example 8.8 is 0.518. The turbine efficiency of 0.87 lowers the cycle efficiency to

$$\eta_{\text{combined cycle}} = \eta_{\text{ideal cycle}} \times \eta_{\text{turbine}} = 0.518 \times 0.87 = 0.45$$

Including the additional efficiencies described in this section, the overall plant efficiency is estimated to be

$$\begin{aligned}\eta_{\text{plant}} &= \eta_{\text{combined cycle}} \times \eta_{\text{combustion}} \times \eta_{\text{electrical}} \times \eta_{\text{auxiliaries}} \\ &= 0.45 \times 0.9 \times 0.92 \times 0.94 = 0.35 \text{ or } \underline{35\%}\end{aligned}$$

If the lost heat from the condenser and process heat could be utilized, a utilization factor of possibly 80% could be realized, with only 20% of the energy contained in the fuel going to waste. Perhaps future power plants could be located close to industrial complexes with condenser waste heat piped to population centers.

☑ **You have completed Learning Outcome** **(4)**

8.6 Summary

The basic Rankine cycle, which forms the foundation for all steam power plants, was presented in this chapter. The ideal cycle, in which the turbine and pump are considered to be 100% efficient, was considered first. Two major modifications of the ideal cycle were analyzed: the reheater and the feedwater heater. Both of these improve cycle efficiency. Examples included combinations of reheaters and feedwater heaters that substantially increased cycle efficiency. In modern-day power plants, several reheaters and perhaps eight feedwater heaters may be used.

Losses were then considered in some detail in the various components of the Rankine cycle. In addition, use of the high-pressure steam for industrial processes and the condenser waste heat for both heating and cooling of buildings was began with the introduction of a utilization factor, which is a measure of the increase in the effectiveness of a power plant. Such power plants exist on many university campuses and near industrial plants.

The following additional terms were introduced or emphasized in this chapter:

Boiler (steam generator): *A heat exchanger used to transfer heat from the fuel to the working fluid.*

Cogeneration cycle: *A Rankine cycle that utilizes the output of the boiler to produce more than one useful form of energy.*

Combined Rankine cycle: *A modified Rankine cycle that includes both reheat and regeneration.*

Condensate: *The water exiting the condenser.*

Condenser: *A heat exchanger that condenses the steam into a saturated liquid.*

Feedwater heater: *A heat exchanger in which superheated steam from the turbine is used to preheat the condensate before it enters the boiler.*

Feedwater pump: *The device used to produce the high pressure required at the turbine inlet.*

Ideal Rankine cycle: *It provides an upper limit of the cycle efficiency since there are no losses.*

Plant efficiency: *The energy output of the plant divided by the energy input in the fuel burned.*

Rankine cycle efficiency: *The net work output of the cycle divided by the energy input to the cycle.*

Reheat Rankine cycle: *The steam in the turbine is returned to the boiler, reheated, and returned to the turbine.*

Regenerative Rankine cycle: *Superheated steam from the turbine is used to preheat the condensate from the condenser before it enters the boiler.*

Steam turbine: *The device that creates mechanical power from the superheated steam that leaves the boiler.*

Several important equations were utilized in this chapter:

Cycle efficiency: $$\eta_{\text{cycle}} = \frac{w_T - w_P}{q_B} = \frac{\dot{W}_T - \dot{W}_P}{\dot{Q}_B} \cong \frac{\dot{W}_T}{\dot{Q}_B}$$

Boiler heat transfer rate: $$\dot{Q}_B = \dot{m}(h_{\text{in}} - h_{\text{out}})$$

Turbine power: $$\dot{W}_T = \dot{m}(h_{\text{in}} - h_{\text{out}})$$

Pump power: $$\dot{W}_P = \dot{m}\frac{P_{\text{out}} - P_{\text{in}}}{\rho}$$

Condenser heat loss rate: $$\dot{Q}_C = \dot{m}(h_{\text{in}} - h_{\text{out}})$$

Utilization factor: $$\varepsilon = \frac{\dot{W}_T + \dot{Q}_{PH} + \dot{Q}_{\text{Heat}}}{\dot{Q}_B}$$

Problems

FE Exam Practice Questions (afternoon M.E. discipline exam)

8.1 The component that results in a relatively low Rankine cycle efficiency is:

- **(A)** The condenser
- **(B)** The turbine
- **(C)** The boiler
- **(D)** The pump

8.2 Regeneration allows:

- **(A)** The condenser to accept steam at a lower quality
- **(B)** The turbine energy to be restored
- **(C)** The boiler water to be preheated
- **(D)** The pump to pressurize the water before it enters the boiler

8.3 Cogeneration occurs when:

- **(A)** The condenser's lost heat is utilized
- **(B)** The turbine's large decrease in energy content is used for heating purposes
- **(C)** The boiler's heat transfer is used for heating purposes rather than generating electricity
- **(D)** The pump creates a larger pressure than needed by the turbine

Problems 8.4–8.12 will refer to the ideal Rankine cycle of Figures 8.29 and 8.30. The high and low pressures are 15 MPa and 10 kPa, respectively, and the high temperature is 600°C. The water mass flow rate is 8 kg/s.

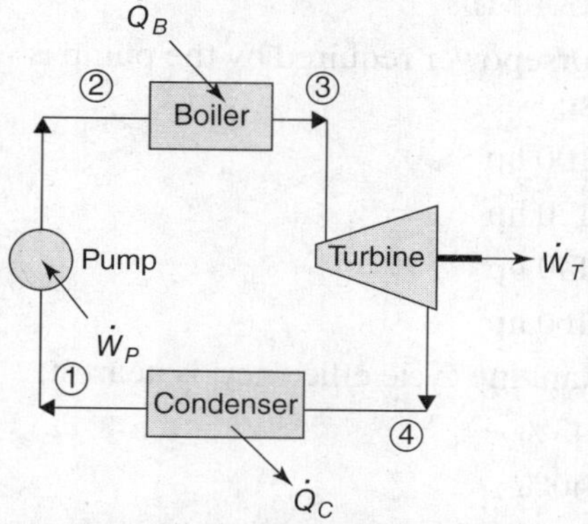

Figure 8.29

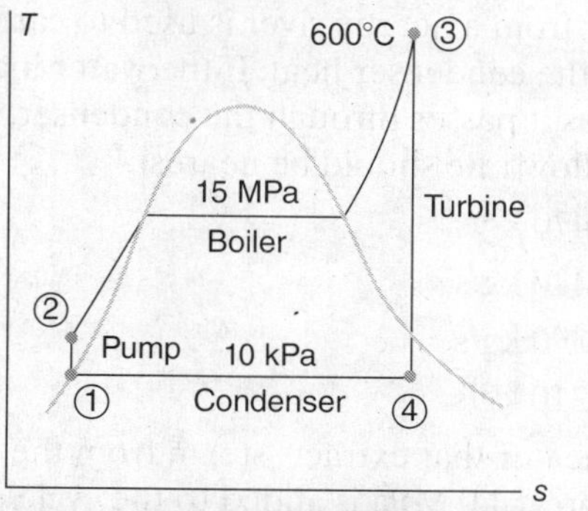

Figure 8.30

8.4 The rate of heat transfer to the steam in the boiler is nearest:

- **(A)** 20.6 MJ/s
- **(B)** 27.1 MJ/s
- **(C)** 31.9 MJ/s
- **(D)** 43.2 MJ/s

8.5 The power produced by the turbine is nearest:

(A) 16 MW

(B) 14 MW

(C) 12 MW

(D) 10 MW

8.6 The rate of heat transfer from the condenser is nearest:

(A) 18.1 MJ/s

(B) 16.9 MJ/s

(C) 15.4 MJ/s

(D) 13.1 MJ/s

8.7 The horsepower required by the pump is nearest:

(A) 100 hp

(B) 120 hp

(C) 140 hp

(D) 160 hp

8.8 The Rankine cycle efficiency is nearest:

(A) 43%

(B) 40%

(C) 38%

(D) 35%

8.9 Water from a nearby river is used to carry away the condenser heat. If the water increases 12°C as it passes through the condenser, its mass-flow rate should be nearest:

(A) 470 kg/s

(B) 420 kg/s

(C) 390 kg/s

(D) 310 kg/s

8.10 A reheater that extracts steam from the turbine at 1.0 MPa is added to the cycle of Fig. 8.29 and reheats it to 600°C at constant pressure. Such a device increases the cycle efficiency to:

(A) 51%

(B) 49%

(C) 47%

(D) 45%

8.11 If 50% of the heat from the condenser of Fig. 8.29 is used to heat buildings, the utilization factor ε of the cycle is nearest:

(A) 71%

(B) 67%

(C) 64%

(D) 61%

8.12 An open feedwater heater extracts steam from the turbine of Fig. 8.29 at 1.0 MPa and 200°C. Saturated liquid water leaves the heater. The mass flux of the extracted steam should be nearest:

(A) 2.1 kg/s

(B) 1.9 kg/s

(C) 1.7 kg/s

(D) 1.5 kg/s

Simple Rankine Cycles

8.13 An ideal Rankine power cycle is shown in Fig. 8.31. Steam leaves the steam generator at 4 MPa and 500°C with a mass flow rate of 8 kg/s. The steam leaves the turbine at 40 kPa. Sketch the cycle on a *T-s* diagram; then calculate:

i) The power output of the turbine

ii) The rate of heat loss from the condenser

iii) The pump horsepower requirement

iv) The rate of heat addition by the boiler

v) The cycle efficiency

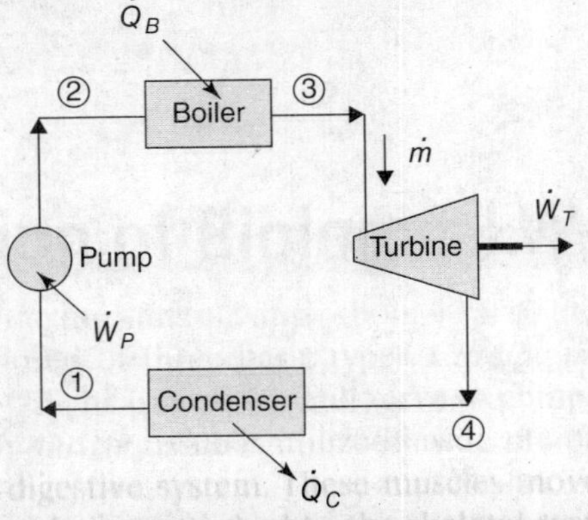

Figure 8.31

8.14 Work Problem 8.13, retaining all quantities except with the following turbine exit pressure:

a) 50 kPa *b*) 30 kPa
c) 20 kPa *d*) 10 kPa

Comment

Near an actual turbine exit, $x_4 \geq 1$, so droplets do not form on the last rows of the turbine blades.

8.15 Work Problem 8.13, retaining all quantities except with the following turbine inlet conditions:

a) 4 MPa, 400°C *b*) 4 MPa, 600°C
c) 4 MPa, 700°C *d*) 4 MPa, 800°C

Comment

Metal turbine blades limit the entering temperature to just over 600°C; ceramic and cooled blades in the leading rows allow higher temperatures.

8.16 Work Problem 8.13, retaining all quantities except with the following turbine inlet conditions:

a) 6 MPa, 600°C *b*) 10 MPa, 600°C
c) 20 MPa, 600°C *d*) 30 MPa, 600°C

Comment

In some power plants, the maximum pressure exceeds the critical pressure of 22 MPa.

8.17 An ideal Rankine power cycle shown in Fig. 8.32 has a boiler that produces steam at 4 MPa and 500°C with a mass flux of 8 kg/s. The steam leaving the turbine has a pressure of 70 kPa. Sketch this cycle on a *T-s* diagram and then calculate:

i) The power output of the turbine
ii) The rate of heat loss from the condenser
iii) The pump horsepower requirement
iv) The rate of heat addition by the boiler
v) The cycle efficiency

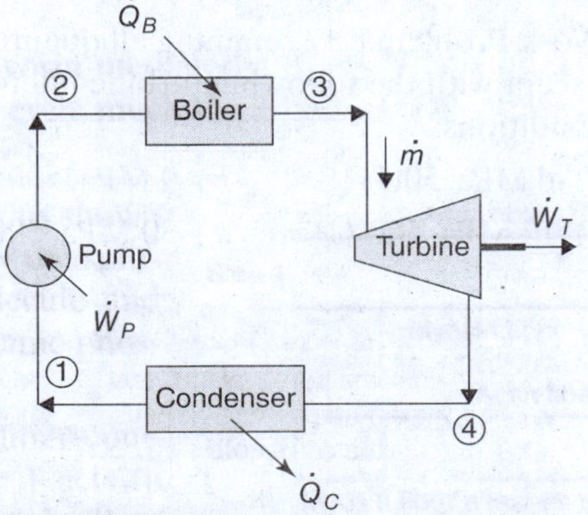

Figure 8.32

8.18 Work Problem 8.17, retaining all quantities except with the following turbine exit pressure:

a) 60 kPa *b*) 55 kPa
c) 30 kPa *d*) 15 kPa

Comment

The quality near the turbine exit should be at least 1.0, so droplets do not form on the last rows of the turbine blades.

8.19 Work Problem 8.17, retaining all quantities except with the following turbine inlet conditions:

a) 4 MPa, 350°C *b*) 4 MPa, 500°C
c) 4 MPa, 600°C *d*) 4 MPa, 800°C

Comment

Metal turbine blades limit the entering temperature to just over 600°C; ceramic and cooled blades in the leading rows allow higher temperatures.

8.20 Work Problem 8.17, retaining all quantities except with the following turbine inlet conditions:

a) 6 MPa, 500°C *b*) 9 MPa, 500°C
c) 12 MPa, 500°C *d*) 30 MPa, 500°C

Comment

In some power plants, the maximum pressure exceeds the critical pressure of 22 MPa.

8.21 For Problem 8.13 assume that the turbine has an efficiency of 90% and the pump has an efficiency of 85%. Calculate the power output and efficiency of the cycle.

8.22 For turbine efficiencies of i) 0.82, ii) 0.84, iii) 0.86, and iv) 0.88 in Problem 8.13 with a pump efficiency of 80%, calculate the cycle efficiency. State your observation of the thermodynamic efficiency of the cycle as a function of the turbine efficiency.

8.23 A geothermal energy source is to be used in the operation of a Rankine power cycle using R134a as the working fluid. The R134a is to leave the heater at 400 kPa and 90°C and the ideal turbine at 14 kPa. Sketch a *T*-s diagram. For a mass flow rate of 2.2 kg/s, calculate:

i) The maximum power to be produced by the turbine
ii) The heat loss in the condenser
iii) The pump horsepower required
iv) The rate of heat addition to the R134a
v) The efficiency of the ideal cycle

Modified Rankine Cycles

8.24 An ideal Rankine power cycle with reheat is shown in Fig. 8.33. Steam leaves the boiler at 8 MPa and 700°C with a mass flux of 20 kg/s. It leaves the high-pressure turbine (HPT) at 1 MPa and is reheated to 700°C before being sent back to the low-pressure turbine (LPT). The steam then enters the condenser at 20 kPa. Cooling water carries away the heat from the condenser. Refer to the *T*-s diagram of Fig. 8.17 and calculate:

i) The heat transfer rate by the boiler from state 2 to state 3
ii) The heat transfer rate in the reheat process
iii) The total power output of the turbine
iv) The rate of heat loss from the condenser
v) The state of the steam leaving the turbine
vi) The pump horsepower
vii) The efficiency of the cycle
viii) The mass flux of cooling water as it passes through the condenser if a 40°C increase in the cooling water is allowed

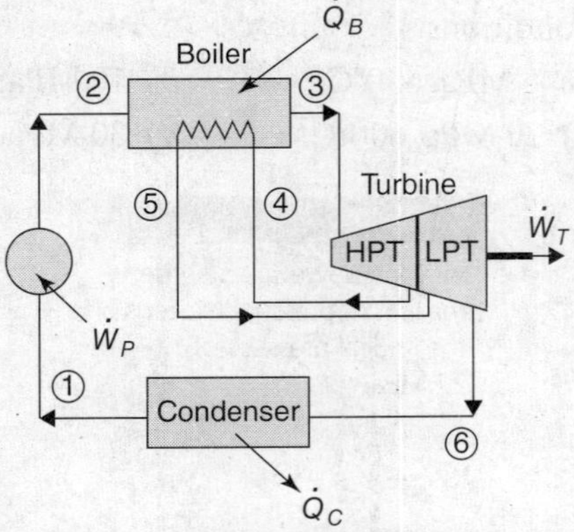

Figure 8.33

Warning: The problems in this section can take a considerable amount of time to solve. Take time to organize your information so that it is easily available for the calculations required. Always check your work. Ask yourself "Does this make sense?" Use of the IRC Calculator will allow you to go to a party after you finish your homework!

8.25 Work Problem 8.24 with the following turbine reheat intercept pressure: *a*) 200 kPa and *b*) 600 kPa.

8.26 Work Problem 8.24 with a condenser pressure of *a*) 10 kPa and *b*) 5 kPa.

8.27 Determine the state (quality, or temperature if superheat) of the steam exiting the turbine of Problem 8.24 if the efficiency of both stages of the turbine is *a*) 92% and *b*) 86%.

8.28 In the ideal Rankine power cycle with reheat of Fig. 8.33, steam leaves the steam generator at 9 MPa and 600°C with a mass flow rate of 22 kg/s. It leaves the high-pressure turbine at 1 MPa and is reheated to 600°C at this pressure. The steam leaves the low-pressure turbine at 40 kPa. The condenser heat is carried away by cooling water. Refer to the *T-s* diagram of Fig. 8.17 and calculate:

i) The heat transfer rate by the boiler from state 2 to state 3
ii) The heat transfer rate in the reheat process
iii) The total power output of the turbine
iv) The rate of heat loss from the condenser
v) The state of the steam leaving the turbine
vi) The pump horsepower
vii) The efficiency of the cycle
viii) The mass flux of cooling water as it passes through the condenser if a 40°C increase in the cooling water is allowed

8.29 Work Problem 8.28 for the following turbine reheat intercept pressure: *a*) 200 kPa and *b*) 600 kPa.

8.30 An ideal Rankine cycle with two reheaters is shown in Fig. 8.34. Low pressure is 10 kPa, high pressure is 10 MPa, and the high temperature is 600°C. The first reheater heats 2 MPa steam to 500°C, and the second reheater heats 200 kPa steam to 400°C. For a mass flux of 20 kg/s, show the cycle on a *T-s* diagram and then determine:

i) The heat added during the first reheat process
ii) The heat added during the second reheat process
iii) The total heat added to the power cycle
iv) The work output of the turbine
v) The state of the steam leaving the turbine
vi) The pump horsepower
vii) The cycle efficiency

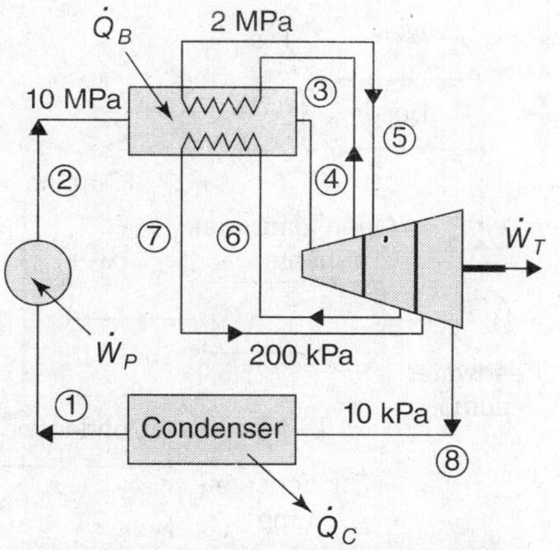

Figure 8.34

8.31 An ideal regenerative Rankine cycle with an open feedwater heater is shown in Fig. 8.35. Steam leaves the boiler at 8 MPa and 600°C with a mass flow rate of 20 kg/s. It is extracted from the turbine at 1 MPa and sent to the open feedwater heater. The remaining steam leaves the turbine at 20 kPa. Assume the turbine and pumps are isentropic. Sketch this cycle on a *T-s* diagram and then determine:

i) The heat flow rate into the steam generator
ii) The mass flow rate diverted to the feedwater heater
iii) The net power output of the turbine
iv) The net power input to the pumps
v) The quality of the steam exiting the turbine
vi) The cycle efficiency

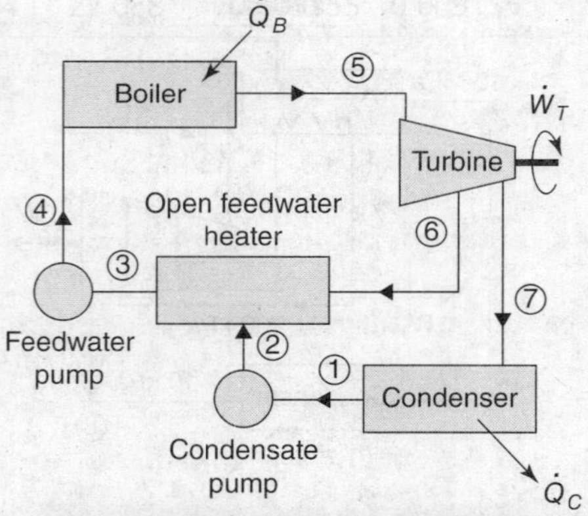

Figure 8.35

Note: The pressures at states 2, 3, and 6 are all equal.

8.32 Work Problem 8.31, retaining all quantities except the feedwater heater pressure at state 6 is *a*) 1200 kPa and *b*) 800 kPa.

8.33 The regenerative Rankine cycle shown in Fig. 8.35 operates with a turbine efficiency of 90% and a mass flow rate of 22 kg/s. Steam leaves the steam generator at 9 MPa and 550°C. Steam is removed from the turbine at 1.2 MPa and sent to the open feedwater heater. The remaining steam leaves the turbine at 30 kPa. Sketch this cycle on a *T-s* diagram and calculate:

i) The heat flow rate into the steam generator
ii) The mass flow rate diverted to the feedwater heater
iii) The power output of the turbine
iv) The net power input to the pumps
v) The quality of the steam exiting the turbine
vi) The thermodynamic efficiency of the cycle

8.34 Work Problem 8.33, retaining all quantities except the preheater intercept pressure which at Section 6 is *a*) 1.4 MPa and *b*) 1.0 MPa.

8.35 An ideal regenerative Rankine cycle contains two open reheaters as shown in the *T-s* diagram of Fig. 8.36. Steam with a flow rate of 20 kg/s leaves the boiler at 8 MPa and 600°C. Steam is removed from the turbine at 2 MPa and at 400 kPa. The remaining steam leaves the turbine at 20 kPa. Calculate:

i) The heat flow rate into the boiler
ii) The mass flow rate diverted to each feedwater heater
iii) The power output of the turbine
iv) The quality of the steam exiting the turbine
v) The net power input to the three pumps
vi) The heat loss from the condenser
vii) The thermodynamic efficiency of the cycle

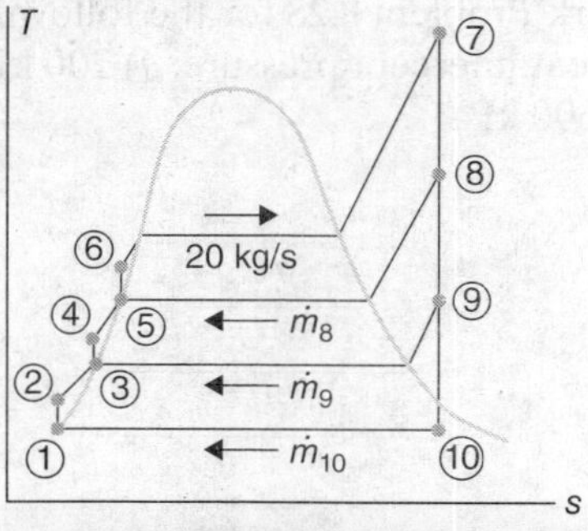

Figure 8.36

8.36 Figure 8.37 shows an ideal Rankine power cycle with regeneration and reheat. Steam leaves the boiler at 8 MPa and 600°C with a mass flow rate of 20 kg/s. Some steam is removed from the high-pressure turbine at 500 kPa and sent to an open feedwater heater. The remaining steam flow leaves the high-pressure turbine at 200 kPa, where it is reheated to 500°C and sent to the low-pressure turbine. The steam then leaves the low-pressure turbine at 20 kPa. Sketch this cycle on a *T-s* diagram and determine:

i) The steam flux extracted to go to the feedwater heater
ii) The total heat addition by the boiler
iii) The power output of the turbine
iv) The quality, or temperature, of the steam leaving the turbine
v) The heat loss from the condenser
vi) The total horsepower of the two pumps
vii) The thermodynamic efficiency of the cycle

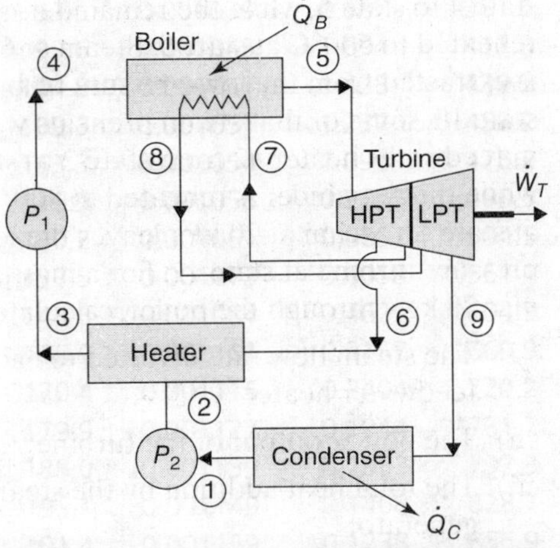

Figure 8.37

8.37 For the ideal power cycle shown in Fig. 8.37, assume the steam leaving the boiler at state 5 is at 8 MPa and 550°C with a mass flux of 20 kg/s. Steam is removed at state 6 at 2 MPa and sent to an open feedwater heater. The remaining steam leaves the high-pressure turbine at state 7 at 1.4 MPa and is reheated to 500°C. This steam is sent back to the low-pressure turbine and leaves this turbine at 30 kPa. Plot this cycle on a *T-s* diagram and determine:

i) The steam flux extracted to go to the feedwater heater
ii) The total heat addition by the boiler
iii) The power output of the turbine
iv) The quality, or temperature, of the steam leaving the turbine
v) The heat loss from the condenser
vi) The total horsepower of the two pumps
vii) The thermodynamic efficiency of the cycle

8.38 For the power cycle described in Problem 8.37, determine the cycle efficiency if the pressure at state 6 is 4 MPa and at state 7 it is 0.8 MPa.

8.39 An ideal regenerative Rankine power cycle is shown in Fig. 8.38. Steam leaves the boiler at 15 MPa and 600°C with a mass flow rate of 10 kg/s. Steam is removed from the turbine at 1.6 MPa and directed to a closed feedwater heater. The remaining steam leaves the turbine at 10 kPa. Sketch the cycle on a *T-s* diagram and calculate:

i) The heat flow rate into the steam generator
ii) The mass flow rate diverted to the feedwater heater
iii) The net power output of the turbine
iv) The cycle efficiency

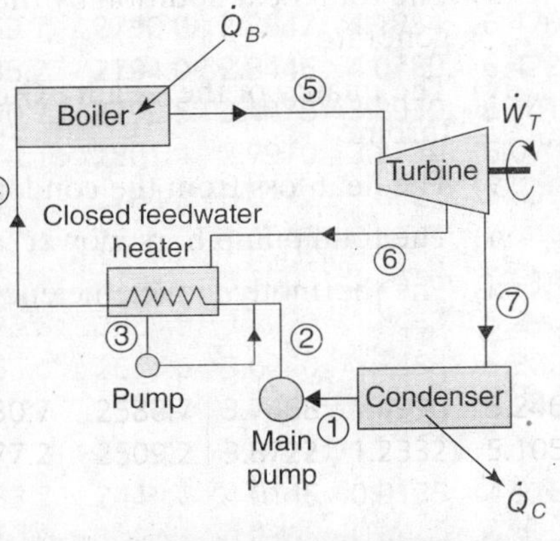

Figure 8.38

8.40 Repeat Problem 8.39 for a turbine efficiency of 85%.

8.41 For the ideal regenerative Rankine cycle shown in Fig. 8.38, steam leaves the steam generator at 15 MPa and 500°C, with a mass flux of 12 kg/s. Steam at 3 MPa is removed from the turbine and directed to a closed feedwater heater. The remaining steam leaves the turbine at 30 kPa. Sketch the cycle on a *T-s* diagram and determine:

i) The heat flow rate into the steam generator
ii) The mass flow rate diverted to the feedwater heater.
iii) The net power output of the turbine
iv) The cycle efficiency

8.42 The *T-s* diagram of an ideal Rankine power cycle with two reheat cycles and two open feedwater heaters is shown in Fig. 8.39. Steam leaves the boiler at state 7 at 15 MPa and 600°C. Steam at 4 MPa leaves the high-pressure turbine at state 8; some of the steam preheats water in feedwater heater #1 from state 4 to state 5, while the remainder is reheated to 500°C at state 9. Steam at 600 kPa is extracted from the low-pressure turbine at state 10; some of that steam preheats water in feedwater heater #2 from state 2 to state 3, while the remainder is reheated to 400°C at state 11. Steam at 10 kPa leaves the low-pressure turbine at state 12. For a mass flux $\dot{m}$ = 20 kg/s through the boiler, calculate:

i) The steam flow rate diverted to each feedwater heater
ii) The power output of the turbine
iii) The total heat addition by the steam generator
iv) The quality of the steam exiting the turbine
v) The heat loss from the condenser
vi) The total pump horsepower
vii) The thermodynamic efficiency of the cycle

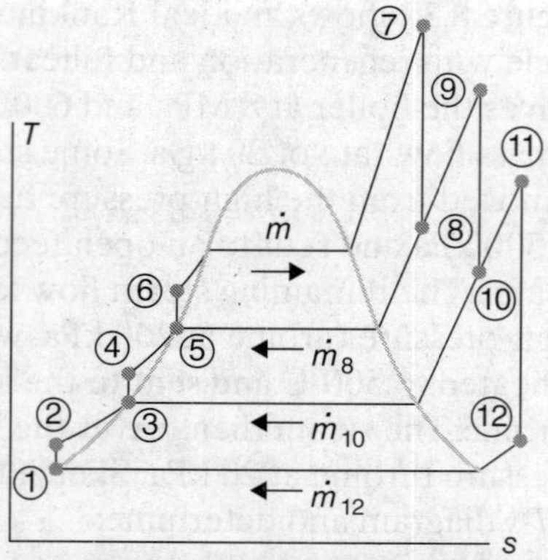

Figure 8.39

8.43 Repeat Problem 8.42, retaining all quantities except that each reheater reheats the steam back to 600°C, that is, $T_9 = T_{11} = 600$°C.

8.44 The *T-s* diagram of an ideal Rankine power cycle with two reheat cycles and two open feedwater heaters is shown in Fig. 8.39. Steam leaves the boiler at state 7 at 20 MPa and 600°C. Steam at 3 MPa leaves the high-pressure turbine at state 8; some of the steam preheats water in feedwater heater #1 from state 4 to state 5, while the remainder is reheated to 500°C at state 9. Steam at 600 kPa is extracted from the low-pressure turbine at state 10; some of that steam preheats water in feedwater heater #2 from state 2 to state 3, while the remainder is reheated to 400°C at state 11. Steam at 20 kPa leaves the low-pressure turbine at state 12. For a mass flux $\dot{m}$ = 20 kg/s through the boiler, calculate:

i) The steam flow rate diverted to each feedwater heater
ii) The power output of the turbine
iii) The total heat addition by the steam generator
iv) The temperature of the steam exiting the turbine
v) The heat loss from the condenser
vi) The total pump horsepower
vii) The thermodynamic efficiency of the cycle

Cogeneration Cycles

8.45 Steam is produced in the ideal cogeneration cycle, shown in Fig. 8.40, at 8 MPa and 600°C at a mass flow rate of 20 kg/s. Thirty percent of the steam is extracted from the turbine at 800 kPa and diverted to a process heater. This steam leaves the process heater at 800 kPa as a saturated liquid. The remaining steam leaves the turbine at 30 kPa, but 20% of the condenser heat transfer is used for heating and cooling buildings. Sketch this process on a *T-s* diagram and calculate:

i) The rate of heat input to the boiler
ii) The power output of the turbine
iii) The rate of heat removal from the process heater
iv) The heat rate used to heat and cool
v) The utilization factor for the cycle

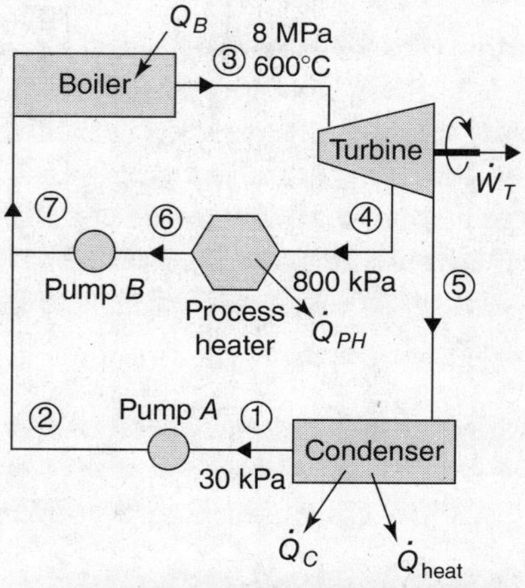

Figure 8.40

8.46 Steam leaves the boiler of the ideal cogeneration cycle shown in Fig. 8.41 at 4 MPa and 500°C. Forty percent of the steam flow is extracted from the turbine by a process heater at 2 MPa. The remaining 60% leaves the turbine at 20 kPa. Assume the turbine and pumps to have 100% efficiencies. Sketch this process on a *T-s* diagram and calculate:

i) The specific heat addition in the steam generator
ii) The net specific pwork output of the turbine
iii) The specific heat removal in the process heater
iv) The specific heat removal in the condenser
v) The utilization factor for the cycle

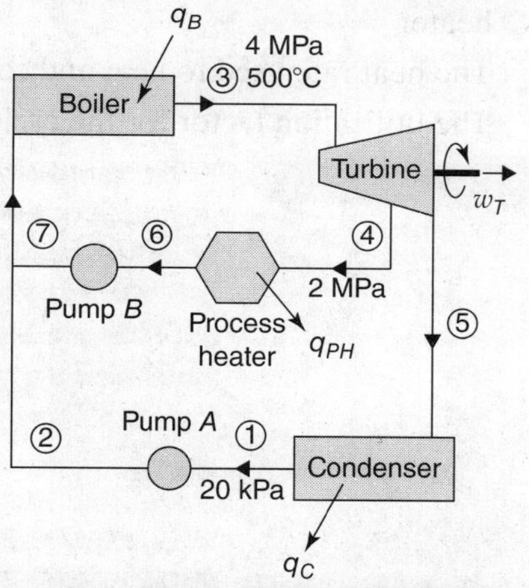

Figure 8.41

8.47 Rework Problem 8.46 assuming that the total steam mass flow rate is 25 kg/s and that the steam is reheated back to 500°C at 2 MPa. The specific quantities will change to rates.

8.48 For the ideal cogeneration cycle shown in Fig. 8.42, steam leaves the boiler at 6 MPa and 600°C with a mass flux of 40 kg/s. The valve is equipped with a pressure regulator that removes 20% of the steam flow and sends it to the process heater at 800 kPa. The remaining 80% of the steam travels to the turbine where 30% of this flow is removed from the turbine at 800 kPa and is also sent to the process heater. The remaining steam is reheated back to 600°C at 800 kPa. Fifty percent of the steam that leaves the turbine at 50 kPa is used for heating and cooling buildings. Sketch this process on a *T-s* diagram and determine:

i) The rate of heat addition in the steam generator
ii) The power output of the turbine
iii) The rate of heat removal in the process heater
iv) The heat rate used to heat and cool
v) The utilization factor for the cycle

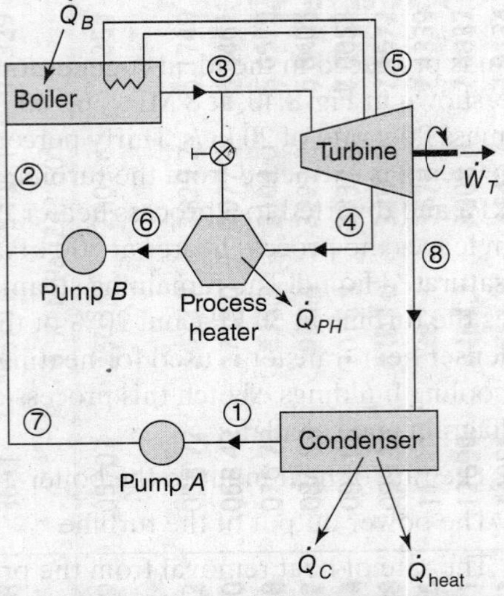

Figure 8.42

CHAPTER

9

Gas Power Cycles

Gas Power Cycle - Jet Propulsion Technology, A Case Study

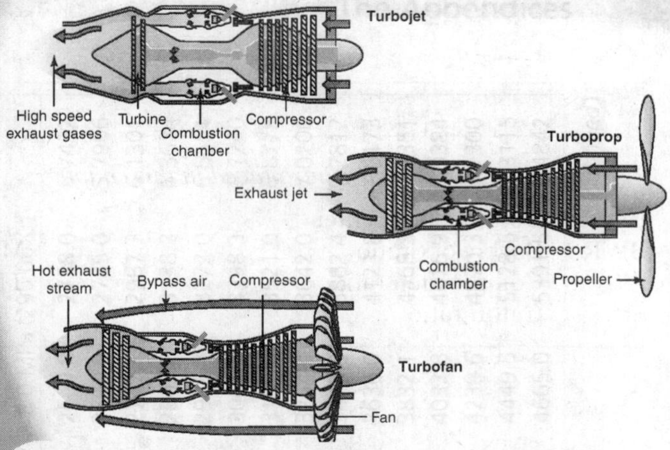

Based on a Microsoft Illustration

Outline

Nomenclature

The following variables are introduced in this chapter:

BDC	Bottom dead center	r_p	Pressure ratio
BWR	Back-work ratio	MEP	Mean effective pressure
r	Compression ratio	TDC	Top dead center
r_c	Cutoff ratio		

Learning Outcomes

- Describe gas power cycles
- Analyze the Otto cycle
- Analyze the diesel cycle
- Analyze the dual, Stirling, and Ericsson cycles
- Analyze the Brayton cycle
- Combine the Brayton and Rankine cycles

Motivational Example—The Variable-Compression-Ratio Engine

The compression ratio of a car engine is the ratio of the largest volume occupied by the air inside one of the cylinders to the smallest volume of the air in that cylinder. Its value for a gas engine is usually around 8 and for a diesel engine around 18. The efficiency of an engine is a direct function of this compression ratio. It would be desirable to vary the compression ratio in a car engine to allow it to perform better under different operating conditions, such as uphill driving or high-speed driving. The first practical variable-compression-ratio engine was developed by Saab in 2000. This engine has a split engine block that is hinged on one side. As the two halves of the engine block rotate away from each other, the compression ratio gets smaller. A hydraulic actuator is used to move the blocks. Paul Corina developed a variable-compression-ratio engine where the two halves of the engine block slide relative to each other. This allows for better sealing of the engine block and the alignment of all moving parts. A four-bar linkage (to be studied in a future design course for mechanical engineering students) is used in Fig. 9.1 to vary the compression ratio, with the highest ratio on the right. These

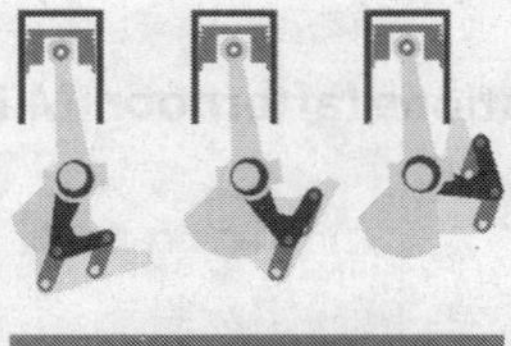

Figure 9.1
A four-bar linkage used to vary the compression ratio.

engines represent a significant development in engine design by allowing the performance of the engine to vary with the demands placed on the engine.

In this chapter we will study power cycles that use an ideal gas, most often air, as the working fluid. For this reason they are called *gas power cycles*. These engines do not have the advantage of using phase change to transfer large amounts of heat. Instead, a gas-powered engine moves large quantities of air to produce power. The engines of primary interest in our study are called *internal combustion engines* since the fuel burns inside the engines. Engines that operate on the Rankine power cycle presented in the last chapter are *external combustion engines* since the combustion of fuel takes place outside the power-producing turbine. Another example of an external combustion engine is the old steam locomotive. Combustion took place in a furnace where wood or coal was used to create steam. The steam was then delivered to a reciprocating-piston device that converted the thermal energy in the steam to mechanical energy. External combustion engines that operate on a gas do not have sufficient application to be of interest in our study, although a short description will be presented of two such cycles.

The power cycles to be analyzed in this chapter mostly operate as *open cycles* in that the working fluid does not return to its original state and location. A good example of an open power cycle is that of the automobile engine. Air at ambient conditions enters the engine, and products of combustion at high temperature leave the engine.

Gas power cycles: Power cycles that use an ideal gas as the working fluid.

Internal combustion engine: The fuel burns inside the engine.

External combustion engine: The combustion of fuel takes place outside the power-producing device.

Open cycle: The working fluid does not return to its original state and location.

9.1 Air-Standard Analysis

The engines to be analyzed in this chapter use air as the working fluid. Air from the environment will enter the engine prior to compression and be vented from the engine immediately after the mechanical power is produced, as happens in the engines of cars, trucks, locomotives, and tanks. The rapid succession of processes in these open cycles will be simplified by a series of assumptions used in the *air-standard analysis*, which makes use of the following assumptions:

1. The working fluid is air that will be treated as an ideal gas.
2. The compression and expansion processes in the cycle will be assumed to be adiabatic and reversible, resulting in two isentropic processes.
3. The combustion process in the engine will be modeled as a simple heat transfer process, with heat transferred to the air from an external source equal to that which the combustion process would supply. The process will depend on the type of engine being analyzed: a gas engine, a diesel engine, or a gas turbine engine.
4. The heat rejection in the exhaust will be accomplished by transferring heat to the surroundings with a process that returns the air to the initial state while doing no work to complete the cycle.

The *cold-air standard* adds another assumption to the air-standard assumptions: The constant specific heats listed in Table B-2, which are at 27°C, will be assumed for all processes throughout a cycle.

Comment

The small amount of fuel that mixes with the air can be ignored when analyzing the processes in these cycles. The fuel simply introduces heat into the combustion process.

Air-standard assumptions:

1. Air is an ideal gas.
2. Compression and power processes are isentropic.
3. A heat transfer process replaces combustion.
4. Exhaust is a process that transfers heat while doing no work.

Cold-air standard: It adds to the air-standard analysis (assumptions 1 through 4), requiring that the specific heats be constant as given in Table B-2.

Some of these assumptions simplify the analyses compared to using the ideal-gas tables or including real-gas effects; even though they do introduce some error, they provide estimates of the power output and efficiency of actual cycles. They also allow quick estimates of the effects of modifications made to cycles of interest. The cycles presented in this chapter utilize processes and devices introduced in Chapters 3 and 4.

Example 9.1 Air-standard analysis

All three processes in the fictitious cycle of Fig. 9.2 are assumed to be quasi-equilibrium processes. The pressure and temperature of the air at the beginning of the compression stroke, from state 1 to state 2, are 100 kPa and 25°C. Assume a piston-cylinder arrangement with a volume ratio of 4 to 1. Determine:

i) The maximum temperature and pressure
ii) The net heat transfer
iii) The net work output
iv) The thermal efficiency of the cycle

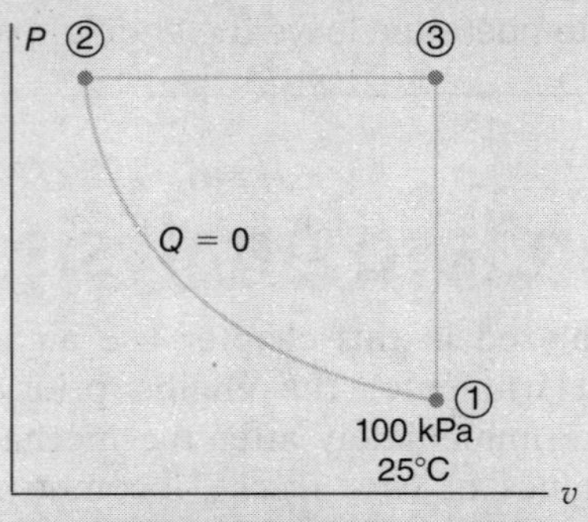

Figure 9.2

Solution

With a piston-cylinder arrangement, the air is contained in the cylinder and treated as a system. So, the first law applied to various processes in Section 3.5 is applicable.

i) At state 1, $P_1 = 100$ kPa and $T_1 = 25°C = 298$ K. The compression stroke is an isentropic process (it's a quasi-equilibrium adiabatic process) for which we use Eqs. 3.44 and 3.46. They give the temperature at state 2 and the maximum pressure in the cycle at state 2, using $k = 1.4$:

$$T_2 = T_1\left(\frac{v_1}{v_2}\right)^{k-1} = 298 \times 4^{0.4} = 519 \text{ K}$$

$$P_2 = P_1\left(\frac{v_1}{v_2}\right)^{k} = 100 \times 4^{1.4} = \underline{696 \text{ kPa}} = P_3$$

The highest temperature in the cycle occurs at state 3. It is found for this constant-pressure process, using the ideal-gas law, to be

$$T_3 = T_2 \frac{v_3}{v_2} = 519 \times 4 = 2075\ \text{K} \quad \text{or} \quad \underline{1800°\text{C}}$$

ii) The heat transfer for the constant-pressure process is

$$q_{2\text{-}3} = \Delta h = C_p(T_3 - T_2)$$
$$= 1.0 \times (2075 - 519) = 1556\ \text{kJ/kg}$$

The heat transfer for the constant-volume process (work is zero) is

$$q_{3\text{-}1} = \Delta u = C_v(T_1 - T_3)$$
$$= 0.717 \times (298 - 2075) = -1274\ \text{kJ/kg}$$

The net heat transfer is

$$q_{\text{net}} = q_{2\text{-}3} + q_{3\text{-}1}$$
$$= 1556 - 1275 = \underline{282\ \text{kJ/kg}}$$

Note: The ideal-gas law $v = RT/P$ is used.

iii) Work occurs for the first two processes, and the net work is calculated as follows:

$$w_{1\text{-}2} = -\Delta u = -C_v(T_2 - T_1) = -0.717 \times (519 - 298)$$
$$= -158\ \text{kJ/kg}$$

$$w_{2\text{-}3} = P(v_3 - v_2) = 696\left(\frac{0.287 \times 2075}{696} - \frac{0.287 \times 519}{696}\right)$$
$$= 447\ \text{kJ/kg}$$

$$\therefore w_{\text{net}} = w_{1\text{-}2} + w_{2\text{-}3} + \not{w}_{3\text{-}1} = -158 + 447 = \underline{289\ \text{kJ/kg}}$$

Comment

The first law for a cycle is $w_{\text{net}} = q_{\text{net}}$ (see Eq. 3.21). The small difference (about 2%) between w_{net} and q_{net} is not significant.

iv) Assuming $q_{\text{in}} = q_{2\text{-}3}$ and $q_{\text{out}} = q_{3\text{-}1}$ the thermal efficiency is

$$\eta = \frac{w_{\text{net}}}{q_{\text{in}}} = \frac{289}{1556} = 0.186 \quad \text{or} \quad \underline{18.6\%}$$

This is not a cycle used in any particular engine, but it is a cycle, and if you can find a use for it, please do!

A reminder is in order to emphasize the difference between a quasi-equilibrium process and a reversible process. The working fluid, the air in the above quasi-equilibrium processes, experiences heat transfer due to processes in which the temperature changes slowly (thermodynamically). The high-temperature reservoir would have to be at 2075 K or above for the heat transfer process from state 2 to state 3 to occur. Such a process would be highly irreversible since it would be across a large temperature difference. The efficiency of 92% of a Carnot cycle between the high and low temperatures would require all processes to be reversible. Some texts may refer to the processes being internally reversible but externally irreversible.

9.2 Reciprocating Engine Terminology

Many unique terms and symbols are used in describing internal combustion engines that use reciprocating pistons in cylinders. The following are some basic definitions. Refer to Fig. 9.3.

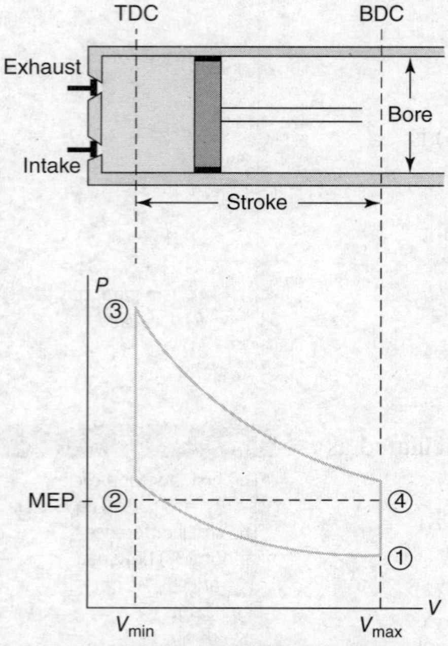

Figure 9.3

Piston-engine terminology.

Top dead center (TDC): The position of the piston in the cylinder that provides the minimum volume.

Bottom dead center (BDC): The position of the piston in the cylinder that provides the maximum volume.

Bore: The inside diameter of the cylinder, which is assumed to be the outside diameter of the piston.

Stroke: The distance of travel of the piston from BDC to TDC.

Displacement: A volume equal to the stroke times the cross-sectional area of the piston, which is $V_{max} - V_{min}$ in Fig. 9.3:

$$\text{Displacement} = \text{Stroke} \times \pi \frac{\text{Bore}^2}{4} \qquad (9.1)$$

The total engine displacement is this value times the number of cylinders in the engine. The total displacement is often used to describe the power level of a particular engine.

Clearance volume: The volume between the cylinder head and the top of the piston when the piston is at TDC.

Compression ratio: The ratio of the maximum volume (piston at BDC) in the cylinder to the minimum volume (piston at TDC) in the cylinder. A higher compression ratio produces more engine power. The compression ratio r is

$$r = \frac{V_{TDC}}{V_{BDC}} = \frac{V_{max}}{V_{min}} = \frac{v_1}{v_2} \qquad (9.2)$$

Intake valve: Allows the air or air–fuel mixture to enter the cylinder.

Exhaust valve: Allows the combustion products to exit the cylinder.

Spark ignition: Occurs when a spark plug ignites the air–fuel mixture.

Compression ignition: Occurs when the high temperature during compression ignites the air–fuel mixture.

Mean effective pressure (MEP): Refers to the ratio of the net work done during the cycle to the displacement. If the MEP acted on the piston during the power stroke, it would produce the same work produced during the actual stroke; that is, the rectangular area under the MEP dashed line in Fig. 9.3 is equal to the area under the *P-V* diagram encompassed by the solid lines. The MEP is used to rate the effectiveness of an engine compared to one with the same displacement: the higher the MEP, the better the performance. The MEP (kPa or psia) can be expressed as

$$\text{MEP} = \frac{W_{cycle}}{V_{max} - V_{min}} = \frac{w_{cycle}}{v_1 - v_2} \qquad (9.3)$$

Comment

The higher the MEP, the better the performance for an engine with the same displacement.

Units:

$$\frac{\text{kN} \cdot \text{m}}{\text{m}^3} = \frac{\text{kN}}{\text{m}^2} = \text{kPa}$$

Several of the preceding terms were presented for information purposes only. The compression ratio r and the MEP are of particular interest. An analysis of the Otto cycle will utilize these terms.

Reciprocating engine terminology — Example 9.2

The dimensions of a piston cylinder arrangement are displayed in Fig. 9.4. The work for one cycle is 400 J. Determine the displacement, the clearance volume, the compression ratio, and the MEP.

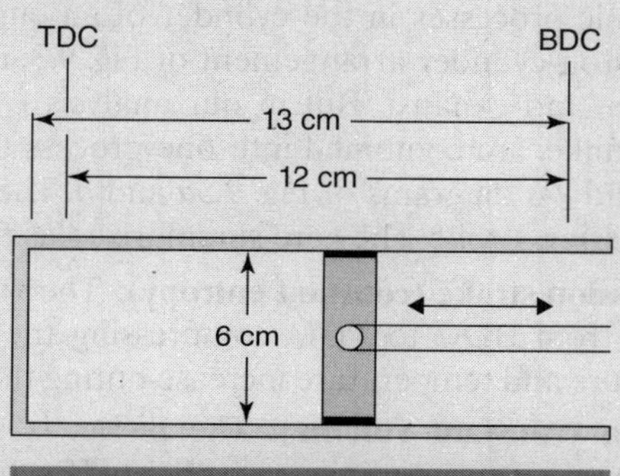

Figure 9.4

Solution

The displacement is the volume displaced while the piston moves from TDC to BDC. It is

$$\text{Displacement} = \text{stroke} \times \pi R^2 = 12 \times \pi \times 3^2 = \underline{339.3 \text{ cm}^3}$$

The clearance volume is the volume in the cylinder when the piston is at TDC:

$$\text{Clearance} = \pi R^2 t = \pi \times 3^2 \times 1 = \underline{28.3 \text{ cm}^3}$$

The compression ratio, a ratio of great importance in engine parameters, is found to be

$$r = \frac{V_{max}}{V_{min}} = \frac{\pi 3^2 \times 13}{\pi 3^2 \times 1} = \underline{13}$$

The MEP, using $W_{cycle} = 0.400$ kJ, is

$$\text{MEP} = \frac{W_{cycle}}{V_{max} - V_{min}} = \frac{0.400 \text{ kN}\cdot\text{m}}{\pi \times 0.03^2 \times (0.13 - 0.01) \text{ m}^3} = \underline{1180 \text{ kPa}}$$

☑ **You have completed Learning Outcome (1)**

9.3 The Otto Cycle

9.3.1 The four-stroke Otto cycle

The four-stroke Otto cycle engine was developed by Nikolaus Otto and was first tested in 1876. This is a spark-ignition engine and is the type of engine used in almost all automobiles around the world. Since it has been in use for over 130 years, there have been significant improvements on this engine. We use it extensively since the engines are so reliable. The engine starts and runs well in all weather conditions, over a wide range of temperatures. It provides a considerable amount of power despite being a compact engine.

Five thermodynamic processes in the cylinder of an engine operate on the Otto cycle with the piston-cylinder arrangement of Fig. 9.3: intake, compression, combustion, expansion, and exhaust. But in our analysis two of the processes, the exhaust and the intake, are combined into one process from state 4 to state 1, shown on the *P-v* and *T-s* diagrams of Fig. 9.5*a* and *b*. The processes typically begin with the compression stroke. They are summarized as follows.

Approximation: The fuel burns in an extremely short time increment, so in the ideal cycle the piston doesn't move.

Note: The fictitious constant-volume process from state 4 to state 1 completes the cycle while emitting heat but doing no work. Isn't that great!

Four strokes: Compression, power, exhaust, and intake.

Process 1, the compression stroke (constant entropy): The valves are closed, and the piston moves from BDC to TDC, compressing the air from state 1 to state 2. Both pressure and temperature increase during this isentropic process.

Process 2, combustion (constant volume): The piston is at top dead center, with the air compressed to its minimum volume. If a fuel injector is being used, the fuel is injected into the cylinder by the time the piston reaches state 2. The spark plug fires and the air–fuel mixture burns, providing heat Q in an instantaneous constant-volume process with a simultaneous rise in the pressure and temperature to state 3. The piston does not move as the heat transfer Q is added to the air.

Process 3, the power stroke (constant entropy): The high- pressure air (the products of combustion) forces the piston from state 3 at TDC to BDC at state 4, with an isentropic process resulting in decreasing pressure and temperature.

Process 4, exhaust and intake (constant volume): This fictitious process replaces the exhaust stroke and the intake stroke. A complete rotation of the

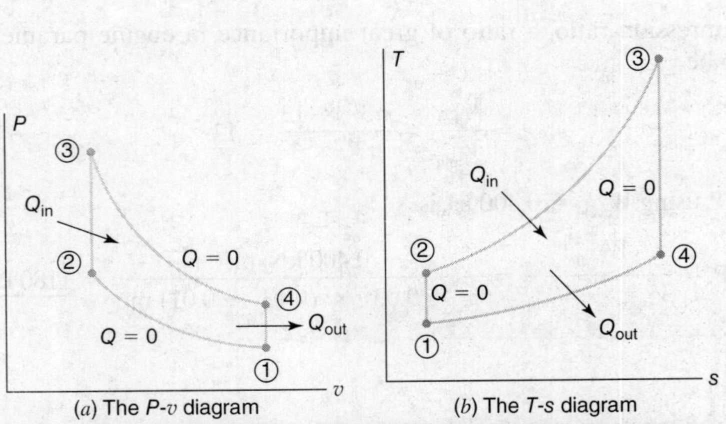

Figure 9.5

The ideal Otto cycle.

crankshaft moves the piston from BDC to TDC forcing the hot exhaust gases out (energy leaves the cylinder) and from TDC to BDC when new air (and perhaps fuel) is drawn in. There is no work accomplished, but heat transfer is needed to return the cycle to state 1. Hence, the constant-volume, quasi-equilibrium process accomplishes three objectives: it requires no work, it emits heat, and it completes the cycle from state 4 to state 1.

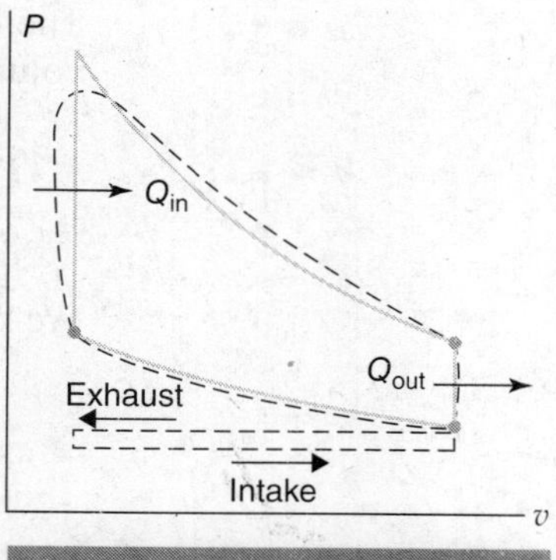

Figure 9.6

The dotted lines form a sketch of an actual P-v diagram. The solid lines show the ideal Otto cycle.

In Fig. 9.6 a pressure-volume diagram of an actual Otto cycle is sketched. The combustion process does not actually take place at constant volume since the piston is continually moving. The intake and exhaust processes do require a small amount of work by the piston. The compression and power strokes are close to being isentropic but vary due to small amounts of friction between the piston and cylinder and the little heat transfer (there is only a very short time during which heat transfer can occur) to the engine block. The ideal cycle does provide a good approximation to the actual cycle.

9.3.2 Otto cycle analysis

Analysis of a reciprocating piston engine is performed using the first law of thermodynamics for a system. Once the intake valve is closed, the air (the system) is completely contained inside the cylinder. This means that the energy level of the air will be represented by its internal energy. We will refer to Fig. 9.5 for the state numbers for this cycle. The thermodynamic efficiency of the cycle is the ratio of the net work output and the heat transfer during combustion. It is

$$\eta = \frac{w_{net}}{q_{in}} = 1 - \frac{q_{out}}{q_{in}} \tag{9.4}$$

Note: We have used $w_{net} = q_{in} - q_{out}$.

The two heat transfer processes from state 2 to state 3 and state 4 to state 1 occur at constant volume, so there is no work done. The two heat transfers (see Eq. 3.26), assuming constant specific heat C_v, are given by

$$\begin{aligned} q_{in} &= u_3 - u_2 = C_v(T_3 - T_2) \\ q_{out} &= u_4 - u_2 = C_v(T_4 - T_1) \end{aligned} \tag{9.5}$$

These expressions allow the efficiency to take the form

$$\eta_{Otto} = 1 - \frac{T_4 - T_1}{T_3 - T_2} \tag{9.6}$$

Comment

Specific heats are assumed constant to allow for quick approximations and easy comparisons. This will usually introduce significant error, but the comparison between cycles will be reasonable. The cold-air standard will be used unless otherwise stated.

Factor out T_1 the numerator and T_2 from the denominator of the ratio of temperature differences. The efficiency is then written as

$$\eta_{Otto} = 1 - \frac{T_1\left(\frac{T_4}{T_1} - 1\right)}{T_2\left(\frac{T_3}{T_2} - 1\right)} \tag{9.7}$$

The two isentropic processes allow the temperature ratios to be related to the volume ratios using Eq. 3.44:

$$\frac{T_2}{T_1} = \left(\frac{v_1}{v_2}\right)^{k-1} \qquad \frac{T_3}{T_4} = \left(\frac{v_4}{v_3}\right)^{k-1} \tag{9.8}$$

We know that $v_1 = v_4$ and $v_2 = v_3$. Algebraic manipulation shows the following:

$$\frac{T_2}{T_1} \cdot \frac{T_4}{T_3} = \left(\frac{v_1}{v_2}\right)^{k-1} \left(\frac{v_3}{v_4}\right)^{k-1} = \left(\frac{v_1}{v_2} \cdot \frac{v_3}{v_4}\right)^{k-1} = 1 \qquad \therefore \frac{T_4}{T_1} = \frac{T_3}{T_2} \tag{9.9}$$

Consequently, the quantities in the parentheses of Eq. 9.7 are equal, and the thermal efficiency of the Otto cycle is expressed as

$$\eta_{\text{Otto}} = 1 - \frac{T_1}{T_2} = 1 - \left(\frac{v_2}{v_1}\right)^{k-1} = 1 - \frac{1}{r^{k-1}} \tag{9.10}$$

Remember, for air:
$R = 0.287$ kJ/kg·K
$C_v = 0.717$ kJ/kg·K
$C_p = 1.0$ kJ/kg·K

recognizing that $v_1/v_2 = r$, where r is the compression ratio.

An example will illustrate the above equations. Examples in this chapter will use the ideal-gas law and assume constant specific heats from Table B-2. Comparisons may also be made using the ideal-gas tables or values from Table B-6.

Example 9.3 The Otto cycle

Note: Assuming constant specific heat introduces significant errors. It is used for comparison and estimating purposes.

The heat input to the Otto cycle shown in Fig. 9.7 is 1000 kJ/kg. The compression ratio is 8. The pressure and temperature of the air at the beginning of the compression stroke are 100 kPa and 15°C. Determine:

i) The maximum temperature and pressure
ii) The thermal efficiency of the cycle
iii) The net work output
iv) The mean effective pressure

First, solve this example assuming an ideal gas with constant specific heats, and then use the air tables from the Appendix.

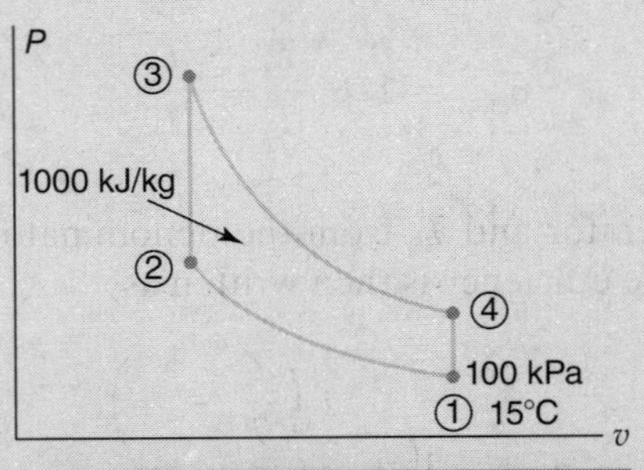

Figure 9.7

First Solution

i) At state 1, $P_1 = 100$ kPa and $T_1 = 15°C = 288$ K. The compression stroke is an isentropic process for which we use Eqs. 3.44 and 3.46. There results

$$T_2 = T_1\left(\frac{v_1}{v_2}\right)^{k-1} = 288 \times 8^{0.4} = 661.7 \text{ K}$$

$$P_2 = P_1\left(\frac{v_1}{v_2}\right)^{k} = 100 \times 8^{1.4} = 1838 \text{ kPa}$$

The combustion process is simulated by constant-volume heat addition. No work is done during the constant-volume process, so the first law allows T_3 to be determined as follows:

$$q_{2\text{-}3} = u_3 - u_2 = C_v(T_3 - T_2)$$

$$1000 = 0.717(T_3 - 661.7) \qquad \therefore T_3 = 2056 \text{ K} \text{ or } \underline{1783°\text{C}}$$

The ideal-gas law for this constant-volume process provides

$$P_3 = P_2\left(\frac{T_3}{T_2}\right) = 1838\left(\frac{2056}{661.7}\right) = \underline{5711 \text{ kPa}}$$

Remember: Temperatures and pressures in this equation must be absolute.

The pressure and temperature at state 3 are the maximum values for the cycle.

ii) We can calculate the thermal efficiency of the cycle using Eq. 9.10 with the given compression ratio:

$$\eta_{\text{Otto}} = 1 - \frac{1}{r^{k-1}} = 1 - \frac{1}{8^{0.4}} = 0.565 \text{ or } \underline{56.5\%}$$

iii) The net work output of the cycle equals the product of the thermal efficiency of the cycle and heat input. It is

$$w_{\text{net}} = \eta_{\text{Otto}} \times q_{\text{in}} = 0.565 \times 1000 = \underline{565 \text{ kJ/kg}}$$

iv) The MEP, the mean effective pressure, is the ratio of the net work to the displacement (see Eq. 9.3). The specific volumes at top dead center and bottom dead center are

$$v_1 = \frac{RT_1}{P_1} = \frac{0.287 \times 288}{100} = 0.8266 \text{ m}^3/\text{kg}$$

$$v_2 = \frac{v_1}{r} = \frac{0.8266}{8} = 0.1033 \text{ m}^3/\text{kg}$$

The MEP is then

$$\text{MEP} = \frac{w_{\text{cycle}}}{v_1 - v_2} = \frac{565 \text{ kN·m/kg}}{(0.8266 - 0.1033) \text{ m}^3/\text{kg}} = \underline{781 \text{ kPa}}$$

(Continued)

Example 9.3 (*Continued*)

Second Solution

i) Table F-1 in the Appendix will be used. At $T_1 = 288$ K the internal energy is interpolated to be $u_1 = 206$ kJ/kg and $v_{r1} = 688$. The isentropic process (review Eq. 6.34) from state 1 to state 2 allows

$$v_{r2} = v_{r1} \times \frac{v_2}{v_1} = 688 \times \frac{1}{8} = 86$$

At $v_{r2} = 86$, the air tables are interpolated to give $T_2 = 648$ K and $u_2 = 472$ kJ/kg.

From state 2 to state 3 the first law gives

$$q_{2\text{-}3} = u_3 - u_2 \qquad 1000 = u_3 - 472 \qquad \therefore u_3 = 1472 \text{ kJ/kg}$$

At $u_3 = 1472$ kJ/kg, interpolation gives $T_3 = 1784$ K or $\underline{1511°\text{C}}$ and $v_{r3} = 4.07$. Then, using $v_1/v_3 = v_1/v_2 = r$, we find

$$P_3 = P_1 \times \frac{v_1}{v_3} \times \frac{T_3}{T_1} = 100 \times 8 \times \frac{1784}{288} = \underline{4956 \text{ kPa}}$$

ii) and iii) Before the efficiency can be found, we must find the net work since the specific heat is not constant in Table F-1, so Eq. 9.10 cannot be used. The isentropic process from state 3 to state 4 again allows

$$v_{r4} = v_{r3} \times \frac{v_4}{v_3} = v_{r3} \times r = 4.07 \times 8 = 32.56$$

For $v_{r4} = 32.56$, by interpolation $T_4 = 917$ K and $u_4 = 689$ kJ/kg. The work and the efficiency can now be found:

$$w_{\text{net}} = (u_3 - u_4) - (u_2 - u_1)$$

$$= (1472 - 689) - (472 - 206) = \underline{517 \text{ kJ/kg}}$$

$$\eta = \frac{w_{\text{net}}}{q_{\text{in}}} = \frac{517}{1000} = 0.517 \text{ or } \underline{51.7\%}$$

This efficiency compares with 56.5% using the cold-air assumptions.

iv) Using $v_1 = 0.827$ m^3/kg and $v_2 = 0.103$ m^3/kg from the first part (they would be the same), we find that the MEP is

$$\text{MEP} = \frac{w_{\text{cycle}}}{v_1 - v_2} = \frac{517 \text{ kN·m/kg}}{(0.827 - 0.103) \text{ m}^3/\text{kg}} = \underline{714 \text{ kPa}}$$

Observe that the cold-air assumptions result in an error of about 10% in the net work, efficiency, and MEP. That represents significant error, but the cold-air analysis does provide a basis for comparing the performance of cycles and estimating the effects of proposed modifications of a cycle. The problems will indicate which procedure to use.

Note: Using the ideal-gas tables is more accurate than the cold-air analysis, but it does introduce errors compared to assuming real gas effects (Section 7.5).

Comment

When interpolating, it is neither necessary nor advisable to carry the significant digits to five or six figures since the numbers are a curve fit and there is no significance to such accuracy. Three significant digits will suffice, although four are acceptable.

9.3.3 Two-stroke Otto cycle engine

The Otto cycle engine has been modified so that two piston strokes are used to produce power in each cycle instead of four piston strokes in the standard Otto cycle engine. This greatly increases the power output of the engine with respect to its size. The intake and exhaust valves used in the four-stroke engine are eliminated. Instead, air intake and exhaust ports are designed into the wall of the cylinder and are opened and closed by the motion of the piston, as suggested by the sketch in Fig. 9.8. The air–fuel mixture is compressed, and as the piston approaches TDC ignition occurs. The power stroke then begins.

While the piston is moving downward in the power stroke, the exhaust port *B* is exposed. The products of combustion vent from the cylinder during the final portion of the power stroke. This will decrease the force on the piston, reducing the total amount of work produced. The motion of the piston also opens a port at *A* for the air–fuel mixture to enter the cylinder, helping to force out the exhaust gases through the exit port *B*. During this process some fuel is lost through the exhaust port. The piston reaches BDC ready to again start the compression stroke. The intake and exhaust ports are covered by the piston after it has traveled a distance above BDC, at the position sketched. The air–fuel mixture is then compressed, and the power stroke follows. A pressurized air–fuel mixture enters the region below the piston at *C* as the piston travels upward during the compression stroke.

Google: To see many different arrangements, simply Google "Images for two-stroke engines."

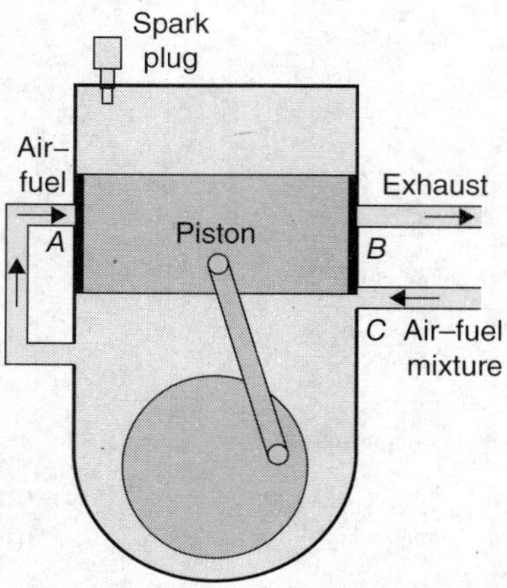

Figure 9.8

A two-stroke cylinder.

Because of the complexities of the intake and exhaust processes, the overall efficiency of an engine operating on the two-stroke Otto cycle is significantly less than that of an engine operating on the four-stroke cycle, even though the *P-v* and *T-s* diagrams, and hence the ideal cycle efficiencies, are the same. The higher power-to-size ratio for the two-stroke engine makes it ideal for the small engines used in such devices as lawn mowers, chain saws, and scooters. It has also been used in small cars.

In a two-stroke engine, lubrication occurs during the compression stroke by mixing the lubricating oil with the fuel. The analysis of a cold-air standard, two-stroke engine is performed in the same manner as the analysis of the four-stroke engine illustrated in Example 9.3.

Comment

The *P-v* and *T-s* diagrams of the four-stroke and two-stroke engines operating on the Otto cycle are identical. It is the actual operation of the engines that is different.

The Carnot cycle is the most efficient cycle operating between a high-temperature and a low-temperature heat reservoir. The next example compares the efficiency of a Carnot cycle with the efficiency of the Otto cycle of Example 9.2 operating between the same high and low temperatures.

The Carnot cycle — Example 9.4

Calculate the work output and the efficiency of a Carnot cycle, shown in Fig. 9.9, operating between the same temperature limits as the Otto cycle of Example 9.3 and accepting the same heat input of 1000 kJ/kg.

Solution

If the efficiency is known, the work can be found. The efficiency of the Carnot cycle can be found if the high and low temperatures are known. From

(Continued)

Example 9.4 *(Continued)*

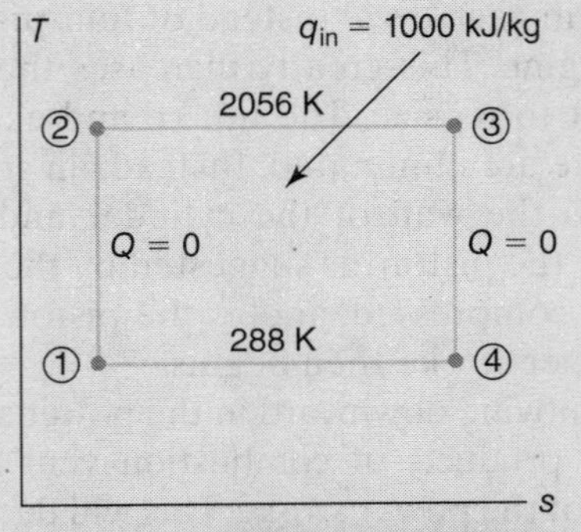

Figure 9.9

Example 9.3, they are $T_1 = 288$ K and $T_3 = 2056$ K. Referring to Eq. 5.15, the maximum possible efficiency (the efficiency of a Carnot cycle) between the high and low temperatures is

$$\eta_{\text{Carnot}} = 1 - \frac{T_L}{T_H} = 1 - \frac{288}{2056} = 0.860 \quad \text{or} \quad \underline{86\%}$$

This compares to the efficiency of the ideal Otto cycle, which was 51.7%. The efficiency of the Otto cycle is 34% below the maximum possible efficiency because the heat transfer occurs across a large temperature difference. It is assumed that the heat is transferred from a constant-temperature heat reservoir, which would have to be at 2056 K, or higher. The temperature at state 2 was 667.1 K, so the temperature difference was at least 1389 K at the beginning of the heat transfer process. That's a very irreversible process. Reversible heat transfer must occur across an infinitesimal temperature difference between the heat reservoir and the system, as in the Carnot cycle, to be considered reversible.

The efficiency can also be expressed as the ratio of net work output to heat input. The net work can then be found:

$$\eta_{\text{Carnot}} = \frac{w_{\text{net}}}{q_{\text{in}}} = \frac{w_{\text{net}}}{1000} = 0.86 \qquad \therefore w_{\text{net}} = \underline{86 \text{ kJ/kg}}$$

The Otto cycle is not a very thermodynamically efficient cycle when compared to the Carnot cycle. The process that transfers heat across the large temperature difference is the culprit, but cycles that attempt to transfer heat over a much smaller temperature difference have not been successful for use in vehicles. Two of these cycles, the Stirling and Ericsson cycles, will be analyzed later in this chapter.

9.3.4 The Wankel engine

The Wankel or rotary engine operates on the Otto cycle but uses a different technique to extract the energy from the combusted fuel. Instead of a reciprocal piston, it uses a rotating mechanism. The engine was first patented by Felix Wankel in

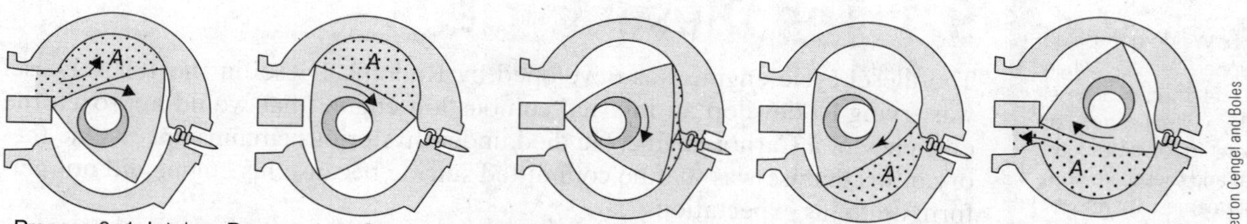

Figure 9.10

The processes of a Wankel engine.

1929. Many auto manufacturers worked on developing the Wankel engine in the 1950s and 1960s. Mazda introduced the first commercial vehicle in 1964 using the Wankel engine. They are still producing rotary engines for their line of pickup trucks. A description of how this engine works is presented in Fig. 9.10.

The "piston" has a triangular shape and rotates in an offset orientation to the central engine shaft. In the first process 0–1, the piston is rotating clockwise and draws an air–fuel mixture into the engine through the intake vent. As the piston continues to rotate, it closes off the intake vent and compresses the air–fuel mixture in an isentropic process 1–2. In process 2–3, a spark plug ignites the fuel and heat is transferred to the air at constant volume. As the piston continues to rotate, the high-pressure products of combustion push against the offset piston, forcing it to rotate. This is the isentropic power stroke. In process 4–1–0, the piston passes the exhaust vent, opening it up. Further rotation of the piston pushes the products of combustion out the exhaust port, and then the air–fuel mixture enters.

The processes that make up the cycle used by the Wankel engine are isentropic compression, constant-volume combustion, isentropic expansion, and constant-volume exhaust/intake. These are the same processes of the Otto cycle, so the cycle analysis would be identical to that of the Otto cycle.

Comment

For every complete rotation of the piston in a Wankel engine, three power strokes occur.

During the power stroke 3–4, another mass of air is being compressed. For every complete rotation of the piston, three power strokes occur. For this reason the Wankel engine is ideal for applications where you need a high power-to-size ratio, as in aircraft propulsion.

The major disadvantage of the Wankel engine is the sealing system at the points of the piston that keep the chambers separated. The floating seals on the piston cause a loss of efficiency and tend to burn oil. The small piston offset requires a very high pressure to produce the required torque. A conventional piston engine has a crankshaft *throw* of half the piston stroke, resulting in about a 50-mm moment arm to create torque. A Wankel engine has only a fraction of that throw, so the torque is inherently relatively low. Remember: Power is torque times angular speed. The three power strokes per revolution of the crankshaft help compensate for the small moment arm.

Throw: The throw on a crankshaft is the moment arm needed to create torque.

Neither an example nor problems are presented for the Wankel engine since it operates on the Otto cycle and the engine has little application.

☑ You have completed Learning Outcome (2)

Note: We do not use an upper-case "D" when referring to a diesel cycle, a diesel engine, or diesel fuel even though Otto and Diesel are both proper names. There is no Otto engine or Otto fuel.

9.4 The Diesel Cycle

The diesel cycle engine was developed by Rudolph Diesel in the 1890s. Diesel was trying to develop an internal combustion engine that would approach the efficiency of a Carnot engine. He died under mysterious circumstances. One theory of his demise was that he committed suicide because his engine did not perform up to his expectations.

Diesel engines originally replaced stationary steam engines, but are now used in submarines, ships, locomotives, trucks, heavy equipment, and electric-generating plants. They are the choice for larger on-road and off-road vehicles as well as automobiles, large and small. About 50% of all new car sales in Europe are diesel because of the lower fuel consumption combined with the high cost of fuel. The largest diesels, up to about 110 000 hp, are used to propel large ships.

The diesel cycle is very similar to the Otto cycle in that it utilizes a reciprocating piston-cylinder arrangement. There are two major differences between the diesel cycle and the Otto cycle. The combustion process takes place at constant pressure in the diesel cycle rather than constant volume; the fuel is injected at top dead center and combustion takes place during the first part of the power stroke. The diesel engine is a *compression ignition engine* in that there is no spark plug used to ignite the fuel. A diesel cycle has a much higher compression ratio than that of the Otto cycle. With a higher compression ratio, both the pressure and the temperature of the air at TDC are very high, the temperature being so high that the injected fuel immediately ignites.

Compression ignition: Ignition occurs when fuel is injected into a high-temperature volume. No spark plug is needed.

Diagrams of the processes of the four-stroke ideal diesel cycle are shown in Fig. 9.11*a* and Fig. 9.11*b*. Process 1–2 is the compression stroke, modeled as an isentropic process. The inlet valve is closed at state 1, and the air is compressed to a very high temperature and pressure (much higher than in an Otto cycle, which would follow the dashed line in Fig. 9.11*a*). During the constant-pressure process 2–3, the fuel is continually injected into the cylinder and immediately ignites due to the high temperature of the air. The piston begins its power stroke at state 2, and fuel injection stops at state 3: the isentropic process 3–4 continues the power stroke. The constant-volume process 4–1 simulates the exhaust stroke and intake stroke, as in the Otto cycle: the piston does no work, yet heat is transferred from the cylinder. Note that the primary difference between the diesel cycle and the Otto cycle is the combustion process: a constant-pressure process in the diesel cycle and a constant-volume process in the Otto cycle.

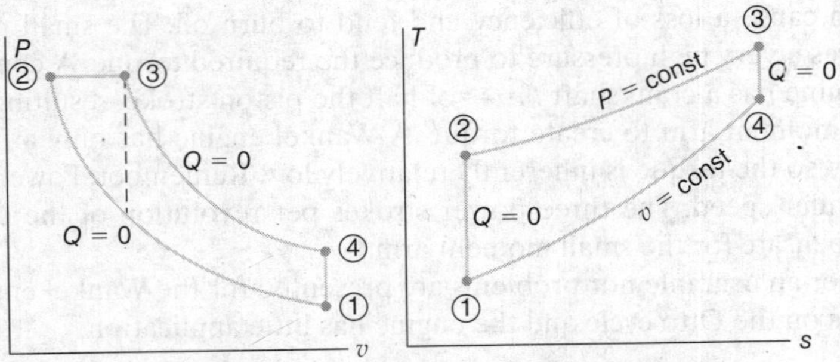

(*a*) The *P*-*v* diagram (*b*) The *T*-*s* diagram

Figure 9.11

The ideal diesel cycle.

The compression ratio of a diesel engine is much higher than that of an Otto cycle engine. An Otto cycle engine might have a compression ratio of up to about 12, whereas that of a diesel engine may be up to about 22. As a result, diesel engines use different fuels than an Otto cycle engine. If a volatile fuel like gasoline was used in a diesel engine, it would auto-ignite before the piston reached TDC. This would cause an effect called *engine knock*, which is the sound of the flame front hitting the oncoming piston. Diesel fuel produces less energy per unit mass than gasoline, but the diesel cycle is significantly more efficient because of its higher compression ratio, resulting in perhaps 50% better fuel mileage.

Engine knock: The sound of a flame front hitting the piston.

The diesel cycle will now be analyzed assuming an ideal gas with constant specific heats, the cold-air standard. The heat input during the constant-pressure combustion process 2–3, assumed to be a quasi-equilibrium heat addition process, is

$$q_{in} = C_p(T_3 - T_2) \tag{9.11}$$

The heat rejection during the constant-volume intake/exhaust process 1–4 is

$$q_{out} = C_v(T_4 - T_1) \tag{9.12}$$

The thermal efficiency of the ideal diesel cycle is then

$$\eta_{diesel} = \frac{w_{net}}{q_{in}} = 1 - \frac{q_{out}}{q_{in}} = 1 - \frac{C_v(T_4 - T_1)}{C_p(T_3 - T_2)}$$

$$= 1 - \frac{(T_4 - T_1)}{k(T_3 - T_2)} \tag{9.13}$$

As was done with the Otto cycle, factor T_1 from the numerator of the ratio on the right-hand side of Eq. 9.13 and T_2 from the denominator of this same ratio. There results

$$\eta_{diesel} = 1 - \frac{T_1\left(\frac{T_4}{T_1} - 1\right)}{k\,T_2\left(\frac{T_3}{T_2} - 1\right)} \tag{9.14}$$

The isentropic compression and power strokes allow the same expressions as in Otto cycle:

$$\frac{T_1}{T_2} = \left(\frac{v_2}{v_1}\right)^{k-1} = \frac{1}{r^{k-1}} \tag{9.15}$$

$$\frac{T_4}{T_3} = \left(\frac{v_3}{v_4}\right)^{k-1} = \frac{1}{r^{k-1}} \tag{9.16}$$

Recognize that $v_4 = v_1$ (see Fig. 9.10*a*) and, for the constant-pressure process, $T_3/T_2 = v_3/v_2$, we can write

$$\frac{T_4}{T_1} = \frac{T_4}{T_3} \times \frac{T_3}{T_2} \times \frac{T_2}{T_1} = \left(\frac{v_3}{v_1}\right)^{k-1}\left(\frac{v_3}{v_2}\right)\left(\frac{v_1}{v_2}\right)^{k-1}$$

$$= \left(\frac{v_3}{v_2}\right)\left(\frac{v_3}{v_2}\right)^{k-1} = \left(\frac{v_3}{v_2}\right)^{k} \tag{9.17}$$

Comment

The high pressure in the diesel cycle requires much heavier components to resist the larger stresses in the bolts and connecting rods. GM converted gasoline engines to diesel engines but did not provide sufficiently for the higher stresses, resulting in engine failure and discontinuation of the engines.

Comment

Constant specific heats provide formulas that allow quick comparisons between cycles. Since the temperatures and pressures are quite high, the errors will be larger than in the Otto cycle analysis when using the cold-air analysis.

Cutoff ratio: The ratio of the volume at the end of combustion to the volume at the start of combustion.

The ratio v_3/v_2 is the ratio of the volume at the end of the combustion process to the volume at the start of the combustion process. For the Otto cycle, the value of this ratio is 1. For the diesel cycle, this number will be greater than 1 since $v_3 > v_2$. This ratio of volumes v_3/v_2 is called the *cutoff ratio* r_c. It is related for this constant-pressure process to the temperature by

$$r_c = \frac{v_3}{v_2} = \frac{T_3}{T_2} \tag{9.18}$$

and is typically about 2. The cycle efficiency, using Eqs. 9.15, 9.17, and 9.18, can now be expressed as

$$\eta_{\text{diesel}} = 1 - \frac{1}{k} \times \frac{T_1}{T_2}\left(\frac{\dfrac{T_4}{T_1} - 1}{\dfrac{T_3}{T_2} - 1}\right) = 1 - \frac{1}{r^{k-1}} \times \frac{r_c^k - 1}{k(r_c - 1)} \tag{9.19}$$

This efficiency of the ideal diesel cycle is a function of two design parameters: the compression ratio r and the cutoff ratio r_c. As the compression ratio increases, the efficiency increases like that of the Otto cycle. As the cutoff ratio increases, the diesel cycle efficiency decreases.

The diesel engine is also manufactured as a two-stroke engine that operates much like the two-stroke Otto cycle engine described in Section 9.3.3. Small two-stroke diesel engines are used in model airplanes, and huge ones, like the one in Fig. 9.12, are used to power big ships.

Courtesy of Wärtsilä

Figure 9.12

A two-stroke diesel engine that produces 108 920 hp built by Aioi Works of Japan to power a container ship.

The diesel cycle — Example 9.5

The compression ratio of a diesel cycle is 18 with a cutoff ratio of 2. The pressure and temperature of the air at the beginning of the compression stroke are 100 kPa and 15°C, as shown in Fig. 9.13. Determine:

i) The maximum temperature and pressure of the air
ii) The thermal efficiency
iii) The required heat input and the net work output
iv) The mean effective pressure

The primary difference between this example and Example 9.3 is the compression ratio.

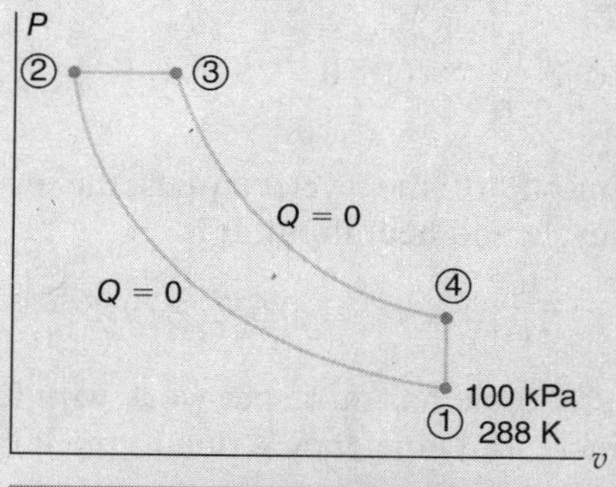

Figure 9.13

Solution

i) At state 1, $P_1 = 100$ kPa and $T_1 = 15°C = 288$ K. The compression stroke is an isentropic process for which we use Eqs. 3.44 and 3.46. They give

$$T_2 = T_1\left(\frac{v_1}{v_2}\right)^{k-1} = 288 \times 18^{0.4} = 915.2 \text{ K}$$

$$P_2 = P_1\left(\frac{v_1}{v_2}\right)^{k} = 100 \times 18^{1.4} = 5720 \text{ kPa}$$

The next process is the combustion process simulated by heat addition, which takes place at constant pressure so $P_3 = P_2 = \underline{5720 \text{ kPa}}$. The cutoff ratio of 2 allows

$$\frac{\cancel{P_3}v_3}{T_3} = \frac{\cancel{P_2}v_2}{T_2} \qquad \frac{T_3}{T_2} = \frac{v_3}{v_2} = 2$$

$$\therefore T_3 = 2T_2 = 2 \times 915 = 1830 \text{ K} \quad \text{or} \quad \underline{1557°C}$$

(Continued)

Example 9.5 *(Continued)*

Note: If $r_c = 3$, $\eta_{diesel} = 58.9\%$.

ii) The thermal efficiency of the cycle, using Eq. 9.19 with the given compression and cutoff ratios, is

$$\eta_{diesel} = 1 - \frac{1}{r^{k-1}} \times \frac{r_c^k - 1}{k(r_c - 1)}$$

$$= 1 - \frac{1}{18^{0.4}} \times \frac{2^{1.4} - 1}{1.4 \times (2 - 1)} = 0.632 \text{ or } \underline{63.2\%}$$

iii) The required heat input by this constant-pressure process (see Eq. 3.31) is

$$q_{in} = h_3 - h_2 = C_p(T_3 - T_2)$$

$$= 1.0 \frac{\text{kJ}}{\text{kg}\cdot\text{K}} \times (1830 - 915.2) \text{ K} = \underline{915 \text{ kJ/kg}}$$

The net work output of the cycle equals the product of the thermal efficiency of the cycle and heat input. It is

$$w_{net} = \eta_{diesel} \times q_{in} = 0.632 \times 915 = \underline{578 \text{ kJ/kg}}$$

Note that the diesel cycle has more net work with less heat input than the Otto cycle. The increased efficiency is due to the larger compression ratio.

iv) The MEP, the mean effective pressure, is the ratio of the net work to the volume difference. The specific volumes at top dead center and bottom dead center are

$$v_1 = \frac{RT_1}{P_1} = \frac{0.287 \times 288}{100} = 0.8266 \text{ m}^3/\text{kg}$$

$$v_2 = \frac{v_1}{r} = \frac{0.8266}{18} = 0.0459 \text{ m}^3/\text{kg}$$

The MEP is found using Eq. 9.3 to be

$$\text{MEP} = \frac{w_{cycle}}{v_1 - v_2} = \frac{578 \text{ kN}\cdot\text{m/kg}}{(0.8266 - 0.0459) \text{ m}^3/\text{kg}} = \underline{740 \text{ kPa}}$$

☑ **You have completed Learning Outcome (3)**

9.5 Other Gas Power Cycles

This section presents three engine cycles that have limited application but are significant to engine development. Processes utilized by cycles that are not actually used in existing engines may provide insight into research on engine development so they have value in being presented.

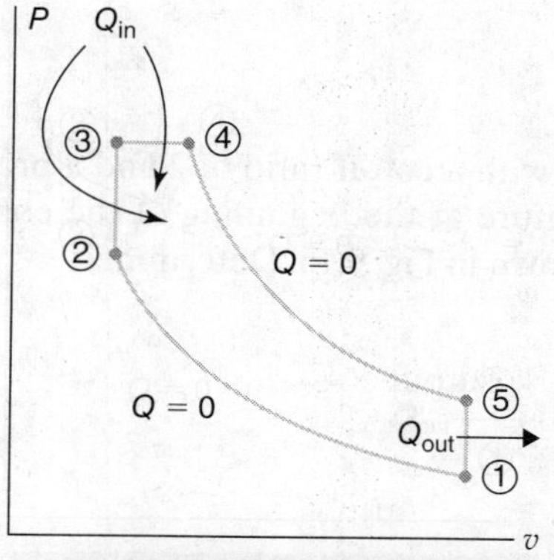

Figure 9.14
The *P-v* diagram of the ideal dual cycle.

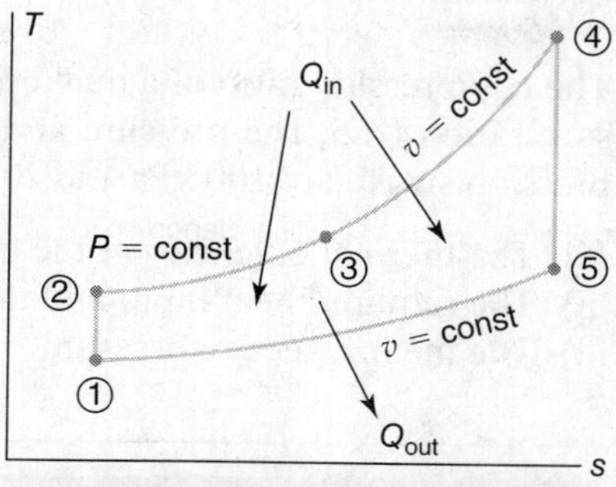

Figure 9.15
The *T-s* diagram of the ideal dual cycle.

9.5.1 The dual cycle

The dual cycle is a combination of the Otto and diesel cycles. The combustion process is divided into two processes, as shown in the diagrams in Figs. 9.14 and 9.15. The first part of the combustion process is done at constant volume, like the Otto cycle. The second part is done at constant pressure, like the diesel cycle. The dual cycle does a fairly realistic job of modeling the combustion process of the Otto cycle, considering the actual Otto cycle curves shown in Fig. 9.6. The analysis of the dual cycle is performed in the same manner as the Otto cycle, with the addition of the constant-pressure combustion process.

The heat addition and heat rejection are, using the cold-air standard,

$$q_{in} = (u_3 - u_2) + (h_4 - h_3)$$
$$= C_v(T_3 - T_2) + C_p(T_4 - T_3) \quad (9.20)$$

$$q_{out} = C_v(T_5 - T_1) \quad (9.21)$$

The thermodynamic efficiency of the cycle is

$$\eta_{dual} = 1 - \frac{q_{out}}{q_{in}} = 1 - \frac{T_5 - T_1}{T_3 - T_2 + k(T_4 - T_3)} \quad (9.22)$$

A *pressure ratio* r_p is defined to be $r_p = P_3/P_2$, allowing the efficiency to be expressed as (the algebra will be an exercise in the problem set)

Pressure ratio: The ratio P_3/P_2 in Fig. 9.14. It is not the ratio of the high and low pressures.

$$\eta_{dual} = 1 - \frac{1}{r^{k-1}} \times \frac{r_p r_c^k - 1}{r_p - 1 + k r_p (r_c - 1)} \quad (9.23)$$

If $r_c = r_p = 1$, the Otto cycle efficiency results; if $r_p = 1$, the diesel cycle efficiency results. Because of the possible higher compression ratio, the efficiency of the dual cycle will exceed that of the Otto cycle. However, for a given compression ratio r, the Otto cycle has the highest efficiency.

Example 9.6 The dual cycle

The compression ratio of a dual cycle is 18, with a cutoff ratio of 2 and a pressure ratio of 1.5. The pressure and temperature at the beginning of the compression stroke are 100 kPa and 20°C, as shown in Fig. 9.16. Determine:

i) The thermal efficiency of the cycle
ii) The required heat input and the net work output
iii) The mean effective pressure

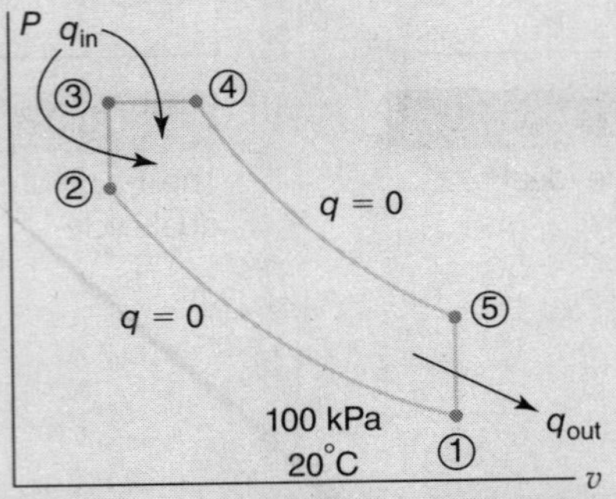

Figure 9.16

Solution

i) Since r, r_c, and r_p are given, the thermodynamic efficiency of the dual cycle is

$$\eta_{dual} = 1 - \frac{1}{r^{k-1}} \times \frac{r_p r_c^k - 1}{r_p - 1 + kr_p(r_c - 1)}$$

$$= 1 - \frac{1}{18^{0.4}} \times \frac{1.5 \times 2^{1.4} - 1}{1.5 - 1 + 1.4 \times 1.5(2 - 1)}$$

$$= 0.642 \text{ or } \underline{64.2\%}$$

Note: This efficiency compares to 63.2% for the diesel cycle with the same compression ratio.

ii) To find the heat input, the temperatures T_2, T_3, and T_4 in Fig. 9.14 are needed. They are found as follows:

$$T_2 = T_1\left(\frac{v_1}{v_2}\right)^{k-1} = 293 \times 18^{0.4} = 931 \text{ K}$$

$$T_3 = T_2\frac{P_3}{P_2} = 931 \times 1.5 = 1396 \text{ K}$$

$$T_4 = T_3\frac{v_4}{v_3} = 1396 \times 2 = 2792 \text{ K}$$

The heat input is, using $C_v = 0.717$ kJ/kg·K and $C_p = 1.0$ kJ/kg·K,

$$q_{in} = C_v(T_3 - T_2) + C_p(T_4 - T_3)$$

$$= 0.717 \times (1396 - 931) + 1.0 \times (2792 - 1396)$$

$$= \underline{1740 \text{ kJ/kg}}$$

The net work output is

$$w_{net} = \eta_{dual} \times q_{in}$$
$$= 0.642 \times 1740 = \underline{1120 \text{ kJ/kg}}$$

iii) The MEP is the ratio of the net work to the displacement. See Eq. 9.3. We find the specific volumes at top dead center and bottom dead center to be

$$v_1 = \frac{RT_1}{P_1} = \frac{0.287 \dfrac{\text{kJ}}{\text{kg·K}} \times 293 \text{ K}}{100 \text{ kPa}} = 0.841 \text{ m}^3\text{/kg}$$

$$v_2 = \frac{v_1}{r} = \frac{0.841}{18} = 0.047 \text{ m}^3\text{/kg}$$

The MEP, using Eq. 9.3, is found to be

$$\text{MEP} = \frac{w_{cycle}}{v_1 - v_2}$$

$$= \frac{1120 \text{ kJ/kg}}{(0.841 - 0.047) \text{ m}^3\text{/kg}} = 1400 \text{ kPa} \quad \text{or} \quad \underline{1.4 \text{ MPa}}$$

9.5.2 The Stirling and Ericsson cycles

The Stirling and Ericsson cycles are composed of processes that are assumed to be reversible in the ideal cycles. Because each ideal cycle is completely reversible, each cycle would have the efficiency of a Carnot cycle. If losses due to friction and the irreversibility of heat transfer across the small but finite temperature differences could be eliminated, the Carnot cycle efficiency would be approached. Stirling engines have been built and tested in research laboratories but have not had significant practical use. Figure 9.17 shows the components of the Stirling and Ericsson cycles. The Stirling engine is typically constructed using pistons, but the sketch in Fig. 9.17 using steady-flow devices demonstrates the cycle.

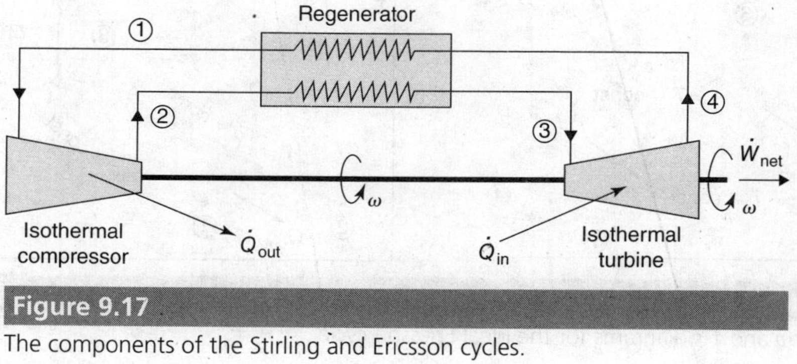

Figure 9.17
The components of the Stirling and Ericsson cycles.

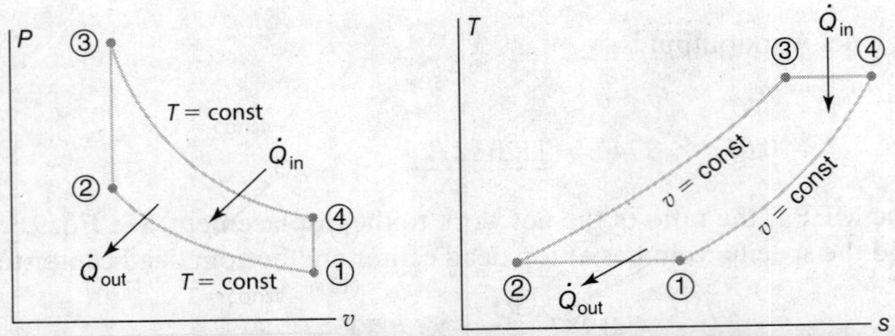

Figure 9.18

The *P-v* and *T-s* diagrams for the ideal Stirling cycle.

Regenerator: A heat exchanger that transfers heat from one fluid to another fluid.

Note: Heat added during an isothermal process can approach reversibility.

The *regenerator* is a heat exchanger that transfers heat from the air leaving the turbine to the air that is leaving the compressor. Air enters the ideal regenerator at temperature T_4 and leaves at temperature T_1. In the other line, the air enters at $T_2 = T_1$ and leaves at $T_3 = T_4$. Inside the regenerator, the temperature at each location is assumed to be essentially the same in each line so that the heat transfer is always across an infinitesimal temperature difference and thus reversible. In an actual regenerator, the temperatures would not be the same at each location, so the heat transfer process would be irreversible but could approach reversibility.

The *P-v* and *T-s* diagrams are displayed in Figs. 9.18 and 9.19 for the two cycles. The process begins at state 1 with air at standard conditions, as in the other cycles that have been analyzed. An isothermal compression process 1–2 then increases the pressure of the air. The temperature of the air is then increased in process 2–3 as it passes through the regenerator. In the Stirling cycle, process 2–3 is a constant-volume process, and in the Ericsson cycle, it is a constant-pressure process, The air then passes through the turbine, where heat is added from an external source by an isothermal process 3–4 and work is accomplished. (The net work is the work produced by the turbine minus the work required to drive the compressor.) The air leaves the turbine at a temperature higher than the temperature at state 1, so thermal energy is transferred to the air about to enter the turbine; regeneration allows a very efficient cycle.

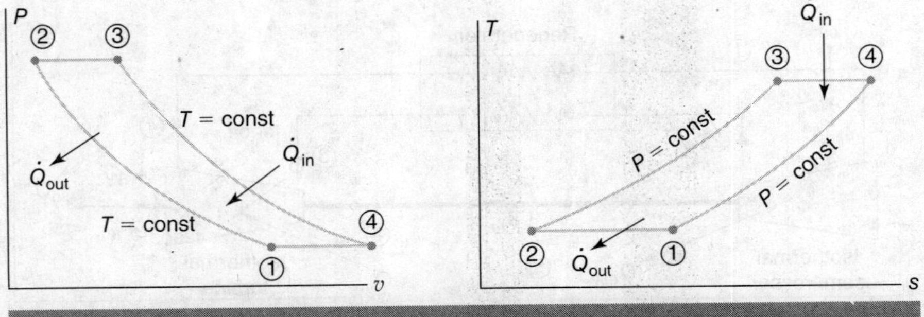

Figure 9.19

The *P-v* and *T-s* diagrams for the ideal Ericsson cycle.

The expression for the work required by the isothermal compressor and provided by the isothermal turbine is found by applying the first law and Eq. 6.18:

$$q - w = \Delta h = C_p \cancel{\Delta T} = 0$$

$$\therefore w = q = T\Delta s = T\left(C_v \ln \cancel{\frac{T_2}{T_1}} + R \ln \frac{v_2}{v_1}\right)$$

$$= TR \ln \frac{v_2}{v_1} \tag{9.24}$$

Because the heat transfer in the ideal cycles is assumed to be transferred at infinitesimal temperature differences, and the other processes are assumed to be reversible, the thermal efficiency of both cycles is given by the Carnot cycle efficiency, which is

$$\eta = 1 - \frac{T_L}{T_H} \tag{9.25}$$

Comment

An engine operating on an ideal Sterling or Ericsson cycle has the same efficiency as the Carnot cycle, the highest efficiency of any possible engine.

The major problems with an engine operating on either of these cycles are the two isothermal heat transfer processes. For reversible heat transfer to occur, the temperature difference between the two surfaces across which the heat transfer occurs must be infinitesimal. A small temperature difference could approach a reversible process, but the time for the heat transfer to occur would be too long for an engine at high rpm. Even with these major problems, significant research is being done on the Stirling engine. Several actual running engines can be observed on the Internet by Googling "Stirling engines." Because of its limited success, this engine does not attract significant engineering interest.

The Stirling and Ericsson cycles — Example 9.7

The compression ratio of both the Stirling and Ericsson cycles of Fig. 9.20 is 12, the low temperature is 90°C, the high temperature is 650°C, and the low pressure is 170 kPa. Determine the efficiency, the power output, and the rate of heat input of each cycle if the mass flow rate is 0.9 kg/s.

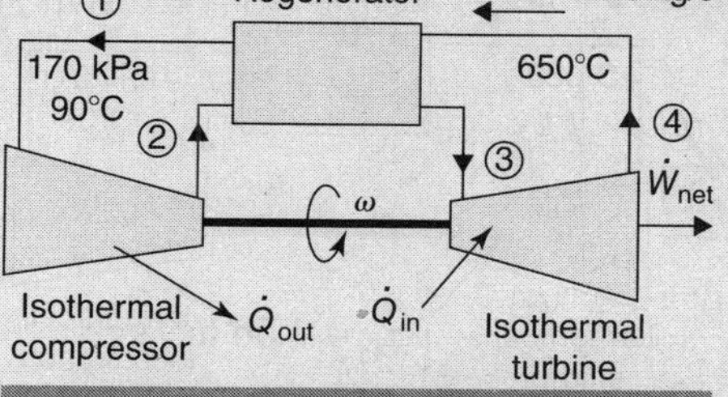

Figure 9.20

(Continued)

Example 9.7 (Continued)

Solution

The efficiency of each ideal cycle is given by Eq. 9.25:

$$\eta = 1 - \frac{T_L}{T_H} = 1 - \frac{363}{923} = 0.607 \quad \text{or} \quad \underline{60.7\%}$$

Note: The compression ratio is a ratio of specific volumes.

The Stirling cycle: The work for the ideal Stirling cycle, using Eq. 9.24, is

$$\begin{aligned} w_{\text{net}} &= w_{1\text{-}2} + w_{3\text{-}4} \\ &= RT_1 \ln \frac{v_2}{v_1} + RT_3 \ln \frac{v_4}{v_3} \\ &= 0.287 \times 363 \ln \frac{1}{12} + 0.287 \times 923 \ln 12 = 399.4 \text{ kJ/kg} \end{aligned}$$

$$\therefore \dot{W}_{\text{net}} = \dot{m}w_{\text{net}} = \frac{0.9 \text{ kg/s} \times 399.4 \text{ kJ/kg}}{0.746 \text{ kw/hp}} = \underline{482 \text{ hp}}$$

The rate of heat input for the Stirling cycle is found using the efficiency. It is

$$\eta = \frac{\dot{W}_{\text{net}}}{\dot{Q}_{\text{in}}} = \frac{482 \times 0.746}{\dot{Q}_{\text{in}}} = 0.607$$

$$\therefore \dot{Q}_{\text{in}} = \underline{592 \text{ kJ/s}}$$

The Ericsson cycle: The efficiency of the ideal Ericsson cycle is the same as that of the ideal Stirling and Carnot cycles at 60.2%. There is work associated with each of the four processes. It is

$$\begin{aligned} w_{\text{net}} &= w_{1\text{-}2} + w_{2\text{-}3} + w_{3\text{-}4} + w_{4\text{-}1} \\ &= RT_1 \ln \frac{v_2}{v_1} + P_2(v_3 - v_2) + RT_3 \ln \frac{v_4}{v_3} + P_1(v_1 - v_4) \end{aligned}$$

Units: Here is the ideal-gas law with English units. Review Example 9.6 if the units are a problem.

The properties P_2, v_1, v_2, v_3, and v_4 are not known. They are found as follows:

Ideal-gas law: $$v_1 = \frac{RT_1}{P_1} = \frac{0.287 \times 363}{170} = 0.613 \text{ m}^3\text{/kg}$$

4→1, P = const: $$\therefore \frac{T_4}{v_4} = \frac{T_1}{v_1}$$

$$\frac{923}{v_4} = \frac{363}{0.613} \qquad \therefore v_4 = 1.56 \text{ m}^3\text{/kg}$$

Compression ratio = 12: $$\therefore v_2 = \frac{v_1}{12} = \frac{0.613}{12} = 0.0511 \text{ m}^3\text{/kg}$$

$$2 \rightarrow 3, P = \text{const}: \therefore P_3 = P_2 = \frac{RT_2}{v_2}$$

$$= \frac{0.287 \times 323}{0.0511} = 1814 \text{ kPa}$$

$$3 \rightarrow 4, T_3 = T_4: \quad v_3 = \frac{RT_3}{P_3} = \frac{0.287 \times 923}{1814} = 0.146 \text{ m}^3/\text{kg}$$

Substituting into the expression above, the work is

$$w_{\text{net}} = 0.287 \times 363 \ln \frac{1}{12} + 1814(0.146 - 0.0511)$$

$$+ 0.287 \times 923 \ln 12 + 170(0.613 - 1.56) = 410.5 \text{ kJ/kg}$$

or $$\dot{W}_{\text{net}} = \dot{m}w_{\text{net}} = 0.9 \frac{\text{kg}}{\text{s}} \times \frac{410.5 \text{ kJ/kg}}{0.746 \text{ hp/kW}} = \underline{495 \text{ hp}}$$

$$\therefore \dot{Q}_{\text{in}} = \frac{495 \times 0.746}{0.607} = \underline{608 \text{ kJ/s}}$$

Note: The work from state 1 to state 2 is the compressor work, which is negative since ln(1/12) is negative.

Remember: 746 w = 1 hp.

☑ **You have completed Learning Outcome (4)**

9.6 The Brayton Cycle

Gas turbine: An engine that operates on the Brayton cycle.

The Brayton cycle was first patented as a piston engine by George Brayton in 1872 and then in the 1920s as an engine that operated on the Brayton cycle using a compressor and a turbine. Because the efficiencies of the compressor and turbine were not sufficiently high, it did not attract significant industrial interest until much later. Now the engine, referred to as a *gas turbine*, is used in all larger commercial and military aircraft as well as power plants, tanks, ships, large trucks, and even in a motorcycle. Its disadvantages compared to piston engines are: high cost, inefficiency at low speeds, longer start-up, and less responsive to speed change. But its power to weight ratio is very attractive.

The gas turbine that operates on a Brayton cycle is an internal combustion engine. It is different from the Otto and diesel cycle engines in that it is a continuous flow engine. Unlike the piston engine, there is a constant flow of air into the engine and a constant flow of products of combustion from the engine, as shown in Fig. 9.21. The compression of air is accomplished by a rotary compressor. Combustion takes place in a separate combustion

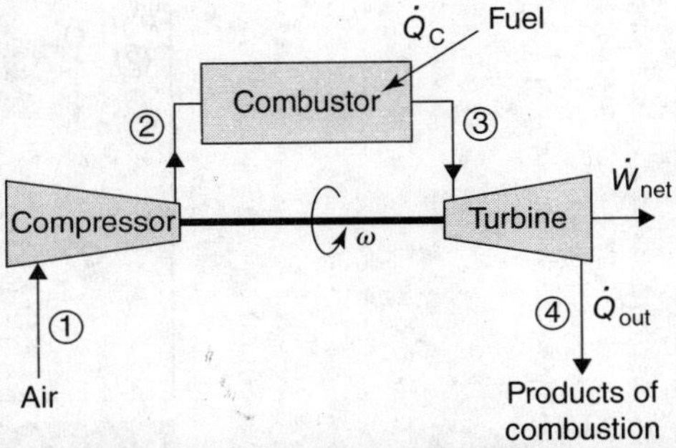

Figure 9.21
The open-cycle gas turbine.

Note: Because it can burn different types of fuels, it may be considered a "green" engine.

chamber, so it can burn many different types of fuel. The high-pressure, high-temperature products of combustion then flow through a gas turbine that produces power. To analyze the Brayton cycle we will use the first law of thermodynamics for control volumes (the compressor, the turbine, and the heat exchangers), as presented in Chapter 4. A diagram of a closed-cycle gas turbine, for analysis purposes, is shown in Fig. 9.22. The top heat exchanger accounts for the energy input that would occur during combustion of the fuel, and the lower heat exchanger emits the heat associated with the products of combustion and does no work; the open-cycle and closed-cycle gas turbine engines are thermodynamically identical.

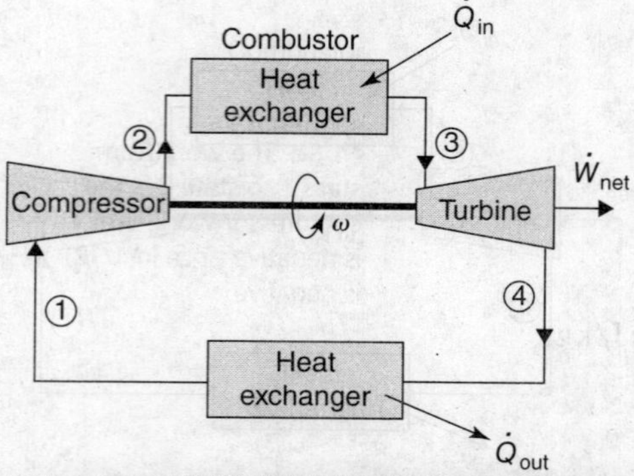

Figure 9.22

The closed-cycle gas turbine.

A necessary function of the turbine is to drive the compressor. Only since the 1940s has the turbine generated sufficient power to both drive the compressor and have additional power to be competitive with piston engines. If the turbine and compressor were only 80% efficient, a gas turbine engine would not be competitive. As efficiencies increased, gas turbine engines have found numerous applications. Perhaps the most common application is the gas turbine engine that powers modern-day commercial aircraft; these engines will be considered in detail in Section 9.6.3.

Let us now analyze the ideal Brayton cycle. The *P-v* and *T-s* diagrams are displayed in Fig. 9.23. Both the heat input and heat removal processes are accomplished with heat exchangers, which utilize constant-pressure processes. The thermal efficiency of the Brayton cycle, assuming constant C_p, is

$$\eta = 1 - \frac{\dot{Q}_{out}}{\dot{Q}_{in}} = 1 - \frac{\dot{m}\, C_p(T_4 - T_1)}{\dot{m}\, C_p(T_3 - T_2)} = 1 - \frac{T_4 - T_1}{T_3 - T_2}$$

$$= 1 - \frac{T_1\left(\dfrac{T_4}{T_1} - 1\right)}{T_2\left(\dfrac{T_3}{T_2} - 1\right)} \tag{9.26}$$

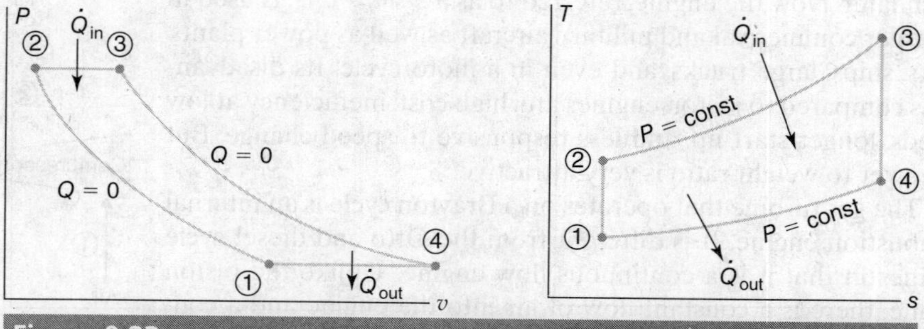

Figure 9.23

The *P-v* and *T-s* diagrams for the ideal Brayton cycle.

For the constant-pressure heat exchangers, $P_3 = P_2$ and $P_4 = P_1$, so

$$\frac{P_3}{P_4} = \frac{P_2}{P_1} \tag{9.27}$$

The compressor and turbine processes are idealized as isentropic processes in the ideal Brayton cycle. Using Eq. 3.45, we obtain

$$\frac{P_2}{P_1} = \left(\frac{T_2}{T_1}\right)^{\frac{k}{k-1}} \quad \text{and} \quad \frac{P_3}{P_4} = \left(\frac{T_3}{T_4}\right)^{\frac{k}{k-1}} \tag{9.28}$$

Equations 9.27 and 9.28 provide

$$\frac{T_2}{T_1} = \frac{T_3}{T_4} \quad \text{or} \quad \frac{T_4}{T_1} = \frac{T_3}{T_2} \tag{9.29}$$

The thermal efficiency equation now takes the rather simple form

$$\eta_{\text{Brayton}} = 1 - \frac{T_1}{T_2} \tag{9.30}$$

In terms of pressures, the efficiency is

$$\eta_{\text{Brayton}} = 1 - \left(\frac{P_1}{P_2}\right)^{(k-1)/k} \tag{9.31}$$

The ratio of the pressure P_2 exiting the compressor to the pressure P_1 entering the compressor, typically atmospheric pressure, is the *pressure ratio* r_p. In terms of this pressure ratio, the efficiency is

$$\eta_{\text{Brayton}} = 1 - r_p^{(1-k)/k} \tag{9.32}$$

Of course, the efficiency is also the ratio of the net work to the heat input.

An important term associated with the gas turbine cycle is the *back-work ratio* (BWR). It is the ratio of the work required to drive the compressor to the total work produced by the turbine. Until the 1940s, too much of the turbine work was required to drive the compressor, but with increased efficiencies of the compressors and the turbines due to better blade design and higher inlet temperature to the turbines (better heat-resistant materials), gas turbine engines enjoy numerous applications.

Pressure ratio r_p: The ratio of the high pressure to the low pressure in the Brayton cycle. Refer to the *P-v* diagram of Figure 9.23.

Back-work ratio (BWR): The ratio of the work required to drive the compressor to the total work produced by the turbine. It is often expressed as a percent, much like an efficiency.

The ideal Brayton cycle — Example 9.8

Air enters the compressor of a gas turbine of Fig. 9.24 at 300 K and 100 kPa where it is compressed to 700 kPa. The combustor increases the temperature to 1000 K. Determine i) the thermal efficiency, ii) the net work, iii) the BWR, and iv) the heat input. Assume ideal processes with constant specific heats.

Solution

The gas turbine engine is a continuous flow device, and since the mass flow rate of air through the engine is not given in the problem statement, this problem will be solved in terms of specific properties.

(Continued)

Example 9.8 (*Continued*)

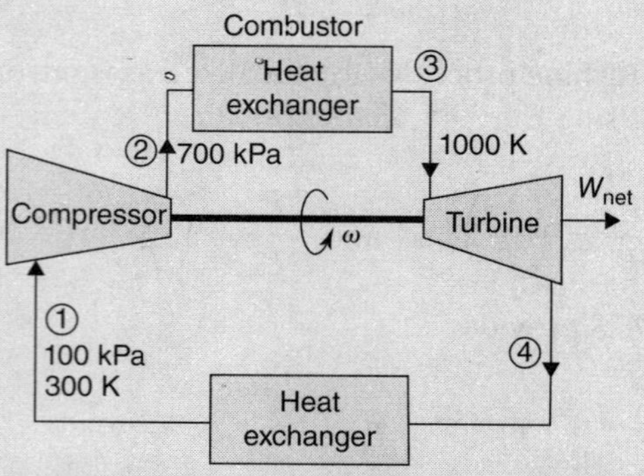

Figure 9.24

i) The thermodynamic efficiency of this ideal cycle is found, using Eq. 9.32, to be

$$\eta = 1 - r_p^{(1-k)/k} = 1 - \left(\frac{700}{100}\right)^{-0.4/1.4} = 0.426 \quad \text{or} \quad \underline{42.6\%}$$

Note: The cold-air standard is assumed.

Equation 9.30 was not used since the temperature T_2 is not known.

ii) The temperatures at states 2 and 4 are

$$T_2 = T_1\left(\frac{P_2}{P_1}\right)^{k-1/k} = 300 \times 7^{0.2857} = 523 \text{ K}$$

$$T_4 = T_3\left(\frac{P_4}{P_3}\right)^{k-1/k} = 1000 \times \left(\frac{1}{7}\right)^{0.2857} = 574 \text{ K}$$

The net work is then

$$\begin{aligned} w_{net} &= w_{turb} - w_{comp} = (h_3 - h_4) - (h_2 - h_1) \\ &= C_p(T_3 - T_4) - C_p(T_2 - T_1) \\ &= 1.0 \times (1000 - 574 - 523 + 300) \\ &= 426 - 223 = \underline{203 \text{ kJ/kg}} \end{aligned}$$

iii) Observe from the above equation that $w_{turb} = 426$ kJ/kg and $w_{comp} = 223$ kJ/kg, so the BWR is

$$\text{BWR} = \frac{w_{comp}}{w_{turb}} = \frac{223}{426} = 0.523 \quad \text{or} \quad \underline{52.3\%}$$

Over half of the work produced by the turbine is needed to drive the compressor in this ideal cycle.

iv) The heat needed by the combustor is found to be

$$q_{in} = C_p(T_3 - T_2) = 1.0 \times (1000 - 523) = \underline{477 \text{ kJ/kg}}$$

Turbine and compressor losses — Example 9.9

This example demonstrates the effect of compressor and turbine efficiencies on the effectiveness of the Brayton cycle. Rework Example 9.8 but assume the adiabatic compressor to be 85% efficient and the turbine to be 89% efficient. Air enters the compressor of a gas turbine engine at 300 K and 100 kPa where it is compressed to 700 kPa and exits the combustor at 1000 K. Determine i) the net work, ii) the BWR, iii) the heat input, and iv) the thermal efficiency. Assume constant specific heats.

Solution

i) The isentropic work required by the compressor of Example 9.8 was 223 kJ/kg. The efficiency of the compressor is the isentropic work input divided by the actual work input. Hence, the actual work requirement is

$$w_{comp,a} = \frac{w_{comp,s}}{\eta_{comp}} = \frac{223}{0.85} = 262 \text{ kJ/kg}$$

The isentropic work produced by the turbine of Example 9.8 was 426 kJ/kg. For a turbine, the efficiency is the actual work output divided by the isentropic work output, so the actual work produced is

$$w_{turb,a} = \eta_{turb} \times w_{turb,s} = 0.87 \times 426 = 371 \text{ kJ/kg}$$

The net work is found to decrease by almost 50% to

$$w_{net,a} = w_{turb,a} - w_{comp,a} = 371 - 262 = \underline{109 \text{ kJ/kg}}$$

ii) The BWR has markedly increased by 35% to

$$\text{BWR} = \frac{w_{comp,a}}{w_{turb,a}} = \frac{262}{371} = 0.706 \quad \text{or} \quad \underline{70.6\%}$$

Observe that almost three-quarters of the work produced by the turbine is needed to drive the compressor in this more realistic cycle.

iii), iv) To find the efficiency, the heat input must be determined. Return to the adiabatic compressor and find its exit temperature T_2:

$$w_{comp,a} = h_2 - h_1 = C_p(T_2 - T_1)$$

$$262 = 1.0 \times (T_2 - 300) \quad \therefore T_2 = 562 \text{ K}$$

The heat input and efficiency are then

$$q_{in} = C_p(T_3 - T_2) = 1.0 \times (1000 - 562) = \underline{438 \text{ kJ/kg}}$$

$$\eta = \frac{w_{net}}{q_{in}} = \frac{109}{438} = 0.249 \quad \text{or} \quad \underline{24.9\%}$$

The cycle efficiency has decreased over 40%. It is now obvious why the gas turbine engine is a relatively recent source of power. The compressor and turbine efficiencies had to be well above 80% before the cycle efficiency was sufficiently high to merit its use. Much research and development was needed to accomplish that feat.

Note: The thermal efficiency is often referred to as the thermodynamic efficiency or simply the efficiency.

9.6.1 The Brayton cycle with regenerative heating

Regenerator: A heat exchanger that preheats the air before combustion.

Regenerative Brayton cycle: The air is preheated before it enters the combustor.

The exhaust gasses from the turbine can be used to preheat the air using a *regenerator* before it enters the combustion chamber. If the gases leaving the turbine have a higher temperature than the temperature of the air leaving the compressor, a regenerator significantly improves the cycle thermal efficiency. The open *regenerative Brayton cycle* is shown in Fig. 9.25, with its *T-s* diagram for the ideal cycle shown in Fig. 9.26, assuming $T_2 < T_5$.

The cycle could be drawn as a closed cycle if a heat exchanger was placed between state 6 and state 1, as sketched in the *T-s* diagram of Fig. 9.26. It would have no effect on the following analysis.

The regenerator is a gas-to-gas heat exchanger. For reversible regeneration, $T_3 = T_5$, $T_2 = T_6$, and there must be only an infinitesimal temperature difference between the two fluids at each location inside the regenerator. Let's determine the efficiency of this ideal regenerative Brayton cycle.

Note: The cold-air standard, which assumes constant specific heats, will be used unless otherwise requested.

The first law for a control volume allows the heat furnished by the combustor, the turbine work, and the compressor work to be determined:

$$\begin{aligned} q_{\text{in}} &= h_4 - h_3 = C_p(T_4 - T_3) \\ w_{\text{turb}} &= h_4 - h_5 = C_p(T_4 - T_5) \\ w_{\text{comp}} &= h_2 - h_1 = C_p(T_2 - T_1) \end{aligned} \tag{9.33}$$

The net work produced by the cycle is then

$$w_{\text{net}} = w_{\text{turb}} - w_{\text{comp}} = C_p(T_4 - T_5 - T_2 + T_1) \tag{9.34}$$

The ideal regenerator allows $T_5 = T_3$, so we observe from Eqs. 9.33 that $q_{\text{in}} = w_{\text{turb}}$, a rather interesting result. The cycle efficiency takes the form

$$\eta = \frac{w_{\text{net}}}{q_{\text{in}}} = \frac{w_{\text{turb}} - w_{\text{comp}}}{w_{turb}} = 1 - \frac{w_{\text{comp}}}{w_{\text{turb}}}$$

$$= 1 - \frac{T_2 - T_1}{T_4 - T_5} = 1 - \frac{T_1}{T_4} \times \frac{\dfrac{T_2}{T_1} - 1}{1 - \dfrac{T_5}{T_4}} \tag{9.35}$$

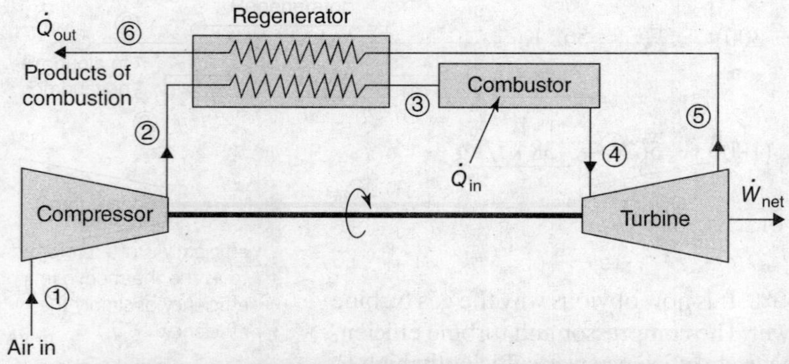

Figure 9.25

The open regenerative Brayton cycle.

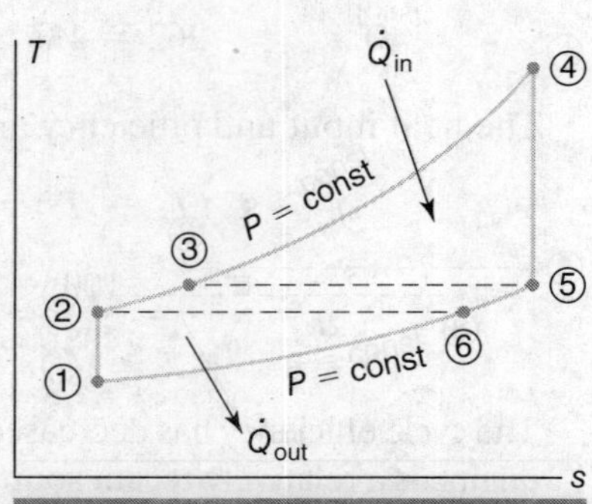

Figure 9.26

The *T-s* diagram for the closed ideal regenerative Brayton cycle.

Recognize that on the combustor side $P_2 = P_3 = P_4$ and on the exhaust side $P_1 = P_5 = P_6$, since there is no pressure drop across an ideal heat exchanger. The respective isentropic processes across the compressor and the turbine then provide

$$\frac{T_2}{T_1} = \left(\frac{P_2}{P_1}\right)^{(k-1)/k} \quad \text{and} \quad \frac{T_5}{T_4} = \left(\frac{P_5}{P_4}\right)^{(k-1)/k} = \left(\frac{P_1}{P_2}\right)^{(k-1)/k} \qquad (9.36)$$

This allows Eq. 9.35, after some algebra with $P_2/P_1 = r_p$, to be expressed as

$$\eta = 1 - \frac{T_1}{T_4} r_p^{(k-1)/k} \qquad \text{(Ideal regenerative Brayton cycle)} \qquad (9.37)$$

This expression for the efficiency is rather surprising in that the efficiency decreases as the pressure ratio r_p increases, an opposite effect for the efficiency of the ideal Brayton cycle given by Eq. 9.32. The sketch in Fig. 9.27 shows this graphically. That means there is a pressure ratio where the efficiencies of the two cycles are equal, corresponding to the condition where $T_2 = T_5$ in Fig. 9.26. If $T_2 > T_5$, a regenerator would actually have a negative effect. If $T_1/T_4 = 0.25$, the pressure ratio where the efficiencies are equal is 11.32.

The following example will include a regenerator in the ideal Brayton cycle of Example 9.8 so that the effect of the regenerator can be assessed.

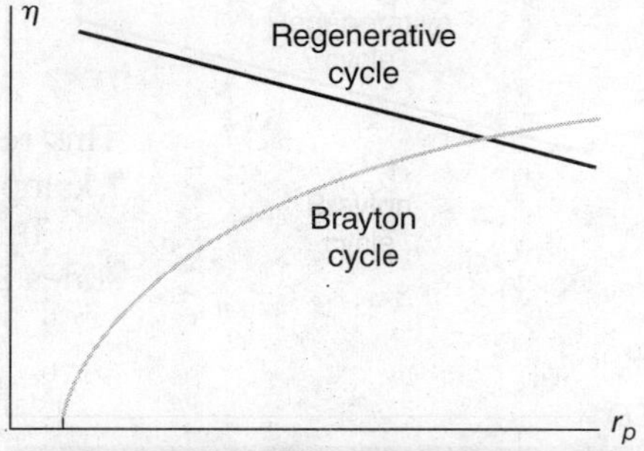

Figure 9.27
The effect of r_p on cycle efficiency for the ideal Brayton cycle and the ideal regenerative Brayton cycle.

The ideal regenerative Brayton cycle — Example 9.10

Air enters the compressor of a gas turbine engine at 300 K and 100 kPa where it is compressed to 700 kPa with a maximum temperature of 1000 K, as in Example 9.8. A regenerator is inserted into the cycle, as sketched in Fig. 9.28. Determine the thermal efficiency and the BWR, assuming ideal processes with constant specific heats.

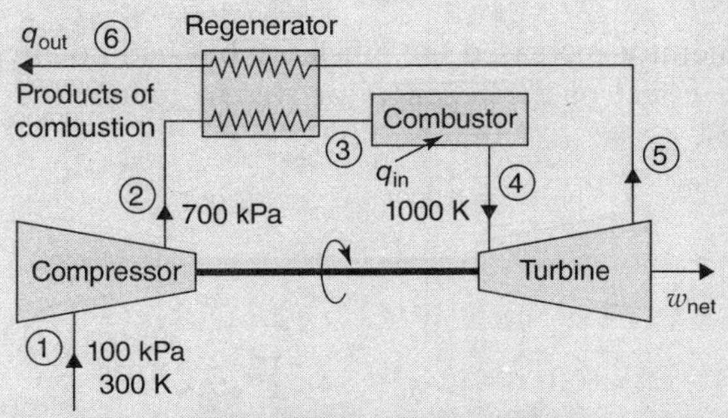

Figure 9.28

(Continued)

Example 9.10 (*Continued*)

Solution

The *T-s* diagram for the ideal regenerative Brayton cycle is identical to that of the ideal Brayton cycle, with the exception that two additional states are included. The minimum and maximum temperatures are not affected. The efficiency of the regenerative cycle is

$$\eta = 1 - \frac{T_1}{T_4} r_p^{(k-1)/k}$$

$$= 1 - \frac{300}{1000} \times 7^{0.2857} = 0.477 \text{ or } \underline{47.7\%}$$

This represents a 12% increase in cycle efficiency compared to the cycle of Example 9.8.

To find the BWR, the temperatures must be known. The temperatures at states 2 and 5 are

$$T_2 = T_1\left(\frac{P_2}{P_1}\right)^{k-1/k} = 300 \times 7^{0.2857} = 523 \text{ K}$$

$$T_5 = T_4\left(\frac{P_5}{P_4}\right)^{k-1/k} = 1000 \times \left(\frac{1}{7}\right)^{0.2857} = 574 \text{ K}$$

The turbine and compressor work are then

$$w_{\text{turb}} = (h_4 - h_5) = C_p(T_4 - T_5)$$

$$= 1.0 \times (1000 - 574) = 426 \text{ kJ/kg}$$

$$w_{\text{comp}} = (h_2 - h_1) = C_p(T_2 - T_1)$$

$$= 1.0 \times (523 - 300) = 223 \text{ kJ/kg}$$

so the BWR of this ideal regenerative Brayton cycle is

$$\text{BWR} = \frac{w_{\text{comp}}}{w_{\text{turb}}} = \frac{223}{426} = 0.523 \text{ or } \underline{52.3\%}$$

The regenerator increased the efficiency but had no effect on the back-work ratio. The effect of the regenerator was to reduce the energy input to the combustor.

9.6.2 The Brayton cycle with regeneration, intercooling, and reheat

Intercooler: A device used to reduce compressor work input.

In addition to the regenerator, two other devices can be added to the Brayton cycle to increase its efficiency. An *intercooler* allows compression to occur in two stages, with the intercooler located between the stages. It reduces the compressor work and the temperature at the compressor outlet, thereby controlling the maximum

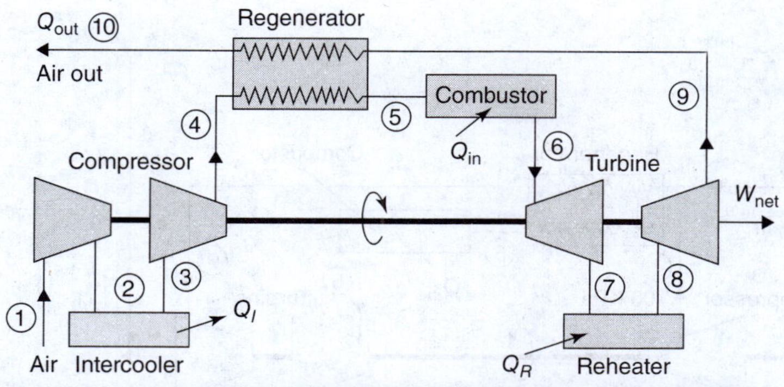

Figure 9.29

The regenerative Brayton cycle with intercooling and reheating.

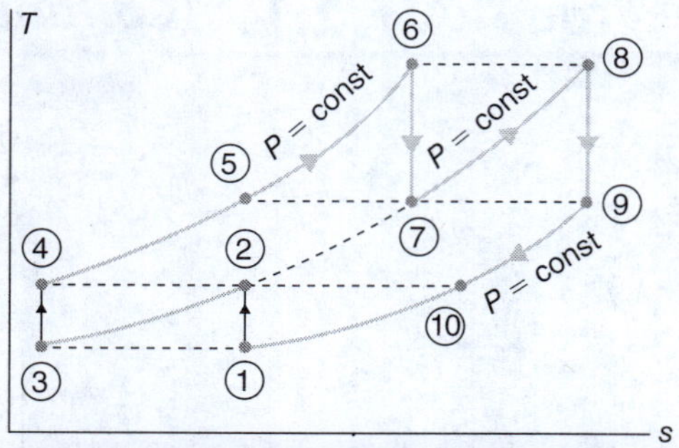

Figure 9.30

The *T-s* diagram for the ideal Brayton cycle of Fig. 9.28.

temperature in the cycle. Also, a *reheater* allows the turbine to produce power in two stages, resulting in increased turbine output. The pressures at which the intercooler and the reheater are inserted in the ideal cycle are selected such that

Reheater: A device used to increase the turbine work output.

$$\frac{P_2}{P_1} = \frac{P_4}{P_3} \quad \text{and} \quad \frac{P_6}{P_7} = \frac{P_8}{P_9} \tag{9.38}$$

with reference to Figs. 9.29 and 9.30. It is assumed in the ideal cycle that there is no pressure drop across a heat exchanger so $P_6 = P_5 = P_4$ and $P_9 = P_{10} = P_1$. Figure 9.30 also indicates that $T_4 = T_2 = T_{10}$ and $T_5 = T_7 = T_9$. The high turbine temperatures T_6 and T_8 are also equal. Intercooling and reheating are effective only if a regenerator is used in the Brayton cycle. If a regenerator is not in the cycle, intercooling and reheating will actually decrease the ideal cycle efficiency. Several stages in both the compressor and turbine may be used in the larger gas turbines.

Example 9.11 The ideal regenerative Brayton cycle with intercooling and reheating

Insert an intercooler and a reheater in the regenerative Brayton cycle of Example 9.10 and calculate the efficiency, assuming ideal components and constant specific heats. Air enters the compressor at 300 K and 100 kPa; it is compressed to a maximum pressure of 700 kPa, and the maximum cycle temperature is 1000 K, as shown in Fig. 9.31.

Solution

Recognizing that $P_2 = P_3$, using Eq. 9.38 the intermediate compressor pressure is found to be

$$P_2^2 = P_1P_4 = 100 \times 700 \quad \therefore P_2 = 265 \text{ kPa}$$

(Continued)

Example 9.11 (Continued)

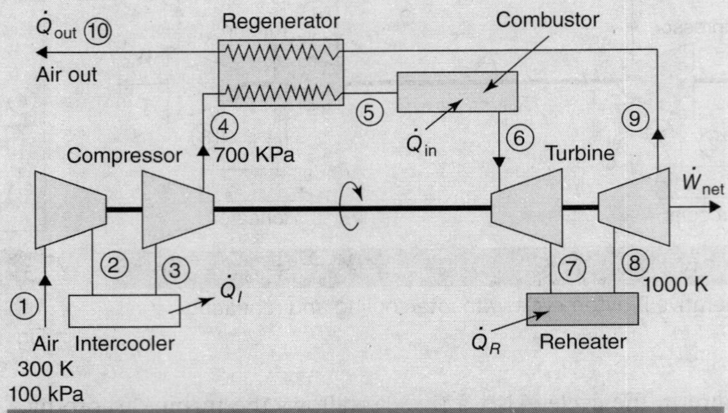

Figure 9.31

The isentropic compressor provides

$$T_2 = T_1\left(\frac{P_2}{P_1}\right)^{(k-1)/k} = 300 \times \left(\frac{265}{100}\right)^{0.2857} = 396 \text{ K}$$

From the maximum temperature of $T_6 = 1000$ K, the temperature at the reheater inlet, assuming an isentropic turbine, is

$$T_7 = T_6\left(\frac{P_7}{P_6}\right)^{(k-1)/k} = 1000 \times \left(\frac{265}{700}\right)^{0.2857} = 758 \text{ K}$$

The temperatures are all known, so the thermal efficiency is found to be

$$\eta = \frac{w_{net}}{q_{in,total}} = \frac{w_{turb} - w_{comp}}{q_{in} + q_R}$$

$$= \frac{(T_8 - T_9) + (T_6 - T_7) - (T_4 - T_3) - (T_2 - T_1)}{(T_6 - T_5) + (T_8 - T_7)}$$

$$= \frac{242 + 242 - 96 - 96}{242 + 242} = 0.603 \quad \text{or} \quad \underline{60.3\%}$$

This represents a 26% increase in efficiency over the ideal regenerative Brayton cycle and a 41% increase over the ideal Brayton cycle. These numbers would change somewhat if losses were considered and the ideal-gas tables were used (i.e., specific heats were not assumed constant).

9.6.3 The turbojet engine

Perhaps the primary use of the gas turbine that operates on the Brayton cycle has been in aircraft propulsion. Turbofan and turbojet engines were first developed for use in large commercial aircraft and in military jet planes since a gas turbine has a relatively large power to weight ratio. The turbofan engine is primarily used in aircraft designed to fly at subsonic speeds and the turbojet engine is used in aircraft designed to fly at supersonic speeds; however, both can fly in both

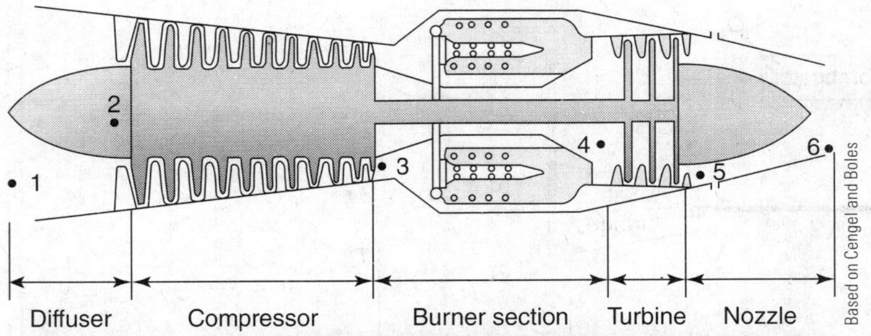

Figure 9.32

A schematic of a turbojet engine.

speed ranges. The basic configuration of the *turbojet engine* is shown in Fig. 9.32. A diffuser[1] slows down the incoming air flow from state 1 to state 2. The compressor sends pressurized air at state 3 into the burner section (the combustor) where fuel is injected into the air; the fuel burns at constant pressure. Products of combustion leave the burner at state 4 and travel through the gas turbine to state 5 and accelerate through the diverging supersonic nozzle to state 6, providing the thrust required to propel the aircraft. The primary function of the turbine is to produce the power needed to drive the compressor. The turbine also produces auxiliary power for the aircraft. The primary purpose of the turbojet engine is to provide the thrust developed by the diverging nozzle.

A *turboprop engine* uses the turbine power to drive a propeller. These engines are used mainly in smaller commuter aircraft. A gear system is used to change the high-speed turbine rotation into a lower-speed, high-torque power that drives the propellers.

The analysis of the air as it flows through a turbojet engine described above involves the air flow through a diffuser and a nozzle as well as the compressor and turbine and is the subject of a course in aerodynamics. It was described here to indicate that a jet engine operates on the Brayton cycle.

☑ **You have completed Learning Outcome (5)**

9.7 The Combined Brayton–Rankine Cycle

The high-temperature exhaust gases from a gas turbine operating on the Brayton cycle can be used as the heat source for a power plant that operates on the Rankine cycle. This *combined-cycle power* plant is an excellent example of cogeneration; rather than discard the energy-laden exhaust gases from a gas turbine engine, they can be used to generate additional power by generating steam in the boiler of Fig. 9.33. The key element in this system is the gas-to-steam heat exchanger (the boiler for the

[1]In a supersonic flow, a diffuser has a decreasing area, whereas a nozzle has an increasing area.

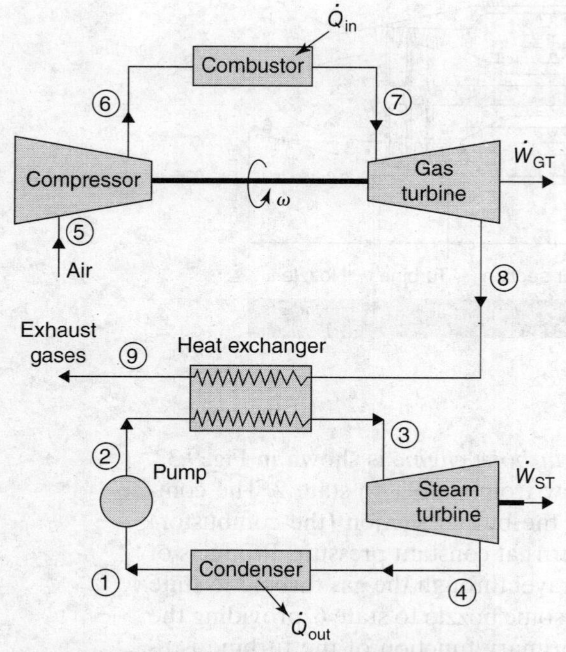

Figure 9.33

The combined Brayton–Rankine cycle.

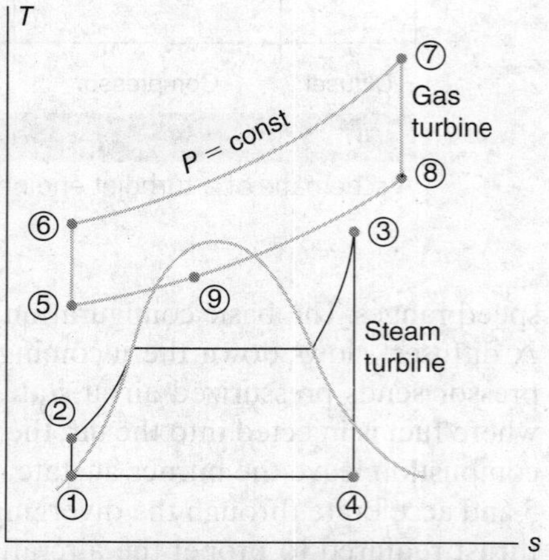

Figure 9.34

The *T-s* diagram for the ideal combined Brayton–Rankine cycle.

Rankine cycle) used to transfer heat from the products of combustion leaving the gas turbine to the water that leaves the pump. The *T-s* diagram is displayed in Fig. 9.34.

This combined-cycle power plant is more sustainable than either of the plants operating separately. Since Brayton cycle engines can use a low heating-value fuel in the external combustion chamber, they are compatible with the use of biofuels that could be made from used vegetable oil or cooking oil. This is especially useful in remote locations where small Brayton cycle engines can be located. Most combined-cycle power plants, however, burn natural gas in the combustor of the Brayton cycle.

Conventional coal-fired power plants range in size to over 2000 MW, while combined cycle Brayton–Rankine power plants approach the 1500 MW size. The efficiency of actual combined-cycle power plants may now exceed 60%, an unheard of efficiency for a coal-fired plant, so it has become a very attractive form of power production. To make it even more attractive, the rejected heat can be used for the various uses mentioned in Section 8.4 on cogeneration.

An example will illustrate an ideal combined Brayton–Rankine power cycle.

Example 9.12 The ideal combined Brayton–Rankine cycle

Air enters the compressor of the Brayton cycle of Fig. 9.35 at 300 K and 100 kPa, where it is compressed to 600 kPa. It exits the combustor at 1200 K. The exhaust gases exit the heat exchanger at 350 K. The pressure in the Rankine cycle is raised by the pump from 20 kPa to 4 MPa. The heat exchanger provides the

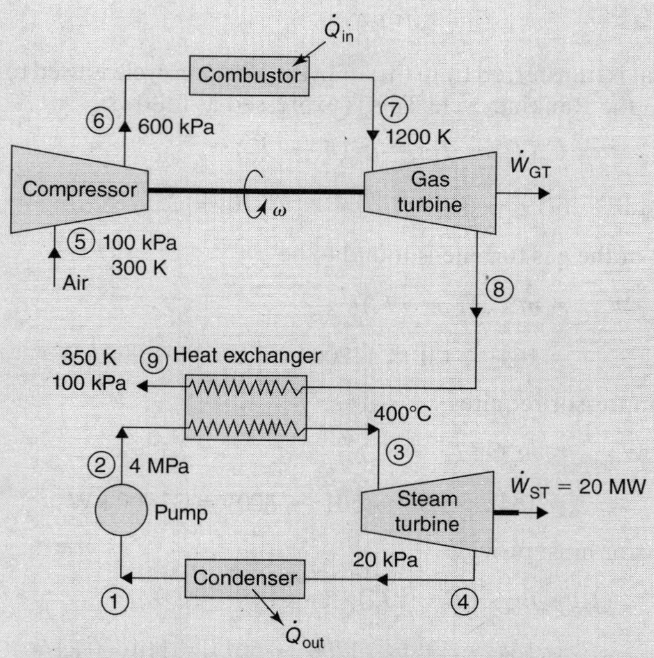

Figure 9.35

steam turbine with steam at 400°C. Determine the efficiency of the combined cycle if the steam turbine power output is 20 MW.

Solution

As usual, the pump work will be neglected in the analysis of the Rankine cycle. Refer to Figs. 9.34 and 9.35 for the appropriate state numbers. The enthalpy at the pump outlet is $h_2 \cong h_1 = 251$ kJ/kg. At the inlet to the steam turbine $h_3 = 3210$ kJ/kg and $s_3 = 6.77$ kJ/kg·K. The isentropic turbine allows the properties at state 4 to be determined using $s_3 = s_4$ and $P_4 = 20$ kPa. We find $h_4 = 2230$ kJ/kg with $x_4 = 0.839$. The mass flow rate passing through the steam turbine is then calculated, using the specified power output:

Note: The IRC Calculator has been used.

$$\dot{W}_{ST} = \dot{m}_S(h_3 - h_4)$$

$$20000 = \dot{m}_S(3210 - 2230) \quad \therefore \dot{m}_S = 20.4 \text{ kg/s}$$

Now, focus on the Brayton cycle. The temperature at the outlet of the compressor is

$$T_6 = T_5\left(\frac{P_6}{P_5}\right)^{(k-1)/k} = 300 \times \left(\frac{600}{100}\right)^{0.2857} = 501 \text{ K}$$

The temperature that exits the gas turbine and enters the heat exchanger (the steam boiler) is

$$T_8 = T_7\left(\frac{P_8}{P_7}\right)^{(k-1)/k} = 1200 \times \left(\frac{100}{600}\right)^{0.2857} = 719 \text{ K}$$

(Continued)

Example 9.12 (*Continued*)

The heat that is transferred from the air in the Brayton cycle is used to superheat the water in the Rankine cycle. This is expressed as follows:

$$\dot{m}_a C_p (T_8 - T_9) = \dot{m}_s (h_3 - h_2)$$

$$\dot{m}_a \times 1.0 \times (719 - 350) = 20.4 \times (3210 - 251) \quad \therefore \dot{m}_a = 164 \text{ kg/s}$$

The output of the gas turbine is found to be

$$\dot{W}_{GT} = \dot{m}_a C_p (T_7 - T_8)$$

$$= 164 \times 1.0 \times (1200 - 719) = 78\,900 \text{ kW}$$

The gas compressor requires

$$\dot{W}_{\text{comp}} = \dot{m}_a C_p (T_6 - T_5)$$

$$= 164 \times 1.0 \times (501 - 300) = 32\,900 \text{ kW}$$

The combustor must provide

$$\dot{Q}_{\text{in}} = \dot{m}_a C_p (T_7 - T_6)$$

$$= 164 \times 1.0 \times (1200 - 501) = 114\,600 \text{ kJ/s}$$

The combined cycle efficiency can now be determined. It is

$$\eta = \frac{\dot{W}_{\text{net}}}{\dot{Q}_{\text{in}}} = \frac{\dot{W}_{ST} + \dot{W}_{GT} - \dot{W}_{\text{comp}}}{\dot{Q}_{\text{in}}}$$

$$= \frac{20\,000 + 78\,900 - 32\,900}{114\,600} = 0.576 \quad \text{or} \quad \underline{57.6\%}$$

Compare this efficiency with that of the Brayton cycle of Example 9.8 of 42.6% and that of the Rankine cycle of Example 8.1 of 38%. Even though a direct comparison cannot be made because of the different pressure ratios and high temperatures, the efficiency of 57.6% is much higher than the individual efficiencies of the Brayton and Rankine cycles. The addition of reheaters, regenerators, and intercoolers would increase the efficiency well above the 57.6% value.

☑ You have completed Learning Outcome (6)

9.8 Summary

The three basic cycles of most importance presented in this chapter were the Otto cycle, the diesel cycle, and the Brayton cycle. Engines that produce power that use a gas, usually air, as the working fluid invariably operate using one of these three basic cycles. The ideal cycles were considered initially in which isentropic processes were used for the compressions and expansions in all cycles. The dual cycle modified the diesel cycle. The very efficient Stirling and Ericsson cycles, which have minimal applications, were also presented. The Brayton cycle was then introduced showing the magnified effects of losses in the compressors and turbines. The Brayton cycle was then modified to include regenerators, intercoolers, and reheaters. Finally, the combined Brayton–Rankine cycle with its relatively high efficiency was presented.

The following additional terms were introduced or emphasized in this chapter:

Air-standard analysis: *Air is an ideal gas, compression and power processes are isentropic, a heat transfer process replaces combustion, and exhaust is a heat transfer process.*

Back-work ratio: *The ratio of the work required to drive the compressor to the total work produced by the turbine.*

Bottom dead center: *The position of the piston where the volume in the cylinder is a maximum.*

Bore: *The diameter of the piston, which is assumed to be the inside diameter of the cylinder.*

Clearance volume: *The volume between the cylinder head and the top of the piston when the piston is at TDC.*

Cold-air standard: *Air enters a cycle at standard conditions with constant specific heats assumed throughout the cycle.*

Compression ignition: *The high temperature after compression ignites the air–fuel mixture.*

Compression ratio: *The ratio of the maximum volume in the cylinder to the minimum volume in the cylinder.*

Cutoff ratio: *The ratio of the volume at the end of combustion to the volume at the start of combustion.*

Displacement: *A volume equal to the stroke times the cross-sectional area of the piston.*

Engine knock: *The sound of the flame front hitting the oncoming piston.*

Exhaust valve: *Allows the combustion products to exit the cylinder.*

Intake valve: *Allows the air or air–fuel mixture to enter the cylinder.*

Intercooler: *Reduces the compressor work and the temperature at the compressor outlet.*

Mean effective pressure: *The ratio of the net work done during the cycle to the change in volume (the displacement) of the cycle.*

Regenerative Brayton cycle: *The air is preheated before it enters the combustor.*

Regenerator: *A heat exchanger that transfers heat from one fluid to another fluid; most often it preheats air.*

Reheater: *Increases the turbine output.*

Spark ignition: *A spark plug ignites the air–fuel mixture.*

Stroke: *The total linear distance of travel of the piston.*

Throw: *The throw on a crankshaft is the moment arm that creates the torque.*

Top dead center: *The position of the piston where the volume in the cylinder is a minimum.*

Several important equations were introduced in this chapter:

Compression ratio: $r = \frac{V_{max}}{V_{min}}$

Mean effective pressure: $\text{MEP} = \frac{w_{cycle}}{v_{max} - v_{min}}$

Otto cycle efficiency: $\eta_{Otto} = 1 - r^{1-k}$

Diesel cycle efficiency: $\eta_{diesel} = 1 - r^{1-k} \times \frac{r_c^k - 1}{k(r_c - 1)}$

Dual cycle efficiency: $\eta_{dual} = 1 - r^{1-k} \times \frac{r_p r_c^k - 1}{r_p - 1 + kr_p(r_c - 1)}$

Brayton cycle efficiency: $\eta_{Brayton} = 1 - r_p^{(1-k)/k}$

Back-work ratio: $\text{BWR} = \frac{W_{Comp}}{W_{Turb}}$

Regenerative Brayton cycle efficiency: $\eta = 1 - \frac{T_1}{T_4} r_p^{(k-1)/k}$

Combined cycle efficiency: $\eta = \frac{\dot{W}_{ST} + \dot{W}_{GT} - \dot{W}_{Comp}}{\dot{Q}_{in}}$

Problems

FE Exam Practice Questions (afternoon M.E. discipline exam)

Problems 9.1–9.3 refer to Fig. 9.36. Assume air with constant specific heats.

9.1 The ideal Otto cycle operates with a compression ratio of 8 and inlet conditions of 20°C and 100 kPa. The high temperature is 1200°C. The work output of the cycle is nearest:

(A) 325 kJ/kg

(B) 350 kJ/kg

(C) 400 kJ/kg

(D) 425 kJ/kg

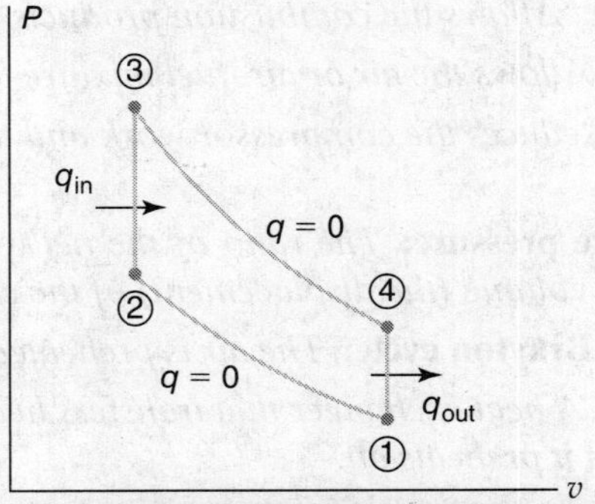

Figure 9.36

9.2 The MEP of the Otto cycle of Problem 9.1 is nearest:

(A) 370 kPa

(B) 410 kPa

(C) 440 kPa

(C) 475 kPa

9.3 The efficiency of the Otto cycle of Problem 9.1 is nearest:

(A) 37%

(B) 42%

(C) 51%

(D) 56%

Problems 9.4–9.6 refer to Fig. 9.37. Assume air with constant specific heats.

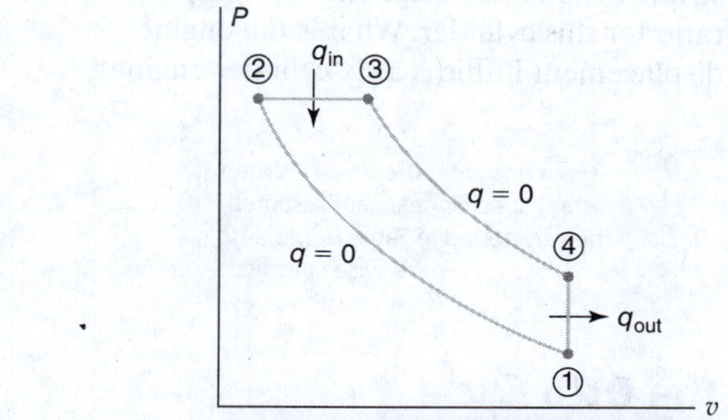

Figure 9.37

9.4 The ideal diesel cycle operates with a compression ratio of 20 and inlet conditions of 20°C and 100 kPa. If the cutoff ratio is 2, the high temperature in the cycle is nearest:

(A) 1710°C

(B) 1670°C

(C) 1580°C

(D) 1430°C

9.5 The work output of the diesel cycle of Problem 9.4 is nearest:

(A) 560 kJ/kg

(B) 490 kJ/kg

(C) 450 kJ/kg

(D) 420 kJ/kg

9.6 The efficiency of the diesel cycle of Problem 9.4 is nearest:

(A) 41%

(B) 50%

(C) 56%

(D) 65%

Problems 9.7–9.10 refer to Fig. 9.38. Assume air with constant specific heats and an adiabatic compressor.

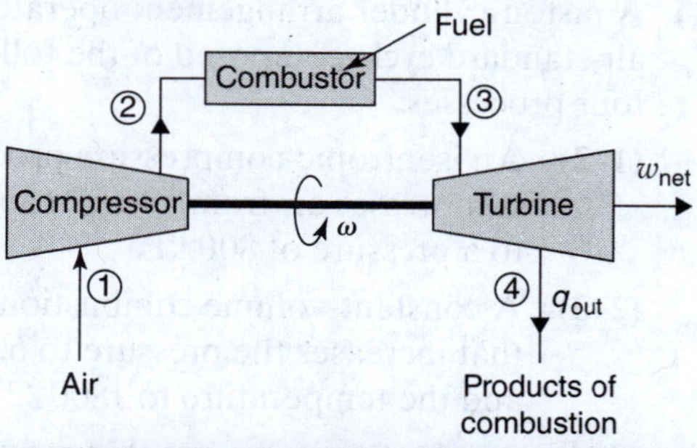

Figure 9.38

9.7 The ideal Brayton cycle operates with a pressure ratio of 6 and inlet conditions of 20°C and 100 kPa. The high temperature is 1800°C. The heat input to the cycle is nearest :

(A) 1210 kJ/kg

(B) 1380 kJ/kg

(C) 1420 kJ/kg

(D) 1580 kJ/kg

9.8 The efficiency of the Brayton cycle of Problem 9.7 is nearest:

(A) 40%

(B) 45%

(C) 50%

(D) 55%

9.9 If both the compressor and the turbine of Problem 9.7 are 80% efficient, the cycle efficiency reduces to:

(A) 27%

(B) 31%

(C) 35%

(D) 39%

9.10 If an ideal regenerator is added to the ideal Brayton cycle of Problem 9.7, the efficiency of the regenerative Brayton cycle is nearest:

(A) 45%

(B) 50%

(C) 55%

(D) 60%

Air-Standard Analysis

9.11 A piston-cylinder arrangement operates on an air-standard cycle composed of the following four processes:

(1–2) An isentropic-compression process that compresses air from 100 kPa and 20°C to a pressure of 800 kPa

(2–3) A constant-volume combustion process that increases the pressure to 6.4 MPa and the temperature to 1800°C

(3–4) An isentropic process that reduces the pressure to 800 kPa

(4–1) A constant-volume process that reduces the pressure back to 100 kPa to complete the cycle

Sketch the cycle on *P-v* and *T-s* diagrams. Using the ideal-gas relationships developed in Chapter 3 assuming constant specific heats, calculate the work output and heat input for this cycle. What is its efficiency?

9.12 A piston-cylinder arrangement operates on an air-standard cycle composed of the following four processes:

(1–2) An isentropic-compression process that takes air from 100 kPa and 10°C to a pressure of 2.1 MPa

(2–3) A constant-pressure combustion process that occurs at 2.1 MPa and increases the temperature to 1650°C

(3–4) An isentropic process that reduces the pressure to 700 kPa

(4–1) A constant-volume process that further reduces the pressure to 100 kPa to complete the cycle

Sketch the cycle on *P-v* and *T-s* diagrams. Using the ideal-gas relationships developed in Chapter 3 assuming constant specific heats, calculate the net work output and heat input for this cycle. What is its efficiency?

Note: If neither the volume nor the mass is given, specific quantities are calculated.

Reciprocating Engine Terminology

9.13 An engine cylinder has a diameter of 6 cm. The piston has a clearance of 5 mm and a stroke of 14 cm. Calculate the compression ratio for this cylinder. What is the engine displacement if this is an eight-cylinder engine?

9.14 An engine cylinder has a diameter of 100 mm. The piston has a clearance of 15 mm and a stroke of 150 mm. Calculate the compression ratio for this cylinder. What is the engine displacement if this is a six-cylinder engine?

Constant specific heats: When constant specific heats are assumed, they are found in Table *B-2*, unless otherwise specified.

The Otto Cycle

9.15 The Otto cycle of Fig. 9.39 has a compression ratio of 8 and a maximum temperature of 2000°C. What is the efficiency of this cycle? Determine the heat transfer in and the net work of the cycle if the air enters the cylinder at *a*) 25°C, *b*) 120°C, and *c*) 200°C. Assume the cold air-standard applies.

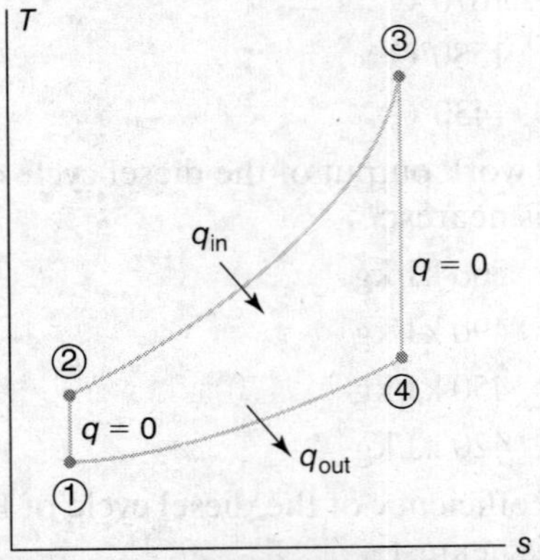

Figure 9.39

9.16 An Otto cycle has an efficiency of 70% and a maximum temperature of 1900 K. What is the compression ratio of this cycle? Determine the heat transfer in and the net work of the cycle if the air enters the cylinder at *a*) 290 K, *b*) 340 K, and *c*) 390 K. Assume the cold air-standard applies.

9.17 Heat is added to the air in an Otto cycle in the amount of 1200 kJ/kg. The compression ratio for this engine is 10. Air enters the cycle at 100 kPa and 20°C. Determine, using a cold air-standard analysis,

i) The temperature and pressure of the air at the end of the combustion process
ii) The cycle efficiency
iii) The net work output
iv) The mean effective pressure

Remember: All parts of problems with i), ii), iii), etc., are to be worked.

Comment

The combustion products are assumed to have the properties of air, a reasonable assumption.

9.18 Rework Problem 9.17 except that the compression ratio is *a*) 6 and *b*) 8.

9.19 Using the air tables in Appendix F-1, find the net work, the efficiency, and the MEP for the cycle of *a*) Problem 9.17 and *b*) Problem 9.18*a*.

Remember: Parts with lower-case italic letters are separate problems.

9.20 Heat is added to the air in an Otto cycle in the amount of 1160 kJ/kg. The compression ratio for this engine is 10. Air enters the cycle at 100 kPa and 290 K. Determine, using a cold-air analysis,

i) The temperature and pressure at the end of the combustion process
ii) The thermal efficiency of the cycle
iii) The work output of the cycle in kJ
iv) The mean effective pressure

9.21 Rework Problem 9.20 except that the compression ratio is *a*) 6 and *b*) 8.

9.22 Using the air tables in Appendix F-1, determine the work, the efficiency, and the MEP for the cycle of *a*) Problem 9.20 and *b*) Problem 9.21*a*.

The Diesel Cycle

9.23 The diesel cycle of Fig. 9.40 has a compression ratio of 16. Calculate the thermal efficiency of the cycle for cutoff ratios of 2, 2.5, and 3.0.

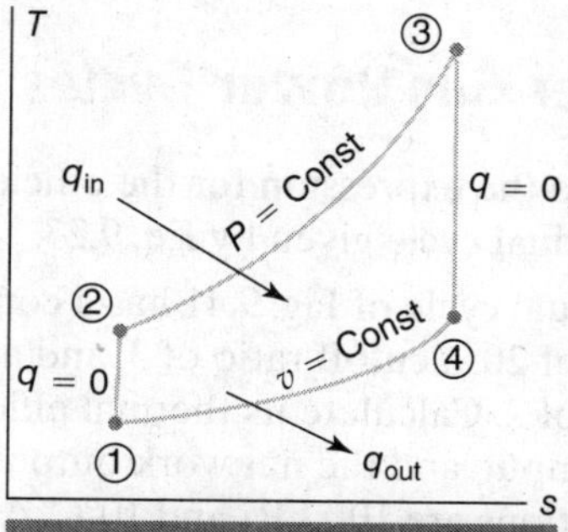

Figure 9.40

9.24 A diesel cycle is designed to have a thermal efficiency of 70%. What must the compression ratio be if the cutoff ratio is 2 and 3?

9.25 Heat is added in the amount of 800 kJ/kg of air to a diesel cycle with a compression ratio of 19 and a cutoff ratio of 1.8. Air enters the engine at 100 kPa and 27°C. Determine, using a cold-air analysis:

i) The temperature and pressure of the products of combustion at the end of the combustion process
ii) The thermal efficiency of the engine
iii) The specific work output of the cycle
iv) The mean effective pressure

9.26 Rework Problem 9.25 except that the compression ratio and cutoff ratio are, respectively, *a*) 16 and 2 and *b*) 20 and 1.75.

9.27 Using the air tables in Appendix F-1, determine the maximum cycle temperature, the net work, the efficiency, and the MEP for the cycle of *a*) Problem 9.25, and *b*) Problem 9.26a.

9.28 Heat is added in the amount of 700 kJ/kg of air to a diesel cycle, with a compression ratio of 22 and a cutoff ratio of 2. Air enters the engine at 100 kPa and 300 K. Determine, using a cold-air analysis:

i) The temperature and pressure of the products of combustion at the end of the combustion process
ii) The specific work output of the cycle
iii) The mean effective pressure
iv) The thermal efficiency of the cycle

9.29 Using the air tables in Appendix F-1, calculate the maximum temperature, the net work, the cycle efficiency, and the MEP for the cycle described in Problem 9.28.

Other Gas Power Cycles

9.30 Derive the expression for the efficiency of the ideal dual cycle given by Eq. 9.23.

9.31 The dual cycle of Fig. 9.41 has a compression ratio of 20, a cutoff ratio of 3, and a pressure ratio of 2. Calculate its thermal efficiency, the heat input, and the net work output if the inlet conditions are 100 kPa and 10ºC. Assume a cold-air analysis.

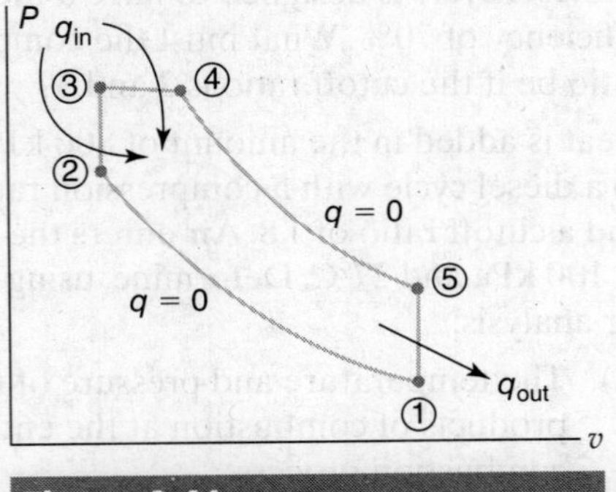

Figure 9.41

9.32 Air enters a dual cycle at 100 kPa and 20°C. Using a cold-air analysis, determine the thermal efficiency, the required specific heat input, and the net work output of this engine if:

a) $r = 18, r_c = 2, \text{and } r_p = 2$

b) $r = 20, r_c = 2, \text{and } r_p = 2$

c) $r = 18, r_c = 1.8, \text{and } r_p = 2.2$

9.33 An air-standard Stirling cycle shown in Fig. 9.42 operates with a minimum pressure of 75 kPa, a compression ratio of 15, and a high temperature of 1000ºC. The heat is rejected at 40°C. Using constant specific heats for air, calculate i) the net work produced, ii) the heat addition, and iii) the heat rejection.

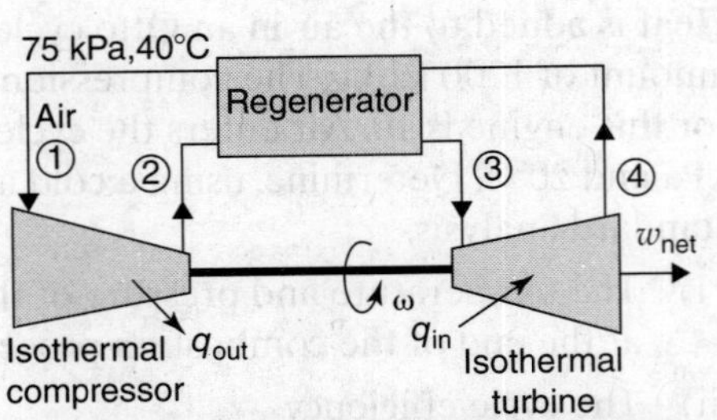

Figure 9.42

9.34 The *P-v* diagram of an air-standard Ericsson cycle, shown in Fig. 9.43, operates with a minimum pressure of 60 kPa and a maximum pressure of 780 kPa. Heat is rejected at a constant temperature of 400°C, and heat is added at 1200°C. Using constant specific heats for air, calculate i) the thermal efficiency of the cycle, ii) the net work, iii) the heat addition, and iv) the heat rejection.

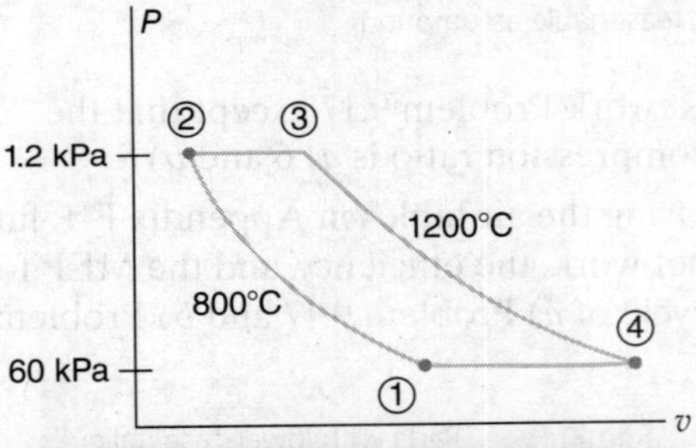

Figure 9.43

The Brayton Cycle

9.35 Calculate the thermal efficiency of an ideal Brayton cycle operating with air if the pressure ratio is i) 6, ii) 8, and iii) 10.

9.36 What pressure ratio is needed for an ideal Brayton cycle operating with air to have an efficiency of i) 50%, ii) 60%, and iii) 70%?

9.37 Air enters the gas turbine cycle of Fig. 9.44 at 25°C and 100 kPa with a flow rate of 1.2 kg/s. It is compressed so that the temperature at state 2 is 400°C. The temperature entering the turbine is 1400ºC. Assuming constant specific heats for the air and ideal conditions in all components, calculate:

i) The thermal efficiency

ii) The heat added in the combustor

iii) The net power produced by the cycle
iv) The back-work ratio

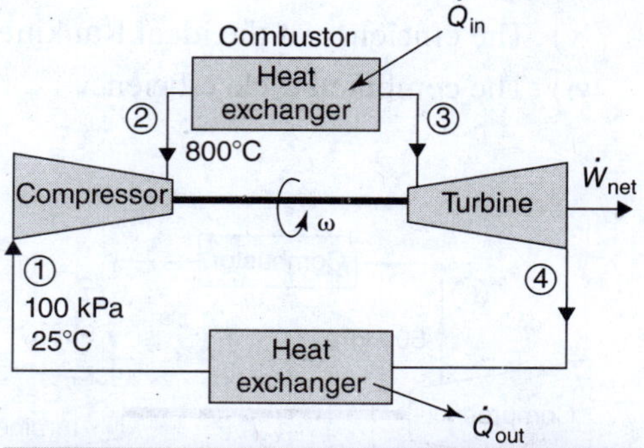

Figure 9.44

9.38 Rework Problem 9.37 except that the compressor exit temperature is *a*) 600°C and *b*) 800°C.

9.39 Determine the thermal efficiency of the Brayton cycle of Problem 9.37 if the efficiency of the compressor and the efficiency of the turbine are both *a*) 85%, *b*) 80%, and *c*) 70%. Maintain the same temperatures.

Remember: The specific heats are found in Table B-2.

9.40 Assume that the efficiencies of the compressor and turbine of the ideal Brayton cycle shown in Fig. 9.45 are the same. What efficiency would cause the net work output to be zero for a high temperature of 1000°C?

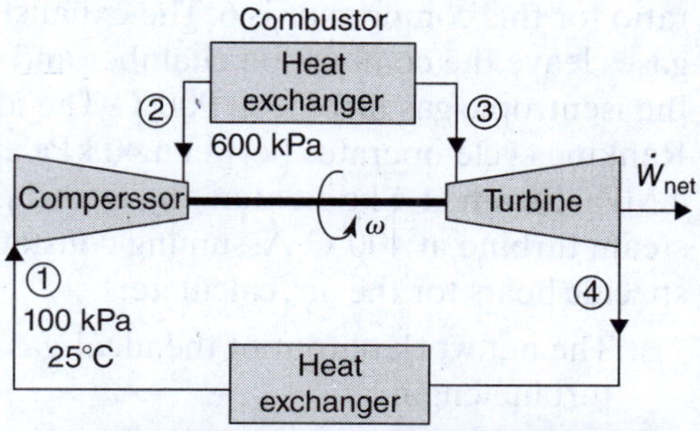

Figure 9.45

9.41 Air enters a gas turbine cycle at 265 K and 100 kPa. It is compressed to 700 kPa. The high temperature is 1250 K. Assuming constant specific heats for the air and ideal conditions in all components, calculate:

i) The thermal efficiency
ii) The heat added in the combustor
iii) The work produced by the turbine
iv) The back-work ratio

9.42 For the cycle of Problem 9.41, find the cycle thermal efficiency if the compressor and turbine have respective isentropic efficiencies of *a*) 85% and 80%, and *b*) 80% and 75%. Determine the percentage decrease in the thermal efficiency of the cycle from that of the ideal cycle.

9.43 Air enters the ideal regenerative Brayton cycle of Fig. 9.46 at 25°C and 100 kPa. The compressor raises the pressure to 500 kPa. The products of combustion leave the combustion chamber at 1200°C. Assuming constant specific heats and an air mass flow rate of 6 kg/s and an ideal cycle, calculate:

i) The cycle efficiency
ii) The power produced by the turbine
iii) The power used by the compressor
iv) The back-work ratio
v) The rate of heat exchange in the regenerator

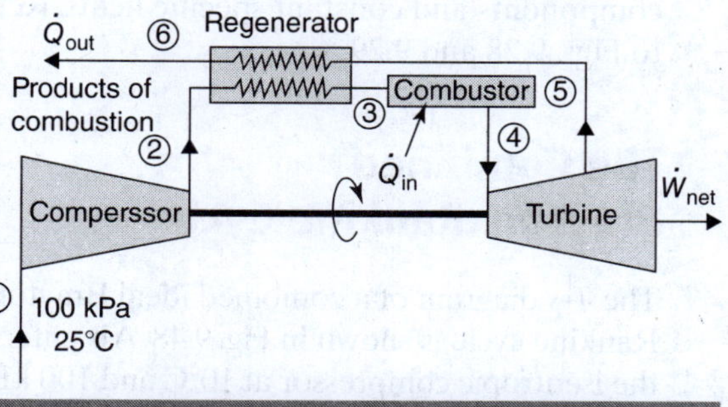

Figure 9.46

9.44 Rework Problem 9.43 except that the pressure ratio and maximum temperature are, respectively, *a*) 4 and 1200°C, *b*) 5 and 1400°C, and *c*) 6 and 1600°C.

9.45 For the cycle of Problem 9.43, find the cycle thermal efficiency using respective turbine and compressor efficiencies of *a*) 92% and 85%, *b*) 87% and 80%, and *c*) 83% and 75%.

9.46 Air enters the regenerative Brayton cycle of Fig. 9.47 cycle at 100 kPa and 265 K. The pressure ratio is 4 and the maximum temperature is 1250 K. Assuming an ideal cycle with constant specific heats, calculate the cycle efficiency and the back-work ratio. Compare with the results of Problem 9.41 if that problem was previously assigned.

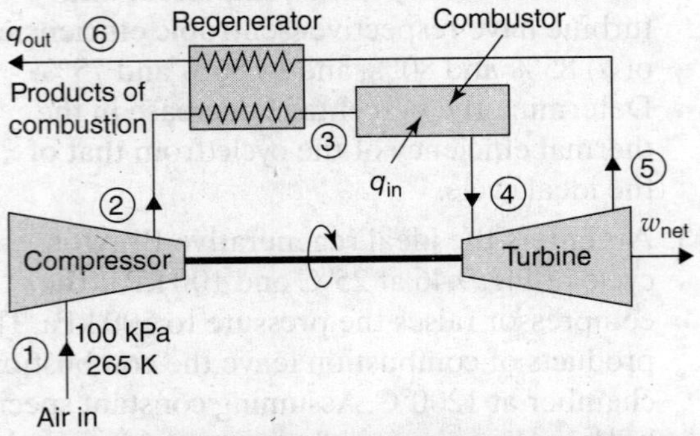

Figure 9.47

9.47 Repeat Problem 9.46 using turbine and compressor efficiencies of 0.85 and 75%, respectively.

9.48 Insert two intercoolers in the regenerative ideal Brayton cycles of *a*) Problem 9.43, *b*) Problem 9.44b, and *c*) Problem 9.44c. Calculate the cycle efficiency assuming ideal components and constant specific heats. Refer to Figs. 9.28 and 9.29.

The Combined Brayton–Rankine Cycle

9.49 The *T-s* diagram of a combined ideal Brayton–Rankine cycle is shown in Fig. 9.48. Air enters the isentropic compressor at 10°C and 100 kPa at state 5, with a mass flux of 8 kg/s. The pressure ratio for the compressor is 5. The combustor heats the air to 950°C. The air leaves the gas turbine and enters the "boiler" of an ideal Rankine cycle that operates between 10 kPa and 4 MPa. The air leaves the boiler at 325°C, and the steam leaves the boiler at 400°C. Assuming constant specific heats for the air, calculate:

i) The net power output of the ideal-gas turbine

ii) The efficiency of the ideal-gas turbine cycle

iii) The mass flux of the steam

iv) The power output of the ideal Rankine cycle

v) The efficiency of the ideal Rankine cycle

vi) The combined cycle efficiency

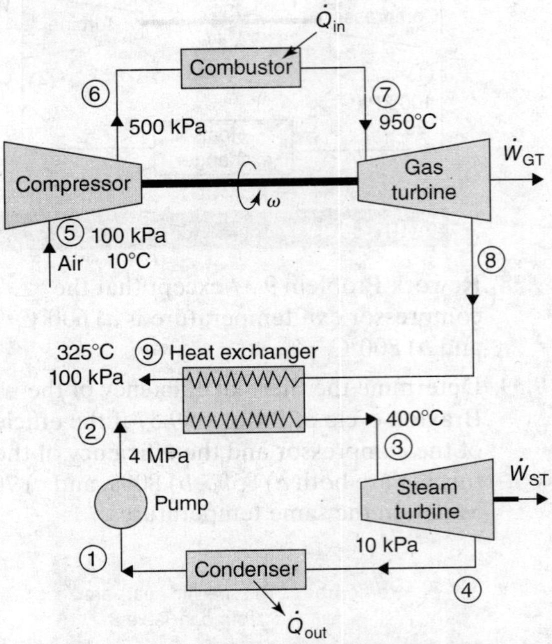

Figure 9.48

9.50 The basic layout of a combined Brayton–Rankine cycle is shown in Fig. 9.49. Air enters the isentropic compressor at 10°C and 100 kPa, with a mass flux of 30 kg/s. The pressure ratio for this compressor is 6. The exhaust gases leave the combustion chamber and enter the isentropic gas turbine at 900°C. The ideal Rankine cycle operates between 40 kPa and 4 MPa. Steam at 4 kg/s enters the isentropic steam turbine at 400°C. Assuming constant specific heats for the air, calculate:

i) The net work output of the ideal-gas turbine engine

ii) The efficiency of the ideal-gas turbine cycle

iii) The heat exchanged in the heat exchanger

iv) The work output of the ideal Rankine cycle

v) The efficiency of the ideal Rankine cycle
vi) The combined cycle efficiency

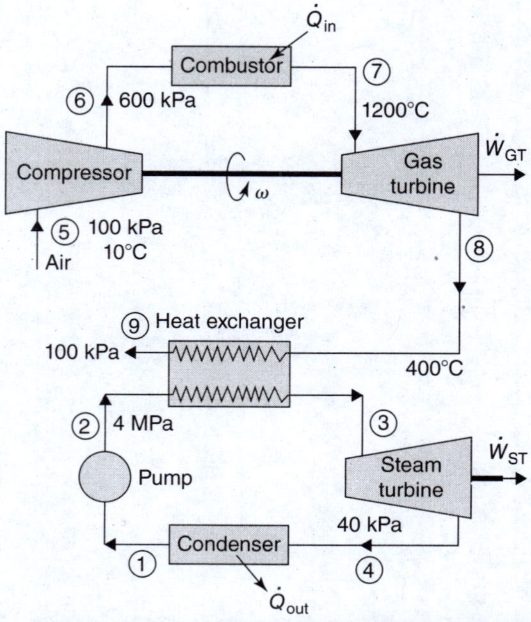

Figure 9.49

9.51 Repeat Problem 9.50 with the only changes being the maximum gas temperature and the maximum steam temperature, which are, respectively, *a*) 1000°C and 450°C and *b*) 1400°C and 500°C.

9.52 Repeat Problem 9.50 with the following information:

a) A gas compressor efficiency of 80%, a gas turbine efficiency of 86%, a steam turbine efficiency of 90%

b) A gas compressor efficiency of 82%, a gas turbine efficiency of 88%, a steam turbine efficiency of 92%

Ignore the water pump power, as usual.

9.53 Air enters the ideal Brayton–Rankine combined cycle of Fig. 9.50 at 300 K and an atmospheric pressure of 100 kPa. The compressor pressure ratio is 6. Exhaust leaves the gas turbine at 1380 K and leaves the heat exchanger at 580 K. The mass flux of air through the Brayton cycle is 16 kg/s. Steam leaves the Rankine side of the heat exchanger at 6 MPa and 500°C and the steam turbine at 30 kPa. Assuming constant specific heats for the air, calculate:

i) The net horsepower produced by the gas turbine
ii) The mass flux of the steam through the Rankine cycle
iii) The horsepower produced by the steam turbine
iv) The combined cycle efficiency

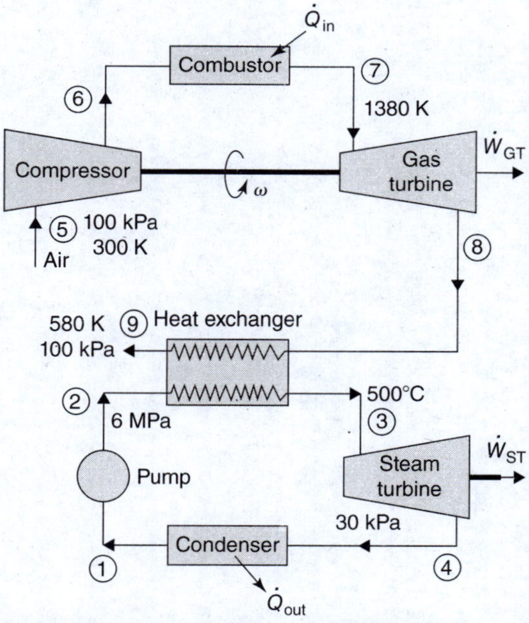

Figure 9.50

9.54 Repeat Problem 9.53 except change the maximum gas temperature and the maximum steam temperature, respectively, to *a*) 1000°C and 400°C and *b*) 1200°C and 550°C.

CHAPTER

10

Refrigeration Cycles

Outline

Nomenclature

No new nomenclature is introduced in this chapter.

Learning Outcomes

- ❑ **Understand the basic cycles and components used in refrigeration**
- ❑ **Analyze refrigeration systems**
- ❑ **Analyze the cascade refrigeration system**
- ❑ **Introduce absorption refrigeration**
- ❑ **Analyze gas refrigeration**

Motivational Example—Evaporative Coolers

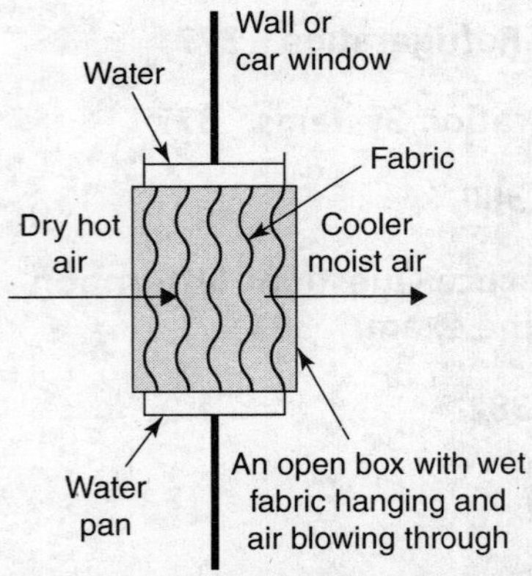

Figure 10.1

An evaporative cooler. Water flows down the fabric layers.

A serious problem in Third World countries and in remote locations is how to provide cooling without the availability of electricity to run a conventional refrigerator. One solution is the evaporative cooler sketched in Fig.10.1. These coolers have been in use for many decades by covering old-style wooden canteens with fabric. A small amount of water would seep from inside the canteen through the wood body and be absorbed by the fabric on the exterior of the canteen. The water in the fabric would evaporate slowly, and the heat of evaporation would keep the canteen cool. A modern version of this idea uses an inside waterproof container that holds food or medicine and an outside container made of a porous material. The space between the containers is filled with a material like fiber or sand that can absorb water. In use, the sun heats the outer container, causing heat to be transferred to the filler material. Water evaporates from the filler material and travels through the outside container. The heat of evaporation removed from the system will keep the inside container at about 6°C.

10.1 The Vapor Compression-Refrigeration Cycle

10.1.1 Refrigeration cycle terminology

The vapor compression-refrigeration cycle resembles a reverse Rankine cycle. Use is made of the phase-change characteristics of the working fluid, the refrigerant, to exchange large quantities of heat. Refrigeration cycles, on which refrigerators, air conditioners, freezers, and heat pumps operate, consist of the following four components, as shown in Fig. 10.2. An ordinary refrigerator with the components identified is sketched in Fig. 10.3.

Comment

Refrigerators, air conditioners, freezers, and heat pumps operate on refrigeration cycles.

1. A *compressor* is used to raise the pressure of the refrigerant. The compressor is usually powered by an electric motor that provides the input power that drives the refrigeration system.
2. A *condenser*, a heat exchanger that removes heat from the refrigerant, transfers heat from the refrigerant to the outside environment, usually as a refrigerant-to-air heat exchanger. In a home central air-conditioning system, this is the large unit that sits outside the house.
3. A *throttle* is then used to suddenly reduce the pressure and temperature of the refrigerant, making it very cold. This is usually a fixed throttle, though an adjustable throttling valve can be used.
4. Finally, the cycle is completed with an *evaporator*, another heat exchanger that transfers heat from the volume being cooled to the refrigerant. In a refrigerator or freezer, this heat exchanger would be built into the inside wall to transfer heat from the interior.

The compression process is ideally an isentropic process. Since it is accomplished with a compressor, the refrigerant entering and leaving this device must

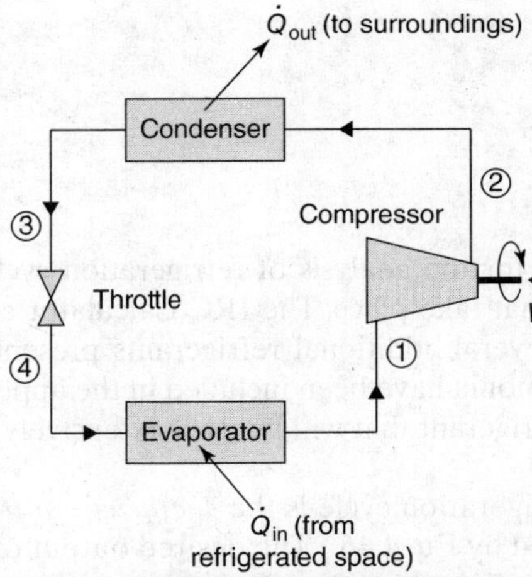

Figure 10.2

The basic refrigeration cycle.

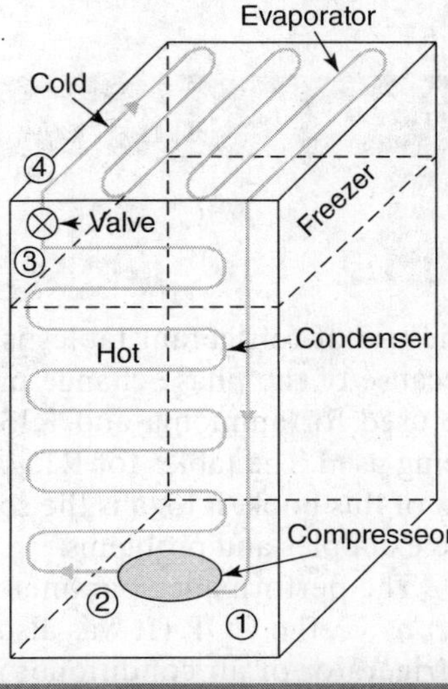

Figure 10.3

A sketch of an ordinary refrigerator (the hot coils are on the back).

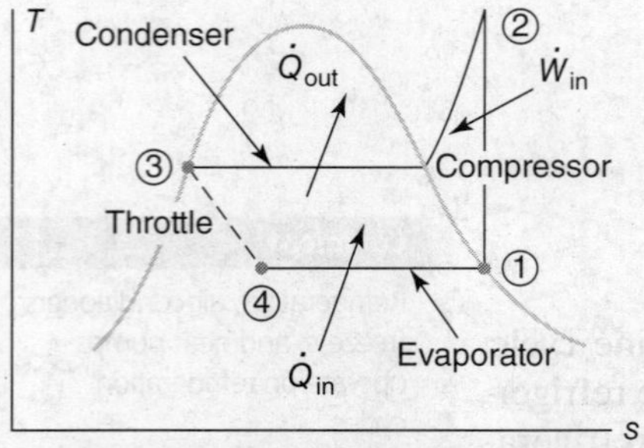

Figure 10.4

The *T-s* diagram for the ideal refrigeration cycle.

be in the gaseous state since droplets on the compressor blades cannot be tolerated. Both the condenser and evaporator are heat exchangers that operate as constant-pressure processes. The throttle is a constant-enthalpy process. Figure 10.4 shows a *T-s* diagram for the ideal refrigeration cycle.

In the ideal refrigeration cycle, the refrigerant enters the compressor at state 1 in Figs. 10.2 and 10.4 as a saturated vapor ($x = 1$) and is compressed to state 2 using an isentropic process. The compressor is powered by an electric motor and accounts for the purchased energy in the definition of the cycle's performance.

At state 2 the refrigerant exits the compressor and enters the condenser as a superheated vapor. The temperature of the refrigerant from state 2 to state 3 must be sufficiently high that the condenser transfers heat to the surroundings. If the device is a refrigerator, the refrigerant must be significantly hotter than the air in the kitchen; the condenser is made up of the coils on the back of a refrigerator. If the device is an air conditioner, the refrigerant must be hotter than the outside air.

Comment

In the ideal refrigeration cycle, the compressor inlet is saturated vapor ($x = 1$) and the condenser exit is saturated liquid ($x = 0$).

At state 3 in the ideal cycle at the condenser exit, the refrigerant is a saturated liquid ($x = 0$). Because the condenser is a heat exchanger, the pressure at state 3 will be the same as the pressure of state 2 in this ideal cycle. The purpose of the condenser is to cool the refrigerant so that it undergoes a phase change from a superheated vapor to a saturated liquid.

Comment

A refrigerant must have a large temperature drop across a throttle while the enthalpy remains constant.

From state 3 to state 4, the refrigerant is forced through a throttle, a valve, or a set of capillary tubes (very small diameter tubes). There is a large drop in the pressure and the temperature, and some of the refrigerant vaporizes while the enthalpy remains constant, as described in Section 4.2.3. State 4 is a mixture of liquid and vapor. The type of refrigerant determines the temperature drop across the throttle.

Comment

The evaporator performs the desired effect for a refrigerator.

The refrigerant enters the evaporator at state 4. It absorbs heat from the volume to be cooled, so it must be at a temperature significantly below the temperature of the cooled space. The cycle is completed, and the refrigerant again enters the compressor at state 1 as a saturated vapor ($x = 1$) in the ideal cycle.

☑ You have completed Learning Outcome (1)

10.1.2 The ideal refrigeration cycle

The use of refrigerant tables is required for the analysis of refrigeration cycles because of the phase-change processes that take place. The IRC Calculator can be used for ammonia and R134a and several additional refrigerants presently being used. The tables for R134a and ammonia have been included in the appendix of this book; R134a is the selected refrigerant that will be used extensively in the examples and problems.

Coefficient of performance: The measure of the performance of a refrigeration system. It's often greater than 1 so it is not called efficiency, but it has the same definition.

The performance parameter for refrigeration cycle is the *coefficient of performance*, the COP. (It was also presented by Eq. 4.46.) The desired output of a refrigerator or air conditioner is the rate of heat removal $\dot{Q}_{in}$ by the evaporator from the region being kept cold. This heat transfer is called the cooling load. The necessary input to force the refrigerant through the cycle is the input power $\dot{W}_{in}$ to the compressor. The COP is most often greater than 1, so it is not called efficiency, but it has the same definition. For a refrigerator or an air conditioner, it is

$$\mathrm{COP}_R = \frac{\text{desired effect}}{\text{input energy}} = \frac{\dot{Q}_{\text{in}}}{\dot{W}_{\text{in}}} \quad \textbf{(10.1)}$$

An example will illustrate the calculations required to analyze the ideal refrigeration cycle.

The ideal refrigeration cycle — Example 10.1

Refrigerant R134a enters the compressor of an ideal refrigeration cycle as a saturated vapor at 120 kPa and enters the condenser at 900 kPa, as shown in Fig. 10.5. Determine the lowest temperature, the cooling load, the necessary compressor horsepower, and the coefficient of performance for this ideal refrigeration cycle for a refrigerant mass flow rate of 0.15 kg/s.

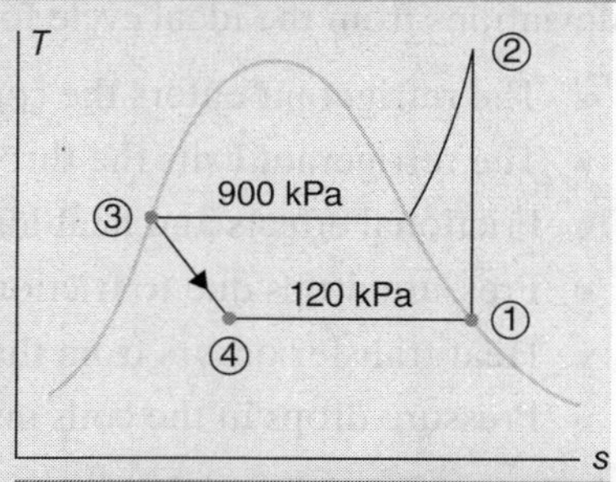

Figure 10.5

Solution

The IRC Calculator will be used to obtain the refrigerant properties. If, for some reason, it is not available, Appendix D can be used. The properties of interest at the four states are listed below. The enthalpies are of primary interest for these control volumes.

> **Remark:** The IRC Calculator provides values that are slightly different from those in Appendix D.

State 1 (saturated vapor): $P_1 = P_4 = 120$ kPa, $x = 1$,
$h_1 = 237$ kJ/kg, $s_1 = 0.948$ kJ/kg·K

State 2 (superheated vapor): $P_2 = 900$ kPa, $s_2 = 0.948$ kJ/kg·K,
$h_2 = 279$ kJ/kg

State 3 (saturated liquid): $P_3 = P_2 = 900$ kPa, $x_3 = 0$,
$h_3 = 102$ kJ/kg, $T_3 = 35.5°\text{C}$

State 4 (quality region): $P_4 = P_1 = 120$ kPa, $T_2 = \underline{-22.3°\text{C}}$,
$h_4 = h_3 = 102$ kJ/kg

> **Comment**
> The temperature dropped substantially across the throttle from 35.5°C to −22.3°C, a necessary property of a refrigerant.

Using the above enthalpies, we can determine the quantities of interest. The heat transferred from the condenser is

$$\dot{Q}_{\text{out}} = \dot{m}(h_2 - h_3)$$

$$= 0.15 \text{ kg/s} \times (279 - 102) \text{ kJ/kg} = 26.6 \text{ kJ/s}$$

> **Note:** The condenser heat transfer is negative since it leaves the refrigerant, but we state the heat transfer from the condenser as a positive number.

The cooling load provided by the evaporator is

$$\dot{Q}_{\text{in}} = \dot{m}(h_1 - h_4) = 0.15(237 - 102) = \underline{20.2 \text{ kJ/s}}$$

The compressor work requirement is

$$\dot{W}_{\text{in}} = \dot{m}(h_2 - h_1) = 0.15 \times (279 - 237) = 6.3 \text{ kW} \quad \text{or} \quad \underline{8.4 \text{ hp}}$$

The coefficient of performance for this ideal cycle is

$$\mathrm{COP}_R = \frac{\dot{Q}_{\text{in}}}{\dot{W}_{\text{in}}} = \frac{20.2}{6.3} = \underline{3.2}$$

> **Note:** The IRC Calculator uses only three significant digits, which is more reasonable than the five or six digits found in tables. One digit subtracts off so only two digits remain.

10.1.3 An actual refrigeration cycle

An actual refrigeration cycle deviates from the ideal cycle in several ways. Several deviations are intentional, while others cannot be avoided. The refrigerant enters the compressor with $x > 1$, which avoids any condensation on the compressor blades; an actual cycle will be designed so that the refrigerant enters slightly superheated. Also, optimum operation requires a compressed liquid to enter the throttle, so state 3 in Fig. 10.4 is designed to be several degrees in the compressed liquid region. Losses occur in the compressor, causing an entropy increase, but the compressor may also lose heat, causing an entropy decrease. A list of the deviations from the ideal cycle follow:

- The refrigerant enters the compressor slightly superheated.
- The refrigerant exits the throttle slightly subcooled.
- Frictional effects and heat loss exist in the compressor.
- Pressure drops due to friction in the connecting pipes between components.
- Heat transfer occurs from the connecting pipes.
- Pressure drops in the coils inside the condenser and evaporator.

Comment

Sufficient heat may cause an entropy decrease across the compressor.

An actual cycle with eight states is shown with the various components in Fig. 10.6. The *T-s* diagram in Fig. 10.7 shows four actual states, assuming short pipes between components with negligible losses. An example illustrates an actual cycle.

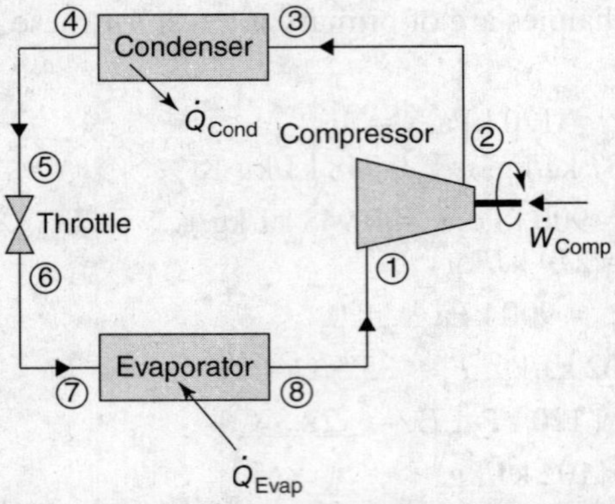

Figure 10.6

The actual states of a refrigeration cycle.

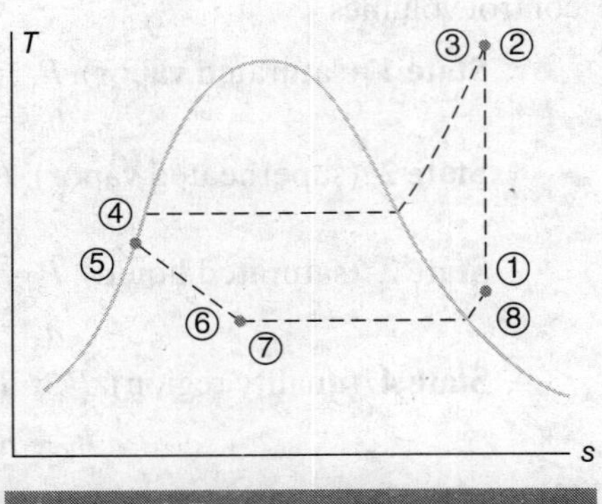

Figure 10.7

The *T-s* diagram for an actual refrigeration cycle with short pipe sections between components.

Example 10.2 An actual refrigeration cycle

Comment

The pipes between components are assumed short, so only four states are needed since losses between components are neglected in the given information.

The ideal refrigeration cycle of Example 10.1 experiences the following real effects:

- The refrigerant leaves the evaporator and enters the compressor at 120 kPa and −18°C.
- The refrigerant leaves the compressor and enters the condenser at 920 kPa and 52°C.

- The refrigerant leaves the condenser and enters the throttle at 900 kPa and 34°C.
- The refrigerant enters the evaporator at −14°C and 125 kPa.

Determine the lowest temperature, the cooling load, and the coefficient of performance for this refrigeration cycle for a refrigerant mass flow rate of 0.15 kg/s.

Solution

The IRC Property Calculator will be used to obtain the refrigerant properties, if it is available on the Web. Appendix D could also be used, but intricate interpolations would be required. The properties of interest at the various states identified in Figs. 10.6 and 10.7 are as follows.

States 8 and 1 (compressor inlet): $P_1 = P_8 = 120$ kPa,
$T_1 = -18°C$, so $h_1 = 240$ kJ/kg

States 2 and 3 (condenser inlet): $P_2 = P_3 = 920$ kPa,
$T_2 = 52°C$, so $h_2 = 286$ kJ/kg

States 4 and 5 (throttle inlet): $P_4 = P_5 = 900$ kPa,
$T_4 = 34°C$, so $h_4 = 99.4$ kJ/kg

States 6 and 7 (evaporator inlet): $P_6 = P_7 = 125$ kPa,
$T_6 = -21.4°C$, so $h_6 = 99.4$ kJ/kg

The lowest temperature in the cycle is at the evaporator inlet, where $T_6 = \underline{-21.4°C}$. The temperature would increase slightly to −18°C as the refrigerant passed through the evaporator.

Using the above enthalpies, we can determine the additional quantities of interest. The cooling load is

$$\dot{Q}_{in} = \dot{m}(h_1 - h_6) = 0.15(240 - 99.4) = \underline{21.1 \text{ kJ/s}}$$

Observe that the real effects have actually increased the cooling load, a rather interesting result.

To find the COP, the work input must be known. The compressor power requirement is

$$\dot{W}_{in} = \dot{m}(h_2 - h_1) = 0.15 \times (286 - 240) = 6.9 \text{ kW}$$

The coefficient of performance for this actual cycle has decreased about 4% because of the increased compressor power requirement. It is

$$\text{COP}_R = \frac{\dot{Q}_{in}}{\dot{W}_{in}} = \frac{21.1}{6.9} = \underline{3.06}$$

Comment

If a compressor efficiency of, say 85%, were included, the COP would decrease significantly more.

10.1.4 Heat pumps

A *heat pump* is a refrigeration system in which the condenser is located inside the space to be heated and the evaporator is located outside the space so that the heat rejected by the condenser warms the space. Motels often use such a system; it cools the room when it's too warm and heats it when it's too cool. Valves are used to reverse the flow of refrigerant through the heat exchangers and throttle. The heat exchanger that was the condenser on the air conditioner is now the evaporator. The air-conditioner condenser is now the evaporator for the heat pump.

Heat pump: A refrigeration system in which the condenser is located inside the space to be heated.

Comment

In a heat pump the condenser performs the desired effect.

Comment

Heat pumps are used primarily in states that are not too hot and not too cold. If they are used in northern states, heating is supplied by electric resistance heaters during the colder weather; that's too expensive.

Heat pumps work well in temperate regions where outside temperatures do not rise above about 40°C or drop below about −7°C, typically the midwestern United States. These systems are not advised for extreme temperatures, hot and cold. If the heat pump cannot supply enough heat in winter using the vapor cycle, an internal electric heater is used to maintain the desired temperature in the heated space. The electric heater is quite expensive to operate, so some systems use a separate natural gas backup heater, which is much more economical. In regions with hot climates (e.g., Florida and Arizona), only air conditioners are needed, with small electric resistance heaters to provide occasional heating.

The COP of a heat pump can be expressed as

$$COP_{HP} = \frac{\text{desired effect}}{\text{purchased energy}} = \frac{\dot{Q}_{out}}{\dot{W}_{in}} \tag{10.2}$$

It is related to the COP_R as follows (let $\dot{Q}_{out}/\dot{Q}_{in} = y$ for simplicity in the algebra):

$$COP_R = \frac{\dot{Q}_{in}}{\dot{W}_{in}} = \frac{\dot{Q}_{in}}{\dot{Q}_{out} - \dot{Q}_{in}} = \frac{1}{\dfrac{\dot{Q}_{out}}{\dot{Q}_{in}} - 1} = \frac{1}{y - 1}$$

$$COP_{HP} = \frac{\dot{Q}_{out}}{\dot{W}_{in}} = \frac{\dot{Q}_{out}}{\dot{Q}_{out} - \dot{Q}_{in}} = \frac{1}{1 - \dfrac{\dot{Q}_{in}}{\dot{Q}_{out}}} = \frac{y}{y - 1} \tag{10.3}$$

$$\therefore COP_{HP} = COP_R + 1$$

Example 10.3 A heat pump

A refrigeration cycle, using R134a, operates as shown in Fig. 10.8. The mass flow rate of refrigerant is 0.3 kg/s. The power input to the compressor, which is 85% efficient, is 20 kW. Determine the coefficient of performance for this system operating in both the cooling mode and heating mode.

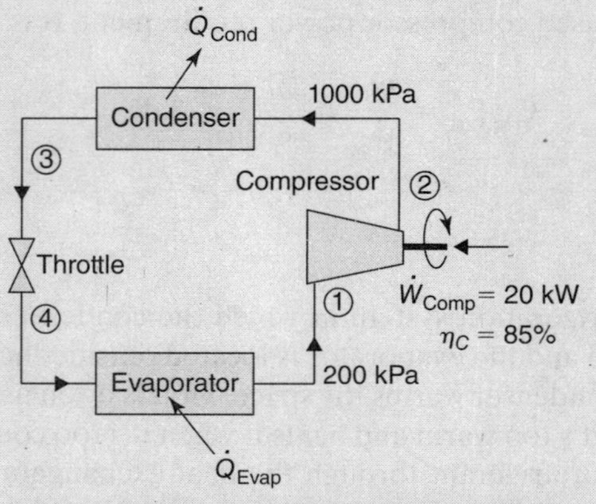

Figure 10.8

Solution

The known properties at the four states are:

State 1: $P_1 = 200$ kPa, $x_1 = 1$: $h_1 = h_g = 244$ kJ/kg, $T_1 = -10.1°C$
State 2: $P_2 = 1000$ kPa
State 3: $P_3 = 1000$ kPa, $x_3 = 0$: $h_3 = 107$ kJ/kg
State 4: $P_4 = 200$ kPa: $h_4 = h_3 = 107$ kJ/kg

The enthalpy leaving the compressor is found using the power input as follows:

$$\dot{W}_{\text{comp}} = \frac{\dot{m}(h_2 - h_1)}{\eta_{\text{comp}}} \qquad 20 = \frac{0.3 \times (h_2 - 244)}{0.85} \qquad \therefore h_2 = 301 \text{ kJ/kg}$$

Using the above enthalpies, the rate of heat transfer from the condenser and evaporator is, respectively,

$$\dot{Q}_{\text{cond}} = \dot{m}(h_2 - h_3)$$
$$= 0.3 \times (301 - 107) = 58 \text{ kJ/s}$$
$$\dot{Q}_{\text{evap}} = \dot{m}(h_1 - h_4)$$
$$= 0.3(244 - 107) = 41.1 \text{ kJ/s}$$

The energy that is input to the refrigerant is

$$\dot{W}_{\text{in}} = \eta \dot{W}_{\text{comp}} = 0.85 \times 20 = 17 \text{ kJ/s}$$

Comment

Because the compressor is 85% efficient, only 17 kJ/s of the 20 kJ/s input power actually enters the refrigerant.

When operating in the cooling mode, the COP is

$$\text{COP}_R = \frac{\dot{Q}_{\text{evap}}}{\dot{W}_{\text{in}}} = \frac{41.1}{17} = 2.4$$

When operating in the heating mode, the COP is

$$\text{COP}_{HP} = \frac{\dot{Q}_{\text{cond}}}{\dot{W}_{\text{in}}} = \frac{58}{17} = 3.4$$

Note: $\text{COP}_{HP} = \text{COP}_R + 1$

10.1.5 Refrigerants

A *refrigerant* is the working fluid used in a refrigeration cycle, a cycle used to either cool (e.g., an air conditioner) or heat (a heat pump) a space. A refrigerant must experience a substantial temperature drop when it passes through a throttle, a device that provides a sudden reduction in pressure. Review Section 2.3, where properties of refrigerants were introduced, and Section 4.2.2, where throttles were analyzed.

Refrigerant: The working fluid used in a refrigeration cycle. It must experience a large temperature drop when it passes through a throttle.

In the early days of refrigeration, the main chemicals used as refrigerants were ammonia, sulfur dioxide, and carbon dioxide. Ammonia was used the most and is still in use in some refrigeration systems today; however, ammonia is very toxic and was first replaced with Freon refrigerants. Refrigerants R12 and R22 were used for several decades under various brand names. In the 1970s, it was concluded that

Comment

The release of fluorocarbons into the atmosphere damages the ozone layer.

Comment

HFO-1234yf may replace R134a, which replaced R12.

Freon refrigerants released into the environment negatively affected the ozone layer. This prompted the development of more environmentally benign refrigerants. Refrigerant R134a was developed to replace the Freons in commercial and domestic refrigeration and air-conditioning systems. It has been in use since the early 1990s and is currently the only refrigerant allowed for domestic use. R134a has a negligible effect on the ozone layer. Refrigerant HFO-1234yf has been proposed as a replacement for R134a in automotive air conditioners because of its low flammability. It is being resisted by air-conditioning specialists because of the high cost with little return, according to the specialists (read comments on the Internet.)

The use of refrigerants in air conditioning and refrigeration systems may be contributing to a pattern of global warming that can cause significant damage to many ecosystems. Traditionally, refrigerants were vented into the atmosphere when the systems were serviced. It was determined that the release of these fluorocarbons into the atmosphere contributed to the damage of the ozone layer, which protects plants from dangerous solar radiation. Two engineering solutions were developed for this problem. First, refrigerants are now collected and destroyed, as required by law, when refrigeration and air-conditioning systems are serviced. The second solution is to switch from the refrigerants that damage the environment to refrigerants such as HFO-1234yf that are safer and more environmentally friendly.

☑ You have completed Learning Outcome (2)

10.2 Cascade Refrigeration Systems

Comment

We refer to a cascade cycle or a cascade system. They have the same meaning.

Comment

A cascade system can achieve the low temperatures needed to create liquid nitrogen.

Note: If the refrigerant is different in each cycle, two different *T-s* diagrams will be required.

The simple refrigeration system of Fig. 10.2 is adequate for most applications. For large industrial air-conditioning or refrigeration needs, however, the simple system would place too much load on a large high pressure-ratio compressor. The efficiency of a large compressor to power large refrigeration requirements would be relatively low. One alternative configuration is the *cascade system*, in which a cascade refrigerator is actually two or more separate refrigeration cycles working in series. Figure 10.9 shows the layout of a two-stage cascade system. A three-stage system would add another cycle below the bottom cycle. The two cycles of the two-stage system are joined by a heat exchanger that serves as the condenser for cycle #1 and the evaporator for cycle #2. Since the flows in the two systems never mix, the mass flow rates and even the type of refrigerant can be different. In addition to allowing multiple compressors to operate under more acceptable conditions resulting in decreased work input, another advantage of a cascade system is to allow a lower temperature in the evaporator of cycle #1. If a very low temperature is desired, several stages may be needed. Increased refrigeration also occurs in the cascade system, as illustrated in Fig. 10.10. A single-stage system with the same low temperature would result in state 9 exiting the compressor and state 10 exiting the throttle.

The optimal value for the intermediate pressure P_I is given by

$$P_I = \sqrt{P_H P_L} \tag{10.4}$$

where P_H and P_L are the high and low pressures shown in Fig. 10.10. The same refrigerant is assumed in both cycles.

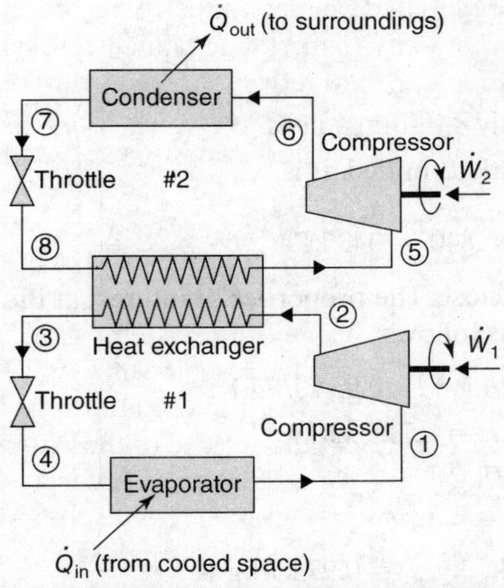

Figure 10.9

A two-stage refrigeration system.

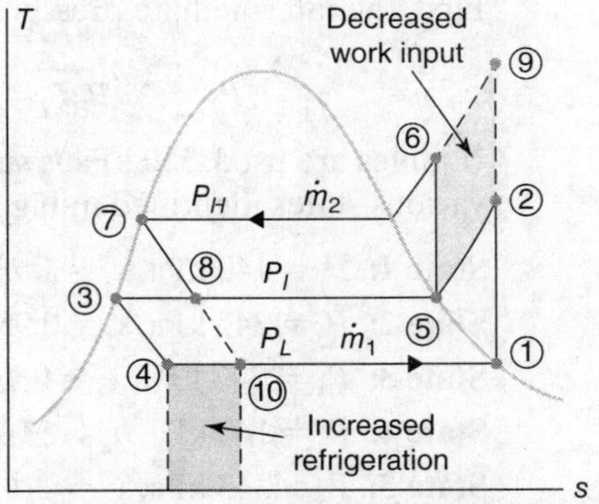

Figure 10.10

The *T-s* diagram for the ideal two-stage cycle. The shaded areas show the decreased work and increased refrigeration when compared to a single-stage system.

To determine the relationship between the mass flow rates in the two cycles, apply the first law to the heat exchanger. It is

$$\dot{m}_1(h_2 - h_3) = \dot{m}_2(h_5 - h_8) \tag{10.5}$$

where $\dot{m}_1$ is the mass flux in low-pressure cycle #1 and $\dot{m}_2$ is the mass flux in high-pressure cycle #2. The ratio of mass fluxes is then

$$\frac{\dot{m}_2}{\dot{m}_1} = \frac{h_2 - h_3}{h_5 - h_8} \tag{10.6}$$

Typically, the low-pressure mass flux $\dot{m}_1$ is known since it is cycle #1 that determines the required refrigeration. If X tons of refrigeration is desired, then, in the SI system,

$$3.52X = \dot{m}_1(h_1 - h_4) \tag{10.7}$$

This equation allows the mass flux in cycle #1 to be determined.

The overall coefficient of performance would be determined by using $\dot{Q}_{in}$ divided by the total compressor work input.

Comment

Historically, in the United States. refrigeration has been measured in tons, a rather unusual unit. It was considered to be the cooling effect of one ton of ice per day. Now, we define:

1 ton = 3.52 kW

A two-stage cascade refrigeration system — Example 10.4

The ideal refrigeration cycle that operates between high and low pressures of 140 kPa and 840 kPa is to be replaced with a two-stage system. Determine the cooling load for a refrigerant (R134a) mass flow rate of 0.07 kg/s and the COP.

(*Continued*)

Example 10.4 (*Continued*)

Solution (The tables used may have slightly different values.)

First, the intermediate pressure must be determined. It is

$$P_I = \sqrt{P_H P_L} = \sqrt{140 \times 840} = 343 \text{ kPa}$$

If tables are used, 350 kPa is sufficiently close. The properties of interest at the various states identified in Fig. 10.9 are as follows:

State 1: $P_1 = 140$ kPa, $x_1 = 1$, $h_1 = 240$ kJ/kg, $s_1 = 0.95$ kJ/kg·K
State 2: $P_2 = 343$ kPa, $s_2 = 0.95$ kJ/kg·K, $h_2 = 258.5$ kJ/kg
State 3: $P_3 = 343$ kPa, $x_3 = 0$, $h_3 = 57.3$ kJ/kg
State 4: $P_4 = 140$ kPa, $h_4 = 57.3$ kJ/kg, $T_4 = -19$°C
State 5: $P_5 = 343$ kPa, $x_5 = 1$, $h_5 = 253.8$ kJ/kg, $s_5 = 0.93$ kJ/kg·K
State 6: $P_6 = 840$ kPa, $s_6 = 0.93$ kJ/kg·K, $h_6 = 272.5$ kJ/kg
State 7: $P_7 = 840$ kPa, $x_7 = 0$, $h_7 = 97.4$ kJ/kg
State 8: $P_8 = 343$ kPa, $h_8 = 97.4$ kJ/kg

Using the above enthalpies, we can determine the quantities of interest. The cooling load is

Units: 1 ton = 3.52 kW

$$\dot{Q}_{\text{in}} = \dot{m}(h_1 - h_4) = 0.07(240 - 57.3) = \underline{12.8 \text{ kW}} \quad \text{or} \quad 3.63 \text{ tons}$$

This would compare with 9.7 kW for a single-stage cycle, a 31% increase in cooling load.

To find the COP, the compressor power must be known. It is the sum of the power required by each cycle. The mass flux $\dot{m}_2$ is found, using Eq. 10.6, to be

$$\dot{m}_2 = \dot{m}_1 \times \frac{h_2 - h_3}{h_5 - h_8} = 0.07 \times \frac{258.5 - 57.3}{253.8 - 97.4} = 0.09 \text{ kg/s}$$

The compressor power can now be calculated. It is

$$\dot{W}_{\text{in}} = \dot{m}_1(h_2 - h_1) + \dot{m}_2(h_6 - h_5)$$
$$= 0.07 \times (258.5 - 240) + 0.09 \times (272.5 - 253.8) = 2.98 \text{ kW}$$

The COP is then found to be

$$\text{COP}_R = \frac{\dot{Q}_{\text{in}}}{\dot{W}_{\text{in}}} = \frac{12.8}{2.98} = \underline{4.3}$$

This would compare with 3.82 using an ideal single-stage system, a 13% increase.

This example illustrates that a two-stage system increases the cooling load and decreases the compressor power for the same pressure difference across the compressor. It was not intended to demonstrate the lower temperature that can be attained using a multistage system. The problems at the end of the chapter will do that.

☑ You have completed Learning Outcome (3)

10.3 Absorption Refrigeration

Absorption refrigeration is a refrigeration system that uses heat to power the system. It is an attractive alternative to compression refrigeration if a source of relatively cheap heat is available, such as the heat from the condenser of a power plant or even from a solar collector. The higher the temperature of the heat source, the more efficient is the absorption refrigeration unit. A pump is used in some systems, so electricity may be required, but that is minimal when compared to the power to operate a compressor. So, electricity from a small dam on a stream and bottled propane can operate an air-conditioning unit or a refrigerator, or both, at a remote cabin in the mountains. Or, the "free" heat from the power plant on a university campus can be used to cool all the buildings on campus, providing huge savings when compared to using electricity to drive compressors.

Absorption refrigeration: A refrigeration system that uses heat to power the system. Refrigerators that run on propane use such a system.

Figure 10.11 shows the basic configuration of an absorption refrigeration system. The most common refrigerant used in such a system is ammonia. The ammonia in the generator exists as a mixture with water. Heat $\dot{Q}_G$ is supplied to the generator by using, for example, a natural gas flame unless there is a source of relatively inexpensive heat. The ammonia is heated to a superheated vapor and separates from the water allowing the generator to perform the same function as a compressor. The water is drained directly to the absorber. The ammonia then flows through a condenser and a throttle the same way it would in a conventional refrigerator, providing cooling $\dot{Q}_E$ in the evaporator. The refrigerant then flows into an absorber, where heat is removed from the refrigerant. In the absorber the ammonia recombines with the water. The mixture is then pumped back to the generator.

Because of the toxicity of ammonia, a lithium bromide–water combination is often used.

Absorption refrigerators are commonly used in camping facilities to provide refrigeration when sufficient electricity to operate a compressor is not available. Alternative sources of heat for absorption refrigeration systems would include solar energy, geothermal energy, or waste heat from a power plant. Absorption refrigerators have great potential as energy sustainable devices because of the low electrical energy requirement. Detailed problems will not be included in this book. This subject is introduced to inform the reader of its existence.

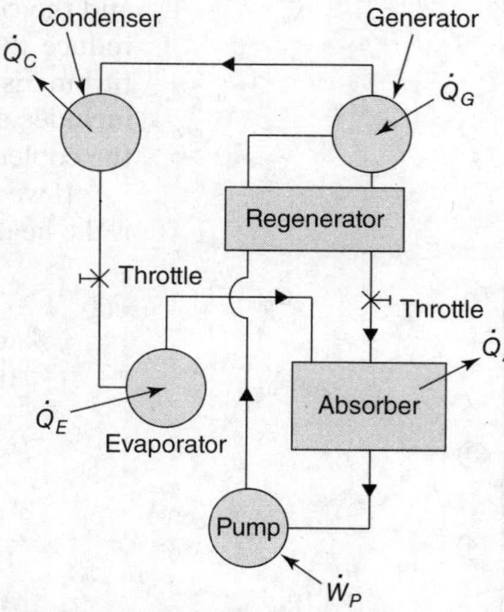

Figure 10.11
An absorption refrigeration system.

☑ **You have completed Learning Outcome (4)**

10.4 Gas Refrigeration Systems

When a supply of compressed air is available, it can be used to power a gas refrigeration system. The most common example of this application is the open-cycle air-conditioning system used in aircraft. The gas refrigeration system is essentially a reverse Brayton cycle, as shown in Fig. 10.12. An airplane moving at high speed can bleed air from the engines at high pressure and divert this flow into a compressor. The air leaving from the compressor is at high temperature

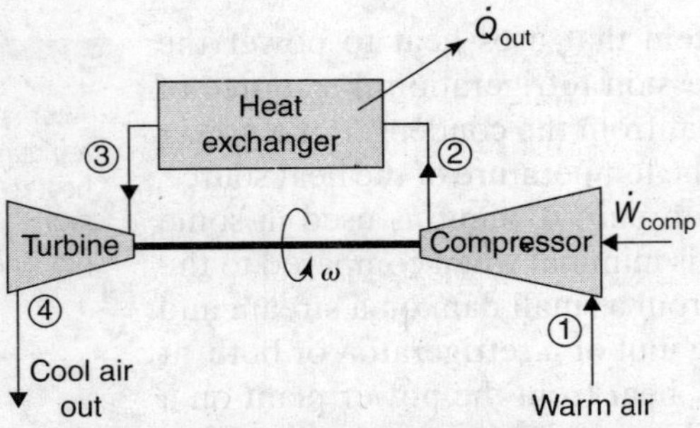

Figure 10.12

The open air-refrigeration cycle.

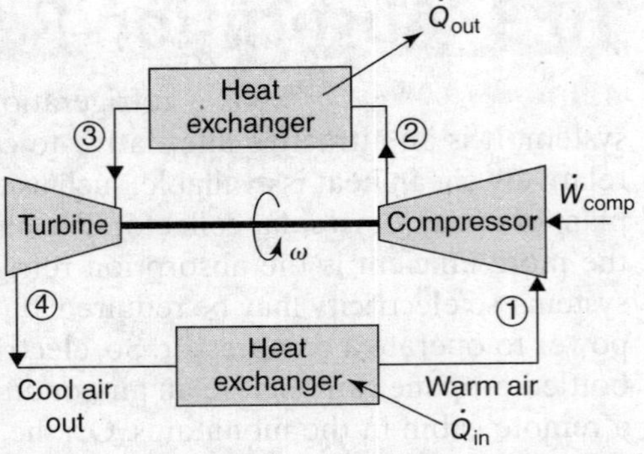

Figure 10.13

The closed air-refrigeration cycle.

and pressure. Heat is removed from the air in a high-temperature heat exchanger, and the cooler air then flows through a turbine. The function of the turbine is to reduce the air temperature and help power the compressor. Cool air from the turbine is then directed into the airplane cabin. The closed system of Fig. 10.13 includes a low-temperature heat exchanger that shows the heat extracted from the cooled space. Its *T-s* diagram is shown in Fig. 10.14 for the ideal cycle.

If we consider the air-conditioning system shown in Fig. 10.13, the cooling load is the heat entering the heat exchanger between states 4 and 1. It is expressed as

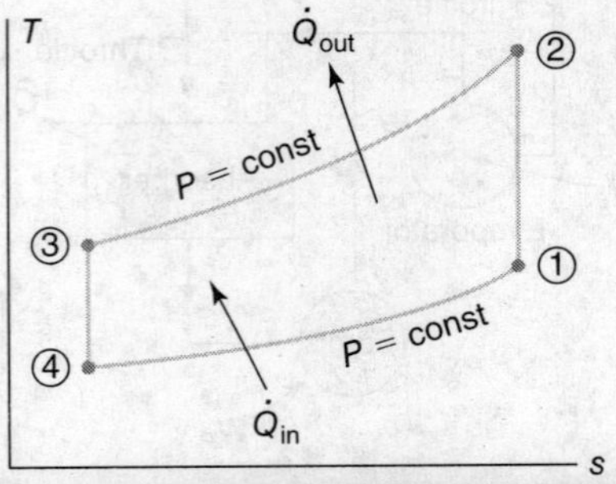

Figure 10.14

The *T-s* diagram for an ideal air-refrigeration cycle.

$$\dot{Q}_{in} = \dot{m}(h_1 - h_4) \tag{10.8}$$

where $\dot{m}$ is the mass flow rate of the air. The isentropic compressor and turbine work rates are

$$\dot{W}_{comp} = \dot{m}(h_2 - h_1) \tag{10.9}$$

$$\dot{W}_{turb} = \dot{m}(h_3 - h_4) \tag{10.10}$$

Because of the low temperature at the turbine exit, the specific heats are usually not assumed constant. Either the ideal-gas tables are used, or use could be made of the IRC Calculator. For a gas refrigeration system, the performance decreases as the compressor pressure ratio increases. An example will illustrate the low temperature that can exist at the turbine exit.

Example 10.5 An ideal-gas refrigeration cycle

Air from an aircraft jet engine[1] is diverted into an air-conditioning system at 85 kPa and −7°C at a flow rate of 0.54 m³/s. The compressor has a rated pressure ratio of 3. The inlet temperature to the turbine is 38°C. Determine the cooling load, the net work input, and the coefficient of performance for this ideal cycle.

[1] At an elevation where commercial aircraft usually fly, the outside temperature is about −50°F.

Solution

Air is an ideal gas, and the IRC Calculator will be used to determine the properties, or the air tables in Appendix F-1 could be used, or constant specific heats could be assumed. The IRC Calculator provides:

State 1: $P_1 = 85$ kPa, $T_1 = -7°C$
$h_1 = 268$ kJ/kg, $s_1 = 6.83$ kJ/kg·K

State 2: $P_2 = 3P_1 = 255$ kPa, $s_2 = s_1 = 6.83$ kJ/kg·K
$h_2 = 372$ kJ/kg

State 3: $P_3 = P_2 = 255$ kPa, $T_3 = 38°C$
$h_3 = 312$ kJ/kg, $s_3 = 6.66$ kJ/kg·K

State 4: $P_4 = 85$ kPa, $s_4 = s_3 = 6.66$ kJ/kg·K
$h_4 = 230$ kJ/kg, $T_4 = -43.3°C$

The mass flux is found, using $\rho_1 = 1.113$ kg/m³ (from the IRC Calculator or using $\rho_1 = P_1/RT_1 = 85/0.287 \times 266$), to be

$$\dot{m} = \rho_1(AV)_1 = 1.113\frac{\text{kg}}{\text{m}^3} \times 0.54\frac{\text{m}^3}{\text{s}} = 0.6 \text{ kg/s}$$

Comment

The flow rate of 0.54 m³/s is not constant. The mass flux is constant.

The cooling load $\dot{Q}_{in}$ can now be calculated. It is

$$\dot{Q}_{in} = \dot{m}(h_1 - h_4)$$
$$= 0.6 \text{ kg/s} \times (268 - 230) \text{ kJ/kg} = \underline{22.7 \text{ kW}}$$

The net work input is

$$\dot{W}_{net} = \dot{W}_{comp} - \dot{W}_{turb}$$
$$= \dot{m}(h_2 - h_1) - \dot{m}(h_3 - h_4)$$
$$= 0.6 \times [(372 - 268) - (312 - 230)] = \underline{13.2 \text{ kW}}$$

The COP is found to be

$$\text{COP} = \frac{\dot{Q}_{in}}{\dot{W}_{net}} = \frac{22.7}{13.2} = \underline{1.72}$$

Note the low temperature at state 4 of −43.3°C. The low turbine outlet temperature is a primary feature of the gas refrigeration cycle. The relatively low COP means the gas cycle requires significantly more energy input than a comparable vapor cycle, so it is used in special situations such as on aircraft.

☑ **You have completed Learning Outcome (5)**

10.5 Summary

This chapter presented the thermodynamic cycles that form the basis for refrigeration systems. Included were refrigerators, heat pumps, and air conditioners. A cascade system that allows for extremely low temperatures was also presented. Absorption refrigeration systems that operate using heat as the energy input and ammonia as the refrigerant were described; absorption refrigeration is used on university campuses as well as in refrigerators that run on natural gas. Finally, systems that operate with a gas, such as those used in commercial aircraft that use air, were analyzed.

Terms that were used in this chapter include:

Absorption refrigeration: *A refrigeration system that uses heat as a source of energy to power the system.*

Cascade refrigeration: *Two or more separate refrigeration cycles working in series.*

Coefficient of performance: *The measure of the performance of a refrigeration system, the ratio of the desired output to the power input.*

Compressor: *A device used to raise the pressure of the refrigerant.*

Condenser: *A heat exchanger that removes heat from the refrigerant until it's a saturated liquid.*

Cooling load: *The heat transfer from the space to be cooled.*

Evaporator: *A heat exchanger that transfers heat from the volume being cooled to the refrigerant.*

Gas refrigeration: *A reverse Brayton cycle.*

Ideal cycle: *A cycle in which the compression and expansion processes are isentropic processes.*

Refrigerant: *The working fluid used in a refrigeration cycle.*

Throttle: *A device used to suddenly reduce the pressure and hence the temperature of the refrigerant.*

Ton: *A measure of refrigeration equal to 3.52 kW.*

Three important equations were utilized:

Coefficient of performance, refrigerator: $\text{COP}_R = \dfrac{\dot{Q}_{\text{in}}}{\dot{W}_{\text{in}}}$

Coefficient of performance, heat pump: $\text{COP}_{HP} = \dfrac{\dot{Q}_{\text{out}}}{\dot{W}_{\text{in}}}$

Intermediate cascade pressure: $P_I = \sqrt{P_H P_L}$

Problems

FE Exam Practice Questions (afternoon M.E. discipline exam)

10.1 The device that is responsible for cooling a space with an air conditioner is the:

(A) Evaporator
(B) Regenerator
(C) Condenser
(D) Absorber

10.2 The device that is responsible for heating a space with a heat pump is the:

(A) Evaporator
(B) Regenerator
(C) Condenser
(D) Absorber

10.3 The *T-s* diagram in Fig. 10.15 indicates an ideal refrigeration cycle operating between 120 kPa and 1200 kPa. If R134a is the refrigerant, the compression work is nearest:

(A) 42 kJ/kg
(B) 47 kJ/kg
(C) 56 kJ/kg
(D) 67 kJ/kg

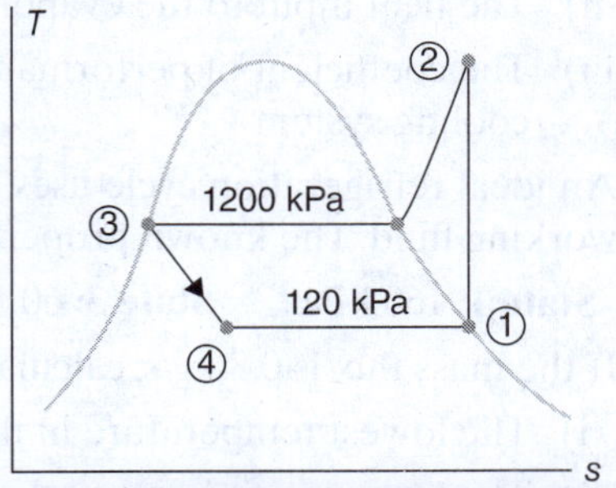

Figure 10.15

10.4 The cooling load of the refrigeration cycle of Problem 10.3 is nearest:

(A) 118 kJ/kg
(B) 98 kJ/kg
(C) 83 kJ/kg
(D) 67 kJ/kg

10.5 If the cycle of Problem 10.3 were used to heat a space, the heating load would be nearest:

(A) 120 kJ/kg
(B) 139 kJ/kg
(C) 146 kJ/kg
(D) 165 kJ/kg

10.6 The compressor of the cycle of Problem 10.3 is 85% efficient, and if that is the only loss considered, the percentage drop in the COP_R would be nearest:

(A) 8%
(B) 11%
(C) 13%
(D) 15%

10.7 The ideal gas air-conditioning cycle, shown in Fig. 10.16, operates with air as the refrigerant. The compressor inlets air at 80 kPa and 0°C and operates with a pressure ratio of 7. If the temperature of the air entering the turbine is 100°C, the lowest temperature in the cycle is nearest:

(A) −38°C
(B) −46°C
(C) −59°C
(D) −67°C

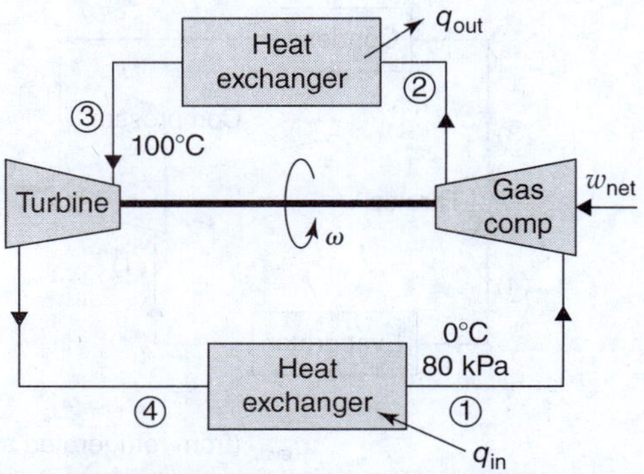

Figure 10.16

10.8 The cooling load for the gas air-conditioning cycle of Problem 10.7 is nearest:

(A) 59 kJ/kg

(B) 66 kJ/kg

(C) 77 kJ/kg

(D) 83 kJ/kg

10.9 The COP of the gas air-conditioning cycle of Problem 10.7 is nearest:

(A) 0.29

(B) 1.05

(C) 1.34

(D) 1.98

The Basic Vapor Refrigeration Cycle

10.10 An ideal refrigeration cycle uses R134a as the working fluid. The refrigerant enters the compressor at state 1, shown in Fig. 10.17, at 200 kPa as a saturated vapor. It leaves the condenser as a saturated liquid at 1 MPa. Calculate:

i) The work input to the compressor

ii) The cooling load from the evaporator

iii) The heat rejection from the condenser

iv) The coefficient of performance for the system

v) The Carnot coefficient of performance

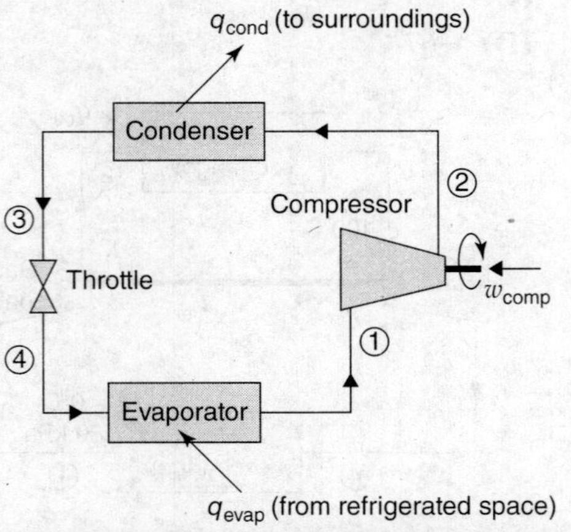

Figure 10.17

10.11 An ideal refrigeration cycle uses ammonia as the working fluid. The refrigerant enters the compressor at 120 kPa and the throttle at 1200 kPa. If it produces 12 tons of refrigeration, calculate:

i) The power input to the compressor

ii) The heat rejection from the condenser

iii) The coefficient of performance for the system

iv) The volume rate of flow entering the compressor

10.12 An ideal refrigeration cycle uses R134a as the working fluid. The refrigerant leaves the evaporator at −24°C and enters the throttle at 45°C. If it produces 40 kJ/s of heating, calculate:

i) The power input to the compressor

ii) The heat input to the evaporator

iii) The coefficient of performance for the heat pump

iv) The volume rate of flow entering the compressor

10.13 An ideal refrigeration cycle, using R134a as the working fluid, operates between saturation temperatures of −20°C and 50°C. If the compressor requires 10 hp, calculate:

i) The mass flow rate

ii) The heat input to the evaporator

iii) The coefficient of performance for a cooling system

10.14 An ideal refrigeration cycle uses R134a as the working fluid. The known properties are:

State 1: 160 kPa **State 3:** 60°C

If the mass flux is 0.8 kg/s, calculate:

i) The lowest temperature in the cycle

ii) The horsepower input to the compressor

iii) The cooling load

iv) The coefficient of performance for the system

10.15 An ideal air-conditioning system uses R134a as the refrigerant. The compressor has an inlet pressure of 140 kPa and an exit pressure of 1.6 MPa with a mass flow rate of 0.27 kg/s. Refer to Fig. 10.17 and determine:

i) The lowest cycle temperature
ii) The highest cycle temperature
iii) The cooling load from the evaporator
iv) The heat rejection from the condenser
v) The compressor horsepower
vi) The coefficient of performance for the system

10.16 The compressor in an ideal refrigeration cycle using R134a has an inlet pressure of 120 kPa. Referring to Fig. 10.4, state 1 has a quality of 1.0 and state 4 has a quality of 0.4. For a cooling load of 8 kJ/s, calculate:

i) The necessary mass flow rate of refrigerant
ii) The power input to the compressor
iii) The coefficient of performance for this system

10.17 The ideal air-conditioning system in Problem 10.16 is operated as a heat pump. Determine the heating load and the coefficient of performance.

10.18 An air conditioner using R134a as the refrigerant has a compressor exit temperature of 70°C and an exit pressure of 1.4 MPa. The refrigerant enters the compressor at 140 kPa and −18°C. The condenser cools the refrigerant to 50°C and 1.3 MPa. For a cooling load of 15 kJ/s, determine:

i) The mass flow rate
ii) The isentropic efficiency of the compressor
iii) The power input to the compressor
iv) The rate of heat rejection by the condenser
v) The coefficient of performance for this system

10.19 A large refrigeration system uses ammonia as the refrigerant. The compressor, with an isentropic efficiency of 90%, requires 150 kW. The ammonia enters the compressor at 120 kPa as a saturated vapor and leaves the compressor at 800 kPa. It leaves the condenser very close to the saturated liquid state. If the cooling load is 600 kJ/s, determine:

i) The mass flow rate of ammonia
ii) The rate of heat rejection by the condenser
iii) The coefficient of performance for this system

10.20 The refrigeration cycle shown in Fig. 10.18 used R134a as the working fluid. The following data were measured at each state:

State 1: 120 kPa, $x_1 = 1.0$
State 2: 1200 kPa, 60°C
State 3: 1190 kPa, 58°C
State 4: 1170 kPa, 44°C
State 5: 1160 kPa, 43°C
State 6: 130 kPa
State 7: 128 kPa, $x_7 = 0.42$
State 8: 122 kPa, $x_8 = 0.95$

Assuming an adiabatic compressor, determine the compressor efficiency and the coefficient of performance of a refrigerator.

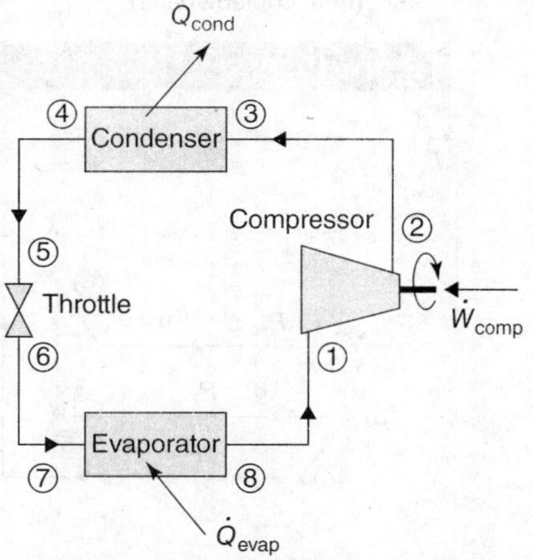

Figure 10.18

Cascade Refrigeration Systems

10.21 A two-stage, ideal cascade refrigeration system shown in Fig. 10.19 and Fig. 10.20 uses R134a as the working fluid in both cycles. The mass flow rate through the first stage is 8 kg/min. The known properties are:

State 1: 120 kPa

State 7: 40°C

Determine:

i) The cooling load
ii) The mass flow rate in system #2
iii) The net compressor power required
iv) The overall coefficient of performance

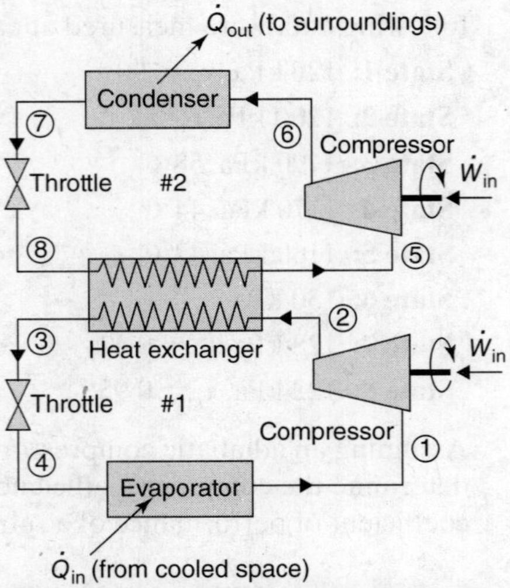

Figure 10.19

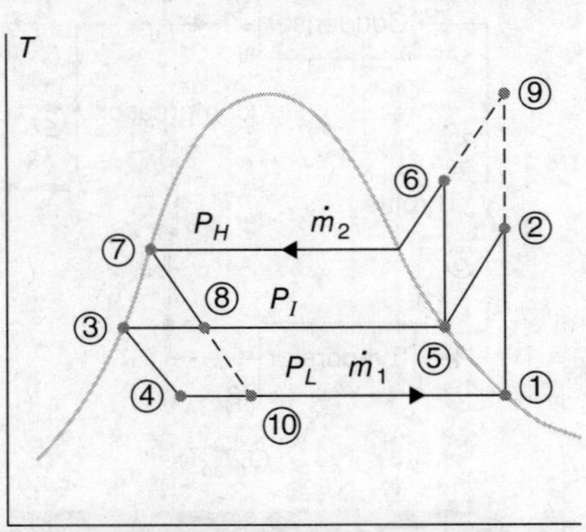

Figure 10.20

10.22 Replace the two-stage system of Problem 10.21 with an ideal single-stage system that produces the same cooling load. Determine the minimum cycle temperature, the mass flow rate of the R134a, the compressor horsepower, and the COP. Compare with the results of Problem 10.21 and state your observations.

10.23 A two-stage, ideal cascade refrigeration system shown in Fig. 10.19 uses R134a as the working fluid in both cycles and replaces the ideal single-stage cycle of Problem 10.14. The mass flow rate through the first stage is 0.8 kg/s. The known properties are:

State 1: 160 kPa

State 7: 60°C

Determine:

i) The cooling load
ii) The mass flow rate in system #2
iii) The net compressor power
iv) The coefficient of performance
v) The percentage increase in the COP and the percentage decrease in compressor power compared with that of Problem 10.14

10.24 Replace the two-stage system of Problem 10.23 with an ideal single-stage cycle that uses R134a and produces the same cooling load. Determine the maximum cycle temperature, the mass flow rate of the R134a, the compressor horsepower, and the COP.

10.25 An ideal two-stage cascade refrigeration system shown in Fig. 10.19 uses 0.8 kg/s of R134a as the working fluid in the first cycle and ammonia as the working fluid in the second cycle. Use Eq. 10.4 for the intermediate pressure. The known properties are:

State 1: R134a at 160 kPa

State 7: Ammonia at 60°C

Determine:

i) The cooling load
ii) The mass flow rate of the ammonia
iii) The total compressor horsepower
iv) The coefficient of performance for this system

10.26 An ideal three-stage cascade system uses R134a in all three cycles with a low pressure of 100 kPa and a high pressure of 1000 kPa and with intermediate pressures of 200 kPa and 600 kPa. The mass flux in the low-pressure cycle is 0.8 kg/s. Sketch the cycle on a *T-s* diagram and determine:

i) The cooling load
ii) The mass flux in each cycle
iii) The total compressor power
iv) The COP of this cascade system
v) Also, for a single-stage cycle between the same high and low pressures, find the compressor power and the COP for the same cooling load.

Gas Refrigeration Systems

10.27 Carbon dioxide is used in the ideal-gas refrigeration system of Fig. 10.21. The gas enters the compressor at 150 kPa and −15°C. The gas leaves the compressor at 1.2 MPa. The CO_2 enters the turbine at 40°C. Calculate the temperature at state 4 and the coefficient of performance for this system. Assume constant specific heats and ideal components.

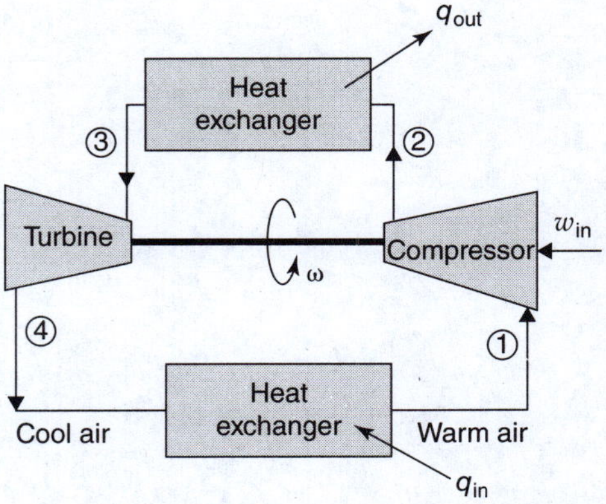

Figure 10.21

10.28 Air from a jet engine intake is diverted to an ideal-gas air-conditioning system at 100 kPa and −20°C, as shown in Fig. 10.21. The air leaves the compressor at 600 kPa. The air enters the turbine at 25°C. Assuming constant specific heats and ideal components, calculate the low-cycle temperature and the cycle COP. For a cabin cooling load of 3000 kJ/min, calculate the necessary mass flow rate of air.

10.29 Air enters an ideal reverse Brayton cycle air conditioner at 100 kPa and 2°C with a mass flux of 1.8 kg/s. The compressor has a pressure ratio of 4. The temperature at the inlet of the turbine is 38°C. Assuming constant specific heats and ideal components, calculate:

i) The low-cycle temperature
ii) The power requirement of the cycle
iii) The rate of heat flow from the heat exchanger
iv) The coefficient of performance for this system

CHAPTER

11

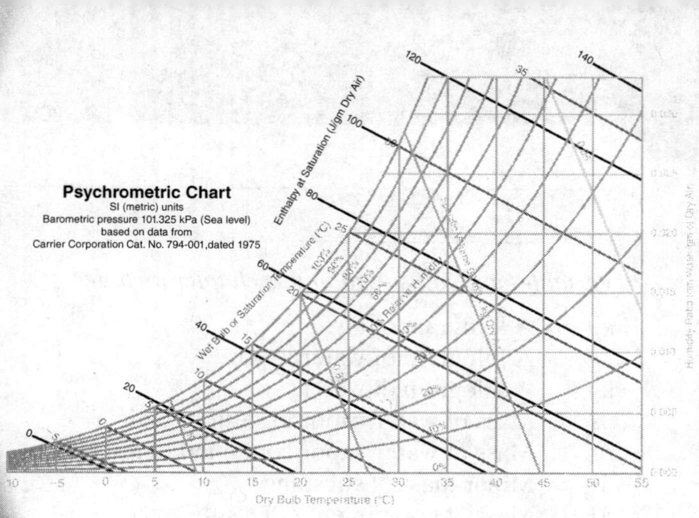

ArthurOgawa

Mixtures and Psychrometrics

Outline

Nomenclature

Variables and units used in this chapter include:

kg_a	A kilogram of air	P_i	Partial pressure of component *i*
kg_v	A kilogram of water vapor	R_u	Universal gas constant
kg_w	A kilogram of water	P_v	Partial pressure of water vapor
m_g	Mass of water vapor in saturated air	T_{db}	Dry-bulb temperature
m_v	Mass of water vapor contained in the air	T_{dp}	Dew-point temperature
M_i	Molar mass of substance	T_{wb}	Wet-bulb temperature
M_i	Mass of component *i* of a substance	x_i	The mass fraction of a component
N	Number of moles in a mixture	y_i	The mole fraction
N_i	Number of moles of component *i*	ϕ	Relative humidity
P_a	Partial pressure of the dry air	ω	Specific humidity, or humidity ratio

Learning Outcomes

- **Become familiar with the terminology and definitions used in gas mixtures and gas-vapor mixtures**
- **Solve first-law problems for gas and gas-vapor mixtures**
- **Become familiar with the definitions and terminology in the field of psychrometrics**
- **Solve air-water vapor problems using a psychrometric chart**
- **Solve problems involving the conditioning of air**

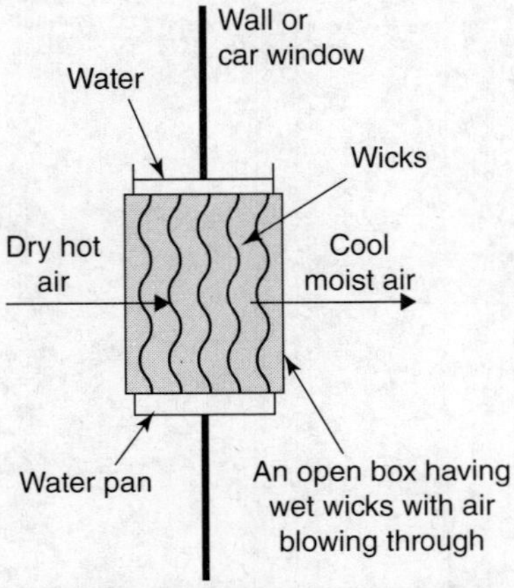

Figure 11.1

A simple swamp cooler.

Motivational Example—The Swamp Cooler

In the last chapter we studied vapor refrigeration and air-conditioning systems. The swamp cooler is a technique used to cool air without the use of an expensive and energy-hungry air conditioner. The technical term for this device is an *evaporative cooler*. It is also called a *swamp cooler*. It works best when the humidity is relatively low. A fine mist of water is sprayed, or water evaporates from a wick, into an air flow, usually caused by a fan. When the water evaporates in the low-humidity air, it absorbs heat as it changes from a liquid to a vapor, with the result that the air is cooled with increased humidity. The name "swamp cooler" comes from the odor associated with the evaporating water. A swamp cooler can be attached to a house or a car window, resulting in a cool air flow into the passenger compartment, as sketched in Fig. 11.1.

11.1 Gas Mixtures

Up to this point in this text, the substances of interest have been assumed to be pure substances composed of only one chemical component. Even when air was used, which we know to be a mixture of gases, it was treated as a pure substance with averaged properties. In this chapter, gases that are mixtures of two or more chemical substances will now be considered. The water vapor that exists in air will be a primary focus of our study.

11.1.1 Definitions and terminology

A *mixture* contains two or more unique components. In this chapter, each of the components will be labeled with a number subscript. For example, if air is considered to be a mixture of nitrogen, oxygen, argon, and carbon dioxide, the total mass m of the mixture is

$$m = m_1 + m_2 + m_3 + m_4 = \sum_i^4 m_i \tag{11.1}$$

In this equation, nitrogen could be identified by subscript 1, oxygen by 2, argon by 3, and carbon dioxide by 4. The number N_i *of moles* of each component is defined by

$$N_i = \frac{m_i}{M_i} \tag{11.2}$$

where M_i is the *molar mass* of substance i, which can be found in Table B-2 in the Appendix. Molar mass has units of kg/kmol in the SI system and lbm/lbmol in the English system of units. One kmol of a substance has a mass in kilograms equal to its molar mass, and one lbmol of a substance has a mass in lbm equal to its molar mass, so, for oxygen, $M = 32$ kg/kmol.

The total number N of moles in a mixture can be found by adding the number of moles of all the components as

$$N = \sum_i^n N_i \tag{11.3}$$

where n is the number of components that make up the mixture. The *mass fraction* x_i of an individual component is defined as

$$x_i = \frac{m_i}{m} \tag{11.4}$$

The *mole fraction* y_i is defined to be the ratio of the moles of a component to the total moles:

$$y_i = \frac{N_i}{N} \tag{11.5}$$

The following equations are satisfied for the fractional properties x_i and y_i:

$$\sum_i^n x_i = 1 \quad \text{and} \quad \sum_i^n y_i = 1 \tag{11.6}$$

Mixture: A substance that contains two or more unique components.

Mole: A unit of measurement that expresses the amount of a substance.

Santia/Shutterstock.com

Here's what a mole really looks like!

Remember:

$\sum_i^n N_i = N_1 + N_2 + \cdots + N_n$

Mass fraction: The ratio of the mass of a component to the total mass.

Mole fraction: The ratio of the moles of a component to the total moles. It does not represent a ratio of volumes or masses.

Note: A bar may be used to denote the average, that is, $\overline{M}$ or M_{avg}, for a mixture, but the notation is unnecessary since its meaning is obvious. Air is a mixture of gases for which we use M = 28.97 kg/kmol as included in Tables B-2 and B-3.

The average molar mass for a mixture, such as air, is

$$M = \frac{m}{N} = \frac{1}{N}\sum_i^n m_i = \sum_i^n y_i M_i \tag{11.7}$$

The mass and mole fractions are related by

$$x_i = \frac{m_i}{m} = \frac{N_i M_i}{NM} = y_i \frac{M_i}{M} \tag{11.8}$$

Gravimetric analysis ⟶ Mass specified

Molar analysis ⟶ Moles specified

Volumetric analysis ⟶ Volume specified (same as the molar analysis for ideal gases)

If the composition of a mixture is described by specifying the mass of each component, it is referred to as a *gravimetric analysis.* If the number of moles of each component is specified, it is a *molar analysis.* If the volume of each component is specified, it is a *volumetric analysis.* In this chapter, only ideal gases will be analyzed, and for an ideal gas, the molar analysis and the volumetric analysis are identical. The type of analysis will be stated or be obvious in each example and problem.

Example 11.1 The primary components of air

Comment

Air is assumed to contain no water vapor unless otherwise stated. The water vapor in air, even moist air, is so small that it has an insignificant influence on the quantities of interest in this section and in the preceding chapters. It will be considered in the next section.

Calculate the average molar mass of air if the molar fractions of the four main components are as follows:

Substance	Mole fraction y_i	Molar mass (kg/kmol)
Nitrogen (N_2)	0.7808	28
Oxygen (O_2)	0.2095	32
Argon	0.0093	40
Carbon dioxide (CO_2)	0.0004	44

Also, calculate the mass of each of the four components for a 10-kg mixture.

Solution

The average molar mass for the air can be found using Eq. 11.7 since the mole fractions are given. It is

$$M_{air} = 0.7808 \times 28 + 0.2095 \times 32 + 0.0093 \times 40 + 0.0004 \times 44$$
$$= \underline{28.96 \text{ kg/kmol}}$$

If air is assumed to be a mixture composed of 79% nitrogen and 21% oxygen, as it often is, the average molar mass is

$$M_{air} = 0.79 \times 28 + 0.21 \times 32 = \underline{28.8 \text{ kg/kmol}}$$

Table B-2 lists the value as 28.97, less than 1% difference compared to 28.8. The argon and carbon dioxide do not make a significant contribution to air as a mixture. However, carbon dioxide does play an important role in other ways, as does a small amount of water vapor, which has been ignored in this example.

Next, let's analyze the air mixture on the basis of mass in order to determine the mass of each component. The mass fractions are calculated using

$$x_i = y_i \frac{M_i}{M}$$

They are listed in the following table along with the mole fractions in the example statement. The calculation for the mass fraction of nitrogen, for example, is

$$x_1 = y_1 \frac{M_1}{M} = 0.7808 \times \frac{28}{28.96} = 0.755$$

Substance	Mole Fraction y_i	Mass Fraction x_i
Nitrogen (N_2)	0.7808	0.755
Oxygen (O_2)	0.2095	0.231
Argon	0.0093	0.0128
Carbon dioxide (CO_2)	0.0004	0.0006

The mass of nitrogen in 10 kg of air is found to be

$$m_1 = x_1 m = 0.755 \times 10 = 7.55 \text{ kg}$$

The mass of the other components follows. Hence, 10 kg of air contains 7.55 kg of nitrogen, 2.31 kg of oxygen, 0.128 kg of argon, and 0.006 kg of carbon dioxide.

Comment

Air can be modeled using only nitrogen and oxygen with acceptable accuracy.

☑ You have completed Learning Outcome (1)

11.1.2 The Amagat and Dalton laws

Let us analyze mixtures of ideal gases in greater detail. Two laws describe their behavior: the Amagat law and the Dalton law. The *Amagat law* treats each component as though it exists independently of the others but at the same pressure and temperature as the mixture. The total volume of the mixture is the sum of the volumes of the components, as shown in Fig. 11.2. The *Dalton law*, which will be used in this text, states that each component occupies the same volume and exists at the same temperature as the mixture, as shown in Fig. 11.3. The total pressure P of the mixture is the sum of the *partial pressures* P_i of the components:

$$P = \sum_i P_i \tag{11.9}$$

Partial pressures: The total pressure is the sum of the partial pressures.

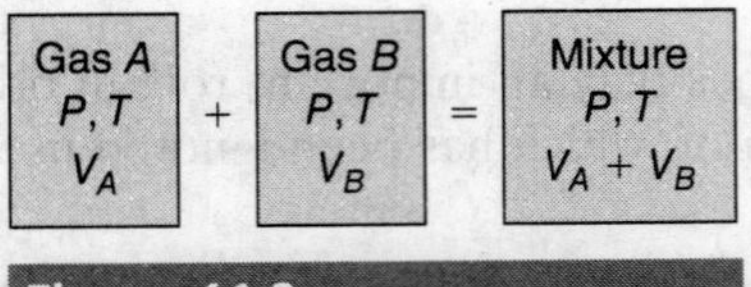

Figure 11.2

The Amagat law: Each component is at the same pressure and temperature as the mixture.

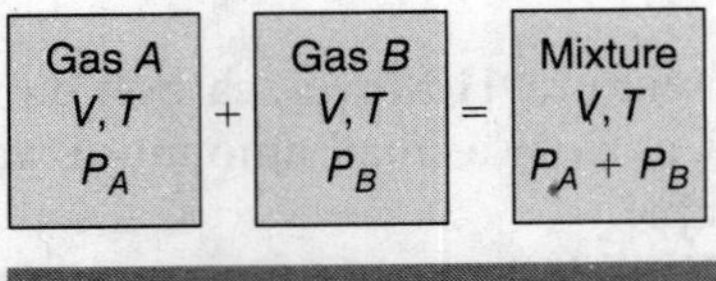

Figure 11.3

The Dalton law: Each component occupies the same volume and temperature as the mixture.

The partial pressure of a component is found, using the ideal-gas law (see Eq. 2.14), to be

Recall: R_u= 8.314 kJ/kmol·K

$$P_i = \frac{N_i R_u T}{V} \tag{11.10}$$

In this equation N_i is the number of moles of gas i, R_u is the universal gas constant, T is the absolute temperature of the mixture, and V is the volume of the mixture. For the complete mixture, the ideal-gas law is

$$P = \frac{N R_u T}{V} \tag{11.11}$$

The ratio of the above two equations provides

Note: With the Dalton law, the volume, as well as the temperature, is the same for all components.

$$\frac{P_i}{P} = \frac{N_i}{N} = y_i \tag{11.12}$$

The properties of mixtures of gases are typically determined on a molar basis. The internal energy, enthalpy, and entropy can be calculated based on the mole fraction of each component and the molar specific values of each property. The molar properties for several common gases are presented in the Appendix in Tables F-2 through F-5. The properties of water vapor treated as an ideal gas are presented in Appendix F-6. The equations for calculating the average properties for a gas mixture are

Units: kJ/kmol and kJ/kg

$$\overline{u} = \sum_i y_i \overline{u}_i \qquad u = \sum_i x_i u_i \qquad C_v = \sum_i x_i C_{v,i} \tag{11.13}$$

Units: kJ/kmol and kJ/kg

$$\overline{h} = \sum_i y_i \overline{h}_i \qquad h = \sum_i x_i h_i \qquad C_p = \sum_i x_i C_{p,i} \tag{11.14}$$

Units: kJ/kmol·K and kJ/kg·K

$$\overline{s} = \sum_i y_i \overline{s}_i \qquad s = \sum_i x_i s_i \tag{11.15}$$

where the overbar indicates a molar basis for the property. Molar specific internal energy and molar specific enthalpy have units of kJ/kmol in the SI system. Molar specific entropy has units of kJ/kmol·K in the SI system.

Molar properties can be related to the mass-based properties by

$$\overline{u}_i = M_i u_i \qquad \overline{h}_i = M_i h_i \qquad \overline{s}_i = M_i s_i \qquad \overline{C}_{p,i} = M_i C_{p,i} \tag{11.16}$$

Partial pressures — Example 11.2

A rigid tank contains 6 kg of oxygen and 4 kg of nitrogen at 800 kPa and 60°C. Calculate the partial pressure of each component and the gas constant of the mixture; see Fig. 11.4.

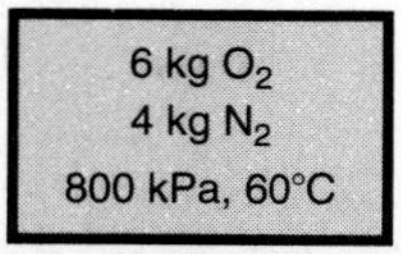

Figure 11.4

Note: Dalton's law is used exclusively in this text.

Solution

The partial pressures are related to the mole fractions by Eq. 11.12. To find the mole fractions, the number of moles of each gas must be known. They are, respectively,

$$\left.\begin{aligned} N_{O_2} &= \frac{m_1}{M_1} = \frac{6\text{ kg}}{32\text{ kg/kmol}} = 0.1875 \\ N_{N_2} &= \frac{m_2}{M_2} = \frac{4\text{ kg}}{28\text{ kg/kmol}} = 0.1429 \end{aligned}\right\} \quad \therefore N = N_{O_2} + N_{N_2} = 0.330\text{ kmol}$$

The mole fractions are then

$$y_1 = \frac{0.1875\text{ kmol}}{0.330\text{ kmol}} = 0.568 \qquad y_2 = \frac{0.1429}{0.330} = 0.433$$

The partial pressures, using Eq. 11.12, are

$$P_{O_2} = P_1 = y_1 P = 0.568 \times 800 = \underline{454\text{ kPa}}$$
$$P_{N_2} = P_2 = y_2 P = 0.433 \times 800 = \underline{346\text{ kPa}}$$

The molar mass is

$$M = y_1 M_1 + y_2 M_2 = 0.568 \times 32 + 0.433 \times 28 = 30.3\text{ kg/kmol}$$

The mixture's gas constant is found to be

$$R = \frac{R_u}{M} = \frac{8.314}{30.3} = \underline{0.274\text{ kJ/kg}\cdot\text{K}}$$

It is in between the gas constants of oxygen and nitrogen.

Properties of a mixture — Example 11.3

A gas mixture containing 5 kg of nitrogen and 8 kg of carbon dioxide exists at 300 kPa and 300 K; see Fig. 11.5. Determine:

i) The mass fraction of each component
ii) The mole fraction of each component
iii) The amount of heat required to raise the temperature of the mixture to 340 K at constant pressure
iv) The change in the entropy of the mixture for the process of (iii)

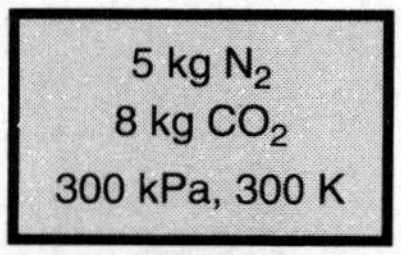

Figure 11.5

(Continued)

Example 11.3 (*Continued*)

Solution

$m_1 = 5$ kg, $m_2 = 8$ kg $\therefore m = 13$ kg

i) $x_1 = \dfrac{m_1}{m} = \dfrac{5}{13} = \underline{0.385}$ $x_2 = \dfrac{m_2}{m} = \dfrac{8}{13} = \underline{0.615}$

ii) $N_1 = \dfrac{m_1}{M_1} = \dfrac{5 \text{ kg}}{28 \text{ kg/kmol}} = 0.179 \text{ kmol}$

$$N_2 = \frac{m_2}{M_2} = \frac{8}{44} = 0.182 \text{ kmol}$$

$$N = N_1 + N_2 = 0.179 + 0.182 = 0.361 \text{ kmol}$$

$$y_1 = \frac{0.179}{0.361} = \underline{0.50} \qquad y_2 = \frac{0.182}{0.361} = \underline{0.50}$$

Note: Mass fractions are used since $C_{p,i}$ is known on a mass basis. If $\bar{C}_{p,i}$ were known, the mole fractions would be used.

iii) The first law (for P = const): $Q = \Delta H = mC_p\Delta T$. Using

$$C_p = \sum x_i C_{p,i}$$

$$C_p = x_1 \cdot C_{p,1} + x_2 \cdot C_{p,2}$$

$$= 0.385 \times 1.04 + 0.615 \times 0.86 = 0.929 \text{ kJ/kg·K}$$

$$\therefore Q = mC_p\Delta T = 13 \times 0.929(340 - 300) = \underline{483 \text{ kJ}}$$

iv) From Eq. 6.19, the entropy change in an ideal gas for this constant-pressure process is, using C_p from part (iii),

Remember: The temperatures must be absolute.

$$\Delta S = m\left(C_p \ln\frac{T_2}{T_1} - R\ln\frac{P_2}{P_1}\right) = mC_p \ln\frac{T_2}{T_1}$$

$$= 13 \times 0.929 \times \ln\frac{340}{300} = \underline{1.511 \text{ kJ/kg·K}}$$

Example 11.4 A mixture flowing through a turbine

A gas mixture, with the mole fractions listed below, enters the turbine of Fig. 11.6 at 1200 kPa and 600 K with an entering volume flow rate of 1.2 m³/s. The mixture leaves the turbine at 300 K. Determine the power produced by the turbine.

Gas	**Mole Fraction y_i**
Carbon dioxide	0.20
Oxygen	0.02
Nitrogen	0.78

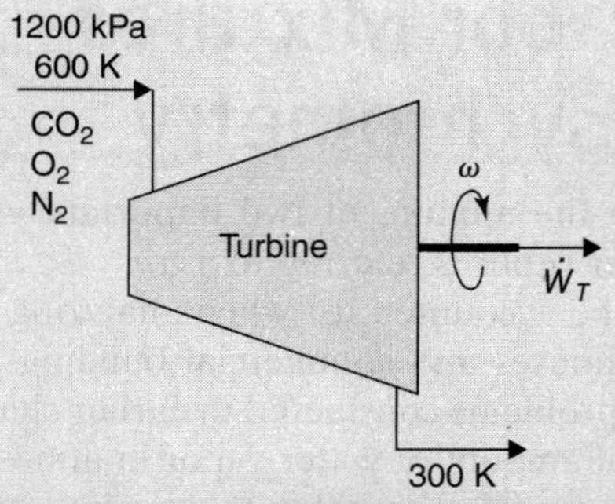

Figure 11.6

Solution

First, calculate the average molar mass of the mixture:

$$M = \sum_i y_i M_i = 0.2 \times 44 + 0.02 \times 32 + 0.78 \times 28 = 31.3 \text{ kg/kmol}$$

Next, the mass flow rate of the mixture is found, using Eq. 4.10, to be

$$\dot{m} = \frac{\dot{V}}{v} = \dot{V} \cdot \frac{P_1}{(R_u/M)T_1}$$
$$= 1.2 \frac{\text{m}^3}{\text{s}} \times \left(\frac{1200 \times 31.3}{8.314 \times 600}\right)\frac{\text{kg}}{\text{m}^3} = 9.04 \text{ kg/s}$$

Units: We know v, which the ideal-gas law gives as $v = RT/P$, has units of m^3/kg.

Now we can calculate the power output using the molar enthalpies found in Tables F-2, F-3, and F-4. Using Eq. 11.14, we have

Units: Using the ideal-gas tables in the Appendix provides $\bar{h}$ in kJ/kmol, so mole fractions are necessary.

$$\dot{W} = \frac{\dot{m}}{M}\sum_i y_i\left[\bar{h}_i(600) - \bar{h}_i(300)\right]$$

$$= \frac{9.04 \dfrac{\text{kg}}{\text{s}}}{31.3 \dfrac{\text{kg}}{\text{kmol}}}\left[(0.2)(22\,280 - 9431) + (0.02)(17\,929 - 8736)\right.$$
$$\left. + (0.78)(17\,563 - 8723)\right]\frac{\text{kJ}}{\text{kmol}}$$

$$= \left(0.289 \frac{\text{kmol}}{\text{s}}\right)\left(9649 \frac{\text{kJ}}{\text{kmol}}\right) = \underline{2790 \text{ kW}}$$

☑ **You have completed Learning Outcome (2)**

11.2 Air–Vapor Mixtures and Psychrometry

Dry air: Air that contains no water vapor.

Atmospheric air: Air that contains dry air and water vapor.

This section deals with the mixture of two important gases: dry air and water vapor. Air with no water vapor is referred to as *dry air*. *Atmospheric air* contains water vapor that must be accounted for when analyzing problems involving the conditioning of air in houses and commercial buildings, and in the analysis of combustion. The air in problems considered in earlier chapters was considered to be dry air, but the small amount of water vapor in atmospheric air does not significantly influence heat transfer, enthalpy change, internal energy change, or the other properties of interest for a process. However, when considering comfort in a building or the combustion process, the amount of water vapor must be included. Combustion will be considered in Chapter 12.

Note: Water vapor in air is assumed to be an ideal gas at relatively low temperature and near atmospheric pressure.

Water vapor exists in the atmosphere as a superheated gas. In fact, it can be modeled as an ideal gas, providing the pressure is near atmospheric pressure and the amount of superheat is not extensive. For example, Table C-1 indicates that saturated water vapor has a specific volume of 57.8 m^3/kg at 20°C, whereas the ideal-gas law provides $v = RT/P = 0.462 \times 293/2.338 = 57.9$ m^3/kg, using the gas constant for water vapor (steam) from Table B-2 and the partial pressure of the water vapor presented next to the temperature of 20°C in Table C-1. Consequently, atmospheric air can be treated as a mixture of two ideal gases: dry air and water vapor. The Dalton law is again used so that

Observation: Atmospheric air can be treated as a mixture of two ideal gases: dry air and water vapor.

$$P = P_a + P_v \tag{11.17}$$

where P_a is the partial pressure of the dry air and P_v is the partial pressure of the vapor, both pressures being absolute. In the above example the partial pressure of the vapor is the saturation pressure of water at 20°C, which is $P_v = 2.338$ kPa from Table C-1. So, if $P_{atm} = 100$ kPa, the partial pressure of the dry air in the atmosphere would be $P_a = 100 - 2.338 = 97.66$ kPa.

Note: h_g is taken to be zero at 0°C by the heating and air-conditioning community. That's quite different from the h_g value listed in Tables C-1 and C-2. It's the enthalpy change that is of interest, so where the 0-datum is selected is not significant.

The enthalpy and internal energy of the water vapor depends on temperature only (being an ideal gas), and thus it is assumed that

$$h_v(T) = h_g(T) \qquad u_v(T) = u_g(T) \tag{11.18}$$

The ideal-gas assumption for the vapor in atmospheric air is acceptable from about 0°C to 50°C.

A key element in the field of heating, ventilating, and air conditioning (HVAC) is how to maintain comfortable levels of humidity in our homes and workplaces. Too much water in the atmosphere makes it uncomfortable, that is, makes it difficult to lose heat by perspiration. Too little moisture irritates the skin and tissue in our respiratory systems. The remainder of this chapter is a detailed analysis of HVAC.

Remember: Water vapor in atmospheric air can be treated as an ideal gas up to about 50°C.

Psychrometry: The study of air–water vapor mixtures.

11.2.1 Terminology and definitions

The study of air–water vapor mixtures is called *psychrometry*. There are two primary quantities that specify the amount of water vapor in the air. The *specific humidity* ω, or *humidity ratio*, is the mass of water vapor per unit mass of dry air, defined by

$$\omega = \frac{m_{\text{vapor}}}{m_{\text{air}}} \tag{11.19}$$

This is not the definition of humidity with which we are familiar. The humidity that the weather channel presents is the "relative humidity," defined below. The humidity ratio is more useful to engineers because it clearly defines the ratio of water mass to air mass. The humidity ratio will be required for engineering calculations.

Relative humidity ϕ is the ratio of the mass m_v of the water vapor actually contained in the air to the mass m_g of water vapor required for the air to be completely saturated at the same temperature. It is defined to be

$$\phi = \frac{m_v}{m_g} \tag{11.20}$$

For dry air that contains no water vapor, the relative humidity is zero, as is the humidity ratio. For humid air in which the water vapor is at the saturation condition, the relative humidity is 1 (or, often stated as 100%).

In the following analysis, the humidity ratio is related to the relative humidity. In doing this, the water vapor, both m_v and m_g, will be treated as an ideal gas. Consequently, the relative humidity is related to the ratio of partial pressures by

$$\phi = \frac{m_v}{m_g} = \frac{P_v V/RT}{P_g V/RT} = \frac{P_v}{P_g} \tag{11.21}$$

The volume and temperature are assumed the same for both conditions: when the air contains some water vapor and when it is saturated.

The humidity ratio takes the form

$$\omega = \frac{m_v}{m_a} = \frac{P_v V/R_v T}{P_a V/R_a T} = \frac{R_a P_v}{R_v P_a} = 0.622\frac{P_v}{P_a} \tag{11.22}$$

where $R_a/R_v = 0.622$. A combination of Eqs. 11.21 and 11.22 relates the two functions:

$$\phi = \frac{\omega P}{(0.622 + \omega)P_g} \quad \text{or} \quad \omega = \frac{0.622\phi P_g}{P - \phi P_g} \tag{11.23}$$

The pressure P is related to P_v and P_a by Eq. 11.17.

Humidity ratio: The mass of water vapor in a unit mass of air. It has units of kg vapor/kg dry air.

Comment

The term "humidity" refers to the relative humidity.

Note: The quantity m_g is the maximum amount of water vapor air can hold. If the amount of water vapor exceeds this value, it condenses out, like dew in the morning (in Michigan) or on the outside of a glass of ice water.

Comment

The gas constant for water vapor (steam) is given in Appendix B.2. We have not used this gas constant very often up to now since the properties of water vapor have been found in Appendix C.

Comment

The humidity ratio is used more often than relative humidity in calculations.

The condensation of water vapor — Example 11.5

Air at 100 kPa, 25°C, and a relative humidity of 75% is compressed isothermally until water droplets begin to appear, as in Fig. 11.7. Determine the pressure at this final state assuming no change in the amount of water in the air.

(Continued)

Example 11.5 (*Continued*)

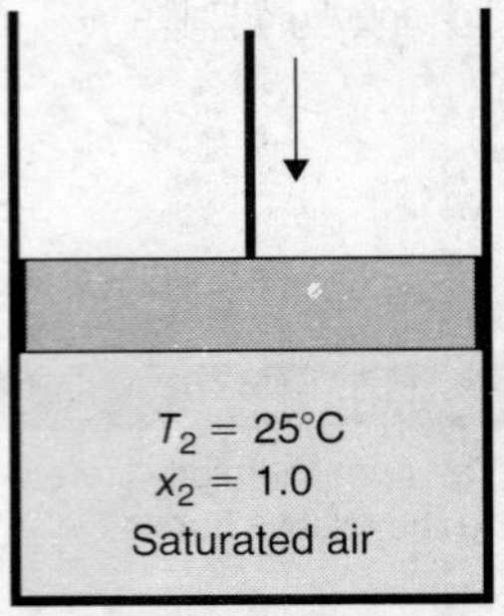

Figure 11.7

Comment

Rather than write kg vapor/kg dry air, we will write kg_v/kg_a.

Solution

The states of the water are:

State 1: $P_1 = 100$ kPa, $T_1 = 25°C$, $\phi = 0.75$
State 2: $T_2 = T_1 = 25°C$, $x_2 = 1$, $P_v = P_g = 3.17$ kPa

The initial pressure of 100 kPa is a total pressure equal to the sum of the partial pressure P_a of the air and the partial pressure P_v of the water vapor. From Table C-1, the saturation pressure of water at 25°C is 3.17 kPa. Equation 11.21 provides the partial pressure of the water vapor at state 1:

$$P_v = \phi \times P_g = 0.75 \times 3.17 \text{ kPa} = 2.38 \text{ kPa}$$

The humidity ratio follows from Eq. 11.23 to be

$$\omega_1 = \frac{0.622\phi P_g}{P - \phi P_g} = \frac{0.622 \times 0.75 \times 3.17}{100 - 0.75 \times 3.17} = 0.0152 \text{ kg}_v/\text{kg}_a$$

where we will use kg_v to be the mass of the water vapor and kg_a to be the mass of dry air.

At state 2, $\omega_2 = \omega_1 = 0.0152 \text{ kg}_v/\text{kg}_a$ since the amount of vapor in the air remains unchanged. Using Eq. 11.22, with $P_g = 3.17$ kPa since $\phi = 1$,

$$P_a = \frac{0.622 \times P_g}{\omega} = \frac{0.622 \times 3.17 \text{ kPa}}{0.0152} = 129.7 \text{ kPa}$$

$$\therefore P_2 = P_v + P_a = 3.17 + 129.7 = \underline{132.9 \text{ kPa}}$$

Example 11.6 The drying of clothes

In an industrial drying process, wet fabric enters an air dryer with a mass flow rate of 0.5 kg/s and a moisture content of 80% by mass. It leaves the dryer with a moisture content of 10%. Air enters the dryer at 80°C with a relative humidity of 20% and leaves the dryer at 60°C with a relative humidity of 70%. See the schematic of Fig. 11.8. Estimate the mass flow rate of air needed to accomplish this process. Assume the vapor to be an ideal gas.

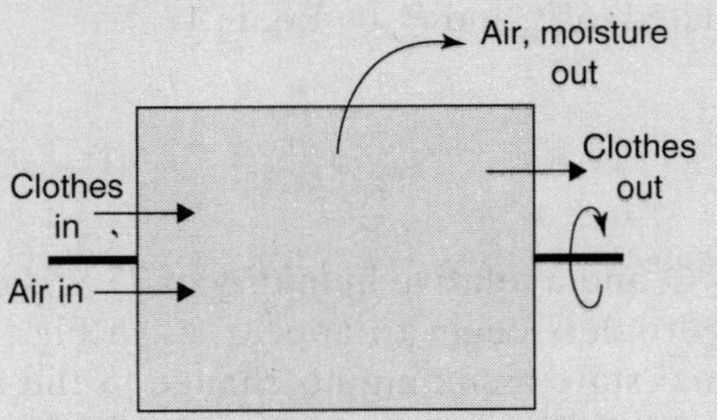

Figure 11.8

Solution

For the clothes entering the dryer:

$$\dot{m}_{\text{in}} = 0.5 \text{ kg/s} \qquad \text{wet clothes}$$

$$\dot{m}_{w,\text{in}} = 0.8 \times 0.5 = 0.4 \text{ kg/s} \qquad \text{water in fabric}$$

$$\therefore \dot{m}_{f,\text{in}} = 0.1 \text{ kg/s} \qquad \text{fabric}$$

Entering air: $T_{a,\text{in}} = 80°\text{C}$ and $\phi_{\text{in}} = 0.2$. $\therefore P_g = 50$ kPa. The humidity ratio is given by Eq. 11.23 to be

$$\omega_{\text{in}} = \frac{0.622\phi P_g}{P - \phi P_g} = \frac{0.622 \times 0.2 \times 50}{101.3 - 0.2 \times 50} = 0.068 \frac{\text{kg}_v}{\text{kg}_a}$$

Fabric and air leaving the dryer:

$$\dot{m}_{f,\text{out}} = \dot{m}_{f,\text{in}} = 0.1 \text{ kg/s}$$

$$T_{\text{air,out}} = 60°\text{C} \quad \text{and} \quad \phi = 0.7 \quad \therefore P_g = 20 \text{ kPa}$$

The humidity ratio leaving is then

$$\omega_{\text{out}} = \frac{0.622\phi P_g}{P - \phi P_g} = \frac{0.622 \times 0.7 \times 20}{101.3 - 0.7 \times 20} = 0.094 \frac{\text{kg}_v}{\text{kg}_a}$$

For the water leaving in the fabric, 10% of the total mass is water. So, in the fabric only, a mass balance of the water takes the form

$$(\dot{m}_w)_{\text{out}} = 0.1\left[(\dot{m}_w)_{\text{out}} + (\dot{m}_f)_{\text{out}}\right] = 0.1\left[(\dot{m}_w)_{\text{out}} + 0.1 \text{ kg/s}\right]$$

$$\therefore \dot{m}_{w,\text{out}} = 0.011 \text{ kg/s in fabric}$$

The total conservation of mass for the water in the air and in the fabric requires

$$0.4 + 0.068 \times \dot{m}_{\text{air,in}} = 0.011 + 0.094 \times \dot{m}_{\text{air,out}}$$

Recognizing that $\dot{m}_{\text{air,in}} = \dot{m}_{\text{air,out}}$, this equation provides the answer:

$$\dot{m}_{\text{air}} = \underline{14.9 \text{ kg/s}}$$

Conservation of mass:

Fabric: $\dot{m}_{f,\text{in}} = \dot{m}_{f,\text{out}}$

Water: $\dot{m}_{w,\text{in}} = \dot{m}_{w,\text{out}}$

Air: $\dot{m}_{\text{air,in}} = \dot{m}_{\text{air,out}}$

Note: kg_v = kgs of water vapour; kg_a = kgs of air

11.2.2 Adiabatic saturation temperature

Instruments that measure relative humidity are called *hygrometers*. Electronic hygrometers use temperature of condensation, or changes in electrical capacitance or resistance, to measure changes in humidity. Hygrometers do not provide accurate measurements of the relative humidity; in some engineering design situations, an accurate value for the humidity ratio is necessary. A setup that can provide an accurate value for humidity ratio ω, and thus the relative humidity ϕ, is sketched in Fig. 11.9 with the *T-s* diagram shown in Fig. 11.10. The channel must be sufficiently long to ensure that the relative humidity at the exit is 100%, or very close to 100%. The *T-s* diagram shows the

Comment

Whenever the term "humidity" is used, it always refers to the relative humidity ϕ, never the humidity ratio ω.

Note: An accurate measurement of the humidity is quite difficult.

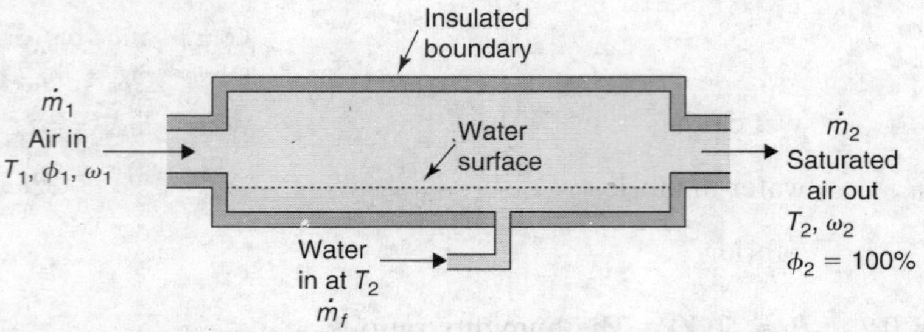

Figure 11.9

The setup used to determine the humidity ratio.

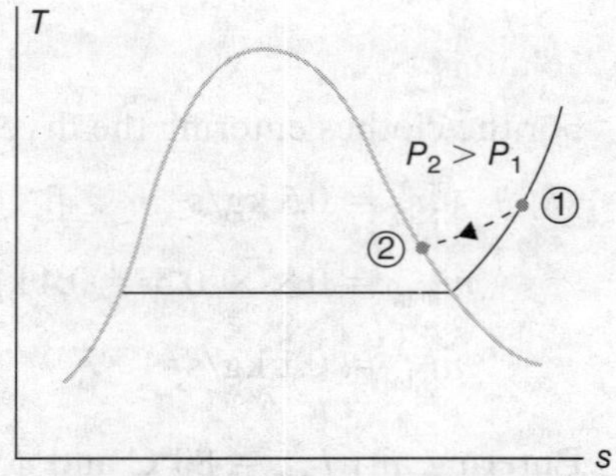

Figure 11.10

T-s diagram for Fig. 11.9. P_1 and P_2 are the partial pressures of the water vapor.

increase in the partial pressure of the water vapor as the humidity increases and the temperature decreases.

The objective of this analysis is to find an expression for the humidity ratio ω_1 in terms of measurable properties. The exit temperature is the *adiabatic saturation temperature*—"adiabatic" because the channel is insulated so that no heat transfer occurs and "saturation" because the air is saturated ($x_2 = 1$), as indicated by state 2 in Fig. 11.10.

Note: As for any steady flow for which $Q = 0$ and $W = 0$, the enthalpy is constant if kinetic and potential energy changes are negligible.

The first law, an enthalpy balance for this steady-flow process, applied to the control volume with two inlets and one exit takes the form

$$\dot{m}_{v1}h_{v1} + \dot{m}_{a1}h_{a1} + \dot{m}_f h_{f2} = \dot{m}_{v2}h_{v2} + \dot{m}_{a2}h_{a2} \tag{11.24}$$

where the temperature of the makeup water is assumed to be the same as the temperature at the exit, so h_{f2} was used for the enthalpy. The mass balances, for both the air and the vapor, take the forms

$$\dot{m}_{a1} = \dot{m}_{a2} = \dot{m}_a \quad \text{and} \quad \dot{m}_{v1} + \dot{m}_f = \dot{m}_{v2} \tag{11.25}$$

The definition $\omega = m_v/m_a$ allows the second equation of Eqs. 11.25 to be written as

$$\dot{m}_f = \dot{m}_a(\omega_2 - \omega_1) \tag{11.26}$$

Using $\phi_2 = 1.0$, we can calculate the humidity ratio ω_2 using Eq. 11.23:

$$\omega_2 = 0.622\frac{P_{g2}}{P - P_{g2}} \tag{11.27}$$

Note: State 1 is quite close to the saturated vapor curve so that $h_{v1} \cong h_{g1}$.

Substitute Eqs. 11.25 and 11.26 into Eq. 11.24 and, using $h_v \cong h_g$ along with some algebra, an expression for the entering humidity ratio is found to be

$$\omega_1 = \frac{\omega_2 h_{fg2} + C_p(T_2 - T_1)}{h_{g1} - h_{f2}} \tag{11.28}$$

Note: The measurements can be made with an acceptable degree of accuracy, so the calculated humidity ω_1 is quite accurate.

We used $h_{a2} - h_{a1} = C_p(T_2 - T_1)$ for the dry air component and $h_{fg2} = h_{g2} - h_{f2}$.

An expression for the humidity ratio in atmospheric air is now available using the measurable properties T_1, T_2, and P, with ω_2 calculated from Eq. 11.27

and with the remaining properties found in Appendix C. Before we apply the equations developed in this section, some terms should be defined. Then Example 11.7 will apply the equations of this section.

☑ **You have completed Learning Outcome (3)**

11.2.3 Psychrometrics

Psychrometrics is the term used to describe the analysis of engineering problems that involve determining the physical properties of air considered as a gas-vapor mixture. It is also referred to as psychrometry. Much attention is focused on relative humidity, which, rather than using the complicated setup of Fig. 11.9, or an inaccurate hygrometer, can be determined with sufficient accuracy using a *psychrometer*, a device that utilizes two thermometers. An accepted technique for determining humidity uses an instrument called a *sling psychrometer*. This device has two thermometers mounted parallel on a rather narrow board, as shown in Fig. 11.11, attached by a bearing to a handle. The user holds the handle and swings the device in a circle. One of the thermometers is exposed to the atmosphere and reads the atmospheric temperature, which in psychrometrics is called the *dry-bulb temperature* T_{db}. The bulb of the other thermometer is enclosed in a cloth-wicking material that is saturated with water. The temperature measured with this bulb is the temperature at which the water evaporates from the wick, which is the saturation temperature called the *wet-bulb temperature* T_{wb}; it is approximately equal to the adiabatic saturation temperature T_2 of Fig. 11.9.

Before the operation of the sling psychrometer is described, a third temperature, the *dew point temperature* T_{dp}, in the study of psychrometrics is defined; it is the temperature at which moisture begins to condense at atmospheric pressure. A good example of the effect of the dew point can be observed with a glass of ice water. If the outside of the glass is at or below the dew point for the existing atmospheric conditions, water will condense on the outside surface of the glass. If the glass stays dry, the temperature of the exterior of the glass is above the dew point. An insulated glass usually keeps the outside surface above the dew point.

The procedure for operating the sling psychrometer is to first soak the wick on the wet-bulb thermometer. Then the psychrometer is swung in a circle for about 30 seconds. The dry-bulb and wet-bulb temperatures are read, and the relative humidity is determined using a *psychrometric chart*. An SI version of the psychrometric chart, developed by the Carrier Corporation, is shown in the Appendix as Table G-1. This chart may appear confusing but can be read with some instruction. The sketch in Fig. 11.12 may be helpful. The dry-bulb temperature T_{db} is plotted on the horizontal axis and the humidity ratio ω (moisture content: kg_{vapor}/kg_{air}) on the right-side vertical axis. The wet-bulb temperatures T_{wb} are plotted as straight lines that slope downward from left to right. Dew-point temperatures T_{dp} are horizontal lines with values the same as the wet-bulb temperature values on the heavy left curved line. Relative humidity is plotted on curves that extend from the lower left-hand corner to the top right-hand side of the chart. State A can be located by knowing any of the two properties, for example, T_{db} and T_{wb}, provided by the psychrometer. Other properties are then read from the chart. It should be emphasized that the chart in Appendix G is for 100 kPa only. At elevations

Psychrometer: A device that measures the humidity. This device is much simpler than the setup of Fig. 11.9, but it is not as accurate.

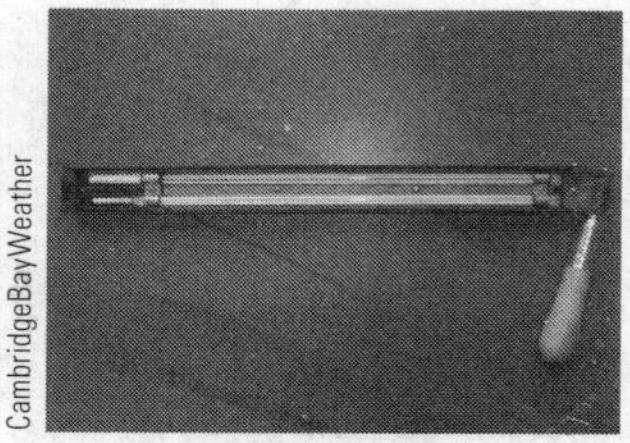

Figure 11.11
A sling psychrometer used to measure humidity.

Dry-bulb temperature T_{db}: The temperature of the air measured with an ordinary thermometer.

Wet-bulb temperature T_{wb}: The temperature reached by a wet material in an air stream, approximately equal to the adiabatic saturation temperature.

Dew point temperature, T_{dp}: The temperature at which water vapor just begins to condense in atmospheric air.

Psychrometric chart. A chart that relates the dry-bulb temperature, wet-bulb temperature, dew-point temperature, humidity, specific humidity, and enthalpy at a particular atmospheric pressure.

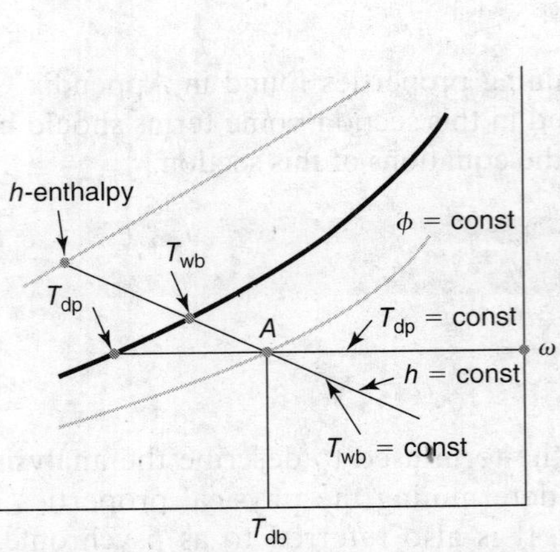

Figure 11.12

A sketch of the psychrometric chart.

Warning: The psychrometric chart in Appendix G is for 100 kPa only. It can be used for reasonable elevations but for significant elevations, charts for those elevations should be used.

where the atmospheric pressures are significantly different from 100 kPa, charts for those elevations must be used.

Humidity curves ranging from 10% to 90% are drawn on the charts in Appendix G. The top-most curved line is not labeled but represents the 100% relative-humidity line. The specific enthalpy of the mixture is plotted on sloped lines that are parallel to the wet-bulb temperature lines. It has units of kJ/kg of dry air. The specific volume of the dry air is also plotted on sloped lines somewhat steeper than the specific-enthalpy lines but is of only occasional use in our study.

Example 11.7 The determination of ω and ϕ

The dry-bulb and wet-bulb temperatures are measured using a sling psychrometer in atmospheric air to be 30°C and 25°C, respectively. Use the atmospheric pressure as 100 kPa and calculate the humidity ratio, the relative humidity, and the specific enthalpy entering the channel of Fig. 11.9.

Solution

The entering and exiting states are:

State 1: $T_1 = 30°\text{C},\ h_{g1} = 2556\ \text{kJ/kg}$
State 2: $T_2 = 25°\text{C},\ h_{f2} = 105\ \text{kJ/kg},\ h_{fg2} = 2442\ \text{kJ/kg}$

The exiting temperature from Fig. 11.9 would be the wet-bulb temperature since $\phi_2 = 1.0$. The properties above were found in Table C-1. The humidity ratio at the exit, obtaining $P_{g2} = 3.169$ kPa from Table C-1, is

$$\omega_2 = 0.622\frac{P_{g2}}{P - P_{g2}} = 0.622 \times \frac{3.169}{100 - 3.169} = 0.0204\ \frac{\text{kg}_v}{\text{kg}_a}$$

The humidity ratio can now be calculated. It is

$$\omega_1 = \frac{\omega_2 h_{fg2} + C_p(T_2 - T_1)}{h_{g1} - h_{f2}}$$

$$= \frac{0.0204 \times 2442 + 1.0 \times (25 - 30)}{2556 - 105} = \underline{0.0183 \text{ kg}_v/\text{kg}_a}$$

The relative humidity is calculated as follows:

$$\omega_1 = 0.622\frac{P_{v1}}{P_{a1}} \qquad 0.0183 = 0.622\frac{P_{v1}}{100 - P_{v1}}$$

$$\therefore P_{v1} = 2.86 \text{ kPa and } \phi_1 = \frac{P_{v1}}{P_{g1}} = \frac{2.86}{4.246} = 0.673 \text{ or } \underline{67.3\%}$$

where P_{g1} was found in Table C-1 at 30°C.

The specific enthalpy for this entering two-component mixture of dry air and water vapor is found to be

$$h_1 = h_{a1} + h_{v1} \cong C_p T_1 + \omega_1 h_{g1}$$

$$= 1.0 \times 30\frac{\text{kJ}}{\text{kg}_a} + 0.0183\frac{\text{kg}_v}{\text{kg}_a} \times 2556\frac{\text{kJ}}{\text{kg}_v} = \underline{76.8 \text{ kJ/kg}_a}$$

Comment

The temperature is used as degrees celcius as it is on the psychrometric chart, which is acceptable since it is the difference $h_2 - h_1$ that's of interest.

Example 11.8 Check the results of Example 11.7

Use the psychrometric chart and check the results of Example 11.7 using the dry-bulb and wet-bulb temperatures measured in atmospheric air to be 30°C and 25°C, respectively, as shown in Fig. 11.13. Find the humidity ratio, the relative humidity, and the specific enthalpy of state 1 of that example.

Solution

The psychrometric chart of Fig. G-1 in the Appendix at T_{db} = 30°C and T_{wb} = 25°C provides the following properties:

$$\omega_1 = \underline{0.018 \text{ kg}_v/\text{kg}_a}, \quad \phi_1 = \underline{67\%}, \quad \text{and} \quad h_1 = \underline{76.5 \text{ kJ/kg}_a}$$

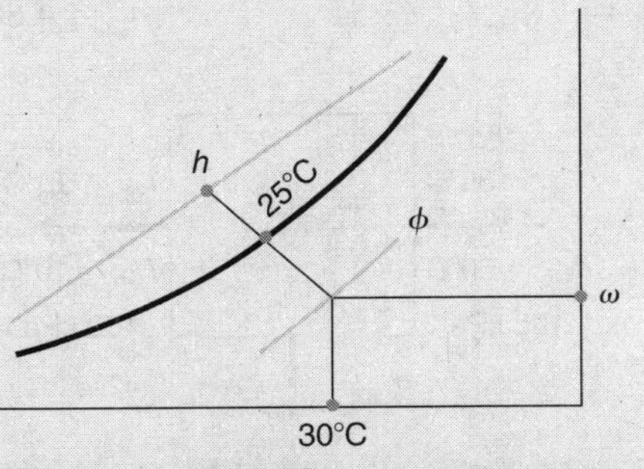

Figure 11.13

(*Continued*)

Example 11.8 (*Continued*)

Comment

The psychrometric chart provides quick results that are sufficiently accurate for engineering calculations.

Obviously, using the chart is much easier than solving the equations, and it is quite accurate. It has an error of only 1.6% for ω_1 and 0.4% for ϕ_1 and h_1. A large chart would allow for even increased accuracy, although most often the properties using the relatively small charts in Appendix G are sufficiently accurate.

Example 11.9 The removal of water vapor by an air conditioner

A large commercial air-conditioning unit of Fig. 11.14 processes 5 kg/s of atmospheric air. The air enters at a pressure of 105 kPa. Its dry-bulb temperature is 35°C, and its wet-bulb temperature is 30°C. The conditioned air leaves the unit at a dry-bulb temperature of 20°C, a wet-bulb temperature of 10°C, and a pressure of 100 kPa. How much water is being removed from the air in one hour? Convert the answer to liters/hr.

Solution

The psychometric chart in Fig. G-1 will be used to determine the relative humidity and the humidity ratio for the air entering and leaving the air conditioner.

Note: The pressure of 105 kPa is sufficiently close to 100 kPa, so the chart provides sufficiently accurate properties.

Inlet Air: For $T_{db} = 35°C$ and $T_{wb} = 30°C$, the psychrometric chart provides $\phi_{in} = 0.70$ and $\omega_{in} = 0.0253\ kg_v/kg_a$.

Exit Air: For $T_{db} = 20°C$ and $T_{wb} = 10°C$, the psychrometric chart is read to obtain $\phi = 0.26$ and $\omega_{out} = 0.0036\ kg_v/kg_a$.

The mass flux of atmospheric air entering the air conditioner is

$$\dot{m}_a + \dot{m}_v = 5\ \text{kg/s}$$

Using Eq. 11.19, this allows the mass flux of dry air to be determined. It is

$$\dot{m}_a + \omega\dot{m}_a = \dot{m}_{in}$$

$$\dot{m}_a + 0.0253\dot{m}_a = 5\ \text{kg}_a/\text{s} \qquad \therefore \dot{m}_a = 4.88\ \text{kg}_a/\text{s}$$

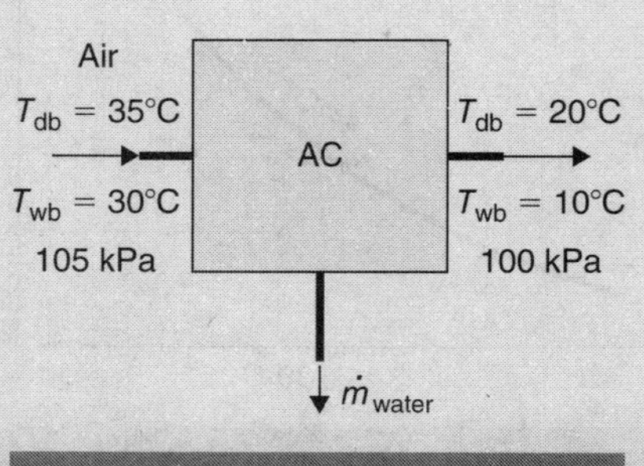

Figure 11.14

The mass flux of water vapor entering the air conditioner is

$$\dot{m}_{v,\text{in}} = 5 - 4.88 = 0.12 \text{ kg}_v/\text{s}$$

> **Note:** The chart was read to two significant digits, so the answer is expressed to two significant digits.

The mass flux of dry air is conserved, so the mass flux exiting the air conditioner is also 4.88 kg_a/s. This allows the mass flux of the water vapor leaving the air conditioner to be found:

$$\dot{m}_{v,\text{out}} = \omega_{\text{out}}\dot{m}_a = 0.0036 \times 4.88 = 0.0176 \text{ kg}_v/\text{s}$$

The conservation of the water vapor component leads to

$$\dot{m}_{\text{water}} = \dot{m}_{v,\text{in}} - \dot{m}_{v,\text{out}}$$
$$= 0.12 - 0.0176 = 0.102 \text{ kg}_v/\text{s} \quad \text{or} \quad \underline{370 \text{ kg}_v/\text{hr}}$$

☑ **You have completed Learning Outcome (4)**

11.3 Air-Conditioning Processes

The air in a space where people either reside or work must be in a specified *"comfort zone"* if they are to be comfortable. The zone may be larger or smaller depending on the time of year (in the summer the temperatures would be somewhat higher than in the winter) and the activity of the people. A reasonable comfort zone is inside the dashed lines on the psychrometric chart sketched in Fig. 11.15. The objective of an *air-conditioning process* is to move the air from an uncomfortable condition outside the comfort zone into a comfortable condition inside the comfort zone. The most expensive move is from a high-humidity, high-temperature condition, such as state F into the comfort zone at state I, a Florida situation that decreases moisture. From a dry, hot condition, a less expensive move would be from state J to state K, an Arizona situation that adds moisture. The more common moves are summarized as follows:

> **Comfort zone:** The zone in which the temperature and humidity allow for reasonable comfort.
>
> **Air-conditioning process:** A process that moves air from an uncomfortable condition into the comfort zone.

- **Heating**—Air is too cold: Heat is added to move from state A to state C. Observe that the humidity decreases without removing moisture.
- **Cooling**—Air is too hot: Heat is removed to move from state B to state C. Observe that the humidity increases and moisture is not added.
- **Heating and humidifying**—Air is too cold, and the humidity is too low: Heat is added to move from state D to state E, and then moisture is added to move to state C. Observe that moisture must be added to enter the comfort zone.
- **Cooling and dehumidifying**—Air is too hot and too humid: First, cool the air from state F to state G. Then continue to cool the air and remove moisture

> **Sensible heat:** When the temperature is changed with no change in moisture content, as from state D to state E. It is represented by $Q = mC_p\Delta T$.

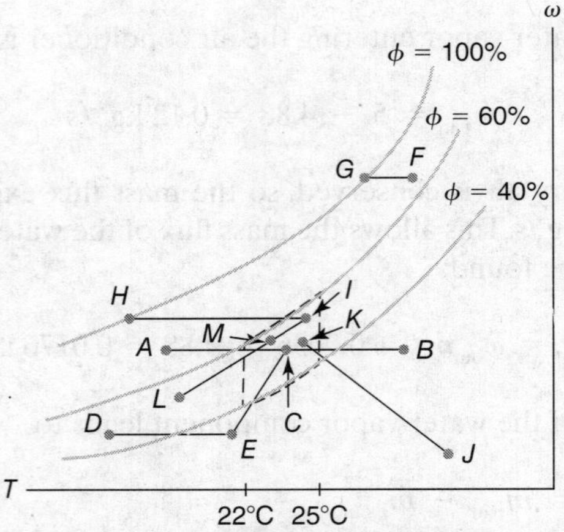

Figure 11.15

Air-conditioning processes.

Latent heat cooling: When the temperature remains unchanged but moisture is removed. A dehumidifier can remove moisture and move a state from directly above the comfort zone into the comfort zone. This process is not shown.

until state H is reached. Finally, heat the air into the comfort zone at state I. This is typical in very humid locations, such as in Orlando, Florida.

- **Evaporative cooling**—Air is too hot, and the humidity is too low: Moisture is added while the air moves from state J to state K. If a state in the comfort zone is not reached, heat can be added. This is a swamp cooler used in very dry locations, such as Phoenix, Arizona. Observe that if moisture is not added when the hot dry air is cooled, it cannot enter the comfort zone.
- **Mixing outside and inside air**—Air is too warm inside and too cool outside: A not-too-humid air stream at state L from outside is mixed with an air stream at state I from inside, with state M representing the final state. Often, state M is in the comfort zone, so no conditioning of the air is needed. Energy and mass balances of the air streams would show that state M lies on a straight line connecting states L and I.

Note: The mass and energy equations would show that the ratio of the distances $d_{I\text{-}M}/d_{L\text{-}M}$ is equal to the ratio of the mass fluxes $\dot{m}_{aL}/\dot{m}_{aI}$.

Several examples will illustrate the above processes. No additional equations are needed.

Example 11.10 Heating

Outside air at 8°C and 80% humidity is to be heated to 22°C, as shown in Fig. 11.16. For a mass flux of 2 kg$_a$/s, estimate the required rate of heat transfer and the exiting relative humidity.

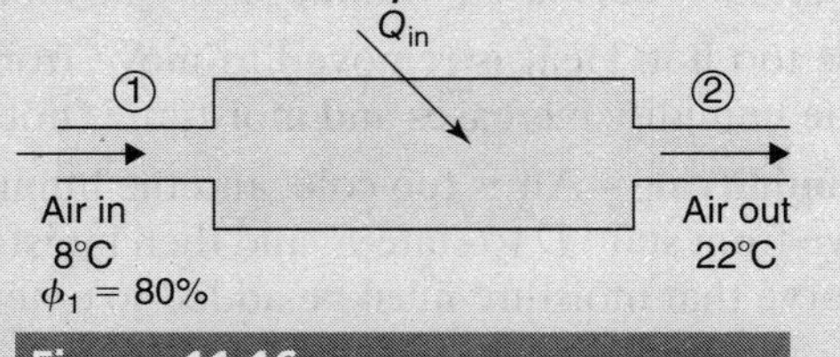

Figure 11.16

Solution

The rate of heat transfer is found by applying the first law. There results

$$\dot{Q}_{in} = \dot{m}(h_2 - h_1) = 2(36 - 21) = \underline{30 \text{ kJ/s}}$$

where the enthalpies were found on the psychrometric chart in Fig. G-1. The process follows path A-C on the chart. The relative humidity is read to be

$$\phi_{out} = \underline{33\%}$$

Observe the significant decrease in humidity when the temperature of the air is increased. When air is heated (this is sensible heat), as in homes in the winter, if humidification is not available, the dry air can lead to health problems, dried-out wooden joints on furniture and musical instruments, cracks in the walls, and so on.

Note: The chart is quite small, so it can be read to only two significant digits. A larger chart would allow three digits.

Heating and humidifying — Example 11.11

Outside air at 10°C and 20% humidity is to be heated to 22°C and humidified to 50% humidity, as shown in Fig. 11.17. For a mass flux of outside air of 2 kg_a/s, find the required heat transfer rate and the amount of water added in one hour. Assume that the small amount of added moisture does not significantly affect the first law.

Solution

The process follows path D-E-C in Fig. 11.15. The rate of heat transfer is

$$\dot{Q}_{in} = \dot{m}(h_2 - h_1) = 2(43 - 15) = \underline{56 \text{ kJ/s}}$$

where h_1 was extrapolated on the psychrometric chart.

The amount of water that must be added is found by using the difference in the humidity ratios multiplied by the mass flux of air. It is found to be

$$\begin{aligned} \dot{m}_{water} &= \dot{m}_a(\omega_2 - \omega_1) \\ &= 2\frac{\text{kg}_a}{\text{s}} \times (0.0082 - 0.0016)\frac{\text{kg}_v}{\text{kg}_a} = 0.0132 \text{ kg}_v\text{/s} \\ &\text{or} \quad \underline{48 \text{ kg}_v\text{/hr}} \end{aligned}$$

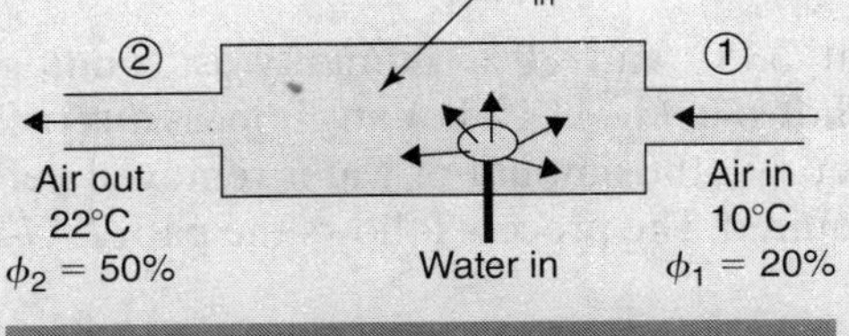

Figure 11.17

Example 11.12 Evaporative cooling

Dry air at 37°C and 10% humidity is passed through a series of water-laden wicks of Fig. 11.18 until the temperature reaches 24°C. Estimate the exiting humidity and the minimum mass of water each hour passing through the wicks if the air flows at 900 kg_a/hr.

Solution

The air follows path *J-K* in Fig. 11.15. No heat is added or extracted as the air flows through the device. From the first law, the enthalpy remains constant. On the psychrometric chart, follow the $h_2 = h_1 = 48$ kJ/kg_a line (the intersection of $T = 37$°C and $\phi = 10\%$) until the vertical line at 24°C is intersected. The humidity reads

$$\underline{\phi = 50\%}$$

The amount of water that is absorbed in the air is

$$\dot{m}_{\text{water}} = \dot{m}_a(\omega_2 - \omega_1)$$

$$= 900\,\frac{\text{kg}_a}{\text{hr}} \times (0.0093 - 0.0039)\,\frac{\text{kg}_w}{\text{kg}_a}$$

$$= \underline{4.9\ \text{kg}_w/\text{hr}}$$

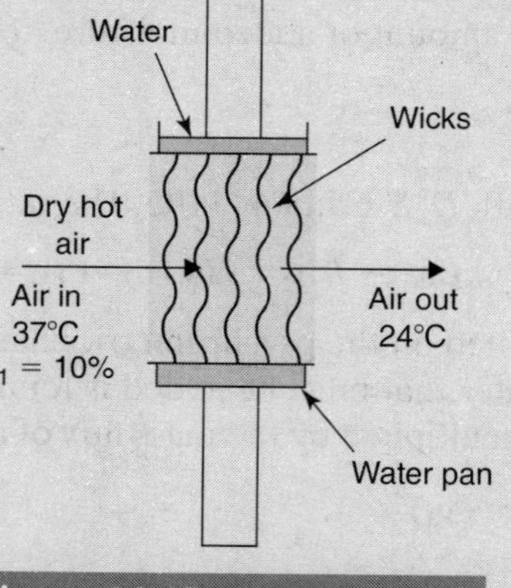

Figure 11.18

Example 11.13 Conditioning hot, humid air

Hot, humid air at 33°C and 80% humidity is conditioned to 22°C and 50% humidity, as shown in Fig. 11.19. For an air mass flux of 2 kg_a/s, estimate the rate of heat transfer out, the amount of water removed per hour, and the rate of heat transfer required. The process follows the path *F-G-H-I* in Fig. 11.15.

Solution

The first step in conditioning hot, humid air is to wring the moisture out of the air. That requires heat to be removed as the air is cooled from state *F* until the

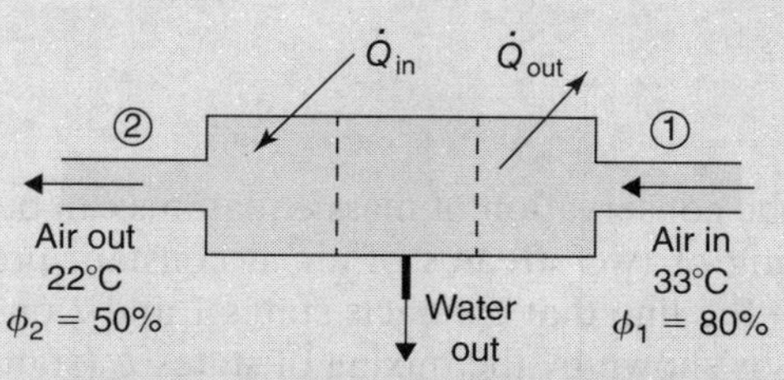

Figure 11.19

dew point is reached at state G and then to continue to state H. The amount of heat removed each second is

$$\dot{Q}_{out} = \dot{m}_a(h_F - h_H) = 2\frac{\text{kg}_a}{\text{s}} \times (99 - 32)\frac{\text{kJ}}{\text{kg}_a} = \underline{134 \text{ kJ/s}}$$

The amount of water removed per hour is

$$\dot{m}_{water} = \dot{m}_a(\omega_G - \omega_H)$$

$$= 2\frac{\text{kg}_a}{\text{s}} \times (0.0258 - 0.0082)\frac{\text{kg}_v}{\text{kg}_a} = 0.0352 \text{ kg}_v\text{/s}$$

$$\text{or} \quad \underline{127 \text{ kg}_v\text{/hr}}$$

Heat is then required to condition the air to the desired state. It is

$$\dot{Q}_{in} = \dot{m}_a(h_I - h_H) = 2\frac{\text{kg}_a}{\text{s}} \times (43 - 32)\frac{\text{kJ}}{\text{kg}_a} = \underline{22 \text{ kJ/s}}$$

Mixing of air streams Example 11.14

Cool air at 10°C and 20% humidity is mixed with inside air taken from near the ceiling at 30°C and 65% humidity. The mass flow rates are shown in Fig. 11.20. Estimate the temperature, the relative humidity, and the volume flow rate of the exiting stream at state 3. Assume the process occurs at 100 kPa.

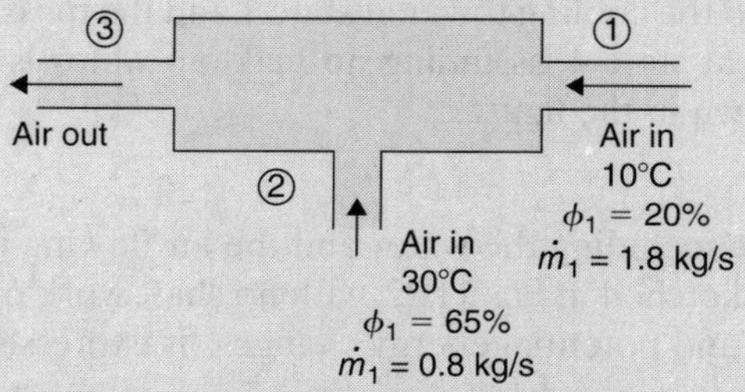

Figure 11.20

(Continued)

Example 11.14 (Continued)

Solution

The first law and the conservation of mass equations can be used to show that the adiabatic mixing of two streams of air at constant pressure occurs such that state 3 lies on the line that connects states 1 and 2 on the psychrometric chart. This process is shown by the mixing of states L (state 1) and I (state 2), resulting in state M (state 3) on the chart sketched in Fig. 11.15. The equations would also show that the ratio of the distances $d_{2\text{-}3}/d_{1\text{-}3}$ is equal to the ratio of the mass fluxes $\dot{m}_{a1}/\dot{m}_{a2}$.

Since states 1 and 2 are given in Fig. 11.20, state 3 can be located by physically measuring the distances on the psychrometric chart. There results

$$\frac{d_{2\text{-}3}}{d_{1\text{-}3}} = \frac{\dot{m}_1}{\dot{m}_2} = \frac{1.8}{0.8} = 2.25$$

The distance measured on Fig. G-1 was 81 mm. So,

$$\left.\begin{aligned} d_{2\text{-}3} &= 2.25d_{1\text{-}3} \\ d_{2\text{-}3} + d_{1\text{-}3} &= 81 \end{aligned}\right\} \quad \therefore d_{2\text{-}3} = 56 \text{ mm} \quad \text{and} \quad d_{1\text{-}3} = 25 \text{ mm}$$

Locate state 3 on the chart by connecting state 1 and state 2 with a straight line and measure 25 mm from state 1 to locate state 3. At state 3 we find $T = \underline{16.3°\text{C}}$ and $\phi = \underline{57\%}$.

Using Eqs. 4.3 and 4.8, we find that the volume flow rate, using $\dot{m}_{a3} = 0.8 + 1.8 = 2.6$ kg/s, is

$$\dot{V} = \dot{m}_{a3}v_{a3} = 2.6 \times 0.83 = \underline{2.16 \text{ m}^3/\text{s}}$$

where v_{a3} was estimated on the psychrometric chart.

Example 11.15 A cooling tower

The heat that is removed by a condenser ($\dot{Q}_C$ of Fig. 8.5) in a Rankine power cycle is often done by the hot water from the condenser being sprayed in a cooling tower, as sketched in Fig. 11.21. The cooler water then returns to the heat exchanger in the condenser to complete the cycle. Calculate the required mass flux of air into the cooling tower at state 1 and the mass flux of water from the cooling tower at state 4, assuming no makeup water is available. Needed properties are shown in the figure.

Comment

A cooling tower uses the concept of evaporative cooling. The water mixing with the air tends to cool the air. There is also an enthalpy exchange between the warm water and the cooler air.

Solution

An energy balance including the water and the air flowing into and out of the cooling tower, as sketched in Fig. 11.22, with no shaft work or heat transfer and neglecting kinetic and potential energy changes, is expressed as

$$\dot{E}_{\text{in}} = \dot{E}_{\text{out}}$$

$$\dot{m}_{a1}h_1 + \dot{m}_{w3}h_3 = \dot{m}_{w4}h_4 + \dot{m}_{a2}h_2$$

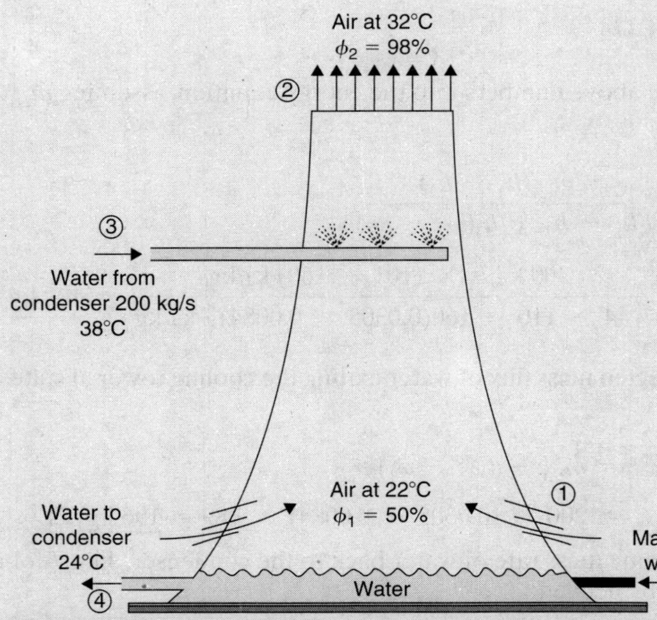

Figure 11.21

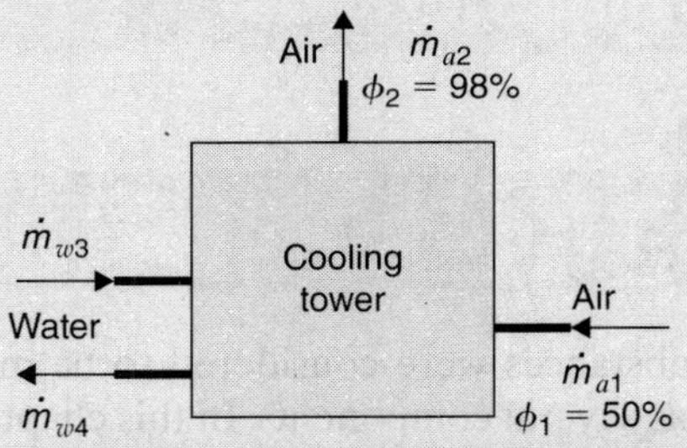

Figure 11.22

Comment

The mass flux of water exiting at section 4 is less than the mass flux that enters at section 3 because more moisture is leaving through the top of the cooling tower at section 2 than enters the vents at section 1.

The mass flux of dry air in equals the mass flux of dry air out, as does the mass flux of water in and out. These conservation of mass equations are

$$\dot{m}_{a1} = \dot{m}_{a2} = \dot{m}_a \qquad \text{for dry air}$$

$$\dot{m}_{w3} = \dot{m}_{w4} + (\omega_2 - \omega_1)\dot{m}_a \qquad \text{for water}$$

From the psychrometric chart and the steam tables we find

$$h_1 = 44 \text{ kJ/kg}_a \qquad \omega_1 = 0.0084 \text{ kg}_v/\text{kg}_a$$

$$h_2 = 110 \text{ kJ/kg}_a \qquad \omega_2 = 0.0305 \text{ kg}_v/\text{kg}_a$$

$$h_3 = h_{f@38°C} = 160 \text{ kJ/kg}_a$$

$$h_4 = h_{f@24°C} = 101 \text{ kJ/kg}_a$$

Note: We use kg_a to represent the mass of dry air.

(*Continued*)

Example **11.15** *(Continued)*

Substitute the above numbers into the energy equation, assuming $\dot{m}_{w4} \cong \dot{m}_{w3}$, to obtain

$$\dot{m}_a = \frac{\dot{m}_{w4}(h_4 - h_3)}{h_1 - h_2 + h_3(\omega_2 - \omega_1)}$$

$$= \frac{200\ \text{kg}_a/\text{s} \times (101 - 160)\ \text{kJ/kg}_a}{[44 - 110 + 160(0.0305 - 0.0084)]\ \text{kJ/kg}_a} = \underline{189\ \text{kg}_a/\text{s}}$$

The corrected mass flux of water exiting the cooling tower at state 4 is now found to be

$$\dot{m}_{w4} = \dot{m}_{w3} - (\omega_2 - \omega_1)\dot{m}_a$$

$$= 200 - (0.0305 - 0.0084) \times 189 = \underline{196\ \text{kg}_w/\text{s}}$$

To send the same mass rate of water back to the condenser, 4 kg_w/s of makeup water would be needed.

Note: We use kg_w to represent the mass of water.

☑ You have completed Learning Outcome (5)

11.4 Summary

Before this chapter, substances were considered to be made up of one component, even air, which has several components. In this chapter, the analysis of gases made up of several components was presented, with attention focused on the water vapor contained in air. Because air contains so little water vapor, it could be ignored in the problems of previous chapters with little if any consequences. But when air-conditioning processes (or combustion in Chapter 12) are concerned, the water vapor component becomes extremely significant. It must be accounted for in engineering calculations. The analysis and equations presented in this chapter allow a basis for those calculations.

The following additional terms were introduced or emphasized in this chapter:

Adiabatic saturation temperature: *The temperature of atmospheric air after it has been saturated adiabatically.*

Amagat law: *Each component is at the same pressure and temperature as the mixture.*

Atmospheric air: *Air that contains water vapor.*

Comfort zone: *The zone in which the temperature and humidity allow for reasonable comfort.*

Dalton law: *Each component has the same volume and temperature as the mixture.*

Dew point: *The temperature at which water vapor just begins to condense in atmospheric air.*

Dry air: *Air that contains no water vapor.*

Dry-bulb temperature: *The atmospheric temperature measured by a conventional thermometer.*

Humidity ratio: *The mass of water vapor in a unit mass of air.*

Hygrometer: *An instrument that measures relative humidity.*

Mass fraction: *The ratio of the mass of a component to the total mass.*

Mixture: *A substance that contains two or more unique components.*

Molar mass: *One kmol of a substance has a mass in kilograms equal to its molar mass.*

Mole: *A unit of measurement that expresses the amount of a substance.*

Mole fraction: *The ratio of the moles of a component to the total moles.*

Partial pressures: *The sum of the partial pressures is the total pressure.*

Psychrometer: *A device that measures the relative humidity.*

Psychrometrics: *The term used when it is necessary to consider air as a gas-vapor mixture.*

Relative humidity: *The ratio of the mass of the vapor to the mass of vapor in saturated air at the same temperature.*

Wet-bulb temperature: *The temperature reached by a wet material in an air stream, approximately the adiabatic saturation temperature.*

Several important equations were introduced in this chapter:

Component moles: $$N_i = \frac{m_i}{M_i}$$

Mass fraction: $$x_i = \frac{m_i}{m}$$

Mole fraction: $$y_i = \frac{N_i}{N}$$

Fraction properties: $$\sum_i^n x_i = 1 \qquad \sum_i^n y_i = 1$$

Average molar mass: $$M = \sum_i^n y_i M_i$$

Fraction relation: $$x_i = y_i \frac{M_i}{M}$$

Partial pressures: $$P = \sum_i^n P_i$$

Partial pressure: $$\frac{P_i}{P} = y_i$$

Mixture property:	$\overline{h} = \sum_i^n y_i \overline{h}_i, \quad h = \sum_i^n y_i h_i$
Molar properties:	$\overline{h}_i = M_i h_i \quad \overline{C}_{p,i} = M_i C_{p,i}$
Total pressure:	$P = P_a + P_v$
Humidity ratio:	$\omega = \dfrac{m_{vapor}}{m_{air}} = 0.622\dfrac{P_v}{P_a}$
Relative humidity:	$\phi = \dfrac{m_v}{m_g} = \dfrac{P_v}{P_g}$
Humidity relations:	$\phi = \dfrac{\omega P}{(0.622 + \omega)P_g} \quad \omega = \dfrac{0.622\phi P_g}{P - \phi P_g}$
Humidity ratio:	$\omega_1 = \dfrac{\omega_2 h_{fg2} + C_p(T_2 - T_1)}{h_{g1} - h_{f2}}$

Problems

FE Exam-Type Questions

11.1 A gas is composed of N_2, O_2, and CO_2. The mole fractions are $y_{N_2} = 0.6$, $y_{O_2} = 0.3$, and $y_{CO_2} = 0.1$. Its molar mass is nearest:

(A) 29.6

(B) 30.8

(C) 33.5

(D) 35.1

11.2 The mass of oxygen in 5 kg of the gas described in Problem 11.1 is nearest:

(A) 1.35 kg

(B) 1.42 kg

(C) 1.56 kg

(D) 1.68 kg

11.3 A tank at 400 kPa and 25°C contains 10 kg of nitrogen and 5 kg of oxygen. The partial pressure of the nitrogen is nearest:

(A) 251 kPa

(B) 266 kPa

(C) 278 kPa

(D) 291 kPa

11.4 In what nation would there least likely be dew on the grass in a summer morning?

(A) France

(B) South Africa

(C) England

(D) Morocco

11.5 A glass of soda-pop with ice in it is sitting on the table of Figure 11.23. What would you expect if the humidity is 40%, the outside of the glass is at 15°C, and the temperature in the room is 22°C? You are near sea level.

(A) Moisture would collect on the glass.

(B) The glass would remain dry.

(C) Additional information is needed.

(D) If I drink the soda-pop, then moisture would form.

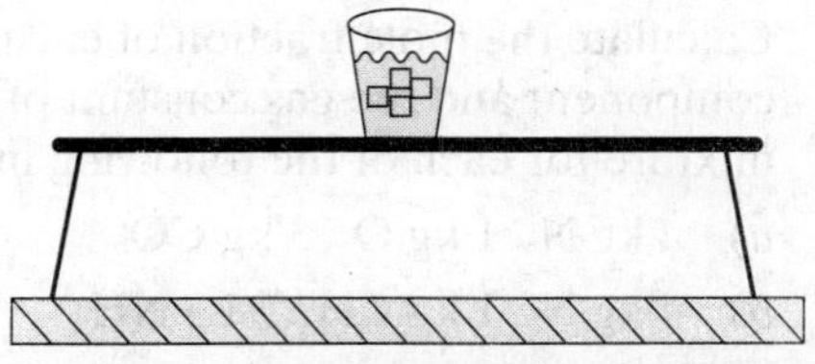

Figure 11.23

Remember: "Humidity" refers to "relative humidity."

11.6 The dry-bulb and wet-bulb temperatures are measured in atmospheric air to be 25°C and 20°C, respectively. If the atmospheric pressure is 100 kPa, the humidity ratio is nearest:

(A) 0.0129 kg vapor/kg dry air
(B) 0.0128 kg vapor/kg dry air
(C) 0.0127 kg vapor/kg dry air
(D) 0.0126 kg vapor/kg dry air

Hint: The equations are needed to obtain the desired accuracy.

11.7 Outside air at 5°C and 80% humidity is heated to 22°C. The humidity of the heated air is nearest:

(A) 27%
(B) 31%
(C) 35%
(D) 38%

Notes: The psychrometric chart provides results that are sufficiently accurate. Pressure is assumed to be 100 kPa if not stated otherwise.

11.8 Dry air at 44°C and 10% humidity is cooled by the swamp cooler of Fig. 11.24. The lowest possible temperature at the exit of the swamp cooler is nearest:

(A) 34°C
(B) 29°C
(C) 25°C
(D) 21°C

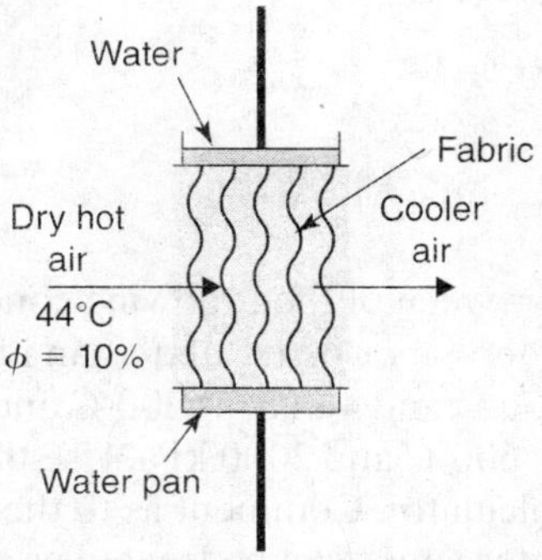

Figure 11.24

11.9 A flow rate of 40 m^3/s of outside air at 10°C and 40% humidity is mixed with 20 m^3/s of inside air from near the ceiling of an industrial plant at 36°C and 70% humidity. The temperature of the mixed air stream is nearest:

(A) 16°C
(B) 18°C
(C) 21°C
(D) 23°C

11.10 A classroom measures 30 m by 10 m and is 3 m in height. If the temperature is 20°C and the humidity is 60%, the amount of water in the air is nearest:

(A) 6.9 kg
(B) 7.1 kg
(C) 8.7 kg
(D) 9.4 kg

11.11 The temperature is 20°C and the humidity is 80% in a classroom that measures 30 m by 10 m by 3 m in height. If the temperature is reduced to 10°C with an airconditioner, the amount of water that will condense out is nearest:

(A) 5 kg
(B) 4 kg
(C) 3 kg
(D) 2 kg

Gas Mixtures

11.12 For which of the following conditions can water vapor be treated as an ideal gas? i) 30°C and 4 kPa, ii) 200°C and 120 kPa, and iii) 600°C and 2000 kPa. Use the IRC Calculator. Comment as to the suggestion that water vapor can be treated as an ideal gas.

Work all parts of problems identified by i), ii), iii), iv), etc.

11.13 A mixture of 4 kg of water vapor and 96 kg of dry air exists at 100 kPa and 25°C in a 10-m^3 volume. Determine the molar mass of the mixture and the volume occupied by the vapor. This represents very humid air. Would a golf ball travel further if the air were significantly less humid?

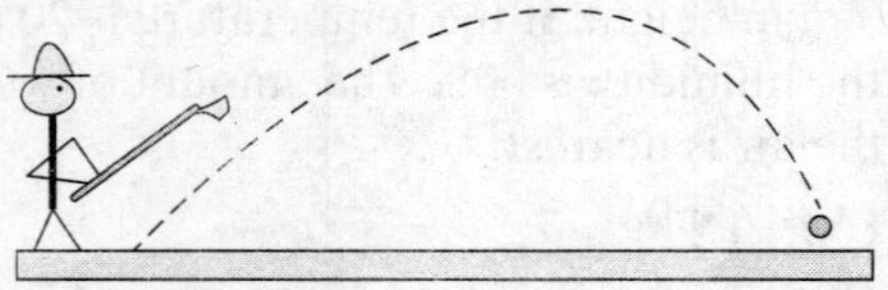

Hint: The drag on the ball is directly proportional to the air density.

11.14 Calculate the mole fraction of each component and the gas constant of the mixture for each of the following mixtures:

a) 4 kg N_2, 1 kg O_2, 3 kg CO_2
b) 4 kg N_2, 1 kg CH_4, 3 kg NH_3
c) 5 kg air, 3 kg CO_2
d) 3 kmol CH_4, 4 kmol N_2, 3 kmol NH_3
e) 5 kmol air, 3 kmol CO_2

Remember: This isn't multiple choice anymore. Each part is a different problem!

11.15 Calculate the mass fraction and mole fraction of each component and the gas constant of the mixture for each of the following mixtures:

a) 2 kg N_2, 3 kg O_2, 5 kg CO_2
b) 3 kmol CH_4, 4 kmol N_2, 3 kmol NH_3
c) 5 kmol air, 3 kmol CO_2
d) 4 kg CH_4, 1 kg N_2, 3 kg NH_3
e) 5 kg air, 3 kg CO_2

11.16 An analysis of a mixture of gases indicates 50% N_2, 40% O_2, and 10% CO_2 at 120 kPa and 40°C. Determine the mixture's gas constant and how many kilograms would be contained in 4 m^3 if the analysis is *a*) gravimetric, and *b*) molar.

11.17 Atmospheric air is assumed to be composed of two ideal gases: dry air and water vapor. If 100 kg of humid atmospheric air contains 97 kg of dry air and 3 kg of water vapor (an extremely moist condition), determine its molar mass. Comment as to the wisdom of assuming atmospheric air to be dry air, as was done in the first 10 chapters.

11.18 Mars has the following molar composition per mole of its atmosphere:

Carbon dioxide	0.955
Nitrogen	0.027
Argon	0.016
Oxygen	0.002

Calculate the mass fraction of each component and the molar mass of the Martian atmosphere.

11.19 A tank contains 50 kg of nitrogen and 50 kg of carbon dioxide. Calculate the mole fraction of each component and the molecular mass of the mixture. What is the gas constant for this mixture?

11.20 The average air pressure on Mars is 600 Pa. For the composition described in Problem 11.18, determine the partial pressure of each constituent.

11.21 If the tank in Problem 11.19 has a volume of 40 m^3 and the contents are at 20°C, calculate the total pressure in the tank. What are the partial pressures of the nitrogen and carbon dioxide?

11.22 A gas mixture of 4 kg of oxygen and 6 kg of carbon dioxide are contained in the cylinder of Fig. 11.25 at a pressure of 420 kPa. Calculate:

i) The mass fraction of each component
ii) The mole fraction of each component
iii) The gas constant for this mixture
iv) The amount of heat needed to raise the mixture temperature from 26°C to 48°C at constant pressure

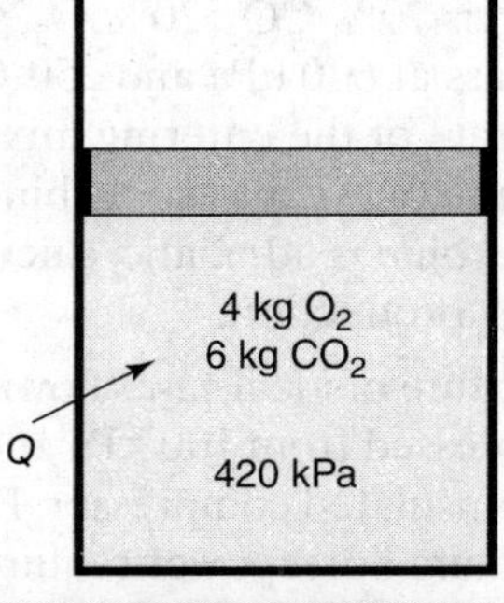

Figure 11.25

Comment

The specific heats are assumed constant unless otherwise stated.

11.23 A gas mixture of 4 kg of O_2, 6 kg of N_2, and 8 kg of CO_2 is contained at a pressure of 120 kPa and 20°C in a rigid container. Calculate:

i) The mass fraction of each component
ii) The mole fraction of each component
iii) The gas constant for this mixture
iv) The amount of heat needed to raise the mixture temperature from 20°C to 100°C

11.24 The partial pressures in a mixture of N_2 and O_2 are 40 kPa and 60 kPa, respectively. If the temperature is 30°C, find the volume of 10 kg of the mixture. How much heat would be required to raise the temperature to 100°C if the volume is held constant? The initial state is shown in Fig. 11.26.

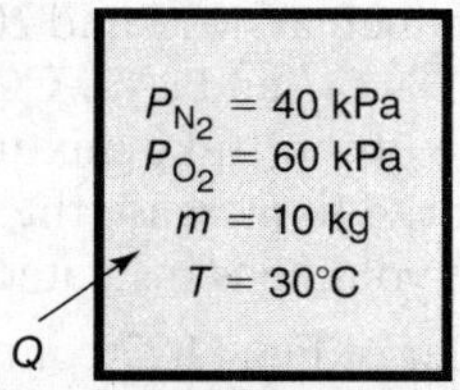

Figure 11.26

11.25 A mixture contains 4 moles of N_2 and 6 moles of O_2. If the temperature is 30°C, find the volume of the mixture if the pressure is 200 kPa. How much heat would be needed to raise the temperature to 100°C at constant pressure?

11.26 A gas mixture of 4 moles of O_2, 6 moles of N_2, and 8 moles of CO_2 are contained at 26°C. Calculate:

i) The mole fraction of each component
ii) The mass fraction of each component
iii) The gas constant for this mixture
iv) The amount of heat needed to raise the mixture temperature to 94°C at constant pressure

11.27 Show that the water vapor in atmospheric air does not significantly influence the change in enthalpy if air with water vapor undergoes a temperature change from 20°C to 400°C at constant pressure. Consider atmospheric air to be composed of 22.5% oxygen, 76% nitrogen, and 1.5% water vapor (all by mass). Compare the results to air with C_p = 1.0 kJ/kg·K and R = 0.287 kJ/kg·K. Calculate the error in Δh if the water vapor is ignored.

Note: The 1.5% water vapor represents 100% humidity.

11.28 A molar analysis of a mixture of ideal gases contained in an 8-m^3 rigid volume at 40°C shows 60% N_2, 30% O_2, and 10% H_2. Determine the pressure of the 10-kg mixture and the heat transfer needed to raise the temperature to 400°C.

11.29 A gravimetric analysis of a mixture of ideal gases contained at 40°C and 200 kPa shows 40% N_2, 35% O_2, and 25% CO_2. Determine the volume of the 10-kg mixture and the work required to increase the temperature to 350°C. The volume is insulated.

11.30 The cylinder of Fig. 11.27 contains a 4-kg mixture of ideal gases at 40°C and 200 kPa, with a molar analysis of 30% N_2, 15% O_2, and 55% H_2. It undergoes a constant-pressure expansion until the volume doubles. Determine the heat transfer and the entropy change.

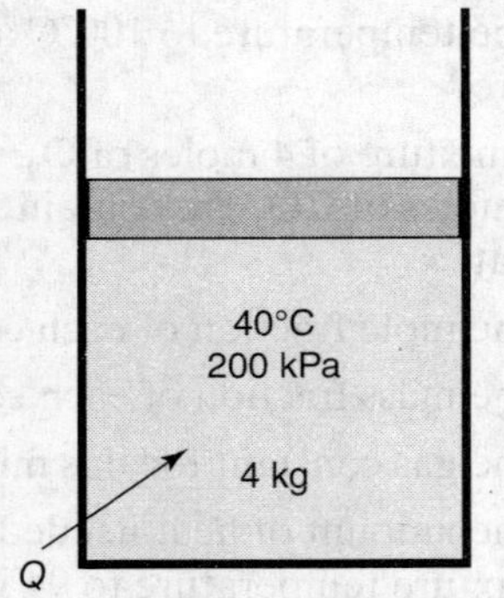

Figure 11.27

11.31 A cylinder contains a 10-kg mixture of ideal gases at 27°C and 280 kPa with a molar analysis of 40% N_2, 25% O_2, and 35% H_2. It undergoes a constant-pressure expansion until the volume doubles to 1.6 m^3. Determine the heat transfer and the entropy change.

11.32 A mixture of gases at 40°C and 400 kPa is contained in a cylinder with an initial volume of 0.2 m^3. The mixture contains 2 kg of CO_2 and 4 kg of air. The pressure is reduced during an isothermal expansion to 100 kPa. Determine the heat transfer and the entropy change for that process.

11.33 The insulated rigid mixing chamber of Fig. 11.28 accepts a 4 kg/min flow of O_2 at 120 kPa and 40°C in one inlet and a 6 kg/min flow of N_2 at 120 kPa and 120°C. Estimate the temperature of the exiting mixed flow, which exits at 120 kPa.

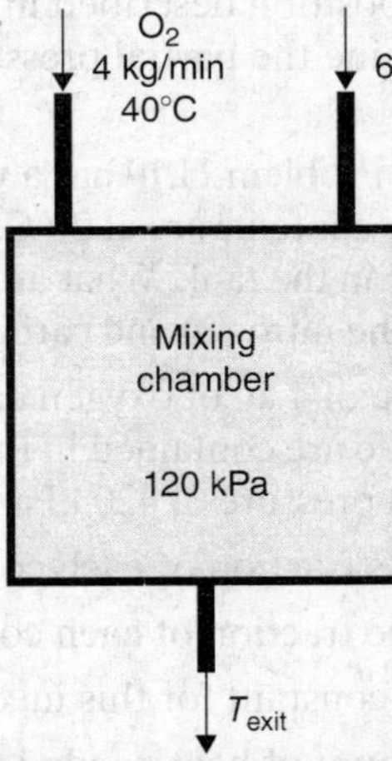

Figure 11.28

11.34 An ideal-gas turbine receives a gas mixture which is 70% CO_2, 20% O_2, and 10% N_2 by mass at 600 kPa and 250°C. The volume flow rate of the entering mixture is 1.5 m^3/s. This mixture exits the turbine at 20°C. If the turbine is adiabatic, calculate the rate of power production.

11.35 A mixture of ideal gases flowing at 50 kg/min is compressed from 100 kPa and 20°C to 600 kPa in an insulated compressor. Determine the minimum horsepower requirement if the gravimetric analysis of the mixture is:

a) 80% N_2 and 20% O_2

b) 20% N_2 and 80% O_2

c) 60% N_2, 30% O_2, and 10% CO_2

d) 40% N_2, 30% O_2, and 30% CO_2

11.36 A mixture of 40% N_2 and 60% O_2 by mass enters the insulated nozzle of Fig. 11.29 at 180°C and 200 kPa with a velocity of 20 m/s. If it exits the nozzle at 60°C, determine the exiting pressure and velocity.

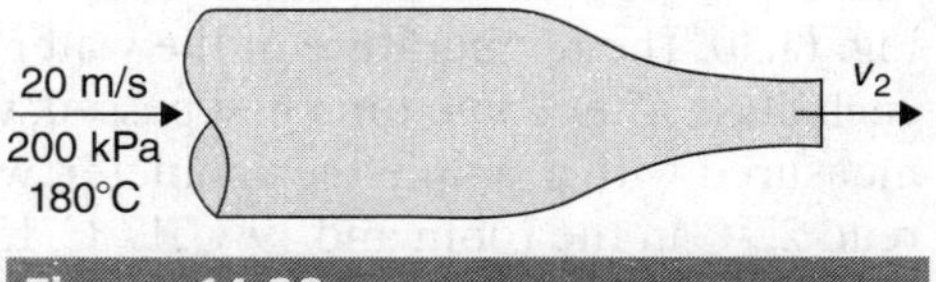

Figure 11.29

11.37 A mixture of 60% N_2 and 40% O_2 by mass enters an insulated nozzle at 90°C and 200 kPa with a velocity of 3 m/s. If it exits the nozzle at 27°C, determine the exiting pressure and velocity.

Air–Vapor Mixtures and Psychrometry

11.38 Calculate the relative humidity, the humidity ratio, and the partial pressure of the dry air if, for the atmospheric air:

a) $T_{db} = 20°C, T_{dp} = 15°C, P_{atm} = 100$ kPa
b) $T_{db} = 30°C, T_{dp} = 20°C, P_{atm} = 104$ kPa

11.39 Calculate the relative humidity, the humidity ratio, and the partial pressure of the dry air if, for the atmospheric air:

a) $T_{db} = 32°C, T_{dp} = 16°C, P_{atm} = 101.3$ kPa
b) $T_{db} = 21°C, T_{dp} = 16°C, P_{atm} = 101.3$ kPa

11.40 Determine the humidity ratio and the partial pressure of the dry air if, for the atmospheric air:

a) $T_{db} = 20°C, P_{atm} = 100$ kPa, and $\phi = 50\%$
b) $T_{db} = 40°C, P_{atm} = 105$ kPa, and $\phi = 20\%$

11.41 Determine the humidity ratio and the partial pressure of the dry air if, for the atmospheric air:

a) $T_{db} = 15°C, P_{atm} = 101$ kPa, and $\phi = 60\%$
b) $T_{db} = 50°C, P_{atm} = 99$ kPa, and $\phi = 15\%$

11.42 Determine the relative humidity and the partial pressure of the dry air if, for the atmospheric air:

a) $T_{db} = 20°C, P_{atm} = 95$ kPa, and $\omega = 0.01\ kg_v/kg_a$
b) $T_{db} = 40°C, P_{atm} = 98$ kPa, and $\omega = 0.02\ kg_v/kg_a$

11.43 Determine the relative humidity and the partial pressure of the dry air if, for the atmospheric air:

a) $T_{db} = 15°C, P_{atm} = 98$ kPa, and $\omega = 0.01\ kg_v/kg_a$
b) $T_{db} = 35°C, P_{atm} = 97$ kPa, and $\omega = 0.02\ kg_v/kg_a$

Note: If the atmospheric pressure is not given, we assume it to be the standard atmospheric pressure of 100 kPa.

11.44 Determine the dew-point temperature of atmospheric air if:

a) $T_{db} = 40°C, P_{atm} = 100$ kPa, and $\phi = 30\%$
b) $T_{db} = 20°C, P_{atm} = 100$ kPa, and $\phi = 60\%$

11.45 Determine the dew-point temperature of atmospheric air if:

a) $T_{db} = 15°C, P_{atm} = 100$ kPa, and $\phi = 70\%$
b) $T_{db} = 32°C, P_{atm} = 100$ kPa, and $\phi = 40\%$

11.46 Air at 20°C and 60% relative humidity enters a heater and exits at 70°C. Calculate the exiting relative humidity assuming a constant pressure of 100 kPa through the heater.

11.47 Air at 100 kPa and 21°C has a relative humidity of 70%. The air is cooled in a constant-pressure process until the water begins to condense. Determine the final temperature of this mixture.

11.48 The air outside is at 20°C. Calculate the humidity ratio and the surface temperature at which water begins to condense for a humidity of *a)* 50%, *b)* 65%, and *c)* 90%.

11.49 Verify that Eq. 11.28 follows from the preceding equations.

11.50 A psychrometer is used to measure a dry-bulb temperature of 22°C and a wet-bulb temperature of 15°C in a 95 kPa atmosphere. Determine the relative humidity and the humidity ratio for this air. What is the specific enthalpy of this air? Do not use the psychrometric chart.

The Psychrometric Chart

11.51 Rework the following problems using the psychrometric chart where helpful:

a) Problem 11.38*a* *b)* Problem 11.38*b*
c) Problem 11.40*a* *d)* Problem 11.42*a*
e) Problem 11.42*b* *f)* Problem 11.50

11.52 Rework the following problems using the psychrometric chart where helpful:

a) Problem 11.39*a* *b)* Problem 11.39*b*
c) Problem 11.41*a* *d)* Problem 11.43*a*
e) Problem 11.43*b* *f)* Problem 11.45*a*

11.53 Use the psychrometric chart to determine the dew point and humidity ratio for air at atmospheric pressure, a temperature of 18°C, and a relative humidity of i) 50%, ii) 70%, and iii) 90%.

11.54 Use the psychrometric chart to provide the missing entries in the following table.

Table 11.1

	T_{db}	T_{wb}	ϕ	ω	T_{dp}	h
a)	20°C		60%			
b)	40°C			0.02		
c)		25°C	60%			
d)	30°C				20°C	

11.55 Two engineering students wanted to determine the humidity in their dorm room. They decided to fill several glasses with tap water and add a little ice to the first glass, a little more to the second glass, and the most to the last of six glasses, as shown in Fig. 11.30. The temperature of the water in each glass, after some time had passed, was measured with a simple thermometer, which read 22°C in the room and 19°C, 17°C, 15°C, 12.5°C, 10°C, and 7°C in the respective glasses. They noticed that moisture was collecting on the outside of each of the last four glasses. Approximate the humidity.

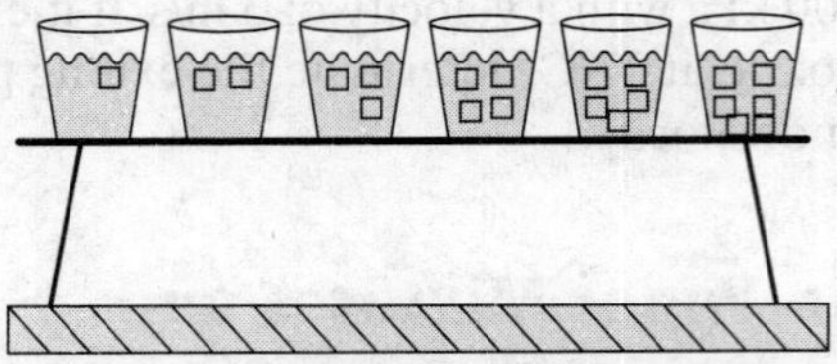

Figure 11.30

11.56 Twenty kilograms of air at 30°C and a humidity of 90% enter a window air-conditioning unit each minute, illustrated in Fig. 11.31. The air leaves the air conditioner at 18°C with a humidity of 40%. What is the change in the humidity ratio for this process? What mass of water is removed from this system each hour? Use the psychrometric chart.

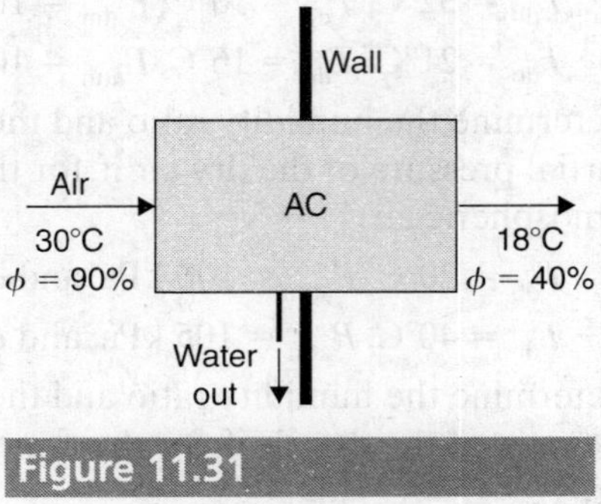

Figure 11.31

11.57 For the data given in Problem 11.56, calculate the rate of change in enthalpy of the air traveling through the air conditioner. Use the psychrometric chart.

11.58 Moist air at 20°C and 100 kPa flows into a dehumidifier at a mass flow rate of 75 kg/min, as shown in Fig. 11.32. If the air is completely dehumidified, how much water will drain from the dehumidifier per hour if the inlet relative humidity is *a*) 50%, *b*) 70%, and *c*) 90%? Use the psychrometric chart.

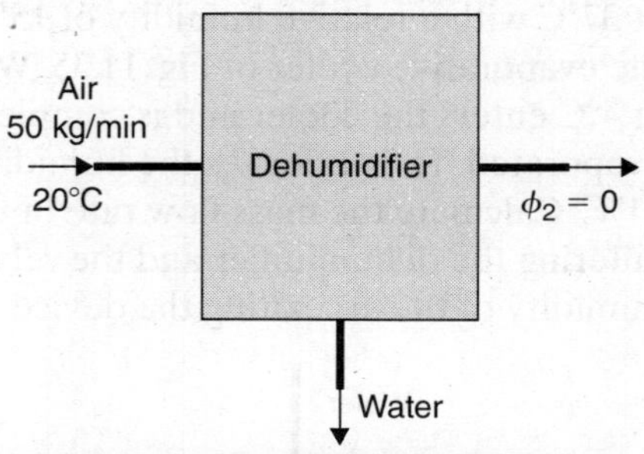

Figure 11.32

11.59 Air with a mass flux of 2 kg/s enters a humidifier at 100 kPa and 20°C with a wet-bulb temperature of 15°C. It leaves the humidifier at 100 kPa and 18°C with a wet-bulb temperature of 17°C. What is the humidity ratio at each of these states? How much water is being added per hour? Use the psychrometric chart.

11.60 Dry air flows in an insulated duct at 21°C and 100 kPa at a flow rate of 2.7 m^3/min. Superheated steam with a mass flux of 9 kg/hr is injected into the air flow so that the downstream humidity is 40%. What is the temperature of the downstream moist air? Use the psychrometric chart.

11.61 A 100-m^3 room contains air at 100 kPa and 20°C. What is the mass fraction and the mole fraction of water in the room if the relative humidity is *a*) 40%, *b*) 60%, and *c*) 90%? Use the psychrometric chart.

Comment

The pressure is assumed to be 100 kPa unless otherwise stated.

Air-Conditioning Processes (use the psychrometric chart)

11.62 An air-conditioning system reduces the temperature of the air in a basement area from 27°C to 20°C. No significant amount of moisture is removed from the basement air. If the initial humidity in the basement is 60%, estimate the humidity after the temperature is lowered.

11.63 Air in a typical house (see Fig. 11.33) is completely exchanged with outside air about every two hours unless care is taken to seal cracks. If the outside air is at 0°C with 60% humidity and the inside air is maintained at 22°C, estimate the humidity of the inside air if no humidifier is operating and other moisture inputs are ignored.

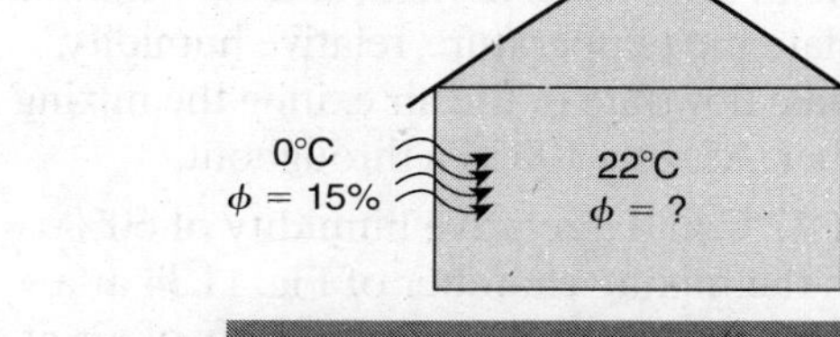

Figure 11.33

11.64 Air in a typical house is completely exchanged with outside air about every two hours unless care is taken to seal cracks. If the outside air is at −7°C with 80% humidity and the inside air is maintained at 22°C, estimate the humidity of the inside air if no humidifier is operating and other moisture inputs are ignored.

11.65 One type of ultrasonic humidifier uses a vibrating piezoelectric crystal actuator to produce tiny water droplets in the air. If 9 kg/min of air at 21°C enters the humidifier with a relative humidity of 20% and leaves at the same temperature with a relative humidity of 60%, how much water is being added each hour?

11.66 A large air-conditioning unit processes 5 kg/s of air. The air enters with a dry-bulb temperature of 35°C and a wet-bulb temperature of 25°C. How much condensate is being removed each hour if the conditioned air leaves the unit with respective dry-bulb and a wet-bulb temperatures of *a*) 20°C and 10°C, *b*) 23°C and 12°C, and *c*) 22°C and 8°C?

11.67 Forty kilograms per minute of air at 10°C and a relative humidity of *a*) 30%, *b*) 50%, and *c*) 80% pass through a heater and are heated to 30°C. How much heat must be added to accomplish this? What is the relative humidity of the air leaving this heater?

11.68 Air leaves the research lab at a university at the rate of 100 m³/min. The same volume rate of outside air at 5°C and 65% humidity enters and is heated to the exiting temperature of 25°C. Water vapor is added to the heated air, resulting in an exiting humidity of 60%. Determine the heat transfer rate and mass flux of water vapor required.

11.69 Air at 15°C and a relative humidity of 70% enters a mixing chamber at a flow rate of 80 m³/min. Another flow of air at 40°C and a relative humidity of 30% enters at a flow rate of *a*) 40 m³/min, *b*) 60 m³/min, and *c*) 80 m³/min. Calculate the temperature, relative humidity, and mass flow rate of the air exiting the mixing chamber. Assume 100 kPa throughout.

11.70 Air at 27°C and a relative humidity of 60% enters the mixing chamber of Fig. 11.34 at a flow rate of 54 m³/min. Another flow of air at −1°C and a relative humidity of 30% enters at a flow rate of 27 m³/min. Calculate the temperature, relative humidity, and mass flow rate of the air exiting the mixing chamber. Assume total pressure in the mixing chamber as 100 kPa throughout.

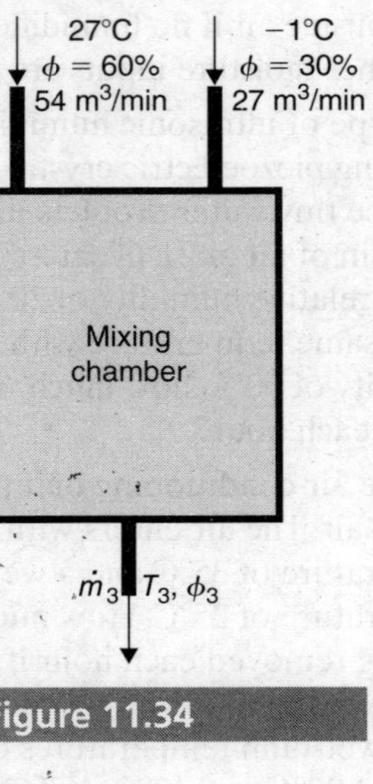

Figure 11.34

11.71 A conditioner takes in 40 m³/min of air at 20°C and a relative humidity of 75%. The air is cooled 12°C and 60% of the moisture is removed. The air is then heated back to 20°C. What is the relative humidity of the exiting air, and what heat rate is required?

11.72 Two hundred cubic meter per minute of air at 37°C with a relative humidity of 15% enter the evaporative cooler of Fig. 11.35. Water at 4°C enters the cooler and is completely evaporated. The air leaves the humidifier at 21°C. Calculate the mass flow rate of water entering the dehumidifier and the relative humidity of the air exiting the device.

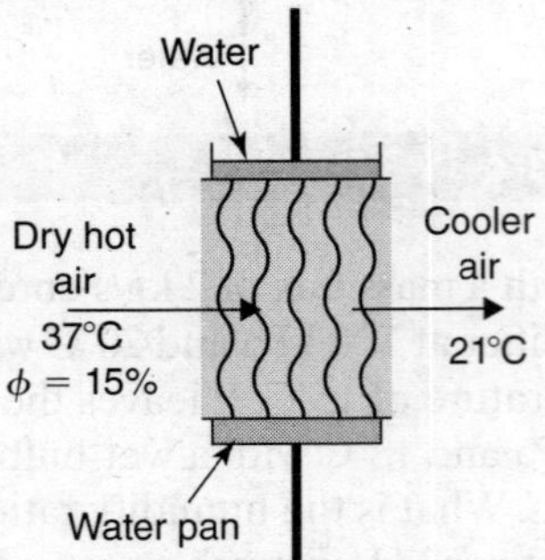

Figure 11.35

11.73 The cooling tower of Fig. 11.36 receives 45°C water from the condenser of a power plant. The water is cooled to 25°C by atmospheric air that enters at 22°C and 60% humidity. Saturated air exits the top of the cooling tower at 30°C. Estimate the volume flow rate of air into the cooling tower and mass flux of the makeup water if the mass flow rate of the condenser water is *a*) 50 kg/s, *b*) 75 kg/s, and *c*) 100 kg/s.

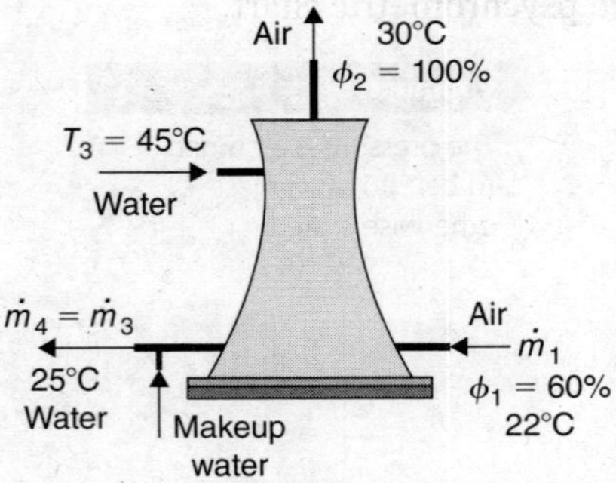

Figure 11.36

CHAPTER

12

Combustion

Outline

Nomenclature

The following variables are used in this chapter:

AF	Air–fuel mass ratio	HHV	Higher heating value
FA	Fuel–air mass ratio	K_P	Equilibrium constant
G, g	Gibbs function	LHV	Lower heating value
$\overline{H}_P, H_P$	Enthalpy of the products	$\overline{U}_P, U_P$	Internal energy of products
$\overline{H}_R, H_R$	Enthalpy of the reactants	$\overline{U}_R, U_R$	Internal energy of reactants
$\overline{h}^\circ$	Enthalpy at the reference state	v	Stoichiometric coefficient
$\overline{h}_f^\circ$	Enthalpy of formation	ϕ	Equivalence ratio
$\overline{h}_{fg}$	Enthalpy of vaporization		

Learning Outcomes

- ❑ **Balance combustion chemical equations**
- ❑ **Determine how much heat is produced in a combustion reaction**
- ❑ **Calculate adiabatic flame temperatures**
- ❑ **Predict the effects of high-temperature dissociation reactions on the combustion process**

Motivational Example—Common Rail Fuel Injector

One of the major drawbacks of the diesel engine is the amount of smoke that is created. The fuel injectors used in traditional diesel engines were driven by cams powered by the engine itself. This resulted in a fuel pressure that was proportional to the engine speed, resulting in much of the smoke. A device has recently been developed that will significantly reduce the smoke. A rail fuel-injection system has been developed that uses a high-pressure pump to supply fuel to a common distribution pipe. This pipe supplies fuel to piezoelectric solenoid valves that inject fuel into the engine cylinders at high pressure independent of the engine speed. The result is that the fuel burns much better, providing a "smokeless" diesel engine that performs well at all speeds.

12.1 Introduction

Combustion is the extremely rapid chemical oxidation of a fuel. It provides the energy for major power plants and internal combustion engines. In this chapter the chemical reactions will be formulated, and the heat provided will be determined when a fuel is burned. The fuels considered are hydrocarbon fuels, which may exist as a solid, liquid, or gas: for example, coal, gasoline, or propane (our focus will be on the liquid and gaseous forms of fuels). Fuels burn in the presence of oxygen to produce heat and products of combustion, as illustrated in Fig. 12.1 and presented as the generalized combustion equation

Combustion: The extremely rapid chemical oxidation of a fuel.

$$\text{Fuel} + \text{Oxidizer} \xrightarrow{\text{Energy}} \text{Products of combustion} \qquad (12.1)$$

The oxidizer can be either pure oxygen, as used with a cutting torch, or air, as used in an engine. In Chapter 11 air was considered to be a mixture of gases, primarily nitrogen and oxygen. In this chapter dry air will be treated as 79% nitrogen, N_2, and 21% oxygen, O_2. Thus, when using dry air, every mole of oxygen used in the combustion process requires 0.79/0.21 = 3.76 moles of nitrogen. The nitrogen in air will reduce combustion temperatures by absorbing heat from the process. Burning a fuel in pure oxygen will create much higher combustion temperatures, which is why care is taken to keep flames away from a pure oxygen environment.

Air: There are 3.76 moles of N_2 for every mole of O_2. (Always assume dry air unless otherwise stated.) Moisture in the air will influence combustion. See Example 12.4.

The fuels to be studied are primarily hydrocarbon fuels, which are complex organic compounds of hydrogen and carbon. Table 12.1 shows a list of a variety of common fuels along with their chemical formulas. Gasoline and diesel fuel are assumed to be C_8H_{18} and $C_{12}H_{26}$, respectively, for simplicity. Some hydrocarbon fuels are mixtures of these simple fuels. A common example is natural gas, which is a mixture of methane, ethane, and propane. Biofuels are alcohol-based compounds, created from substances such as vegetable oil and animal fats; they are used to produce biodiesel fuel. The primary component of a vegetable-based biofuel is ethanol, C_2H_5OH.

Coal: Each variety has a different chemical formula. It is composed primarily of carbon.

Table 12.1 Common Hydrocarbon Fuels

Methane	CH_4
Benzene	C_6H_6
Acetylene	C_2H_2
Octane	C_8H_{18} (gasoline)
Propane	C_3H_8
Dodecane	$C_{12}H_{26}$ (diesel fuel)
Ethane	C_2H_6
Coal	C (primarily)

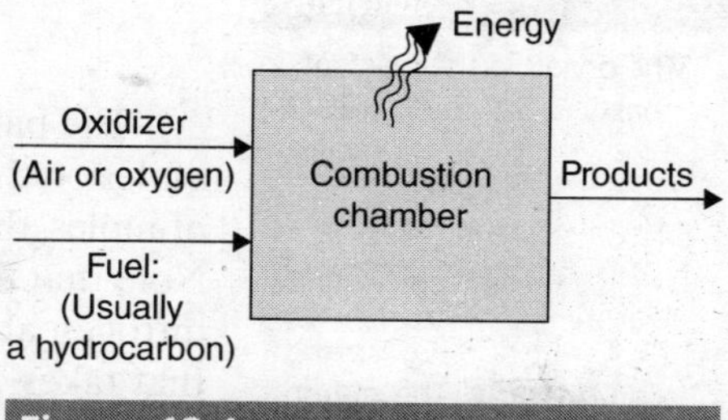

Figure 12.1
The combustion chamber.

12.2 Combustion Reactions

Stoichiometric reaction: Exactly enough oxidizer exists to combust all the fuel.

Reactants: The terms typically on the left of the reaction equation.

Products of combustion: The terms typically on the right of the reaction equation.

The first step in establishing the chemical equation for a combustion process is to determine the chemical equation of the reaction assuming that there is exactly enough oxidizer to combust all of the fuel. This is called a *stoichiometric reaction*. This reaction equation will be modified if too much or too little oxidizer is available.

As an example of this process, let's consider the stoichiometric combustion of propane in dry air. The *reactants*, the terms that are typically on the left of the reaction equation, are propane (C_3H_8) and air ($O_2 + 3.76N_2$). The *products of combustion*, the terms typically on the right, will be carbon dioxide CO_2, water (H_2O), and N_2. The chemical reaction equation (assuming one mole of propane) is written, with unknowns *A, B, C,* and *D*, as

$$C_3H_8 + A(O_2 + 3.76N_2) \Rightarrow B\,CO_2 + C\,H_2O + D\,N_2 \tag{12.2}$$

Note: A compound C_3H_8 has 3 carbon atoms and 8 hydrogen atoms; it represents 1 mole. Moles do not balance; atoms balance.

The equation is balanced by determining the coefficients *A, B, C,* and *D,* so that the mass (i.e., the number) of the atoms of each element is the same in the reactants on the left-hand side and the products on the right-hand side. This is done as follows:

$$\begin{array}{lll} \textbf{C:} & 3 = B & \therefore B = 3 \\ \textbf{H:} & 8 = 2C & \therefore C = 4 \\ \textbf{O:} & 2A = 2B + C & \therefore A = 5 \\ \textbf{N:} & 3.76A = D & \therefore D = 18.8 \end{array} \tag{12.3}$$

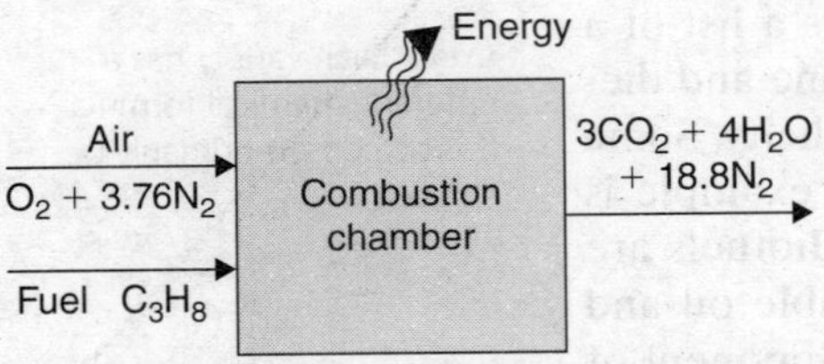

Figure 12.2
A stoichiometric reaction.

The reaction represented by Eq. 12.2 is demonstrated by Fig. 12.2 (the energy emitted will be determined in a later section) and now is written as

$$C_3H_8 + 5(O_2 + 3.76N_2) \Rightarrow 3CO_2 + 4H_2O + 18.8N_2 \tag{12.4}$$

Comment

The dry air in the reactants has $5 + 5 \times 3.76 = 23.8$ moles.

The combustion equation represented by Eq. 12.4 is an example of *complete combustion*: The carbon burns to CO_2, and the hydrogen[1] burns to H_2O (if air is used, the nitrogen does not react, but simply tags along with 18.8 moles on each side of the equation). *Incomplete combustion* occurs if the products contain CO, C, H_2, or HO.

Complete combustion: The carbon burns to CO_2, and the hydrogen burns to H_2O.

The balance just performed was based on atoms, that is, mass. The number of moles on the left is 24.8, and the number of moles on the right is 25.8. The number of moles, that is, the volume, is not necessarily conserved in a chemical reaction.[2] Note that the nitrogen does not actively participate in the chemical reaction. The nitrogen absorbs heat from the reaction and provides mass for the work process that takes place in an engine after the combustion process occurs. It adds mass and reduces the maximum temperature, so from an energy consideration, it does somewhat more than just going along for the ride.

[1]Any sulfur would burn to SO_2.
[2]The simple reaction $2H_2 + O_2 \rightarrow 2H_2O$ has 3 moles on the left and 2 moles on the right.

Dry air will be assumed to be the reactant unless moist air is specified. The moisture content, even though it's a small percentage of the reactants, is quite influential in an energy balance, so when specified, it must be included. An example will illustrate its role in the reaction equation. Its contribution to the energy balance will be considered in a subsequent section.

We can choose to burn fuel with more or less air than is needed for a stoichiometric reaction. When just enough air is provided, it is *stoichiometric air* or *theoretical air*. If more air than is necessary is used in a reaction, oxygen that is not used in the combustion process is also in the products. This unused oxygen is pushed out of an engine in the exhaust stroke. When more air is used than the stoichiometric air, an engine is said to be *running lean*.

Theoretical air: Just enough air for a stoichiometric reaction.

Running lean: When more air is used than is necessary for a stoichiometric reaction.

When more air is present than is needed for stoichiometric combustion, it is expressed as *percent theoretical air* or *percent excess air*. This takes the form

$$\%\ \text{theoretical air} = 100\% + \%\ \text{excess air} \tag{12.5}$$

The air–fuel mixture may not be mixed uniformly in the combustion chamber, so excess air allows for more complete combustion.

If less air is used than the stoichiometric air, not enough oxygen is present to burn all of the fuel and CO, and possibly some carbons, exist in the products. If insufficient oxygen is present to burn all the fuel, an engine is said to be *running rich*. Otto cycle engines tend to perform more smoothly when the engine is running rich. There are, however, serious environmental hazards to exhausting unburned hydrocarbon fuels into the atmosphere. These fumes combine with sulfur from coal burned in power plants, the burning of fuel in vehicles, and volcanoes to produce sulfuric acid in the atmosphere. The law in some states requires that gas stations have gasoline nozzles that draw gasoline fumes back into the fuel storage tank rather than let them escape into the atmosphere every time people "fill up."

Running rich: When insufficient air is present to burn all the fuel.

The *air–fuel ratio (AF)* is defined as the ratio of the mass of air used to the mass of fuel used. Note that this term is defined on a mass basis rather than a molar basis. The *fuel–air ratio (FA)* is simply the ratio of the mass of fuel used to the mass of air used and is the inverse of the air–fuel ratio. The relationships are

Air–fuel ratio (AF): The ratio of the mass of air to the mass of fuel.

Fuel–air ratio (FA): The ratio of the mass of fuel to the mass of air.

$$AF = \frac{\text{mass of air}}{\text{mass of fuel}} \qquad FA = \frac{\text{mass of fuel}}{\text{mass of air}} = \frac{1}{AF} \tag{12.6}$$

The *equivalence ratio* ϕ (not to be confused with relative humidity) is defined as the air–fuel ratio using stoichiometric air divided by the actual air–fuel ratio as

Equivalence ratio: The air–fuel ratio using stoichiometric air to the actual air–fuel ratio.

$$\phi = \frac{AF_{\text{stoichiometric}}}{AF_{\text{actual}}} = \frac{FA_{\text{actual}}}{FA_{\text{stoichiometric}}} \tag{12.7}$$

$\phi = 1$ stoichiometric or complete combustion

$\phi < 1$ engine is running lean

$\phi > 1$ engine is running rich

If the equivalence ratio is 1, the reaction is stoichiometric. If it is less than one, the engine is running lean; if it is greater than one, the engine is running rich.

You will observe that the products of combustion for the fuels considered contain H_2O as one of the products. It is initially in the form of water vapor since the combustion process produces significant amounts of heat, but the products

Comment

It is important that the temperature remains above the dew point when exhausting the products of combustion so that acidic condensate does not damage pipes or chimneys.

cool down as they flow out a chimney or out a tailpipe. If they cool down too much, and the temperature falls below the dew point, the water vapor will condense and may result in damage from the acids that are contained in the condensate. So, it's important that the temperature remains above the dew point all the way to the exit. In high-efficiency furnaces, the exhaust temperature from the furnace is indeed below the dew point. Therefore, the products of combustion are not sent up a chimney but are condensed and sent down a drain. An example will illustrate how the dew point is calculated.

Example 12.1 Stoichiometric combustion of natural gas

Natural gas is 70% methane, 15% ethane, and 15% propane, by volume. Write a chemical equation for the stoichiometric combustion of natural gas in air, as demonstrated by Fig. 12.3.

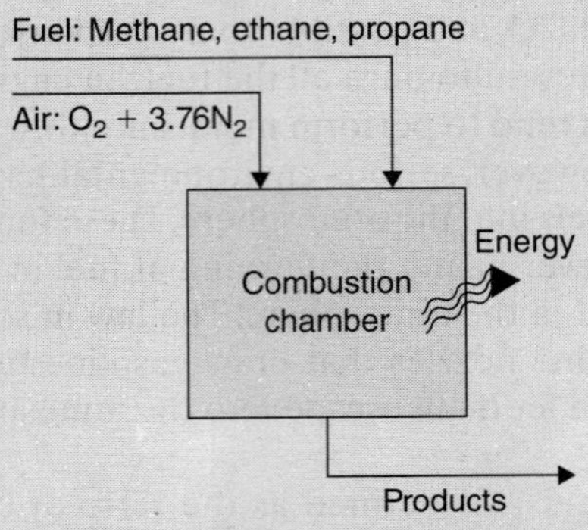

Figure 12.3

Solution

The first objective is to determine the stoichiometric combustion equations for each constituent of natural gas. The equations are, following the procedure for propane outlined above with Eqs. 12.3:

$$\textbf{Methane:}\quad CH_4 + 2(O_2 + 3.76N_2) \Rightarrow CO_2 + 2H_2O + 7.52N_2$$
$$\textbf{Ethane:}\quad C_2H_6 + 3.5(O_2 + 3.76N_2) \Rightarrow 2CO_2 + 3H_2O + 13.16N_2$$
$$\textbf{Propane:}\quad C_3H_8 + 5(O_2 + 3.76N_2) \Rightarrow 3CO_2 + 4H_2O + 18.8N_2$$

Next, multiply the equations by their percent content in the natural gas and add them together, that is,

$$\text{gas} = 0.15[\text{propane}] + 0.15[\text{ethane}] + 0.70[\text{methane}]$$

The combustion equation results for 1 mole of propane:

$$0.15C_3H_8 + 0.15C_2H_6 + 0.7CH_4 + 2.675(O_2 + 3.76N_2)$$
$$\Rightarrow 1.45CO_2 + 2.45H_2O + 10.06N_2$$

Combustion with excess air Example 12.2

Propane is burned with 20% excess air. Determine i) the air–fuel ratio, ii) the volume percentage of water vapor in the products, and iii) the dew-point temperature of the products. Assume complete combustion; that is, there is no CO in the products.

Solution

The stoichiometric reaction was presented in Eq. 12.4 to be

$$C_3H_8 + 5(O_2 + 3.76N_2) \Rightarrow 3CO_2 + 4H_2O + 18.8N_2$$

With 20% excess air, the "5" is multiplied by 1.2 to give 20% more air. The combustion equation is then

$$C_3H_8 + 1.2 \times 5(O_2 + 3.76N_2) \Rightarrow A\,CO_2 + B\,H_2O + C\,N_2 + D\,O_2$$

Balance the atoms:

C:	$3 = A$	$\therefore A = 3$
H:	$8 = 2B$	$\therefore B = 4$
N:	$6 \times 3.75 \times 2 = 2C$	$\therefore C = 22.56$
O:	$6 \times 2 = 2A + B + 2D$	$\therefore D = 1$

The combustion, expressed schematically in Fig. 12.4, is then represented by

$$C_3H_8 + 6(O_2 + 3.76N_2) \Rightarrow 3CO_2 + 4H_2O + 22.56N_2 + O_2$$

i) The air–fuel ratio is

$$AF = \frac{\text{mass air}}{\text{mass fuel}} = \frac{6 \times 4.76 \times 29}{1 \times (36 + 8)}$$

$$= \frac{(6 \times 4.76)\ \text{kmol}_{\text{air}} \times 29\ \text{kg}_{\text{air}}/\text{kmol}_{\text{air}}}{1\ \text{kmol}_{\text{fuel}} \times (36 + 8)\ \text{kg}_{\text{fuel}}/\text{kmol}_{\text{fuel}}} = \underline{18.82\ \text{kg}_{\text{air}}/\text{kg}_{\text{fuel}}}$$

ii) The volume percentage is a ratio of moles. It is

$$\%\ \text{water} = \frac{\text{moles water}}{\text{moles of products}} \times 100 = \frac{4}{30.56} \times 100 = \underline{13.1\%}$$

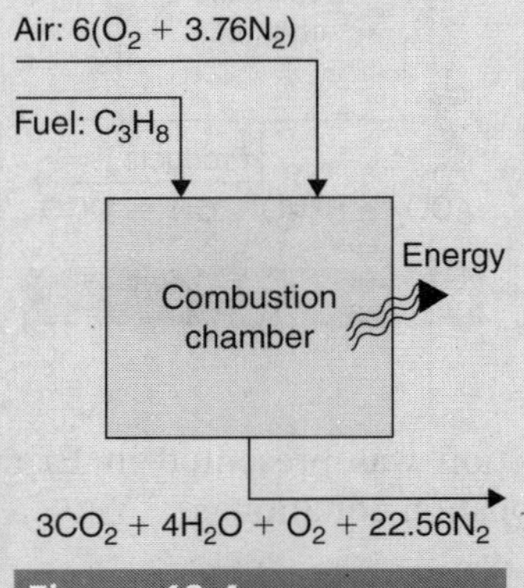

Figure 12.4

(Continued)

Example 12.2 (Continued)

iii) The partial pressure P_v of the water vapor is found using its mole fraction (see Eq. 11.12) of the products. It is

Comment

The atmospheric pressure is assumed to be 100 kPa unless otherwise stated.

$$P_v = y_{vapor} P_{atm} = \frac{4}{30.56} \times 100 = 13.1 \text{ kPa}$$

The dew-point temperature is interpolated in Table C-2, or is found, using the IRC Calculator (use quality = 1.0), to be

$$T_{dp} = \underline{51.2°C}$$

If the temperature of the products, such as the temperature of the exhaust leaving the exhaust pipe of a vehicle, is less than T_{dp}, the water vapor would condense, along with some nasty acids that are always present in combustion products (in negligibly small but corrosive amounts).

Example 12.3 Combustion with less than theoretical air

Ethane is burned with 85% theoretical air, as shown in Fig. 12.5. Calculate the mass percentage of carbon monoxide in the products and the air–fuel ratio. Assume that CO is the only additional compound in the products due to the lack of sufficient oxygen.

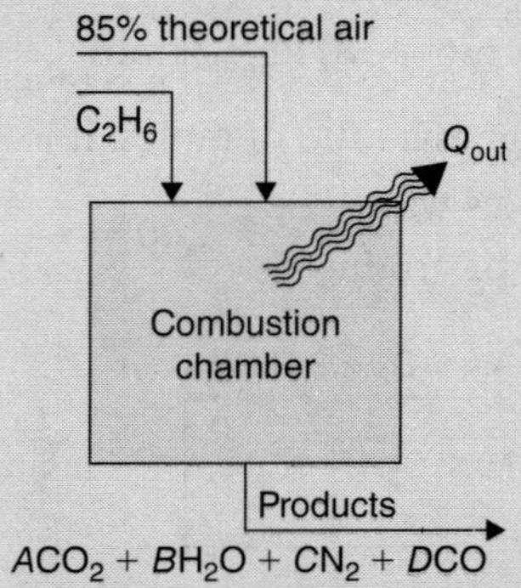

Figure 12.5

Solution

The stoichiometric reaction was presented in Example 12.1 so, using 85% theoretical air, the combustion equation is

$$C_2H_6 + 0.85 \times 3.5(O_2 + 3.76N_2)$$
$$\Rightarrow A\,CO_2 + B\,H_2O + C\,N_2 + D\,CO$$

Balance the atoms:

$$\begin{aligned} &\textbf{H:} && 6 = 2B && \therefore B = 3 \\ &\textbf{N:} && 0.85 \times 3.5 \times 3.76 \times 2 = 2C && \therefore C = 11.2 \\ &\textbf{C:} && 2 = A + D \\ &\textbf{O:} && 5.95 = 2A + B + D && \therefore A = 0.95, \quad D = 1.05 \end{aligned}$$

The combustion equation is then

$$C_2H_6 + 0.85 \times 3.5(O_2 + 3.76N_2) \Rightarrow 0.95\,CO_2 + 3\,H_2O + 11.2\,N_2 + 1.05\,CO$$

The mass percentage of CO is calculated to be

$$\%\ CO \text{ by mass} = \frac{1.05 \times 28}{0.95 \times 44 + 3 \times 18 + 11.2 \times 28 + 1.05 \times 28} \times 100 = \underline{6.70\%}$$

The air–fuel ratio is

$$AF = \frac{\text{air mass}}{\text{fuel mass}} = \frac{0.85 \times 3.5 \times 4.76 \times 29}{1 \times 30} = \underline{13.7\ kg_{air}/kg_{fuel}}$$

Combustion with moist air — Example 12.4

Determine the chemical equation for propane, burned in the combustor of Fig. 12.6, with 40% excess of air that has a specific humidity of 0.01. Assume a constant pressure of 100 kPa and also determine the dew point.

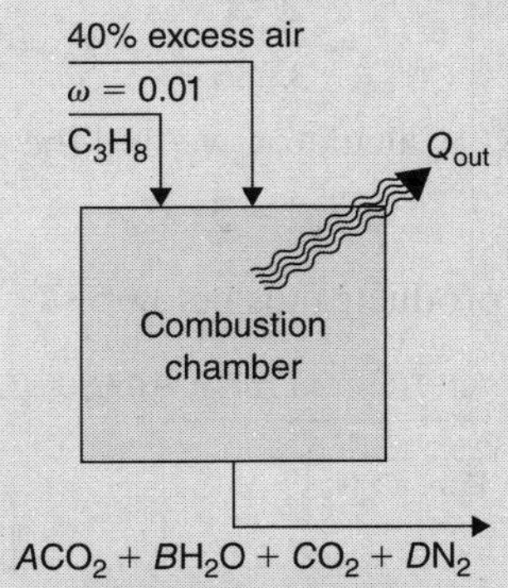

Figure 12.6

Solution

We use the stoichiometric combustion equation with dry air presented by Eq. 12.4:

$$C_3H_8 + 5(O_2 + 3.76N_2) \Rightarrow 3CO_2 + 4H_2O + 18.8N_2$$

(Continued)

Example 12.4 (Continued)

Using 40% excess air, the 5 is multiplied by 1.4, and the combustion equation with dry air becomes

$$C_3H_8 + 1.4 \times 5(O_2 + 3.76N_2) \Rightarrow 3CO_2 + 4H_2O + 2O_2 + 26.32N_2$$

For every mole of propane burned, 2 moles of oxygen are not used and we are now using 26.32 moles of nitrogen.

Comment

When moisture is present in the air, we ignore it and balance the reaction equation with dry air. The moisture is then added since it simply tags along, like nitrogen. It doesn't affect the other terms in the equation (except to add to the water), but it does alter the heat transfer from the combustion chamber. (Heat transfer will be considered in Section 12.3.)

The specific humidity was defined in Chapter 11 as the ratio of the mass of water in the air to the mass of air. For the purposes of this analysis, we need to change this to a molar ratio. The molar mass of water is 18 kg/kmol, and the molar mass of air is 29 kg/kmol. The molar specific humidity $\overline{\omega}$ is determined to be

$$\overline{\omega} = \omega\left(\frac{M_{\text{air}}}{M_{\text{water}}}\right) = 0.01\left(\frac{29}{18}\right) = 0.0161\,\frac{\text{moles of water}}{\text{mole of air}}$$

where the bar indicates a molar basis. The number of moles of water that enters the reaction with the air is now calculated to be

$$\begin{aligned} N_{\text{water}} &= \overline{\omega} N_{\text{air}} \\ &= 0.0161 \times 1.4 \times 5 \times (4.76) = 0.536 \text{ mol of water} \end{aligned}$$

The complete chemical reaction for combustion with moist air is shown in Fig. 12.7 and is represented by

$$\begin{aligned} &C_3H_8 + 7(O_2 + 3.76N_2) + 0.536H_2O \\ &\qquad \Rightarrow 3CO_2 + 4.536H_2O + 2O_2 + 26.32N_2 \end{aligned}$$

Note that the water vapor in the air, like the nitrogen, simply tags along. Its dew point can be found by determining its partial pressure (see Eq. 11.12), which is

$$P_v = y_{H_2O} P_{\text{atm}} = \frac{4.54}{35.86} \times 100 = 12.7 \text{ kPa}$$

Using Table C-2 or the IRC Calculator, we find the dew point to be

$$T_{\text{dp}} = \underline{51°\text{C}}$$

If the temperature of the products falls below 53°C, condensation will occur.

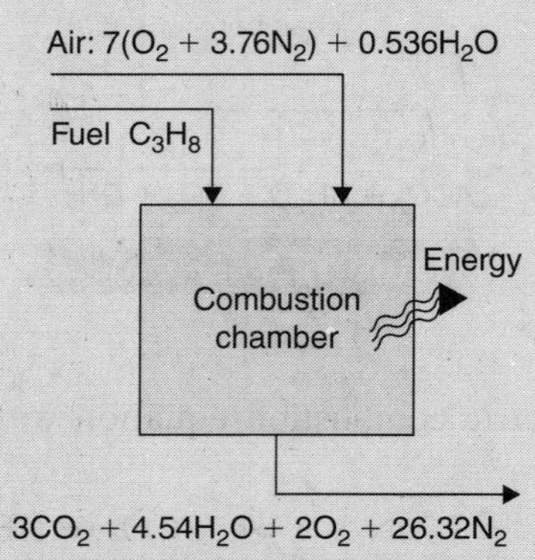

Figure 12.7

Determine the *AF* from the products — Example 12.5

Benzene is burned with dry air, and a volumetric analysis is performed[3] on the dry products (no water vapor is measured) of combustion. It is found that the dry products contain 80.69% N_2, 8.58% CO_2, and 10.73% O_2. The moles of each gas are shown in Fig. 12.8. Determine the composition of the air–fuel ratio and the percent excess air.

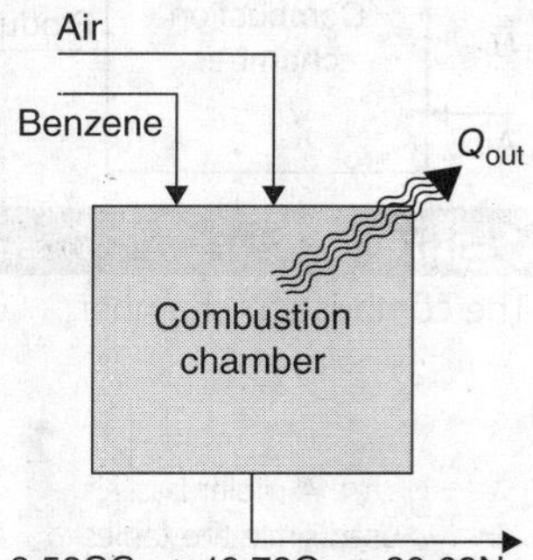

Figure 12.8

Note: A "dry analysis" of the products means that water is not included in the analysis, but it is always present.

Solution

Let us assume that an unknown supply of fuel provides 100 moles of dry products so that the reaction equation is

$$AC_6H_6 + B(O_2 + 3.76N_2) \Rightarrow 8.58CO_2 + CH_2O + 10.73O_2 + 80.69N_2$$

in which the moles of fuel, air, and water are not known. The products must contain water so that term is included, but the number of moles is unknown since it was a dry analysis. The atoms are balanced:

C:	$6A = 8.58$	$\therefore A = 1.43$
N:	$3.76B = 80.69$	$\therefore B = 21.46$
H:	$6A = 2C$	$\therefore C = 4.29$
O:	$2B = 17.16 + C + 21.46$	$42.92 = 42.91$

Observe: The oxygen balance is a check on the values obtained for A, B, and C.

Divide all terms in the reaction equation by A so the equation is in standard form:

$$C_6H_6 + 15(O_2 + 3.76N_2) \Rightarrow 6CO_2 + 3H_2O + 7.5O_2 + 56.43N_2$$

The air–fuel ratio is the ratio of the mass of air to the mass of fuel using the molar mass of each. There results the following:

$$AF = \frac{\text{air mass}}{\text{fuel mass}} = \frac{15 \times 4.76 \times 29}{1 \times (6 \times 12 + 6 \times 1)} = 26.55\ \text{kg}_{\text{air}}/\text{kg}_{\text{fuel}}$$

Molar mass:
$M_{fuel} = 6 \times 12 + 6 \times 1 = 78$ kg/kmol

For theoretical air, the reaction equation can be shown to be

$$C_6H_6 + 7.5(O_2 + 3.76N_2) \Rightarrow 6CO_2 + 3H_2O + 28.2N_2$$

Thus, 15/7.5 = 2, yielding 200% theoretical air. The percent excess air, referring to Eq. 12.5, is

$$\%\ \text{Excess air} = 200 - 100 = \underline{100\%}$$

☑ You have completed Learning Outcome (1)

[3] An Orsat gas analyzer, used to analyze the products, does not measure the water vapor content.

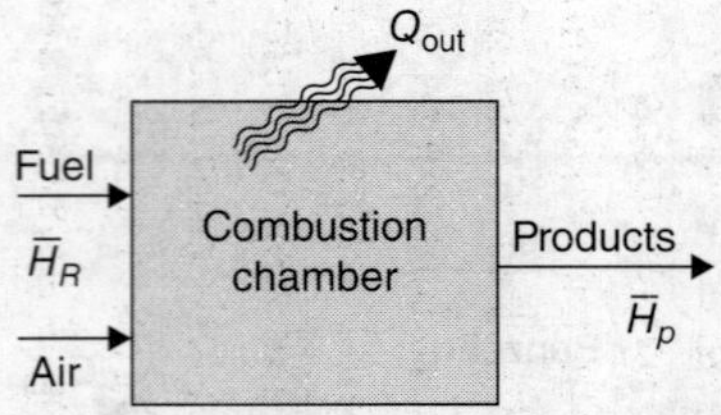

Figure 12.9

The combustion chamber.

Note: A molar basis is used since the tables present enthalpy on a molar basis.

Comment

The symbol "ℓ" after a compound signifies the liquid phase, and the symbol "g" signifies the gaseous phase, which is assumed if the "ℓ" is not present.

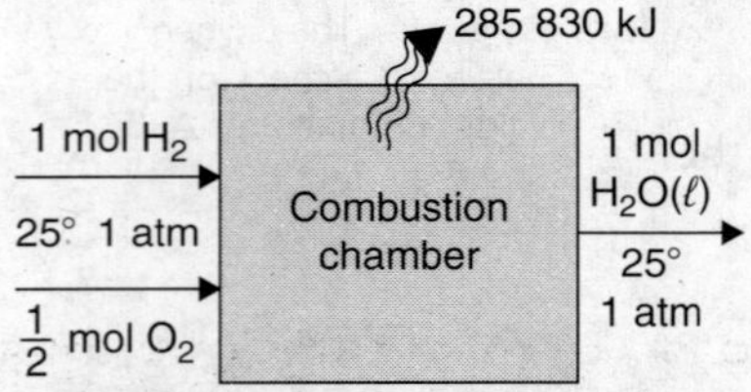

Figure 12.10

The combustion of hydrogen and oxygen.

Enthalpy of formation: The energy emitted or required to form a compound from its stable elements.

Exothermic reaction: A reaction that emits heat, such as the formation of H_2O.

Endothermic reaction: A reaction that requires heat.

12.3 The Enthalpy of Formation and the Enthalpy of Combustion

In Section 12.2 we considered only the chemical reaction and did not attempt to predict the amount of heat emitted by a particular reaction. In this section, the heat emitted by a particular reaction, which is the primary objective in our study of combustion, will be determined.

The amount of heat Q_{out} given off in the combustion reaction shown in Fig. 12.9 is determined by applying the first law of thermodynamics to the process. This is represented by

$$Q_{\text{out}} = \overline{H}_R - \overline{H}_P \tag{12.8}$$

where $\overline{H}_R$ is the enthalpy of the reactants and $\overline{H}_P$ is the enthalpy of the products, on a molar basis. The changes in kinetic and potential energy across the combustion chamber (the control volume) are, as usual, assumed to be insignificant, except in certain applications such as a rocket nozzle, which will be illustrated in Example 12.8. Also, it is assumed that there is no work performed. So, to find the heat emitted from the combustion chamber, $\overline{H}_P$ and $\overline{H}_R$ must be determined. To determine the enthalpies, a reference state is established. It is agreed that the reference state shall be 25°C at standard atmospheric pressure and a superscript "o" on a property will signify such a state. For instance, $\overline{h}^\circ_1$ would be the enthalpy on a molar basis of a substance at the reference state, usually at the inlet.

To illustrate the application of Eq. 12.8, consider the constant-pressure combustion of hydrogen and oxygen, represented by

$$2H_2 + O_2 \Rightarrow 2H_2O(\ell) \tag{12.9}$$

with the inlet and exit conditions both being at the reference state of 25°C and a pressure of 1 atm, hence liquid water. In order for the exit temperature to be at 25°C, the heat transfer for this combustion process must be[4] −285 830 kJ for each kmol of $H_2O(\ell)$ formed, as illustrated in Fig. 12.10. The reactants H_2 and O_2 are stable since they exist at standard conditions in the atmosphere in that form. So, at 25°C and 1 atm, it takes no energy to form those two elements. Thus, the heat given off by the reaction of H_2 and O_2 is called the *enthalpy of formation* $\overline{h}^\circ_f$, as listed in Table B-7 for $H_2O(\ell)$. The enthalpy of formation for H_2 is zero since it is a stable element. The enthalpy of H is listed as a positive 218 000 kJ/kmol because it takes energy to break apart H_2. Observe in Table B-7 that some compounds have a negative $\overline{h}^\circ_f$, which indicates that they emit heat (an *exothermic reaction*), while others have a positive $\overline{h}^\circ_f$, which indicates that they require heat to form (an *endothermic reaction*).

The enthalpy of a particular compound or element is the sum of two enthalpies. The enthalpy of formation $\overline{h}^\circ_f$ is the energy needed to take the constituent atoms and form them into a compound at the standard conditions of 25°C and 1 atm. Elements such as nitrogen, hydrogen, oxygen, and carbon that exist naturally

[4]The minus sign is due to our sign convention: Heat transfer leaving a control volume is negative.

have an enthalpy of formation of zero. The other component of the enthalpy is the amount of energy needed to bring the material to a temperature different from the standard state. For example, the enthalpy of nitrogen at 127°C is $\overline{h}_{N_2} = \overline{h}^{\circ}_{N_2} + \Delta\overline{h}_{N_2} = 0 + 11\,640 - 8669 = 2971$ kJ/kmol, using values at 298 K and 400 K in Table F-2. Tables F-2 through F-6 show the enthalpies of formation and enthalpy increments for common gases in SI units, all on a molar basis.

> **Comment**
> A compound consists of two or more elements. H_2 is an element since it consists of only hydrogen, as are O_2 and N_2.

Table B-7 lists the enthalpies of formation for fuels and common compounds. It must be noted that the enthalpies of formation in Table B-7 are referenced to a temperature of 25°C rather than 0 K, as is done in Tables F-2 through F-6.

> **Comment**
> A good approximation: $\Delta\overline{h}_{N_2} = \overline{C}_p\Delta T = (1.042 \times 28)(400 - 298) = 2976$ kJ/kmol, less than a 0.2% error. For high temperatures, the error would be unacceptable. The tables will be used exclusively.

If the enthalpy of formation of propane is desired, Table B-7 gives a value of $\overline{h}^{\circ}_f = -103\,850$ kJ/kmol. This is the energy needed to combine three atoms of carbon and eight atoms of hydrogen to form one mole of propane at 25°C at a pressure of one atmosphere. Enthalpy changes to other temperatures are not given for propane or other fuels in the Appendix.

If the molar enthalpy of carbon dioxide at 700 K (427°C) is desired, the enthalpy of formation is added to the enthalpy increment from 298 K to 700 K from Table F-4. This is expressed, for carbon dioxide, as

$$\begin{aligned}\overline{h}_{700\,\text{K}} &= \overline{h}^{\circ}_f + \overline{h} - \overline{h}_0 \qquad (\text{for } CO_2)\\ &= -393\,520 + 27\,125 - 6601 = -372\,996 \text{ kJ/kmol} \qquad \textbf{(12.10)}\end{aligned}$$

> **Comment**
> All the digits may be used in the terms of an equation, but the answer should be rounded off to three or four significant digits, never five or more. We're engineers!

The above number could be rounded off to −373 MJ/kmol when used in engineering calculations. If units of kJ/kg are desired, simply divide by the molar mass, which for the CO_2 above would be −372 996 kJ/kmol/44 kg/kmol = −8477 kJ/kg.

The enthalpy of formation is the energy required to form a compound. The energy emitted when a compound undergoes stoichiometric combustion at constant temperature and atmospheric pressure is the *enthalpy of combustion*. In Fig. 12.10 the heat transfer is −285 830 kJ/kmol for the combustion of hydrogen with oxygen at 25°C and atmospheric pressure. It is listed in Table B-8 and is referred to as the *higher heating value* (HHV) since water in the products at 25°C is in liquid form. If the water is in gaseous form, the *lower heating value* (LHV) will result. To make sure that the water is a vapor, the LHV uses an exit temperature of perhaps 150°C. Because the exit temperature is not uniformly accepted to be 150°C, we most often use the HHV. The enthalpy of vaporization, which for water at 25°C is $\overline{h}_{fg} = 44\,010$ kJ/kmol (2445 kJ/kg), is also listed in Table B-7. The enthalpies of combustion (the HHVs) for various compounds and fuels are listed in Table B-8 in the Appendix.

> **Enthalpy of combustion:** The energy emitted when a compound undergoes stoichiometric combustion at constant temperature and atmospheric pressure.
>
> **Higher heating value (HHV):** When the water in the products is in liquid form.
>
> **Lower heating value (LHV):** When the water in the products is in gaseous form.

The first law, stated in Eq. 12.8, can be stated for any number of components making up the reactants and products. In an expanded form, neglecting kinetic and potential energy changes and shaft work, the first law, for a steady-flow control volume, takes the form

$$\begin{aligned}Q &= \overline{H}_P - \overline{H}_R\\ &= \sum_{\text{prod}} N_i(\overline{h}^{\circ}_f + \overline{h} - \overline{h}^{\circ})_i - \sum_{\text{react}} N_i(\overline{h}^{\circ}_f + \overline{h} - \overline{h}^{\circ})_i \qquad \textbf{(12.11)}\end{aligned}$$

> **Note:** The HHV and the LHV are listed as positive heat transfers, so they are negative the enthalpy of combustion.

where N_i represents the number of moles of substance i, with i summed over all substances in the products and the reactants. If the shaft work or kinetic energy

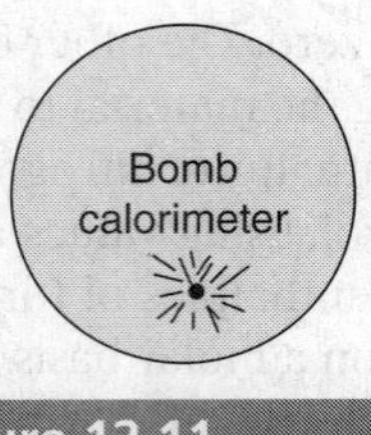

Figure 12.11
A bomb calorimeter.

change is not zero, as in a combustion jet turbine, it is simply included appropriately in the equation.

Note: This is the first law applied to a steady flow through a control volume with no shaft work.

If the control volume has no inlet or outlet, as in the bomb calorimeter of Fig. 12.11, it is a system and the first law becomes

$$Q = U_P - U_R$$
$$= \sum_{\text{prod}} N_i(\overline{h}_f^\circ + \overline{h} - \overline{h}^\circ - Pv)_i - \sum_{\text{react}} N_i(\overline{h}_f^\circ + \overline{h} - \overline{h}^\circ - Pv)_i \qquad (12.12)$$

Recall: $u = h - Pv$

where enthalpy $h = u + Pv$ has been used since tables list h and not u. Because the volume of any solid or liquid is very small when compared to the volume of a gas, we retain the Pv term for a gas only. Assuming an ideal gas, Eq. 12.12 for a rigid volume takes the form

Note: This is the first law applied to a rigid volume, such as a bomb calorimeter.

$$Q = \sum_{\text{prod}} N_i(\overline{h}_f^\circ + \overline{h} - \overline{h}^\circ - R_u T)_i - \sum_{\text{react}} N_i(\overline{h}_f^\circ + \overline{h} - \overline{h}^\circ - R_u T)_i \qquad (12.13)$$

In the above three equations, an ideal gas will be assumed to find $\overline{h} - \overline{h}^\circ$ using the tables in Appendix F. For vapors, tables such as the steam tables will be used. For solids and liquids, we will use $\overline{h} - \overline{h}^\circ = \overline{C}_p \Delta T$.

Example 12.6 Enthalpy of formation of CO_2

One mole of carbon and one mole of oxygen O_2 enter a combustion chamber at 25°C and 1 atm, combustion occurs, and the CO_2 that is formed exits at 25°C and 1 atm. What is the enthalpy of formation of CO_2? Refer to the steady-flow combustion chamber sketched in Fig. 12.12.

Note: Two moles enter, one mole leaves.

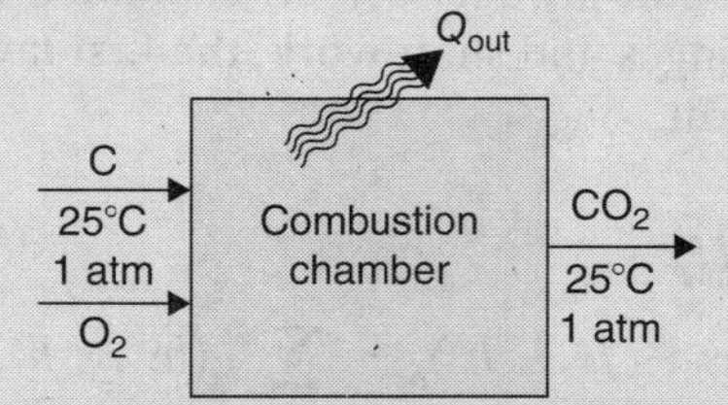

Figure 12.12

Solution

The combustion occurs according to the reaction

$$C + O_2 \Rightarrow CO_2$$

The steady-flow combustion chamber is flow through a control volume, so Eq. 12.11, using entries from Table F-4, is applied as follows:

$$Q = \sum_{\text{prod}} N_i(\bar{h}_f^\circ + \bar{h} - \bar{h}^\circ)_i - \sum_{\text{react}} N_i(\bar{h}_f^\circ + \bar{h} - \bar{h}^\circ)_i$$

$$= \left[1 \times (\bar{h}_f^\circ + \bar{h}_{25^\circ\text{C}} - \bar{h}^\circ)_{CO_2}\right]_{\text{prod}}$$

$$-\left[1 \times \cancel{(\bar{h}_f^\circ + \bar{h} - \bar{h}^\circ)}_C + 1 \times \cancel{(\bar{h}_f^\circ + \bar{h} - \bar{h}^\circ)}_{O_2}\right]_{\text{react}}$$

$$= 1 \times (-393\,520 + 9364 - 9364) - 0 - 0 = -393\,520 \text{ kJ/kmol}$$

where the two zeros are for the reactants C and O_2 since $\bar{h}_f^\circ = 0$ and $(\bar{h} - \bar{h}^\circ) = 0$ at 25°C for each reactant. The heat transfer, <u>−393 500 kJ/kmol</u>, is the enthalpy of formation of CO_2, since that is the heat emitted when the compound is formed. The heat transfer is negative, so it is leaving the control volume. We could say that <u>393 500 kJ/kmol</u> left the control volume, in which case Q_{out} would be written as a positive quantity.

Repeated comment: The numbers listed in tables are often given to five or more significant digits. Material properties are typically temperature dependent, and such properties are not known to four significant digits in most, if not all, problems. Data is typically collected, a best-curve fit is attached to the data, and then numbers are generated to as many digits as the researcher cares to provide, as is done to generate the steam tables. So, to give answers to five or more significant digits in an engineering problem involving material properties and dimensions is simply misleading and wrong! Usually three digits, or possibly even only two, may have significance.

Enthalpy of combustion of liquid propane — Example 12.7

Using the enthalpies of formation from Table B-7, determine the enthalpy of combustion of liquid propane. The reactants and products are all at 25°C and 1 atm of pressure, as shown in Fig. 12.13.

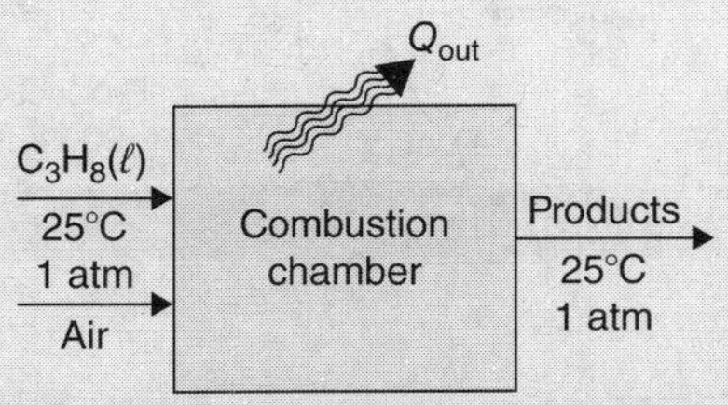

Figure 12.13

Solution

First, determine the reaction equation. It is found to be

$$C_3H_8(l) + 5(O_2 + 3.76N_2) \Rightarrow 3CO_2 + 4H_2O(\ell) + 18.8N_2$$

The water must be in liquid form since it is at 25°C and 1 atm. The combustion chamber is assumed to be a steady flow through a control volume, so Eq. 12.11

(Continued)

Note: When burned with air, it is the same as burning with oxygen since the nitrogen simply tags along. If the temperatures are different, the nitrogen does influence the heat transfer. So, the nitrogen is simply ignored in this equation.

Example 12.7 *(Continued)*

Comment

$\bar{h}_f^\circ$ for air is zero since it is composed of oxygen and nitrogen.

Comment

The LHV would result if for H_2O, −241 820 (vapor) was used rather than −258 830.

Note: The $\bar{h}_c$ for propane gas would not include $\bar{h}_{fg}$ = 15 060 kJ/kmol, so the LHV would be −2 205 000 kJ/kmol.

is applied as follows:

$$Q = \sum_{\text{prod}} N_i(\bar{h}_f^\circ + \cancel{\bar{h} - \bar{h}^\circ})_i - \sum_{\text{react}} N_i(\bar{h}_f^\circ + \cancel{\bar{h} - \bar{h}^\circ})_i$$

$$= [N\bar{h}_f^\circ]_{CO_2} + [N\bar{h}_f^\circ]_{H_2O} - [N\bar{h}_f^\circ]_{C_3H_8(l)} - \cancel{[N\bar{h}_f^\circ]_{O_2}} - \cancel{[N\bar{h}_f^\circ]_{N_2}}$$

$$= 3 \times (-393\,520) + 4 \times (-285\,830) - 1 \times (-103\,850)$$

$$= -2\,220\,000 \text{ kJ/kmol of } C_3H_8(g)$$

The enthalpy of combustion $\bar{h}_c$ of $C_3H_8(\ell)$ is then found to be

$$\bar{h}_c = -Q_{\text{out}} = \underline{-2\,220\,000 \text{ kJ/kmol}}$$

This is the HHV since the water is in liquid form. It could also be written as −2220/44 = <u>−50.45 MJ/kg</u>, common units for the enthalpy of combustion.

Example 12.8 Combustion in a jet engine

Air with negligible kinetic energy (compared to the exiting kinetic energy) and liquid octane, both at 25°C and 1 atm, enter an insulated engine. The exhaust products exit at 900 K and 1 atm of pressure, as shown in Fig. 12.14. Estimate the velocity of the exhaust products assuming no shaft work. Assume stoichiometric combustion.

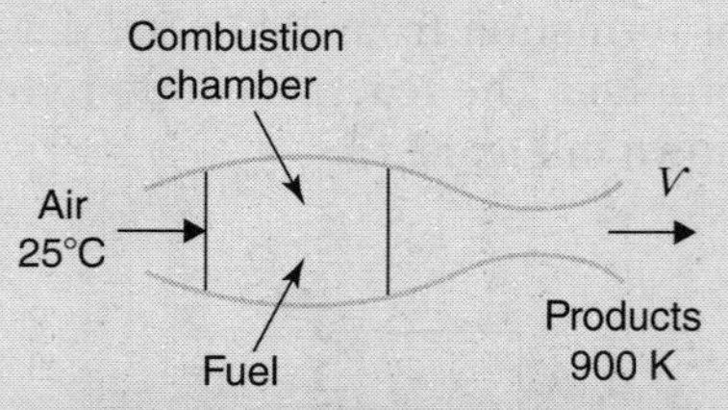

Figure 12.14

Solution

Stoichiometric combustion occurs according to the reaction

$$C_8H_{18}(\ell) + 12.5(O_2 + 3.76N_2) \Rightarrow 8CO_2 + 9H_2O + 47N_2$$

The water is obviously in vapor form since the products are at 900 K. The insulated ($Q = 0$) combustion chamber is assumed to involve steady flow through a control volume, so Eq. 12.8 takes the form

$$0 = H_P + m_P\frac{V^2}{2} - H_R$$

where $m_P V^2/2$ is the kinetic energy of the products per mole of fuel. The mass of the products per mole of fuel is found to be

$$m_P = 8 \times 44 + 9 \times 18 + 47 \times 28 = 1830 \text{ kg/kmol fuel}$$

Let's find H_P. It is

$$H_P = H_{CO_2} + H_{H_2O} + H_{N_2}$$
$$= 8(-393\,520 + 37\,405 - 9364) + 9(-241\,810 + 31\,828 - 9904) + 47(0 + 26\,890 - 8669) = -4\,046\,000 \text{ kJ/kmol fuel}$$

Next, H_R is found to be

$$H_R = H_{C_8H_{18}(\ell)} + H_{O_2} + H_{N_2}$$
$$= 1(-249\,910) + 0 + 0 = -249\,910 \text{ kJ/kmol fuel}$$

The first law, represented by Eq. 12.8, now allows the velocity to be found. It is

$$V^2 = \frac{2}{m_P}(H_R - H_P)$$
$$= \frac{2}{1830 \text{ kg/kmol}}(-249\,910 + 4\,046\,000)\frac{\text{kJ}}{\text{kmol}} \times 1000\frac{\text{J}}{\text{kJ}}$$
$$= 4.149 \times 10^6 \text{ m}^2/\text{s}^2 \qquad \therefore V = \underline{2040 \text{ m/s}}$$

Enthalpy of formation for liquid octane:
$\bar{h}^\circ_f = -208\,450 - 41\,460 = -249\,910$ kJ/kmol

Units: $\frac{\text{J}}{\text{kg}} = \frac{\text{N} \cdot \text{m}}{\text{N} \cdot \text{s}^2/\text{m}} = \frac{\text{m}^2}{\text{s}^2}$

Combustion with excess air — Example 12.9

Propane gas, flowing at 0.04 kg/min, is burned with 40% excess air in the steady-flow combustion chamber of Fig. 12.15, both of which enter at standard conditions of 25°C and 1 atm of pressure. The air has a relative humidity of 70%. Determine the rate of heat transfer from the chamber if the products exit at 800 K and 1 atm.

Solution

Stoichiometric combustion occurs according to the reaction

$$C_3H_8 + 5(O_2 + 3.76N_2) \Rightarrow 3CO_2 + 4H_2O(g) + 18.8N_2$$

Recall: Dry air is assumed unless otherwise stated. Moisture in the air would definitely influence the result in this example.

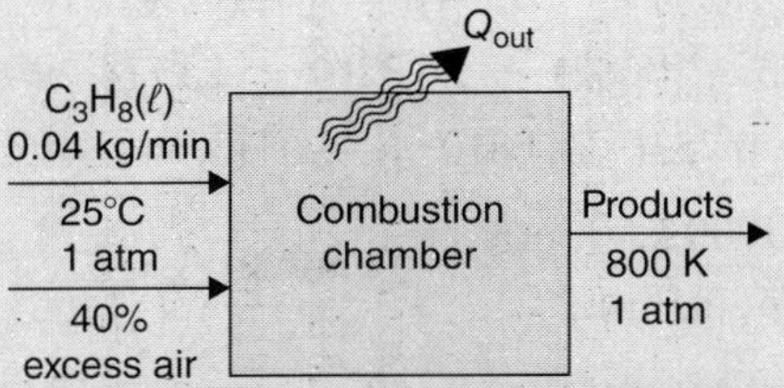

Figure 12.15

(*Continued*)

Example 12.9 (*Continued*)

With 40% excess air, the moles of oxygen are multiplied by 1.4 (see Eq. 12.4), leading to the equation

$$C_3H_8 + 7(O_2 + 3.76N_2) \Rightarrow 3CO_2 + 4H_2O + 2O_2 + 26.32N_2$$

To find the mass flux of the air into the chamber, first AF is found to be

$$AF = \frac{\text{mass air}}{\text{mass fuel}} = \frac{7 \times 4.76 \times 29\ \text{kg}_{\text{air}}}{1 \times 44\ \text{kg}_{\text{fuel}}} = 21.96\ \text{kg air/kg fuel}$$

Then, the mass flux of the air can be found knowing the mass flux of the fuel. It is

Note: The mass of water vapor in the air has been ignored since it is negligibly small.

$$\dot{m}_{\text{air}} = AF \times \dot{m}_{\text{fuel}}$$

$$= 21.96 \frac{\text{kg}_{\text{air}}}{\text{kg}_{\text{fuel}}} \times 0.04 \frac{\text{kg}_{\text{fuel}}}{\text{min}} = \underline{0.878\ \text{kg}_{\text{air}}/\text{min}}$$

From the psychrometric chart in Appendix G for air at 25°C and a relative humidity of 70%, the humidity ratio is 0.014. The mass flow rate of water entering the combustion chamber can be calculated to be

$$\dot{m}_{\text{water}} = \dot{m}_{\text{air}} \times \omega \times \frac{M_{\text{water}}}{M_{\text{air}}}$$

$$= 0.878 \times 0.014 \times \frac{18}{29} = 0.012 \frac{\text{kg}_{\text{vapor}}}{\text{min}}$$

The number of moles of water in the air per mole of dry air is:

$$\frac{N_{\text{water}}}{N_{\text{air}}} = \omega \times \frac{M_{\text{air}}}{M_{\text{water}}} = 0.014 \times \frac{29}{18} = 0.0226$$

The combustion equation can now be modified to include the moisture content of the air.

$$C_3H_8 + 7(O_2 + 3.76N_2) + (7 \times 4.76 \times 0.0226)H_2O$$
$$\Rightarrow 3CO_2 + 4.75H_2O + 2O_2 + 26.32N_2$$

The first law, Eq. 12.8, applied to the combustion chamber takes the form

$$Q = H_P - H_R$$
$$= \{3(-393\,520 + 32\,179 - 9364) + 4.75(-241\,810 + 27\,896 - 9904)$$
$$+ 2(0 + 24\,523 - 8682) + 26.32(0 + 23\,714 - 8669)\}$$
$$- \{-103\,850 + 0.753(-241\,810 + 9904)\} = -1\,470\,000\ \text{kJ/kmol fuel}$$

$$\text{or} \quad Q_{\text{out}} = \frac{1\,470\,000\ \text{kJ/kmol}}{44\ \text{kg/kmol}} = 33\,400\ \text{kJ/kg}_{\text{fuel}}$$

The rate of heat transfer from the chamber is

$$\dot{Q}_{\text{out}} = 33\,400 \frac{\text{kJ}}{\text{kg}_{\text{fuel}}} \times 0.04 \frac{\text{kg}_{\text{fuel}}}{\text{min}} \times \frac{1\ \text{min}}{60\ \text{s}} = \underline{22.3\ \text{kJ/s}}$$

Combustion in a rigid bomb Example 12.10

A rigid bomb, shown in Fig. 12.16, contains propane and oxygen at 25°C and a pressure of 100 kPa. Stoichiometric combustion occurs in an instant. The final temperature in the bomb some time after combustion is measured to be 1727°C. Determine the final pressure of the products and the heat transfer that occurred from the bomb.

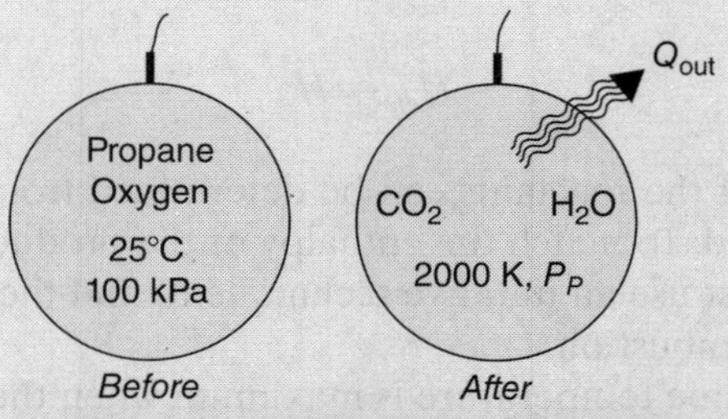

Figure 12.16

Solution

Stoichiometric combustion occurs according to the reaction

$$C_3H_8 + 5O_2 \Rightarrow 3CO_2 + 4H_2O$$

The final pressure can be found using the ideal-gas law. It takes the forms

$$\left.\begin{aligned} P_R V_R &= N_R R_u T_R \\ P_P V_P &= N_P R_u T_P \end{aligned}\right\} \quad \frac{P_R \cancel{V_R}}{P_P \cancel{V_P}} = \frac{N_R \cancel{R_u} T_R}{N_P \cancel{R_u} T_P}$$

$$\frac{100}{P_P} = \frac{6 \times 298}{7 \times 2000} \qquad \therefore P_P = \underline{783 \text{ kPa}}$$

The heat transfer can be found by applying the first law, represented by Eq. 12.13. Using the ideal-gas tables in the Appendix, using $T_P = 2000°C$, the first law provides

$$Q = \sum_{\text{prod}} N_i(\bar{h}_f^\circ + \bar{h} - \bar{h}^\circ - R_u T)_i - \sum_{\text{react}} N_i(\bar{h}_f^\circ + \bar{h} - \bar{h}^\circ - R_u T)_i$$

$$= \{3(-393\,520 + 100\,804 - 9364 - 8.314 \times 2000)$$
$$+4(-241\,810 + 82\,593 - 9904 - 8.314 \times 2000))\}$$
$$-\{1(-103\,850 - 8.314 \times 298) + 5(0 - 8.314 \times 298)\}$$

$$= -1\,580\,000 \text{ kJ/kmol fuel}$$

The heat transfer from the bomb is then

$$Q_{out} = -Q = \frac{1\,580\,000 \text{ kJ/kmol}}{44 \text{ kg/kmol}} = \underline{35\,920 \text{ kJ/kg fuel}}$$

Note: Our sign convention is heat transfer in is positive and out is negative.

☑ You have completed Learning Outcome (2)

12.4 Flame Temperature

Adiabatic flame temperature: The temperature after combustion of a constant-pressure flow through an insulated combustion chamber.

If all of the heat developed during combustion remains in the products of combustion and if kinetic and potential energy is negligible for a constant-pressure flow through the combustion chamber, the temperature of the products is called the *adiabatic flame temperature* since no heat is removed from the control volume. To determine the adiabatic flame temperature, we apply the first law of thermodynamics to the control volume knowing that $Q = 0$. This modifies Eq. 12.8 to the simpler form

$$H_R = H_P \tag{12.14}$$

The total enthalpy of the reactants can be determined from information on the fuel and oxidizer used. To match the enthalpy of the products to the enthalpy of the reactants, we must use an iterative technique to find the correct temperature of the products of combustion.

The adiabatic flame temperature is maximum when theoretical air is mixed with the fuel in the insulated combustor. The adiabatic flame temperature can be reduced by adding excess air to the mixture. This addition of air does not alter the first-law equation expressed by Eq. 12.14. The temperature of the products of combustion could also be lowered to a specified flame temperature by transferring heat from the combustor; an example of that calculation was given by Example 12.9. If a specified amount of heat is transferred but the temperature is not known, a trial-and-error procedure is also required. Examples will illustrate.

Example 12.11 Maximum adiabatic flame temperature

Liquid octane (gasoline) and stoichiometric air enter an insulated combustion chamber at 25°C and 100 kPa, as shown in Fig. 12.17. Find the temperature after combustion in the exiting gases, neglecting kinetic and potential energy changes and assuming no shaft work.

Solution

The first step is to determine the stoichiometric equation for the combustion of octane (C_8H_{18}) in air. It is

Comment

Dry air is assumed unless otherwise stated.

$$C_8H_{18}(\ell) + 12.5(O_2 + 3.76N_2) \Rightarrow 8CO_2 + 9H_2O + 47N_2$$

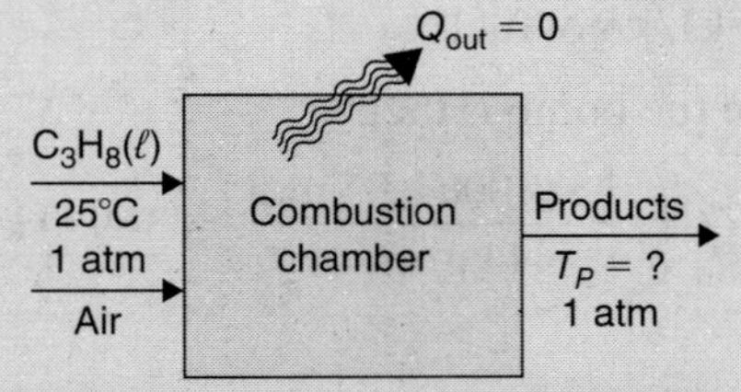

Figure 12.17

One mole of gasoline (octane) produces 8 moles of carbon dioxide and 9 moles of water. The 47 moles of nitrogen heat up, so it affects the exiting temperature T_P of the products.

Let's determine the molar enthalpy of the reactants. From Table B-7 the enthalpy of formation for liquid octane is listed as −208 450 − 41 460 = −249 910 kJ/kmol. The oxygen and nitrogen are at 298 K, so $\bar{h}_f^\circ + (\bar{h} - \bar{h}^\circ) = 0$ for both of those reactants. Hence, the total molar enthalpy of the reactants is

$$\bar{h}_R = \bar{h}^\circ_{f,C_8H_{18}} = -249\,910 \text{ kJ/kmol}$$

Next, the sum of the enthalpies of the three components that make up the products at the unknown final temperature T_P forms $\bar{h}_P$. The enthalpy for each component is listed as follows:

$$CO_2:\ \bar{h}_{CO_2} = \bar{h}_f^\circ + \bar{h}_{CO_2} - \bar{h}_{298\text{ K}} = -393\,520 + \bar{h}_{CO_2} - 9364$$

$$H_2O:\ \bar{h}_{H_2O} = \bar{h}_f^\circ + \bar{h}_{H_2O} - \bar{h}_{298\text{ K}} = -241\,810 + \bar{h}_{H_2O} - 9904$$

$$N_2:\ \bar{h}_{N_2} = \bar{h}_f^\circ + \bar{h}_{N_2} - \bar{h}_{298\text{ K}} = \bar{h}_{N_2} - 8669$$

Using $\bar{h}_R = \bar{h}_P$, the above are combined into the first law which states that the energy contained in the fuel and air at 25°C must equal the energy contained in the products at T_P, expressed by Eq. 12.14. The equation simplifies to

$$5\,650\,000 = 8\bar{h}_{CO_2} + 9\bar{h}_{H_2O} + 47\bar{h}_{N_2}$$

The objective is to find T_P that balances the above equation. It is a trial-and-error procedure. Guess a value for T_P and check the equation. The first guess will undoubtedly be incorrect. But with two or three guesses, an answer can usually be obtained. Two of our guesses follow:

2200 K: $5650000 \stackrel{?}{=} 8 \times 112900 + 9 \times 92900 + 47 \times 72000 = 5120000$

2600 K: $5650000 \stackrel{?}{=} 8 \times 137400 + 9 \times 114300 + 47 \times 86600 = 6200000$

The temperature is in between 2200 K and 2600 K and is interpolated to be (see Fig. 12.18)

$$T_P = \frac{5650 - 5120}{6200 - 5120}(2600 - 2200) + 2200 = \underline{2400\text{ K}}$$

This answer represents the maximum adiabatic flame temperature since stoichiometric air was used. Excess air can be used to decrease the exiting temperature, a desired result if the temperature cannot be as high as 2400 K because of certain materials exposed to the products of combustion. This will be illustrated in the following example.

Observe: At 2200 K, the rhs is too low. At 2600 K, the rhs is too high. It must be between the two temperatures.

Hint: A sketch of the interpolation may be helpful, as in Fig. 12.18.

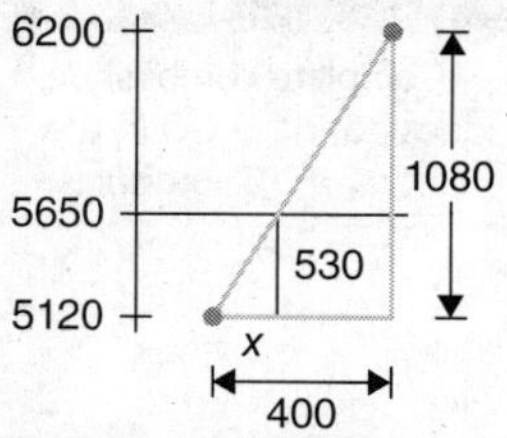

Figure 12.18

Note: The answer would be 2396 K, but that's four significant digits that are not supported by the three significant digits in the numbers used to generate the interpolation. We should not use more significant digits in an answer than are allowed by the numbers used in the calculations.

Example 12.12 Adiabatic flame temperature with excess air

The octane liquid of Example 12.11 undergoes complete combustion with 200% excess air in the insulated combustion chamber, using the same inlet conditions and the same assumptions. Estimate the temperature of the exhaust gases leaving the chamber of Fig. 12.19. The lower temperature gases would have less damaging effects on the metallic parts of an engine located after the combustion process.

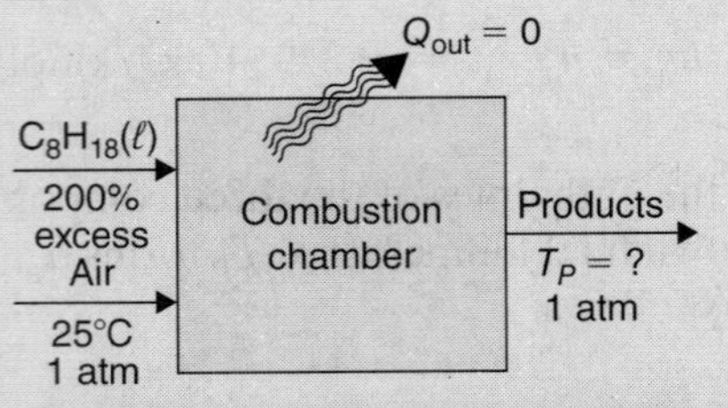

Figure 12.19

Solution:

The stoichiometric equation for the combustion of octane (C_3H_8) in air is

$$C_8H_{18}(\ell) + 12.5(O_2 + 3.76N_2) \Rightarrow 8CO_2 + 9H_2O + 47N_2$$

Note: For 200% excess air, the moles of air are tripled (see Eq. 12.5).

From the definition of excess air, 300% theoretical air enters the combustion chamber, so the reaction equation with the excess air becomes

$$C_8H_{18} + 37.5(O_2 + 3.76N_2) \Rightarrow 8\,CO_2 + 9H_2O + 25\,O_2 + 141\,N_2$$

Comment

Complete combustion does not have CO in the products.

Observe the extra oxygen in the products.

Let's determine the molar enthalpy of the reactants. From Table B-7 the enthalpy of formation for liquid octane is −208 450 − 41 460 = −249 910 kJ/kmol. The oxygen and nitrogen are at 298 K, so $\bar{h}_f^\circ + (\bar{h} - \bar{h}^\circ) = 0$ for both of those reactants. Hence, the total molar enthalpy of the reactants is

$$\bar{h}_R = \bar{h}^o_{f,C_8H_{18}} = -249\,910 \text{ kJ/kmol fuel}$$

Next, the enthalpy of each product at the unknown final temperature is listed as follows:

$$CO_2\text{:}\quad \bar{h}_{CO_2} = \bar{h}_f^\circ + \bar{h}_{CO_2} - \bar{h}_{298\text{ K}} = -393\,520 + \bar{h}_{CO_2} - 9364$$

$$H_2O\text{:}\quad \bar{h}_{H_2O} = \bar{h}_f^\circ + \bar{h}_{H_2O} - \bar{h}_{298\text{ K}} = -241\,810 + \bar{h}_{H_2O} - 9904$$

$$O_2\text{:}\quad \bar{h}_{O_2} = \bar{h}_f^\circ + \bar{h}_{O_2} - \bar{h}_{298\text{ K}} = \bar{h}_{O_2} - 8682$$

$$N_2\text{:}\quad \bar{h}_{N_2} = \bar{h}_f^\circ + \bar{h}_{N_2} - \bar{h}_{298\text{ K}} = \bar{h}_{N_2} - 8669$$

The sum of the four terms forms $\bar{h}_P$. Using $\bar{h}_R = \bar{h}_P$, the above are combined into Eq. 12.14, which states that the energy contained in the fuel and air at 25°C must equal the energy contained in the products at T_P. The equation is

$$6\,673\,000 = 8\bar{h}_{CO_2} + 9\bar{h}_{H_2O} + 50\bar{h}_{O_2} + 141\bar{h}_{N_2}$$

The objective is to find T_P that balances the preceding equation. Guess a value for T_P and check the equation. With two or three guesses, an answer can usually be obtained. Two guesses, one above $\bar{h}_R$ and one below $\bar{h}_R$, are

1200 K: $6\,673\,000 \stackrel{?}{=} 8 \times 53\,800 + 9 \times 43\,400 + 50 \times 38\,400 + 141 \times 36\,800 = 7\,930\,000$

1000 K: $6\,673\,000 \stackrel{?}{=} 8 \times 42\,800 + 9 \times 35\,900 + 50 \times 31\,400 + 141 \times 30\,100 = 6\,480\,000$

The temperature is closer to 1000 K than 1200 K and is interpolated to be

$$T_P = \frac{6673 - 6480}{7930 - 6480}(1200 - 1000) + 1000 = \underline{1030\text{ K}}$$

The exit temperature has been substantially reduced, which is one reason why excess air is often used. Without excess air, the exit temperature may be too high for the materials encountered by the products.

Flame temperature with heat transfer — Example 12.13

Gaseous propane reacts with 100% theoretical air both at 25°C and atmospheric pressure in the steady-flow combustion process displayed in Fig. 12.20. Determine the flame temperature if heat is transferred from the combustor at the rate of 20 MJ/kg of fuel. Neglect kinetic and potential energy changes as well as any shaft work.

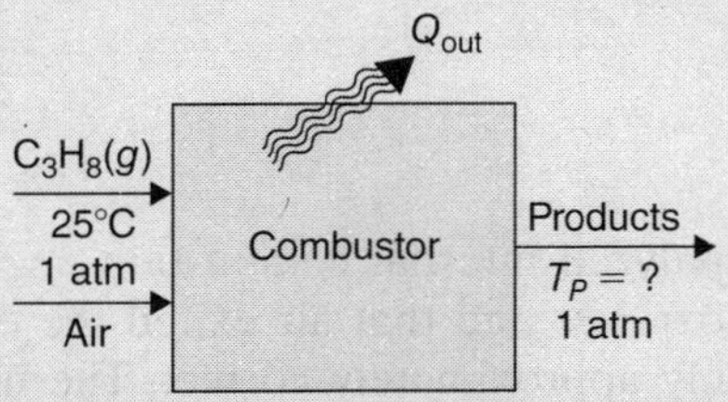

Figure 12.20

Solution

The stoichiometric equation for the combustion of propane (C_3H_8) in air is

$$C_3H_8 + 5(O_2 + 3.76N_2) \Rightarrow 3CO_2 + 4H_2O + 18.8N_2$$

One mole of gaseous propane produces 3 moles of carbon dioxide and 4 moles of water. The inert nitrogen does require some heat to reach the exiting temperature.

(Continued)

Example 12.13 (*Continued*)

From Table B-7 the enthalpy of formation for gaseous propane is listed as −103850 kJ/kmol. The oxygen and nitrogen are at 298 K so $\bar{h}_f^\circ + (\bar{h} - \bar{h}^\circ) = 0$ for both of those reactants. Hence, the total molar enthalpy of the reactants is

$$\bar{h}_R = \bar{h}^\circ_{f,C_3H_8} = -103850 \text{ kJ/kmol}$$

Next, the enthalpy of each product at the final temperature is listed as follows:

$$CO_2: \ \bar{h}_{CO_2} = \bar{h}_f^\circ + \bar{h}_{CO_2} - \bar{h}_{298\,K} = -393520 + \bar{h}_{CO_2} - 9364$$

$$H_2O: \ \bar{h}_{H_2O} = \bar{h}_f^\circ + \bar{h}_{H_2O} - \bar{h}_{298\,K} = -241810 + \bar{h}_{H_2O} - 9904$$

$$N_2: \ \bar{h}_{N_2} = \bar{h}_f^\circ + \bar{h}_{N_2} - \bar{h}_{298\,K} = \bar{h}_{N_2} - 8669$$

Note: The heat transfer $\bar{q}$ is negative since it leaves the combustor.

The energy balance $\bar{q} = \bar{h}_P - \bar{h}_R$, with the heat transfer $\bar{q} = -20000 \times 44 = -880000$ kJ/kmol, simplifies to

$$1400000 = 3\bar{h}_{CO_2} + 4\bar{h}_{H_2O} + 18.8\bar{h}_{N_2}$$

The objective is to find T_P that balances the above equation. Guess a value for T_P and check the equation. Two guesses, one above $\bar{h}_R$ and one below $\bar{h}_R$, are

1400 K: $1400000 \overset{?}{=} 3 \times 65300 + 4 \times 45700 + 18.8 \times 43600 = 1200000$

1800 K: $1400000 \overset{?}{=} 3 \times 88800 + 4 \times 60400 + 18.8 \times 57700 = 1600000$

The temperature is halfway between 1400 K and 1800 and is close to 1600 K.

If the exit temperature is known and the heat transfer is desired, that problem does not require a trial-and-error solution.

Observation: The two vapor trails made by those high-flying airplanes are due to the water vapor from the engines and the atmosphere that gets collected into the two extremely strong vortices created by the two wings of the aircraft.

In Chapter 9 we analyzed internal combustion engines assuming that the engine used air as the oxidizer and that air exited the engine as exhaust. Now we know that this is only approximately correct. The nitrogen passes through the engine unchanged, but the oxygen is converted to carbon dioxide and water. Knowing this and using the tools developed in this chapter, we could perform a more accurate analysis of engine cycles. The error that would be found using the more accurate analysis would be relatively small since the water and carbon dioxide make up a small percentage of the products. So, the comparisons made in Chapter 9 are valid. However, it is now obvious why water vapor is always present in the products of combustion and can be observed under certain conditions from the tailpipe of a motor vehicle, or the contrails behind a high-flying aircraft.

☑ You have completed Learning Outcome (3)

12.5 Equilibrium Reactions

This chapter has demonstrated that the products created in the stoichiometric combustion of a hydrocarbon fuel are carbon dioxide and water. So where do the pollutants come from that exist in automobile exhaust? At very high temperatures, other reactions are taking place during the combustion process. These reactions are different from the combustion reactions previously in that the reaction is a two-way process. In these equilibrium reactions, a compound will *dissociate* into smaller compounds or atoms at high temperature. At the same time, these smaller compounds will re-associate back into the original compound. Both the dissociation and re-association processes take place simultaneously. The relative amounts of the various compounds will be a function of the temperature of the substances.

Note: The nitrogen isn't created when air is the reactant. It just tags along.

To describe these equilibrium reactions, use will be made of the Gibbs function defined in Chapter 7:

$$G = H - TS \tag{12.15}$$

The differential of this equation is

$$dG = dH - TdS - SdT \tag{12.16}$$

From Chapter 6 we have

$$TdS = dH - VdP \tag{12.17}$$

Combining these two equations yields the following relationship:

$$dG = VdP - SdT \tag{12.18}$$

Under equilibrium conditions, both pressure and temperature are assumed constant; Eq. 12.18 then provides

$$dG = 0 \quad \text{so that} \quad G = \text{const} \tag{12.19}$$

An equilibrium reaction is a process during which the Gibbs function is constant. It can be expressed as

$$\text{Reactants} \Leftrightarrow \text{Products} \tag{12.20}$$

For an equilibrium reaction, Eq. 12.19 shows the conservation of Gibbs function for this process:

$$\sum \nu_R g_R = \sum \nu_P g_P \tag{12.21}$$

where ν (the Greek letter nu) represents a *stoichiometric coefficient*. The specific Gibbs function g can be calculated from properties using Eq. 12.15:

$$g = h - Ts \tag{12.22}$$

If an equilibrium reaction occurs during which chemicals A and B create chemicals C and D, the reverse reaction will occur simultaneously. Chemicals C and D will re-form chemicals A and B. For a constant Gibbs function process, this is represented by

$$\nu_A A + \nu_B B \Leftrightarrow \nu_C C + \nu_D D \tag{12.23}$$

$$\nu_A g_A + \nu_B g_B = \nu_C g_C + \nu_D g_D \tag{12.24}$$

The double arrows show that the equation works in both directions.

For a constant-temperature process, Eq. 12.16 can be simplified:

$$\begin{aligned} dG &= VdP - SdT \\ &= VdP \end{aligned} \tag{12.25}$$

Since the ideal-gas law states that $PV = NR_uT$, we can insert the volume V into Eq. 12.25 and define the differential of the Gibbs function in terms of pressure to be

$$dG = NR_uT\frac{dP}{P} \tag{12.26}$$

Divide this equation by the number of moles N and the differential of the molar Gibbs function is

$$d\bar{g} = R_uT\frac{dP}{P} \tag{12.27}$$

Integrate this equation between the limits of a reference pressure of one atmosphere and an arbitrary upper limit P:

$$\int_{\text{ref}}^{\bar{g}} d\bar{g} = R_uT\int_{1\,\text{atm}}^{P}\frac{dP}{P} \tag{12.28}$$

This integration yields an expression for the molar Gibbs function at a pressure P in terms of a reference molar Gibbs function at one atmosphere, expressed as

$$\bar{g}_p = \bar{g}_{\text{ref}} + R_uT\ln P \tag{12.29}$$

For the equilibrium reaction $A + B \Leftrightarrow C + D$ we can now express the following:

$$\begin{aligned} \nu_A\bar{g}_A &= \nu_A\bar{g}_{A_{\text{ref},A}} + \nu_AR_uT\ln P_A \\ \nu_B\bar{g}_B &= \nu_B\bar{g}_{B_{\text{ref},B}} + \nu_BR_uT\ln P_B \\ \nu_C\bar{g}_C &= \nu_C\bar{g}_{C_{\text{ref},C}} + \nu_CR_uT\ln P_C \\ \nu_D\bar{g}_D &= \nu_D\bar{g}_{D_{\text{ref},D}} + \nu_DR_uT\ln P_D \end{aligned} \tag{12.30}$$

In these equations P_A, P_B, P_C, and P_D are the partial pressures of constituents A, B, C, and D in atmospheres, as required by Eq. 12.28. The total pressure is

$$P = P_A + P_B + P_C + P_D \tag{12.31}$$

When expressions (12.30) are substituted back into Eq. 12.24, the following reaction equation results:

Equilibrium constant: A dimensionless constant used to determine the equilibrium composition of reacting gases.

$$\nu_C\bar{g}_{\text{ref}_C} + \nu_D\bar{g}_{\text{ref}_D} - \nu_A\bar{g}_{\text{ref}_A} - \nu_B\bar{g}_{\text{ref}_B} = R_uT\ln\left[\frac{P_A^{\nu_A}P_B^{\nu_B}}{P_C^{\nu_C}P_D^{\nu_D}}\right] \tag{12.32}$$

An *equilibrium constant* K_P, a dimensionless constant used to determine the equilibrium composition of reacting gases, is defined to be

$$K_P = \frac{P_C^{\nu_C}P_D^{\nu_D}}{P_A^{\nu_A}P_B^{\nu_B}} \tag{12.33}$$

Equation 12.32 now takes the form

Note: We have used $\ln(K_P)^{-1} = -\ln K_P$

$$\ln K_P = \frac{\nu_A\bar{g}_{\text{ref}_A} + \nu_B\bar{g}_{\text{ref}_B} - \nu_C\bar{g}_{\text{ref}_C} - \nu_D\bar{g}_{\text{ref}_D}}{R_uT} \tag{12.34}$$

Table B-9 in the Appendix lists the natural logarithms of the equilibrium constants for a variety of equilibrium reactions as a function of the absolute temperature of a gas.

The equilibrium composition of products Example 12.14

A mixture of 50% carbon dioxide and 50% oxygen, by volume, is heated in a steady-flow process at 3000 K and 1 atm of pressure. Determine the equilibrium composition of the products.

Solution

Carbon dioxide and oxygen enter the reaction. Some of the carbon dioxide dissociates into carbon monoxide and oxygen. The reaction equation is

$$CO_2 + O_2 \Rightarrow aCO_2 + bO_2 + cCO$$

A carbon balance demands $1 = a + c$. An oxygen balance demands $4 = 2a + 2b + c$. The dissociation reaction is written as

$$CO_2 \Leftrightarrow CO + \frac{1}{2}O_2$$

The natural log of the equilibrium constant for this dissociation reaction at 3000 K is obtained from Table B-9:

$$\ln K_P = -1.111 \qquad \therefore K_P = e^{-1.111} = 0.329$$

Expressions are written for the partial pressures of each component:

$$P_{O_2} = \frac{b}{a+b+c}P \qquad P_{CO_2} = \frac{a}{a+b+c}P \qquad P_{CO} = \frac{c}{a+b+c}P$$

The equation for K_P is written in terms of these partial pressures using Eq. 12.33. It takes the form

$$K_P = \frac{P_{CO}P_{O_2}^{1/2}}{P_{CO_2}} = 0.329$$

$$K_P = \frac{\dfrac{c}{a+b+c}P\left(\dfrac{b}{a+b+c}P\right)^{1/2}}{\dfrac{a}{a+b+c}P} = \frac{c}{a}\left(\frac{b}{a+b+c}P\right)^{1/2} = 0.329$$

We know from the chemical balances that

$$a = 1 - c \quad \text{and} \quad b = \frac{4 - 2a - c}{2} = \frac{4 - 2(1-c) - c}{2} = \frac{2+c}{2}$$

(Continued)

Example 12.14 (Continued)

This gives

$$K_P = \frac{c}{a}\left(\frac{b}{a+b+c}P\right)^{1/2} = 0.329 = \frac{c}{1-c}\left[\frac{2-c}{4-c}\right]^{1/2}$$

using $P = 1$ atm. This equation cannot be solved directly for a value of c. By trial and error, we can get a value of $c = 0.328$.

$$c = 0.328 \qquad a = 1 - c = 0.672 \qquad b = \frac{2+c}{2} = 1.164$$

The full reaction equation is now written as

$$CO_2 + O_2 \Leftrightarrow 0.672\,CO_2 + 1.164\,O_2 + 0.328\,CO$$

☑ **You have completed Learning Outcome (4)**

12.6 Summary

In this chapter, attention was focused on the chemical reaction and the energy emitted when combustion occurred, since that is of considerable interest to engineers. In previous chapters, should a mixture occur, it was homogeneous throughout the process, and one component did not react with another component. In this chapter, attention was focused on steady-flow processes, often at constant pressure because of its common occurrence, but a few constant-volume processes were also considered. Neither isothermal nor isentropic processes were of interest. Equilibrium reactions that occur during the combustion process were also analyzed.

The following terms were introduced or emphasized in this chapter:

Adiabatic flame temperature: *The temperature after combustion from an insulated combustor.*

Air–fuel ratio: *The ratio of the mass of air to the mass of fuel.*

Combustion: *The extremely rapid chemical oxidation of a fuel.*

Complete combustion: *The carbon burns to CO_2 and the hydrogen burns to H_2O.*

Endothermic reaction: *A reaction that requires heat to form.*

Enthalpy of combustion: *The energy emitted when a compound undergoes stoichiometric combustion.*

Enthalpy of formation: *The energy emitted or required to form a compound from its stable elements.*

Equivalence ratio: *The air/fuel ratio using stoichiometric air divided by the actual air–fuel ratio.*

Exothermic reaction: *A reaction that emits heat when formed.*

Flame temperature: *The temperature after a combustor that is not insulated.*
Fuel–air ratio: *The ratio of the mass of fuel to the mass of air.*
Higher heating value: *When the water in the products is in liquid form.*
Lower heating value: *When the water in the products is in gaseous form.*
Products of combustion: *The terms typically on the right of the reaction equation.*
Reactants: *The terms typically on the left of the reaction equation.*
Running lean: *When more air is used than is necessary for a stoichiometric reaction.*
Running rich: *When insufficient air is present to burn all the fuel.*
Stoichiometric coefficients: *Coefficients used in a stoichiometric reaction.*
Stoichiometric reaction: *Exactly enough oxidizer to combust all the fuel.*
Theoretical air: *Just enough air for a stoichiometric reaction.*

Several important equations were introduced in this chapter:

A sample chemical reaction: $CH_4 + 2(O_2 + 3.76N_2) \Rightarrow CO_2 + 2H_2O + 7.52N_2$

Air–fuel ratio: $AF = \dfrac{\text{mass of air}}{\text{mass of fuel}}$

Fuel–air ratio: $FA = \dfrac{\text{mass of fuel}}{\text{mass of air}}$

Equivalence ratio: $\phi = \dfrac{AF_{\text{stoichiometric}}}{AF_{\text{actual}}}$

Emitted heat (first law): $Q = \overline{H}_P - \overline{H}_R$

Equilibrium constant $K_P = \dfrac{P_C^{\nu_C} P_D^{\nu_D}}{P_A^{\nu_A} P_B^{\nu_B}}$

Problems

FE-Style Problems

(Combustion problems are not usually on the FE Exam.)

12.1 Which of the following is always present after combustion of a hydrocarbon fuel?

(A) H_2O and CO_2
(B) H_2O and N_2
(C) N_2 and CO_2
(D) H_2O, N_2, and CO_2

12.2 Benzene (C_6H_6) is burned with stoichiometric air. The number of nitrogen moles in the products of combustion for each mole of fuel is:

(A) 5.64
(B) 11.28
(C) 16.92
(D) 28.2

12.3 The air–fuel ratio in the reaction

$$CH_4 + 2(O_2 + 3.76N_2) \Rightarrow CO_2 + 2H_2O + 7.52N_2$$

is nearest:

(A) 2.0

(B) 3.6

(C) 17.3

(D) 23.2

12.4 Propane (C_3H_8) is burned with 200% excess air, resulting in complete combustion. The number of moles of oxygen in the products of combustion is:

(A) 5

(B) 10

(C) 20

(D) 30

Note: Dry air is always assumed unless otherwise noted.

12.5 Octane (C_8H_{18}) is burned with 80% stoichiometric air, resulting in incomplete combustion with CO and 4 moles of unburned carbon being the only additional components in the products. The number of moles of CO_2 in the products of combustion is:

(A) 2.2

(B) 2.8

(C) 3.4

(D) 3.6

12.6 The following reaction equation represents complete combustion of a fuel. Calculate the percent excess air.

$$C_2H_6 + 6.5(O_2 + 3.76N_2) \Rightarrow 2CO_2 + 3H_2O + 3O_2 + 24.44N_2$$

(A) 56%

(B) 86%

(C) 136%

(D) 176%

12.7 One mole of hydrogen and dry air enter a combustion chamber at 25°C and 1 atm and leave at 800 K. Calculate the heat transfer from the chamber for stoichiometric air.

(A) 200 MJ

(B) 150 MJ

(C) 100 MJ

(D) 50 MJ

12.8 Hydrogen and theoretical dry air combust in an insulated combustion chamber. The adiabatic flame temperature is nearest:

(A) 1200 K

(B) 1600 K

(C) 2000 K

(D) 2400 K

Chemical Equations for Combustion

12.9 Develop the stoichiometric equation for the combustion in air of *a*) methane, *b*) benzene, *c*) octane, *d*) propane, and *e*) ethane.

Remember: Each part with a lower-case italic letter is a separate problem.

12.10 Develop the chemical equation for the combustion in 50% excess air of *a*) methane, *b*) benzene, *c*) octane, *d*) propane, and *e*) ethane. Assume complete combustion.

Note: Dry air is always assumed unless otherwise noted.

12.11 Develop the chemical equation for the combustion with 90% theoretical air of *a*) methane, *b*) benzene, *c*) octane, *d*) propane, and *e*) ethane. Assume that CO is the only additional compound in the products due to the lack of sufficient oxygen.

12.12 Develop the stoichiometric equation for the burning in air of:

a) one mole of propane and one mole of methane

b) one mole of methane and one mole of octane

c) one mole of octane and one mole of dodecane

d) one mole of methane, one mole of propane, and one mole of benzene

12.13 A fuel mixture of 20% ethane, 50% methane, and 30% propane, by volume, undergoes stoichiometric combustion. If the fuel enters the combustion chamber of Fig. 12.21 at 4 kg/hr, determine the volume flow rate of air required if the air is at 24°C and 100 kPa.

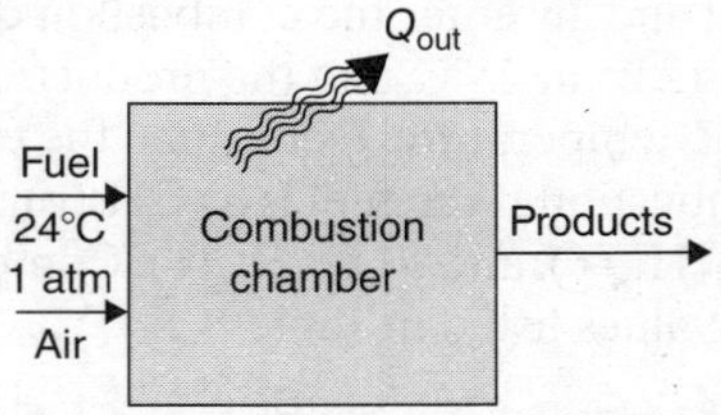

Figure 12.21

Remember: Assume standard atmospheric pressure unless otherwise stated. A combustion chamber is a heat exchanger, so the pressure is assumed constant throughout except in a control volume that may include a nozzle. A rigid volume is also an obvious exception.

12.14 Develop the chemical equation for burning a 50/50 (by volume) mixture of benzene and propane with 90% theoretical air. Calculate the air-fuel ratio, the volume percentage of CO_2 in the products, and the dew-point temperature of the products. Assume that CO is the only additional compound in the products.

12.15 Air enters a combustion chamber at 30°C with volume flow rate of 25 m^3/min. Determine the mass flux of fuel for stoichiometric combustion if the fuel is *a*) methane, *b*) benzene, *c*) octane, *d*) propane, and *e*) ethane.

12.16 Develop the chemical equation for the stoichiometric combustion of acetylene with pure oxygen. Calculate the oxygen-fuel ratio, the volume percentage of CO_2 in the products, and the dew-point temperature of the products.

12.17 Calculate the air-fuel ratio for the stoichiometric combustion of dodecane with 10% excess air, as shown in Fig. 12.22. Also determine the volume percentage of CO_2 in the products, and the dew-point temperature of the products.

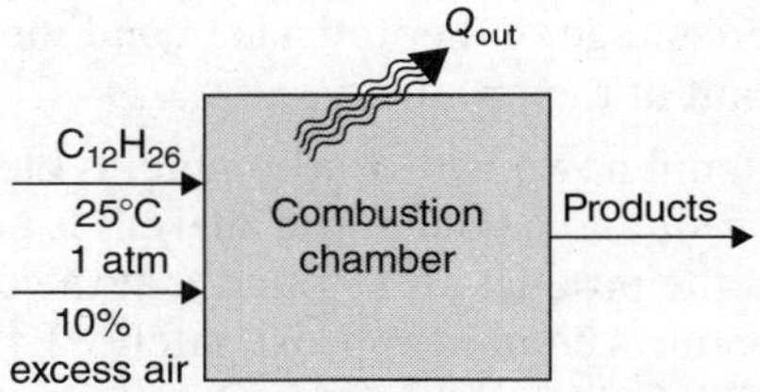

Figure 12.22

12.18 Calculate the air-fuel ratio for the combustion of kerosene (assumed to be $C_{12}H_{26}$) with 200% excess air in a gas turbine engine. Also, determine the volume percentage of CO_2 in the products, and the dew-point temperature of the products.

12.19 A fuel is composed, by volume, of 70% CH_4, 12% H_2, 8% O_2, and 10% CO_2. If complete combustion occurs with 75% humidity air at 20°C, determine the dew point of the products.

12.20 Methane is combusted with dry air, as displayed in Fig. 12.23. A dry volumetric analysis of the products indicates 6.2% CO_2, 9.9 % O_2, and 83.9% N_2. Estimate the air-fuel ratio and the percentage of theoretical air.

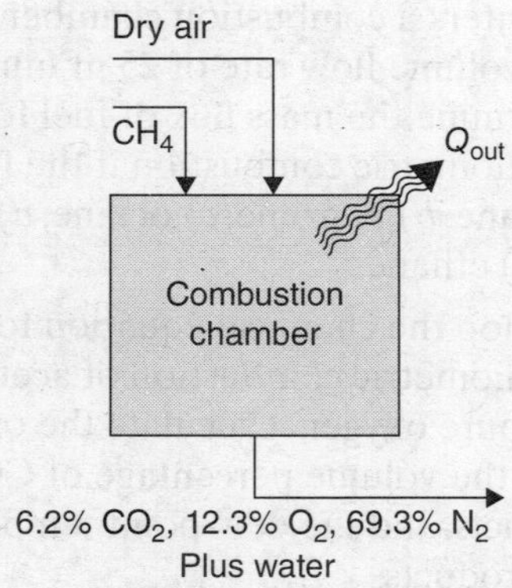

Figure 12.23

12.21 Octane and dry air enter a combustion chamber. A dry volumetric analysis of the products indicates 10.6% CO_2, 5.8% O_2, and 83.6% N_2. Determine the air–fuel ratio, the percentage of theoretical air, and the dew point of the products.

12.22 An unknown hydrocarbon fuel is burned with dry air and a volumetric analysis is performed on the products. It is found that the dry products contain approximately 83.4% N_2, 8.3% CO_2, and 8.3% O_2. Determine the probable composition of the hydrocarbon fuel (whole number atoms). Assume initially that one mole of the fuel is burned to produce 100 moles of dry products.

12.23 A hydrocarbon fuel C_aH_b is burned with dry air. Determine the composition of the fuel combusted (whole number atoms) and the percent theoretical air if a dry volumetric analysis of the products indicates

a) 10.37% CO_2, 5.93% O_2, and 83.7% N_2

b) 9.97 % CO_2, 6.64% O_2, and 83.4% N_2

c) 11.14% CO_2, 5.02% O_2, and 83.84% N_2

Assume initially that one mole of the fuel is burned to produce 100 moles of dry products.

12.24 A producer gas, possibly created from coal, is composed of 3.5% CH_4, 4.5% CO_2, 27% CO, 14% H_2, and 51% N_2. If complete combustion occurs in the combustion chamber of Fig. 12.24 with *a*) 120% theoretical air, *b*) 150% theoretical air, and *c*) 200% theoretical air, determine the air–fuel ratio and the dew-point temperature.

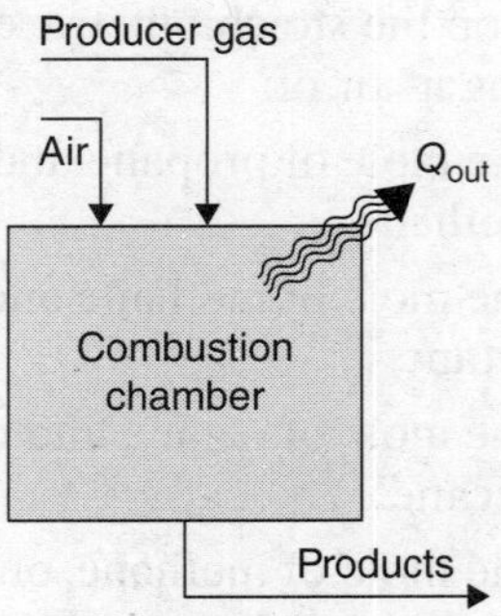

Figure 12.24

Enthalpy of Formation and Enthalpy of Combustion

12.25 Fuel and air enter the combustion chamber of Fig. 12.25 at 25°C, and the products exit at the same temperature. Determine the enthalpy of combustion if the fuel is *a*) $CH_4(g)$, *b*) $C_3H_8(g)$, *c*) $C_8H_{18}(\ell)$, and *d*) $C_{12}H_{26}(\ell)$. Compare with the values listed in Table B-8.

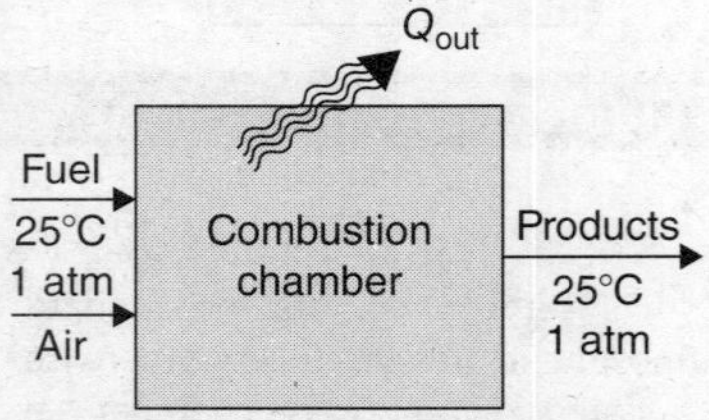

Figure 12.25

Comment

Temperatures at the inlet and the exit are important when calculating heat transfer. Standard atmospheric pressure will be assumed unless otherwise stated (as in a nozzle) throughout the combustion chamber.

Assumption: The air is assumed to be dry air unless otherwise stated.

12.26 A fuel is burned with 100% theoretical air. Both the fuel and air are at 25°C. The products of combustion exit the combustion process at 1200 K. Determine the heat transfer from the combustion chamber if the fuel is one mole of *a*) $CH_4(g)$, *b*) $C_3H_8(g)$, *c*) $C_8H_{18}(\ell)$, and *d*) $C_{12}H_{26}(\ell)$.

12.27 Liquid octane and 40% excess air enter a combustion chamber at 25°C, and the products exit at 827°C. What is the rate of heat transfer from the chamber if the flow rate of fuel is 4 kg/min and the relative humidity is *a*) 0%, *b*) 50%, *c*) 70%, and *d*) 85%? Assume complete combustion.

12.28 A fuel enters the combustion chamber of Fig. 12.26 at 25°C and is completely burned with 180% theoretical air possessing 60% relative humidity. The products of combustion leave the combustor at 200°C. Determine the rate of heat transfer for one mole of fuel if the fuel is *a*) $CH_4(g)$, *b*) $C_3H_8(g)$, *c*) $C_8H_{18}(\ell)$, *d*) $C_{12}H_{26}(g)$, and *e*) $H_2(g)$.

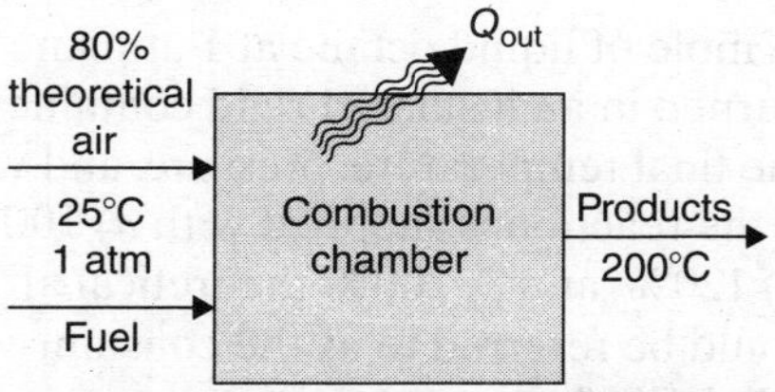

Figure 12.26

12.29 Ten pounds of powdered coal burns with 100% excess dry air each second. The reactants enter the combustion chamber at 25°C. The products of combustion leave the combustion chamber at 500°C. Calculate the rate of heat transfer per hour from the walls of the combustion chamber. Assume the coal to be 100% carbon (the better coal is about 90% carbon, so this represents an upper limit).

12.30 In a jet engine, liquid $C_{12}H_{26}$ at 25°C is added to air at 25°C in the insulated engine of Fig. 12.27. If the fuel is completely burned with 100% excess air and the products of combustion leave the nozzle at 1000 K, estimate the velocity of the products leaving the nozzle.

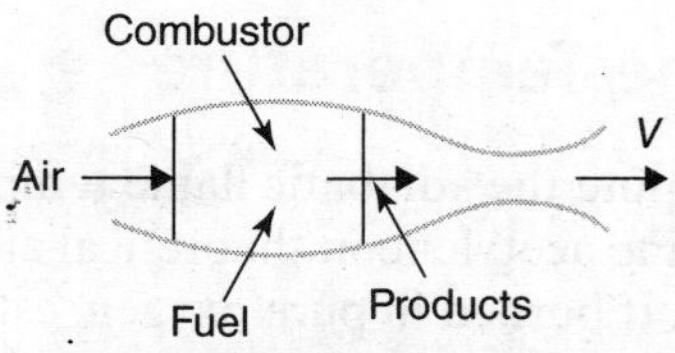

Figure 12.27

12.31 In a jet engine, liquid $C_{12}H_{26}$ at 25°C is added to air at 25°C in the insulated engine of Fig. 12.27 and burned. If the fuel is burned with 80% excess air and the products of combustion leave the nozzle at 1100 K, estimate the velocity of the products leaving the nozzle.

12.32 A rigid volume contains 0.1 kg of fuel and stoichiometric oxygen at 25°C and 100 kPa. Ignition occurs and the fuel undergoes complete combustion, resulting in a final temperature of 1000 K. Determine the heat transfer from the volume if the fuel is *a*) $CH_4(g)$, *b*) $C_3H_8(g)$, *c*) $C_8H_{18}(\ell)$, and *d*) $H_2(g)$.

12.33 The insulated, rigid 1-m³ volume in Fig. 12.28 contains 0.1 kg of fuel and 120% theoretical air at 25°C and 100 kPa. Ignition occurs and the fuel undergoes complete combustion. Determine the final pressure and temperature if the fuel is *a*) $CH_4(g)$, *b*) $C_3H_8(g)$, and *c*) $C_8H_{18}(\ell)$.

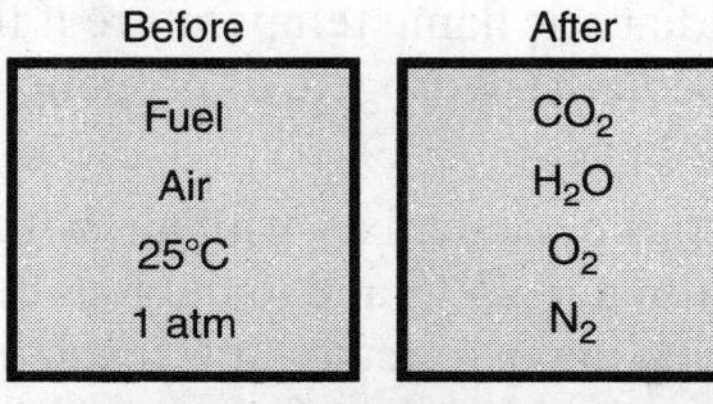

Figure 12.28

12.34 An insulated, rigid volume contains 0.4 kg of fuel that is 20% methane, 30% propane, and 50% butane. Stoichiometric air and the fuel are at 25°C and 100 kPa. Ignition occurs, and the fuel undergoes complete combustion with a final temperature of 660 K. Determine the final pressure, the volume, and the heat transfer.

Flame Temperature

12.35 Compute the adiabatic flame temperature for burning acetylene in theoretical air, both at 25°C. If burned in pure oxygen, estimate the temperature that could be expected.

Assumptions: Kinetic and potential energy changes are negligible unless otherwise stated.

12.36 Propane mixes with air, both initially at 1 atm and 25°C, and undergoes complete combustion in the steady-flow insulated combustor of Fig. 12.29. Calculate the adiabatic flame temperature for *a*) theoretical air, *b*) 50% excess air, and *c*) 200% theoretical air.

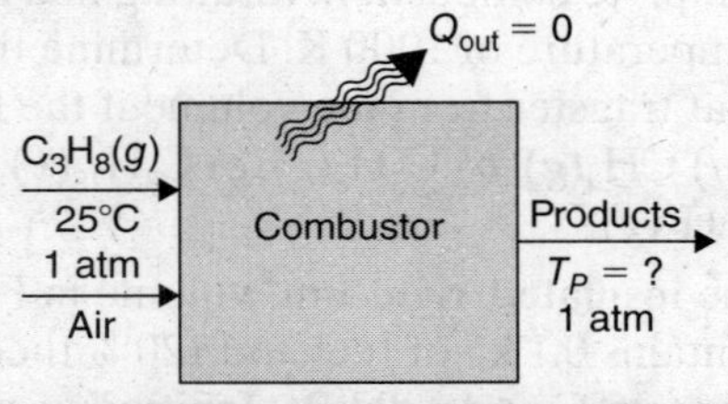

Figure 12.29

12.37 Stoichiometric air mixes with a fuel both of which are at 25°C and 100 kPa. Calculate the adiabatic flame temperature if the fuel is gaseous *a*) hydrogen, *b*) propane, and *c*) methane.

12.38 Gaseous octane mixes with air, both entering at 1 atm and 25°C, and undergoes complete combustion in a steady-flow insulated combustion chamber. Calculate the adiabatic flame temperature for *a*) theoretical air, *b*) 50% excess air, and *c*) 200% theoretical air.

12.39 Propane mixes with air, both entering at 1 atm and 25°C, and undergoes complete combustion in a steady-flow insulated combustion chamber. Calculate the adiabatic flame temperature for *a*) theoretical air, *b*) 50% excess air, and *c*) 200% theoretical air.

12.40 A 50/50 (by volume) mixture of ethane and methane is burned with 150% theoretical air, all gases at 25°C, as shown in Fig. 12.30. Calculate the adiabatic flame temperature.

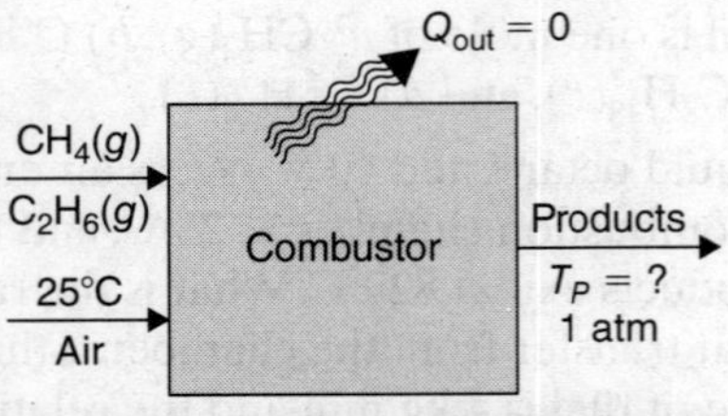

Figure 12.30

12.41 Liquid kerosene (essentially dodecane) is burned in air at 25°C and 50% relative humidity with an air–fuel ratio of 20. Calculate the adiabatic flame temperature.

12.42 Propane is burned with 100% excess dry air at 25°C, but only 75% of the fuel is burned in the reaction. Determine the adiabatic flame temperature for this reaction.

12.43 A mole of liquid octane at 1 atm and 25°C is burned in an insulated rigid container. Find the final temperature, pressure, and volume if the reaction is supplied with *a*) 100%, *b*) 120%, and *c*) 200% theoretical air. This could be referred to as the constant-volume adiabatic flame temperature.

12.44 Gaseous propane and theoretical air both at 25°C undergo combustion in a steady-flow process. Determine the flame temperature if heat transfer of *a*) 800 MJ/kmol of fuel, *b*) 1200 MJ/kmol of fuel, and *c*) 1600 MJ/kmol of fuel leaves the combustor of Fig. 12.31.

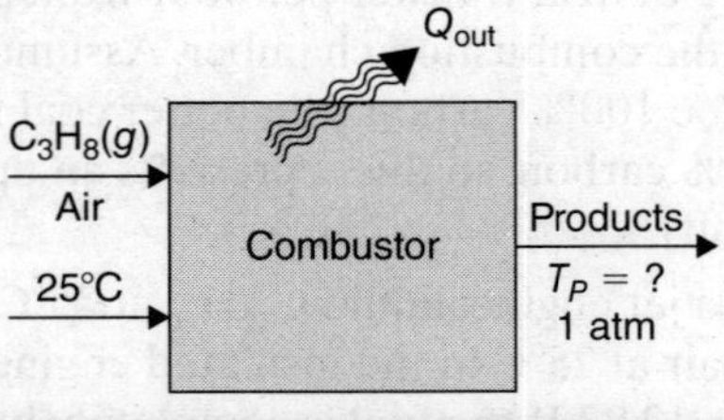

Figure 12.31

12.45 The liquid octane in a gas turbine is burned with 400% excess air that enters the combustion chamber at 7°C. The products of combustion leave the constant-pressure combustion chamber at 707°C. The fuel flow rate is 0.5 kg/s. Calculate the heat flow rate from the combustion chamber.

12.46 A rocket engine uses liquid hydrazine (N_2H_4) as a fuel and pure oxygen at 25°C as the oxidizer. Hydrazine has an enthalpy of formation of −50 410 kJ/kmol. The oxygen is supplied to the engine with an oxygen–fuel ratio of 3. The heat loss from the combustor is 200 kJ/kg of N_2H_4 burned. Calculate the flame temperature just before the nozzle assuming a constant-pressure combustion chamber.

Equilibrium Reactions

12.47 Oxygen exists at 500 kPa and 3000 K. In this state, a dissociation reaction occurs forming single oxygen atoms. Determine the partial pressure of the diatomic (O_2) and monatomic (O) oxygen.

12.48 For the dissociation reaction of water separating into hydrogen (H_2) and hydroxyl (OH) ions, plot the partial pressure of the hydrogen as a function of temperature for the temperature range of 1000 K to 5000 K.

12.49 One mole of water and one mole of CO_2 are placed in a reactor at 3000 K and a total pressure of 400 kPa. The water dissociates into hydrogen and oxygen, and the carbon dioxide dissociates into carbon monoxide and oxygen. Determine the partial pressures of the H_2O, the CO_2, the O_2, and the CO.

Part III

Contemporary Topics

Several alternative energy sources are presented in Chapter 13, sources that will aid in our nation's goal of energy sustainability. This is followed in Chapter 14 by a short introduction to the thermodynamics of living organisms. These two chapters are not intended to provide equations, examples, and home problems that would allow detailed calculations. They are primarily a description of sources of energy that do not have large-scale potential but, added together, are having increased importance to our sustainable energy production.

CHAPTER

13

Alternative Energy Conversion

Ashiga/Shutterstock.com

Outline

Learning Outcomes

- **Understand the many ways in which energy can be produced**
- **Develop an understanding of the environmental and sociological implications of power-plant design**
- **Develop an understanding of the economical factors involved in creating power**

These learning outcomes will be provided throughout the chapter and are not concentrated in particular articles. This chapter is not intended to provide the details of equations and worked out examples; it is for informational purposes only.

Motivational Example—Heat-Induced Magnetism

The University of Minnesota has developed a multiferroic alloy that undergoes a reversible phase transformation when heated. During this phase change, the alloy goes from being nonmagnetic to being highly magnetic. If the heated alloy is placed near a ceramic magnet, the increase in the magnetic field of the alloy causes a significant increase in the magnetic field of the magnet. If this device is surrounded by a wire coil, the change in the magnetic field of the core will induce an electric current in the coil. This device has potential to replace conventional power plants by directly creating electricity from heat. The Rankine cycle would no longer be needed.

13.1 Biofuels

Biofuels: Fuels made from organisms that were recently living (edible plants, inedible plants, animal waste, and algae).

The fuels used to provide the energy needed by the power plants of Chapter 8 and the fuels that power the engines of Chapter 9 are primarily fossil fuels. These include coal and petroleum products, such as gasoline, diesel fuel, and natural gas. Fossil fuels are fuels composed of organisms that have been dead for eons. However, the supply of fossil fuels is finite. We continue to discover new coal, oil, and natural gas fields, but this will stop at some point in time, although new discoveries are allowing that time to be extended. Fossil fuels are considered to be an unsustainable source of fuels over the hundreds of years to come.

Biofuels are fuels made from organisms that were recently living. These organisms vary from vegetable matter to animal by-products to algae. Biofuel sources can be divided into four categories: edible plants, inedible plants, animal waste, and algae.

Biofuels represent a sustainable alternative to fossil fuels. Following is a list of concerns about the safety and economics of using biofuels.

1. Can biofuels be produced on a scale necessary to replace fossil fuels? In 2011 less than 2% of the fuels being used were biofuels.
2. Biofuels are sustainable in that the plants, animals, and algae used to produce them are renewable. But what are the environmental and socioeconomic consequences of developing biofuels? How will the mass production and use of biofuels affect the environment? Does the production of biofuels from plants mean that impoverished people will not get the food they need?
3. What is the cost of mass-producing biofuels compared to the cost of producing fossil fuels?
4. Can biofuel technology be developed in a timely manner to replace diminishing supplies of fossil fuels?

Biofuels are the only source of alternative energy that produces greenhouse gasses.

13.1.1 Ethanol

Ethanol, also known as ethyl alcohol, pure alcohol, or grain alcohol, has been used as an additive to gasoline for many years. Internal combustion engines can be designed to run on pure ethanol. The original Model-T Fords ran on pure ethanol until 1908. There are plans to develop an ethanol engine to be used in hybrid electric cars. Some cities around the world use ethanol to power their public transportation fleets.

Ethanol (CH_3CH_2OH) is produced from agricultural vegetable matter. It can be produced from edible vegetables such as wheat, corn, sugar cane, sugar beets, barley, sweet potatoes, potatoes, sunflowers, and sorghum, to name just a few. It can also be produced from inedible plants such as hemp, switchgrass, kenaf, and common straw. There are three techniques for producing ethanol from vegetable matter.

- *Fermentation* of sugar is the process by which microbes convert plant sugars into ethanol. This process is used only with sugar cane, which is easily grown in the tropics.
- *Distillation* is the separation of ethanol and water from a biomass. This vaporization process will remove about 95% of the water. The other 5% must be removed by other treatments if the ethanol is to be mixed with gasoline.
- *Dehydration*, or azeotropic distillation, is a process where benzene or cyclohexane is added to the biomass, which is then distilled. These chemicals attract the water to the top of the distillation column, leaving pure ethanol at the bottom.

Ethanol: A fuel produced from agricultural vegetable matter.

Fermentation: The process by which microbes convert plant sugars into ethanol.

Distillation: The separation of ethanol and water from a biomass.

Dehydration: A process where a chemical is added to the biomass, which is then distilled.

The chemical equation for the combustion of ethanol in air is shown by the reaction

$$CH_3CH_2OH + 3(O_2 + 3.76N_2) \Rightarrow 2CO_2 + 3H_2O + 11.3N_2 \quad (13.1)$$

The major problem in using gasoline–ethanol blended fuels is that ethanol has difficulty vaporizing. This makes it difficult to start an automobile engine at temperatures below 11°C. Ethanol has a heating value that is about 35% lower than that of gasoline; hence fuel consumption increases when using an ethanol blend. Engine performance using gasoline–ethanol blends can be improved by using higher compression ratios or turbochargers.

13.1.2 Biodiesel

Biodiesel: Diesel fuel based on vegetable oil or animal fat.

B60: 60% biodiesel and 40% petrodiesel.

Biodiesel fuel is produced completely or partially from vegetable oils or animal fats. Used vegetable oil is often processed into biodiesel fuel. Most diesel engines can run on a blend of petrodiesel fuel mixed with biodiesel fuel. Some companies (e.g., Volkswagen) have developed diesel-cycle engines that will run on pure biodiesel fuel. Fuel that is 40% petrodiesel and 60% biodiesel is referred to as B60, whereas B100 is pure biodiesel fuel.

A biodiesel production experiment

Many processes can be used to convert vegetable oil into biodiesel fuel. Some techniques are best for small-scale production for personal use, whereas others are better suited to mass production. The process described below converts the triglycerides in used vegetable oil to methyl esters (a biodiesel fuel) and glycerol. These esters are the usable biodiesel fuel. The glycerol can be used to manufacture soap. The process described here is an esterification process that can be performed in any chemistry lab. It will provide insight into the processes needed to run a batch operation to produce small quantities of fuel. This process can then be scaled up to determine what might be necessary to operate a large-scale industrial biodiesel plant.

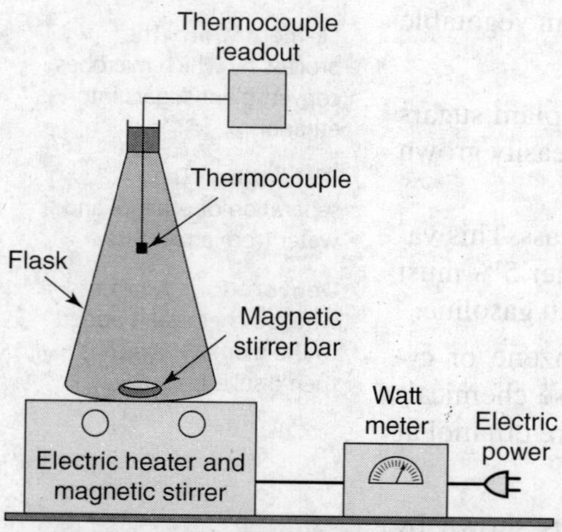

Figure 13.1

A biodiesel production apparatus.

1. **The equipment**
   The process is simple and requires only a few basic pieces of lab equipment. Figure 13.1 shows the experimental apparatus for producing biofuel from used or pure cooking oil. A hot plate and stirrer are needed to heat the chemicals used in the process. Canola oil can be converted into biofuel in one hour if the temperature of the reactants is brought to and maintained at 65°C while stirring the reactants. A watt meter measures the energy used in the process. A laboratory flask holds the reactants, and a thermocouple monitors the temperature of the reactants.

2. **The process**
   The process uses 200 g of cooking oil along with 55 mL of methanol (CH_4OH). Used canola oil may have to be filtered to remove any particulates. The process uses twice as much methanol as needed for the stoichiometric reaction in order to be sure that all of the canola oil reacts. The excess methanol will be removed after the reaction. Two grams of sodium hydroxide (NaOH) are used as a catalyst to improve the rate of reaction and the yield of biofuel. The catalyst is mixed with the methanol before adding it to the canola oil. The fatty acids in the cooking oil react with the

methanol to form the biodiesel methyl ester and glycerol as shown by the biofuel reaction

$$1 \text{ triglyceride} + 3 \text{ methanol} \Rightarrow 1 \text{ glycerol} + 3 \text{ methyl esters} \quad (13.2)$$

which describes the chemical reaction. The two main products are the methyl ester biofuel and glycerol.

The canola oil and methanol–sodium hydroxide mixture are placed in the flask. The top of the flask is plugged with a stopper, and the thermocouple is inserted through the stopper into the reactants. The stirrer operates at 450 rpm, and the watt meter is read to determine how much energy is being used by the stirrer. The stirrer operates continuously throughout the process. The electric heater is then turned on to heat the reactants to 65°C. During this transient process, the temperature of the reactants should be recorded along with watt-meter readings to determine the amount of energy used. This process should take five to ten minutes. When 65°C is reached, the heater is reduced so as not to overheat the reactants. Figure 13.2 shows typical temperature and power profiles as a function of time. At the end of one hour the heater and stirrer are turned off. The stir bar is removed from the flask, and the flask is stored to allow the methyl ester and glycerol to separate. This process can take from two hours to several days, depending on how much separation is desired. By allowing the products to rest in the flask, a decanting process is used to separate the products. The glycerol has a higher density and will sink to the bottom of the flask while the biofuel rises to the top.

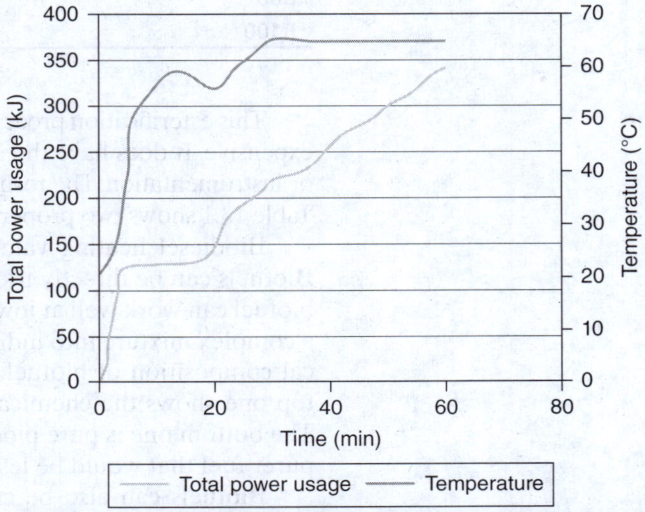

Figure 13.2
Power usage and temperature as a function of time.

3. Separating and washing the fuel

At the end of the decanting process, the biofuel and glycerol are carefully separated and weighed. Both the fuel and the glycerol must be washed to remove unused chemicals and contaminants. This is done by adding water to the separated products and either shaking the containers or bubbling air through the material. The water is then allowed to settle and separate from the products. The waste water is contaminated and must be processed as such.

4. Testing the biofuel

The fuel produced can be tested by using a bomb calorimeter or by powering a diesel engine with a dynamometer. The biofuel should have a heating value very similar to dodecane (petrodiesel fuel). The results of the watt-meter readings will indicate how much power was actually used to process the biofuel. If the mass of each reactant is multiplied by its specific heat and the temperature difference between 65°C and the ambient room temperature, the theoretical energy needed to complete the process will be obtained. Most of the heat loss occurs from the outer surface of the flask and any exposed surface on the electric heater. The ratio of the energy needed to run the process to the actual energy produced defines the efficiency of the process.

Table 13.1 Biodiesel Mixture Properties

Blend	% Diesel Heating Value	Freezing Temperature (°C)
Diesel	100	−10
B20	99	−16
B40	99	−10
B60	94	−3
B80	85	−1
B100	84	4

This esterification process is time consuming, and some of the chemicals used are expensive. It does have the advantage of not needing special processing equipment or instrumentation. The resulting fuel can be mixed with diesel fuel or used by itself. Table 13.1 shows two properties of the fuel in different mixtures with diesel fuel.

Biodiesel heating values compare well with petroleum-based diesel fuels. Biofuels can be directly used in common diesel engines. A mixture of diesel and biofuel can work well at low temperatures. A gas chromatograph, which separates a complex mixture into individual components, was used to determine the chemical composition of biofuel. Figure 13.3 compares two gas chromatographs. The top one shows the chemical constituent distribution for commercial diesel fuel. The bottom one is pure biodiesel fuel. It is clearly seen that the biofuel is a much purer fuel that would be less likely to produce harmful exhaust fumes.

Biofuels can also be created from animal waste products and animal fats. These alternatives are much more expensive than producing energy from vegetable matter.

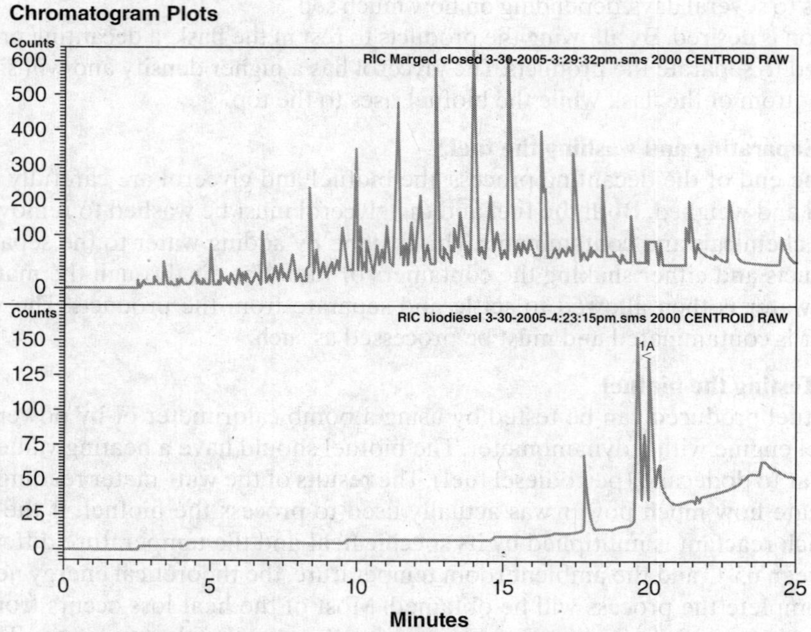

Figure 13.3

Gas chromatographs for diesel (top) and biofuel.

13.1.3 Algae fuel

Various types of algae can be processed into biofuel using the same techniques that convert vegetable matter into biodiesel fuel. The advantage to using algae is that 10 to 100 times the amount of energy per unit area can be produced compared to that produced with vegetable biofuels. Algae can be grown in open ponds or in closed containers. Open-pond algae farms are inexpensive to build and operate. They are, however, subject to contamination from viruses and bacteria. The algae in open ponds are also very susceptible to changes in water temperature and pH values. Closed containers eliminate all of these risks but are very expensive to build and maintain. Considerable research is being applied to developing economical closed-container algae farms.

Algae fuel: Fuel, primarily biodiesel, produced from algae.

Algae can be processed into many types of fuels. The most common is biodiesel fuel. Algae can also be made into biobutanol, biomethane, ethanol, and biogasoline.

13.2 Solar Energy

13.2.1 Photovoltaic cells

Solar energy can be converted directly into electricity, or solar radiation can be used as a power source. *Photovoltaic cells* use photons from the sun to create direct-current electricity by energizing semiconductor material to release electrons. A p-n semiconductor junction consists of a positive material rich in electrons and a negative semiconductor rich in electron "holes." The energy from a photon raises the energy level of an electron in the p-type semiconductor and allows it to jump to the n-type semiconductor, creating an electric current. The semiconductor materials currently in use in the production of solar cells are monocrystalline silicon, polycrystalline silicon, cadmium telluride, and copper indium gallium selenide/sulfide.

Photovoltaic cells: Cells that use photons from the sun to create direct-current electricity.

Solar cells can be used for small applications such as powering traffic warning signs, pumps, calculators, parking meters, and outdoor lighting. Large photovoltaic power stations have been built in Canada and Europe which produce power in the range of 50 MW to 90 MW. Figure 13.4 shows the 72 MW Lieberose Photovoltaic Park in Germany. The most efficient mass-produced solar panels provide electrical power of about 140 W/m^2.

ICHAEL URBAN/Staff/AFP/ Getty Images

Figure 13.4

The Lieberose Photovoltaic Park in Germany has 700 000 panels.

Advantages of solar cells:

1. Photovoltaic solar power is pollution free.
2. Solar radiation on the Earth's surface is plentiful and sustainable.
3. Photovoltaic cells can operate for several years with little maintenance.
4. The direct current produced by photovoltaic cells can be converted to alternating current compatible with the electric grid serving the public.

Disadvantages of solar cells:

1. Solar cells are expensive to produce. Considerable research and development is being applied to develop low-cost solar cells.
2. Solar cells work only when the sun is shining. This requires an electric storage facility for night-time power.
3. The manufacture of solar cells requires the use of toxic materials.

Photovoltaic cells have been used very effectively in satellites, where the effects of atmosphere, weather, and night are nonexistent. Large photovoltaic power plants that would be very efficient could be built in space. The problem is transferring the electrical power back to the Earth's surface. The next generation of engineers can figure that one out!

Light-sensitive dyes can be used to create flexible solar cells. Ruthenium metalorganic dye is used as a light-absorbing material. The electrons set free in the dye are transferred to a titanium–dioxide layer to complete the electric circuit. The main difficulty with these solar cells is that the dye degrades over time when exposed to heat and ultraviolet light. Organic polymers such as polyphenylene vinylene and copper phthalocyanine have been used to create thin-film solar cells with efficiencies as high as 8%. Dye-based and polymer solar cells are cheaper and less toxic to manufacture than silicon solar cells.

13.2.2 Active solar heating

Active solar heating: Solar radiation that is used directly as a heat source.

Solar radiation can be used directly as a heat source. Solar hot-water heaters are used to provide hot water in many households. These systems consist of a solar collector, a pump, a control system, and a storage tank, as sketched in Fig. 13.5. The solar collector is a flat panel consisting of three layers. The top layer is glass that allows solar radiation to pass through to the second layer, which is a dark radiation-absorbing metal panel. The dead air between the glass and the absorption panel provides some insulation from convective heat loss. The bottom layer is a water channel. The pump moves water through the solar collector and into the water tank. Water passing under the absorption panel is heated and stored in the water tank.

Water heated while the sun shines must be sufficient for the needs of the house for a 24-hour period. Backup energy is needed for times of sunless hours. A control box monitors the temperatures of the solar collector and the water tank and uses this information to operate the pump. In some cases, the pump can be eliminated using free convection of the water to circulate through the system. In a cold environment, a two-fluid system can be used. Ethylene glycol (antifreeze) can be pumped through the solar collector to absorb heat. The hot ethylene glycol will then travel through a heat exchanger in the water tank to heat the water.

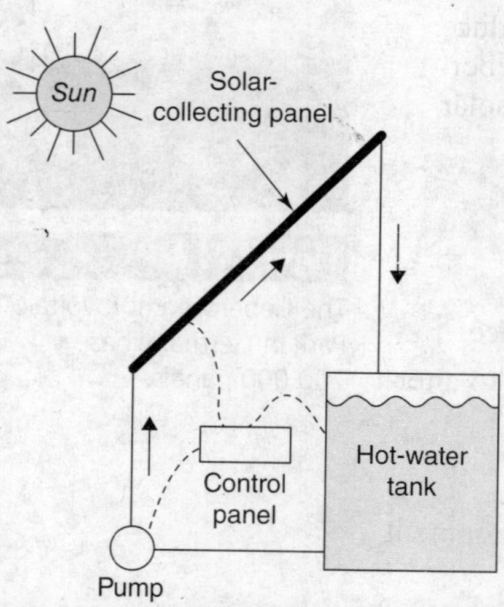

Figure 13.5

Residential solar hot-water system.

Many factors influence the cost effectiveness of solar hot-water heaters. The climate, the type of collector, the installation costs, and the local cost of electricity will greatly influence the profitability of solar heating.

The Solar City Tower shown in Fig 13.6, proposed for the 2016 Olympic Games, is a concept envisaged by a Zurich-based architecture and design studio, RAFAA. The solar design was created for a competition in 2009. The image shows an artist's impression of what the proposed tower might look like if constructed. However, the project is just a proposal, and the 105-meter tower may never actually be built. It would operate with only solar energy. The solar energy is meant to supply energy for all of the Olympic city, as well as for part of Rio. It pumps up water from the ocean during the daytime that flows through turbines to produce energy during the night from the stored potential energy.

Solar power can also be concentrated using mirrors to operate a Rankine cycle power system. Massive solar farms, like the one shown in

Courtesy of Rafael Schmidt Architekten

Figure 13.6

The Solar City Tower proposed for the 2016 Olympic Games in Rio. (Google "Rio's solar tower" for details.)

Fig. 13.7, utilize direction-controlled mirror arrays to focus large amounts of solar radiation on a tower containing a boiler.

The major difficulty in operating such a station is directing each individual mirror, a heliostat, to focus on the boiler. Solar farms constructed in Spain can produce up to 20 MW of power.

© Kevin Burke/Corbis

Figure 13.7

A solar farm.

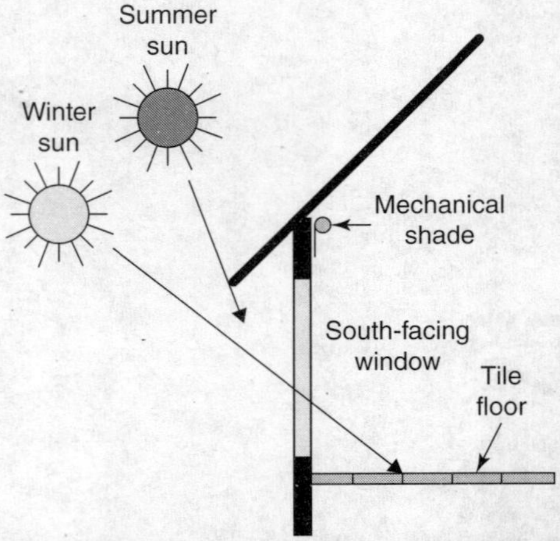

Figure 13.8

A sketch of a solar wall.

13.2.3 Passive solar heating

Passive solar heating is the use of solar radiation to heat water or the inside of a house without forcibly moving the water or air, using only natural free convection for circulation. Many architectural solutions have been formulated for using available exposure to the sun to heat residences. These designs can be inexpensive to construct and very cheap to maintain.

In passive solar-building design, the windows, walls, and floors store and distribute solar energy in the form of heat in the winter and reject solar heat in the summer, all without the use of mechanical and electrical devices. An insulated curtain is lowered at night in the winter over the large south-facing windows to retain heat (this could be manually operated, or a small motor could be used). The overhang is properly dimensioned to shade the tile floor in the summer and expose the tile to the sun in the winter, as sketched in Fig. 13.8.

13.3 Fuel Cells

Fuel cells: Direct energy conversion devices that use chemical reactions to create electric power.

Fuel cells are direct energy conversion devices that use chemical reactions to create electric power. In these reactions, a fuel reacts with an oxidizer to directly create electricity. Unlike batteries, fuel cells do not store their fuel within the device. Instead, the fuel and oxidizer continuously flow from external sources into the fuel cell. Fuel cells generally contain three parts: an anode (the negatively charged electrode), a cathode (the positively charged electrode), and an electrolyte that allows for charge transmission between the two electrodes.

The operation of a typical fuel cell is shown in Fig. 13.9. Hydrogen fuel is drawn from an external tank into the anode, while an oxidant is drawn from an

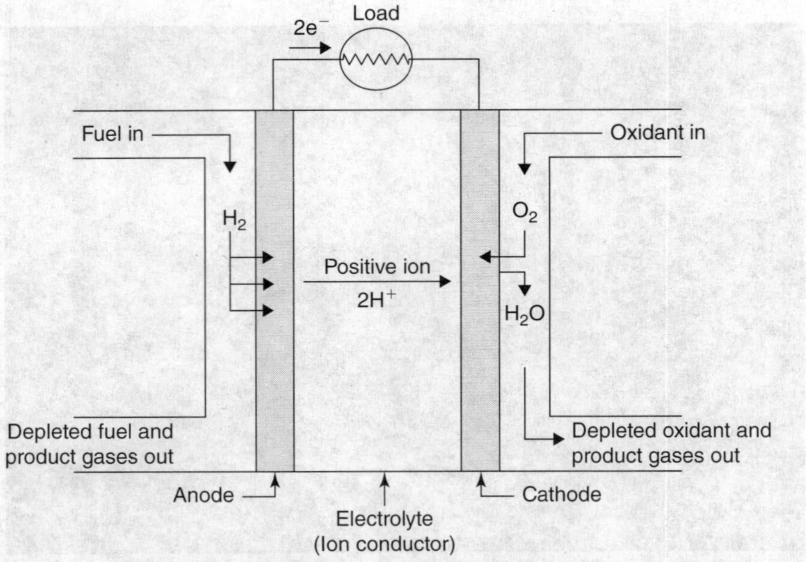

Figure 13.9

A fuel-cell schematic.

external source into the cathode. This oxidant is usually oxygen from the ambient air. At the anode, hydrogen is catalyzed with platinum to form positively charged hydrogen ions and free electrons. The electrons are conducted by direct current to the electrical load, and the hydrogen ions are transported across the electrolyte layer. The hydrogen ions are catalyzed, using nickel, with the oxygen ions in the cathode to form water. Fuel-cell research has focused on transportation because hydrogen fuel cells emit no pollution, producing only power and water.

Paul Kane/Stringer/Getty Images News/Getty Images

Figure 13.10

Hydrogen fuel-cell bus.

Space applications are particularly attractive since fuel cells provide both power and drinking water for the crew. Today, fuel cells are used by some buses and in space shuttles. Figure 13.10 shows a fuel-cell-powered bus. A possible configuration for an automotive fuel-cell application is displayed in Fig. 13.11. Hydrogen tanks are mounted in the rear of the vehicle, and the fuel cell is mounted under the front seats. An electric motor is used to power the car.

Fuel cells provide an opportunity for a green energy source since a non-hydrocarbon is used as the fuel. However, it must be mentioned that the most common current hydrogen production technology uses steam re-forming of methane. Still, other hydrogen production methods such as electrolysis allow for a hydrocarbon-free energy source.

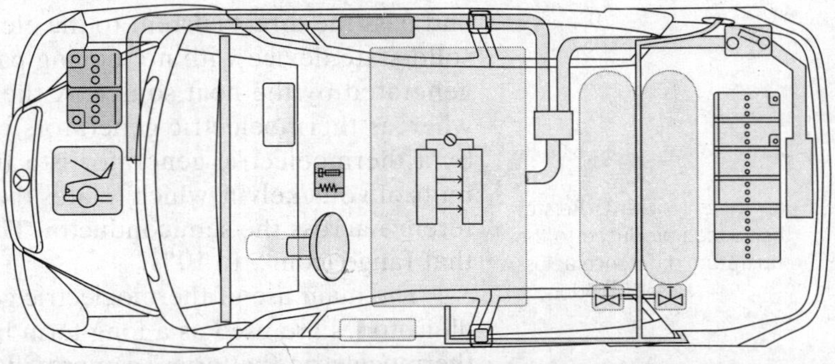

Figure 13.11

Schematic of a fuel-cell system in an automobile.

In general, fuel cells come in a stack, where a stack contains several fuel cells connected electrically in series. Each fuel cell not only contains an anode, cathode, and electrolyte per the diagram in Fig. 13.11, but also separator plates to provide both a gas barrier and a series electrical connection between cells. The configuration of these stacks may be comprised of flat fuel cells or concentric tubular fuel cells. Tubular fuel cells are generally only used in high-temperature fuel cells such as the solid oxide fuel cell.

A number of fuel-cell types are available, and they are categorized by electrolyte type:

- Polymer electrolyte fuel cell (PEFC)
- Alkaline fuel cell (AFC)
- Phosphoric acid fuel cell (PAFC)
- Molten carbonate fuel cell (MCFC)
- Solid oxide fuel cell (SOFC)

Fuel cells have a limited lifetime of about 5000 hours. Efficiencies will vary from 30 to 50%, depending on operating conditions. There are concerns about being able to store a hydrogen fuel safely, but these issues are being addressed. Concerns have also been raised about being able to supply hydrogen in the quantities needed to replace gasoline as our main automotive fuel. Currently, hydrogen is produced by steam methane re-forming, which produces carbon dioxide. Hydrogen could be produced by electrolysis if a sufficient supply of electricity were available. Research is being conducted to reduce the cost of producing fuel cells; one alternative being studied is to not use platinum as the anode catalyst.

13.4 Thermoelectric Generators

Thermoelectric generator: A device that converts heat directly into electricity.

A *thermoelectric generator* is a device that will convert heat directly into electricity. It works on the same principle as a thermocouple but on a much larger scale. It utilizes two pieces of semiconductor material. An n-type semiconductor has an excess of electrons, while a p-type semiconductor has an excess of electron "holes." The two semiconductors are joined by a conductive material, as shown in Fig. 13.12. Heat from a high-temperature source enters the hot shoe and is conducted through to the semiconductors. The hot shoe has a high thermal conductivity and a high electrical conductivity. Electrons are released from the n-type semiconductor and establish a current through the p-type semiconductor. Cold shoes are electrically and thermally conductive and pass the direct current to the electric load. A thermoelectric generator is a solid-state device with no moving parts. The only pollutants created would be generated by the heat source. A thermocouple creates millivolts of potential, whereas thermoelectric generators create several volts. The voltage produced by a thermoelectric generator is a function of the *Seebeck coefficient S*, with units of volts/kelvin, which relates the voltage produced to the temperature difference across the semiconductor. Thermoelectric generators have efficiencies that range from 5 to 10%.

Seebeck coefficient: Relates the voltage produced to the temperature difference across a semiconductor.

The main use of thermoelectric generators has been to power satellites. Radioisotopes are used as a long-term heat source to power the generators. These thermoelectric "batteries" can provide useful power without maintenance for up to 10 years. Thermoelectric generators also have the potential to provide power in places that have a steady supply of heat, such as geothermal sites.

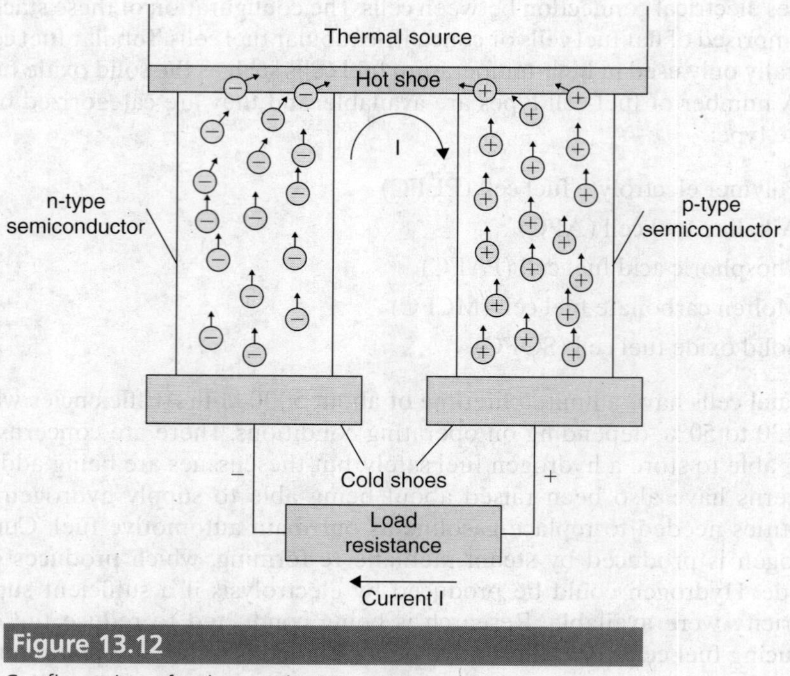

Figure 13.12

Configuration of a thermoelectric generator.

13.5 Geothermal Energy

Geothermal energy is obtained from energy stored in the Earth. The Earth's core is an almost unlimited heat source with highest temperatures at the core of about 5500°C. At a depth of 28 to 55 km, the temperature ranges from 200°C to 1000°C. Unfortunately, we have the ability to drill to only a depth of 10 km. Fissures that exist in the crust of the Earth allow high-temperature steam to permeate to the surface at geothermal sources. These locations are usually found in areas of volcanic activity. Steam from these geothermal sources can be used to generate electricity using either of two techniques. In a *dry steam plant*, high-pressure steam at 150°C or higher is taken directly from the geothermal source and used to operate a Rankine cycle power plant. In a *flash steam plant*, sketched in Fig. 13.13, steam at 180°C or higher is passed through a flash tank, which separates liquid water from steam by forcing it into the low-pressure tank. Liquid water is drained from the flash tank, and the dry steam is used to provide energy to a Rankine cycle power plant. Geothermal power has the advantage of being a free energy source.

Geothermal power is not totally without environmental hazards. Some geothermal-steam sources also vent toxic gases. The rate of flow of gases, such as carbon dioxide, from geothermal sources is, however, small compared to the rate of production of carbon dioxide in a coal-fired power plant. Iceland, a volcanic island in the North Atlantic Ocean, gets 30% of its electricity from geothermal power production and 70% from hydroelectric plants.

Geothermal energy: Energy obtained from internal energy stored in the Earth.

Dry steam plant: High-pressure steam that provides energy directly to a Rankine cycle power plant.

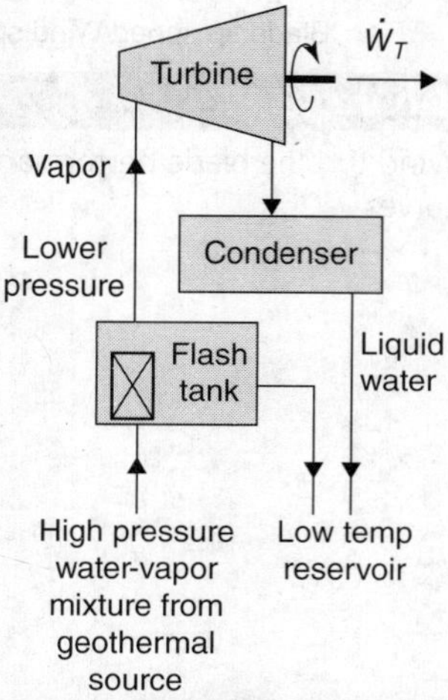

Figure 13.13

Flash steam geothermal power plant.

13.6 Wind Energy

Wind, another sustainable source of energy, accounts for 2.5% of the solar energy received by the Earth. Wind turbines have been used since ancient times to pump water and grind grain. Figure 13.14 shows the famous windmills of Mykonos, dating from the 16th century.

Kenneth Kroos

Figure 13.14

Windmills of Mykonos.

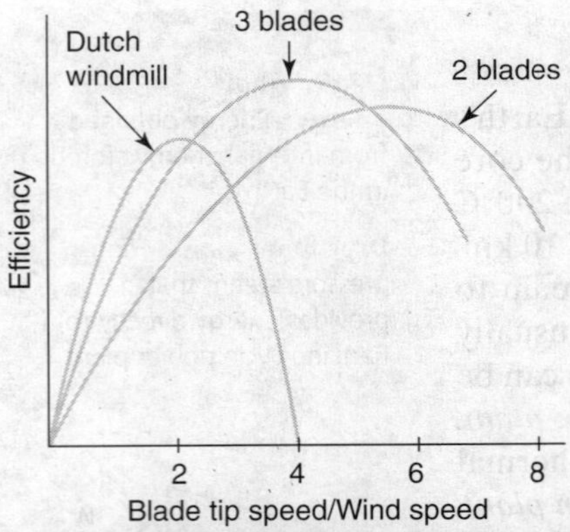

Figure 13.15

Wind turbine blade performance curves.

The maximum possible power produced by a wind turbine is related to the density ρ of the air, the swept area A of the blades, and the approach velocity V by the relation

$$\dot{W}_{\text{turbine}} = \frac{1}{2}\rho A V^3 \quad \textbf{(13.3)}$$

Analysis shows that the maximum possible wind turbine efficiency is 60%, but 50% is probably not attainable. The maximum efficiency for a particular turbine is rated at a particular wind speed, all other speeds having a lower efficiency.

The number of blades on a wind turbine will affect the efficiency of the turbine, as shown in Fig. 13.15. This figure shows that two or three blades on the turbine produce higher efficiencies. Obviously, major power-producing wind turbines only have two or three blades.

Wind-power-generating plants use no fuel; however, the stations are very expensive to build and must be regularly maintained. The primary negative by-product of wind turbines is the noise produced as the blades "whoosh" by. They also require acreage, and some feel they are an eyesore. There is considerable interest in building offshore wind turbine farms. The wind offshore is more regular, and the noise is irrelevant. An unexpected negative by-product is the decommissioning of a wind farm. Some wind farms have been abandoned, and landowners and residents have no apparent recourse to force companies to spend millions of dollars to correct the situation.

13.7 Ocean and Hydroelectric Energy

13.7.1 Wave energy

Wave power: Power derived directly or indirectly from the kinetic energy in waves.

The motion of waves can be used to generate power. *Wave power* has been in use since 1890 but has not been widely employed. The Agucadoura Wave Park in Portugal began operation on 2008 as an experimental attempt to produce 2.2 MW of electric power. The Agucadoura facility uses a series of devices called Pelamis machines. These consist of a series of partially submerged cylindrical tanks connected end to end. As a wave passes the line of Pelamis machines, the rise of the wave raises each cylinder in succession. The Pelamis machines are connected by hydraulic cylinders that use the relative motion of the machines to drive generators. The output of these hydraulic motors is used to generate electricity. Figure 13.16 shows a conceptual drawing of a wave energy facility using Pelamis machines. A new generation of Pelamis machines started testing in Scotland in 2010.

Another application of wave power uses the motion of an incoming wave to move water through a hydroelectric-generating station, sketched in Fig. 13.17. A wave pushes against a wall. The motion of the wall pushes a piston that moves water through a hydraulic line into a hydroelectric conversion plant. The moving water flows through a water turbine that drives an electric generator. This type of power generation is different from normal hydroelectric plants in that the power production is intermittent, depending on the wave intensity and frequency.

Figure 13.16

Wave energy facility.

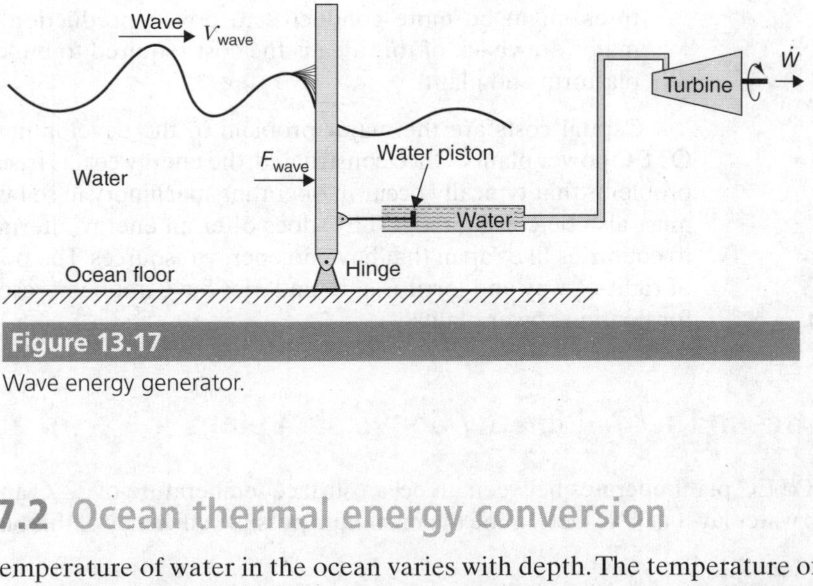

Figure 13.17

Wave energy generator.

13.7.2 Ocean thermal energy conversion

The temperature of water in the ocean varies with depth. The temperature of sea water is highest near the surface, where it is heated by the sun, and it then decreases with depth. Figure 13.18 shows how the temperature in the ocean typically varies with depth at low to moderate latitudes. A surface layer has a uniform temperature up to a depth of about 100 m. This layer is heated by the sun, and the wave action mixes the water, keeping the temperature uniform. Below the surface layer is a thermocline where the water temperature steadily drops. Below 1000 m the water temperature stabilizes, and the difference in temperature

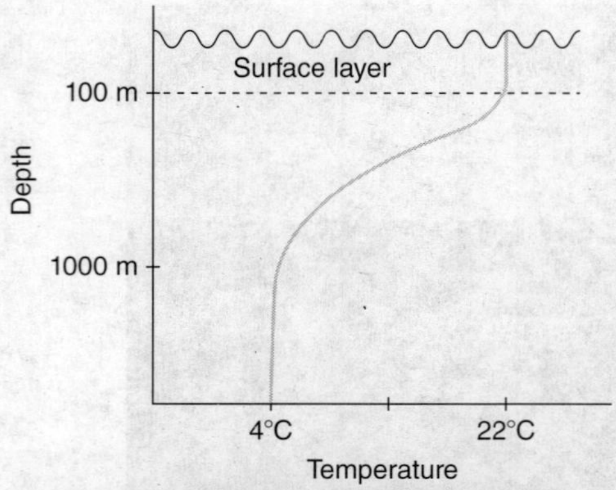

Figure 13.18

Ocean water temperature as a function of depth.

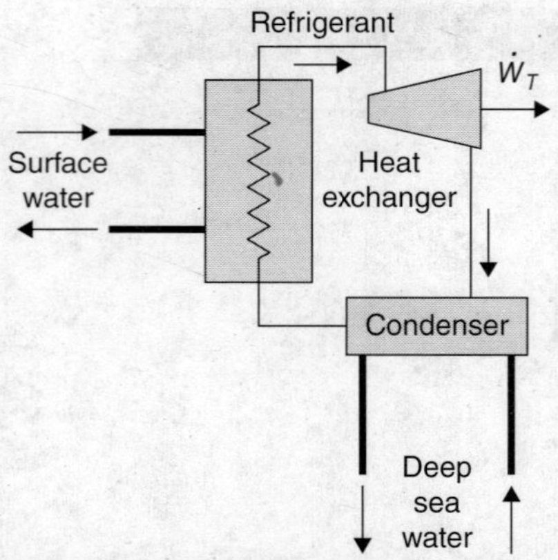

Figure 13.19

Ocean thermal energy conversion plant.

between the surface layer and the deep layer is about 18°C. It is possible to operate a Rankine power cycle using this temperature difference.

The configuration of an Ocean Thermal Energy Conversion (OTEC) plant is shown in Fig. 13.19. Water is drawn from the surface to heat and vaporize the working fluid, which is typically ammonia. A refrigerant is used instead of steam since refrigerants will vaporize at lower temperatures. The working fluid drives a turbine-generator unit to produce electricity and is cooled in a condenser using water from the deep layer.

With a maximum efficiency of 6%, an OTEC plant would have to be massive to produce the power necessary to support a city or industry. There are three ways to build these plants:

1. Land-based plants would pump hot and cold water from the ocean to a land-based power plant. The logistics of moving that much water over the distance required makes this approach questionable.
2. Shore-based plants would be built in the ocean water on shelves along the shore. This reduces the distance the water must be transported but leaves the plant susceptible to damage from wave action and storms.
3. Floating plants would be mounted on floating platforms in the ocean. This concept has considerable merit since the platform could be moved to a location where water temperatures might be more conducive to power production. The major drawback of this idea is the cost required to build the platform and plant.

Capital costs are the major problem in the development of OTEC power plants. Once constructed, the energy cost is free. The problems that typically occur in operating machinery in sea water must also be considered. OTEC does offer an energy alternative to countries like Japan that have no energy resources. The politics of right-of-way and territorial boundaries have not been resolved for floating power plants.

Example 13.1 An Ocean Thermal Energy Conversion plant

An OTEC plant operates between an ocean surface temperature of 22°C and a deep water layer at 4°C. Determine the maximum possible efficiency of this plant.

Solution

The maximum possible efficiency of a heat engine is the efficiency of a Carnot engine operating between the two heat reservoir temperatures. From Chapter 5, the Carnot efficiency is given by

$$\eta = 1 - \frac{T_L}{T_H}$$

For this problem $T_H = 22°C = 295\,K$ and $T_L = 4°C = 277\,K$, so the maximum possible efficiency of the thermal plant is

$$\eta = 1 - \frac{T_L}{T_H} = 1 - \frac{277}{295} = 0.061 \quad \text{or} \quad \underline{6.1\%}$$

An actual plant would have an efficiency of 3% or less.

13.7.3 Hydroelectric power

Hydroelectric power is produced from falling or flowing water. These power plants can vary in size from 100 000 MW to small units along fast-running streams producing electric power in the kilowatt range. Hydroelectric power is dependent on the steady flow of water. River water is easily accessible, but flow rates vary with the seasons and rainfall amounts. Most power plants use a reservoir system to store enough water to avoid supply problems. There are several ways to accomplish this. Five will be suggested.

Hydroelectric power: Power produced from falling or flowing water.

1. **Dams:** Positioning a dam across a river creates a reservoir sufficient to supply a hydroelectric power plant. The dam will also house the generating plant. Water is directed from the reservoir through a turbine and continues on down the river.
2. **Pumped storage:** This technique moves water between reservoirs at different elevations. When the demand on power is low in the middle of the night, water is pumped from the low-elevation reservoir back to the high-elevation reservoir, such as the one shown in Fig. 13.20 between Lake Michigan and a human-made reservoir. The water at the higher elevation then passes through the pump/turbine to generate power during peak demand.

Comment

Power plants are difficult to turn on and off, so it's desirable to store the energy during the off hours.

Courtesy of Consumers Energy

Figure 13.20

Consumers Energy pumped-storage 842-acre reservoir near Ludington, Michigan: It can store up to 27 billion gallons of water and produce 1.9 MW of energy during peak demand.

3. **Run of the river:** This technique is used for small-power production. A turbine is mounted directly in a river or channel.
4. **Tidal power:** This method makes use of daily tidal action to supply a flow of water to a generating station.
5. **Underground supply:** This method uses two water supplies that exist at different elevations such as a waterfall or a mountain lake. Water is channeled underground from the high-elevation source to run a power-generating station.

Hydroelectric power generation depends on having the natural water resources necessary to operate such a plant. Brazil, which has the Amazon River basin as a source of water, produces 85% of its electricity from hydroelectric power. Hydroelectric power accounts for 19% of the total electricity production in the world and about 7% in the United States.

Two distinct types of water turbines generate electricity. *Reaction turbines* operate on a pressure change along the flow of water. *Impulse turbines* convert the kinetic energy of the water to mechanical energy transported by a turbine shaft. Impulse turbines, unlike reaction turbines, do not have to be sealed to contain the water.

Using hydroelectric power has several advantages, especially these three:

1. No fuel burned, so no pollution
2. Water provided by nature
3. Lower operation and maintenance costs

Comment

The Marmot Dam in Oregon, with a generating capacity of 22 MW, was decommissioned due primarily to the migration of salmon.

There are also disadvantages to hydroelectric power, some of which have led to the decommissioning of many dams.

1. High capital construction costs
2. Variable water supply
3. Modification of fish habitats and migration routes
4. Displacement of local population and wildlife

Comment

The fishing industry on the Columbia River has been severely damaged by several large dams, with the loss of over 20 000 jobs.

See http://www.pcffa.org/fn-oct97.htm.

When the Aswan Dam was built on the Nile River in southern Egypt, large numbers of people from lower Numidia had to be relocated to provide space for the reservoir needed to supply the dam. The power supply was highly beneficial to Egypt, as were additional irrigation and fishing. The dam also served to prevent flooding of the Nile in downstream areas, especially in the Fertile Delta, which would lose most of its crops during extreme floods. Some of Egypt's archaeological sites were also damaged, but many were relocated. The evaporation of water from the surface of the reservoir reduced the volume of flow to the sea so that the Fertile Delta was reduced significantly in size, decreasing the amount of fertile farming acreage. In addition to adding to the effects of the Mediterranean Sea's higher level, the Fertile Delta is experiencing grave problems that will only become more severe with time. Engineers must always weigh what is to be gained against what will be lost.

13.8 Osmotic Power Generation

Osmotic power generation is a method of generating either electricity or fluid power using the difference in chemical potential between salty sea water and fresh river water. The best place to establish these generating stations is where a fresh-water river empties into a sea or an ocean. Areas such as the Netherlands have excellent potential to use this type of power generation since they continuously pump fresh water into the North Sea to keep farmland from flooding.

Many techniques may be used to produce power using osmosis, but the two most common ones are reverse electrodialysis (RED) and pressure-retarded osmosis (PRO). Reverse electrodialysis is essentially a salt-water battery. Differences in the ion concentration between the fresh and sea water establish an electric potential or voltage. Anode and cathode membranes are stacked much in the same way as fuel cells are stacked to produce electric power. This type of energy production can produce 1 W of power per square meter of membrane. A 250-kW power station would be about the size of a shipping container. This technology is still in its infancy, and environmental impact studies are required to determine its effect on the local ecology.

Pressure-retarded osmosis uses the difference in salinity between sea water and fresh water to physically move the fresh water. A semipermeable membrane separates the sea water and river water. This membrane has small openings that will allow fresh water to move into the salt-water region as shown in Fig. 13.21 The membrane is impermeable to salt water moving in the opposite direction. The fresh water is driven by the chemical potential between it and the sea water to move through the membrane and increase the level of the sea water. This system has the potential to raise the sea water over 200 m in height. This increased level of sea water is then channeled through a water turbine to produce power.

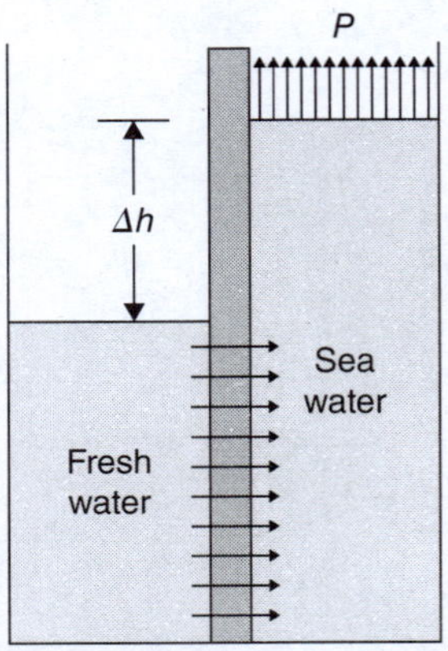

Figure 13.21

A pressure-retarded osmosis system.

Experimental osmotic generating stations have been built in the Netherlands and in Norway. The ecologies of sea-water and fresh-water regions are very different. Changes in salinity caused by osmotic power stations could have a harmful impact on both ecologies. There are also concerns about the longevity of the membranes used in these devices.

13.9 Summary

In the previous chapter, we considered the energy that comes from combustible substances, especially hydrocarbons, which are not sustainable over the centuries that lie ahead. In this chapter, we reviewed the sources of energy that are sustainable since they come from the variety of sources found in nature that are replenished continually. Cost is the primary factor that does not allow these sustainable energy sources to be used to generate a substantial percentage of our electrical power. With time, however, as the supply of hydrocarbons diminishes, making costs rise, the development of alternative sources will continue to contribute more and more to the energy supply in the United States and the world. Renewable energy, energy that comes from natural resources (e.g., sunlight, wind, rain, tides, geothermal), provides about 16% of global energy consumption, with 10% coming from traditional biomass, which is used mainly for heating, and 3.4% from hydroelectricity. Small hydro, biomass, wind, solar, geothermal, and biofuels account for another 3% and are growing rapidly. Renewables contribute about 19% in global electricity generation, with 16% of electricity coming from large hydro plants.

The following additional terms were introduced in this chapter:

Active solar heating: *Solar radiation that is used directly as a heat source.*

Algae fuel: *Fuel, primarily biodiesel, produced from algae.*

Biodiesel: *Diesel fuel based on vegetable oil or animal fat.*

Biofuels: *Fuels made from organisms that were recently living (edible plants, inedible plants, animal waste, and algae).*

Dehydration: *A process whereby benzene is added to the biomass, which is then distilled.*

Distillation: *The separation of ethanol and water from a biomass.*

Dry steam plant: *High-pressure steam that provides energy directly to a Rankine cycle power plant.*

Ethanol: *Energy produced from agricultural vegetables.*

Fermentation: *The process by which microbes convert plant sugars into ethanol.*

Fuel cells: *Direct energy conversion devices that use chemical reactions to create electric power.*

Geothermal energy: *Energy obtained from internal energy stored in the Earth.*

Hydroelectric power: *Power produced from falling or flowing water.*

Ocean thermal energy: *The energy generated using the temperature difference between the warm surface layer and the deeper cool layers.*

Passive solar heating: *The use of solar radiation to provide space heating using only gravity for circulation.*

Photovoltaic cells: *Cells that use photons from the sun to create direct-current electricity.*

Seebeck coefficient: *Relates the voltage produced to the temperature difference across a semiconductor.*

Thermoelectric generator: *A device that converts heat directly into electricity.*

Wave power: *Power derived directly or indirectly from the kinetic energy in waves.*

Wet steam plant: *Steam passes through a flash evaporator and powers a Rankine cycle power plant.*

An equation utilized in this chapter:

Maximum wind power: $$\dot{W}_{\text{turbine}} = \frac{1}{2}\rho A V^3$$

Problems

13.1 Estimate a typical student's personal carbon footprint using the EPA Individual Carbon Footprint Calculator at: www.epa.gov/climatechange/emissions/ind_calculator2.html State your assumptions.

13.2 Use the Carbon Footprint Calculator from the previous problem to propose ways to reduce a student's carbon emissions. Which ways do you anticipate being most effective?

13.3 Use the business Carbon Footprint Calculator at: www.carbonfootprint.com/calculator1.html to estimate the carbon emissions from your university.

13.4 Which of the alternative energy sources described in this chapter have a carbon footprint?

13.5 Which of the alternative energy sources described in this chapter are dependent on weather conditions? What percentage of a nation's energy production do you think could be on weather-dependent conditions? (Argue your case based on scientific reasoning.)

Biofuels

13.6 One of the major by-products of the biodiesel conversion process is glycerol. Develop a list of five commercial uses for this substance.

13.7 Develop a list of inedible crops of vegetable matter that can be used to create biofuel. From what part of the world do these crops come?

13.8 Follow the steps outlined in Section 13.1.2 and actually perform the biofuel production process using both waste cooking oil and fresh cooking oil and compare the resulting fuels. What is the difference in the economics of using waste oil as opposed to fresh oil?

Solar Energy

13.9 Perform an online search of the commercial production of photovoltaic solar cells and develop a list of the toxic materials produced in this process.

13.10 Perform an online search of commercially available photovoltaic solar cells and select a system that would provide the power needs for a typical 180 m^2 home in your location. Develop a cost estimate for installing this system.

13.11 Design a solar hot-water heating system in your location that would replace a 750 liters conventional water heater. Estimate the cost of such a system.

Fuel Cells

13.12 Design a system that would use 110-V AC power to produce enough hydrogen to power a personal automobile. Estimate the cost of such a system to purchase and install.

13.13 Perform an online search to determine how hydrogen fuel tanks in automobiles are developed to ensure safety in the event of an accident.

13.14 Perform an online search and develop a list of models and prices for electric vehicles that are commercially available today.

Thermoelectric Generators

13.15 Develop a list of satellites that use radioactive thermoelectric generators.

13.16 Develop a list of commercially available systems that use a thermoelectric process for cooling and refrigeration.

Geothermal Energy

13.17 Develop a list of the regions on each continent where geothermal energy is readily available.

13.18 Perform an online search to determine if geothermal energy poses any health or environmental risks.

Wind Energy

13.19 How could an array of wind turbines affect the local ecology?

13.20 Develop a list of the wind energy facilities in your state.

13.21 A hilltop has a steady wind velocity of 30 kmph. What percent increase in power would result from using 3-m-radius turbine blades rather than 1.8-m-radius blades?

Ocean and Hydroelectric Energy

13.22 Develop a list of places on each continent that have the potential to produce hydroelectric power on a commercial scale.

13.23 How could ocean wave energy affect offshore ecology? Consider both positive and negative aspects.

13.24 Where is the largest reverse electrodialysis power plant in the world? How much power does it produce?

13.25 Where is the largest pressure-retarded osmosis plant in the world? How much power does it produce?

CHAPTER

14

Thermodynamics of Living Organisms

Outline

Nomenclature

Nomenclature introduced in this chapter:

c	Speed of light: 299.8×10^6 m/s
h	Planck's constant: 6.626×10^{-34} J·s
ν	Light frequency: 405 THz to 790 THz

Learning Outcomes

- ❑ **Become familiar with the mechanism by which plants use energy**
- ❑ **Understand the mechanism by which animals use energy**
- ❑ **Become familiar with the energy and nutritional values of foods**
- ❑ **Solve calorimetry problems**
- ❑ **Understand how muscles work in animals**
- ❑ **Study how living organisms can regulate energy production and body temperature**

Motivational Example—The Bionic Ankle

The field of prosthetics has experienced great achievements in recent years. The development of controllable robotic arms and hands has had a significant impact on the quality of life for amputees. In prosthetic design, the new fields of mechatronics and bioengineering have combined to produce some very innovative solutions. One of the more recent developments has been the robotic-controlled ankle. In earlier prosthetic designs, the ankle mechanism was modeled using a simple cantilever beam made of carbon composite materials. While this provided an ankle that would simulate many aspects of walking or running, it was not controllable. Researchers have developed dynamic robotic ankles that have the ability to provide the thrusting power of an ankle as well as the motion.

The field of bioengineering has grown rapidly over the past decade. Bioengineering is the solution of biological problems using the problem-solving skills developed in engineering. This discipline has found a home in all the traditional engineering fields. Electrical engineers develop electronic measuring devices ranging from simple blood pressure meters to high-resolution CAT scan technology. Civil engineers work in the field of water supply and environmental resource management. Chemical engineers use nanotechnology to solve biological problems. Mechanical engineers work in both the structural aspects of biology such as the skeletal system and in the thermofluid area studying the respiratory and circulatory systems. Figure 14.1 shows a robotic ankle developed at MIT's biomechatronics lab. This device uses motor-controlled springs to provide a thrusting action that produces a more fluid gait and reduces fatigue in the leg.

Figure 14.1

The MIT bionic ankle.

14.1 Energy Conversion in Plants

Plants and algae convert sunlight into usable energy by a process called *photosynthesis*, in which the plant is considered to be an open thermodynamic system. Sunlight, water, and carbon dioxide are absorbed by the plant, and sugar and oxygen are produced, represented by

$$CO_2 + H_2O + \text{Light} \Rightarrow CH_2O + O_2 \tag{14.1}$$

This is a one-way reaction. If any of the three input constituents are not present, the process does not occur. A pigment called *chlorophyll* absorbs the sunlight. The photons break electrons loose from the chlorophyll molecules and allow the carbon dioxide to react with the water to form sugar. This process takes place in, for example, the leaves of plants; a leaf cross section is quite complex, as displayed in Fig. 14.2. The rate of sugar production is a function of the surrounding temperature, the carbon dioxide level, and the type of light available. The photosynthesis reaction increases with temperature and with the amount of sunlight radiation, but reaches a plateau at high-radiation levels.

Photosynthesis: The process by which plants and algae convert sunlight into usable energy.

Chlorophyll: The pigment that absorbs the sunlight.

Note: CH_2O is a carbohydrate.

The photosynthesis reaction takes place in two stages. In the first stage, a light-dependent reaction uses the energy of the sunlight to create energy-storing molecules called adenosine triphosphate (ATP) and nicotine adenine dinucleotide phosphate (NADPH). In the second stage, another light-dependent reaction uses these energy-storing molecules to capture and react with the carbon dioxide to produce sugars (carbohydrates).

Organisms using the photosynthesis process are thought to have been in existence for over 3.5 billion years. Before higher life-forms appeared, the atmosphere of the Earth had high levels of carbon dioxide. Hydrogen was used instead of oxygen to produce electrons in the photosynthesis process. It is possible

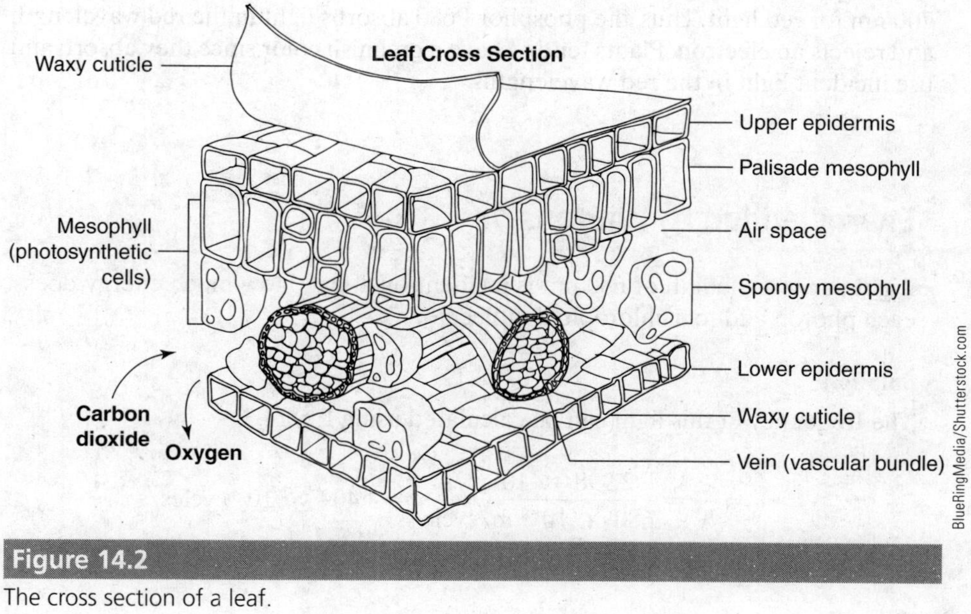

Figure 14.2

The cross section of a leaf.

that oxygen-producing bacteria appeared on Earth about 3 billion years ago and began to transform the atmosphere into a more oxygen-rich environment. This would have allowed the development of more advanced unicellular organisms. As time passed, the relative concentrations of carbon dioxide and oxygen in the atmosphere stabilized.

Comment

Photosynthesis transfers mass and converts energy while chemical reactions occur.

Photosynthesis is an endothermic (energy-absorbing) process. Light energy in the form of photons is converted to chemical energy that is used to convert carbon dioxide into carbohydrates. Carbon dioxide and light are supplied from the atmosphere to the cellular membranes of the leaf, while water is supplied through the roots and stem of the plant. Thus, photosynthesis is a complex process of mass transfer, energy conversion, and chemical reactions.

The thermodynamics of photosynthesis enters when attempting to understand the process by which sunlight breaks loose an electron in the chlorophyll pigment. Light exhibits the characteristics of both particles and waves. Light particles (photons) carry a discrete amount of energy. Light also acts like a wave in that it will have distinct frequencies. The amount of energy contained by a photon is given by

$$E = h\nu \tag{14.2}$$

where Planck's constant $h = 6.626 \times 10^{-34}$ J·s; the light frequency ν has units of cycles/s (hertz). The frequency of light is related to the wavelength λ (m/cycle) of the light by

ν (the Greek letter nu)
λ (the Greek letter lambda)

$$\nu = \frac{c}{\lambda} \tag{14.3}$$

Note: Physicists state that all nine digits are significant.

The speed of light c is 299 792 458 m/s. The light-reactive phosphors in chlorophyll are labeled according to the wavelength of the light with which they react. The visible spectrum of light has wavelengths varying from 400 nm for purple light to 700 nm for red light. Thus, the phosphor P680 absorbs light in the red wavelength and rejects an electron. Plants tend to have a greenish color since they absorb and use incident light in the red wavelengths.

Example 14.1 Energy added by photons

Light at a wavelength of 680 nm is incident on a leaf. How much energy does each photon add to a chlorophyll molecule?

Solution

The frequency of this light can be calculated using Eq. 14.3:

$$\nu = \frac{c}{\lambda} = \frac{2.998 \times 10^{8}\ \text{m/s}}{680 \times 10^{-9}\ \text{m/cycle}} = 4.409 \times 10^{14}\ \text{cycles/s}$$

The energy level of the photon is calculated, using Eq. 14.2, to be

$$E = h\nu = 6.626 \times 10^{-34} \frac{\text{J}\cdot\text{s}}{\text{cycle}} \times 4.409 \times 10^{14} \frac{\text{cycles}}{\text{s}} = \underline{2.92 \times 10^{-19}\ \text{J}}$$

Some texts omit "cycle" in this equation because it is actually dimensionless, but it does give a better understanding of the equation since wavelength is measured in m/cycle.

☑ **You have completed Learning Outcome (1)**

14.2 Energy Conversion in Animals

The life-supporting processes in animals are almost the reverse of photosynthesis. In animals, oxygen is taken into the respiratory system, and food is taken into the digestive system. Food is processed in the digestive system, where it is transformed by chemical reactions to a form of fuel that can be absorbed into the circulatory system and distributed throughout the body. Air is taken into the respiratory system, where oxygen in the air is absorbed into the circulatory system. The processed food and oxygen combine in an exothermic (energy-emitting) reaction to produce the energy needed to maintain body temperature and to do work.

Digestion of food takes place through several organs. Sugars begin to digest in the mouth, and the remainder of the food then travels through the esophagus to the stomach. The stomach serves three functions. First, it is a receptacle for storing predigested food. Second, the food is mixed with liquids and enzymes that will be used to process the food. Third, the stomach causes food to move through the intestinal system, where the process of breaking down the food into nutrients takes place. Several complex reactions take place in the small intestine.

Carbohydrates, such as starch and sugar, are broken down by enzymes beginning in the mouth. They are converted to glucose, primarily in the small intestines, and absorbed into the blood, which carries the glucose to the liver, where it is either used for energy or stored for future use. Proteins are broken down by enzymes in the small intestine to create amino acids, which are absorbed into the bloodstream and used throughout the body to create cells. The digestion of fats uses bile from the liver to dissolve the fat into small particles. Enzymes from the small intestine and pancreatic juices break large fat molecules into smaller ones, which are absorbed into the bloodstream. The blood carries the fat to different parts of the body where it is stored. Salt is absorbed in the small intestine by being dissolved in water that has been consumed. Some fibers can be dissolved by enzymes and processes, but insoluble fibers are simply passed through the digestive system.

A key process in the digestive system is the conversion of food to usable energy by oxidation. We can determine how much energy can be obtained from

Calorimetry: An experimental method for determining the energy of oxidation, or heating value.

food by the use of *calorimetry*, an experimental method for determining the energy of oxidation, or heating value. The device used to determine the energy (calories) obtained from both solid and liquid fuels is the bomb calorimeter, displayed in Fig. 14.3. The outside container is a Dewar flask that allows a negligible amount of heat transfer from the device. Inside the container is a large water bath. The amount of water used is precisely measured. Suspended in the water bath are a stirring propeller and a precision thermometer. The stirrer agitates the water to maintain a uniform temperature, which is measured to the nearest 1/10th of a degree Celsius.

Also suspended in the water bath is the bomb, which is a thick, stainless steel, sealed container used to house the oxidation reaction. The substance to be oxidized is precisely weighed and placed in a cup suspended in the middle of the bomb. A fuse wire (usually a nickel alloy called nichrome) is attached to a solid fuel or dipped in a liquid fuel. The wire is attached to electrical leads that run through the top cover of the bomb to the power supply used to ignite the fuel. The nichrome wire used is precisely weighed so that the energy of vaporization in any wire destroyed in the process can be accounted for. The top cover of the bomb is screwed on, and the bomb is charged with pure oxygen. This will make sure that more than enough oxygen is available to oxidize the entire amount of fuel and the specimen.

When the device is assembled, the stirrer is turned on, and the electricity is connected to the fuse wire. When the wire is ignited, the fuel and specimen in the bomb burn in the pure oxygen environment and are completely oxidized. The energy released is absorbed by the case of the bomb and the water bath. The temperature of the water bath is observed to increase until it reaches its maximum temperature, which is used to calculate the amount of energy absorbed by the system from the reaction. The gaseous products of combustion are released from the bomb, and it is carefully opened. The remaining fuse wire is collected and weighed to determine how much was vaporized. Calculations are made to determine the amount of heat given to the water bath and the bomb, and to determine the energy used to vaporize the missing fuse wire. The total energy given off is divided by the mass of the specimen to obtain its heating value (HV). Proper units of heating value are calories/gram (cal/g) or joules per gram (J/g). One calorie is the equivalent of 4.184 joules of energy.

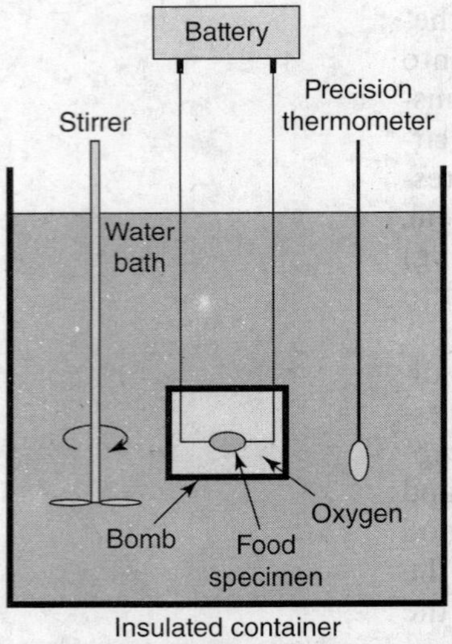

Figure 14.3

A bomb calorimeter.

Example 14.2 A bomb calorimeter

A bomb calorimeter is used to burn 14 g of gasoline. The water bath contains 2.08 kg of water initially at 21.24°C. The stainless steel bomb has a mass of 2.95 kg. The heat capacity of stainless steel is 461 J/kg·°C. Nichrome wire weighing 0.072 g is used to assemble the bomb. The bomb is charged with oxygen, submersed in the water bath, and the fuse wire is ignited. The water bath temperature is observed to rise to a maximum temperature of 26.14°C. After allowing the bomb and the water bath to cool, the bomb is removed.

The pressurized products of combustion are vented from the bomb, and it is disassembled. The remaining nichrome fuse wire is collected and weighed, yielding 0.051 g of wire. The nichrome wire has a heat of combustion of 5858 J/g. Calculate the heating value of the gasoline.

Solution

The heat produced by the fuse wire is

$$HV_{wire} = (0.092 - 0.051) \times 5858\ \text{J/g} = 240.18\ \text{J} = 0.240\ \text{kJ}$$

Comment

Three significant digits are stated because of the accuracy of the numbers used.

Water in this temperature range has a specific heat of 4.177 kJ/kg·°C, which is found in Appendix B-4. The amount of heat absorbed by the water bath is calculated to be

$$\begin{aligned} Q_{water} &= mC_p\Delta T \\ &= 2.08\ \text{kg} \times 4.177\ \text{kJ/kg·°C} \times (26.14°\text{C} - 21.24°\text{C}) = 42.6\ \text{kJ} \end{aligned}$$

The amount of heat absorbed by the body of the bomb is now calculated:

$$\begin{aligned} Q_{bomb} &= mC_p\Delta T \\ &= 2.96\ \text{kg} \times 0.461\ \text{kJ/kg·°C} \times (26.14°\text{C} - 21.24°\text{C}) = 6.69\ \text{kJ} \end{aligned}$$

The total heat released in this reaction is the sum

$$Q = 0.240\ \text{kJ} + 42.6\ \text{kJ} + 6.69\ \text{kJ} = 49.5\ \text{kJ}$$

The heating value of the gasoline is the amount of heat released by the combustion process divided by the mass of the fuel. It is

$$HV = \frac{49.5\ \text{kJ}}{0.014\ \text{kg}} = \underline{3530\ \text{kJ/kg of gasoline}}$$

One published value of the heating value of gasoline is about 3670 kJ/kg. Heating values will vary with the octane rating of the fuel and whether or not the water produced in the combustion reaction stays as steam or condenses into liquid water.

If the proper procedure is followed and the constituents are carefully weighed, the bomb calorimeter will yield excellent results for the heating value of food, as listed in Table 14.1. The heating value is the amount of energy released per unit mass when the food is completely oxidized. The heating value is given in units of joules per gram of food. Also listed are the percentages by mass amounts of fat, carbohydrate, and protein. It is not only the type of food, but the manner of preparation that determines the nutritional value. Meat that is roasted or baked will have much less fat content. The traditional unit of food energy is the *food calorie* or the kilogram calorie. One food calorie is equal to 4.184 kJ of energy.

Food calorie: The amount of energy needed to raise 1 kg of water 1°C.

Table 14.1 Nutritional Table of Common Foods

Food	Cooking Method	HV (kJ/kg)	% Fat	% Carbohydrate	% Protein
Almonds	Whole	24.4	53	21	21
Apple Juice	Canned	1.9	0	12	0
Apricots	Raw	2	0	11	0.01
Avocado	California	26	11	7	4
Banana		3.9	0.8	24	2
Beef Steak	Sirloin—Broiled	11.8	18	0	27
Blueberries	Raw	2.3	0.7	14	0.7
Broccoli	Raw	1.1	0.7	5	3
Butter	Salted	3.3	79	0	0
Catsup		4.4	0.4	25	2
Chicken	Roasted Breast	6.8	3	0	31
Egg	Raw	1.5	0	0	12
Flounder	Baked	5.9	7	0	19
Ground Beef	Fried	12.1	21	0	24
Italian Bread		11.9	0	56	9
Macaroni	Tender	4.6	1	23	4
Marshmallow		13.3	0	81	4
Milk	Whole	2.2	3	5	3
Milk	Skim	1.5	0	5	3
Oysters	Raw	2.8	2	3	8
Peanut Butter		24.8	50	19	31
Popcorn	Air Popped	15.7	0	75	13
Potato	Baked w. Skin	4.6	0	25	2.5
Raisins		12.6	1	79	3
White Rice	Instant	4.6	0	24	2
Salt		0	0	0	0
Sugar	Granulated	16.1	0	100	0
Turkey	Roasted	6.6	4	0	29

☑ **You have completed Learning Outcome (2)**

14.3 The Generation of Biological Work

Muscle tissue creates motion by contracting and relaxing. About 40% of the mass of the human body is muscle, composed of three basic types. *Cardiac muscles* control the contraction and release of the heart muscles and serve to pump blood throughout the human body. *Smooth muscle* tissue is utilized inside the body to produce motion in systems like the digestive system. These muscles move food along the digestive tract. *Skeletal muscles* are attached to the skeletal system to produce body motion.

Muscle motion is caused by the relative motion of two types of fiber in the muscle. Myosin fibers lie parallel to actin fibers and are connected. The chemical process by which the myosin fibers pull themselves along the actin fibers is called actomyosin-mediated adenosine triphosphate (ATP) hydrolysis. ATP is the same organic compound that converts light energy to chemical energy in photosynthesis. It acts as an organic energy converter. In this complex reaction, the myosin fiber relocates the base of the attached actin fiber by about 6 nm (6×10^{-9} m). Bulk muscle motion is achieved through billions of such reactions in a single muscle. A muscle is, in effect, a combination of many little motors each made up of the actin and myosin fibers. The relative motion between the actin and myosin fibers causes the muscle to act like a "linear" motor. The muscle is contracted in a fraction of a second, and the fibers are released to return to their original positions. We are used to thinking of motors as rotational devices, like electric motors and automobile engines. Muscles function more like linear actuators which produce reciprocal motion, similar to that of a piston.

The thermodynamics of muscle motion is based in the actomyosin-mediated ATP hydrolysis reaction. This process is a four-step, cyclic power cycle much the same as the Rankine power generation cycle described in Chapter 8.

Step 1: The myosin fiber is in position to attach to the actin fiber, as shown in Fig. 14.4. The myosin fiber is allowed to move freely by the splitting of ATP. The ATP kicks loose one phosphate molecule and breaks into adenosine diphosphate (ADP) and an inorganic phosphate P_i.

Step 2: The myosin fiber attaches to the actin fiber. The actin fiber combines with the myosin complex. The P_i released in Step 1 acts to strengthen the bonds between the actin fiber and the myosin fiber. This must be a reversible process if these bonds are to be released to later allow relaxation of the muscle.

Step 3: The myosin fiber moves the actin fiber. Products of the combination process in Step 2 are displaced. The released P_i causes an energy release, which is converted into a force moving the actin fiber.

Step 4: The myosin fiber releases its hold on the actin fiber. ATP is bound, and there is a dissociation of actomyosin. ADP is replaced by ATP, which breaks down the actomyosin complex and releases the bond between the fibers.

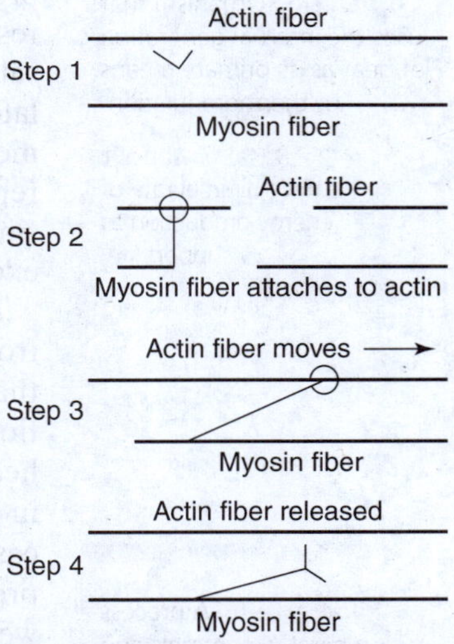

Figure 14.4
Muscle fiber actions.

☑ You have completed Learning Outcomes (3) and (4)

14.4 Temperature Regulation in Biological Systems

Living organisms live in a temperature range that extends from 0°C at the low end to 42°C at the high end. Life cannot exist at temperatures below 0°C without cryopreservation techniques because water in the body will freeze. Above 42°C,

Thermoregulation: Balancing the heat to and from a body with that generated within the body.

Metabolism: All the chemical reactions that produce energy in the body.

Homeothermic: Maintaining a relatively constant body temperature.

Poikilothermic: Having a varying body temperature.

proteins undergo a denaturing process that alters the structure of the protein. The subject of maintaining acceptable temperatures in living organisms is called *thermoregulation*; which is achieved by balancing the rate of energy production within the body and the rate of heat intake from outside the body with the rate of heat loss from the body. The term *metabolism* addresses all the chemical reactions that produce energy in the body. Some biological systems, like the digestive system, produce energy. Others produce energy in the form of work. Living organisms use a wide variety of processes to obtain and remove body heat.

Organisms that maintain a relatively constant body temperature are called *homeothermic*: mammals and birds. Organisms that have body temperatures that can vary considerably are called *poikilothermic*: lizards and fish.

14.4.1 Endothermic organisms

Endothermic organism: Organism that uses the internal generation of heat as its primary means of thermoregulation.

Basal metabolic rate: Minimal rate of energy production to support life.

Organisms that use the internal generation of heat as their primary means of thermoregulation are *endothermic organisms*. Mammals fall into this classification. The minimal rate of energy production needed to support life in an endothermic organism is called the *basal metabolic rate*. This assumes that the organism is at rest and that no digestion is occurring when its temperature is measured. The basal metabolic rate accounts for the energy needed to support breathing, to circulate blood, and to maintain electrical activity to support the nervous system. The metabolic rate in an organism is affected by physical activity, eating, body growth, reproduction, and heat production. The metabolic rate is lowest when sleeping since nonessential muscles are not working. The metabolic rate during extreme exercising can be 10 to 20 times the basal metabolic rate.

Heat loss in endothermic organisms occurs by convection and evaporation from the skin to the surrounding atmosphere. If your skin temperature is greater than the atmospheric temperature around you, you lose heat primarily by convection heat transfer from the surface of the skin. Endothermic organisms can gain heat by immersing themselves in warm water or cool off by immersing themselves in cold water. Mammals and birds can reduce heat loss in cold weather by a process called *piloerection*, where hair or feathers are lifted away from the skin to produce additional insulation near the skin. Since humans do not have much hair, we notice this phenomenon as "goose bumps." If your skin temperature is less than the atmospheric temperature, you lose heat by sweat evaporating from the surface of your skin. The phase change releases large amounts of heat. Bull frogs achieve this same effect by secreting mucus from their skin.

Piloerection: A process whereby mammals and birds reduce heat loss.

Hypothermia: When the human body gets below 35°C.

Hyperthermia: When the human body gets above 38°C.

If the cooling rate or heating rate of the body is insufficient to maintain a constant body temperature, the temperature of the body will change, with serious consequences. Table 14.2 lists the physical effect of either increasing or decreasing the body temperature. *Hypothermia* occurs when the human body temperature gets below 35°C. *Hyperthermia* occurs when the human body temperature reaches 38°C or higher.

14.4.2 Ectothermic organisms

Ectothermic organism: The organism is heated from sources external to the body.

An *ectothermic organism* is heated from sources external to the body. Lizards will absorb energy by convection from the surrounding air, by conduction from laying on warm rocks or gravel, or from direct solar radiation. Fish absorb heat energy from the surrounding water. Fish tend to be homeothermic since they live

Table 14.2 Physical Effects of Changing the Human Body Temperature

Body Temp	Symptoms
44°C	Death
43°C	Brain damage and possibly death
41°C	Fainting, dehydration, vomiting, headaches, or dizziness
38°C	Sweating and uncomfortable feelings
37°C	Normal body temperature
36°C	Shivering
34°C	Severe shivering, loss of movement in the extremities, and skin color changes
33°C	Confusion, drowsiness, slow heartbeat, and shallow breathing
31°C	Loss of consciousness and change in heart rhythm
26°C	Death

in environments where the water temperature is stable. Both fish and lizards are poikilothermic. Since their body temperature will vary and they receive external heat, their metabolic rates are very low. The body temperature of an ectothermic organism will largely be a function of the environmental temperature. Cold weather will decrease the body temperature and thus reduce physical activity, whereas warm weather will increase body temperature and physical activity. Ectothermic organisms will seek to regulate body temperature by altering their physical environment: Lizards sit on warm, sunny rocks to gain energy and move into shady areas when they are too warm; fish move to warmer or colder water depending on their individual needs. Bees and other insects can raise their body temperature in anticipation of flying by exercising their flying muscles.

14.4.3 Temperature regulation in plants

Some plants have the ability to raise their temperature above the ambient temperature by a process called *thermogenesis*. This can make them resistant to frost. Thermogenesis is the development of thermal energy as a part of a chemical reaction in the cellular respiratory process. The sacred lotus plant shown in Fig. 14.5, which is native to Southeast Asia, can remain up to 20°C above the ambient temperature. In this plant, heat is generated by breaking down starches that are stored in its roots.

Figure 14.5

Sacred lotus plant.

14.5 Summary

In this chapter, the energy that can be generated from the sun through the action of photosynthesis in plants and algae was examined. Then the generation of energy from the foods eaten by both animals and humans was presented. Finally, the energy required by the muscular systems in the bodies of animals and humans was investigated, along with the regulation of temperature in biological systems.

The following additional terms were introduced or emphasized in this chapter:

Calorimetry: *An experimental method for determining the energy of oxidation, or heating value.*

Chlorophyll: *The pigment that absorbs sunlight.*

Ectothermic: *Energy gained from an external source.*

Ectothermic organism: *An organism heated from sources external to the body.*

Endothermic: *Energy generated internally.*

Endothermic organism: *An organism that uses the internal generation of heat.*

Homeothermic: *An organism that has a constant body temperature.*

Metabolism: *The chemical reactions that produce energy in the body.*

Photosynthesis: *The process by which sunlight converts plants and algae into usable energy.*

Poikilothermic: *An organism that has a varying body temperature.*

Thermogenesis: *The production of heat in living organisms.*

Thermoregulation: *Processes that are used to control body temperature.*

Two equations were utilized in this chapter:

Photon energy: $E = h\nu$

Frequency of light: $\nu = c/\lambda$

Problems

Energy Conversion in Plants

14.1 Plot the energy of a photon as a function of the wavelength of light for the range of visible wavelengths.

14.2 If a plant could be engineered that would convert gamma rays ($\lambda = 0.01$ nm) into chemical energy, what would be the percent increase in energy over plants that use light with a wavelength of 660 nm?

14.3 Using Eq. 14.1, determine how many grams of sugar are produced if one kilogram of carbon dioxide is used in the photosynthesis process. How much water is needed?

Energy Conversion in Animals

14.4 For the bomb calorimeter described in Example 14.2, determine the maximum temperature the bomb would achieve if 20 g of apple juice and 0.005 g of fuse wire were burned.

14.5 For one day, record what types of food you ate and roughly how much of each type you consumed. Calculate your energy intake for that day in joules and in food calories.

14.6 Design a menu that will achieve the following caloric intake:

a) 1000 calories per day for a diet

b) 1500 calories per day for a smaller person

c) 2000 calories per day for a larger person

d) 6000 calories per day for a Tour-de-France bike rider

14.7 Compare the daily caloric intake for breakfast, lunch, and dinner at a university dining hall to the typical meals you would eat at home. Which diet has a higher fat content? Carbohydrate content?

14.8 Develop a daily diet that would reduce your amount of fat intake.

14.9 A very popular diet encourages people to eat a high-protein diet. Develop a daily high-protein diet and compare it to your normal daily diet. Did you increase or decrease the fat and carbohydrate intake?

The Generation of Biological Work

14.10 Most joint actions are the combination of one or more muscles tightening and other muscles relaxing at the same time. Describe the action of three human joints that act in this manner.

14.11 Give one example of smooth muscle tissue in the human body and describe how it functions.

14.12 Perform an online search and report on what happens to cardiac muscle tissue during a heart attack.

Temperature Regulation in Biological Systems

14.13 If the ambient temperature is 38°C, estimate the heat transferred from your body by the evaporation of 5 g of sweat.

14.14 List three types of fish that live in cold water and three types of fish that live in warm water.

14.15 List three plants that have the ability to regulate their temperatures.

14.16 During the first expeditions to the Antarctic, explorers found out that short, stocky men had a much better survival rate than tall athletic men. Why is this?

14.17 List five ways to cool the human body in case of overheating. What is the best method in an industrial plant?

14.18 Find an online Basic Metabolic Rate calculator and use it to determine your BMR.

14.19 Using the BMR calculator from Problem 14.18, determine the percent drop in your BMR if you lost 10 pounds on a diet.

THE APPENDICES

List of Appendices

A Conversions of Units

Length	Force	Mass	Velocity
1 cm = 0.3937 in	1 lbf = 0.4448 × 10^6 dyne	1 oz = 28.35 g	1 mph = 1.467 ft/s
1 m = 3.281 ft	1 dyne = 2.248 × 10^{-6} lbf	1 lbm = 0.4536 kg	1 mph = 0.8684 knot
1 km = 0.6214 mi	1 kip = 1000 lbf	1 slug = 32.17 lbm	1 ft/s = 0.3048 m/s
1 in = 2.54 cm	1 N = 0.2248 lbf	1 slug = 14.59 kg	1 km/h = 0.2778 m/s
1 ft = 0.3048 m	1 N = 10^5 dyne	1 kg = 2.205 lbm	1 knot = 1.688 ft/s
1 mi = 1.609 km			
1 mi = 5280 ft			
1 mi = 1760 yd			

Work and Heat	Power	Pressure	Volume
1 J = 10^7 ergs	1 hp = 550 ft-lbf/s	1 psi = 2.036 in Hg	1 ft^3 = 7.481 gal (U.S.)
1 ft-lbf = 1.356 J	1 hp = 2545 Btu/hr	1 psi = 27.7 in H_2O	1 ft^3 = 0.02832 m^3
1 Cal = 3.088 ft-lbf	1 hp = 0.7455 kW	1 atm = 29.92 in Hg	1 gal (U.S.) = 231 in^3
1 Cal = 0.003968 Btu	1 W = 1 J/s	1 atm = 33.93 ft H_2O	1 gal (Brit.) = 1.2 gal (U.S.)
1 Btu = 1055 J	1 W = 10^7 dyne·cm/s	1 atm = 101.3 kPa	1 L = 10^{-3} m^3
1 Btu = 0.2930 W·hr	1 W = 3.412 Btu/hr	1 atm = 1.0133 bar	1 L = 0.03531 ft^3
1 Btu = 778 ft-lbf	1 kW = 1.341 hp	1 atm = 14.7 psi	1 L = 0.2642 gal
1 kWh = 3412 Btu	1 ton = 12,000 Btu/hr	1 in Hg = 0.4912 psi	1 m^3 = 264.2 gal
1 therm = 10^5 Btu	1 ton = 3.517 kW	1 ft H_2O = 0.4331 psi	1 m^3 = 35.31 ft^3
1 quad = 10^{15} Btu	1 W = 0.7375 ft-lbf/s	1 psi = 6.895 kPa	1 ft^3 = 28.32 L
		1 kPa = 0.145 psi	1 in^3 = 16.387 cm^3
		1 Torr = 1 mm Hg	

B Material Properties

Table B-I Properties of the U.S. Standard Atmosphere

$P_0 = 101.3$ kPa $\rho_0 = 1.225$ kg/m^3

Altitude m	Temperature °C	Pressure P/P_0	Density ρ/ρ_0
0	15.2	1.000	1.000
1 000	9.7	0.8870	0.9075
2 000	2.2	0.7846	0.8217
3 000	−4.3	0.6920	0.7423
4 000	−10.8	0.6085	0.6689
5 000	−17.3	0.5334	0.6012
6 000	−23.8	0.4660	0.5389
7 000	−30.3	0.4057	0.4817
8 000	−36.8	0.3519	0.4292
10 000	−49.7	0.2615	0.3376
12 000	−56.3	0.1915	0.2546
14 000	−56.3	0.1399	0.1860
16 000	−56.3	0.1022	0.1359
18 000	−56.3	0.07466	0.09930
20 000	−56.3	0.05457	0.07258
30 000	−46.5	0.01181	0.01503
40 000	−26.6	0.2834×10^{-2}	0.3262×10^{-2}
50 000	−2.3	0.7874×10^{-3}	0.8383×10^{-3}
60 000	−17.2	0.2217×10^{-3}	0.2497×10^{-3}
70 000	−53.3	0.5448×10^{-4}	0.7146×10^{-4}

Table B-2 Properties of Various Ideal Gases (at 25°C)

Gas	Chemical Formula	Molar Mass	R		C_p		C_v		k
			kJ/kg·K	ft-lbf/lbm-°R	kJ/kg·K	Btu/lbm-°R	kJ/kg·K	Btu/lbm-°R	k
Air	–	28.97	0.2870	53.34	1.003	0.240	0.717	0.171	1.400
Argon	Ar	39.95	0.2081	38.68	0.520	0.1253	0.312	0.0756	1.667
Butane	C_4H_{10}	58.12	0.1430	26.58	1.716	0.415	1.573	0.381	1.091
Carbon dioxide	CO_2	44.01	0.1889	35.10	0.842	0.203	0.653	0.158	1.289
Carbon monoxide	CO	28.01	0.2968	55.16	1.041	0.249	0.744	0.178	1.400
Ethane	C_2H_6	30.07	0.2765	51.38	1.766	0.427	1.490	0.361	1.186
Ethylene	C_2H_4	28.05	0.2964	55.07	1.548	0.411	1.252	0.340	1.237
Helium	He	4.00	2.0770	386.0	5.198	1.25	3.116	0.753	1.667
Hydrogen	H_2	2.02	4.1242	766.4	14.209	3.43	10.085	2.44	1.409
Methane	CH_4	16.04	0.5184	96.35	2.254	0.532	1.735	0.403	1.299
Neon	Ne	20.18	0.4120	76.55	1.020	0.246	0.618	0.1477	1.667
Nitrogen	N_2	28.01	0.2968	55.15	1.042	0.248	0.745	0.177	1.400
Octane	C_8H_{18}	114.23	0.0728	13.53	1.711	0.409	1.638	0.392	1.044
Oxygen	O_2	32.00	0.2598	48.28	0.922	0.219	0.662	0.157	1.393
Propane	C_3H_8	44.10	0.1886	35.04	1.679	0.407	1.491	0.362	1.126
Steam	H_2O	18.02	0.4615	85.76	1.872	0.445	1.411	0.33.5	1.327

Note: C_p, C_v, and k are at 298 K. Also, kJ/kg·K is the same as kJ/kg·°C.

Based on data from G. J. Van Wylen and R. E. Sonntag, Fundamentals of Classical Thermodynamics, Wiley, New York, 1976.

Table B-3 Critical-Point Constants

Substance	Formula	Molar Mass	Temperature		Pressure		Volume		Z_{cr}
			K	°R	MPa	psia	m^3/kmol	ft^3/lbmol	
Air	–	28.97	133	239	3.77	547	0.0883	1.41	0.30
Ammonia	NH_4	17.03	405.6	729.8	11.28	1636	0.0724	1.16	0.243
Argon	Ar	39.94	151	272	4.86	705	0.0749	1.20	0.290
Benzene	C_6H_6	78.11	562	1012	4.92	714	0.2603	4.17	0.274
Butane	C_4H_{10}	58.12	425.2	765.2	3.80	551	0.2547	4.08	0.274
Carbon dioxide	CO_2	44.01	304.2	547.5	7.39	1070	0.0943	1.51	0.275
Carbon monoxide	CO	28.01	133	240	3.50	507	0.0930	1.49	0.294
Carbon tetrachloride	CC_{14}	153.84	556.4	1001.5	4.56	661	0.2759	4.42	0.272
Ethane	C_2H_6	30.07	305.5	549.8	4.88	708	0.148	2.37	0.284
Ethylene	C_2H_4	28.05	282.4	508.3	5.12	742	0.1242	1.99	0.271
Helium	He	4.00	5.3	9.5	0.23	33.2	0.0578	0.926	0.302
Hydrogen	H_2	2.02	33.3	59.9	1.30	188	0.0649	1.04	0.304
Methane	CH_4	16.04	191.1	343.9	4.64	673	0.0993	1.59	0.290
Neon	Ne	20.18	44.5	80.1	2.73	395	0.0417	0.668	0.308
Nitrogen	N_2	28.02	126.2	227.1	3.39	492	0.0899	1.44	0.291
Oxygen	O_2	32.00	154.8	278.6	5.08	736	0.078	1.25	0.308
Propane	C_6H_6	44.09	370.0	665.9	4.26	617	0.1998	3.20	0.277
Propylene	C_3H_6	42.08	365.0	656.9	4.62	670	0.1810	2.90	0.276
R134a	CF_3CH_2F	102.03	374.3	613.7	4.07	596	0.2478	2.96	0.324
Sulfur dioxide	SO_2	64.06	430.7	775.2	7.88	1143	0.1217	1.95	0.269
Water	H_2O	18.02	647.4	1165.3	22.1	3204	0.0568	0.90	0.233

Based on data from K. A. Kobe and R. E. Lynn, Jr., Chem. Rev., 52: 117–236 (1953).

Table B-4 Specific Heats and Densities of Liquids and Solids

C_p kJ/kg·°C ρ kg/m^3

Liquids

Substance	State	C_p	Density ρ	Substance	State	C_p	Density ρ
Water	1 atm, 25°C	4.177	1000	Glycerin	1 atm, 10°C	2.32	1261
Ammonia	1 atm, −20°C	4.52	665				
	1 atm, 25°C	4.80	602	Mercury	1 atm, 10°C	0.138	13 600
R-134a	1 atm, −50°C	1.23	1440				
	1 atm, 25°C	1.42	1210	Propane	1 atm, 0°C	2.41	529
Benzene	1 atm, 15°C	1.80	879	Ethyl Alcohol	1 atm, 25°C	2.43	789

Solids

Substance	T °C	C_p	Density ρ	Substance	T °C	C_p	Density ρ
Ice	−20	2.033	920	Lead	0	0.124	11 310
	0	2.10	917				
Aluminum	−100	0.699	2700				
	0	0.870	2700	Copper	−100	0.328	8940
	100	0.941	2700		0	0.381	8940
Iron	20	0.448	7800		100	0.393	8940
Silver	20	0.233	10 500				

Based on data from Kenneth Wark, Thermodynamics, 3rd ed., McGraw-Hill, New York, 1981.

Table B-5 Ideal-Gas Specific Heat of Several Common Gases as a Cubic Function of Temperature

$\bar{C}_p = a + bT + cT^2 + dT^3$ with T in K and $\bar{C}_p$ in kJ/kmol·K

Gas	a	b	c	d
Air	28.11	0.001967	0.4802×10^{-5}	-1.966×10^{-9}
N_2	28.9	−0.001571	0.8081×10^{-5}	-2.873×10^{-9}
O_2	25.48	0.0152	-0.7155×10^{-5}	1.312×10^{-9}
H_2	29.11	−0.001916	0.4003×10^{-5}	-0.8704×10^{-9}
CO	28.16	0.001675	0.5372×10^{-5}	-2.222×10^{-9}
NO	29.34	−0.0009395	0.9747×10^{-5}	-4.187×10^{-9}
H_2O	32.24	0.001923	1.055×10^{-5}	-3.595×10^{-9}
CO_2	22.26	0.05981	-3.501×10^{-5}	7.469×10^{-9}
NO_2	22.9	0.05715	-3.52×10^{-5}	7.87×10^{-9}

Based on data from G. J. Van Wylen and R. E. Sonntag, Fundamentals of Classical Thermodynamics, Wiley, New York, 1976.

Table B-6 Ideal-Gas Specific Heats of Several Common Gases at Different Temperatures

	c_p	c_v	k	c_p	c_v	k	c_p	c_v	k
	kJ/kg·K			kJ/kg·K			kJ/kg·K		
Temperature, K	**Air**			**Carbon dioxide, CO_2**			**Carbon monoxide, CO**		
250	1.003	0.716	1.401	0.791	0.602	1.314	1.039	0.743	1.400
300	1.005	0.718	1.400	0.846	0.657	1.288	1.040	0.744	1.399
350	1.008	0.721	1.398	0.895	0.706	1.268	1.043	0.746	1.398
400	1.013	0.726	1.395	0.939	0.750	1.252	1.047	0.751	1.395
450	1.020	0.733	1.391	0.978	0.790	1.239	1.054	0.757	1.392
500	1.029	0.742	1.387	1.014	0.825	1.229	1.063	0.767	1.387
550	1.040	0.753	1.381	1.046	0.857	1.220	1.075	0.778	1.382
600	1.051	0.764	1.376	1.075	0.886	1.213	1.087	0.790	1.376
650	1.063	0.776	1.370	1.102	0.913	1.207	1.100	0.803	1.370
700	1.075	0.788	1.364	1.126	0.937	1.202	1.113	0.816	1.364
750	1.087	0.800	1.359	1.148	0.959	1.197	1.126	0.829	1.358
800	1.099	0.812	1.354	1.169	0.980	1.193	1.139	0.842	1.353
900	1.121	0.834	1.344	1.204	1.015	1.186	1.163	0.866	1.343
1000	1.142	0.855	1.336	1.234	1.045	1.181	1.185	0.888	1.335
Temperature, K	**Hydrogen, H_2**			**Nitrogen, N_2**			**Oxygen, O_2**		
250	14.051	9.927	1.416	1.039	0.742	1.400	0.913	0.653	1.398
300	14.307	10.183	1.405	1.039	0.743	1.400	0.918	0.658	1.395
350	14.427	10.302	1.400	1.041	0.744	1,399	0.928	0.668	1.389
400	14.476	10.352	1.398	1.044	0.747	1.397	0.941	0.681	1.382
450	14.501	10.377	1.398	1.049	0.752	1.395	0.956	0.696	1.373
500	14.513	10.389	1.397	1.056	0.759	1.391	0.972	0.712	1.365
550	14.530	10.405	1.396	1.065	0.768	1.387	0.988	0.728	1.358
600	14.546	10.422	1.396	1.075	0.778	1.382	1.003	0.743	1.350
650	14.571	10.447	1.395	1.086	0.789	1.376	1.017	0.758	1.343
700	14.604	10.480	1.394	1.098	0.801	1.371	1.031	0.771	1.337
750	14.645	10.521	1.392	1.110	0.813	1.365	1.043	0.783	1.332
800	14.695	10.570	1.390	1.121	0.825	1.360	1.054	0.794	1.327
900	14.822	10.698	1.385	1.145	0.849	1.349	1.074	0.814	1.319
1000	14.983	10.859	1.380	1.167	0.870	1.341	1.090	0.830	1.313

Based on data from Kenneth Wark, Thermodynamics, 4th ed. (New York: McGraw-Hill, 1983), p. 783, Table A-4M. Originally published in Tables of Thermal Properties of Gases, NBS Circular 564, 1955.

Table B-7 Enthalpy of Formation and Enthalpy of Vaporization

25°C (77°F), 1 atm

Substance	Formula	$\bar{h}_f^\circ$ kJ/kmol	$\bar{h}_{fg}$ kJ/kmol	$\bar{h}_f^\circ$ Btu/lbmol	$\bar{h}_{fg}$ Btu/lbmol
Carbon	C(s)	0		0	
Hydrogen	$H_2(g)$	0		0	
Nitrogen	$N_2(g)$	0		0	
Oxygen	$O_2(g)$	0		0	
Carbon monoxide	$CO(g)$	−110 500		−47,540	
Carbon dioxide	$CO_2(g)$	−393 500		−169,300	
Water	$H_2O(g)$	−241 800		−104,000	
Water	$H_2O(\ell)$	−285 800	44 010	−123,000	
Hydrogen peroxide	$H_2O_2(g)$	−136 300	61 090	−58,640	26,260
Ammonia	$NH_3(g)$	−46 190		−19,750	
Oxygen	$O(g)$	249 200		107,200	
Hydrogen	$H(g)$	218 000		93,780	
Nitrogen	$N(g)$	472 700		203,300	
Hydroxyl	$OH(g)$	39 040		16,790	
Methane	$CH_4(g)$	−74 850		−32,210	
Acetylene (Ethyne)	$C_2H_2(g)$	226 700		97,540	
Ethylene (Ethene)	$C_2H_4(g)$	52 280		22,490	
Ethane	$C_2H_6(g)$	−84 680		−36,420	
Propylene (Propene)	$C_3H_6(g)$	20 410		8,790	
Propane	$C_3H_8(g)$	−103 800	15 060	−44,680	6,480
n-Butane	$C_4H_{10}(g)$	−126 200	21 060	−54,270	9,060
n-Pentane	$C_5H_{12}(g)$	−146 400	31 410	−62,980	11,360
n-Octane	$C_8H_{18}(g)$	−208 400	41 460	−89,680	17,840
n-dodecane	$C_{12}H_{26}(\ell)$	−291 000	44 000	−125,200	18,900
Benzene	$C_5H_6(g)$	82 930	33 830	35,680	14,550
Methyl alcohol	$CH_3OH(g)$	−200 900	37 900	−86,540	16,090
Ethyl alcohol	$C_2H_5OH(g)$	−235 300	42 340	−101,200	18,220

Based on data from JANAF Thermochemical Tables, NSRDS-NBS-37, 1971; Selected Values of Chemical Thermodynamic Properties, NBS Technical Note 270-3, 1963; and API Res. Project 44, Carnegie Institute of Technology, Pittsburg, 1953.

Table B-8 Enthalpy of Combustion and Enthalpy of Vaporization

25°C (77°F), 1 atm

Substance	Formula	HHV, kJ/kmol	$\bar{h}_{fg}$, kJ/kmol	HHV, Btu/lbmol	$\bar{h}_{fg}$, Btu/lbmol
Hydrogen	$H_2(g)$	285 800		123,000	
Carbon	C(s)	393 500		169,300	
Carbon monoxide	CO(g)	283 000		121,800	
Methane	$CH_4(g)$	890 400		383,000	
Acetylene	$C_2H_2(g)$	1 300 000		559,000	
Ethylene	$C_2H_4(g)$	1 411 000		607,000	
Ethane	$C_2H_6(g)$	1 560 000		671,000	
Propylene	$C_3H_6(g)$	2 058 000		885,600	
Propane	$C_3H_8(g)$	2 220 000	15 060	955,000	6,480
n-Butane	$C_4H_{10}(g)$	2 877 000	21 060	1,238,000	9,060
n-Pentane	$C_5H_{12}(g)$	3 536 000	26 410	1,521,000	11,360
n-Hexane	$C_6H_{14}(g)$	4 195 000	31 530	1,805,000	13,560
n-Heptane	$C_7H_{16}(g)$	4 854 000	36 520	2,090,000	15,710
n-Octane	$C_8H_{18}(g)$	5 512 000	41 460	2,370,000	17,840
n-dodecane	$C_{12}H_{26}(\ell)$	8 126 000	44 000	3,370,000	18,900
Benzene	$C_6H_6(g)$	3 302 000	33 830	1,420,000	14,550
Toluene	$C_7H_8(g)$	3 948 900	39 920	1,700,000	17,180
Methyl alcohol	$CH_3OH(g)$	764 500	37 900	328,700	16,090
Ethyl alcohol	$C_2H_5OH(g)$	1 409 000	42 340	606,200	18,220

Note: Water appears as a liquid in the products of combustion.

Based on data from Kenneth Wark, Thermodynamics, 3rd ed. McGraw-Hill, New York, 1981, Table A 23M.

Table B-9 Natural Logarithms of the Equilibrium Constant K_p

The equilibrium constant K_p for the reaction $\nu_A A + \nu_B B \Leftrightarrow \nu_C C + \nu_D D$ is defined as $K_p = \dfrac{P_C^{\nu_C} P_D^{\nu_D}}{P_A^{\nu_A} P_B^{\nu_B}}$

Temp., K	$H_2 \Leftrightarrow 2H$	$O_2 \Leftrightarrow 2O$	$N_2 \Leftrightarrow 2N$	$H_2O \Leftrightarrow H_2 + ½O_2$	$H_2O \Leftrightarrow ½H_2 + HO$	$CO_2 \Leftrightarrow CO + ½O_2$	$½N_2 + ½O_2 \Leftrightarrow NO$
298	−164.005	−186.975	−367.480	−92.208	−106.208	−103.762	−35.052
500	−92.827	−105.630	−213.373	−52.691	−60.281	−57.616	−20.295
1000	−39.803	−45.150	−99.127	−23.163	−26.034	−23.529	−9.388
1200	−30.874	−35,605	−80.611	−18.182	−20.283	−17.871	−7.569
1400	−24.463	−27.742	−66.329	−14.609	−16.099	−13.842	−6.270
1600	−19.637	−22.285	−56.055	−11.921	−13.066	−10.830	−5.294
1800	−15.866	−18.030	−48.051	−9.826	−10.657	−8.497	−4.536
2000	−12.840	−14.622	−41.645	−8.145	−8.728	−6.635	−3.931
2200	−10.353	−11.827	−36.391	−6.768	−7.148	−5.120	−3.433
2406	−8.276	−9.497	−32,011	−5.619	−5.832	−3.860	−3.019
2600	−6.517	−7.521	−28.304	−4.648	−4.719	−2.801	−2.671
2800	−5.002	−5.826	−25.117	−3.812	−3.763	−1.894	−2.372
3000	−3.685	−4357	−22.359	−3.086	−2.937	−1.111	−2.114
3200	−2.534	−3.072	−19.937	−2.451	−2.212	−0.429	−1.888
3400	−1.516	−1.935	−17.800	−1.891	−1.576	0.169	−1.690
3600	−0.609	−0.926	−15.898	−1.392	−1.088	0.701	−1.513
3800	0.202	−0.019	−14.199	−0.945	−0.501	1.176	−1.356
4000	0.934	0.796	−12.660	−0.542	−0.044	1.599	−1.216
4500	2.486	2.513	−9.414	0.312	0.920	2.490	−0.921
5000	3.725	3.895	−6.807	0.996	1.689	3.197	−0.686
5500	4.743	5.023	−4.665	1.560	2.318	3.771	−0.497
6000	5.590	5.963	−2.865	2.032	2.843	4.245	−0.341

Based on data from Gordon J. Van Wylen and Richard E. Sonntag, Fundamentals of Classical Thermodynamics, English/SI Version. 3rd ed. (New York: John Wiley & Sons, 1986), p 723, Table A.14. Based on thermodynamic data given in JANAF, Thermochemical Tables (Midland, MI: Thermal Research Laboratory, The Dow Chemical Company, 1971).

Table B-10 Constants for the van der Waals Equation of State

	a, Pa·m^6/kg^2	b, m^3/kg	a, lbf-ft^4/lbm^2	b, ft^3/lbm
Air	163.0	0.00127	870	0.0202
Ammonia	1468	0.00220	7850	0.0351
Carbon dioxide	188.3	0.000972	1010	0.0156
Carbon monoxide	188.0	0.00141	1010	0.0227
Freon 12	71.8	0.000803	394	0.0132
Helium	214	0.00587	1190	0.0959
Hydrogen	6083	0.0132	32,800	0.212
Methane	888	0.00266	4780	0.0427
Nitrogen	174.7	0.00138	934	0.0221
Oxygen	134.4	0.000993	720	0.0159
Propane	481	0.00204	2580	0.0328
Water	1703	0.00169	9130	0.0271

C Steam Tables

Table C-1 Properties of Saturated H_2O—Temperature Table

T °C	P MPa	v m³/kg v_f	v_g	u kJ/kg u_f	u_g	h kJ/kg h_f	h_{fg}	h_g	s kJ/kg·K s_f	s_{fg}	s_g
0.01	00.000611	0.001000	206.1	0.0	2375.3	0.0	2501.3	2501.3	0.0000	9.1571	9.1571
2	0.0007056	0.001000	179.9	8.4	2378.1	8.4	2496.6	2505.0	0.0305	9.0738	9.1043
5	0.0008721	0.001000	147.1	21.0	2382.2	21.0	2489.5	2510.5	0.0761	8.9505	9.0266
10	0.001228	0.001000	106.4	42.0	2389.2	42.0	2477.7	2519.7	0.1510	8.7506	8.9016
15	0.001705	0.001001	77.93	63.0	2396.0	63.0	2465.9	2528.9	0.2244	8.5578	8.8.7822
20	0.002338	0.001002	57.79	83.9	2402.9	83.9	2454.2	2538.1	0.2965	8.3715	8.6680
25	0.003169	0.001003	43.36	104.9	2409.8	104.9	2442.3	2547.2	0.3672	8.1916	8.5588
30	0.004246	0.001004	32.90	125.8	2416.6	125.8	2430.4	2556.2	0.4367	8.0174	8.4541
35	0.005628	0.001006	25.22	146.7	2423.4	146.7	2418.6	2565.3	0.5051	7.8488	8.3539
40	0.007383	0.001008	19.52	167.5	2430.1	167.5	2406.8	2574.3	0.5723	7.6855	8.2578
45	0.009593	0.001010	15.26	188.4	2436.8	188.4	2394.8	2583.2	0.6385	7.5271	8.1656
50	0.01235	0.001012	12.03	209.3	2443.5	209.3	2382.8	2592.1	0.7036	7.3735	8.0771
55	0.01576	0.001015	9.569	230.2	2450.1	230.2	2370.7	2600.9	0.7678	7.2243	7.9921
60	0.01994	0.001017	7.671	251.1	2456.6	251.1	2358.5	2609.6	0.8310	7.0794	7.9104
65	0.02503	0.001020	6.197	272.0	2463.1	293.0	2333.8	2626.8	0.9549	6.8012	7.7561
70	0.03119	0.001023	5.042	292.9	2469.5	293.0	2333.8	2626.8	0.9549	6.8012	7.7561
75	0.03858	0.001026	4.131	313.9	2475.9	313.9	2321.4	2635.3	1.0155	6.6678	7.6833
80	0.04739	0.001029	3.407	334.8	2482.2	334.9	2308.8	2643.7	1.0754	6.5376	7.6130
85	0.05783	0.001032	2.828	355.8	2488.4	355.9	2296.0	2651.9	1.1344	6.4109	7.5453
90	0.07013	0.001036	2.361	376.8	2494.5	376.9	2283.2	2660.1	1.1927	6.2872	7.4799
95	0.08455	0.001040	1.982	397.9	2500.6	397.9	2270.2	2668.1	1.2503	6.1664	7.4167
100	0.1013	0.001044	1.673	418.9	2506.5	419.0	2257.0	2676.0	1.3071	6.0486	7.3557
110	0.1433	0.001052	1.210	461.1	2518.1	461.3	2230.2	2691.5	1.4188	5.8207	7.2395
120	0.1985	0.001060	0.8919	503.5	2529.2	503.7	2202.6	2706.3	1.5280	5.6024	7.1304
130	0.2701	0.001070	0.6685	546.0	2539.9	546.3	2174.2	2720.5	1.6348	5.3929	7.0277
140	0.3613	0.001080	0.5089	588.7	2550.0	589.1	2144.8	2733.9	1.7395	5.1912	6.9307
150	0.4758	0.001090	0.3928	631.7	2559.5	632.2	2114.2	2746.4	1.8422	4.9965	6.8387
160	0.6178	0.001102	0.3071	674.9	2568.4	675.5	2082.6	2758.1	1.9431	4.8079	6.7510
170	0.7916	0.001114	0.2428	718.3	2576.5	719.2	2049.5	2768.7	2.0423	4.6249	6.6672
180	1.002	0.001127	0.1941	762.1	2583.7	763.2	2015.0	2778.2	2.1400	4.4466	6.5866
190	1.254	0.001141	0.1565	806.2	2590.0	807.5	1978.8	2786.4	2.2363	4.2724	6.5087
200	1.554	0.001156	0.1274	850.6	2595.3	852.4	1940.8	2793.2	2.3313	4.1018	6.4331
210	1.906	0.001173	0.1044	895.5	2599.4	897.7	1900.8	2798.5	2.4253	3.9340	6.3593
220	2.318	0.001190	0.08620	940.9	2602.4	943.6	1858.5	2802.1	2.5183	3.7686	6.2869

(*Continued*)

Based on data from Keenan, Keyes, Hill, and Moore, Steam Tables, Wiley, New York, 1969; G. J. Van Wylen and R. E. Sonntag, Fundamentals of Classical Thermodynamics, Wiley, New York, 1973.

Table C-1 *(Continued)*

		v m³/kg		*u* kJ/kg		*h* kJ/kg			*s* kJ/kg·K		
T °C	*P* MPa	v_f	v_g	u_f	u_g	h_f	h_{fg}	h_g	s_f	s_{fg}	s_g
230	2.795	0.001209	0.07159	986.7	2603.9	990.1	1813.9	2804.0	2.6105	3.6050	6.2155
240	3.344	0.001229	0.05977	1033.2	2604.0	1037.3	1766.5	2803.8	2.7021	3.4425	6.1446
250	3.973	0.001251	0.05013	1080.4	2602.4	1085.3	1716.2	2801.5	2.7933	3.2805	6.0738
260	4.688	0.001276	0.04221	1128.4	2599.0	1134.4	1662.5	2796.9	2.8844	3.1184	6.0028
270	5.498	0.001302	0.03565	1177.3	2593.7	1184.5	1605.2	2789.7	2.9757	2.9553	5.9310
280	6.411	0.001332	0.03017	1227.4	2586.1	1236.0	1543.6	2779.6	3.0674	2.7905	5.8579
290	7.436	0.001366	0.02557	1278.9	2576.0	1289.0	1477.2	2766.2	3.1600	2.6230	5.7830
300	8.580	0.001404	0.02168	1332.0	2563.0	1344.0	1405.0	2749.0	3.2540	2.4513	5.7053
310	9.856	0.001447	0.01835	1387.0	2546.4	1401.3	1326.0	2727.3	3.3500	2.2739	5.6239
320	11.27	0.001499	0.01549	1444.6	2525.5	1461.4	1238.7	2700.1	3.4487	2.0883	5.5370
330	12.84	0.001561	0.01300	1505.2	2499.0	1525.3	1140.6	2665.9	3.5514	1.8911	5.4425
340	14.59	0.001638	0.01080	1570.3	2464.6	1594.2	1027.9	2622.1	3.6601	1.6765	5.3366
350	16.51	0.001740	0.008815	1641.8	2418.5	1670.6	893.4	2564.0	3.7784	1.4338	5.2122
360	18.65	0.001892	0.006947	1725.2	2351.6	1760.5	720.7	2481.2	3.9154	1.1382	5.0536
370	21.03	0.002213	0.004931	1844.0	2229.0	1890.5	442.2	2332.7	4.1114	0.6876	4.7990
374.14	22.088	0.003155	0.003155	2029.6	2029.6	2099.3	0.0	2099.3	4.4305	0.0000	4.4305

Table C-2 Properties of Saturated H_2O—Pressure Table

		v m³/kg		u kJ/kg		h kJ/kg			s kJ/kg·K		
P MPa	T °C	v_f	v_g	u_f	u_g	h_f	h_{fg}	h_g	s_f	s_{fg}	s_g
0.0006	0.01	0.001000	206.1	0.0	2375.3	0.0	2501.3	2501.3	0.0000	9.1571	9.1571
0.0008	3.8	0.001000	159.7	15.8	2380.5	15.8	2492.5	2508.3	0.0575	9.0007	9.0582
0.001	7.0	0.001000	129.2	29.3	2385.0	29.3	2484.9	2514.2	0.1059	8.8706	8.9765
0.0012	9.7	0.001000	108.7	40.6	2388.7	40.6	2478.5	2519.1	0.1460	8.7639	8.9099
0.0014	12.0	0.001001	93.92	50.3	2391.9	50.3	2473.1	2523.4	0.1802	8.6736	8.8538
0.0016	14.0	0.001001	82.76	58.9	2394.7	58.9	2468.2	2527.1	0.2101	8.5952	8.8053
0.002	17.5	0.001001	67.00	73.5	2399.5	73.5	2460.0	2533.5	0.2606	8.4639	8.7245
0.003	24.1	0.001003	45.67	101.0	2408.5	101.0	2444.5	2545.5	0.3544	8.2240	8.5784
0.004	29.0	0.001004	34.80	121.4	2415.2	121.4	2433.0	2554.4	0.4225	8.0529	8.4754
0.006	36.2	0.001006	23.74	151.5	2424.9	151.5	2415.9	2567.4	0.5208	7.8104	8.3312
0.008	41.5	0.001008	18.10	173.9	2432.1	173.9	2403.1	2577.0	0.5924	7.6371	8.2295
0.01	45.8	0.001010	14.67	191.8	2437.9	191.8	2392.8	2584.6	0.6491	7.5019	8.1510
0.012	49.4	0.001012	12.36	206.9	2442.7	206.9	2384.1	2591.0	0.6961	7.3910	8.0871
0.014	52.6	0.001013	10.69	220.0	2446.9	220.0	2376.6	2596.6	0.7365	7.2968	8.0333
0.016	55.3	0.001015	9.433	231.5	2450.5	231.5	2369.9	2601.4	0.7719	7.2149	7.9868
0.018	57.8	0.001016	8.445	241.9	2453.8	241.9	2363.9	2605.8	0.8034	7.1425	7.9459
0.02	60.1	0.001017	7.649	251.4	2456.7	251.4	2358.3	2609.7	0.8319	7.0774	7.9093
0.03	69.1	0.001022	5.229	289.2	2468.4	289.2	2336.1	2625.3	0.9439	6.8256	7.7695
0.04	75.9	0.001026	3.993	317.5	2477.0	317.6	2319.1	2636.7	1.0260	6.6449	7.6709
0.06	85.9	0.001033	2.732	359.8	2489.6	359.8	2293.7	2653.5	1.1455	6.3873	7.5328
0.08	93.5	0.001039	2.087	391.6	2498.8	391.6	2274.1	2665.7	1.2331	6.2023	7.4354
0.1	99.6	0.001043	1.694	417.3	2506.1	417.4	2258.1	2675.5	1.3029	6.0573	7.3602
0.12	104.8	0.001047	1.428	439.2	2512.1	439.3	2244.2	2683.5	1.3611	5.9378	7.2980
0.14	109.3	0.001051	1.237	458.2	2517.3	458.4	2232.0	2690.4	1.4112	5.8360	7.2472
0.16	113.3	0.001054	1.091	475.2	2521.8	475.3	2221.2	2696.5	1.4553	5.7472	7.2025
0.18	116.9	0.001058	0.9775	490.5	2525.9	490.7	2211.1	2701.8	1.4948	5.6683	7.1631
0.2	120.2	0.001061	0.8857	504.5	2529.5	504.7	2201.9	2706.6	1.5305	5.5975	7.1280
0.3	133.5	0.001073	0.6058	561.1	2543.6	561.5	2163.8	2725.3	1.6722	5.3205	6.9927
0.4	143.6	0.001084	0.4625	604.3	2553.6	604.7	2133.8	2738.5	1.7770	5.1197	6.8967
0.6	158.9	0.001101	0.3157	669.9	2567.4	670.6	2086.2	2756.8	1.9316	4.8293	6.7609
0.8	170.4	0.001115	0.2404	720.2	2576.8	721.1	2048.0	2769.1	2.0466	4.6170	6.6636
1	179.9	0.001127	0.1944	761.7	2583.6	762.8	2015.3	2778.1	2.1391	4.4482	6.5873
1.2	188.0	0.001139	0.1633	797.3	2588.8	798.6	1986.2	2784.8	.2170	4.3072	6.5242
1.4	195.1	0.001149	0.1408	828.7	2592.8	830.3	1959.7	2790.0	2.2847	4.1854	6.4701
1.6	201.4	0.001159	0.1238	856.9	2596.0	858.8	1935.2	2794.0	2.3446	4.0780	6.4226
2	212.4	0.001177	0.09963	906.4	2600.3	908.8	1890.7	2799.5	2.4478	3.8939	6.3417
4	250.4	0.001252	0.04978	1082.3	2602.3	1087.3	1714.1	2801.4	2.7970	3.2739	6.0709
6	275.6	0.001319	0.03244	1205.4	2589.7	1213.3	1571.0	2784.3	3.0273	2.8627	5.8900
8	295.1	0.001384	0.02352	1305.6	2569.8	1316.6	1441.4	2758.0	3.2075	2.5365	5.7440
10	311.1	0.001452	0.01803	1393.0	2544.4	1407.6	1317.1	2724.7	3.3603	2.2546	5.6149
12	324.8	0.001527	0.01426	1472.9	2513.7	1491.3	1193.6	2684.9	3.4970	1.9963	5.4933
14	336.8	0.001611	0.01149	1548.6	2476.8	1571.1	1066.5	2637.6	3.6240	1.7486	5.3726
16	347.4	0.001711	0.00931	1622.7	2431.8	1650.0	930.7	2580.7	3.7468	1.4996	5.2464
18	357.1	0.001840	0.00749	1698.9	2374.4	1732.0	777.2	2509.2	3.8722	1.2332	5.1054
20	365.8	0.002036	0.00583	1785.6	2293.2	1826.3	583.7	2410.0	4.0146	0.9135	4.9281
22.09	374.14	0.00316	0.00316	2029.6	2029.6	2099.3	0.0	2099.3	4.4305	0.0	4.4305

Based on data from Keenan, Keyes, Hill, and Moore, Steam Tables, Wiley, New York, 1969; G. J. Van Wylen and R. E. Sonntag, Fundamentals of Classical Thermodynamics, Wiley, New York, 1973.

Table C-3 Superheated Steam

T °C	v m³/kg	u kJ/kg	h kJ/kg	s kJ/kg·K	v m³/kg	u kJ/kg	h kJ/kg	s kJ/kg·K	v m³/kg	u kJ/kg	h kJ/kg	s kJ/kg·K
	P = 0.010 MPa (45.81°C)				P = 0.050 MPa (81.33°C)				P = 0.10 MPa (99.63°C)			
Sat.	14.674	2437.9	2584.7	8.1502	3.240	2483.9	2645.9	7.5939	1.6940	2506.1	2675.5	7.3594
50	14.869	2443.9	2592.6	8.1749								
100	17.196	2515.5	2687.5	8.4479	3.418	2511.6	2682.5	7.6947	1.6958	2506.7	2676.2	7.3614
150	19.512	2587.9	2783.0	8.6882	3.889	2585.6	2780.1	7.9401	1.9364	2582.8	2776.4	7.6134
200	21.825	2661.3	2879.5	8.9038	4.356	2659.9	2877.7	8.1580	2.172	2658.1	2875.3	7.8343
250	24.136	2736.0	2977.3	9.1002	4.820	2735.0	2976.0	8.3556	2.406	2733.7	2974.3	8.0333
300	26.445	2812.1	3076.5	9.2813	5.284	2811.3	3075.5	8.5373	2.639	2810.4	3074.3	8.2158
400	31.063	2968.9	3279.6	9.6077	6.209	2968.5	3278.9	8.8642	3.103	2967.9	3278.2	8.5435
500	35.679	3132.3	3489.1	9.8978	7.134	3132.0	3488.7	9.1546	3.565	3131.6	3483.1	8.8342
600	40.295	3302.5	3705.4	10.1608	8.057	3302.2	3705.1	9.4178	4.028	3301.9	3704.7	9.0976
700	44.911	3479.6	3928.7	10.4028	8.981	3479.4	3928.5	9.6599	4.490	3479.2	3928.2	9.3398
800	49.526	3663.8	4159.0	10.6281	9.904	3663.6	4158.9	9.8852	4.952	3663.5	4158.6	9.5652
900	54.141	3855.0	4396.4	10.8396	10.828	3854.9	4396.3	10.0967	5.414	3854.8	4396.1	9.7767
1000	58.757	4053.0	4640.6	11.0393	11.751	4052.9	4640.5	10.2964	5.875	4052.8	4640.3	9.9764
1100	63.372	4257.5	4891.2	11.2287	12.674	4257.4	4891.1	10.4859	6.337	4257.3	4891.0	10.1659
1200	67.987	4467.9	5147.8	11.4091	13.597	4467.8	5147.7	10.6662	6.799	4467.7	5147.6	10.3463
1300	72.602	4683.7	5409.7	11.5811	14.521	4683.6	5409.6	10.8382	7.260	4683.5	5409.5	10.5183
	P = 0.20 MPa (120.23°C)				P = 0.30 MPa (133.55°C)				P = 0.40 MPa (143.63°C)			
Sat.	.8857	2529.5	2706.7	7.1272	.6058	2543.6	2725.3	6.9919	.4625	2553.6	2738.6	6.8959
150	.9596	2576.9	2768.8	7.2795	.6339	2570.8	2761.0	7.0778	.4708	2564.5	2752.8	6.9299
200	1.0803	2654.4	2870.5	7.5066	.7163	2650.7	2865.6	7.3115	.5342	2646.8	2860.5	7.1706
250	1.1988	2731.2	2971.0	7.7086	.7964	2728.7	2967.6	7.5166	.5951	2726.1	2964.2	7.3789
300	1.3162	2808.6	3071.8	7.8926	.8753	2806.7	3069.3	7.7022	.6548	2804.8	3066.8	7.5662
400	1.5493	2966.7	3276.6	8.2218	1.0315	2965.6	3275.6	8.0330	.7726	2964.4	3273.4	7.8985
500	1.7814	3130.8	3487.1	8.5133	1.1867	3130.0	3486.0	8.3251	.8893	3129.2	3484.9	8.1913
600	2.013	3301.4	3704.0	8.7770	1.3414	3300.8	3703.2	8.5892	1.0055	3300.2	3702.4	8.4558
700	2.244	3478.8	3927.6	9.0194	1.4957	3478.4	3927.1	8.8319	1.1215	3477.9	3926.5	8.6987
800	2.475	3663.1	4158.2	9.2449	1.6499	3662.9	4157.8	9.0576	1.2372	3662.4	4157.3	8.9244
900	2.706	3854.5	4395.8	9.4566	1.8041	3854.2	4395.4	9.2692	1.3529	3853.9	4395.1	9.1362
1000	2.937	4052.5	4640.0	9.6563	1.9581	4052.3	4639.7	9.4690	1.4685	4052.0	4639.4	9.3360
1100	3.168	4257.0	4890.7	9.8458	2.1121	4256.5	4890.4	9.6585	1.5840	4256.5	4890.2	9.5256
1200	3.399	4467.5	5147.3	10.0262	2.2661	4467.2	5147.1	9.8389	1.6996	4467.0	5146.8	9.7060
1300	3.630	4683.2	5409.3	10.1982	2.4201	4683.0	5409.0	10.0110	1.8151	4682.8	5408.8	9.8780

	P = 0.50 MPa (151.86°C)				P = 0.60 MPa (158.85°C)				P = 0.80 MPa (170.43°C)			
Sat.	.3749	2561.2	2748.7	6.8213	.3157	2567.4	2756.8	6.7600	.2404	2576.8	2769.1	6.6628
200	.4249	2642.9	2855.4	7.0592	.3520	2638.9	2850.1	6.9665	.2608	2630.6	2839.3	6.8158
250	.4744	2723.5	2960.7	7.2709	.3938	2720.9	2957.2	7.1816	.2931	2715.5	2950.0	7.0384
300	.5226	2802.9	3064.2	7.4599	.4344	2801.0	3061.6	7.3724	.3241	2797.2	3056.5	7.2328
350	.5701	2882.6	3167.7	7.6329	.4742	2881.2	3165.7	7.5464	.3544	2878.2	3161.7	7.4089
400	.6173	2963.2	3271.9	7.7938	.5137	2962.1	3270.3	7.7079	.3843	2959.7	3267.1	7.5716
500	.7109	3128.4	3483.9	8.0873	.5920	3127.6	3482.8	8.0021	.4433	3126.0	3480.6	7.8673
600	.8041	3299.6	3701.7	8.3522	.6697	3299.1	3700.9	8.2674	.5018	3297.9	3699.4	8.1333
700	.8969	3477.5	3925.9	8.5952	.7472	3477.0	3925.3	8.5107	.5601	3476.2	3924.2	8.3770
800	.9896	3662.1	4156.9	8.8211	.8245	3661.8	4156.5	8.7367	.6181	3661.1	4155.6	8.6033
900	1.0822	3853.6	4394.7	9.0329	.9017	3853.4	4394.4	8.9486	.6761	3852.8	4393.7	8.8153
1000	1.1747	4051.8	4639.1	9.2328	.9788	4051.5	4638.8	9.1485	.7340	4051.0	4638.2	9.0153
1100	1.2672	4256.3	4889.9	9.4224	1.0559	4256.1	4889.6	9.3381	.7919	4255.6	4889.1	9.2050
1200	1.3596	4466.8	5146.6	9.6029	1.1330	4466.5	5146.3	9.5185	.8497	4466.1	5145.9	9.3855
1300	1.4521	4682.5	5408.6	9.7749	1.2101	4682.3	5408.3	9.6906	.9076	4681.8	5407.9	9.5575
	P = 1.00 MPa (179.91°C)				P = 1.20 MPa (187.99°C)				P = 1.40 MPa (195.07°C)			
Sat.	.194 44	2583.6	2778.1	6.5865	.163 33	2588.8	2784.8	6.5233	.140 84	2592.8	2790.0	6.4693
200	.2060	2621.9	2827.9	6.6940	.169 30	2612.8	2815.9	6.5898	.143 02	2603.1	2803.3	6.4975
250	.2327	2709.9	2942.6	6.9247	.192 34	2704.2	2935.0	6.8294	.163 50	2698.3	2927.2	6.7467
300	.2579	2793.2	3051.2	7.1229	.2138	2789.2	3045.8	7.0317	.182 28	2785.2	3040.4	6.9534
350	.2825	2875.2	3157.7	7.3011	.2345	2872.2	3153.6	7.2121	.2003	2869.2	3149.5	7.1360
400	.3066	2957.3	3263.9	7.4651	.2548	2954.9	3260.7	7.3774	.2178	2952.5	3257.5	7.3026
500	.3541	3124.4	3478.5	7.7622	.2946	3122.8	3476.3	7.6759	.2521	3321.1	3474.1	7.6027
600	.4011	3296.8	3697.9	8.0290	.3339	3295.6	3696.3	7.9435	.2860	3294.4	3694.8	7.8710
700	.4478	3475.3	3923.1	8.2731	.3729	3474.4	3922.0	8.1881	.3195	3473.6	3920.8	8.1160
800	.4943	3660.4	4154.7	8.4996	.4118	3659.7	4153.8	8.4148	.3528	3659.0	4153.0	8.8431
900	.5407	3852.2	4392.9	8.7118	.4505	3851.6	4392.2	8.6272	.3861	3851.1	4391.5	8.5556
1000	.5871	4050.5	4637.6	8.9119	.4892	4050.0	4637.0	8.8274	.4192	4049.5	4636.4	8.7559
1100	.6335	4255.1	4888.6	9.1017	.5278	4254.6	4888.0	9.0172	.4524	4254.1	4887.5	8.9457
1200	.6798	4465.6	5145.4	9.2822	.5665	4465.1	5144.9	9.1977	.4855	4464.7	5144.4	9.1262
1300	.7261	4681.3	5407.4	9.4543	.6051	4680.9	5407.0	9.3698	.5186	4680.4	5406.5	9.2984

(Continued)

Table C-3 (Continued)

T °C	v m³/kg	u kJ/kg	h kJ/kg	s kJ/kg·K	v m³/kg	u kJ/kg	h kJ/kg	s kJ/kg·K	v m³/kg	u kJ/kg	h kJ/kg	s kJ/kg·K
	P = 1.60 MPa (201.41°C)				P = 1.80 MPa (207.15°C)				P = 2.0 MPa (212.42°C)			
Sat.	.123 80	2596.0	2794.0	6.4218	.11042	2598.4	2797.1	6.3794	.099 63	2600.3	2799.5	6.3409
225	.13287	2644.7	2857.3	6.5518	.13673	2636.6	2846.7	6.4808	.103 77	2628.3	2835.8	6.4147
250	.141 84	2692.3	2919.2	6.6732	.12497	2686.0	2911.0	6.6066	.11144	2679.6	2902.5	6.5453
300	.15862	2781.1	3034.8	6.8844	.14021	2776.9	3029.2	6.8226	.125 47	2772.6	3023.5	6.7664
350	.17456	2866.1	3145.4	7.0694	.15457	2863.0	3141.2	7.0100	.13857	2859.8	3137.0	6.9563
400	.19005	2950.1	3254.2	7.2374	.16847	2947.7	3250.9	7.1794	.151 20	2945.2	3247.6	7.1271
500	.2203	3119.5	3472.0	7.5390	.19550	3117.9	3469.8	7.4825	.175 68	3116.2	3467.6	7.4317
600	.2500	3293.3	3693.2	7.8080	.2220	3292.1	3691.7	7.7523	.19960	3290.9	3690.1	7.7024
700	.2794	3472.7	3919.7	8.0535	.2482	3471.8	3918.5	7.9983	.2232	3470.9	3917.4	7.9487
800	.3086	3658.3	4152.1	8.2808	.2742	3657.6	4151.2	8.2258	.2467	3657.0	4150.3	8.1765
900	.3377	3850.5	4390.8	8.4935	.3001	3849.9	4390.1	8.4386	.2700	3849.3	4389.4	8.3895
1000	.5668	4049.0	4635.8	8.6938	.3260	4048.5	4635.2	8.6391	.2933	4048.0	4634.6	8.5901
1100	.3958	4253.7	4887.0	8.8837	.3518	4253.2	4886.4	8.8290	.3166	4252.7	4885.9	8.7800
1200	.4248	4464.2	5143.9	9.0643	.3776	4463.7	5143.4	9.0096	.3398	4463.3	5142.9	8.9607
1300	.4538	4679.9	5406.0	9.2364	.4034	4679.5	5405.6	9.1818	.3631	4679.0	5405.1	9.1329
	P = 2.50 MPa (233.99°C)				P = 3.00 MPa (233.90°C)				P = 3.50 MPa (242.60°C)			
Sat.	.079 98	2603.1	2803.1	6.2575	.066 68	2604.1	2804.2	6.1869	.05707	2603.7	2803.4	6.1253
225	.080 27	2605.6	2806.3	6.2639								
250	.087 00	2662.6	2880.1	6.4085	.070 58	2644.0	2855.8	6.2872	.058 72	2623.7	2829.2	6.1749
300	.098 90	2761.6	3008.8	6.6438	.081 14	2750.1	2993.5	6.5390	.068 42	2738.0	2977.5	6.4461
350	.109 76	2851.9	3126.3	6.8403	.090 53	2843.7	3115.3	6.7428	.076 78	2835.3	3104,0	6.6579
400	.120 10	2939.1	3239.3	7.0148	.099 36	2932.8	3230.9	6.9212	.084 53	2926.4	3222.3	6.8405
450	.130 14	3025.5	3350.8	7.1746	.107 87	3020.4	3344.0	7.0834	.091 96	3015.3	3337.2	7.0052
500	.13998	3112.1	3462.1	7.3234	.11619	3108.0	3456.5	7.2338	.099 18	3103.0	3450.9	7.1572
600	.15930	3288.0	3686.3	7.5960	.13243	3285.0	3682.3	7.5085	.11324	3282.1	3678.4	7.4339
700	.17832	3468.7	3914.5	7.8435	.14838	3466.5	3911.7	7.7571	.12699	3464.3	3908.8	7.6837
800	.197 16	3655.3	4148.2	8.0720	.164 14	3653.5	4145.9	7.9862	.14056	3651.8	4143.7	7.9134
900	.215 90	3847.9	4387.6	8.2853	.179 80	3846.5	4385.9	8.1999	.15402	3845.0	4384.1	8.1276
1000	.2346	4046.7	4633.1	8.4861	.195 41	4045.4	4631.6	8.4009	.167 43	4044.1	4630.1	8.3288
1100	.2532	4251.5	4884.6	8.6762	.21098	4250.3	4883.3	8.5912	.18080	4249.2	4881.9	8.5192
1200	.2718	4462.1	5141.7	8.8569	.226 52	4460.9	5140.5	8.7720	.194 15	4459.8	5139.3	8.7000
1300	.2905	4677.8	5404.0	9.0291	.242 06	4676.6	5402.8	8.9442	.207 49	4675.5	5401.7	8.8723

	P = 4.0 MPa (250.40°C)				P = 4.5 MPa (257.49°C)				P = 5.0 MPa (263.99°C)			
Sat.	.04978	2602.3	2801.4	6.0701	.04406	2600.1	2798.3	6.0198	.03944	2597.1	2794.3	5.9734
275	.054 57	2667.9	2886.2	6.2285	.04730	2650.3	2863.2	6.1401	.04141	2631.3	2838.3	6.0544
300	.058 84	2725.3	2960.7	6.3615	.051 35	2712.0	2943.1	6.2828	.04532	2698.0	2924.5	6.2084
350	.06645	2826.7	3092.5	6.5821	.05840	2817.8	3080.6	6.5131	.051 94	2808.7	3068.4	6.4493
400	.073 41	2919.9	3213.6	6.7690	.06475	2913.3	3204.7	6.7047	.057 81	2906.6	3195.7	6.6459
450	.080 02	3010.2	3330.3	6.9363	.070 74	3005.0	3323.3	6.8746	.06330	2999.7	3316.2	6.8186
500	.086 43	3099.5	3445.3	7.0901	.07651	3095.3	3439.6	7.0301	.06857	3091.0	3433.8	6.9759
600	.09885	3279.1	3674.4	7.3688	.087 65	3276.0	3670.5	7.3110	.078 69	3273.0	3666.5	7.2589
700	.11095	3462.1	3905.9	7.6198	.09847	3459.9	3903.0	7.5631	.088 49	3457.6	3900.1	7.5122
800	.122 87	3650.0	4141.5	7.8502	.10911	3648.3	4139.3	7.7942	.09811	3646.6	4137.1	7.7440
900	.13469	3843.6	4382.3	8.0647	.11965	3842.2	4380.6	8.0091	.10762	3840.7	4378.8	7.9593
1000	.14645	4042.9	4628.7	8.2662	.130 13	4041.6	4627.2	8.2108	.11707	4040.4	4625.7	8.1612
1100	.158 17	4248.0	4880.6	8.4567	.140 56	4246.8	4879.3	8.4015	.12648	4245.6	4878.0	8.3520
1200	.169 87	4458.6	5138.1	8.6376	.150 98	4457.5	5136.9	8.5825	.135 87	4456.3	5135.7	8.5331
1300	.181 56	4674.3	5400.5	8.8100	.161 39	4673.1	5399.4	8.7549	.14526	4672.0	5398.2	8.7055
	P = 6.0 MPa (275.64°C)				P = 7.0 MPa (285.88°C)				P = 8.0 MPa (295.06°C)			
Sat.	.03244	2589.7	2784.3	5.8892	.027 37	2580.5	2772.1	5.8133	.023 52	2569.8	2758.0	5.7432
300	.036 16	2667.2	2884.2	6.0674	.029 47	2632.2	2838.4	3.9305	.024 26	2590.9	2785.0	5.7906
350	.04223	2789.6	3043.0	6.3335	.035 24	2769.4	3016.0	6.2283	.029 95	2747.7	2987.3	6.1301
400	.04739	2892.9	3177.2	6.5408	.039 93	2878.6	3158.1	6.4478	.034 32	2863.8	3138.3	6.3634
450	.052 14	2988.9	3301.8	6.7193	.04416	2978.0	3287.1	6.6327	.038 17	2966.7	3272.0	6.5551
500	.056 65	3082.2	3422.2	6.8803	.04814	3073.4	3410.3	6.7975	.04175	3064.3	3398.3	6.7240
550	.06101	3174.6	3540.6	7.0288	.051 95	3167.2	3530.9	6.9486	.04516	3159.8	3521.0	6.8778
600	.06525	3266.9	3658.4	7.1677	.055 65	3260.7	3650.3	7.0894	.04845	3254.4	3642.0	7.0206
700	.073 52	3453.1	3894.2	7.4234	.06283	3448.5	3888.3	7.3476	.054 81	3443.9	3882.4	7.2812
800	.081 60	3643.1	4132.7	7.6566	.06981	3639.5	4128.2	7.5822	.06097	3636.0	4123.8	7.5173
900	.089 58	3837.8	4375.3	7.8727	.076 69	3835.0	4371.8	7.7991	.06702	3832.1	4368.3	7.7351
1000	.09749	4037.8	4622.7	8.0751	.083 50	4035.3	4619.8	8.0020	.073 01	4032.8	4616.9	7.9384
1100	.105 36	4243.3	4875.4	8.2661	.09027	4240.9	4872.8	8.1933	.078 96	4238.6	4870.3	8.1300
1200	.11321	4454.0	5133.3	8.4474	.097 03	4451.7	5130.9	8.3747	.084 89	4449.5	5128.5	8.3115
1300	.121 06	4669.6	5396.0	8.6199	.10377	4667.3	5393.7	8.5473	.09080	4665.0	5391.5	8.4842

(Continued)

Table C-3 (*Continued*)

T °C	v m³/kg	u kJ/kg	h kJ/kg	s kJ/kg·K	v m³/kg	u kJ/kg	h kJ/kg	s kJ/kg·K	v m³/kg	u kJ/kg	h kJ/kg	s kJ/kg·K
	P = 9.0 MPa (303.40°C)				P = 10.0 MPa (311.06°C)				P = 12.5 MPa (327.89°C)			
Sat.	.020 48	2557.8	2742.1	5.6772	.018026	2544.4	2724.7	5.6141	.013 495	2505.1	2673.8	5.4624
325	.023 27	2646.6	2856.0	5.8712	.019 861	2610.4	2809.1	5.7568				
350	.025 80	2724.4	2956.6	6.0361	.022 42	2699.4	2923.4	5.9443	.016 126	2624.6	2826.2	5.7118
406	.029 93	2848.4	3117.8	6.2854	.026 41	2832.4	3096.5	6.2120	.02000	2789.3	3039.3	6.0417
450	.033 50	2955.2	3256.6	6.4844	.029 75	2943.4	3240.9	6.4190	.022 99	2912.5	3199.8	6.2719
500	.03677	3055.2	3336.1	6.6576	.032 79	3045.8	3373.7	6.5966	.02560	3021.7	3341.8	6.4618
550	.03987	3152.2	3511.0	6.8142	.03564	3144.6	3500.9	6.7561	.028 01	3125.0	3475.2	6.6290
600	.04285	3248.1	3633.7	6.9589	.038 37	3241.7	3625.3	6.9029	.030 29	3225.4	3604.0	6.7810
650	.04574	3343.6	3755.3	7.0943	.04101	3338.2	3748.2	7.0398	.032 48	3324.4	3730.4	6.9218
700	.04857	3439.3	3876.5	7.2221	.04358	3434.7	3870.5	7.1687	.034 60	3422.9	3855.3	7.0536
800	.054 09	3632.5	4119.3	7.4596	.04859	3628.9	4114.8	7.4077	.038 69	3620.0	4103.6	7.2965
900	.05950	3829.2	4364.3	7.6783	.053 49	3826.3	4361.2	7.6272	.042 67	3819.1	4352.5	7.5182
1000	.06485	4030.3	4614.0	7.8821	.058 32	4027.8	4611.0	7.8315	.04658	4021.6	4603.8	7.7237
1100	.07016	4236.3	4867.7	8.0740	.06312	4234.0	4865.1	8.0237	.050 45	4228.2	4858.8	7.9165
1200	.07544	4447.2	5126.2	8.2556	.06789	4444.9	5123.8	8.2055	.054 30	4439.3	5118.0	8.0987
1300	.08072	4662.7	5389.2	8.4284	.072 65	4460.5	5387.0	8.3783	.058 13	4654.8	5381.4	8.2717
	P = 15.0 MPa (342.24°C)				P = 17.5 MPa (354.75°C)				P = 20.0 MPa (365.81°C)			
Sat.	.010 337	2455.5	2610.5	5.3098	.0079 20	2390.2	2528.8	5.1419	.005 834	2293.0	2409.7	4.9269
350	.011470	2520.4	2692.4	5.4421								
400	.015 649	2740.7	2975.5	5.8811	.012 447	2685.0	2902.9	5.7213	.009 942	2619.3	2818.1	5.5540
450	.018 445	2879.5	3156.2	6.1404	.015 174	2844.2	3109.7	6.0184	.012 695	2806.2	3060.1	5.9017
500	.020 80	2996.6	3308.6	6.3443	.017 358	2970.3	3274.1	6.2383	.014 768	2942.9	3238.2	6.1401
550	.022 93	3104.7	3448.6	6.5199	.019 288	3083.9	3421.4	6.4230	.016 555	3062.4	3393.5	6.3348
600	.024 91	3208.6	3582.3	6.6776	.02106	3191.5	3560.2	6.5866	.018 178	3174.0	3537.6	6.5048
650	.026 80	3310.3	3712.3	6.8224	.022 74	3296.0	3693.9	6.7357	.019 693	3281.4	3675.3	6.6582
700	.028 61	3410.9	3840.1	6.9572	.024 34	3398.7	3824.6	6.8736	.021 13	3386.4	3809.0	6.7993
800	.032 10	3610.9	4092.4	7.2040	.027 38	3601.8	4081.1	7.1244	.023 85	3592.7	4069.7	7.0544
900	.035 46	3811.9	4343.8	7.4279	.030 31	3804.7	4335.1	7.3507	.026 45	3797.5	4326.4	7.2830
1000	.038 75	4015.4	4596.6	7.6348	.033 16	4009.3	4589.5	7.5589	.028 97	4003.1	4582.5	7.4925
1100	.04200	4222.6	4852.6	7.8283	.035 97	4216.9	4846.4	7.7531	.031 45	4211.3	4840.2	7.6874
1200	.04523	4433.8	5112.3	8.0108	.038 76	4428.3	5106.6	7.9360	.033 91	4422.8	5101.0	7.8707
1300	.04845	4649.1	5376.0	8.1840	.041 54	4643.5	5370.5	8.1093	.036 36	4638.0	5365.1	8.0442

	P = 25.0 MPa				P = 30.0 MPa				P = 40.0 MPa			
375	.001 9731	1798.7	1848.0	4.0320	.001 789	1737.8	1791.5	3.9305	.001 640 7	1677.1	1742.8	3.8290
400	.006 004	2430.1	2580.2	5.1418	.002 790	2067.4	2151.1	4.4728	.001 907 7	1854.6	1930.9	4.1135
425	.007 881	2609.2	2806.3	5.4723	.005 303	2455.1	2614.2	5.1504	.002 532	2096.9	2198.1	4.5029
450	.009 162	2720.7	2949.7	5.6744	.006 735	2619.3	2821.4	5.4424	.003 693	2365.1	2512.8	4.9459
500	.011 123	2884.3	3162.4	5.9592	.008 678	2820.7	3081.1	5.7905	.005 622	2678.4	2903.3	5.4700
550	.012 724	3017.5	3335.6	6.1765	.010 168	2970.3	3275.4	6.0342	.006 984	2869.7	3149.1	5.7785
600	.014 137	3137.9	3491.4	6.3602	.011446	3100.5	3443.9	6.2331	.008 094	3022.6	3346.4	6.0114
650	.015 433	3251.6	3637.4	6.5229	.012 596	3221.0	3598.9	6.4058	.009 063	3158.0	3520.6	6.2054
700	.016 646	3361.3	3777.5	6.6707	.013 661	3335.8	3745.6	6.5606	.009 941	3283.6	3681.2	6.3750
800	.018 912	3574.3	4047.1	6.9345	.015 623	3555.5	4024.2	6.8332	.011523	3517.8	3978.7	6.6662
900	.021 045	3783.0	4309.1	7.1680	.017 448	3768.5	4291.9	7.0718	.012 962	3739.4	4257.9	6.9150
1000	.023 10	3990.9	4568.5	7.3802	.019 196	3978.8	4554.7	7.2867	.014 324	3954.6	4527.6	7.1356
1100	.025 12	4200.2	4828.2	7.5765	.020 903	4189.2	4816.3	7.4845	.015 642	4167.4	4793.1	7.3364
1200	.027 11	4412.0	5089.9	7.7605	.022 589	4401.3	5079.0	7.6692	.016 940	4380.1	5057.7	7.5224
1300	.029 10	4626.9	5354.4	7.9342	.024 266	4616.0	5344.0	7.8432	.018 229	4594.3	5323.5	7.6969

Table C-4 Compressed Liquid

T °C	v m³/kg	u kJ/kg	h kJ/kg	s kJ/kg·K	v m³/kg	u kJ/kg	h kJ/kg	s kJ/kg·K	v m³/kg	u kJ/kg	h kJ/kg	s kJ/kg·K
	P = 5 MPa (264.0°C)				P = 10 MPa (311.1°C)				P = 15 MPa (342.4°C)			
0	0.000 998	0.04	5.04	0.0001	0.000 995	0.09	10.04	0.0002	0.000 993	0.15	15.05	0.0004
20	0.001 000	83.65	88.65	0.296	0.000 997	83.36	93.33	0.2945	0.000 995	83.06	97.99	0.2934
40	0.001 006	167.0	172.0	0.570	0.001 003	166.4	176.4	0.5686	0.001 001	165.8	180.78	0.5666
60	0.001 015	250.2	255.3	0.828	0.001 013	249.4	259.5	0.8258	0.001 010	248.5	263.67	0.8232
80	0.001 027	333.7	338.8	1.072	0.001 024	332.6	342.8	1.0688	0.001 022	331.5	346.81	1.0656
100	0.001 041	417.5	422.7	1.303	0.001 038	416.1	426.5	1.2992	0.001 036	414.7	430.28	1.2955
120	0.001 058	501.8	507.1	1.523	0.001 055	500.1	510.6	1.5189	0.001 052	498.4	514.19	1.5145
140	0.001 077	586.8	592.2	1.734	0.001 074	584.7	595.4	1.7292	0.001 071	582.7	598.72	1.7242
160	0.001 099	672.6	678.1	1.938	0.001 095	670.1	681.1	1.9317	0.001 092	667.7	684.09	1.9260
180	0.001 124	759.6	765.2	2.134	0.001 120	756.6	767.8	2.1275	0.001 116	753.8	770.50	2.1210
200	0.001 153	848.1	853.9	2.326	0.001 148	844.5	856.0	2.3178	0.001 143	841.0	858.2	2.3104
	P = 20 MPa (365.8°C)				P = 30 MPa				P = 50 MPa			
0	0.000 990	0.19	20.01	0.0004	0.000 986	0.25	29.82	0.0001	0.000 977	0.20	49.03	0.0014
20	0.000 993	82.77	102.6	0.2923	0.000 989	82.17	111.8	0.2899	0.000 980	81.00	130.02	0.2848
40	0.000 999	165.2	185.2	0.5646	0.000 995	164.0	193.9	0.5607	0.000 987	161.9	211.21	0.5527
60	0.001 008	247.7	267.8	0.8206	0.001 004	246.1	276.2	0.8154	0.000 996	243.0	292.79	0.8052
80	0.001 020	330.4	350.8	1.0624	0.001 016	328.3	358.8	1.0561	0.001 007	324.3	374.70	1.0440
100	0.001 034	413.4	434.1	1.2917	0.001 029	410.8	441.7	1.2844	0.001 020	405.9	456.89	1.2703
120	0.001 050	496.8	517.8	1.5102	0.001 044	493.6	524.9	1.5018	0.001 035	487.6	539.39	1.4857
140	0.001 068	580.7	602.0	1.7193	0.001 062	576.9	608.8	1.7098	0.001 052	569.8	622.35	1.6915
160	0.001 088	665.4	687.1	1.9204	0.001 082	660.8	693.3	1.9096	0.001 070	652.4	705.92	1.8891
180	0.001 112	751.0	773.2	2.1147	0.001 105	745.6	778.7	2.1024	0.001 091	735.7	790.25	2.0794
200	0.001 139	837.7	860.5	2.3031	0.001 130	831.4	865.3	2.2893	0.001 115	819.7	875.5	2.2634

Table C-5 Saturated Solid–Vapor

T °C	P kPa	v m³/kg: Sat. Solid $v_i \times 10^3$	Sat. Vapor v_g	u kJ/kg: Sat. Solid u_i	Subl. u_{ig}	Sat. Vapor u_g	h kJ/kg: Sat. Solid h_i	Subl. h_{ig}	Sat. Vapor h_g	s kJ/kg·K: Sat. Solid s_i	Subl. s_{ig}	Sat. Vapor s_g
0.01		1.091	206.1	−333.4	2709	2375	−333.4	2835	2501	−1.221	10.378	9.156
0	0.611	1.091	206.3	−333.4	2709	2375	−333.4	2835	2501	−1.221	10.378	9.157
−2	0.518	1.090	241.7	−337.6	2710	2373	−337.6	2835	2498	−1.237	10.456	9.219
−4	0.438	1.090	283.8	−341.8	2712	2370	−341.8	2836	2494	−1.253	10.536	9.283
−6	0.369	1.090	334.2	−345.9	2713	2367	−345.9	2836	2490	−1.268	10.616	9.348
−8	0.611	1.089	394.4	−350.0	2714	2364	−350.0	2837	2487	−1.284	10.698	9.414
−10	0.260	1.089	466.7	−354.1	2716	2361	−354.1	3837	2483	−1.299	10.781	9.481
−12	0.218	1.089	553.7	−358.1	2717	2359	−358.1	2837	2479	−1.315	10.865	9.550
−14	0.182	1.088	658.8	−362.2	2718	2356	−362.2	2838	2476	−1.331	10.950	9.619
−16	0.151	1.088	786.0	−366.1	2719	2353	−366.1	2838	2472	−1.346	11.036	9.690
−20	0.104	1.087	1129	−374.0	2722	2348	−374.0	2838	2464	−1.377	11.212	9.835
−24	0.070	1.087	1640	−381.8	2724	2342	−381.8	2839	2457	−1.408	11.394	9.985
−28	0.047	1.086	2414	−389.4	2726	2336	−389.4	2839	2450	−1.439	11.580	10.141

Based on data from Keenan, Keyes, Hill, and Moore, Steam Tables, Wiley, New York, 1969; G. J. Van Wylen and R. E. Sonntag, Fundamentals of Classical Thermodynamics, Wiley, New York, 1973.

D Properties of R134a

Table D-l Saturated R134a—Temperature Table

Temp °C	Pressure kPa	Specific Volume m^3/kg Sat. Liquid $v_f \times 10^3$	Sat. Vapor v_g	Internal Energy kJ/kg Sat. Liquid u_f	Sat. Vapor u_g	Enthalpy kJ/kg Sat. Liquid h_f	Evap. h_{fg}	Sat. Vapor h_g	Entropy kJ/kg·K Sat. Liquid s_f	Sat. Vapor s_g
−40	51.64	0.7055	0.3569	−0.04	204.45	0.00	222.88	222.88	0.0000	0.9560
−36	63.32	0.7113	0.2947	4.68	206.73	4.73	220.67	225.40	0.02001	0.9506
−32	77.04	0.7172	0.2451	9.47	209.01	9.52	218.37	227.90	0.0401	0.9456
−28	93.05	0.7233	0.2052	14.31	211.29	14.37	216.01	230.38	0.0600	0.9411
−26	101.99	0.7265	0.1882	16.75	212.43	16.82	214.80	231.62	0.0699	0.9390
−24	111.60	0.7296	0.1728	19.21	213.57	19.29	213.57	232.85	0.0798	0.9370
−22	121.92	0.7328	0.1590	21.68	214.70	21.77	212.32	234.08	0.0897	0.9351
−20	132.99	0.7361	0.1464	24.17	215.84	24.26	211.05	235.31	0.0996	0.9332
−18	144.83	0.7395	0.1350	26.67	216.97	26.77	209.76	236.53	0.1094	0.9315
−16	157.48	0.7428	0.1247	29.18	218.10	29.30	208.45	237.74	0.1192	0.9298
−12	185.40	0.7498	0.1068	34.25	220.36	34.39	205.77	240.15	0.1388	0.9267
−8	217.04	0.7569	0.0919	39.38	222.60	39.54	203.00	242.54	0.1583	0.9239
−4	252.74	0.7644	0.0794	44.56	224.84	44.75	200.15	244.90	0.1777	0.9213
0	292.82	0.7721	0.0689	49.79	227.06	50.02	197.21	247.23	0.1970	0.9190
4	337.65	0.7801	0.0600	55.08	229.27	55.35	194.19	249.53	0.2162	0.9169
8	387.56	0.7884	0.0525	60.43	231.46	60.73	191.07	251.80	0.2354	0.9150
12	442.94	0.7971	0.0460	65.83	233.63	66.18	187.85	254.03	0.2545	0.9132
16	504.16	0.8062	0.0405	71.29	235.78	71.69	184.52	256.22	0.2735	0.9116
20	571.60	0.8157	0.0358	76.80	237.91	77.26	181.09	258.36	0.2924	0.9102
24	645.66	0.8257	0.0317	82.37	240.01	82.90	177.55	260.45	0.3113	0.9089
26	685.30	0.8309	0.0298	85.18	241.05	85.75	175.73	261.48	0.3208	0.9082
28	726.75	0.8362	0.0281	88.00	242.08	88.61	173.89	262.50	0.3302	0.9076
30	770.06	0.8417	0.0265	90.84	243.10	91.49	172.00	263.50	0.3396	0.9070
32	815.28	0.8473	0.0250	93.70	244.12	94.39	170.09	264.48	0.3490	0.9064
34	862.47	0.8530	0.0236	96.58	245.12	97.31	168.14	265.45	0.3584	0.9058
36	911.68	0.8590	0.0223	99.47	246.11	100.25	166.15	266.40	0.3678	0.9053
38	962.98	0.8651	0.0210	102.38	247.09	103.21	164.12	267.33	0.3772	0.9047
40	1016.4	0.8714	0.0199	105.30	248.06	106.19	162.05	268.24	0.3866	0.9041
42	1072.0	0.8780	0.0188	108.25	249.02	109.19	159.94	269.14	0.3960	0.9035
44	1129.9	0.8847	0.0177	111.22	249.96	112.22	157.79	270.01	0.4054	0.9030

(Continued)

Table D-1 through D-3 are based on equations from D. P. Wilson and R. S. Basu, "Thermodynamic Properties of a New Stratospherically Safe Working Fluid–Refrigerant 134a," ASHRAE Trans., Vol. 94, Pt. 2, 1988, pp. 2095–2118.

Table D-I (*Continued*)

Temp °C	Pressure kPa	Specific Volume m³/kg Sat. Liquid $v_f \times 10^3$	Sat. Vapor v_g	Internal Energy kJ/kg Sat. Liquid u_f	Sat. Vapor u_g	Enthalpy kJ/kg Sat. Liquid h_f	Evap. h_{fg}	Sat. Vapor h_g	Entropy kJ/kg·K Sat. Liquid s_f	Sat. Vapor s_g
48	1252.6	0.8989	0.0159	117.22	251.79	118.35	153.33	271.68	0.4243	0.9017
52	1385.1	0.9142	0.0142	123.31	253.55	124.58	148.66	273.24	0.4432	0.9004
56	1527.8	0.9308	0.0127	129.51	255.23	130.93	143.75	274.68	0.4622	0.8990
60	1681.3	0.9488	0.0114	135.82	256.81	137.42	138.57	275.99	0.4814	0.8973
70	2116.2	1.0027	0.0086	152.22	260.15	154.34	124.08	278.43	0.5302	0.8918
80	2632.4	1.0766	0.0064	169.88	262.14	172.71	106.41	279.12	0.5814	0.8827
90	3243.5	1.1949	0.0046	189.82	261.34	193.69	82.63	276.32	0.6380	0.8655
100	3974.2	1.5443	0.0027	218.60	248.49	224.74	34.40	259.13	0.7196	0.8117

Table D-2 Saturated R134a—Pressure Table

		Specific Volume m³/kg		Internal Energy kJ/kg		Enthalpy kJ/kg			Entropy kJ/kg·K	
Pressure kPa	Temp. °C	Sat. Liquid $v_f \times 10^3$	Sat. Vapor v_g	Sat. Liquid u_f	Sat. Vapor u_g	Sat. Liquid h_f	Evap. h_{fg}	Sat. Vapor h_g	Sat. Liquid s_f	Sat. Vapor s_g
60	−37.07	0.7097	0.3100	3.14	206.12	3.46	221.27	224.72	0.0147	0.9520
80	−31.21	0.7184	0.2366	10.41	209.46	10.47	217.92	228.39	0.0440	0.9447
100	−26.43	0.7258	0.1917	16.22	212.1.8	16.29	215.06	231.35	0.0678	0.9395
120	−22.36	0.7323	0.1614	21.23	214.50	21.32	212.54	233.86	0.0879	0.9354
140	−18.80	0.7381	0.1395	25.66	216.52	25.77	210.27	236.04	0.1055	0.9322
160	−15.62	0.7435	0.1229	29.66	218.32	29.78	208.19	237.97	0.1211	0.9295
180	−12.73	0.7485	0.1098	33.31	219.94	33.45	206.26	239.71	0.1352	0.9273
200	−10.09	0.7532	0.0993	36.69	221.43	36.84	204.46	241.30	0.1481	0.9253
240	−5.37	0.7618	0.0834	42.77	224.07	42.95	201.14	244.09	0.1710	0.9222
280	−1.23	0.7697	0.0719	48.18	226.38	48.39	198.13	246.52	0.1911	0.9197
320	2.48	0.7770	0.0632	53.06	228.43	53.31	195.35	248.66	0.2089	0.9177
360	5.84	0.7839	0.0564	57.54	230.28	57.82	192.76	250.58	0.2251	0.9160
400	8.93	0.7904	0.0509	61.69	231.97	62.00	190.32	252.32	0.2399	0.9145
500	15.74	0.8056	0.0409	70.93	235.64	71.33	184.74	256.07	0.2723	0.9117
600	21.58	0.8196	0.0341	78.99	238.74	79.48	179.71	259.19	0.2999	0.9097
700	26.72	0.8328	0.0292	86.19	241.42	86.78	175.07	261.85	0.3242	0.9080
800	31.33	0.8454	0.0255	92.75	243.78	93.42	170.73	264.15	0.3459	0.9066
900	35.53	0.8576	0.0226	98.79	245.88	99.56	166.62	266.18	0.3656	0.9054
1000	39.39	0.8695	0.0202	104.42	247.77	105.29	162.68	267.97	0.3838	0.9043
1200	46.32	0.8928	0.0166	114.69	251.03	115.76	155.23	270.99	0.4164	0.9023
1400	52.43	0.9159	0.0140	123.98	253.74	125.26	148.14	273.40	0.4453	0.9003
1600	57.92	0.9392	0.0121	132.52	256.00	134.02	141.31	275.33	0.4714	0.8982
1800	62.91	0.9631	0.0105	140.49	257.88	142.22	134.60	276.83	0.4954	0.8959
2000	67.49	0.9878	0.0093	148.02	259.41	149.99	127.95	277.94	0.5178	0.8934
2500	77.59	1.0562	0.0069	165.48	261.84	168.12	111.06	279.17	0.5687	0.8854
3000	86.22	1.1416	0.0053	181.88	262.16	185.30	92.71	278.01	0.6156	0.8735

Table D-1 through D-3 are based on equations from D. P. Wilson and R. S. Basu, "Thermodynamic Properties of a New Stratospherically Safe Working Fluid–Refrigerant 134a," ASHRAE Trans., Vol. 94, Pt. 2, 1988, pp. 2095–2118.

Table D-3 Superheated R134a

T, °C	v m³/kg	u kJ/kg	h kJ/kg	s kJ/kg·K	v m³/kg	u kJ/kg	h kJ/kg	s kJ/kg·K
	P = 0.06 MPa (−37.07°C)				P = 0.10 MPa (−26.43°C)			
Sat.	0.31003	206.12	224.72	0.9520	0.19170	212.18	231.35	0.9395
−20	0.33536	217.86	237.98	1.0062	0.19770	216.77	236.54	0.9602
−10	0.34992	224.97	245.96	1.0371	0.20686	224.01	244.70	0.9918
0	0.36433	232.24	254.10	1.0675	0.21587	231.41	252.99	1.0227
10	0.37861	239.69	262.41	1.0973	0.22473	238.96	261.43	1.0531
20	0.39279	247.32	270.89	1.1267	0.23349	246.67	270.02	1.0829
30	0.40688	255.12	279.53	1.1557	0.24216	254.54	278.76	1.1122
40	0.42091	263.10	288.35	1.1844	0.25076	262.58	287.66	1.1411
50	0.43487	271.25	297.34	1.2126	0.25930	270.79	296.72	1.1696
60	0.44879	279.58	306.51	1.2405	0.26779	279.16	305.94	1.1977
70	0.46266	288.08	315.84	1.2681	0.27623	287.70	315.32	1.2254
80	0.47650	296.75	325.34	1.2954	0.28464	296.40	324.87	1.2528
90	0.49031	305.58	335.00	1.3224	0.29302	305.27	334.57	1.2799
	P = 0.14 MPa (−18.80°C)				P = 0.18 MPa (−12.73°C)			
Sat.	0.13945	216.52	236.04	0.9322	0.10983	219.94	239.71	0.9273
−10	0.14519	223.03	243.40	0.9606	0.11135	222.02	242.06	0.9362
0	0.15219	230.55	251.86	0.9922	0.11678	229.67	250.69	0.9684
10	0.15875	238.21	260.43	1.0230	0.12207	237.44	259.41	0.9998
20	0.16520	246.01	269.13	1.0532	0.12723	245.33	268.23	1.0304
30	0.17155	253.96	277.97	1.0828	0.13230	253.36	277.17	1.0604
40	0.17783	262.06	286.96	1.1120	0.13730	261.53	286.24	1.0898
50	0.18404	270.32	296.09	1.1407	0.14222	269.85	295.45	1.1187
60	0.19020	278.74	305.37	1.1690	0.14710	278.31	304.79	1.1472
70	0.19633	287.32	314.80	1.1969	0.15193	286.93	314.28	1.1753
80	0.20241	296.06	324.39	1.2244	0.15672	295.71	323.92	1.2030
90	0.20846	304.95	334.14	1.2516	0.16148	304.63	333.70	1.2303
100	0.21449	314.01	344.04	1.2785	0.16622	313.72	343.63	1.2573
	P = 0.20 MPa (−10.09°C)				P = 0.24 MPa (−5.37°C)			
Sat.	0.09933	221.43	241.30	0.9253	0.08343	224.07	244.09	0.9222
−10	0.09938	221.50	241.38	0.9256				
0	0.10438	229.23	250.10	0.9582	0.08574	228.31	248.89	0.9399
10	0.10922	237.05	258.89	0.9898	0.08993	236.26	257.84	0.9721
20	0.11394	244.99	267.78	1.0206	0.09399	244.30	266.85	1.0034
30	0.11856	253.06	276.77	1.0508	0.09794	252.45	275.95	1.0339
40	0.12311	261.26	285.88	1.0804	0.10181	260.72	285.16	1.0637
50	0.12758	269.61	295.12	1.1094	0.10562	269.12	294.47	1.0930
60	0.13201	278.10	304.50	1.1380	0.10937	277.67	303.91	1.1218
70	0.13639	286.74	314.02	1.1661	0.11307	286.35	313.49	1.1501
80	0.14073	295.53	323.68	1.1939	0.11674	295.18	323.19	1.1780
90	0.14504	304.47	333.48	1.2212	0.12037	304.15	333.04	1.2055
100	0.14932	313.57	343.43	1.2483	0.12398	313.27	343.03	1.2326

Table D-1 through D-3 are based on equations from D. P. Wilson and R. S. Basu, "Thermodynamic Properties of a New Stratospherically Safe Working Fluid–Refrigerant 134a," ASHRAE Trans., Vol. 94, Pt. 2, 1988, pp. 2095–2118.

Table D-3 *(Continued)*

T, °C	*v* m³/kg	*u* kJ/kg	*h* kJ/kg	*s* kJ/kg·K	*v* m³/kg	*u* kJ/kg	*h* kJ/kg	*s* kJ/kg·K
	P = 0.28 MPa (−1.23°C)				*P* = 0.32 MPa (2.48°C)			
Sat.	0.07193	226.38	246.52	0.9197	0.06322	228.43	248.66	0.917
0	0.07240	227.37	247.64	0.9238				
10	0.07613	235.44	256.76	0.9566	0.06576	234.61	255.65	0.942
20	0.07972	243.59	265.91	0.9883	0.06901	242.87	264.95	0.974
30	0.08320	251.83	275.12	1.0192	0.07214	251.19	274.28	1.006
40	0.08660	260.17	284.42	1.0494	0.07518	259.61	283.67	1.036
50	0.08992	268.64	293.81	1.0789	0.07815	268.14	293.15	1.066
60	0.09319	277.23	303.32	1.1079	0.08106	276.79	302.72	1.095
70	0.09641	285.96	312.95	1.1364	0.08392	285.56	312.41	1.124
80	0.09960	294.82	322.71	1.1644	0.08674	294.46	322.22	1.152
90	0.10275	303.83	332.60	1.1920	0.08953	303.50	332.15	1.180
100	0.10587	312.98	342.62	1.2193	0.09229	312.68	342.21	1.207
110	0.10897	322.27	352.78	1.2461	0.09503	322.00	352.40	1.234
120	0.11205	331.71	363.08	1.2727	0.09774	331.45	362.73	1.261
	P = 0.40 MPa (8.93°C)				*P* = 0.50 MPa (15.74°C)			
Sat.	0.05089	231.97	252.32	0.9145	0.04086	235.64	256.07	0.911
10	0.05119	232.87	253.35	0.9182				
20	0.05397	241.37	262.96	0.9515	0.04188	239.40	260.34	0.926
30	0.05662	249.89	272.54	0.9837	0.04416	248.20	270.28	0.959
40	0.05917	258.47	282.14	1.0148	0.04633	256.99	280.16	0.991
50	0.06164	267.13	291.79	1.0452	0.04842	265.83	290.04	1.022
60	0.06405	275.89	301.51	1.0748	0.05043	274.73	299.95	1.053
70	0.06641	284.75	311.32	1.1038	0.05240	283.72	309.92	1.082
80	0.06873	293.73	321.23	1.1322	0.05432	292.80	319.96	1.111
90	0.07102	302.84	331.25	1.1602	0.05620	302.00	330.10	1.139
100	0.07327	312.07	341.38	1.1878	0.05805	311.31	340.33	1.167
110	0.07550	321.44	351.64	1.2149	0.05988	320.74	350.68	1.194
120	0.07771	330.94	362.03	1.2417	0.06168	330.30	361.14	1.221
130	0.07991	340.58	372.54	1.2681	0.06347	339.98	371.72	1.248
140	0.08208	350.35	383.18	1.2941	0.06524	349.79	382.42	1.274
	P = 0.60 MPa (21.58°C)				*P* = 0.70 MPa (26.72°C)			
Sat.	0.03408	238.74	259.19	0.9097	0.02918	241.42	261.85	0.9080
30	0.03581	246.41	267.89	0.9388	0.02979	244.51	265.37	0.9197
40	0.03774	255.45	278.09	0.9719	0.03157	253.83	275.93	0.9539
50	0.03958	264.48	288.23	1.0037	0.03324	263.08	286.35	0.9867
60	0.04134	273.54	298.35	1.0346	0.03482	272.31	296.69	1.0182
70	0.04304	282.66	308.48	1.0645	0.03634	281.57	307.01	1.0487
80	0.04469	291.86	318.67	1.0938	0.03781	290.88	317.35	1.0784
90	0.04631	301.14	328.93	1.1225	0.03924	300.27	327.74	1.1074
100	0.04790	310.53	339.27	1.1505	0.04064	309.74	338.19	1.1358
110	0.04946	320.03	349.70	1.1781	0.04201	319.31	348.71	1.1637
120	0.05099	329.64	360.24	1.2053	0.04335	328.98	359.33	1.1910
130	0.05251	339.38	370.88	1.2320	0.04468	338.76	370.04	1.2179
140	0.05402	349.23	381.64	1.2584	0.04599	348.66	380.86	1.2444
150	0.05550	359.21	392.52	1.2844	0.04729	358.68	391.79	1.2706
160	0.05698	369.32	403.51	1.3100	0.04857	368.82	402.82	1.2963

(Continued)

Table D-3 (Continued)

T, °C	v m³/kg	u kJ/kg	h kJ/kg	s kJ/kg·K	v m³/kg	u kJ/kg	h kJ/kg	s kJ/kg·K
	P = 0.80 MPa (31.33°C)				P = 0.90 MPa (35.53°C)			
Sat.	0.02547	243.78	264.15	0.9066	0.02255	245.88	266.18	0.9054
40	0.02691	252.13	273.66	0.9374	0.02325	250.32	271.25	0.9217
50	0.02846	261.62	284.39	0.9711	0.02472	260.09	282.34	0.9566
60	0.02992	271.04	294.98	1.0034	0.02609	269.72	293.21	0.9897
70	0.03131	280.45	305.50	1.0345	0.02738	279.30	303.94	1.0214
80	0.03264	289.89	316.00	1.0647	0.02861	288.87	314.62	1.0521
90	0.03393	299.37	326.52	1.0940	0.02980	298.46	325.28	1.0819
100	0.03519	308.93	337.08	1.1227	0.03095	308.11	335.96	1.1109
110	0.03642	318.57	347.71	1.1508	0.03207	317.82	346.68	1.1392
120	0.03762	328.31	358.40	1.1784	0.03316	327.62	357.47	1.1670
130	0.03881	338.14	369.19	1.2055	0.03423	337.52	368.33	1.1943
140	0.03997	348.09	380.07	1.2321	0.03529	347.51	379.27	1.2211
150	0.04113	358.15	391.05	1.2584	0.03633	357.61	390.31	1.2475
160	0.04227	368.32	402.14	1.2843	0.03736	367.82	401.44	1.2735
170	0.04340	378.61	413.33	1.3098	0.03838	378.14	412.68	1.2992
180	0.04452	389.02	424.63	1.3351	0.03939	388.57	424.02	1.3245
	P = 1.00 MPa (39.39°C)				P = 1.20 MPa (46.32°C)			
Sat.	0.02020	247.77	267.97	0.9043	0.01663	251.03	270.99	0.9023
40	0.02029	248.39	268.68	0.9066				
50	0.02171	258.48	280.19	0.9428	0.01712	254.98	275.52	0.9164
60	0.02301	268.35	291.36	0.9768	0.01835	265.42	287.44	0.9527
70	0.02423	278.11	302.34	1.0093	0.01947	275.59	298.96	0.9868
80	0.02538	287.82	313.20	1.0405	0.02051	285.62	310.24	1.0192
90	0.02649	297.53	324.01	1.0707	0.02150	295.59	321.39	1.0503
100	0.02755	307.27	334.82	1.1000	0.02244	305.54	332.47	1.0804
110	0.02858	317.06	345.65	1.1286	0,02335	315.50	343.52	1.1096
120	0.02959	326.93	356.52	1.1567	0.02423	325.51	354.58	1.1381
130	0.03058	336.88	367.46	1.1841	0.02508	335.58	365.68	1.1660
140	0.03154	346.92	378.46	1.2111	0.02592	345.73	376.83	1.1933
150	0.03250	357.06	389.56	1.2376	0.02674	355.95	388.04	1.2201
160	0.03344	367.31	400.74	1.2638	0.02754	366.27	399.33	1.2465
170	0.03436	377.66	412.02	1.2895	0.02834	376.69	41070	1.2724
180	0.03528	388.12	423.40	1.3149	0.02912	387.21	422.16	1.2980

Table D-3 *(Continued)*

T, °C	*v* m³/kg	*u* kJ/kg	*h* kJ/kg	*s* kJ/kg·K	*v* m³/kg	*u* kJ/kg	*h* kJ/kg	*s* kJ/kg·K
	P = 1.40 MPa (52.43°C)				*P* = 1.60 MPa (57.92°C)			
Sat.	0.01405	253.74	273.40	0.9003	0.01208	256.00	275.33	0.8982
60	0.01495	262.17	283.10	0.9297	0.01233	258.48	278.20	0.9069
70	0.01603	272.87	295.31	0.9658	0.01340	269.89	291.33	0.9457
80	0.01701	283.29	307.10	0.9997	0.01435	280.78	303.74	0.9813
90	0.01792	293.55	318.63	1.0319	0.01521	291.39	315.72	1.0148
100	0.01878	303.73	330.02	1.0628	0.01601	301.84	327.46	1.0467
110	0.01960	313.88	341.32	1.0927	0.01677	312.20	339.04	1.0773
120	0.02039	324.05	352.59	1.1218	0.01750	322.53	350.53	1.1069
130	0.02115	334.25	363.86	1.1501	0.01820	332.87	361.99	1.1357
140	0.02189	344.50	375.15	1.1777	0.01887	343.24	373.44	1.1638
150	0.02262	354.82	386.49	1.2048	0.01953	353.66	384.91	1.1912
160	0.02333	365.22	397.89	1.2315	0.02017	364.15	396.43	1.2181
170	0.02403	375.71	409.36	1.2576	0.02080	374.71	407.99	1.2445
180	0.02472	386.29	420.90	1.2834	0.02142	385.35	419.62	1.2704
190	0.02541	396.96	432.53	1.3088	0.02203	396.08	431.33	1.2960
200	0.02608	407.73	444.24	1.3338	0.02263	406.90	443.11	1.3212

E Properties of Ammo

Table E-I Saturated Ammonia

		Specific volume m^3/kg		Enthalpy kJ/kg			Entropy kJ/kg·K		
T, °C	P, kPa	v_f	v_g	h_f	h_{fg}	h_g	s_f	s_{fg}	s_g
−50	40.88	0.001424	2.6254	−44.3	1416.7	1372.4	−0.1942	6.3502	6.1561
−46	51.55	0.001434	2.1140	−26.6	1405.8	1379.2	−0.1156	6.1902	6,0746
−44	57.69	0.001439	1.9032	−17.8	1400.3	1382.5	−0.0768	6.1120	6.0352
−42	64.42	0.001444	1.7170	−8.9	1394.7	1385.8	−0.0382	6.0349	5.9967
−40	71.77	0.001449	1.5521	0.0	1389.0	1389.0	0.0000	5.9589	5.9589
−38	79.80	0.001454	1.4085	8.9	1383.3	1392.2	0.0380	5.8840	5.9220
−36	88.54	0.001460	1.2757	17.8	1377.6	1395.4	0.0757	5.8101	5.8858
−34	98.05	0.001465	1.1597	26.8	1371.8	1398.5	0.1132	5.7372	5.8504
−32	108.37	0.001470	1.0562	35.7	1365.9	1401.6	0.1504	5.6652	5:8156
−30	119.55	0.001476	0.9635	44.7	1360.0	1404.6	0.1873	5.5942	5.7815
−28	131.64	0.001481	0.8805	53.6	1354.0	1407.6	0.2240	5.5241	5.7481
−26	144.70	0.001487	0.8059	62.6	1347.9	1410.5	0.2605	5.4548	5.7153
−24	158.78	0.001492	0.7388	71.6	1341.8	1413.4	0.2967	5.3864	5.6831
−22	173.93	0.001498	0.6783	80.7	1335.6	1416.2	0.3327	5.3188	5.6515
−20	190.22	0.001504	0.6237	89.7	1329.3	1419.0	0.3684	5.2520	5.6205
−18	207.71	0.001510	0.5743	98.8	1322.9	1421.7	0.4040	5.1860	5.5900
−16	226.45	0.001515	0.5296	107.8	1316.5	1424.4	0.4393	5.1207	5.5600
−14	246.51	0.001521	0.4889	116.9	1310.0	1427.0	0.4744	5.0561	5.5305
−12	267.95	0.001528	0.4520	126.0	1303.5	1429.5	0.5093	4.9922	5.5015
−10	290.85	0.001534	0.4185	135.2	1296.8	1432.0	0.5440	4.9290	5.4730
−8	315.25	0.001540	0.3878	144.3	1290.1	1434.4	0.5785	4.8664	5.4449
−6	341.25	0.001546	0.3599	153.5	1283.3	1436.8	0.6128	4.8045	5.4173
−4	368.90	0.001553	0.3343	162.7	1276.4	1439.1	0.6469	4.7432	5.3901
−2	398.27	0.001559	0.3109	171.9	1269.4	1441.3	0.6808	4.6825	5.3633
0	429.44	0.001566	0.2895	181.1	1262.4	1443.5	0.7145	4.6223	5.3369
2	462.49	0.001573	0.2698	190.4	1255.2	1445.6	0.7481	4.5627	5.3108
4	497.49	0.001580	0.2517	199.6	1248.0	1447.6	0.7815	4.5037	5.2852
6	534.51	0.001587	0.2351	208.9	1240.6	1449.6	0.8148	4.4451	5.2599
8	573.64	0.001594	0.2198	218.3	1233.2	1451.5	0.8479	4.3871	5.2350
10	614.95	0.001601	0.2056	227.6	1225.7	1453.3	0.8808	4.3295	5.2104
12	658.52	0.001608	0.1926	237.0	1218.1	1455.1	0.9136	4.2725	5.1861
14	704.44	0.001616	0.1805	246.4	1210.4	1456.8	0.9463	4.2159	5.1621
16	752.79	0.001623	0.1693	255.9	1202.6	1458.5	0.9788	4.1597	5.1385
18	803.66	0.001631	0.1590	265.4	1194.7	1460.0	1.0112	4.1039	5.1151
20	857.12	0.001639	0.1494	274.9	1186.7	1461.5	1.0434	4.0486	5.0920

(*Continued*)

Based on data from National Bureau of Standards Circular No. 142, Tables of Thermodynamic Properties of Ammonia

Table E-l (*Continued*)

		Specific volume m³/kg		Enthalpy kJ/kg			Entropy kJ/kg·K		
T, °C	P, kPa	v_f	v_g	h_f	h_{fg}	h_g	s_f	s_{fg}	s_g
22	913.27	0.001647	0.1405	284.4	1178.5	1462.9	1.0755	3.9937	5.0692
24	972.19	0.001655	0.1322	294.0	1170.3	1464.3	1.1075	3.9392	5.0467
26	1033.97	0.001663	0.1245	303.6	1162.0	1465.6	1.1394	3.8850	5.0244
28	1098.71	0.001671	0.1173	313.2	1153.6	1466.8	1.1711	3.8312	5.0023
30	1166.49	0.001680	0.1106	322.9	1145.0	1467.9	1.2028	3.7777	4.9805
32	1237.41	0.001689	0.1044	332.6	1136.4	1469.0	1.2343	3.7246	4.9589
34	1311.55	0.001698	0.0986	342.3	1127.6	1469.9	1.2656	3.6718	4.9374
36	1389.03	0.001707	0.0931	352.1	1118.7	1470.8	1.2969	3.6192	4.9161
38	1469.92	0.001716	0.0880	361.9	1109.7	1471.5	1.3281	3.5669	4.8950
40	1554.33	0.001726	0.0833	371.7	1100.5	1472.2	1.3591	3.5148	4.8740
42	1642.35	0.001735	0.0788	381.6	1091.2	1472.8	1.3901	3.4630	4.8530
44	1734.09	0.001745	0.0746	391.5	1081.7	1473.2	1.4209	3.4112	4.8322
46	1829.65	0.001756	0.0707	401.5	1072.0	1473.5	1.4518	3.3595	4.8113
48	1929.13	0.001766	0.0669	411.5	1062.2	1473.7	1.4826	3.3079	4.7905
50	2032.62	0.001777	0.0635	421.7	1052.0	1473.7	1.5135	3.2561	4.7696

Table E-2 Superheated Ammonia

P, kPa (T_{sat}, °C)		Temperature, °C −20	−10	0	10	20	30	40	50	60	70	80	100
	v	2.4474	2.5481	2.6482	2.7479	2.8473	2.9464	3.0453	3.1441	3.2427	3.3413	3.4397	
50	h	1435.8	1457.0	1478.1	1499.2	1520.4	1541.7	1563.0	1584.5	1606.1	1627.8	1649.7	
(−46.54)	s	6.3256	6.4077	6.4865	6.5625	6.6360	6.7073	6.7766	6.8441	6.9099	6.9743	7.0372	
	v	1.6233	1.6915	1.7591	1.8263	1.8932	1.9597	2.0261	2.0923	2.1584	2.2244	2.2903	
75	h	1433.0	1454.7	1476.1	1497.5	1518.9	1540.3	1561.8	1583.4	1605.1	1626.9	1648.9	
(−39.18)	s	6.1190	6.2028	6.2828	6.3597	6.4339	6.5058	6.5756	6.6434	6.7096	6.7742	6.8373	
	v	1.2110	1.2631	1.3145	1.3654	1.4160	1.4664	1.5165	1.5664	1.6163	1.6659	1.7155	1.8145
100	h	1430.1	1452.2	1474.1	1495.7	1517.3	1538.9	1560.5	1582.2	1604.1	1626.0	1648.0	1692.6
(−33.61)	s	5.9695	6.0552	6.1366	6.2144	6.2894	6.3618	6.4321	6.5003	6.5668	6.6316	6.6950	6.8177
	v	0.9635	1.0059	1.0476	1.0889	1.1297	1.1703	1.2107	1.2509	1.2909	1.3309	1.3707	1.4501
125	h	1427.2	1449.8	1472.0	1493.9	1515.7	1537.5	1559.3	1581.1	1603.0	1625.0	1647.2	1691.8
(−29.08)	s	5.8512	5.9389	6.0217	6.1006	6.1763	6.2494	6.3201	6.3887	6.4555	6.5206	6.5842	6.7072
	v	0.7984	0.8344	0.8697	0.9045	0.9388	0.9729	1.0068	1.0405	1.0740	1.1074	1.1408	1.2072
150	h	1424.1	1447.3	1469.8	1492.1	1514.1	1536.1	1558.0	1580.0	1602.0	1624.1	1646.3	1691.1
(−25.23)	s	5.7526	5.8424	5.9266	6.0066	6.0831	6.1568	6.2280	6.2970	6.3641	6.4295	6.4933	6.6167
	v		0.6199	0.6471	0.6738	0.7001	0.7261	0.7519	0.7774	0.8029	0.8282	0.8533	0.9035
200	h		1442.0	1465.5	1488.4	1510.9	1533.2	1555.5	1577.7	1599.9	1622.2	1644.6	1689.6
(−18.86)	s		5.6863	5.7737	5.8559	5.9342	6.0091	6.0813	6.1512	6.2189	6.2849	6.3491	6.4732
	v		0.4910	0.5135	0.5354	0.5568	0.5780	0.5989	0.6196	0.6401	0.6605	0.6809	0.7212
250	h		1436.6	1461.0	1484.5	1507.6	1530.3	1552.9	1575.4	1597.8	1620.3	1642.8	1688.2
(−13.67)	s		5.5609	5.6517	5.7365	5.8165	5.8928	5.9661	6.0368	6.1052	6.1717	6.2365	6.3613
	v			0.4243	0.4430	0.4613	0.4792	0.4968	0.5143	0.5316	0.5488	0.5658	0.5997
300	h			1456.3	1480.6	1504.2	1527.4	1550.3	1573.0	1595.7	1618.4	1641.1	1686.7
(−9.23)	s			5.5493	5.6366	5.7186	5.7963	5.8707	5.9423	6.0114	6.0785	6.1437	6.2693
	v			0.3605	0.3770	0.3929	0.4086	0.4239	0.4391	0.4541	0.4689	0.4837	0.5129
350	h			1451.5	1476.5	1500.7	1524.4	1547.6	1570.7	1593.6	1616.5	1639.3	1685.2
(−5.35)	s			5.4600	5.5502	5.6342	5.7135	5.7890	5.8615	5.9314	5.9990	6.0647	6.1910
	v			0.3125	0.3274	0.3417	0.3556	0.3692	0.3826	0.3959	0.4090	0.4220	0.4478
400	h			1446.5	1472.4	1497.2	1521.3	1544.9	1568.3	1591.5	1614.5	1637.6	1683.7
(−1.89)	s			5.3803	5.4735	5.5597	5.6405	5.7173	5.7907	5.8613	5.9296	5.9957	6.1228
	v			0.2752	0.2887	0.3017	0.3143	0.3266	0.3387	0.3506	0.3624	0.3740	0.3971
450	h			1441.3	1468.1	1493.6	1518.2	1542.2	1565.9	1589.3	1612.6	1635.8	1682.2
(1.26)	s			5.3078	5.4042	5.4926	5.5752	5.6532	5.7275	5.7989	5.8678	5.9345	6.0623

(Continued)

Table E-2 *(Continued)*

P, kPa (T_{sat}, °C)		20	30	40	50	60	70	80	100	120	140	160	180
	v	0.2698	0.2813	0.2926	0.3036	0.3144	0.3251	0.3357	0.3565	0.3771	0.3975		
500	*h*	1489.9	1515.0	1539.5	1563.4	1587.1	1610.6	1634.0	1680.7	1727.5	1774.7		
(4.14)	*s*	5.4314	5.5157	5.5950	5.6704	5.7425	5.8120	5.8793	6.0079	6.1301	6.2472		
	v	0.2217	0.2317	0.2414	0.2508	0.2600	0.2691	0.2781	0.2957	0.3130	0.3302		
600	*h*	1482.4	1508.6	1533.8	1558.5	1582.7	1606.6	1630.4	1677.7	1724.9	1772.4		
(9.29)	*s*	5.3222	5.4102	5.4923	5.5697	5.6436	5.7144	5.7826	5.9129	6.0363	6.1541		
	v	0.1874	0.1963	0.2048	0.2131	0.2212	0.2291	0.2369	0.2522	0.2672	0.2821		
700	*h*	1474.5	1501.9	1528.1	1553.4	1578.2	1602.6	1626.8	1674.6	1722.4	1770.2		
(13.81)	*s*	5.2259	5.3179	5.4029	5.4826	5.5582	5.6303	5.6997	5.8316	5.9562	6.0749		
	v	0.1615	0.1696	0.1773	0.1848	0.1920	0.1991	0.2060	0.2196	0.2329	0.2459	0.2589	
800	*h*	1466.3	1495.0	1522.2	1548.3	1573.7	1598.6	1623.1	1671.6	1719.8	1768.0	1816.4	
(17.86)	*s*	5.1387	5.2351	5.3232	5.4053	5.4827	5.5562	5.6268	5.7603	5.8861	6.0057	6.1202	
	v		0.1488	0.1559	0.1627	0.1693	0.1757	0.1820	0.1942	0.2061	0.2178	0.2294	
900	*h*		1488.0	1516.2	1543.0	1569.1	1594.4	1619.4	1668.5	1717.1	1765.7	1814.4	
(21.54)	*s*		5.1593	5.2508	5.3354	5.4147	5.4897	5.5614	5.6968	5.8237	5.9442	6.0594	
	v		0.1321	0.1388	0.1450	0.1511	0.1570	0.1627	0.1739	0.1847	0.1954	0.2058	0.2162
1000	*h*		1480.6	1510.0	1537.7	1564.4	1590.3	1615.6	1665.4	1714.5	1763.4	1812.4	1861.7
(24.91)	*s*		5.0889	5.1840	5.2713	5.3525	5.4292	5.5021	5.6392	5.7674	5.8888	6.0047	6.1159
	v			0.1129	0.1185	0.1238	0.1289	0.1338	0.1434	0.1526	0.1616	0.1705	0.1792
1200	*h*			1497.1	1526.6	1554.7	1581.7	1608.0	1659.2	1709.2	1758.9	1808.5	1858.2
(30.96)	*s*			5.0629	5.1560	5.2416	5.3215	5.3970	5.5379	5.6687	5.7919	5.9091	6.0214
	v			0.0944	0.0995	0.1042	0.1088	0.1132	0.1216	0.1297	0.1376	0.1452	0.1528
1400	*h*			1483.4	1515.1	1544.7	1573.0	1600.2	1652.8	1703.9	1754.3	1804.5	1854.7
(36.28)	*s*			4.9534	5.0530	5.1434	5.2270	5.3053	5.4501	5.5836	5.7087	5.8273	5.9406
	v				0.0851	0.0895	0.0937	0.0977	0.1053	0.1125	0.1195	0.1263	0.1330
1600	*h*				1502.9	1534.4	1564.0	1592.3	1646.4	1698.5	1749.7	1800.5	1851.2
(41.05)	*s*				4.9584	5.0543	5.1419	5.2232	5.3722	5.5084	5.6355	5.7555	5.8699
	v				0.0739	0.0781	0.0820	0.0856	0.0926	0.0992	0.1055	0.1116	0.1177
1800	*h*				1490.0	1523.5	1554.6	1584.1	1639.8	1693.1	1745.1	1796.5	1847.7
(45.39)	*s*				4.8693	4.9715	5.0635	5.1482	5.3018	5.4409	5.5699	5.6914	5.8069
	v				0.0648	0.0688	0.0725	0.0760	0.0824	0.0885	0.0943	0.0999	0.1054
2000	*h*				1476.1	1512.0	1544.9	1575.6	1633.2	1687.6	1740.4	1792.4	1844.1
(49.38)	*s*				4.7834	4.8930	4.9902	5.0786	5.2371	5.3793	5.5104	5.6333	5.7499

F Ideal-Gas Tables

Table F-I Properties of Air

T K	h kJ/kg	P_r	u kJ/kg	v_r	$s°$ kJ/kg·K	T K	h kJ/kg	P_r	u kJ/kg	v_r	$s°$ kJ/kg·K
200	199.97	0.3363	142.56	1707	1.29559	780	800.03	43.35	576.12	51.64	2.69013
220	219.97	0.4690	156.82	1346	1.39105	820	843.98	52.49	608.59	44.84	2.74504
240	240.02	0.6355	171.13	1084	1.47824	860	888.27	63.09	641.40	39.12	2.79783
260	260.09	0.8405	185.45	887.8	1.55848	900	932.93	75.29	674.58	34.31	2.84856
280	280.13	1.0889	199.75	738.0	1.63279	940	977.92	89.28	708.08	30.22	2.89748
290	290.16	1.2311	206.91	676.1	1.66802	980	1023.25	105.2	741.98	26.73	2.94468
300	300.19	1.3860	214.07	621.2	1.70203	1020	1068.89	123.4	776.10	23.72	2.99034
310	310.24	1.5546	221.25	572.3	1.73498	1060	1114.86	143.9	810.62	21.14	3.03449
320	320.29	1.7375	228.43	528.6	1.76690	1100	1161.07	167.1	845.33	18.896	3.07732
340	340.42	2.149	242.82	454.1	1.82790	1140	1207.57	193.1	880.35	16.946	3.11883
360	360.58	2.626	257.24	393.4	1.88543	1180	1254.34	222.2	915.57	15.241	3.15916
380	380.77	3.176	271.69	343.4	1.94001	1220	1301.31	254.7	951.09	13.747	3.19834
400	400.98	3.806	286.16	301.6	1.99194	1260	1348.55	290.8	986.90	12.435	3.23638
420	421.26	4.522	300.69	266.6	2.04142	1300	1395.97	330.9	1022.82	11.275	3.27345
440	441.61	5.332	315.30	236.8	2.08870	1340	1443.60	375.3	1058.94	10.247	3.30959
460	462.02	6.245	329.97	211.4	2.13407	1380	1491.44	424.2	1095.26	9.337	3.34474
480	482.49	7.268	344.70	189.5	2.17760	1420	1539.44	478.0	1131.77	8.526	3.37901
500	503.02	8.411	359.49	170.6	2.21952	1460	1587.63	537.1	1168.49	7.801	3.41247
520	523.63	9.684	374.36	154.1	2.25997	1500	1635.97	601.9	1205.41	7.152	3.44516
540	544.35	11.10	389.34	139.7	2.29906	1540	1684.51	672.8	1242.43	6.569	3.47712
560	565.17	12.66	404.42	127.0	2.33685	1580	1733.17	750.0	1279.65	6.046	3.50829
580	586.04	14.38	419.55	115.7	2.37348	1620	1782.00	834.1	1316.96	5.574	3.53879
600	607.02	16.28	434.78	105.8	2.40902	1660	1830.96	925.6	1354.48	5.147	3.56867
620	628.07	18.36	450.09	96.92	2.44356	1700	1880.1	1025	1392.7	4.761	3.5979
640	649.22	20.65	465.05	88.99	2.47716	1800	2003.3	1310	1487.2	3.944	3.6684
660	670.47	23.13	481.01	81.89	2.50985	1900	2127.4	1655	1582.6	3.295	3.73541
680	691.82	25.85	496.62	75.50	2.54175	2000	2252.1	2068	1678.7	2.776	3.7994
700	713.27	28.80	512.33	69.76	2.57277	2100	2377.4	2559	1775.3	2.356	3.8605
720	734.82	32.02	528.14	64.53	2.60319	2200	2503.2	3138	1872.4	2.012	3.9191
740	756.44	35.50	544.02	59.82	2.63280						

Based on data from J. H. Keenan and J. Kaye, Gas Tables, Wiley, New York, 1945.

Table F-2 Molar Properties of Nitrogen, N_2

$\overline{h}_f^\circ = 0$ kJ/kmol

T K	*h* kJ/kmol	*u* kJ/kmol	*s*° kJ/kmol·K	*T* K	*h* kJ/kmol	*u* kJ/kmol	*s*° kJ/kmol·K
0	0	0	0	1000	30 129	21 815	228.057
220	6391	4562	182.639	1020	30 784	22 304	228.706
240	6975	4979	185.180	1040	31 442	22 795	229.344
260	7558	5396	187.514	1060	32 101	23 288	229.973
280	8141	5813	189.673	1080	32 762	23 782	230.591
298	8669	6190	191.502	1100	33 426	24 280	231.199
300	8723	6229	191.682	1120	34 092	24 780	231.799
320	9306	6645	193.562	1140	34 760	25 282	232.391
340	9888	7061	195.328	1160	35 430	25 786	232.973
360	10 471	7478	196.995	1180	36 104	26 291	233.549
380	11 055	7895	198.572	1200	36 777	26 799	234.115
400	11 640	8314	200.071	1240	38 129	27 819	235.223
420	12 225	8733	201.499	1260	38 807	28 331	235.766
440	12 811	9153	202.863	1280	39 488	28 845	236.302
460	13 399	9574	204.170	1300	40 170	29 361	236.831
480	13 988	9997	205.424	1320	40 853	29 878	237.353
500	14 581	10 423	206.630	1340	41.539	30 398	237.867
520	15 172	10 848	207.792	1360	42 227	30 919	238.376
540	15 766	11 277	208.914	1380	42 915	31 441	238.878
560	16 363	11 707	209.999	1400	43 605	31 964	239.375
580	16 962	12 139	211.049	1440	44 988	33 014	240.350
600	17 563	12 574	212.066	1480	46 377	34 071	241.301
620	18 166	13 011	213.055	1520	47 771	35 133	242.228
640	18 772	13 450	214.018	1560	49 168	36 197	243.137
660	19 380	13 892	214.954	1600	50 571	37 268	244.028
680	19 991	14 337	215.866	1700	54 099	39 965	246.166
700	20 604	14 784	216.756	1800	57 651	42 685	248.195
720	21 220	15 234	217.624	1900	61 220	45 423	250.128
740	21 839	15 686	218.472	2000	64 810	48 181	251.969
760	22 460	16 141	219.301	2100	68 417	50 957	253.726
780	23 085	16 599	220.113	2200	72 040	53 749	255.412
800	23 714	17 061	220.907	2300	75 676	56 553	257.02
820	24 342	17 524	221.684	2400	79 320	59 366	258.580
840	24 974	17 990	222.447	2500	82 981	62 195	260.073
860	25 610	18 459	223.194	2600	86 650	65 033	261.512
880	26 248	18 931	223.927	2700	90 328	67 880	262.902
900	26 890	19 407	224.647	2800	94 014	70 734	264.241
920	27 532	19 883	225.353	2900	97 705	73 593	265.538
940	28 178	20 362	226.047	3000	101 407	76 464	266.793
960	28 826	20 844	226.728	3100	105 115	79 341	268.007
980	29 476	21 328	227.398	3200	108 830	82 224	269.186

Based on data from JANAF Thermochemical Tables, NSRDS-NBS-37, 1971

Table F-3 Molar Properties of Oxygen, O_2

$\bar{h}_f^\circ = 0$ kJ/kmol

T K	h kJ/kmol	u kJ/kmol	s° kJ/kmol·K	T K	h kJ/kmol	u kJ/kmol	s° kJ/kmol·K
0	0	0	0	1020	32088	23607	244.164
220	6404	4575	196.171	1040	32789	24142	244.844
240	6984	4989	198.696	1060	33490	24677	245.513
260	7566	5405	201.027	1080	34194	25214	246.171
280	8150	5822	203.191	1100	34899	25753	246.818
298	8682	6203	205.033	1120	35606	26294	247.454
300	8736	6242	205.213	1140	36314	26836	248.081
320	9325	6664	207.112	1160	37023	27379	248.698
340	9916	7090	208.904	1180	37734	27923	249.307
360	10511	7518	210.604	1200	38447	28469	249.906
380	11109	7949	212.222	1220	39162	29018	250.497
400	11711	8384	213.765	1240	39877	29568	251.079
420	12314	8822	215.241	1260	40594	30118	251.653
440	12923	9264	216.656	1280	41312	30670	252.219
460	13535	9710	218.016	1300	42033	31224	252.776
480	14151	10160	219.326	1320	42753	31778	253.325
500	14770	10614	220.589	1340	43475	32334	253.868
520	15395	11071	221.812	1360	44198	32891	254.404
540	16022	11533	222.997	1380	44923	33449	254.932
560	16654	11998	224.146	1400	45648	34008	255.454
580	17290	12467	225.262	1440	47102	35129	256.475
600	17929	12940	226.346	1480	48561	36256	257.474
620	18572	13417	227.400	1520	50024	37387	258.450
640	19219	13898	228.429	1540	50756	37952	258.928
660	19870	14383	229.430	1560	51490	38520	259.402
680	20524	14871	230.405	1600	52961	39658	260.333
700	21184	15364	231.358	1700	56652	42517	262.571
720	21845	15859	232.291	1800	60371	45405	264.701
740	22510	16357	233.201	1900	64116	48319	266.722
760	23178	16859	234.091	2000	67881	51253	268.655
780	23850	17364	234.960	2100	71668	54208	270.504
800	24523	17872	235.810	2200	75484	57192	272.278
820	25199	18382	236.644	2300	79316	60193	273.981
840	25877	18893	237.462	2400	83174	63219	275.625
860	26559	19408	238.264	2500	87057	66271	277.207
880	27242	19925	239.051	2600	90956	69339	278.738
900	27928	20445	239.823	2700	94881	72433	280.219
920	28616	20967	240.580	2800	98826	75546	281.654
940	29306	21491	241.323	2900	102793	78682	283.048
960	29999	22017	242.052	3000	106780	81837	284.399
980	30692	22544	242.768	3100	110784	85009	285.713
1000	31389	23075	243.471	3200	114809	88203	286.989

Based on data from JANAF Thermochemical Tables, NSRDS-NBS-37, 1971

Table F-4 Molar Properties of Carbon Dioxide, CO_2

$\bar{h}_f^\circ = -393\ 520$ kJ/kmol

T K	*h* kJ/kmol	*u* kJ/kmol	*s*° kJ/kmol·K	*T* K	*h* kJ/kmol	*u* kJ/kmol	*s*° kJ/kmol·K
0	0	0	0	1020	43859	35378	270.293
220	6601	4772	202.966	1040	44953	36306	271.354
240	7280	5285	205.920	1060	46051	37238	272.400
260	7979	5817	208.717	1080	47153	38174	273.430
280	8697	6369	211.376	1100	48258	39112	274.445
298	9364	6885	213.685	1120	49369	40057	275.444
300	9431	6939	213.915	1140	50484	41006	276.430
320	10186	7526	216.351	1160	51602	41957	277.403
340	10959	8131	218.694	1180	52724	42913	278.361
360	11748	8752	220.948	1200	53848	43871	279.307
380	12552	9392	223.122	1220	54977	44834	280.238
400	13372	10046	225.225	1240	56108	45799	281.158
420	14206	10714	227.258	1260	57244	46768	282.066
440	15054	11393	229.230	1280	58381	47739	282.962
460	15916	12091	231.144	1300	59522	48713	283.847
480	16791	12800	233.004	1320	60666	49691	284.722
500	17678	13521	234.814	1340	61813	50672	285.586
520	18576	14253	236.575	1360	62963	51656	286.439
540	19485	14996	238.292	1380	64116	52643	287.283
560	20407	15751	239.962	1400	65271	53631	288.106
580	21337	16515	241.602	1440	67586	55614	289.743
600	22280	17291	243.199	1480	69911	57606	291.333
620	23231	18076	244.758	1520	72246	59609	292.888
640	24190	18869	246.282	1560	74590	61620	294.411
660	25160	19672	247.773	1600	76944	63741	295.901
680	26138	20484	249.233	1700	82856	68721	299.482
700	27125	21305	250.663	1800	88806	73840	302.884
720	28121	22134	252.065	1900	94793	78996	306.122
740	29124	22972	253.439	2000	100804	84185	309.210
760	30135	23817	254.787	2100	106864	89404	312.160
780	31154	24669	256.110	2200	112939	94648	314.988
800	32179	25527	257.408	2300	119035	99912	317.695
820	33212	26394	258.682	2400	125152	105197	320.302
840	34251	27267	259.934	2500	131290	110504	322.308
860	35296	28125	261.164	2600	137449	115832	325.222
880	36347	29031	262.371	2700	143620	121172	327.549
900	37405	29922	263.559	2800	149808	126528	329.800
920	38467	30818	264.728	2900	156009	131898	331.975
940	39535	31719	265.877	3000	162226	137283	334.084
960	40607	32625	267.007	3100	168456	142681	336.126
980	41685	33537	268.119	3200	174695	148089	338.109
1000	42769	34455	269.215				

Based on data from JANAF Thermochemical Tables, NSRDS-NBS-37, 1971

Table F-5 Molar Properties of Carbon Monoxide, CO

$\bar{h}_f^\circ = -110\ 530$ kJ/kmol

T K	*h* kJ/kmol	*u* kJ/kmol	*s*° kJ/kg·K	*T* K	*h* kJ/kmol	*u* kJ/kmol	*s*° kJ/kg·K
0	0	0	0	1040	31 688	23 041	235.728
220	6391	4562	188.683	1060	32 357	23 544	236.364
240	6975	4979	191.221	1080	33 029	24 049	236.992
260	7558	5396	193.554	1100	33 702	24 557	237.609
280	8140	5812	195.713	1120	34 377	25 065	238.217
300	8723	6229	197.723	1140	35 054	25 575	238.817
320	9306	6645	199.603	1160	35 733	26 088	239.407
340	9889	7062	201.371	1180	36 406	26 602	239.989
360	10 473	7480	203.040	1200	37 095	27 118	240.663
380	11 058	7899	204.622	1220	37 780	27 637	241.128
400	11 644	8319	206.125	1240	38 466	28 426	241.686
420	12 232	8740	207.549	1260	39 154	28 678	242.236
440	12 821	9163	208.929	1280	39 844	29 201	242.780
460	13 412	9587	210.243	1300	40 534	29 725	243.316
480	14 005	10 014	211.504	1320	41 226	30 251	243.844
500	14 600	10 443	212.719	1340	41 919	30 778	244.366
520	15 197	10 874	213.890	1360	42 613	31 306	244.880
540	15 797	11 307	215.020	1380	43 309	31 836	245.388
560	16 399	11 743	216.115	1400	44 007	32 367	245.889
580	17 003	12 181	217.175	1440	45 408	33 434	246.876
600	17 611	12 622	218.204	1480	46 813	34 508	247.839
620	18 221	13 066	219.205	1520	48 222	35 584	248.778
640	18 833	13 512	220.179	1560	49 635	36 665	249.695
660	19 449	13 962	221.127	1600	51 053	37 750	250.592
680	20 068	14 414	222.052	1700	54 609	40 474	252.751
700	20 690	14 870	222.953	1800	58 191	43 225	254.797
720	21 315	15 328	223.833	1900	61 794	45 997	256.743
740	21 943	15 789	224.692	2000	65 408	48 780	258.600
760	22 573	16 255	225.533	2100	69 044	51 584	260.370
780	23 208	16 723	226.357	2200	72 688	54 396	262.065
800	23 844	17 193	227.162	2300	76 345	57 222	263.692
820	24 483	17 665	227.952	2400	80 015	60 060	265.253
840	25 124	18 140	228.724	2500	83 692	62 906	266.755
860	25 768	18 617	229.482	2600	87 383	65 766	268.202
880	26 415	19 099	230.227	2700	91 077	68 628	269.596
900	27 066	19 583	230.957	2800	94 784	71 504	270.943
920	27 719	20 070	231.674	2900	98 495	74 383	272.249
940	28 375	20 559	232.379	3000	102 210	77 267	273.508
960	29 033	21 051	233.072	3100	105 939	80 164	274.730
980	29 693	21 545	233.752	3150	107 802	81 612	275.326
1000	30 355	22 041	234.421	3200	109 667	83 061	275.914
1020	31 020	22 540	235.079				

Based on data from JANAF Thermochemical Tables, NSRDS-NBS-37, 1971

Table F-6 Molar properties of hydrogen, H_2

$\bar{h}_f^\circ = 0$ kJ/kmol

T K	$\bar{h}$ kJ/kmol	$\bar{u}$ kJ/kmol	$\bar{s}^\circ$ kJ/kmol·K	T K	$\bar{h}$ kJ/kmol	$\bar{u}$ kJ/kmol	$\bar{s}^\circ$ kJ/kmol·K
0	0	0	0	1440	42 808	30 835	177.410
260	7370	5209	126.636	1480	44 091	31 786	178.291
270	7657	5412	127.719	1520	45 384	32 746	179.153
280	7945	5617	128.765	1560	46 683	33 713	179.995
290	8233	5822	129.775	1600	47 990	34 687	180.820
298	8468	5989	130.574	1640	49 303	35 668	181.632
300	8522	6027	130.754	1680	50 622	36 654	182.428
320	9100	6440	132.621	1720	51 947	37 646	183.208
340	9680	6853	134.378	1760	53 279	38 645	183.973
360	10 262	7268	136.039	1800	54 618	39 652	184.724
380	10 843	7684	137.612	1840	55 962	40 663	185.463
400	11 426	8100	139.106	1880	57 311	41 680	186.190
420	12 010	8518	140.529	1920	58 668	42 705	186.904
440	12 594	8936	141.888	1960	60 031	43 735	187.607
460	13 179	9355	143.187	2000	61 400	44 771	188.297
480	13 764	9773	144.432	2050	63 119	46 074	189.148
500	14 350	10 193	145.628	2100	64 847	47 386	189.979
520	14 935	10 611	146.775	2150	66 584	48 708	190.796
560	16 107	11 451	148.945	2200	68 328	50 037	191.598
600	17 280	12 291	150.968	2250	70 080	51 373	192.385
640	18 453	13 133	152.863	2300	71 839	52 716	193.159
680	19 630	13 976	154.645	2350	73 608	54 069	193.921
720	20 807	14 821	156.328	2400	75 383	55 429	194.669
760	21 988	15 669	157.923	2450	77 168	56 798	195.403
800	23 171	16 520	159.440	2500	78 960	58 175	196.125
840	24 359	17 375	160.891	2550	80 755	59 554	196.837
880	25 551	18 235	162.277	2600	82 558	60 941	197.539
920	26 747	19 098	163.607	2650	84 368	62 335	198.229
960	27 948	19 966	164.884	2700	86 186	63 737	198.907
1000	29 154	20 839	166.114	2750	88 008	65 144	199.575
1040	30 364	21 717	167.300	2800	89 838	66 558	200.234
1080	31 580	22 601	168.449	2850	91 671	67 976	200.885
1120	32 802	23 490	169.560	2900	93 512	69 401	201.527
1160	34 028	24 384	170.636	2950	95 358	70 831	202.157
1200	35 262	25 284	171.682	3000	97 211	72 268	202.778
1240	36 502	26 192	172.698	3050	99 065	73 707	203.391
1280	37 749	27 106	173.687	3100	100 926	75 152	203.995
1320	39 002	28 027	174.652	3150	102 793	76 604	204.592
1360	40 263	28 955	175.593	3200	104 667	78 061	205.181
1400	41 530	29 889	176.510	3250	106 545	79 523	205.765

Table F-7 Molar Properties of Water, H_2O

$\bar{h}_f^\circ = -241\ 810$ kJ/kmol

T K	$\bar{h}$ kJ/kmol	$\bar{u}$ kJ/kmol	$\bar{s}^\circ$ kJ/kmol·K	T K	$\bar{h}$ kJ/kmol	$\bar{u}$ kJ/kmol	$\bar{s}^\circ$ kJ/kmol·K
0	0	0	0	1020	36 709	28 228	233.415
220	7295	5466	178.576	1040	37 542	28 895	234.223
240	7961	5965	181.471	1060	38 380	29 567	235.020
260	8627	6466	184.139	1080	39 223	30 243	235.806
280	9296	6968	186.616	1100	40 071	30 925	236.584
298	9904	7425	188.720	1120	40 923	31 611	237.352
300	9966	7472	188.928	1140	41 780	32 301	238.110
320	10 639	7978	191.098	1160	42 642	32 997	238.859
340	11 314	8487	193.144	1180	43 509	33 698	239.600
360	11 992	8998	195.081	1200	44 380	34 403	240.333
380	12 672	9513	196.920	1220	45 256	35 112	241.057
400	13 356	10 030	198.673	1240	46 137	35 827	241.773
420	14 043	10 551	200.350	1260	47 022	36 546	242.482
440	14 734	11 075	201.955	1280	47 912	37 270	243.183
460	15 428	11 603	203.497	1300	48 807	38 000	243.877
480	16 126	12 135	204.982	1320	49 707	38 732	244.564
500′	16 828	12 671	206.413	1340	50 612	39 470	245.243
520	17 534	13 211	207.799	1360	51 521	40 213	245.915
540	18 245	13 755	209.139	1400	53 351	41 711	247.241
560	18 959	14 303	210.440	1440	55 198	43 226	248.543
580	19 678	14 856	211.702	1480	57 062	44 756	249.820
600	20 402	15 413	212.920	1520	58 942	46 304	251.074
620	21 130	15 975	214.122	1560	60 838	47 868	252.305
640	21 862	16 541	215.285	1600	62 748	49 445	253.513
660	22 600	17 112	216.419	1700	67 589	53 455	256.450
680	23 342	17 688	217.527	1800	72 513	57 547	259.262
700	24 088	18 268	218.610	1900	77 517	61 720	261.969
720	24 840	18 854	219.668	2000	82 593	65 965	264.571
740	25 597	19 444	220.707	2100	87 735	70 275	267.081
760	26 358	20 039	221.720	2200	92 940	74 649	269.500
780	27 125	20 639	222.717	2300	98 199	79 076	271.839
800	27 896	21 245	223.693	2400	103 508	83 553	274.098
820	28 672	21 855	224.651	2500	108 868	88 082	276.286
840	29 454	22 470	225.592	2600	114 273	92 656	278.407
860	30 240	23 090	226.517	2700	119 717	97 269	280.462
880	31 032	23 715	227.426	2800	125 198	101 917	282.453
900	31 828	24 345	228.321	2900	130 717	106 605	284.390
920	32 629	24 980	229.202	3000	136 264	111 321	286.273
940	33 436	25 621	230.070	3100	141 846	116 072	288.102
960	34 247	26 265	230.924	3150	144 648	118 458	288.9
980	35 061	26 913	231.767	3200	147 457	120 851	289.884
1000	35 882	27 568	232.597	3250	150 250	123 250	290.7

Based on data from JANAF Thermochemical Tables, NSRDS-NBS-37, 1971

G Psychrometric Charts

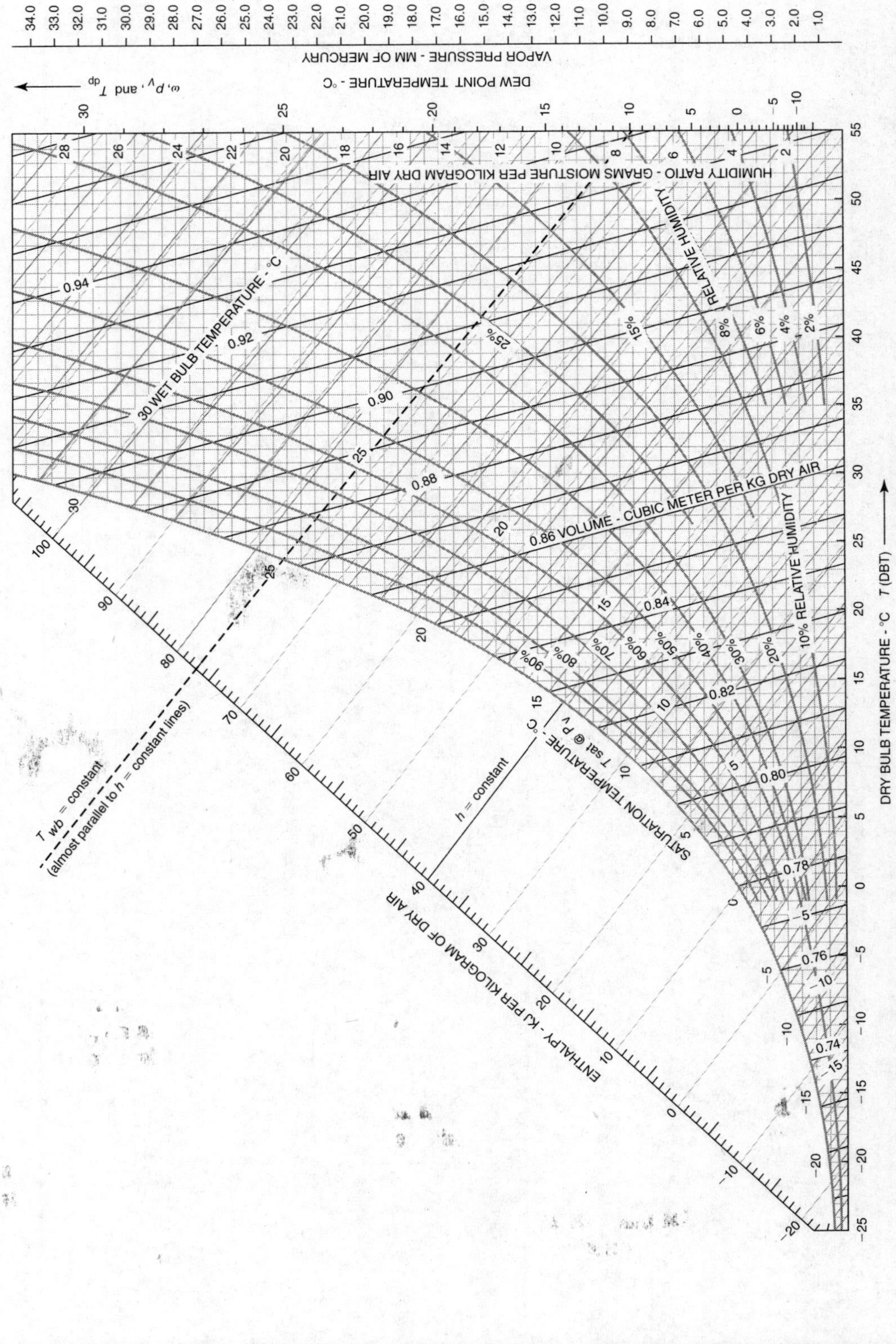

Figure G-1

Psychrometric chart, $P = 1$ atm, SI units.

H Compressibility Ch

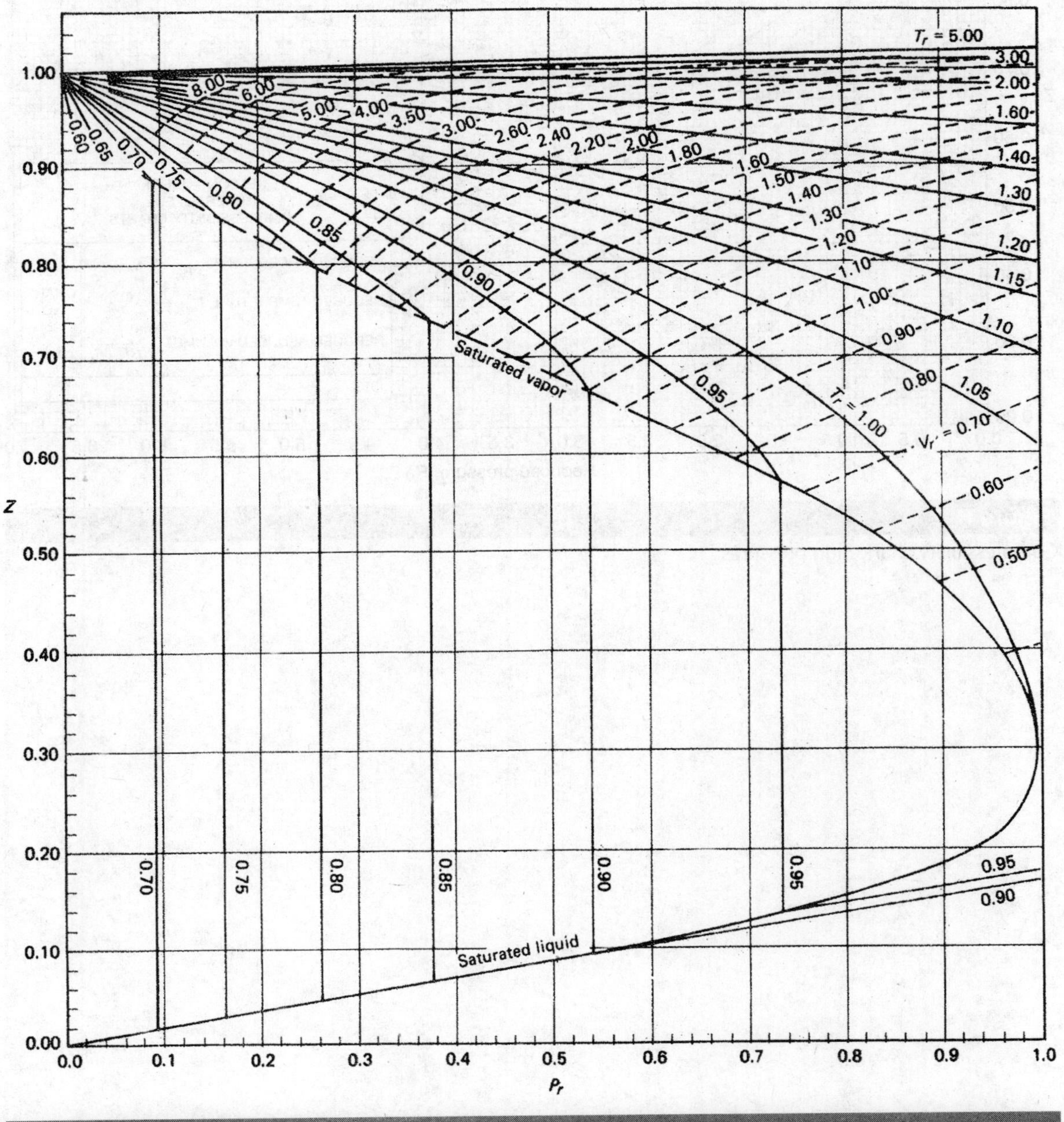

Figure H-1

Compressibility chart, low pressures.

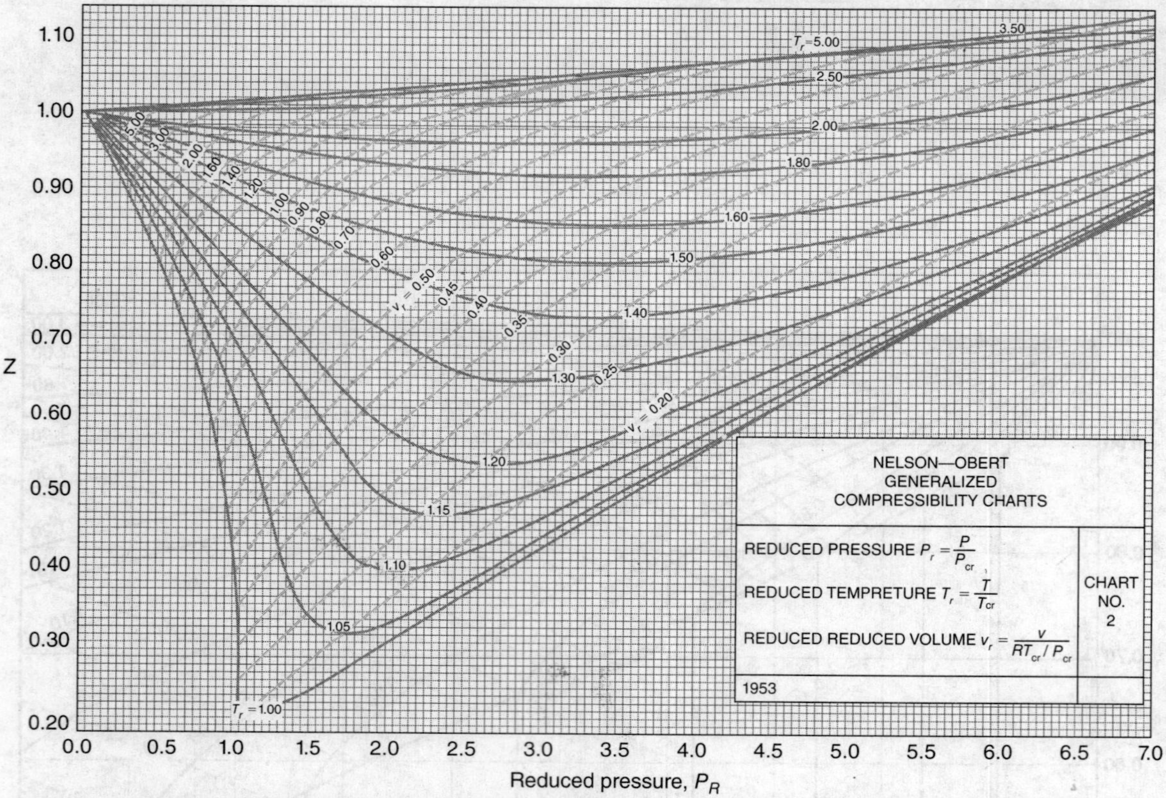

Based on V. M. Faires, Problems on Thermodynamics, Macmillan, New York, 1962. Data from L. C. Nelson and E. P. Obert, Generalized Compressibility Charts, Chem. Eng. 61: 203 (1954)

Figure H-2

Compressibility chart, high pressures.

I Enthalpy Departure Charts

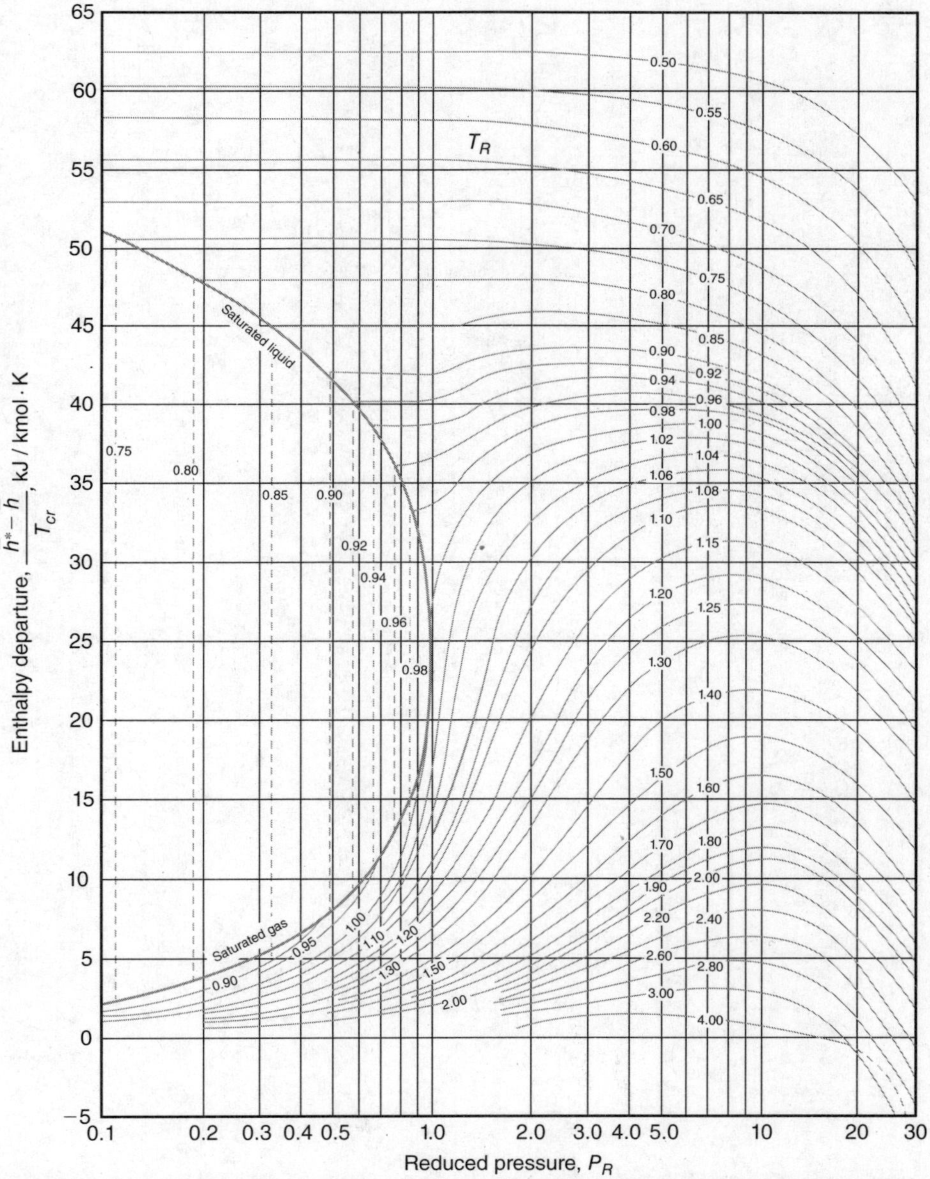

Based on G.J. Van Wylen and R.E. Sonntag, Fundamentals of Classical Thermodynamics, 3rd ed., Wiley, New York.

Figure I-1

Enthalpy departure chart, SI units.

J Entropy Departure Charts

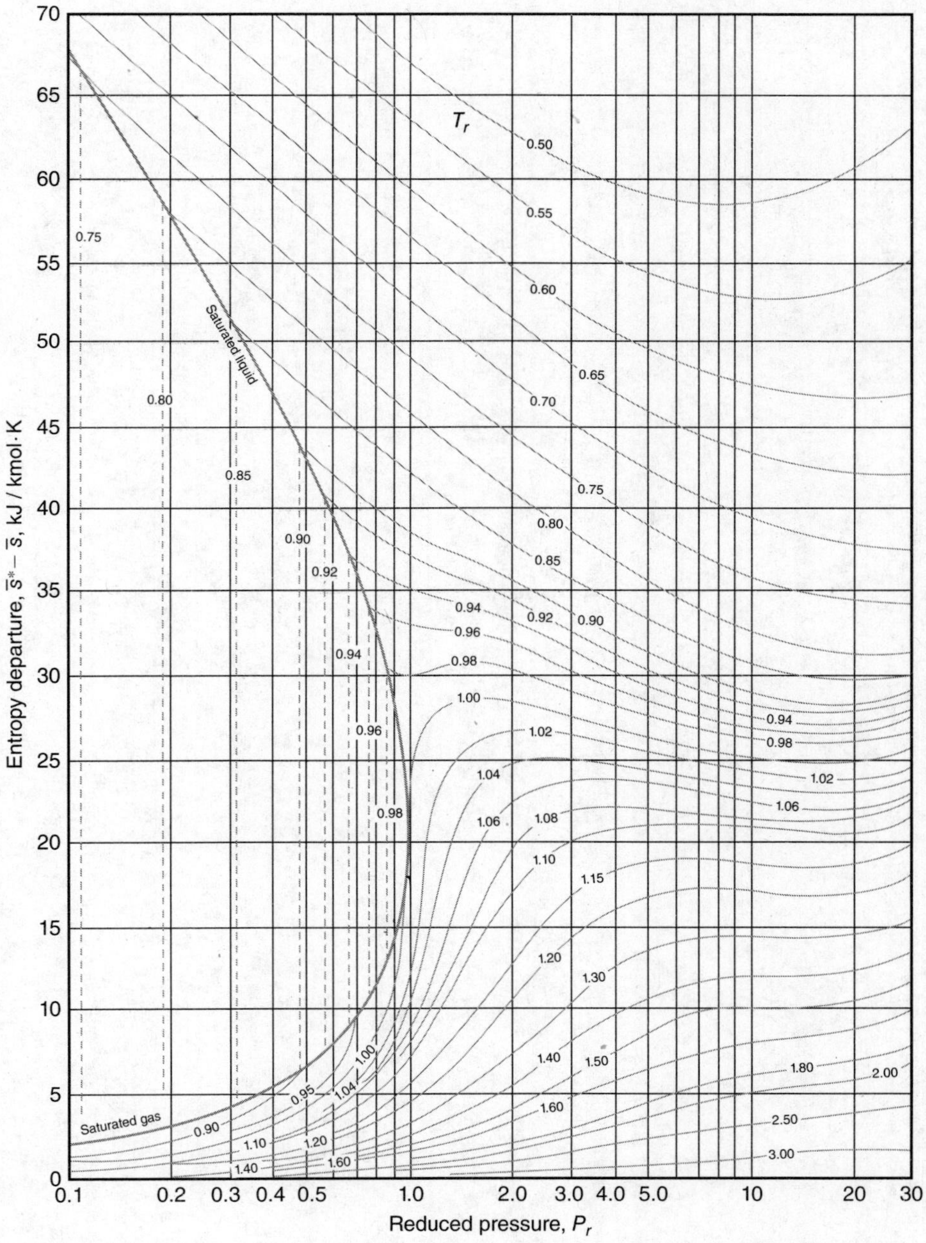

Based on G.J. Van Wylen and R.E. Sonntag, Fundamentals of Classical Thermodynamics, 3rd ed., Wiley, New York.

Figure J-1

Entropy departure chart, SI units.

Answers to Selected Problems

Chapter 1

1.1 **B**
1.2 **C**
1.3 **D**
1.4 **B**
1.5 **C**
1.6 **D**
1.7 **B**
1.8 **D**
1.9 **A**
1.10 **A**
1.11 **D**
1.12 **C**
1.13 **A**
1.14 **C**
1.15 **B**
1.16 **B**
1.20 True
1.26 i) A system, ii) a control volume, iii) a system, iv) a system.
1.28 0.02 mm
1.30 40 kg
1.32 Volume is extensive
1.34 Ice is lighter than water at 0°C
1.36 *b*) 0.5 m^3/kg, 4 kg, 39.2 N
1.36 *d*) 102 kg, 51 kg/m^3, 0.0196 m^3/kg
1.37 *b*) 0.2985 m^3/kg, 1.81 kg, 17.75 N
1.37 *d*) 204 kg, 0.00265 m^3/kg, 377.8 kg/m^3
1.38 0.25 kg/m^3, 0.00025, 2 kg, 19.62 N
1.40 Only (ii) is a quasi-equilibrium process
1.42 307 kPa, 296 kPa, 260 kPa
1.44 13 m
1.46 2.45 kPa gage, 0.724 in. Hg
1.48 *a*) 73.4 kPa or 0.734 bar
1.52 −270°C
1.54 310 K
1.56 *b*) 32 mm
1.58 65 MJ

Chapter 2

2.1 **D**
2.2 **B**
2.3 **B**
2.4 **A**
2.5 **A**
2.6 **D**
2.7 **A**
2.8 **C**
2.9 **A**
2.10 **B**
2.11 **A**
2.12 **C**
2.13 **A**
2.14 **B**
2.15 **B**
2.16 **D**
2.17 **A**
2.18 *b*) 1554 kPa
2.19 *b*) 120.2°C
2.20 2880 m
2.21 *b*) 79.2°C
2.22 0.41

2.24 $0.0042\ m^3$, $3.543\ m^3$
2.26 *b*) $7.528\ m^3$, 120.2°C
2.28 *b*) 180°C
2.31 *b*) 118.5 kg
2.32 $0.005\ m^3$, $6.015\ m^3$
2.33 *b*) 1.24%
2.34 *b*) 45.8%
2.36 $0.000306\ m^3$, $0.000308\ m^3$
2.38 $0.0064\ m^3$
2.40 *b*) 1554 kPa
2.42 *b*) $0.008\ m^3$
2.42 *d*) $0.7849\ m^3$
2.43 *b*) 652°C
2.44 424 kPa
2.48 3294 kJ/kg, 3290 kJ/kg, 3290 kJ/kg
2.49 *b*) 10 600 kJ
2.50 *b*) $0.073\ m^3/kg$, 900 kJ/kg
2.51 *b*) $999\ kg/m^3$, 11 500 kJ, 12 800 kJ
2.52 *a*) 4174 kJ
2.53 *b*) 2518 kJ/kg, 2694 kJ/kg
2.54 4580 kJ
2.56 *b*) Saturated mixture, 0.288%
2.58 24.38 MJ
2.60 225 kJ/kg
2.62 0.0184 kg, 0.636 m
2.64 28.8 kg
2.66 Helium
2.68 ii) 2410 kPa, $12.5\ kg/m^3$, 1230 N
iv) $0.05\ m^3/kg$, 1337 kPa, 1960 N
2.70 $0.241\ kg/m^3$
2.72 $\rho_{\text{dry air}} = 1.61\rho_{\text{humid}}$
2.74 *b*) 662 kPa
2.74 *d*) 664 kPa
2.76 $0.923\ m^3/kg$, $0.921\ m^3/kg$, $0.92\ m^3/kg$, $0.923\ m^3/kg$
2.78 225 kJ/kg, 213 kJ/kg, 210 kJ/kg
2.79 *b*) 370 kJ/kg, 513 kJ/kg
2.80 *b*) 4040 kJ, 5595 kJ
2.80 *d*) 4080 kJ, 5640 kJ
2.82 *b*) 2.20 kJ/kg·°C
2.83 *b*) 2.151 kJ/kg·°C
2.84 12 720 kJ, 12 720 kJ
2.86 *b*) 5820 kJ

Chapter 3

3.1 **C**
3.2 **D**
3.3 **B**
3.4 **B**
3.5 **C**
3.6 **C**
3.7 **B**
3.8 **A**
3.9 **B**
3.10 **A**
3.11 **C**
3.12 **A**
3.13 **D**
3.14 **A**
3.15 **C**
3.16 **B**
3.18 405 J
3.20 8002 kg, 392.5 kJ
3.22 197.7 J
3.24 *b*) −43.84 kJ
3.26 −31.9 kJ
3.28 3125 kJ
3.30 −0.125 J
3.32 −0.1125 kJ
3.34 739 J
3.36 320 J, 76.4 rpm
3.38 $135\ kW/m^2$
3.40 994 kW
3.42 $160\ W/m^2$, $173.6\ W/m^2$
3.46 32°C, no work
3.48 25 MW
3.50 $Q_{1\text{-}2} = -200$ kJ, $W_{3\text{-}4} = 400$ kJ, $Q_{4\text{-}1} = -600$ kJ, $Q_{2\text{-}3} = 1000$ kJ
3.52 *b*) 0.23 kg melts so 0°C
3.54 *b*) 54.5°C

3.56 153°C
3.58 3.776 m^3, 77.4°C, 666 kPa
3.60 *a*) 120 kPa, 0.00877 kg, 104.8ºC, 1100 kPa, 188 J
3.61 *b*) 4.9 cm
3.62 96.9 MJ
3.63 *b*) −17 600 kJ, −4180 kJ
3.64 *b*) 1193 K, 154 kJ, 537 kJ
3.66 *b*) −6480 kJ, −1030 kJ
3.68 42.9°C
3.70 *b*) –2450 kJ
3.72 3.46 kJ, 22.7 kJ
3.74 669°C
3.76 *a*) 0.731, 0, 1130 kJ/kg
3.78 *a*) 630°C
3.80 −30.7 kJ
3.82 400°C, −54.5 kJ
3.83 *b*) −2.27 kJ
3.84 *b*) 0.350 m^3/kg, −86.5 kJ/kg
3.85 *b*) −365 kJ/kg, 1 kJ/kg
3.86 151°C, 78.7 kJ
3.87 *b*) 3750 kJ, −5310 kJ
3.88 *b*) 200 kPa, 0.339 m^3/kg, 0.483 m^3/kg, 200 kJ, 57.6 kJ, 57 kJ
3.89 *b*) 134.6 kJ, 0.13 kJ
3.90 *b*) 1148 kJ, 0, −294 kJ, 854 kJ
3.91 *b*) 0, 148 kJ, −113 kJ, 35 kJ
3.92 *b*) 73 kJ, 182 kJ, −208 kJ, 47 kJ
3.93 *b*) 550 kJ, 550 kJ

Chapter 4

4.1 **A**
4.2 **D**
4.3 **B**
4.4 **D**
4.5 **A**
4.6 **C**
4.7 **B**
4.8 **D**
4.9 **C**
4.10 **A**
4.11 **A**
4.12 **D**
4.13 **D**
4.14 **A**
4.15 **A**
4.16 **C**
4.17 **C**
4.18 **D**
4.19 **A**
4.20 **A**
4.21 **B**
4.22 0.00141 m^3/s, 1.41 kg/s, 17.8 m/s
4.24 0365 kg/s, 0.0337 m^3/s
4.26 *b*) 1.172 m^3/s
4.28 *b*) 2560 kPa gage
4.29 *b*) 2090 kW
4.30 447 kW
4.32 9642 hp
4.34 *b*) 201 kW
4.36 0.0833 kW
4.37 *b*) 0.0957 kW
4.38 *b*) 23.7 kW
4.40 117 hp
4.42 −139 kJ/s
4.44 1.2 hp
4.46 15.3%, 99.6°C
4.47 *b*) 12.3°C
4.48 *b*) 18.1%
4.50 −29°C, 0.946 m^3/s
4.52 27.3 kg/s, 51.6 m/s
4.53 *b*) 0.359
4.54 195°C
4.55 *b*) 143.6°C
4.56 8.85 kg/s
4.58 0.098
4.60 30.9°C
4.62 1008 m/s, 11.57 kg/s, 5.83 cm
4.64 0.963
4.66 0.578 kg/s, 2.33 m/s, 18.4 m^3/s, 28.8 kJ/s
4.67 *b*) 512 m/s

4.68 *b*) 445°C, 61.7 kg
4.69 *b*) −33.6 MJ
4.70 *b*) 127°C, 3.48 kg
4.72 *b*) −129°C
4.73 *b*) 203.5 kPa
4.74 *b*) 133 kPa, −117°C, 7.1 min
4.75 *b*) 7940 kW, 53.4 hp, 26.8%
4.76 *a*) 7106 hp, 66.2 hp, 20.0%
4.77 *b*) 70.8 kJ/s, 2.36
4.78 *b*) 255 kW, 6.83
4.79 *a*) 935 kW, 41.7%, 1550 kJ/s, 37.6%
4.80 *b*) 1206 hp, 42.5%, 1435 kW, 38.5%

Chapter 5

5.1 **C**
5.2 **D**
5.3 **B**
5.4 **C**
5.5 **A**
5.6 **D**
5.7 **C**
5.8 **D**
5.9 **B**
5.10 **A**
5.12 671 W, only (i)
5.13 *b*) 9.89 kJ/s, 67%
5.14 34%, 212 GJ
5.15 *b*) 4440 kW
5.16 957 kJ/s, 68.9 kg/hr
5.18 *b*) 80 kW, 46.6%
5.19 *b*) 103 hp
5.20 *b*) 13.6%
5.22 20.6×10^6 MJ/day
5.23 *b*) 22 700 hp
5.24 6.70 hp
5.25 *b*) 16.7 kJ/s, 2.24
5.26 33.5 kJ/s, 26.8 kJ/s
5.28 33.22 kW, 5.36
5.30 *b*) 19.1 hp
5.34 7.66%
5.36 68.3%
5.38 60%, 236°C
5.40 209 hp
5.42 336 hp
5.44 Impossible
5.45 *b*) 125 kJ/s, 75 kW, 460°C
5.46 0.333
5.48 171°C
5.50 *b*) 57.2 hp, 657 kJ/s
5.52 2.93 hp
5.54 17.4 hp

Chapter 6

6.1 **C**
6.2 **D**
6.3 **B**
6.4 **A**
6.5 **D**
6.6 **C**
6.7 **B**
6.8 **A**
6.9 **A**
6.10 **B**
6.11 **D**
6.12 **A**
6.13 **D**
6.14 **A**
6.15 **A**
6.16 Zero
6.18 −0.200 kJ/kg·K
6.20 0.0711 kJ/kg·K, -0.211 m^3/kg
6.22 *b*) 1.56 kJ/kg·K
6.22 *d*) 19.5 kJ/kg·K
6.23 *c*) 15.3 m^3
6.24 *b*) 357°C, −173 kJ/kg
6.24 *d*) 485°C, −112 kJ/kg
6.25 *b*) 139.8°C, 109.4 kJ/kg
6.26 *b*) 379°C, 3.98 kJ/K
6.27 *b*) 136°C, 0.557 kJ/kg
6.28 *b*) 220 kJ, 220 kJ, 0.623 kJ/kg·K
6.29 *b*) 1838 kPa, 400°C

6.30 *b*) 400°C, 1.18 kJ/K
6.31 *b*) 249°C, 1.81 kJ/K
6.32 *b*) 350°C, −47.3 kJ
6.33 *b*) 102.6°C, 349.5 kJ/kg
6.34 431°C, 1.16 kJ/K
6.36 −0.1066 kJ/K, 0.2396 kJ/K
6.38 9.77°C, 0.170 kJ/K
6.39 *b*) 28.7°C, 0.102 kJ/K
6.40 1.475, 0.6981 kJ/kg·K
6.41 *b*) −488 kJ/kg, −1.19 kJ/kg
6.42 *b*) 75.84 kJ/K
6.43 *b*) 0.346 kJ/K
6.44 1.49 kJ/K
6.46 *b*) 487°C, −446 kJ/kg
6.47 *b*) 4510 kJ
6.48 15.55 MJ, 20.4 kJ/K
6.50 225°C, 232°C, 3.1%
6.52 *b*) 1.49 kJ/kg·K
6.52 *d*) 350°C, −167 kJ/kg
6.52 *f*) 369°C, 4.04 kJ/K
6.54 −30°C, 0.057 kJ/kg·K
6.56 97.5 MJ/s, 147.4 kJ/K·s
6.58 3480 hp
6.59 *b*) 15 MW, 8 kJ/K·s
6.60 *b*) 84 400 hp
6.60 *d*) 88 100 hp
6.61 *b*) 41 hp
6.62 331 hp, 0 kJ/K·s
6.64 2.36 kJ/K·s
6.66 0.37 kJ/K·s
6.67 *b*) 3.5 kJ/K
6.68 10.58 MW, 92.3%
6.70 *b*) 77.4%
6.70 *d*) 77.5%
6.71 *b*) 23 600 hp, 81.9%
6.72 567°C, 98.9%
6.74 *b*) 503 kJ/kg
6.76 79.4%
6.78 113 hp
6.80 208 m/s, 64°C
6.82 194 kJ/kg
6.83 *b*) 237 kJ/kg
6.84 197 hp

Chapter 7

7.1 **C**
7.2 **D**
7.3 **C**
7.4 **B**
7.5 **D**
7.6 **A**
7.7 **A**
7.8 0.0260 m^3/kg, 0.00184 m^3/kg
7.14 −0.8 kJ/kg·K
7.18 40.1°C
7.19 *a*) 397°C
7.20 400°C
7.22 1406 kJ/kg, 2.453 kJ/kg·K, 0.071%, 0.081%
7.24 *b*) −92°C
7.26 $C_v \ln\frac{T_2}{T_1} + R\ln\frac{v_2}{v_1}, C_p \ln\frac{T_2}{T_1} - R\ln\frac{P_2}{P_1}$
7.28 $C_v(T_2 - T_1) + a\left(\frac{1}{v_1} - \frac{1}{v_2}\right) + P_2v_2 - P_1v_1$
7.30 −11.4 kJ/kg, −1.0 kJ/kg·K, −0.976 kJ/kg·K
7.32 $3.86 \times 10^{-4}\ K^{-1}$, 2.33 GPa, 0.109 kJ/kg·K
7.34 $0.26 \times 10^{-4}\ K^{-1}$, 230 MPa, 3.3×10^{-5} kJ/kg·K
7.35 *b*) −25°C
7.36 0.041 °C/kPa, −11°C
7.38 0.0075 °C/kPa, 2.39 kJ/kg·K, 2.38 kJ/kg·K
7.40 *b*) −141 kJ/kg, −173 kJ/kg
7.41 *b*) 0, −48.3 kJ/kg
7.42 *b*) 0.4765 kJ/kg·K, 0.476 kJ/kg·K
7.43 *b*) 0.173 kJ/kg·K, 0.173 kJ/kg·K,
7.44 i) 500 kJ/kg, −0.364 kJ/kg·K
7.46 i) 62.5 kJ/kg, −0.433 kJ/kg·K
7.48 i) 3380 kW
7.50 i) 1600 kPa
7.52 −8.22 kJ/kg, −1.216 kJ/kg·K, 177.8 kJ/kg

Chapter 8

8.1 C

8.2 C

8.3 A

8.4 B

8.5 C

8.6 C

8.7 D

8.8 A

8.9 D

8.10 B

8.11 A

8.12 C

8.14 *b*) 8440 kW, 16.8 MJ/s, 42.6 hp, 25.2 MJ/s, 33.5%

8.14 *d*) 9560 kW, 16.5 MJ/s, 42.8 hp, 26 MJ/s, 36.8%

8.15 *b*) 9150 kW, 17.7 MJ/s, 42.5 hp, 26.8 MJ/s, 34.1%

8.16 *b*) 10 MW, 16.4 MJ/s, 107 hp, 26.4 MJ/s, 37.9%

8.18 *b*) 10 400 hp, 17 000 kJ/s, 42.3 hp, 24 800 kJ/s, 31.3%

8.19 *b*) 11 260 hp, 17 940 Btu/s, 42.14 hp, 26 340 kJ/s, 31.9%

8.20 *b*) 11 050 hp, 15 860 kJ/s, 95.7 hp, 24 100 kJ/s, 34.2%

8.22 26.6%, 27.3% 28.0%, 28.6%

8.24 72.6 MJ/s, 15.7 MJ/s, 38.5 kW, 49.8 MJ/s, 130°C, 214 hp, 43.6%, 298 kg/s

8.25 *b*) 72.6 MJ/s, 18.3 MJ/s, 38.9 kW, 52 MJ/s, 182°C, 214 hp, 42.8%, 311 kg/s

8.26 *b*) 74.9 MJ/s, 15.7 MJ/s, 42.9 kW, 47.6 MJ/s, 98.5%, 214 hp, 47.4%, 285 kg/s

8.27 *b*) 216°C

8.28 72 860 kJ/s, 16 280 kJ/s, 35 200 kJ/s, 53 950 kJ/s, 147°C, 264 hp, 39.5%, 322 kg/s

8.29 *b*) 72 860 kJ/s, 18 700 kJ/s, 35 200 kJ/s, 56 360 kJ/s, 202°C, 264 hp, 38.4%, 336 kg/s

8.30 68.7 MJ/s, 7360 kJ/s, 84.8 kW, 36.4 MJ/s, 58.2°C, 161 hp, 42.9%

8.32 *b*) 58.4 MJ/s, 3.5 kg/s, 24.4 kW, 210 hp, 87.5%, 41.8%

8.34 *b*) 60 400 kJ/s, 3.92 kg/s, 29 800 hp, 259 hp, 90.4%, 36.8%

8.36 3.01 kg/s, 74.1 MJ/s, 30 MW, 185°C, 44.2 MJ/s, 212 hp, 40.5%

8.38 5.34 kg/s, 57 500 kJ/s, 30 530 hp, super, 34 760 kJ/s. 185 hp, 39.6%

8.40 27.2 MJ/s, 2.36 kg/s, 11.1 MW, 40.8%

8.42 2.93 kg/s, 29.1 MW, 59.9 MJ/s, 94.1%, 30.8 MJ/s, 383 hp, 48.6%

8.44 2.94 kg/s and 2.40 kg/s, 39 280 hp, 63 150 kJ/s, 69°C, 33 880 kJ/s, 524 hp, 46.4%

8.46 2936 kJ/kg, 758 kJ/kg, 925 kJ/kg, 1253 kJ/kg, 57.3%

8.48 141.2 MJ/s, 39 MW, 45.6 MJ/s, 28.3 MJ/s, 80%

Chapter 9

9.1 A

9.2 C

9.3 D

9.4 B

9.5 C

9.6 D

9.7 D

9.8 A

9.9 A

9.10 D

9.12 448.6 kJ/kg, 1253 kJ/kg, 35.8%

9.14 11, 1177 cm^3

9.15 *b*) 56.5%, 982.4 kJ/kg, 555 kJ/kg

9.16 *b*) 20.3, 764 kJ/kg, 535 kJ/kg

9.18 *a*) 2001°C and 4100 kPa, 51.2%, 614 kJ/kg, 876 kPa

9.19 *b*) 553 kJ/kg, 46.1%, 731 kPa

9.20 2073°C, 8090 kPa, 60.2%, 698 kJ/kg, 932 kPa

9.21 *b*) 2011°C, 6303 kPa, 56.5%, 655 kJ/kg, 899 kPa

9.22 *b*) 534 kJ/kg, 46.0%, 7760 kPa

9.24 30.1, 39.5

9.26 *b*) 1521°C and 6627 kPa, 65.8%, 527 kJ/kg, 644 kPa

9.27 *b*) 436 kJ/kg, 54.5%, 540 kPa

9.28 1450°C and 7574 kPa, 66.0%, 462 kJ/kg, 566 kPa

9.32 *b*) 66.0%, 2640 kJ/kg, 1740 kJ/kg
9.34 54.3%, 589 kJ/kg, 1080 kJ/kg, 491 kJ/kg
9.36 11.3, 24.7, 67.6
9.38 *b*) 72.2%, 5720 kJ/s, 520 kW, 64.1%
9.39 *b*) 28.7%, 12.6%
9.40 63.4 %
9.42 *b*) 20.7%, 5.14%
9.44 *b*) 71.8%, 3700 kW, 1040 kW, 28.2%, 3500 kJ/s
9.45 *b*) 56.7%
9.46 68.5%, 31.6%
9.48 *b*) 77.6%
9.50 12 MW, 40%, 11.56 MJ/s, 3560 kW, 30.8%, 51.8%
9.51 *b*) 14.4 MW, 40%, 12.5 MJ/s, 4080 kW, 32.6%, 51.4 %
9.52 *b*) 8660 kW, 30.1%, 11.6 MJ/s, 3275 kW, 2\8.3%, 41.4%
9.54 *b*) 9291 hp, 1.744 kg/s, 2737 hp, 51.8%

Chapter 10

10.1 A
10.2 C
10.3 B
10.4 A
10.5 D
10.6 D
10.7 C
10.8 A
10.9 C
10.10 33 kJ/kg, 136 kJ/kg, 169 kJ/kg, 4.12, 5.06
10.12 11.6 kW, 238.4 kJ/s, 3.45, 0.0412 m^3/s
10.14 −15.6ºC, 52 hp, 81.6 kJ/s, 2.1
10.16 0.062 kg/s, 2.79 kW, 2.86
10.18 126 kg/s, 84.2%, 9.6 hp, 22.1 kJ/s, 2.0
10.20 91%, 2.13
10.22 11.1 kg/min, −22.3°C, 11.1 hp, 2.87
10.24 69°C, 1.30 kg/s, 85 hp, 2.1
10.26 156 kJ/s, 1.03 kg/s and 1.23 kg/s, 77 hp, 2.7, 1.26 kg/s, 83 hp, 2.5
10.28 −94°C, 1.48, 0.676 kg/s

Chapter 11

11.1 B
11.2 C
11.3 B
11.4 D
11.5 B
11.6 B
11.7 A
11.8 D
11.9 B
11.10 D
11.11 B
11.13 28.3 kg/kmol, 0.294 kJ/kg·K
11.14 *b*) 0.5, 0.125, 0.375, 0.590, 0.129, 0.281, 0.398 kJ/kg·K
11.14 *d*) 0.3, 0.4, 0.3, 0.227, 0.531, 0.242, 0.394 kJ/kg·K
11.15 *b*) 0.3, 0.4, 0.3, 0.227, 0.531, 0.242, 0.394 kJ/kg·K
11.15 *d*) 0.5, 0.125, 0.375, 0.541, 0.077, 0.382, 0.481 kJ/kg·K
11.16 *b*) 0.266 kJ/kg·K, 5.77 kg
11.18 0.9664, 0.0139, 0.0147, 0.0015, 43.48 kg/kmol
11.20 573 kPa, 16.2 kPa, 9.6 kPa. 1.2 kPa
11.22 0.4, 0.6, 0.478, 0.522, 0.217 kJ/kg·K, 191 kJ
11.24 8.27 m^3, 484 kJ
11.26 0.222, 0.333, 0.444, 128 kg, 168 kg, 352 kg, 0.231 kJ/kg·K, 39.97 kJ
11.28 122 kPa, 3930 kJ
11.30 2530 kJ, 5.6 kJ/K
11.32 661 kJ, 0.352 kJ/K
11.34 1630 kW
11.35 *b*) 152 kW
11.35 *d*) 143 kW
11.36 67.6 kPa, 482 m/s
11.38 *a*) 72.9%, 98.3 kPa, 0.0108 kg_v/kg_a
11.39 *b*) 73.1%, 99.48 kPa, 0.0112 kg_v/kg_a
11.40 *b*) 96.31 kPa, 0.0238 kg_v/kg_a
11.41 *b*) 97.14 kPa, 0.00117 kg_v/kg_a
11.42 *b*) 94.9 kPa, 41.4%
11.43 *b*) 93.0 kPa, 53.1%, 52.9%

11.44 *b*) 12.0°C
11.45 *b*) 16.7°C
11.46 4.5%
11.48 *b*) 13.2°C, 0.0096 kg_v/kg_a
11.50 0.00852 kg_v/kg_a, 48.4%, 43.6 kJ/kg
11.51 *b*) 56%, 0.0147 kg_v/kg_a, 101.7 kPa
11.51 *d*) 69%, 93.4 kPa
11.51 *f*) 48%, 0.0079 kg_v/kg_a, 42 kJ/kg_a
11.52 *b*) 70%, 0.011 kg_v/kg_a, 99.48 kPa
11.52 *d*) 90%, 96.46 kPa
11.52 *f*) 10°C
11.54 *b*) 28.4°C, 44%, 24.8°C, 92 kJ/kg_a
11.56 0.0194 kg_v/kg_a, 23.3 kg_v/hr
11.58 *b*) 50.4 kg_v/hr, 23 kg/hr
11.60 58°C
11.61 *b*) 0.991, 0.0086, 0.986, 0.014
11.62 92%
11.64 10%
11.66 *b*) 205 kg_v/hr
11.67 *b*) 13.3 kJ/s, 15%
11.68 87 kJ/s, 0.017 kg_v/s
11.69 *b*) 25.4°C, 50%, 160 kg/min
11.70 17°C, 73%, 98 kg/min
11.72 80%, 1.6 kg/min
11.73 *b*) 130 kg_a/s, 2.2 kg_w/s

Chapter 12

12.1 **A**
12.2 **D**
12.3 **C**
12.4 **B**
12.5 **B**
12.6 **B**
12.7 **A**
12.8 **D**
12.9 *b*) $C_6H_6 + 7.5(O_2 + 3.76N_2) \Rightarrow 6CO_2 + 3H_2O + 28.2N_2$
12.9 *d*) $C_3H_8 + 5(O_2 + 3.76N_2) \Rightarrow 3CO_2 + 4H_2O + 18.8N_2$
12.10 *b*) $C_6H_6 + 11.25(O_2 + 3.76N_2) \Rightarrow 6CO_2 + 3H_2O + 3.75O_2 + 42.3N_2$
12.10 *d*) $C_3H_8 + 7.5(O_2 + 3.76N_2) \Rightarrow 3CO_2 + 4H_2O + 2.5O_2 + 28.2N_2$
12.11 *b*) $C_6H_6 + 6.75(O_2 + 3.76N_2) \Rightarrow 4.5CO_2 + 3H_2O + 1.5CO + 25.38N_2$
12.11 *d*) $C_3H_8 + 4.5(O_2 + 3.76N_2) \Rightarrow 2CO_2 + 4H_2O + CO + 16.92N_2$
12.12 *b*) $CH_4 + C_8H_{18} + 14.5(O_2 + 3.76N_2) \Rightarrow 9CO_2 + 11H_2O + 54.52N_2$
12.12 *d*) $CH_4 + C_3H_8 + C_6H_6 + 14.5(O_2 + 3.76N_2) \Rightarrow 10CO_2 + 9H_2O + 54.52N_2$
12.14 12.73 kg_{air}/kg_{fuel}, 11.15%, 49°C
12.15 *b*) 2.95 kg_{fuel}/min
12.15 *d*) 1.83kg_{fuel}/min
12.16 3.077 kg_{oxygen}/kg_{fuel}, 66.7%, 72°C
12.18 62.8 kg_{air}/kg_{fuel}, 4.43%, 32°C
12.20 31.0 kg_{air}/kg_{fuel}, 180%
12.22 C_6H_{16}
12.23 *b*) C_9H_{20}
12.24 *b*) 2.31 kg_{air}/kg_{fuel}, 40.5°C
12.25 *b*) −2220 MJ/kmol
12.25 *d*) −8100 MJ/kmol
12.26 *b*) −1245 MJ
12.26 *d*) −4596 MJ
12.27 *b*) 1520 kJ/s
12.27 *d*) 1500 kJ/s
12.28 *b*) −964 MJ/kmol
12.28 *d*) −2945 MJ/kmol
12.29 2720 MJ/hr
12.30 1140 m/s
12.32 *b*) 4280 kJ
12.32 *d*) 11 020 kJ
12.33 *b*) 900 kPa
12.34 −16 450 kJ, 230.5 kPa, 632 m^3
12.36 *b*) 1500°C
12.37 *b*) 2100°C
12.38 *b*) 1580°C
12.39 *b*) 1560°C
12.40 1530°C
12.42 1130°C
12.43 *b*) 1850°C, 746 kPa, 1880 m^3
12.44 *b*) 940°C
12.46 1020°C

Index

PRINCIPAL UNITS USED IN MECHANICS

Quantity	International System (SI)			U.S. Customary System (USCS)		
	Unit	Symbol	Formula	Unit	Symbol	Formula
Acceleration (angular)	radian per second squared		rad/s^2	radian per second squared		rad/s^2
Acceleration (linear)	meter per second squared		m/s^2	foot per second squared		ft/s^2
Area	square meter		m^2	square foot		ft^2
Density (mass) (Specific mass)	kilogram per cubic meter		kg/m^3	slug per cubic foot		slug/ft^3
Density (weight) (Specific weight)	newton per cubic meter		N/m^3	pound per cubic foot	pcf	lb/ft^3
Energy; work	joule	J	N·m	foot-pound		ft-lb
Force	newton	N	kg·m/s^2	pound	lb	(base unit)
Force per unit length (Intensity of force)	newton per meter		N/m	pound per foot		lb/ft
Frequency	hertz	Hz	s^{-1}	hertz	Hz	s^{-1}
Length	meter	m	(base unit)	foot	ft	(base unit)
Mass	kilogram	kg	(base unit)	slug		lb-s^2/ft
Moment of a force; torque	newton meter		N·m	pound-foot		lb-ft
Moment of inertia (area)	meter to fourth power		m^4	inch to fourth power		in.^{4}
Moment of inertia (mass)	kilogram meter squared		kg·m^2	slug foot squared		slug-ft^2
Power	watt	W	J/s (N·m/s)	foot-pound per second		ft-lb/s
Pressure	pascal	Pa	N/m^2	pound per square foot	psf	lb/ft^2
Section modulus	meter to third power		m^3	inch to third power		in.^{3}
Stress	pascal	Pa	N/m^2	pound per square inch	psi	lb/in.^{2}
Time	second	s	(base unit)	second	s	(base unit)
Velocity (angular)	radian per second		rad/s	radian per second		rad/s
Velocity (linear)	meter per second		m/s	foot per second	fps	ft/s
Volume (liquids)	liter	L	10^{-3} m^3	gallon	gal.	231 in.^{3}
Volume (solids)	cubic meter		m^3	cubic foot	cf	ft^3

SELECTED PHYSICAL PROPERTIES

Property	SI	USCS
Water (fresh)		
weight density	9.81 kN/m^3	62.4 lb/ft^3
mass density	1000 kg/m^3	1.94 slugs/ft^3
Sea water		
weight density	10.0 kN/m^3	63.8 lb/ft^3
mass density	1020 kg/m^3	1.98 slugs/ft^3
Aluminum (structural alloys)		
weight density	28 kN/m^3	175 lb/ft^3
mass density	2800 kg/m^3	5.4 slugs/ft^3
Steel		
weight density	77.0 kN/m^3	490 lb/ft^3
mass density	7850 kg/m^3	15.2 slugs/ft^3
Reinforced concrete		
weight density	24 kN/m^3	150 lb/ft^3
mass density	2400 kg/m^3	4.7 slugs/ft^3
Atmospheric pressure (sea level)		
Recommended value	101 kPa	14.7 psi
Standard international value	101.325 kPa	14.6959 psi
Acceleration of gravity (sea level, approx. 45° latitude)		
Recommended value	9.81 m/s^2	32.2 ft/s^2
Standard international value	9.80665 m/s^2	32.1740 ft/s^2

SI PREFIXES

Prefix	Symbol	Multiplication factor		
tera	T	10^{12}	=	1 000 000 000 000
giga	G	10^{9}	=	1 000 000 000
mega	M	10^{6}	=	1 000 000
kilo	k	10^{3}	=	1 000
hecto	h	10^{2}	=	100
deka	da	10^{1}	=	10
deci	d	10^{-1}	=	0.1
centi	c	10^{-2}	=	0.01
milli	m	10^{-3}	=	0.001
micro	μ	10^{-6}	=	0.000 001
nano	n	10^{-9}	=	0.000 000 001
pico	p	10^{-12}	=	0.000 000 000 001

Note: The use of the prefixes hecto, deka, deci, and centi is not recommended in SI.